This book is due for return on or before the last date shown below.

NUMBER THEORY

A Lively Introduction with Proofs, Applications, and Stories

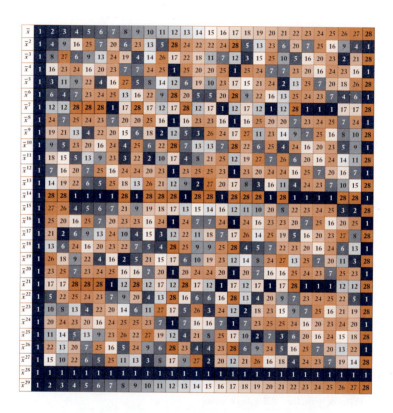

JAMES E. POMMERSHEIM

TIM K. MARKS

ERICA L. FLAPAN

John Wiley & Sons, Inc.

Publisher	Laurie Rosatone
Editor	Shannon Corliss
Marketing Manager	Sarah Davis
Production Manager	Dorothy Sinclair
Senior Production Editor	Sandra Dumas
Designer	James O'Shea
Media Editor	Melissa Edwards
Illustration Editor	Anna Melhorn
Photo Department Manager	Hilary Newman
Photo Editor	Sheena Goldstein
Production Management Services	MPS Limited, A Macmillan Company

This book was typeset in 11/13 Adobe Caslon Pro Regular at MPS Limited, A Macmillan Company and printed and bound by Courier/Kendallville. The cover was printed by Courier/Kendallville.

The paper in this book was manufactured by a mill whose forest management programs include sustained yield harvesting of its timberlands. Sustained yield harvesting principles ensure that the number of trees cut each year does not exceed the amount of new growth.

This book is printed on acid-free paper.

ISBN 13 978-0470-42413-1
Printed in the United States of America.
10 9 8 7 6 5 4 3 2

NUMBER THEORY

A Lively Introduction with Proofs,
Applications, and Stories

Contents

Preface

The goal of this book is to present a rigorous yet leisurely-paced treatment of elementary number theory, together with relevant applications. The book uses an inviting, conversational style throughout, and the authors have put great effort into making difficult concepts and proofs as accessible as possible without sacrificing mathematical depth. Readable discussions motivate new concepts and theorems before their formal definitions and statements are presented. In addition, many applications of number theory are explained in detail throughout the text, including some that have rarely (if ever) appeared in textbooks.

This book is well suited to a one-semester or one-quarter undergraduate number theory course. It is ideal for either a first proofs course or a more advanced course in which students already have experience with proof writing. The book is appropriate for a first proofs course because it introduces students to number theory without assuming familiarity with proof techniques or much experience with mathematical notation or abstraction. In Chapters 1 and 2, students learn how to write formal proofs while building their understanding of the arithmetic properties of the integers. Number theory provides an excellent vehicle for learning to write proofs because the integers are already very familiar to the students, and the book does not require an advanced background in mathematics. This book is also suitable for a more advanced number theory course. Courses in which students are already familiar with proofs can move quickly through the first two chapters, providing extra time to cover more of the advanced topics at the end of the book.

We have found over and over in our teaching that when rigorous mathematics is presented in a friendly, nonthreatening manner, students find it enjoyable rather than intimidating. We have incorporated this idea into every chapter of our book using *math myths*, fictional tales about famous mathematicians that are used to introduce and illustrate mathematical concepts. These stories, together with accompanying cartoons, use a playful style to draw the reader into the fundamental ideas of number theory. The myths provide a concrete representation of more abstract concepts that can aid the student during the transition to general statements of theorems and their proofs. The math myths, as well as the cartoons, exemplify the book's light-hearted approach to serious mathematics. These fictional stories about famous mathematicians also serve to motivate students' curiosity about the real historical figures. To satisfy their curiosity, each chapter begins with an in-depth historical summary of the life of the mathematician who is featured in the myth, as well as a summary of a portion of his or her mathematical work.

The book's approach makes both fundamental and more advanced topics in number theory accessible to a wider audience. We have focused on developing concepts and proving theorems in a way that is understandable to students with a range of backgrounds. This includes even the most technically difficult theorems, such as Quadratic Reciprocity, Fermat's Two Squares Theorem, the Miller-Rabin primality test, and others, for which complete proofs are presented with an emphasis on readability.

In addition to the math myths, use of humor, and conversational style, this book employs a number of other pedagogical techniques to facilitate student understanding. Many theorems are introduced using *Numerical Proof Previews*, specific numerical examples used both to discover the statement of a theorem and to serve as a concrete model for its proof. The book also incorporates a geometric approach when appropriate, so that students can use their geometric understanding as a bridge to a symbolic understanding. Throughout the book, extensive use of color coding helps students follow algebraic manipulations and clarifies the connections between text, equations, tables, and figures. Due to its innovative pedagogy, the book could be particularly appropriate for math education majors.

Structure of the Text

The book is organized into three main parts. Chapters 1 and 2 introduce students to methods of proof and the basic properties of the integers, including mathematical induction. Depending on the background of the students, these two chapters can be covered carefully or quickly skimmed. Instructors who wish to take a more axiomatic approach to the foundations of number theory should use the optional material in the Appendix (available on both the Student Companion Website and the Instructor Companion Website), which presents axioms for the integers and begins to develop their consequences. Written in the style of a chapter, including complete explanations and proofs as well as exercises, the Appendix can be used to augment the material in Chapters 1 and 2.

Chapters 3–9 form the core content of the book. They teach central concepts and theorems in elementary number theory using a developmental approach, in which each chapter builds on material covered in previous chapters. Topics presented in these chapters include basic divisibility properties, prime numbers, the Euclidean Algorithm, modular arithmetic (including applications), Fermat's Little Theorem, Euler's Theorem, and RSA (public-key) encryption.

Chapters 10–15 contain a smorgasbord of additional topics, from which an instructor can select material as time and inclination permit. Topics covered in Chapters 10–15 include primitive roots, primality testing, quadratic reciprocity, Gaussian integers, Fermat's Two Squares Theorem, continued fractions, Pell's Equation, and Fermat's Last Theorem. Numerous real-world applications of number theory are distributed throughout the book. Chapter dependencies are summarized in the dependency diagram.

The exercise sets at the end of each section provide *Numerical Problems* (computational and algorithmic exercises), as well as *Reasoning and Proofs* exercises that develop students' ability to read and

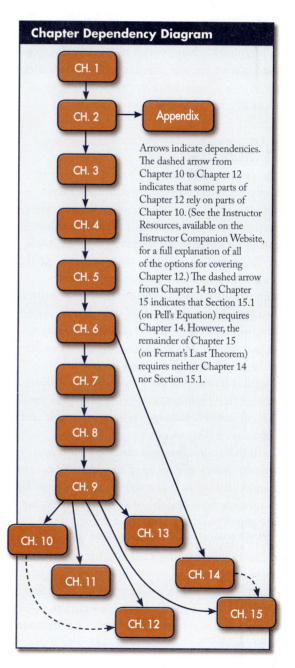

Chapter Dependency Diagram

CH. 1
CH. 2 → Appendix
CH. 3
CH. 4
CH. 5
CH. 6
CH. 7
CH. 8
CH. 9
CH. 10
CH. 11
CH. 12
CH. 13
CH. 14
CH. 15

Arrows indicate dependencies. The dashed arrow from Chapter 10 to Chapter 12 indicates that some parts of Chapter 12 rely on parts of Chapter 10. (See the Instructor Resources, available on the Instructor Companion Website, for a full explanation of all of the options for covering Chapter 12.) The dashed arrow from Chapter 14 to Chapter 15 indicates that Section 15.1 (on Pell's Equation) requires Chapter 14. However, the remainder of Chapter 15 (on Fermat's Last Theorem) requires neither Chapter 14 nor Section 15.1.

write proofs and to reason mathematically. A large number of more advanced exercises are also provided throughout the book, either grouped separately as *Advanced Reasoning and Proofs* exercises or indicated using the symbol * to the left of the exercise number. Many of the exercise sets include in-depth *Explorations*, in which a series of exercises develop a specific topic that is related to the material in the section. The wide range of exercises in each section helps ensure that the book can be used in a variety of different courses and enables the instructor to emphasize proof writing to as little or as great an extent as he or she desires.

To the Student

This book is unlike any math textbook you have ever used. Every chapter includes a *math myth*, which is a fictional story starring a famous mathematician that introduces an important number theory topic in a friendly, inviting manner. Math myths are indicated by the following icon: **Math Myth**. To add to the fun, cartoons are sprinkled throughout the book. The entire book is written in a readable, conversational style, but we don't skimp at all on the serious mathematics. In fact, we wanted to call the book *The Fun Way to Serious Number Theory*, but the publisher wouldn't let us. This book will teach you plenty of high-level mathematics with full rigor—we just do our best to help you be comfortable and have fun in the process.

In this course, you will develop your proof-writing skills while learning the core concepts in number theory as well as advanced topics used in modern applications. You will work hard, but the payoff will be well worth it.

We've written this book because we want you to succeed in learning a subject that we love, and we want to help you enjoy the process of learning it. We've made a concerted effort to explain difficult concepts and proofs in a way that makes them easier to understand. Many of the theorems are preceded by *Naomi's Numerical Proof Previews*,

which are specific numerical examples that will help give you a concrete understanding of both the statements of the theorems and the ideas behind their proofs, before formalizing the statement and proof in more abstract terms. Naomi's Numerical Proof Previews are always accompanied by a picture of Naomi. Where appropriate, we present ideas geometrically as well as algebraically, so that you can use your geometric understanding to aid in your symbolic understanding. We also make liberal use of color coding, both to make it easier to see how an equation follows from the previous ones and to highlight the connections between the text and the corresponding equations, figures, and tables.

Naomi

In addition to Naomi and her Numerical Proof Previews, there are a couple of other characters who pop up at various places in the book. *Put it in Prose, Paul!* is a feature that appears after the statements of some of the theorems. After the theorem is stated in formal mathematical language, Paul restates the theorem in plain English to make it easier to understand and remember. Another character who appears when you least expect him is your friend Phil Lovett. Phil spends too much of his time trying to dream up new mathematical principles but not enough time actually studying number theory. Phil Lovett means well, but he never gets the math quite right, and it will often be your job to figure out where your friend Phil's reasoning went wrong. You will recognize Phil by his half-baked ideas and his picture.

Paul

Phil

If the fictitious stories about famous mathematicians in the math myths get you interested in knowing more about the real historical figures, you will find a brief summary of the featured mathematician's life and work at the beginning of each chapter. The real lives of these mathematicians are fascinating true stories in their own right (often just as weird as the math myths), and the biographies at the beginning of each chapter will take you through the history of number theory and the contributions of many mathematicians, from the ancient Greeks to the present day. As an added bonus, reading the history at the beginning of each chapter will enable you to get many of the inside jokes in the chapter's math myth. We understand that you may not always be in the mood for a math joke (too much groaning can be hard on the throat!), so we have written each chapter so that it can be followed even if you choose not to read the myth.

This book is teeming with exercises. Doing exercises builds mathematical muscle—it helps you test your understanding of the material and further develops your insight and intuition. In order to help you get started on the exercises, the Student Companion Website contains answers to all of the odd-numbered Numerical Problems and hints for all of the odd-numbered Reasoning and Proofs and Advanced Reasoning and Proofs exercises. We encourage you to solve some of the odd-numbered exercises with the help of these student resources before you tackle the rest of the homework assigned by your instructor.

As you learn number theory, keep in mind that mathematics is a human achievement. Like the great works of art and architecture, the mathematical understanding that humans have developed and continue to develop is something we can all be proud of. Everything in this book was discovered by creative and curious people at some point in human history. Each theorem brought to some human being thrill and delight at the moment of first discovery. You can share in this delight as you rediscover number theory and create your own individual understanding of the subject.

If you are starting a course in number theory, you are extremely fortunate. More than 99.9% of people in the world never get this privilege. Enjoy it! This book will be there to help you every step of the way.

To the Instructor

The idea for this book grew out of number theory courses and mathematical reasoning courses that we have taught for many years at the Johns Hopkins University's Center for Talented Youth (CTY), Pomona College, Reed College, New Mexico State University, and the University of California, San Diego. We originally developed the stories in the math myths as humorous skits to engage our students in thinking about some of the ideas of number theory before they were formally introduced. We used Numerical Proof Previews to give the students a concrete way to understand the key ideas of a proof before we presented the abstract proof. Our students have responded so enthusiastically over the years that we decided to write this book, to share our approach to teaching number theory with a wider audience.

Our experience has led us to conclude that this book will work well for many different types of courses. This book is well suited to number theory courses for math majors, whether or not the students are already acquainted with proof writing. As number theory provides a good framework in which to learn to write proofs, the book is also appropriate for mathematical reasoning and introduction to proofs courses. This book can also be used for enrichment courses for motivated first-year college students or nonmajors, and even with exceptionally talented high school students.

For a typical number theory course for majors, you might cover the core material in Chapters 1–9 plus a few of the more advanced topics from Chapters 10–15, as time permits. If your audience is mature enough mathematically and you would like more time to cover advanced topics such as primality testing and the Gaussian integers, you may elect to go very quickly over the material of the first two chapters.

For a mathematical reasoning or introduction to proofs course, the instructor might thoroughly cover Chapters 1 and 2, which include a thoughtful discussion of writing formal proofs, before going on to introduce the more standard number theory material. For an introduction to mathematics course for nonmajors, the instructor might go light on the rigor, perhaps doing some proofs or Numerical Proof Previews in class, while assigning mainly Numerical Problems and some of the easier Reasoning and Proofs exercises. A course for future high school teachers might take a more rigorous approach, focusing on the core material in Chapters 1–9.

The optional Appendix is available on the Web, on both the Student Companion Site and the Instructor Companion Site. The Appendix, which can be used in conjunction with Chapters 1 and 2, provides an axiomatic approach to the integers using the language of rings. You should use the Appendix if you want to incorporate a more rigorous axiomatic approach to the foundations of number theory. The Appendix develops the basic axioms of number theory and their first consequences. After the definitions of binary operation and ring, three axioms for the integers are given: Axiom 1 states that the integers are a ring, Axiom 2 states the order properties (properties of $<$), and Axiom 3 is the Well-Ordering Principle. Next begins the process of deriving consequences of the axioms. Sample proofs are given to illustrate how many familiar results about the integers can be deduced rigorously from the three axioms. The Appendix concludes with a proof that the Principle of Mathematical Induction and the Principle of Strong Induction are each equivalent to the Well-Ordering Principle.

For the instructor who wants to incorporate technology into the course, many of the topics in the book give you the opportunity to introduce the students to mathematical software packages such as Mathematica or Maple. A Maple worksheet containing a library of number-theoretic routines tailored to the course is provided on both the Instructor Companion Website and the Student Companion Website. Organized by chapter, this library allows lecturers or students to explore the computational side of the concepts (sometimes using enormous numbers!) and gives students a feel for how quickly (or slowly) certain number-theoretic functions can be computed. The routines in the worksheet also serve as a

model for students who want to develop their own code for exploring other concepts or for solving the computer exercises that are sprinkled throughout the book.

The book also informally introduces computational complexity in Chapter 4, with a brief discussion of polynomial-time algorithms. Throughout the book, as new material is introduced, we informally discuss whether the new concepts are computable in polynomial time. These discussions can be skipped entirely. However, spending a few minutes discussing computational issues, and perhaps showing a computer demo in class (using the Maple worksheet available in the online resources), will reinforce the idea that almost everything in the early part of the book is efficiently computable, with the notable exception of factoring. If the students understand this well, they will more fully appreciate the RSA algorithm when they get to Chapter 9. It will also help them understand the material on primality testing in Chapter 12, including the section on Pratt primality certificates. Also in Chapter 12, number-theoretic problems are used to motivate a short discussion of the P = NP question.

The *Instructor Resources*, available on the Instructor Companion Website, contain all the information you would receive from a colleague who recently finished teaching the course. You get the full text of the lectures the colleague has given, the descriptions of fun activities, handouts, and games used to engage the students in an interactive class atmosphere, as well as homework assignments, a syllabus, and exam questions. All activities are clearly marked apart from the main lecture notes—they are meant to serve as optional suggestions that can be implemented if they match your teaching style. If you elect to lecture without interruptions, you can use the lecture notes without the activities. The Instructor Resources include complete solutions to all exercises in the book, which may help you to select homework assignments. The Instructor Resources also include a variety of other possible assignments such as student presentations, written projects, and online discussion board topics that can prompt the students to improve one another's understanding of the subject and to learn to distinguish between trustworthy and non-trustworthy Internet sources.

For your convenience, the Instructor Resources are available online on the Instructor Companion Website. The Instructor Resources are not accessible by students. The Student Companion Website provides students with numerical answers for all of the odd-numbered Numerical Problems and hints for all of the odd-numbered Reasoning and Proofs and Advanced Reasoning and Proofs exercises. These hints are designed to provide just enough help to get the student on the right track, without revealing the complete solution.

Acknowledgments

This book has been a long time in the making. When you work on a book for over 15 years, you end up with a long list of people to thank. Numerous people have aided us in this project; it would be impossible to mention all their names here.

We thank all of the students we have taught over the years at Pomona College, Reed College, The University of California San Diego, New Mexico State University, and the Johns Hopkins University's Center for Talented Youth (CTY), most of whom have long since graduated.

Special thanks to Anna Draganova for the amazing job she did writing the Instructor Resources, and thanks to Helen Wong for taking over where Anna left off. We thank "Sideshow" Dave Perry for writing creative and challenging exercises. In addition to Sideshow Dave, we wish to thank Michael "Quimby" Krebs and Igor "Teper" Teper for co-instructing the CTY number theory course over the years and in so doing helping develop the ideas that are presented in this book and in the Instructor Resources. We thank Danielle Potvin Curran for her early support of this effort and for her invaluable advice on the design. We thank Jim Hollan for letting us meet in his lab's conference room. Thanks to Diana Deutsch for checking the section on the missing fundamental auditory illusion. Thanks to Kathy Sheldon for coming to the rescue many times by answering formatting questions when there was no one else to turn to, as well as for all of her other help and support over the years.

For their help writing solutions and/or hints for the exercises, we thank Pomona College students Nicolas Conway, Matt Holden, Mark Junod, Laura Pinzur, Sam Miner, Vincent Earl Selhorst-Jones, Sam Stromberg, Alan Tarr and Will Fletcher; as well as Scripps College students Ariel Bleicher, Elise Hanson, Priya Prasad, and Katherine Shultis; and Harvey Mudd College students Andrea Heald, Gregor Passolt, and Ted Spaide. Daniel Walton, also from Harvey Mudd College, helped write exercises on primality testing. Pomona students Michelle Vijverberg and Anita Cheng helped with the art manuscript, and Jen Hardee, Eita Hatayama, and Nathan Reed helped with formatting and layout. We are grateful to Pomona students Vincent Earl Selhorst-Jones and Will Fletcher for their help with detailed proofreading at the final stage.

We thank the many graduate students who have helped us with the book. In particular, thanks to Dwayne Chambers for writing solutions and proofreading a draft of some of the early chapters, as well as for technical support at various points along the way. Claire Weiss, Lars Bull, Youssef Francis, and Aaron "Dimby" Jones also contributed useful suggestions. We are grateful to America Chambers for writing hints to exercises and to Michelle Manes for writing many exercises and solutions. Thanks to Masood Aryapoor for writing the student hints for the vast majority of odd-numbered problems in the book.

We have also benefited from the input of many of our colleagues, including Art Benjamin, David Meyer, David Pengelley, David Perkinson, Shahriar Shahriari, Jerry Shurman, Irena Swanson, and Chris Towse.

We want to thank our original editor at Key College Publishing, Richard Bonacci, for taking on our project and believing in it and in us despite our slow pace of writing. We thank Key College development editors Allyndreth Cassidy and Annie Mac for their help and patience. We want to thank everyone at Wiley for taking over our project, starting with our editor at Wiley, Shannon Corliss. Thanks to Anna Melhorn for supervising the development of the cartoons, thanks to Sheena Goldstein for finding excellent photos to use in the text, thanks to our production editor Sandra Dumas, and to our designer James O'Shea. Thanks also to Lynn Lustberg at Macmillan Publishing Solutions for handling the copyediting, typesetting, and proofreading of our manuscript. Finally, we want to thank illustrator Jim Haynes for turning our sometimes cryptic textual descriptions into the vibrant cartoons that contribute so much to the book.

On a more personal level, Erica Flapan wants to thank her husband Francis Bonahon and daughter Laure Flapan for their love, support, encouragement, and patience throughout this project. Tim Marks would like to thank his mother and father and his wife and daughter for all of their support and their unconditional love throughout many years of writing this book. Jamie Pommersheim thanks his father for his love and for encouraging his mathematical talents from an early age. Dad, please do not immediately donate your copy of this book to the university in Nairobi; we'll make sure that they get one.

Prologue   Number Theory Through the Ages
or An Anachronistic Assembly

God made the integers;
all the rest is the work of humanity.

Leopold Kronecker

Many years ago, in a college or university much like your own but rather more like Cambridge University, a bell rang at precisely 2:00 pm. A white-haired British gentleman arose, cleared his throat, and began to speak.

G.H. Hardy: Ladies and Gentlemen, welcome to 1938. My esteemed colleagues, throughout history, each of you has made key contributions to number theory, the study of the positive integers—

$$1, 2, 3, \ldots$$

Number theory has developed a great deal since many of you departed, and I thought I would begin today's meeting by bringing you up to speed. You have all been interested in the prime numbers. The numbers

$$2, 3, 5, 7, 11, 13, 17, \ldots$$

are special because none of them can be factored into smaller numbers. But does this sequence follow a pattern? For example . . .

Euclid (interrupting in a nasal voice with an ancient Greek accent): Well, actually, in Book IX, Proposition 20 of my *Elements*, I showed that there are infinitely many primes.

Hardy: Yes, Euclid, my good sir. Your proof is well known to everyone in the room, except possibly the honorable Pythagoras, who is several centuries your senior. But is there a pattern to the infinite sequence of primes? Herr Gauss, perhaps you would like to start off our discussion?

Carl Friedrich Gauss (a mature man, wearing a tall, floppy hat): Well, as you examine larger and larger integers you find that there are fewer and fewer primes. On average, the primes become more spread out the farther you go. Of the numbers less than one hundred, 25% are prime, but of the numbers less than one million, only about 8% are prime. I even found a formula that approximates how many primes there are less than any given number. It's quite remarkable—it has to do with calculus and logarithms and such.

[Two nineteenth-century Frenchmen chime in.]

Jacques Hadamard: You'll be happy to hear that we proved your conjecture about this formula.

Charles de la Vallée Poussin: It's now called the Prime Number Theorem.[1]

Hardy: Thank you, Messieurs Hadamard and de la Vallée Poussin, for bringing Herr Gauss up to date on this matter. This is just the sort of exchange I hoped for. I invite all of you to join in. Let's continue. As you consider larger numbers, the gaps between primes tend to get larger, right?

Pythagoras: Between the primes 113 and 127 is a large gap consisting of 13 consecutive numbers, none of which are prime.

---

[1] The Prime Number Theorem is discussed in Section 3.3.

Euclid (makes a few quick scribbles on a scroll of papyrus, then looks up): And yet soon after that, we find two primes, 137 and 139, that are right next to each other.

Hardy (nodding in agreement): Yes, there is no obvious pattern regarding where there are large gaps and where the primes are close together. It appears to be somewhat random.

Leonhard Euler: Between the primes 2,614,941,710,599 and 2,614,941,711,251 is a large gap consisting of 651 consecutive numbers, none of which are prime.

Gauss (makes a few quick scribbles on a scrap of paper, then looks up): And yet soon after that, we again find two primes—2,614,941,711,749 and 2,614,941,711,751—right next to each other.

Euler (scratching his chin): I have often wondered whether there are infinitely many *twin primes*—prime numbers that differ by 2.

Hardy: Well, Herr Euler, it is still not known whether there are infinitely many twin primes. And I have a feeling this problem will remain unsolved until at least the beginning of the twenty-first century. We must revisit this at future meetings!

Gauss (leaning forward): What I am really curious about is, did anyone ever prove that infernal statement by Fermat?

[Fermat adjusts his white powdered wig as all eyes turn in his direction.]

Hardy: For those of you who predate Monsieur Fermat, I should tell you the story of what has become known as Fermat's Last Theorem[2], which is a bold claim about the following equation. *[Hardy goes to the chalkboard, picks up the chalk, and writes the equation]*

$$x^n + y^n = z^n$$

Pierre de Fermat (chuckling nervously): Ah, *that* equation. One of my favorites. I have *beaucoup* to say on this subject. But first I'd like to hear the story.

Hardy: Of course. Fermat claimed that there are no positive integer solutions to this equation . . .

Pythagoras (interrupting): My friends, when $n = 2$, that equation is $x^2 + y^2 = z^2$, which is my formula for the lengths of the sides of a right triangle. For example, if the two legs of a right triangle have lengths 3 and 4, then the hypotenuse must have length 5, because $3^2 + 4^2 = 5^2$. Similarly, $5^2 + 12^2 = 13^2$, and $8^2 + 15^2 = 17^2$. There are a good number of integer solutions to that equation.

Hardy: That is certainly true, Pythagoras. The equation $x^2 + y^2 = z^2$ has many integer solutions. But what about $x^3 + y^3 = z^3$, or $x^4 + y^4 = z^4$, and so on? Fermat claimed that there are no nonzero integer solutions to the equation $x^n + y^n = z^n$, for any $n > 2$.

---

[2] Fermat's Last Theorem is discussed in Sections 15.2–15.5.

Sophie Germain: In other words, Monsieur Fermat was saying that two cubes cannot add up to a cube. Likewise, two fourth powers cannot add up to a fourth power, and so on.

Hardy: Well put, Madame Germain. To continue the story, legend has it that one night around the year 1630, Pierre de Fermat wrote this claim in the margin of his copy of Diophantus' treatise *Arithmetica*. Fermat went on to write that he had found a remarkable proof, which the margin was too small to contain.

Euler (raising his hand): I proved Fermat's claim for $n = 3$.

Germain: I proved many things about that equation. In particular, my results imply that if there is a nonzero integer solution to the equation $x^n + y^n = z^n$, then n must be greater than 100, or at least one of x, y, and z must be divisible by n.

Ernst Kummer (with a stern expression and a large bushy mustache): I, too, spent many years working on this problem.

Hardy: Yes, Herr Kummer. You succeeded in proving Fermat's Last Theorem for many values of n. And more importantly, in the course of your attempts to solve this problem, you laid the foundations of an important new subject called algebraic number theory.

[Still standing at the chalkboard, Hardy looks with admiration at the roomful of eminent mathematicians.]

Hardy: It is striking how many of you have worked on this problem. Raise your hand if you have tried to prove Fermat's claim. *(Does a quick count)* As I thought—nearly all. Yet here we are in the middle of the twentieth century, and this problem remains unsolved despite all of your valiant efforts.

Fermat (looking dumbfounded): Excusez-moi? You have been working on this problem for over 300 years and you still have not proved it? That's amazing! Especially in light of what I'm about to tell you.

[Voices erupt from all around the room.]

So you did have a proof?

You didn't really prove it, did you?

Enlighten us, Pierre!

[Fermat stands up, adjusts his wig, and walks hesitantly toward the chalkboard. A hush fills the room. The Frenchman takes a deep breath, then begins to speak.]

Fermat: Hmm. . . . Very well, here is the truth. When I wrote that statement, I . . . uh. . . . Well, I. . . .

[Suddenly, a blinding light flashes and the room shakes violently. Out of a cloud of smoke steps a small group of people, dressed casually in the styles of the mid-to-late twentieth century and early 21st century, gasping to catch their breath.]

Julia Robinson: We got here as fast as we could!

Gauss (who hid behind his chair in the commotion, pokes his head out and eyes the newcomers nervously): Who are you?

Robinson: We're from the future. Sorry we're late. We had to make a stop in the early 1940s to help the Allies break the German Enigma code. Ever since then we've been a little behind schedule.

Hardy (steps toward the new arrivals and offers his hand): Welcome, travelers. I am happy you could be with us today. We were just discussing the conjecture known as Fermat's Last Theorem.

[A thin British man emerges from the group of newcomers.]

Andrew Wiles: Obviously, news of my breakthrough has not yet reached your decade. In 1995, I became the first person to give a complete proof of Fermat's Last Theorem.

[The assembled mathematicians burst into applause. Cries of "Bravo!" and "Well done!" are heard.]

Wiles (nods in acknowledgment, then looks around the room): Is Monsieur Fermat here? I would love to meet him.

Hardy: Yes, he's just over. . . . Hey, what happened to Pierre?

[Fermat is nowhere to be seen. In all the smoke and noise, he has vanished.]

Germain (shaking her head): It's a shame we'll never hear Fermat's side of the story, since so many of us worked so hard trying to prove his claim.

Hardy: Indeed, Madame Germain. But we have Mr. Wiles here, and perhaps later we can persuade him to share with us the details of his proof.

Wiles: I would be delighted, especially since my results are built on the mathematical foundations developed by many of the people in this room.

Hardy: Thank you, Andrew. And perhaps some of your colleagues from the future have new discoveries of their own to share with us.

[Three men step forward.]

Ron Rivest, Adi Shamir, and Leonard Adleman (speaking simultaneously): With your permission, Mr. Hardy, we would like to tell you about an important application of number theory that we have discovered.

Hardy: An application? Of number theory? But I always found it charming that my chosen field had no practical use. I studied number theory purely for its inherent beauty. I never imagined that this lofty subject would be applied to worldly pursuits.

Gauss: In my day, I discovered many applications of calculus to the real world. It seems perfectly natural that continuous functions and real numbers can be used to model real-world phenomena, such as the motion of planets and magnetic attraction. But number theory concerns only the integers, and the interesting relations that we find among them. I would be amazed if the study of prime numbers, for instance, had any practical value whatsoever.

Rivest, Shamir, and Adleman (again speaking simultaneously): Then prepare to be amazed, Herr Gauss! We have found a new method for sending secret messages securely[3], based on the properties of prime numbers that you all have discovered over the centuries. This method and others like it are integral to secure communication in the early twenty-first century. In fact, as the use of digital computers became widespread in the second half of the twentieth century, *many* important applications of number theory were developed.

Pythagoras (looking lost and confused): Computers? What are computers? Could someone please explain to me what these people are talking about?

Hardy: There will be plenty of time for explanation later. But now, it is time for tea. I hope you have all enjoyed this brief discussion of number theory through the ages. Please help yourself to port and walnuts.

---

[3]The RSA encryption algorithm is discussed in Section 9.5.

Chapter 1

SPOTLIGHT ON...

Pythagoras (569–ca. 500 BCE)

The ancient Greek mathematician Pythagoras lived in the Greek colony of Croton, which is now part of southern Italy. He traveled to Babylon and Egypt (and perhaps to India), where he learned a great deal about mathematics, philosophy, and religion. Pythagoras founded a secret brotherhood that was devoted to mathematical, philosophical, mystical, and moral studies. The members of the brotherhood were known as the Pythagoreans, and their primary symbol was the pentagram, a symmetric five-pointed star.

The Pythagoreans had strong spiritual beliefs and followed a strict monastic code. They believed that when a person died, his or her soul would transmigrate into an animal. They would not kill animals or wear wool, because the soul of one of their departed friends might be alive in the animal. The Pythagoreans were also known to "abstain from beans." However, what this actually meant is not known for certain. Some historians of mathematics have asserted that the Pythagoreans believed that human souls could also transmigrate to beans, so beans should not be eaten. Other historians have said that since beans were used to vote in Greek elections, the command to "abstain from beans" actually meant to abstain from politics.

Numbers were a key element of the mysticism of the Pythagoreans. Their motto was "All is number," by which they meant that the positive integers are the basis for everything, mathematical as well as nonmathematical. They gave the numbers human attributes, believing that the even numbers had female characteristics and the odd numbers had male

characteristics. Every number had particular characteristics associated with it. For example, 1 represented reason, 2 opinion, 3 harmony, 4 justice, 5 marriage, and so on.

A key element of the Pythagorean philosophy was their belief in the natural harmony of the universe. This idea was exemplified by the musical harmony produced by two vibrating strings, one twice as long as the other (i.e., with lengths in the ratio 2:1). Similarly, other simple ratios of string lengths, such as 3:2 and 4:3, also produce harmonious sounds. The early Pythagoreans' belief in the natural harmony of numbers was interconnected with their fundamental belief that the ratio of any two lengths is equal to a ratio of two integers. In other words, the early Pythagoreans believed that every length is a rational number.

It was well known to the Pythagoreans that $\sqrt{2}$ occurs as a length in geometry (e.g., as the diagonal of a 1×1 square). However, the early Pythagoreans did not know that $\sqrt{2}$ is an irrational number—that is, that $\sqrt{2}$ cannot be written as a quotient of two integers.

While no one knows who first discovered that $\sqrt{2}$ is an irrational number, it is believed that this discovery was made by one of the later Pythagoreans, after Pythagoras's death. Because the assumption that all lengths were rational had been at the heart of the philosophy of the Pythagoreans, the discovery of irrational lengths led to a crisis in the foundations of mathematics that caused the Pythagoreans to question all of their mathematical results. Plato wrote in his *Dialogues* that the entire Greek mathematical establishment was shocked to learn of the existence of irrational numbers.

The Pythagoreans were allied with the aristocracy of Croton. As a result, when a democratic movement arose around 500 BCE, the people of Croton became hostile to Pythagoras and forced him to leave town in haste. Pythagoras moved to Metapontum and died shortly afterward. Later, around 450 BCE, a democratic revolution took place in which the people of Croton attacked the aristocracy. The revolutionaries burned down the Pythagoreans' buildings, and the Pythagoreans scattered.

A Bit of Pythagoras's Math

Because of the secrecy of the Pythagorean brotherhood, it is not possible to know which discoveries Pythagoras made himself and which were made by other Pythagoreans. The most important contribution of Pythagoras was transforming mathematics from a collection of practical numerical methods into an intellectual discipline, worthy of studying for its own sake. While number was everything to the Pythagoreans, their interests were not restricted to number theory. They found close connections between the study of integers and the study of plane geometry.

The Pythagorean Theorem was not discovered by Pythagoras but was already known to the Babylonians. Many historians believe that this theorem became known as the Pythagorean Theorem because Pythagoreans were the first to write down a rigorous proof of it. However, even this is not known for certain. Since the time of Pythagoras, at least 370 different proofs of the Pythagorean Theorem have been given. The following is believed to be the proof of the Pythagorean Theorem that was given by the Pythagoreans.

Consider a right triangle with legs a and b and hypotenuse c. Draw the square with sides of length $a + b$ and the blue triangles as indicated in the following diagrams.

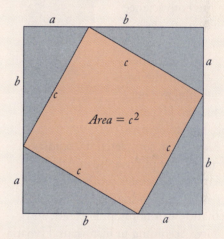

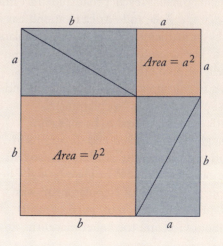

Now compare these two diagrams. Since the total blue area in the top diagram equals the total blue area in the bottom diagram, the total orange area in the top diagram must equal the total orange area in the bottom diagram. We conclude that $a^2 + b^2 = c^2$. (In Exercise 12 of Section 1.1, you will provide the details of this proof.)

Hypatia (370–415 CE)

Hypatia, the first well-known woman mathematician in the historical record, contributed to both science and mathematics as a teacher, a writer, and an inventor. Her father, Theon, was a mathematician and philosopher who educated Hypatia himself. Hypatia grew up in Alexandria, completed her studies of mathematics and philosophy in Athens, and eventually became a professor at the University of Alexandria, where she taught mathematics, astronomy, philosophy, and mechanics. According to historians, Hypatia was such a popular teacher that students came from Europe, Asia, and Africa to attend her lectures on mathematics. In addition, scholars came to her house to discuss science, philosophy, religion, and other intellectual matters.

As a strong supporter of the religion and philosophy known as Neoplatonism, Hypatia was considered to be a pagan. Although Neoplatonism had its origins in the rational philosophy of Plato, it also had a religious aspect that Platonism lacked. The religious philosophy of Neoplatonism was based on a combination of mystical intuition about God and the belief that ultimate truth is unknowable. One of the maxims of Neoplatonism was "The Absolute has its center everywhere but its circumference nowhere." Many Christians of the time saw Neoplatonism as a threat. However, Hypatia had Christians among her philosophy students who regarded her very favorably. In fact, Synesius of Cyrene, who was one of the first scholars to write about the Christian concept of the Trinity, used Neoplatonist ideas that he had learned from Hypatia in his writings about Christianity.

Hypatia lived in a tumultuous time in the Roman Empire when sectarian violence between Christians and pagans was common. The political climate became particularly hostile to Hypatia's views when Cyril became Patriarch of Alexandria in 412. Under Cyril's leadership, the country began persecuting all non-Christians. This included astrologers and numerologists, who claimed to be able to see into the future. Since people generally knew little about mathematics, they were unable to distinguish between mathematicians and numerologists. In fact, the emperor Constantius had made a law stating that "no one may consult a soothsayer or a mathematician." Hypatia was a popular figure who was outspoken about her politics, her Neoplatonism, and her belief in mathematics and the scientific rationalism of the Greeks. She refused to convert to Christianity despite being advised to do so by her close friend Orestes, the prefect of Egypt. In 415, Hypatia was brutally murdered by a mob of fanatical Christian zealots.

A Bit of Hypatia's Math

Hypatia wrote several mathematics textbooks for her students. Her most well-known text was a commentary on the book *Arithmetica*, which had been written around 250 CE by the ancient Greek mathematician Diophantus (see the history of Diophantus at the beginning of Chapter 5). Hypatia's commentary provided many new problems as well as alternative solutions to existing problems, which were incorporated into later versions of *Arithmetica*. She also wrote a book entitled *The Conics of Appolonius*, in which she explained Appolonius's method of using conic sections to understand the orbits of the planets, and she created tables of the movements of the heavenly bodies that she entitled *Astronomical Canon*. Hypatia and her father, Theon, wrote a treatise on Euclid, and it is believed that she helped her father revise Euclid's *Elements* into its current form.

In addition to the books she wrote, Hypatia designed a variety of scientific instruments, including one to measure the positions of the planets and stars, one to distill water, one to measure the level of water, and one to measure the specific gravity of a liquid.

Chapter 1 Numbers, Rational and Irrational
or The Greek System

Called "the queen of mathematics" by Carl Friedrich Gauss, number theory is the study of the natural numbers,

$$1, 2, 3, 4, 5, \ldots$$

the most basic of all number systems. Upon close examination, the apparent simplicity of the natural numbers gives way to startling complexity. The beautiful patterns and theorems that emerge have fascinated many of the greatest mathematical minds throughout the centuries. Yet, a vast array of problems in number theory remain unsolved for current and future generations to explore. In the last few decades, fundamental ideas of number theory have found practical applications that have become essential to the very fabric of modern society. In the chapters that follow, we will delve into many of the great ideas, fundamental theorems, and important applications of number theory.

1.1 Numbers and the Greeks

In our quest to discover the secrets of the natural numbers, we will encounter many other number systems such as the real numbers, the complex numbers, and more. Our modern concept of number includes an amazing variety of number systems, but the ancient Greek notion of number was much more limited. To the ancient Greeks, numbers were only used to count objects and to represent geometric quantities such as length, area, or volume. In the time of Pythagoras, it was believed that any geometric quantity could be represented by a rational number (i.e., as the quotient of two integers). As we will see, this turns out not to be true.

A pair of protracted Panhellenic pathways

The ancient Greek mathematician Pythagoras is credited with the idea that results in mathematics must be proved from a list of axioms using formal reasoning. In addition to his mathematical work, Pythagoras was the leader of a secret society that lived by a strict set of rules based on mathematics, philosophy, and mysticism. What's not so well known about[1] Pythagoras

---

[1]The phrase "what's not so well known about ..." indicates that the information that is about to follow is a *math myth*, a story that has no historical basis but serves to illustrate a mathematical point.

is that he was the founder of the famous Beta Rho Omega fraternity at Croton University. The people of Croton did not like the idea of having fraternities and sororities in their town and had a burning desire to get rid of them. However, the fraternities and sororities felt safe because the Greek system had been in place in Croton for many years.

It was rush week, and Pythagoras had a serious case of senioritis. His grade point average had dropped to delta minus, but he just didn't seem to care. The only thing on his mind these days was Beta Rho Omega (BPΩ). Pythagoras had an acute desire to attract the right crowd to join his fraternity, and he felt that a new walkway in front of the frat house would help. He and his frat brothers constructed a 7-meter-long walkway using seven square tiles that were 1 meter on each side.

Next door to the fraternity house stood the house of the sorority Sigma Iota Sigma (ΣΙΣ). Its president, an ambitious young woman named Hypatia, was working on a quadruple major in math, philosophy, physics, and astronomy. When Hypatia and her sorority sisters saw that the guys next door were building a walkway, they decided to build one of their own. The sisters also used square tiles that were 1 meter on each side. However, they used just five tiles, arranging them point to point to make their walkway more decorative.

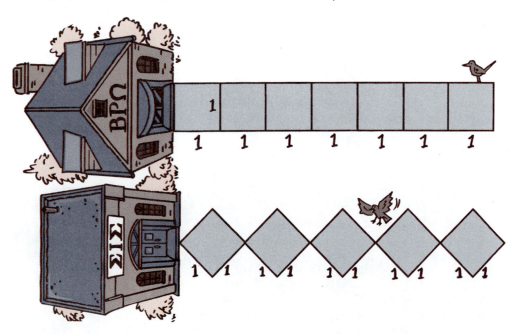

The next morning, Pythagoras looked out his window and noticed an inequality. The sorority's walkway was slightly longer than his fraternity's. He and his frat brothers laid down a few more tiles to make their walkway longer. Hypatia would not stand for this, so the sisters laid down a few more tiles to make their own walkway the longest once again. The dean of the university was not at all pleased with this escalating battle, and he called a meeting of the Greek council that night to resolve the issue once and for all.

The dean entered the assembly hall, and the students followed him in two columns. "I have heard your statements and your reasons," the dean began. "I have decided that the only fair solution is for the two walkways to be of equal length. This ruling encompasses construction of both walkways, and I have hired a contractor to extend both walkways until they reach precisely the same length." Pleased with his solution, the dean smiled broadly as he left the room. He often thought of himself as quite sharp, but as usual the students found him rather obtuse.

Construction began early the next morning. After each new tile was laid, the crew observed that one walkway or the other was longer, and so more tiles had to be added in an attempt to even out the walkways. By the middle of the next week, the walkways from the houses ran across the road to the side opposite, over the adjacent cornfields, and into the center of campus, where the university's main quadrangle was bisected by the construction.

Looking out his office window, the dean felt helpless as the two tiled walkways overran his beautiful campus. He had no idea how long it would take before his decree was fulfilled. Just then Hypatia burst into the room.

"Dean," exclaimed Hypatia, "I was just in the middle of a long walk when I had a sudden realization. No matter how long the construction continues,

the two walkways will never be exactly the same length. And I can prove it!" Hypatia showed the dean her proof, and he ordered an immediate halt to the construction.

"As a result of your brilliant realization, Hypatia, I award to your sorority, Sigma Iota Sigma, the right to have the longest walkway." Thanks to Hypatia's insight, the beautiful campus of Croton University was saved, and the sisters celebrated by inviting all of the frat brothers over for a huge party. They even invited the dean, who showed up with a keg of root beer.

In our myth, Hypatia saved the day by realizing that the two walkways could never be exactly the same length. To see how she figured this out, suppose that the fraternity's walkway had m tiles, and the sorority's walkway had n tiles.

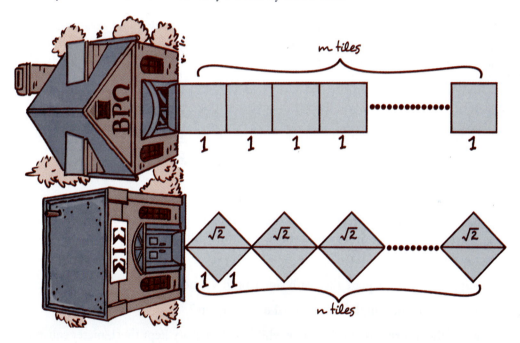

The fraternity's walkway is m meters long, while the sorority's walkway is $n\sqrt{2}$ meters long. In order for the two walkways to have the same length, we would need to have

$$m = n\sqrt{2}.$$

This equation implies that

$$\sqrt{2} = \frac{m}{n},$$

where m and n are both positive integers. But as the ancient Greeks knew (and as we will prove in Section 1.4), the number $\sqrt{2}$ is irrational—it cannot be written as the quotient of two integers. Thus, the two walkways could never have the same length.

Numerical Problems

1. The Pythagorean Theorem asserts that in a right triangle with legs of lengths a and b and hypotenuse of length c, $a^2 + b^2 = c^2$.

    a. Find two examples of right triangles in which a, b, and c are all positive integers.

    b. Find two examples of right triangles in which a and b are both positive integers, but c is not.

    c. Suppose both legs of a right triangle have the same length, a. Write an expression for the length of the hypotenuse, c.

2. At the beginning of our myth, when the fraternity's walkway had seven tiles and the sorority's had five tiles, approximately what was the difference between the lengths of the two walkways?

3. In our myth, if the sorority's walkway had 33 tiles, what is the smallest number of tiles the fraternity would need to have to surpass them in length?

4. In Exercise 2, you found that the lengths of the two walkways are pretty close.

    a. Find positive integers m and n such that the fraternity walkway with m tiles and the sorority walkway with n tiles differ in length by less than 0.05 meter.

    b. Find positive integers m and n such that the fraternity walkway with m tiles and the sorority walkway with n tiles differ in length by less than 0.02 meter.

5. Suppose you had square tiles with side length $\sqrt{2}$ meters.

    a. What is the length of a diagonal of this square?

    b. If the sorority used this square while the fraternity kept the 1-meter squares, as in the myth, could they make walkways of equal lengths? Give an example, or explain why not.

    c. If the fraternity used this square while the sorority kept the 1-meter squares, as in the myth, could they make walkways of equal lengths? Give an example, or explain why not.

Reasoning and Proofs

6. Assume that Pythagoras and his brothers had formed their walkway with equilateral triangles of side length 8, and Hypatia and her sisters had formed their walkway with regular hexagons of side length 14, as shown in the diagram.

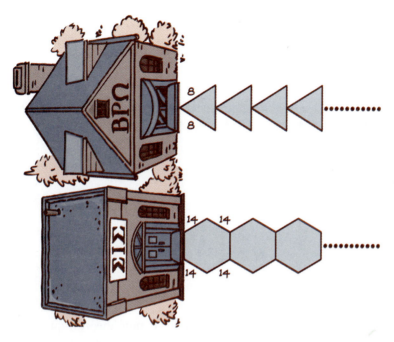

Could the dean's decree be carried out now? If so, what is the shortest possible equal walkway length? Explain your answer.

7. Assume that Hypatia and her sisters had formed their walkway as in our story, but Pythagoras and his brothers had formed their walkway using regular octagons of side length 1.

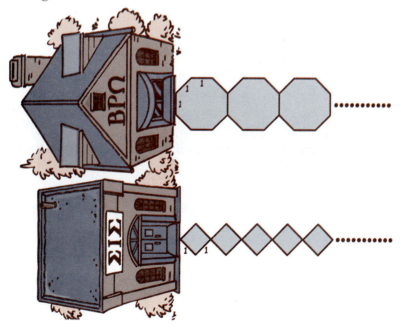

Could the dean's decree be carried out now? Explain.

EXPLORATION Figurate Numbers (Exercises 8–10)

Pythagoras was interested in **figurate numbers**, which are numbers that can be represented as geometric figures, such as the triangular numbers:

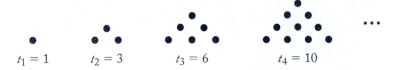

$t_1 = 1$ $t_2 = 3$ $t_3 = 6$ $t_4 = 10$

or the square numbers:

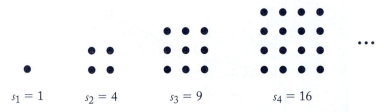

$s_1 = 1$ $s_2 = 4$ $s_3 = 9$ $s_4 = 16$

We let t_n denote the nth triangular number and s_n denote the nth square number.

8. **a.** Calculate t_5, t_6, and t_7.

  b. Given that $t_{1000} = 500500$, calculate t_{1001} as simply as possible.

9. It is easy to see pictorially that a square number can always be decomposed into the sum of two consecutive triangular numbers. For example:

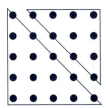

  a. Use this diagram to write a formula for s_5 in terms of triangular numbers.

  b. Generalizing your answer to part a, write a formula for an arbitrary square number s_n (where n is a positive integer) in terms of triangular numbers.

10. Placing two of the same triangular numbers together gives a rectangle that is almost a square. For example:

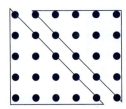

a. Use this diagram to write an equation involving t_5.

b. Generalizing your answer to part a, write an equation involving an arbitrary triangular number t_n (where n is a positive integer).

c. Use your answer to part b to give a formula for t_n.

d. Sometimes we can take points that are arranged in a rectangle and rearrange them into a square. For example, a 4×9 rectangle can be rearranged into a 6×6 square. Rectangles like the one shown in the diagram, however, cannot be rearranged into a square. Prove this fact—that is, prove that for any positive integer n, $2t_n$ is not a perfect square. [*Hint:* Your answer to part c may be helpful.]

11. Assume that you have a rectangular box with length l, width w, height h, and diagonal d (from one corner of the box to the opposite corner of the box), as shown in this diagram:

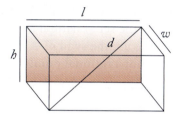

a. Give a formula for d in terms of l, w, and h.

b. Find a box for which l, w, h, and d are all integers. What are the values of l, w, h, and d?

c. Answer the following questions, which exhausted the ancient Greek mathematician Eudoxus:

   i. Given a cube, by what factor must you increase the side length so that the length of the diagonal is doubled?

   ii. By what factor must you increase the side length so that the volume of the cube is doubled?

12. "A Bit of Pythagoras's Math" at the beginning of this chapter offers a sketch of a proof of the Pythagorean Theorem. Write out this geometric proof in detail.

1.2 Numbers You Know

Of all of the types of numbers, the most basic are the numbers we use for counting: 1, 2, 3, 4, 5, ... These numbers, called the **natural numbers**, are the main focus of number theory. The set of all natural numbers is denoted as follows:

$$\mathbf{N} = \{1, 2, 3, 4, 5, \ldots\}. \qquad \leftarrow \textbf{Natural numbers}$$

Those of you with student loans know that not all numbers are positive. The **integers,**

$$\mathbf{Z} = \{\dots, -3, -2, -1, 0, 1, 2, 3, \dots\}, \qquad \leftarrow \textbf{Integers}$$

consist of the natural numbers and their negatives, together with 0. The symbol **Z** stands for *Zahlen*, which is German for "numbers."

The terms *positive integer* and *natural number* are used interchangeably. We often use the term *nonnegative integers* to refer to the set of integers greater than or equal to 0.

Quotients of integers, such as $\frac{3}{5}$ and $-\frac{17}{2}$, are called **rational numbers**. We define the set of rational numbers by

$$\mathbf{Q} = \left\{ \frac{a}{b} \,\middle|\, a, b \in \mathbf{Z}, b \neq 0 \right\}. \qquad \leftarrow \textbf{Rational numbers}$$

For help reading these symbols, look in the following notation notebook.

We use the symbol $\in$ to mean "is an element of." For example, $\mathbf{2 \in N}$ means that the number 2 is an element of the set of natural numbers; in other words, **2 is a natural number**. Similarly, we use the symbol $\notin$ to mean "is not an element of." For example, $\mathbf{\sqrt{2} \notin Q}$ means $\sqrt{2}$ **is not a rational number**.

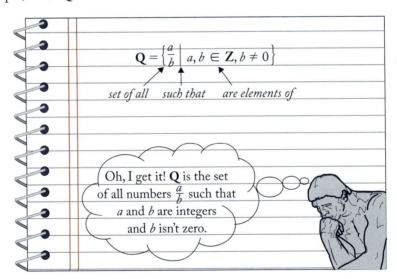

We have now introduced three sets of numbers, **N**, **Z**, and **Q**. Every natural number is an integer, and every integer is a rational number. But there are other numbers that are not rational. The Pythagoreans proved that irrational lengths appear often in geometry. Even the length of the diagonal of a 1×1 square, $\sqrt{2}$, cannot be expressed as the ratio of two integers (as will be proved in Section 1.4). To include all possible lengths, a larger number system is needed.

The set of **real numbers**, denoted by **R**, consists of all lengths and their negatives. Real numbers can be visualized as points on a number line. The following diagram shows the number line with a few of the real numbers labeled.

R is the set of all points on the number line.

The real numbers can also be thought of as the set of all finite and infinite decimals (both repeating and nonrepeating). For example, the finite decimal -47.782, the repeating decimal $0.283838383\ldots$, and the nonrepeating decimal $\sqrt{2} = 1.41421356237309\ldots$ are all real numbers. Rational numbers can always be written using finite or repeating decimals, while irrational numbers can only be written using nonrepeating decimals. In addition to the square roots of some integers, there are plenty of other irrational numbers. For example, the number π is irrational, a fact suspected by the ancient Greeks but not proved until many centuries later. (Johann Heinrich Lambert gave the first proof in 1761.) Defining the real numbers precisely is not an easy task; it belongs to the subject of *real analysis* rather than number theory.

The real numbers can be used to build the **complex numbers**:

$$\mathbf{C} = \{a + bi \mid a, b \in \mathbf{R}\},$$

where i is a square root of -1.

The sets **R** and **C**, as well as other number systems appearing in this book, are useful in number theory. However, the central concern of number theory is the set of natural numbers, **N**, as well as its close cousins **Z** and **Q**.

EXERCISES 1.2

Numerical Problems

1. **a.** Evaluate $(-i)^2$.

   b. Find two different complex numbers each of which square to give -4.

   c. Find a complex number whose square is i.

2. Using the rule $i^2 = -1$ and ordinary arithmetic, practice calculating with complex numbers. The letters a and b represent real numbers.

   a. $(3 + 2i) + (4 - i)$

   b. $(3 + 2i) - (4 - i)$

c. $\dfrac{1}{4-i}$ [*Hint:* Multiply both numerator and denominator by $4 + i$.]

d. $(a + bi) - (a - bi)$

e. $(a + bi) + (a - bi)$

f. $(a + bi)(a - bi)$

g. $\dfrac{1}{a+bi}$ [*Hint:* Multiply both numerator and denominator by $a - bi$.]

3. For each of the following numbers, tell whether it is a natural number, integer, rational number, real number, and/or complex number (list all that apply).

   a. 47.3

   b. $2\sqrt{2}$

   c. -67

   d. 53.53535353535...

   e. 0.101001000100001000001... (continue this pattern, with more and more 0s between each 1 and the next 1)

   f. $\dfrac{23}{5}$

   g. $3i$

4. a. Write $\dfrac{1}{9}$ as a repeating decimal. [*Hint:* Use long division.]

   b. Use your answer to part a to write $\dfrac{2}{9}$ and $\dfrac{5}{9}$ as repeating decimals.

5. Write $\dfrac{1}{7}$ as a repeating decimal. [*Hint:* Use long division.]

6. For this exercise, you may need the formula for the sum of an infinite geometric series:

$$a + ar + ar^2 + ar^3 + \cdots = \frac{a}{1-r}, \text{ if } |r| < 1.$$

   a. Represent the repeating decimal 0.2222... as a ratio of two integers.

   b. Represent the repeating decimal 42.2888... as a ratio of two integers.

Reasoning and Proofs

7. Explain why every integer is a rational number.

8. **a.** Explain why every real number is a complex number.

   b. Explain why every pure imaginary number (i.e., a number of the form ci, where c is real) is a complex number.

9. Prove that $0.99999999\ldots = 1$. [*Hint:* You may need the formula given in Exercise 6.]

1.3 A First Look at Proofs

In the sciences and social sciences, experiments and observations are used as justifications for theories. In mathematics, however, these are not enough to justify a claim. Sometimes a mathematical statement that seems clearly to be true turns out to be false. For example, the early ancient Greeks believed that every length can be represented by a rational number, but this is not true (as we will show in Section 1.4). The only way to know for sure that a mathematical statement is true is to prove the statement rigorously. Furthermore, once a mathematical statement is proven, we know it is true for all eternity.

Not only is mathematical proof fundamental to the notion of mathematical truth, but it is also the heart and soul of mathematical beauty. Learning how to write a mathematical proof is an important part of learning mathematics. There is no algorithm for how to write a proof, just as there is no algorithm for how to write a novel. Experience and creativity each play a role in constructing a proof. The first step in learning to write proofs is to understand some proof techniques.

Direct proof

Often we prove statements *directly*: We assume the hypotheses and follow a logical sequence of steps that culminates in the conclusion. In the course of our argument, we may make use of the hypotheses, the definitions of the terms involved, and any previous results that are relevant.

As a first example of a direct proof, we will prove a familiar fact about odd integers: The product of any two odd integers is an odd integer. An integer n is said to be *odd* if there exists an integer k such that $n = 2k + 1$. An integer n is said to be *even* if there exists an integer k such that $n = 2k$.

LEMMA 1.3.1

Let m and n be integers. If both m and n are odd, then mn is odd.

Like many mathematical statements, this lemma has the form

If P then Q.

This statement can also be expressed as **P implies Q**, which is often written **P ⇒ Q**. A statement of this form is called a *conditional statement*. In the conditional statement **if P then Q**, we say that **P** is the *hypothesis*, and **Q** is the *conclusion*.

How might we go about proving a conditional statement? The most direct way to prove the conditional statement **if P then Q** is to begin by assuming that **P** is true, and then write down a logical sequence of steps that culminates in the assertion that **Q** is true. This is illustrated in the proof of the lemma:

PROOF OF **LEMMA 1.3.1**

Let m and n be odd integers.   ← **Hypothesis**

[**To show:** mn **is odd**]

We know that there exist integers q and r such that

$$m = 2q + 1 \quad \text{and} \quad n = 2r + 1.$$

We multiply m and n together to get

$$mn = (2q + 1)(2r + 1)$$
$$= 4qr + 2q + 2r + 1. \tag{1}$$

Equivalently,

$$mn = 2(2qr + q + r) + 1. \tag{2}$$

Since $2qr + q + r$ is an integer, we have written mn as 2 times an integer plus 1. Thus, mn is odd.   ← **Conclusion**   ■

Lemma 1.3.1 makes an assertion about *any* pair of odd integers. We began our proof by choosing two arbitrary odd integers. We introduced these new numbers using the magic word *Let* in the sentence "Let m and n be odd integers." Since m and n were chosen to be *arbitrary* odd integers, once we showed that mn was odd, we could conclude that the product of *any* two odd integers is odd.

The symbol ■ is used here, and throughout the book, to indicate that a proof is complete. Sometimes, mathematicians end proofs with the initials Q.E.D., which stands for the Latin phrase *quod erat demonstrandum*, meaning "that which was to be shown."

Another technique that is sometimes useful is to consider different cases separately. We use proof by cases in the following example. Here and throughout the chapter, we will assume that every integer is either even or odd, but not both. Although here we assume this property, it can in fact be proved as a result of more basic properties. If this bothers or intrigues you, see the Appendix (on the Student Companion Website) for a more axiomatic approach that includes this proof.

EXAMPLE 1

Prove that for every integer n, the integer $n^2 + n$ is even.

PROOF Let n be an integer. [**To show:** $n^2 + n$ **is even.**]

Either n is even, or n is odd. We consider these two cases separately.

Case 1 Suppose n is even.

Then there exists an integer k such that

$$n = 2k.$$

Multiplying both sides of this equation by $n + 1$ gives

$$n(n + 1) = 2k(n + 1),$$

or, equivalently,

$$n^2 + n = 2k(n + 1).$$

Thus, we have expressed $n^2 + n$ as 2 times the integer $k(n + 1)$. Hence, $n^2 + n$ is even.

Case 2 Suppose n is odd.

Then there exists an integer k such that $n = 2k + 1$. Thus,

$$n + 1 = 2k + 2,$$

or

$$n + 1 = 2(k + 1).$$

Multiplying both sides of this equation by n yields

$$n(n + 1) = n \cdot 2(k + 1),$$

or

$$n^2 + n = 2n(k + 1).$$

Thus, we have expressed $n^2 + n$ as 2 times the integer $n(k + 1)$. It follows that $n^2 + n$ is even.

We conclude that for every integer n, the integer $n^2 + n$ is even. ◼

The result we just proved in Example 1 makes an assertion about *every* integer. We began our proof with the phrase "Let n be an integer," to indicate that we were choosing an arbitrary integer n. Once we had shown that $n^2 + n$ is even for the arbitrary integer n, we could conclude that for every integer n, the number $n^2 + n$ is even.

Indirect proof

While direct proof is the most basic type of proof, it is not the only proof technique that we will use. Some statements are more easily proven *indirectly*. Indirect reasoning is so fundamental that we often use it in our everyday lives without even realizing it. Here's an example.

Professor Josephine Bean is well known at a university on the East Coast for her energetic teaching style. She enthralls students with the joys of number theory, darting from one side of the blackboard to the other with a piece of chalk in her left hand and a mug full of coffee in her right. In fact, people say that from the time Dr. Bean arrives at the Math Department in the morning and takes a mug of coffee, she is never without her coffee until she goes home in the evening.

Last year on Dr. Bean's birthday, her students decided to surprise her by filling her office with balloons. The students snuck in early in the morning with the balloons, hoping to arrive before she did. They wanted to be absolutely certain that Dr. Bean was not there. They went to the Math Department office to ask Pat, the administrative assistant, but Pat had stepped out. While the students waited impatiently for Pat to return, one student decided to help herself to a mug of coffee, and she noticed that every hook on the wall had a mug hanging from it.

Suddenly she turned to her friends and exclaimed, "Dr. Bean has not arrived yet. Let's go put the balloons in her office!"

How did the student know that Dr. Bean was not there? This was the student's reasoning: If Dr. Bean were in her office, she would have a mug of coffee with her, so at least one of the departmental mugs would be gone from its hook. But all of the mugs were on their hooks, so Dr. Bean could not be in her office.

Let's write this indirect argument more formally.

THEOREM. *Dr. Bean has not arrived yet.*

PROOF (By contradiction.)

Assumption: Suppose that Dr. Bean has already arrived.

Then she has already taken a mug of coffee. So, there must be a mug missing from at least one of the hooks. However, every hook has a mug. This is a contradiction. $\Rightarrow\Leftarrow$

Therefore, our Assumption must have been false. We conclude that Dr. Bean has not yet arrived. ●

The method of *indirect proof*, also known as *proof by contradiction*, is an important mathematical technique.[2] In an indirect proof, we make the assumption that the statement we are trying to prove is false and then derive a contradiction. When we reach a contradiction, we know that our original assumption must have been false. We can conclude that the statement we are trying to prove is true. In an indirect proof, we use the symbol $\Rightarrow\Leftarrow$ to indicate that we have reached a contradiction.

---

[2] Proof by contradiction is also known by its Latin name, *reductio ad absurdum*.

The method of proof by contradiction is often used to prove *negative* statements. For example, the statement "Dr. Bean has not arrived yet" is a negative statement because it asserts that something is not the case.

EXAMPLE 2

Prove that there is no smallest positive rational number.

SOLUTION Because this is a negative statement, we may want to prove it by contradiction. ●

PROOF (By contradiction.)

Assumption: Suppose there is a smallest positive rational number, q.

Let $r = \dfrac{q}{2}$. Then $0 < r < q$.

Since q is a rational number, there exist m and n such that $q = \dfrac{m}{n}$ and $n \neq 0$.

Then $r = \dfrac{m}{2n}$. Since m and $2n$ are integers and $2n \neq 0$, by definition r is a rational number.

We have now shown that r is a positive rational number that is less than q. This contradicts our Assumption that q is the smallest positive rational number. $\Rightarrow\Leftarrow$

Hence, there is no smallest positive rational number. ●

Example 1 presented a proof by cases. Sometimes a proof involves cases that are virtually identical. In such a situation, it is sufficient to prove the result for only one of the cases. To do this, we use the phrase *without loss of generality*. The following example illustrates this method.

EXAMPLE 3

Let m and n be integers. Prove that if mn is odd, then both m and n are odd.

PROOF Let m and n be integers such that mn is odd.   ← **Hypothesis**

Assumption: Suppose that m and n are not both odd.
Thus, either m is even, or n is even.   (3)
Without loss of generality, assume m is even.
Then there is some integer k such that

$$m = 2k.$$

Multiplying both sides of this equation by n gives

$$mn = 2kn.$$

Since kn is an integer, this implies that mn is even. This contradicts the hypothesis that mn is odd. $\Rightarrow\Leftarrow$

Since we have reached a contradiction, our Assumption (that m and n are not both odd) must be false.

Hence, both m and n are odd.   ← **Conclusion**   ■

At line (3) in this proof, we knew that either m was even or n was even. At this point, we could have considered one case where m is even and a second case where n is even. However, these two cases would be identical except that the roles of m and n would be switched. After proving the result for Case 1, it would be boring to write a proof for Case 2 and even more boring to read it. Instead, we used the phrase *without loss of generality* to indicate that the two cases are analogous. If we can do the proof for one case, then we can do it for the other case. Using the phrase *without loss of generality* allowed us to prove the result by considering a single case.[3]

The converse

The **converse** of any conditional statement is obtained by exchanging the hypothesis and the conclusion to form a new conditional statement. The converse of the statement

If P then Q   ← **Statement**

is the statement

If Q then P   ← **Converse**

Written using implication arrows, the converse of the statement $\mathbf{P} \Rightarrow \mathbf{Q}$ is the statement $\mathbf{Q} \Rightarrow \mathbf{P}$. Consider the following conditional statement about any number n:

If n is a positive even integer, then $n \geq 2$.   (4)

The converse of this statement is

If $n \geq 2$, then n is a positive even integer.   (5)

Note that a statement and its converse have different meanings and are not, in general, logically equivalent. In this example, statement (4) is true, but its converse, statement (5), is false. (For instance, $3 \geq 2$, but 3 is not a positive even integer.)

We have just seen an example of a statement that is true but whose converse is false. It is also possible for a conditional statement and its converse to both be true. For instance, we proved in Example 3 that for any two integers m and n,

If mn is odd, then both m and n are odd.   (6)

---

[3]Since the phrase *without loss of generality* is used so often in mathematics, it is sometimes abbreviated by the first letter of each word as WLOG.

As we proved in Lemma 1.3.1, the converse of (6) is also true:

If both m and n are odd, then mn is odd. (7)

Statement (6) and its converse, statement (7), can be combined into a single statement:

mn is odd *if and only if* both m and n are odd.

In general, a sentence of the form

P *if and only if* Q

means that both of the following statements are true:

If P then Q and **If Q then P.**

An *if and only if* statement, also called a *biconditional* statement, gives you two conditional statements for the price of one. The statement **P if and only if Q** can be written using implication arrows as **P ⇔ Q**.

To prove **P if and only if Q**, we need to prove two statements: **if P then Q** and **if Q then P**. When proving a statement of the form **P if and only if Q**, we write [⇒] to indicate that we are about to prove **if P then Q**, and we write [⇐] to indicate that we are about to prove **if Q then P**. The following lemma provides an example.

LEMMA 1.3.2

Let n be an integer. Then n^2 is even if and only if n is even.

PROOF Let n be an integer.

[⇒] Suppose that n^2 is even.

[We must show that n is even. We will do this by contradiction.]
Assumption: Suppose that n is not even.

Thus, n must be odd. Then $n^2 = n \cdot n$ is the product of two odd integers; hence, n^2 is odd, by Lemma 1.3.1. This contradicts the hypothesis that n^2 is even. ⇒⇐
Since we reached a contradiction, the Assumption must be false. Hence, n is even.

[⇐] Now suppose that n is even. **[To show: n^2 is even.]**

Since n is even, there exists an integer k such that

$$n = 2k.$$

Multiplying both sides of this equation by n gives

$$n^2 = 2kn.$$

Since kn is an integer, n^2 is even. ∎

The contrapositive

We have just seen how to prove statements by contradiction. A related way to prove a conditional statement is to prove the *contrapositive* of the statement. The **contrapositive** of the statement **if P then Q** is the statement **if (not Q) then (not P)**, which is logically equivalent.

$$P \Rightarrow Q \qquad \leftarrow \textbf{Statement}$$

$$(\textbf{not Q}) \Rightarrow (\textbf{not P}) \qquad \leftarrow \textbf{Contrapositive}$$

For example, consider the following statement about your friend Phil Lovett.

$$\text{If } \underbrace{\text{Phil parties until 3 AM Sunday night}}_{P}, \text{ then } \underbrace{\text{Phil looks tired Monday}}_{Q}. \qquad (8)$$

Here **P** is the statement "Phil parties until 3 AM Sunday night," and **Q** is the statement "Phil looks tired Monday." The contrapositive of (8), **If (not Q) then (not P)**, is the statement

$$\text{If } \underbrace{\text{Phil does not look tired Monday}}_{\textbf{not Q}}, \text{ then } \underbrace{\text{Phil did not party until 3 AM Sunday night}}_{\textbf{not P}}. \ (9)$$

Statement (9) is logically equivalent to statement (8).

EXAMPLE 4

Let x be a positive real number. Prove that if x is irrational, then is $\sqrt{x}$ irrational.

SOLUTION We wish to prove the statement

$$\text{If } x \text{ is irrational, then } \sqrt{x} \text{ is irrational.} \qquad (10)$$

We will prove statement (10) by proving its contrapositive:

$$\text{If } \sqrt{x} \text{ is rational, then } x \text{ is rational.} \qquad (11)$$

Statement (11) is a conditional statement, which we will prove directly. ●

PROOF OF STATEMENT (11)

Let x be a positive real number. Assume that $\sqrt{x}$ is rational.

[To show: x is rational.]

Since $\sqrt{x}$ is rational, there exist integers p and q such that

$$\sqrt{x} = \frac{p}{q}.$$

Squaring this equation yields

$$\sqrt{x} = \frac{p^2}{q^2}.$$

Thus, x can be written as the quotient of the two integers, p^2 and q^2. We conclude that x is a rational number. ∎

We have just proven that for any positive real number x, statement (11) is true. It follows that statement (10) must also be true for any positive real number x.

Understanding why the contrapositive is equivalent

Example 4 illustrates that to prove the conditional statement $\mathbf{P} \Rightarrow \mathbf{Q}$, it is sometimes easier to prove its contrapositive, **(not Q)** $\Rightarrow$ **(not P)**, which is logically equivalent. But how do we know that the statements **(not Q)** $\Rightarrow$ **(not P)** and $\mathbf{P} \Rightarrow \mathbf{Q}$ really are logically equivalent? Let's try to understand more precisely why this is so.

First, let's see why

$$\textbf{(not Q)} \Rightarrow \textbf{(not P)} \quad \text{implies} \quad \mathbf{P} \Rightarrow \mathbf{Q}. \tag{12}$$

To do this, we suppose that

$$\textbf{(not Q)} \Rightarrow \textbf{(not P)}, \tag{13}$$

and we will show that

$$\mathbf{P} \Rightarrow \mathbf{Q}. \tag{14}$$

To prove the conditional statement (14), we assume that P is true.
[To show: Q is true.] We know that Q is either true or false. However, if Q were false, then statement (13) would imply that P is false, which contradicts the assumption that P is true. Hence, Q must be true. We have thus shown that if we assume (13), then (14) must be true. In other words, we have achieved our goal: We have shown that (12) is true.

We can similarly show that $\mathbf{P} \Rightarrow \mathbf{Q}$ implies **(not Q)** $\Rightarrow$ **(not P)**. Therefore, the two statements, $\mathbf{P} \Rightarrow \mathbf{Q}$ and **(not Q)** $\Rightarrow$ **(not P)**, are logically equivalent.

The axiomatic approach

In our proof that the product of any two odd numbers is odd (Lemma 1.3.1), we made a number of assumptions that we did not state explicitly. For example, in order to get from equation (1) to equation (2), we used the distributive property. What gives us the right to blithely make assumptions like this, without stating them? That's a great question. For our development of number theory, we could choose to take a very formal, axiomatic approach, in which we explicitly state all of our assumptions (such as the distributive property) before using them in proofs. In the Appendix (on the Student Companion Website), we take just such an approach, which allows us to build the entire edifice of number theory upon the foundation of a small number of fundamental properties of the integers. If that notion is appealing to you (or to your professor), you should read the Appendix. The main body of this book takes a less axiomatic approach that enables us more quickly to dive into our main business: discovering and proving interesting and exotic theorems about the integers.

EXERCISES 1.3

Numerical Problems

1. Write the hypothesis and conclusion of each conditional statement.

   a. If $m = 10$, then $m^2 = 100$.

   b. If $x = 5$, then $x \geq 3$.

   c. If I eat three hot dogs at lunch, then my stomach is upset before dinner.

2. Write the contrapositive of each statement in Exercise 1.

3. In Example 4, we proved statement (11). Write the hypothesis and conclusion of statement (11). Where does each occur in the proof in Example 4?

Reasoning and Proofs

4. Let n be an integer. In the first part of the proof of Lemma 1.3.2, we proved the statement

$$\text{If } n^2 \text{ is even, then } n \text{ is even.}$$

   What is the contrapositive of this statement?

5. Prove that the sum of any two odd integers is an even integer.

6. Prove that the sum of any two even integers is an even integer.

7. Prove that the sum of an odd integer and an even integer is always odd.

8. Let m and n be integers such that m is even. Prove that the product mn is even.

9. Let n be an integer. Prove that $n^2 \geq 0$. [*Hint:* Consider three cases: $n > 0$, $n = 0$, and $n < 0$.]

10. Here is the beginning of a proof by contradiction:

LEMMA. *Let n be an integer. If n^2 is odd, then n is odd.*

> **PROOF** Let n be an integer such that n^2 is odd.
>
> **[We must show that n is odd. We will do this by contradiction.]**
>
> *Assumption:* Suppose that _____.
>
> . . .

    a. Fill in the blank to complete the Assumption.

    b. Complete the rest of the proof.

11. Prove the following statements about rational numbers:

    a. The sum of any pair of rational numbers is a rational number.

    b. The product of any pair of rational numbers is a rational number.

12. Let a and b be rational numbers with $b \neq 0$. Prove that $\frac{a}{b}$ is a rational number.

13. Prove that the sum of any rational number and any irrational number is irrational. [*Hint:* Use indirect proof.]

14. Prove that if q is an irrational number, then $-q$ is an irrational number.

15. Prove that the product of any nonzero rational number and any irrational number is an irrational number.

16. Prove that the reciprocal of any irrational number is also an irrational number.

17. Suppose that the ordered pairs (x_1, y_1) and (x_2, y_2) each satisfy the equation $y = mx + b$.

    a. Prove the following statement by proving its contrapositive:

$$\text{If } y_1 \neq y_2 \text{, then } x_1 \neq x_2.$$

    b. Now assume that $m \neq 0$. Prove the following statement by proving its contrapositive:

$$\text{If } x_1 \neq x_2 \text{, then } y_1 \neq y_2.$$

Advanced Reasoning and Proofs

18. Let x and y be positive real numbers. Prove that if $\sqrt{xy} \neq \frac{x+y}{2}$, then $x \neq y$.

19. Suppose that x and y are irrational, but $x + y$ is rational. Prove that $x - y$ is irrational.

20. Use proof by contradiction to show that the circle with equation $x^2 + y^2 = 4$ does not intersect the line $y = x + 4$.

21. Omar throws a party, and a total of n of his friends come, where $n > 1$. While all of the guests are friends with Omar, many of them did not know each other prior to the party. For example, Van is only friends with three other guests, while five of Haruko's friends are at the party. Prove that at least two of the guests have the exact same number of friends at the party.

22. Let a, b, and c be nonnegative real numbers. Prove that if $a + b \neq \sqrt{c^2 + 2ab}$, then $a^2 + b^2 \neq c^2$.

23. Prove that the sum of any positive real number and its reciprocal must be at least 2.

24. Prove that there is some irrational number x such that $x^{\sqrt{2}}$ is rational.

1.4 Irrationality of $\sqrt{2}$

Proof that $\sqrt{2}$ is irrational

Recall that in the myth in Section 1.1, Hypatia saved the day by realizing that the fraternity walkway and the sorority walkway would never be the same length. Her argument was based on the fact that $\sqrt{2}$ is irrational. We now prove this fact, using the method of indirect proof. This choice of method should not be surprising, since the assertion that $\sqrt{2}$ is irrational is a negative statement: $\sqrt{2}$ is not a rational number.

Fun Facts

The idea behind our proof that $\sqrt{2}$ is irrational (Theorem 1.4.1) can be found in the writings of Aristotle.

THEOREM 1.4.1

$\sqrt{2}$ *is irrational.*

PROOF (By contradiction.)

Assumption: Suppose $\sqrt{2}$ is rational.

By our Assumption, there are integers a and b such that $\sqrt{2} = \frac{a}{b}$ and $b \neq 0$. Reduce the fraction $\frac{a}{b}$ to lowest terms, to get

$$\sqrt{2} = \frac{p}{q}. \tag{1}$$

where p and q are integers with no common factors. Now square both sides of equation (1) to get rid of the square root:

$$\sqrt{2} = \frac{p^2}{q^2}.$$

Multiplying both sides of this equation by q^2, we get

$$2q^2 = p^2. \tag{2}$$

We have now expressed p^2 as 2 times the integer q^2, so p^2 is an even integer.

Thus, p is an even integer, by Lemma 1.3.2. It follows that there is an integer r such that

$$p = 2r.$$

Substitute this into equation (2) to get

$$2q^2 = (2r)^2;$$

equivalently,

$$2q^2 = 4r^2,$$

and hence,

$$q^2 = 2r^2.$$

This equation implies that q^2 is an even integer; hence, q is even, again by Lemma 1.3.2. But now both p and q are even, and thus they have the common factor 2. This contradicts the fact that p and q have no common factors. $\Rightarrow\Leftarrow$

Since we reached a contradiction, the Assumption must be false. Thus, $\sqrt{2}$ is irrational. ∎

Notice that in this proof, we assumed that any fraction can be reduced to lowest terms. This familiar fact will be proven in Chapter 3 (Corollary 3.4.5).

Incommensurability

As mentioned at the beginning of the chapter, the ancient Greeks saw every number as representing the length, area, or volume of a geometric object. The ancient Greeks defined two lengths as being *commensurable* if both numbers are an integer multiple of some smaller length. For example, the lengths 4 and 6 are commensurable because they are both multiples of the length 2.

The lengths $\frac{1}{3}$ and $\frac{2}{5}$ are commensurable because they are both multiples of the length $\frac{1}{15}$.

$$\frac{1}{3}$$ $$\frac{2}{5}$$

$$5 \cdot \frac{1}{15} \qquad 6 \cdot \frac{1}{15}$$

Are any two segments commensurable? In other words, given two segments, is it always possible to find a small segment that will divide evenly into both? The early ancient Greeks believed that any two segments are commensurable. However, this is not the case. It turns out that even the most basic geometric figures often contain incommensurable segments. Consider the following unit square:

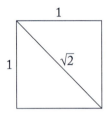

Sometime before 410 BCE, the Greeks discovered that the diagonal of this square is not commensurable with its sides. This discovery shocked the Greeks, whose belief that all lengths were commensurable had been fundamental to their philosophy of natural harmony.

We will now prove the Greeks' surprising result that segments with lengths $\sqrt{2}$ and 1 are not commensurable. This ancient Greek assertion is essentially equivalent to the statement that $\sqrt{2}$ is irrational (Theorem 1.4.1). Proving our assertion is not hard because we have already proven Theorem 1.4.1. A result that can be proved from an earlier result without too much additional work is called a *corollary*.

COROLLARY 1.4.2

The diagonal of the unit square is not commensurable with its sides. In other words, a segment of length $\sqrt{2}$ and a segment of length 1 are incommensurable.

PROOF (By contradiction.)

Assumption: Suppose segments of lengths $\sqrt{2}$ and 1 are commensurable.

Then there is some length x such that $\sqrt{2}$ and 1 are each multiples of x. This means there are positive integers n and m such that

$$nx = \sqrt{2} \qquad \text{and} \qquad mx = 1.$$

Dividing the first equation by the second yields

$$\frac{n}{m} = \sqrt{2}.$$

This contradicts the fact that $\sqrt{2}$ is irrational (Theorem 1.4.1). $\Rightarrow\Leftarrow$

Since we reached a contradiction, the Assumption must be false. Hence, segments of lengths $\sqrt{2}$ and 1 are incommensurable. ∎

EXERCISES 1.4

Numerical Problems

1. Are $\frac{3}{8}$ and 5 commensurable? If so, find a length x such that $\frac{3}{8}$ and 5 are both integer multiples of x.

2. Suppose a and b are integers. Determine whether $\frac{a}{b}$ and 5 are commensurable. If so, find a length x such that $\frac{a}{b}$ and 5 are both integer multiples of x.

3. a. Find two real numbers that are each commensurable with $\sqrt{2}$.

   b. Find two real numbers that are not commensurable with $\sqrt{2}$.

Reasoning and Proofs

4. Prove that $\sqrt{8}$ is irrational.

5. Prove that $\sqrt{6}$ is irrational.

6. a. Prove that $\sqrt{2} + \sqrt{3}$ is irrational. [*Hint:* Use the result of Exercise 5.]

   b. Let a and b be rational numbers such that $\sqrt{ab}$ is irrational. Prove that $\sqrt{a} + \sqrt{b}$ is irrational.

7. Prove that $\sqrt[3]{2}$ is irrational.

8. Prove that if x^2 is irrational, then x is irrational.

9. Suppose that $a, b, c,$ and d are rational, and $a + b\sqrt{2} = c + d\sqrt{2}$. Prove that $a = c$ and $b = d$.

10. Let x and y be any rational numbers. Prove that a segment of length x is commensurable with a segment of length y.

11. Let x be a rational number. Prove that if a segment of length y is commensurable with a segment of length x, then y is a rational number.

12. By the definition given in this section, two segments are commensurable if we can find a smaller segment that goes evenly into both segments. As you will show in this exercise, an equivalent condition is that some number of copies of the first segment will exactly equal some number of copies of the second.

    a. Suppose that a segment of length x and a segment of length y are commensurable. Show that some positive multiple of x is equal to some positive multiple of y. That is, show that there exist natural numbers m and n such that $mx = ny$.

    b. Now suppose that a segment of length x and a segment of length y are incommensurable. Show that no positive multiple of x can equal a positive multiple of y. That is, show that if m and n are natural numbers, then mx cannot equal ny.

13. Prove that if x is a real number, then at least one of the numbers $\sqrt{2} + x$ and $\sqrt{2} - x$ must be an irrational number.

14. Is there any line that passes through the origin $(0, 0)$ of the Cartesian plane but does not pass through any other point whose coordinates are both integers? Explain. [*Hint:* Consider the slope of any line passing through the origin that goes through another point whose coordinates are both integers.]

1.5 Using Quantifiers

The existential and universal quantifiers

When we make a statement involving a variable, we cannot know whether the statement is true or false if we do not know the value of the variable. For example, is the statement

$$x > 2 \tag{1}$$

true or false? You cannot tell unless you know what x represents. You would certainly not want to use statement (1) in a proof unless you had already introduced x. In contrast, consider the statement

$$\text{There exists an integer } x \text{ such that } x > 2. \tag{2}$$

By using the words *there exists* to introduce x, we have created a meaningful statement about the integers. (In this case, the statement happens to be true.)

The phrase "there exists" is called the **existential quantifier**, and statement (2) is an example of an *existential statement*. The symbol $\exists$ (read "there exists") is used to denote the existential quantifier. Thus, we can rewrite statement (2) as follows:

$$\exists x \in \mathbf{Z} \text{ such that } x > 2.$$

We can also assert that a statement is true *for all* values of x in a particular set. For example, while you would not want to write $x^2 \geq x$ without specifying what x is, it makes sense to say that

$$\text{For every natural number } x, x^2 \geq x. \tag{3}$$

The phrase "for every" is called the **universal quantifier**, and statement (3) is an example of a *universal statement*. We may use the symbol $\forall$ (read "for every" or "for all") to denote the universal quantifier. Thus, we can rewrite statement (3) as follows:

$$\forall x \in \mathbf{N}, x^2 \geq x.$$

Quantifiers are often hidden in an English sentence. For example, consider this statement:

$$\text{If } x \text{ and } y \text{ are natural numbers and } x > y, \text{ then } x + 1 > y + 1. \tag{4}$$

This is not a statement about specific natural numbers x and y. It is really a statement about *any* natural numbers x and y. To make the quantifiers in statement (4) explicit, we rewrite it as

$$\text{For all natural numbers } x \text{ and } y, \text{ if } x > y, \text{ then } x + 1 > y + 1.$$

Thus, we could write this statement using symbols as follows:

$$\forall x, y \in \mathbf{N}, \text{ if } x > y, \text{ then } x + 1 > y + 1.$$

Proving statements with quantifiers

Often we want to prove that a statement is true *for every* element of a particular set. In the following example, we wish to prove that a statement is true for every integer greater than 1. The most common way to begin such a proof is to choose an *arbitrary* integer greater than 1, using the phrase

$$\text{Let } n \text{ be an integer such that } n > 1.$$

EXAMPLE 1

Prove that for every integer $n > 1$, $n^2 > n + 1$.

PROOF Let n be an integer such that $n > 1$.

This implies that

$$n \geq 2.$$

Multiply this inequality by n to get

$$n^2 \geq 2n. \tag{5}$$

Starting with the inequality $n > 1$ and adding n to both sides yields

$$2n > n + 1. \tag{6}$$

Applying transitivity to inequalities (5) and (6),

$$n^2 > n + 1. \qquad \blacksquare$$

We began the proof of Example 1 by choosing an arbitrary integer $n > 1$ and then went on to prove the inequality $n^2 > n + 1$. It follows that this inequality must be true for *every* integer $n > 1$.

As you are learning to write proofs, pay special attention to the word *let*, because it is used in two different ways. The first way that *let* is used in proofs is to introduce an *arbitrary* number, as in Example 1. (We saw other examples of this usage in the discussion of direct proof in Section 1.3.) We now describe another way that *let* can be used in a proof: to introduce a *specific* number. Suppose that we have previously defined the numbers a and b, and we want to introduce a new symbol, c, to represent their sum. In a proof, we would do this using the phrase

$$\text{Let } c = a + b.$$

Suppose we want to prove that *there exists* a number x that satisfies a particular statement. Often, the easiest way is simply to provide an example of a specific number that satisfies the statement. This specific number is often introduced in proofs using the word *let*, as demonstrated in Examples 2 and 3.

EXAMPLE 2

Prove that there exists a rational number x such that $x - x^2 > 0$.

PROOF Let $x = \dfrac{1}{2}$.

Then $x - x^2 = \dfrac{1}{2} - \dfrac{1}{4} = \dfrac{1}{4}$, which is greater than 0. $\qquad \blacksquare$

Example 2 demonstrates that an existential statement may be proven by providing a specific number that satisfies the statement. This specific number may also be defined in terms of other variables, as in line (7) of Example 3.

EXAMPLE 3

Let n be an integer greater than 1. Prove that there exists an integer m such that

$$n < m < n^2.$$

PROOF Let n be an integer such that $n > 1$.

By Example 1,

$$n + 1 < n^2.$$

Since $n < n + 1$, we have

$$n < n + 1 < n^2.$$

Let $m = n + 1$.   (7)

Then m is an integer such that

$$n < m < n^2.$$ ∎

In the proof of Example 3, we showed that there is an integer m such that $n < m < n^2$ by giving $m = n + 1$ as an example of such an integer. This specific integer m was introduced in line (7) of the proof using the sentence

Let $m = n + 1$.

The order of quantifiers in a statement

When we use two different quantifiers in a single statement, the order of the quantifiers is very important. For example, consider the sentence

> **For every** integer x, **there exists** an integer y such that $x + y = 0$.   (8)

Symbolically, we would write this as

$$\forall x \in \mathbf{Z}, \exists y \in \mathbf{Z} \text{ such that } x + y = 0.$$

This sentence is telling us that no matter which integer x we pick, we can find an integer y so that x and y add up to 0. This statement is true, since every integer has an additive inverse (e.g., given $x = 5$, we can let $y = -5$).

In contrast with statement (8), consider the following sentence:

> **There exists** an integer y such that **for every** integer x, $x + y = 0$.   (9)

We would write this symbolically as

$$\exists y \in \mathbf{Z} \text{ such that } \forall x \in \mathbf{Z}, x + y = 0.$$

This sentence asserts that there is one particular integer y that is the additive inverse of every integer. This statement is certainly false.

Sentences (8) and (9) differ only in the order of their quantifiers. Yet this change in order completely changes the meaning of the sentence. So, we have to be careful to always put our quantifiers in the correct order.

Negating a statement containing a quantifier

Consider the statement

> Everyone in this class handed in today's homework. (10)

The negation of statement (10) is

> Not everyone in this class handed in today's homework. (11)

Another way to phrase statement (11) is

> There is someone in the class who did not hand in today's homework. (12)

As we see from statements (10) and (12), negating a universal quantifier gives us an existential quantifier followed by the negation of the rest of the statement.

Let's see how this works with a mathematical example. Consider the following statement:

$$\text{For every } x \in \mathbf{Z}, x > 0. \tag{13}$$

The negation of statement (13) is the statement

$$\text{There exists } x \in \mathbf{Z} \text{ such that } x \leq 0.$$

Disproving a universal statement is (at least on the surface) easier than proving that the statement is true. All you need is a single example for which the statement fails. An example that demonstrates that a statement is false is called a **counterexample** to the statement. For instance, the number $x = -3$ is a counterexample to statement (13), which proves that statement (13) is false.

EXAMPLE 4

Disprove the following statement:

$$\text{For every integer } n, n^2 > n. \tag{14}$$

SOLUTION We must demonstrate the negation of statement (14):

$$\text{There exists an integer } n \text{ such that } n^2 \leq n. \tag{15}$$

Any example of statement (15) is a counterexample to statement (14).

Let $n = 1$. Then, since $1^2 \leq 1$, we have found a counterexample to statement (14). This proves that statement (14) is not true. ∎

These examples involve negating universal statements. Now let's look at examples of negating existential statements. The negation of the statement

$$\text{There is somebody in this classroom who is asleep} \tag{16}$$

is the statement

$$\text{Everybody in this classroom is awake.} \tag{17}$$

From statements (16) and (17), we see that negating an existential quantifier gives us a universal quantifier followed by the negation of the rest of the statement.

We also see this when we negate mathematical statements. For instance, the negation of

$$\text{There exists } n \in \mathbf{N} \text{ such that } 2n = 1$$

is the statement

$$\text{For every } n \in \mathbf{N}, 2n \neq 1.$$

EXERCISES 1.5

Numerical Problems

1. Give a few numerical examples to demonstrate the meaning of each statement.

   a. For every integer n, $n^2 \geq n$.

   b. For every pair of integers n and m, if nm is odd, then $n + m$ is even.

   c. For every integer n, $n^2 + n$ is even.

2. Find a counterexample to each statement:

   a. For every integer n, $2n \geq n$.

   b. For every pair of integers n and m, if nm is even, then $n + m$ is even.

3. For each statement, determine whether you believe the statement is true or false. If the statement is true, provide a few examples to demonstrate its

meaning. If the statement is false, provide a counterexample to prove that it is false.

   a.   The sum of any two odd integers is an odd integer.

   b.   Any integer that is divisible by 10 is also divisible by 2.

   c.   If you divide any even integer by 2, the result is an odd integer.

   d.   The square of any real number is a rational number.

   e.   When you square any real number, the result is greater than or equal to the original real number. (That is, if n is a real number, then $n^2 \geq n$.)

   f.   The product of any two consecutive integers is even.

4. Prove that there exist positive integers n and m such that $n^2 + m^2 = 100$.

5. Write the negation of each statement:

   a.   For every real number x, $x^3 \leq x$.

   b.   There exists a natural number x such that $x^2 = 15$.

6. Without using the words *all* or *every*, write the negation of the following statement:

   All the critics loved the movie.

Reasoning and Proofs

7. a.   Prove that for every positive integer n, $n^2 \geq n$.

    b.   Prove that this inequality is also true for every negative integer n.

    c.   Demonstrate the one remaining case needed to show that for all integers n, $n^2 \geq n$.

8. Prove that for every pair of integers n and m, if nm is odd, then $n + m$ is even.

9. Let n be an integer. Prove that there exist integers p and q such that $p - q = n$ and $p + q = n^2$.

10. For each statement, determine whether it is true or false, and explain your reasoning.

    a. $\forall x \in \mathbf{Z}, \exists y \in \mathbf{Z}$ such that $y > x$.

    b. $\exists x \in \mathbf{Z}$ such that $\forall y \in \mathbf{Z}, y > x$.

    c. $\forall x, y \in \mathbf{Z}, y > x$.

    d. $\exists x, y \in \mathbf{Z}$ such that $y > x$.

11. Prove the following statement: For every $x \in \mathbf{N}$, there exists $n \in \mathbf{N}$ such that $n > x$.

12. Phil Lovett You run into your friend Phil Lovett[4] at a party, where he tells you:

"Dude, I discovered this really neat fact about irrational numbers. When you take any two irrational numbers and add 'em together, you always get an irrational number. Like, $\sqrt{2}$ is irrational and so is π, and $\sqrt{2} + \pi$ is totally irrational. It always works. Can't seem to prove it, though."

Is Phil right that the sum of any two irrational numbers is irrational? If so, prove it. If not, provide a counterexample.

13. Is it true that the product of any two irrational numbers is irrational? If so, prove it. If not, provide a counterexample.

---

[4]As you read this book, you will occasionally come across examples and exercises about the brilliant insights of your good friend Phil Lovett. These will always be accompanied by a picture of Phil in the margin. While Phil means well, you should be wary of his claims. He often gets so excited about his results that he presents them to you without fully thinking them through.

Chapter 2

Emmy Noether (1882–1935)

German mathematician Emmy Noether made fundamental and lasting contributions to the field of abstract algebra. Noether was widely recognized for her creativity and her ability to make connections between abstract concepts. In addition, she led the mathematical community as a whole to adopt a more axiomatic approach.

Noether's mathematical contributions are even more impressive in light of the prejudice she encountered throughout her career as a woman in mathematics in early 20th-century Germany. After completing her Ph.D. in 1907, Noether had trouble finding an academic position because most German universities had a policy against hiring women. Without a job, she worked on her research and gave some lectures for her father Max Noether, an algebraist at the University of Erlangen. During World War I, Noether applied for a position as a *Privatdozent* (roughly equivalent to an assistant professor) at the University of Göttingen. However, the philosophers and historians on the faculty opposed her because of her sex. They argued, "What will our soldiers think when they return to the university and find that they are expected to learn at the feet of a woman?" Nonetheless, Noether had the strong support of David Hilbert, one of the leading mathematicians of the time, who asserted at a faculty meeting, "I do not see that the sex of the candidate is

an argument against her admission as Privatdozent. After all, we are a university and not a bathing establishment." Even with Hilbert on her side, Noether was not given the position. Instead, she taught courses that were listed under Hilbert's name. Finally, in 1922, Göttingen gave her a symbolic position as an adjunct professor. The position carried no regular salary and no official duties, but it paid her a small amount to give lectures on abstract algebra.

Noether stayed at Göttingen from 1922 until 1933, doing research and lecturing. When the Nazis came to power in 1933, she and the other Jews at Göttingen lost their positions and were forbidden from participating in any aspect of academic life. As a result, Noether emigrated to the United States and became a professor at Bryn Mawr College. She died suddenly two years later. On the occasion of Noether's death, Albert Einstein wrote in a letter to the *New York Times:*

In the judgement of the most competent living mathematicians, Fräulein Noether was the most significant creative mathematical genius thus far produced since the higher education of women began. In the realm of algebra in which the most gifted mathematicians have been busy for centuries, she discovered methods which have proved of enormous importance in the development of the present day younger generation of mathematicians.

A Bit of Noether's Math

What do we mean by the term *number system*? One way to approach this question is to examine what properties are shared by familiar number systems such as the integers, the rational numbers, the real numbers, and even the complex numbers. Each of these number systems consists of a set, together with two basic operations: addition and multiplication. In all of these systems, the operations satisfy certain fundamental algebraic rules, such as the commutative property of addition and multiplication, and the distributive property of multiplication over addition. In 1914, Abraham Fraenkel wrote down a set of such rules and defined a *ring* as any set with two operations that obeyed these rules. Fraenkel's list of rules was long, however, and contained some rules that could be derived from the others. In 1921, Noether gave a shorter list of rules that is the basis for our modern definition of a ring. For a discussion of rings using the modern definition, see Chapter 8 and the Appendix (on the Student Companion Website).

Noether developed a general theory of rings, proving many theorems that hold in *all* rings. The advantage of having this general theory is that when we concoct or discover a new system, all we need to do is check that our system satisfies the definition of a ring.

If it does, then we know that all of Noether's theorems are true in this new system. For example, because the set of all polynomials is a ring, Noether's theory unifies the study of polynomials with the study of more familiar rings of numbers.

Like most great mathematical achievements, Noether's development of the theory of rings built on the work of her predecessors. A century before Noether and Fraenkel introduced the concept of a ring, the study of abstract algebraic structures (sets with operations) began with a structure known as a *group*. A group is a set of numbers together with a single operation that obeys particular rules of arithmetic. Group theory grew out of the study of solutions to polynomial equations. You are already familiar with the quadratic formula, which expresses the roots of a quadratic polynomial (a polynomial of degree 2). The first formula for solving quadratic equations was developed independently by Brahmagupta in 7th-century India and Al Khwarizmi in 9th-century Persia. It was not until the Italian Renaissance, however, that mathematicians developed cubic and quartic formulas for finding the roots of polynomials of degrees 3 and 4. The ensuing quest for a quintic formula (a formula for finding the roots of any polynomial of degree 5) continued until the early 19th century, when Niels Abel and Evariste Galois used group theory to prove a surprising result: It is impossible to find a quintic formula! In other words, in contrast with lower-degree polynomials, there is no analogous formula for finding the roots of a polynomial of degree 5.

Chapter 2 Mathematical Induction

2.1 The Principle of Mathematical Induction

Emmy Noether is one of the most famous woman mathematicians in history. She is widely recognized for her important contributions to the field of abstract algebra. What's not so well known about Noether is that being a math professor was not her only job. While she enjoyed the challenges of research and teaching at a prestigious university, she found that nothing could compare to the intellectual stimulation provided by her second job: working the night shift as a security guard.

Noether was in charge of security for events at the German government's Galois School of Graduate Group Theory in Göttingen. One event that was particularly popular with the students was the Saturday cinema series at the Arts Center at Hilbert Hall. Noether was assigned the tedious task of making sure that everyone who wanted to see the movie had a student ID. She came up with a system to do this quickly and easily. Every Saturday night, when the line for the movie had formed outside Hilbert Hall, Noether announced over the loudspeaker, "Everyone please hold up your student ID." Before opening the doors, she scanned the crowd to make sure everyone was holding an ID.

Noether's direct method proved quite successful for many years, until one Saturday night saw a record turnout. The university was showing the classic action adventure film *Raiders of the Lost Arctangent*. Hilbert Hall was not big enough to contain the countless students who showed up, and Noether was forced to turn away many disappointed students. During the movie, she investigated and discovered that although all of the people who she had admitted to the Arts Center held student IDs, some of them were not from the Galois School. A sickly group of commuter students from the nearby Abel Academy of Able Algebraists were in the center, while some Galois students had been left out.

Noether resolved that in the future, she would prevent any Abel students from entering the theater. She would need a way to verify that everyone in line was holding an ID from the Galois School, rather than from some other college.

"I've got it!" Noether said to her assistant. "First, we'll check the ID of the first person in line to make sure that person is a Galois student. Then we'll make an announcement over the loudspeaker, asking every Galois student in the line to make sure that the person after them in line is also a Galois student. After this has happened, we'll know that everyone in line is a Galois student, and we can let them all into the movie."

"That sounds like a good idea," replied the assistant. "But I may have trouble remembering all those rules."

"I'm one step ahead of you," Noether replied. "I made this sign so we would remember my principle for admission to the cinema."

Noether's method (from our myth) for checking that everyone in line was a Galois student illustrates a proof technique known as *mathematical induction*. Suppose you want to prove that a statement is true for all natural numbers. The Principle of Mathematical Induction, which we will soon state formally, asserts that you can prove that a statement is true for every natural number n by establishing two things. First, check that the statement is true for the number 1. Second, show that whenever the statement is true for a natural number k, it follows that the statement is true for $k + 1$. Once you have shown both of these things, you may conclude that the statement is true for all natural numbers.

The idea is similar to the principle of falling dominoes. After setting up a long chain of dominoes, you want to be sure that they will all fall down. To guarantee that this will happen, you must make sure of two things:

1. You push over the first domino.
2. Every domino that falls will knock over the next domino in the chain.

If you are sure of both of these, then you can be confident that all of the dominoes in your chain will fall.

We now give a statement of the Principle of Mathematical Induction. We will take this principle as a basic assumption about the natural numbers. (It turns out that the Principle of Mathematical Induction can be proven using another basic assumption, the Well-Ordering Principle, which we discuss in Section 2.2.)

2.1.1   THE PRINCIPLE OF MATHEMATICAL INDUCTION

Suppose that $P(n)$ is a statement about the natural number n.
If it is established that both

Base Case $P(1)$ *is true.*

Inductive Step *For every natural number k, if $P(k)$ is true, then $P(k + 1)$ is also true.*

Then $P(n)$ is true for all natural numbers n.

The Principle of Mathematical Induction is a frequently used technique for proving that a statement is true for all natural numbers. To use this method, we must show that both the base case and the inductive step are satisfied for the statement that we wish to prove. The base case can often be checked directly.

To prove the inductive step, we let k be an arbitrary natural number, and we must prove the statement

$$P(k) \text{ is true} \Rightarrow P(k + 1) \text{ is true.}$$

To prove this conditional statement, we assume that $P(k)$ is true. The assumption that $P(k)$ is true is called the **inductive hypothesis**. Using this assumption, we must prove that $P(k + 1)$ is true.

Once we have proven both the base case and the inductive step, we may invoke the Principle of Mathematical Induction to conclude that $P(n)$ is true for all natural numbers n. The following example illustrates this method.

EXAMPLE 1

Prove that for every natural number n, $2^n > n$.

PROOF We prove this by induction. Let $P(n)$ be the statement $2^n > n$. We want to show that $P(n)$ is true for all positive integers n.

Base Case We must prove that $P(n)$ is true for $n = 1$. In other words, we must check that

$$2^1 > 1.$$

This is clearly true since $2 > 1$.

Inductive Step Let $k \in \mathbf{N}$. Assume that $P(k)$ is true, i.e.,

$$2^k > k. \qquad \leftarrow \textbf{Inductive hypothesis} \qquad (1)$$

[**We must show that $P(k + 1)$ is true, i.e., that $2^{k+1} > k + 1$.**]

Starting with the inductive hypothesis (1), multiply both sides by 2 to get

$$2^{k+1} > 2k. \qquad (2)$$

Since $k \geq 1$, it follows that $k + k \geq k + 1$. In other words,

$$2k \geq k + 1. \qquad (3)$$

Combining inequalities (2) and (3), we get

$$2^{k+1} > k + 1. \qquad (4)$$

Thus, we have shown that $P(k + 1)$ is true. This completes the Inductive Step.

Since we have proven both the base case and the inductive step, the Principle of Mathematical Induction (2.1.1) implies that for every positive integer n, $2^n > n$. ∎

Induction can often be used to prove formulas for the sum of a sequence of numbers. The following example illustrates this use of induction.

EXAMPLE 2

Prove that for every natural number n,

$$1 + 3 + 5 + \cdots + (2n - 1) = n^2.$$

PROOF (By Induction.)

Base Case For $n = 1$, we need to prove that

$$1 = 1^2. \qquad \leftarrow P(1)$$

This is clearly true.

Inductive Step Let $k \in \mathbf{N}$. Assume the inductive hypothesis that

$$1 + 3 + 5 + \cdots + (2k - 1) = k^2. \qquad \leftarrow P(k) \qquad (5)$$

[**To show:** $1 + 3 + 5 + \cdots + (2(k + 1) - 1) = (k + 1)^2, \qquad \leftarrow P(k + 1)$

i.e., $1 + 3 + 5 + \cdots + (2k + 1) \qquad = (k + 1)^2.]$

Add $2k + 1$ to both sides of equation (5) to get

$$1 + 3 + 5 + \cdots + (2k - 1) + (2k + 1) = k^2 + 2k + 1.$$

Now factor the right side to get

$$1 + 3 + 5 + \cdots + (2k - 1) + (2k + 1) = (k + 1)^2, \qquad (6)$$

which is precisely what we needed to prove for the Inductive Step.

Therefore, the Principle of Mathematical Induction implies that for every natural number n, $1 + 3 + 5 + \cdots + (2n - 1) = n^2$. ∎

Induction is often used to prove statements about recursive formulas. A *recursive formula* for a sequence specifies how to calculate each term of the sequence using the term(s) that came before it. For example, here is a recursive[1] definition of a sequence:

$$A_1 = 1$$

$$A_{n+1} = 2A_n + 3, \qquad \text{for } n \geq 1 \qquad (7)$$

The first term in the sequence is 1. After that, each term of the sequence is obtained by doubling the previous term and adding 3. Using this formula, we can calculate the first few terms of the sequence:

n	1	2	3	4	5	6	7	$\cdots$
A_n	1	5	13	29	61	125	253	$\cdots$

Is there a pattern in this sequence? If you stare at the table long enough, you may notice that each number in the sequence is 3 less than a power of 2. We might conjecture that the numbers A_n are given by the following formula:

$$A_n = 2^{n+1} - 3. \qquad (8)$$

---

[1] The word *recursive* contains the prefix *re-*, meaning "back," and the base *curse*, meaning "run." To find the value of a term in this sequence using the recursive definition (7), one has to *run back* to the previous term, double it, and add 3.

This formula is an *explicit formula* for the sequence: It tells us how to compute the value of A_n as a function of n, without the need to calculate any of the previous terms in the sequence.

The following example proves our conjecture that the explicit formula (8) is true for all natural numbers n.

EXAMPLE 3

Let A_n be the sequence defined recursively by formula (7):

$$A_1 = 1$$

$$A_{n+1} = 2A_n + 3, \qquad \text{for } n \geq 1$$

Prove that this sequence satisfies explicit formula (8):

$$A_n = 2^{n+1} - 3.$$

PROOF We prove this by induction.

Base Case We must show that

$$A_1 = 2^{1+1} - 3. \qquad \leftarrow P(1)$$

This is true because by definition, $A_1 = 1$.

Inductive Step Let $k \in \mathbf{N}$. Assume the inductive hypothesis that

$$A_k = 2^{k+1} - 3. \qquad \leftarrow P(k) \tag{9}$$

[To show: $A_{k+1} = 2^{(k+1)+1} - 3$, $\qquad \leftarrow P(k+1)$
i.e., $\qquad A_{k+1} = 2^{k+2} - 3$.]

By the recursive definition of the sequence,

$$A_{k+1} = 2A_k + 3. \tag{10}$$

Substituting equation (9) into equation (10) gives

$$A_{k+1} = 2(2^{k+1} - 3) + 3.$$

$$= 2 \cdot 2^{k+1} - 6 + 3.$$

$$= 2^{k+2} - 3.$$

Thus, $A_{k+1} = 2^{k+2} - 3$, which completes the inductive step.

By the Principle of Mathematical Induction, we conclude that for all $n \in \mathbf{N}$, $A_n = 2^{n+1} - 3$. ∎

The preceding examples illustrate how to use induction to prove statements for all integers $n \geq 1$. Now consider the statement

$$n^2 > n + 1. \tag{11}$$

This statement is not true for $n = 1$ (as you can verify). However, statement (11) *is* true for all $n \geq 2$, as we prove in the following example. To use induction to prove a statement for all $n \geq 2$, our base case will be to prove that $P(2)$ is true, rather than to prove that $P(1)$ is true. Similarly, if we wanted to prove that a statement is true for all $n \geq 100$, we could use induction but take $n = 100$ to be our base case. (For a rigorous justification of this more general use of induction, see Exercise 22.)

EXAMPLE 4

Prove that for every natural number $n \geq 2$, $n^2 > n + 1$.

PROOF We prove this by induction.

Base Case We must show that

$$2^2 > 2 + 1. \qquad \leftarrow P(2)$$

This is clearly true since $4 > 3$.

Inductive Step Let $k \in \mathbf{N}$. Assume the inductive hypothesis that

$$k^2 > k + 1. \qquad \leftarrow P(k) \tag{12}$$

[To show: $(k + 1)^2 > (k + 1) + 1$, $\qquad \leftarrow P(k + 1)$
   i.e., $\quad (k + 1)^2 > k + 2.$]

Adding $2k + 1$ to both sides of inequality (12), we get

$$k^2 + 2k + 1 > 3k + 2. \tag{13}$$

Since $k > 0$, we know that

$$3k + 2 > k + 2. \tag{14}$$

Combining inequalities (13) and (14), we get

$$k^2 + 2k + 1 > k + 2$$

or, equivalently,

$$(k + 1)^2 > k + 2.$$

Hence, the Principle of Mathematical Induction implies that for every natural number $n \geq 2$, $n^2 > n + 1$. ■

EXERCISES 2.1

Numerical Problems

1. Here is a number pattern:

$$2^0 = 2^1 - 1 \tag{15}$$

$$2^0 + 2^1 = 2^2 - 1 \tag{16}$$

$$2^0 + 2^1 + 2^2 = 2^3 - 1 \tag{17}$$

$$2^0 + 2^1 + 2^2 + 2^3 = 2^4 - 1 \tag{18}$$

Let $P(n)$ be the statement $2^0 + 2^1 + 2^2 + \cdots + 2^{n-1} = 2^n - 1$.

a. Fill in the blanks:

Equation (15) is $P(\_)$.

Equation (16) is $P(\_)$.

Equation (17) is $P(\_)$.

Equation (18) is $P(\_)$.

b. Write $P(100)$ and $P(101)$.

c. Check that equation (15) is true.

d. Without verifying equation (18) directly, we can show that equation (17) implies equation (18), as follows:

Add 2^3 to both sides of equation (17) to get

$$2^0 + 2^1 + 2^2 + 2^3 = 2^3 + 2^3 - 1$$

$$= 2 \cdot 2^3 - 1$$

$$= 2^4 - 1,$$

which is equation (18).

Use the same technique to show that equation (16) implies equation (17).

e. Use the technique from part d to show that if $P(100)$ is true, then $P(101)$ is true.

2. Here is a number pattern:

$$2(1) = 1 \cdot 2 \qquad \leftarrow P(1)$$

$$2(1 + 2) = 2 \cdot 3 \qquad \leftarrow P(2)$$

$$2(1 + 2 + 3) = 3 \cdot 4 \qquad \leftarrow P(3) \tag{19}$$

a. Write $P(4)$ and $P(5)$.

b. Write $P(n)$.

c. Check that $P(3)$ is true.

d. Without verifying $P(4)$ directly, we can show that $P(3)$ implies $P(4)$, as follows:

Add $2 \cdot 4$ to both sides of equation (19) to get

$$2(1 + 2 + 3) + 2 \cdot 4 = 3 \cdot 4 + 2 \cdot 4,$$

and simplify to get

$$2(1 + 2 + 3 + 4) = (3 + 2) \cdot 4$$

$$= 5 \cdot 4$$

$$= 4 \cdot 5,$$

which is $P(4)$.

Use the same technique to show that if $P(4)$ is true, then $P(5)$ is true.

e. Write $P(k)$ and $P(k + 1)$.

f. Use the technique from part d to show that if $P(k)$ is true, then $P(k + 1)$ is true.

Reasoning and Proofs

3. a. What is the hypothesis of the Principle of Mathematical Induction (2.1.1)?

b. What is the conclusion of the Principle of Mathematical Induction (2.1.1)?

4. Simplify each of the following sums to express it as a simple fraction:

a. $\dfrac{1}{1 \cdot 2} = \dfrac{?}{?}$

$\dfrac{1}{1 \cdot 2} + \dfrac{1}{2 \cdot 3} = \dfrac{?}{?}$

$\dfrac{1}{1 \cdot 2} + \dfrac{1}{2 \cdot 3} + \dfrac{1}{3 \cdot 4} = \dfrac{?}{?}$

b. Based on part a, make a conjecture and complete the following formula:

$$\frac{1}{1 \cdot 2} + \frac{1}{2 \cdot 3} + \cdots + \frac{1}{n(n+1)} = \frac{?}{?}$$

c. Use mathematical induction to prove that your formula from part b is true for every natural number n.

5. In our myth, suppose that Noether only used her second rule ("Have every Galois student check that the next person in line is a Galois student") but did not check that the first person in line was a Galois student.

    a. What would happen if no one in the line was a Galois student?

    b. Describe a situation in which the line contains both Abel students and Galois students, Noether's second rule (but not her first rule) is followed, yet the Abel students are not detected.

6. Intuitively, we know that every positive integer is either a multiple of 3, one more than a multiple of 3, or two more than a multiple of 3. Stated more formally, for any $n \in \mathbf{N}$, there exists a nonnegative integer s such that $n = 3s$, $n = 3s + 1$, or $n = 3s + 2$. Prove this statement.

7. **a.** Use induction to prove that for all $n \in \mathbf{N}$, $\dfrac{n(n+1)}{2}$ is a natural number.

    b. Use induction to prove that for all $n \in \mathbf{N}$, $\dfrac{n(n+1)(n+2)}{6}$ is a natural number.

8. In Example 4, we used induction to prove that for every natural number $n \geq 2$, $n^2 > n + 1$. This can also be proved without induction. Prove it directly. [*Hint:* Use the fact that $(n - 1)^2 > 0$.]

9. Prove that for all $n \in \mathbf{N}$, $3 + 7 + 11 + 15 + \cdots + (4n - 1) = n(2n + 1)$.

10. **a.** Prove that for all $n \in \mathbf{N}$,

$$1 + 2 + 3 + \cdots + n = \frac{n(n+1)}{2}.$$

    b. Suppose you have a book whose pages are numbered from 1 to 100. Use the formula in part a to find the sum of all of the page numbers in the book.

    c. Suppose that someone has taken the book from part b and torn out the sheet of paper that had page 53 on one side and page 54 on the other side. What is the sum of all of the remaining page numbers in the book?

d. You have another book, in which one of the sheets of paper (containing two consecutive page numbers) is missing. The sum of all the page numbers that are now in the book is 10,000. How many pages were there originally, and what were the two page numbers on the sheet that is missing?

11. Prove that for all $n \in \mathbf{N}$, $1^2 + 2^2 + 3^2 + \cdots + n^2 = \dfrac{n(n+1)(2n+1)}{6}$.

12. a. Evaluate each of the following factorials: $1!$, $2!$, $3!$, $4!$, $5!$

   b. Note that

$$1 \cdot 1! = 1,$$
$$1 \cdot 1! + 2 \cdot 2! = 5,$$
$$1 \cdot 1! + 2 \cdot 2! + 3 \cdot 3! = 23.$$

Calculate

$$1 \cdot 1! + 2 \cdot 2! + 3 \cdot 3! + 4 \cdot 4! = \,?$$

   c. Determine a general formula for the sums in part b. Then use induction to prove that your formula is correct.

13. Prove that for all $n \in \mathbf{N}$, $n! \leq n^n$.

14. a. Expand the following expressions, and look for a pattern:

$$(1 - r)(1) = \underline{\qquad}.$$
$$(1 - r)(1 + r) = \underline{\qquad}.$$
$$(1 - r)(1 + r + r^2) = \underline{\qquad}.$$
$$(1 - r)(1 + r + r^2 + r^3) = \underline{\qquad}.$$

   b. Write an algebraic rule to describe this pattern.

   c. Prove that your rule holds.

15. Let a and r be real numbers, with $r \neq 1$. Prove that for every positive integer n, the following formula (for the sum of the first $n + 1$ terms of a geometric series) is correct.

$$a + ar + ar^2 + \cdots + ar^n = \frac{a(1 - r^{n+1})}{1 - r}.$$

16. Prove that for every nonnegative integer n, there are exactly $n + 1$ pairs of nonnegative integers (a, b) that satisfy the equation $a + 2b = 2n$.

EXPLORATION The Towers of Hanoi (Exercises 17–19)

The Towers of Hanoi game consists of three pegs and n disks that fit over the pegs, each disk a different size. Initially, the disks are stacked on a single peg, with the largest disk on the bottom leading up to the smallest disk on top, such as pictured here for $n = 5$:

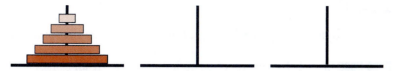

The player is to move the disks from peg to peg, one at a time, with the rule that a disk can never be placed on top of a smaller disk. The object of the game is to move the entire stack of disks to one of the two originally vacant pegs.

17. Play the Towers of Hanoi game with $n = 1, 2, 3,$ and 4 disks, recording the moves. Find a formula for the minimal number of moves it takes to win the game.

18. Prove that no matter what number, n, of disks you start with, this game can always be won, and prove that your formula from Exercise 17 (for the minimal number of moves it takes to win) is correct.

19. According to one legend, monks in a temple had to move a tower of 64 disks, following the rules of the Towers of Hanoi game. The disks were very delicate and could only be moved one at a time, requiring 1 second to move each disk from one peg to another. When the monks finish the task of moving all 64 disks according to the rules of the Towers of Hanoi, the universe will end. If they use the minimal number of moves, how long will it take for them to complete the task?

Advanced Reasoning and Proofs

20. Prove that for every natural number $n > 5$, we have $n! > n^3$.

21. Prove that for every natural number n, the rational expression $\dfrac{n^3 - n}{6}$ is an integer.

22. We can restate the Principle of Mathematical Induction (2.1.1) so that it can be used to prove that a statement is true for all integers $n \geq L$:

> THE PRINCIPLE OF MATHEMATICAL INDUCTION (RESTATED)
>
> *Suppose that $P(n)$ is a statement about the natural number n. Let L be a natural number. If it is established that both*
>
> **Base Case** *$P(L)$ is true.*
>
> **Inductive Step** *For every natural number $k \geq L$, if $P(k)$ is true then $P(k + 1)$ is also true.*
>
> *Then $P(n)$ is true for all natural numbers $n \geq L$.*

Prove that this version of the Principle of Mathematical Induction follows from 2.1.1.

23. Here is a proof that all horses are the same color. Is this a true statement? If not, explain the error in the proof.

> THEOREM. *All horses are the same color.*
>
> **PROOF** Let $P(n)$ be the statement
>
> In any group of n horses, all of the horses are the same color.
>
> **Base Case** $n = 1$. In a group of a single horse, clearly all horses are the same color.
>
> **Inductive Step** Suppose that in any group of k horses, all horses are the same color.
>
> [To show: in any group of $k + 1$ horses, all horses are the same color.]
>
> Consider a group of $k + 1$ horses:

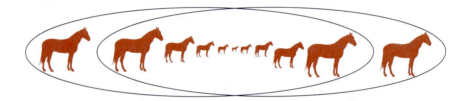

> By the inductive hypothesis, the first k horses are all the same color. Also by the inductive hypothesis, the last k horses are all the same color. Thus, all $k + 1$ horses are the same color. ∎

24. Let m be a natural number. Given $2m$ points placed on a circle, you will label half of the points with the symbol **+** and the other half of the points with the symbol **−**. Show that no matter how you do this, it is always possible to draw m line segments, each connecting a **+** symbol to a **−** symbol, such that none of the line segments intersect.

25. Prove that for any natural number n, if you draw any n lines in a plane, you can shade in each region with either blue or orange so that no two regions that share a border have the same color. (The following diagram shows an example for $n = 4$.)

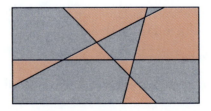

26. Let $a_{n,r}$ be the coefficient of the x^r term in the polynomial $(x + 1)^n$. Use induction on n to prove that for all nonnegative integers $r \leq n$,

$$a_{n,r} = \frac{n!}{r!(n - r)!}$$

27. The game of Moose is played on a checkerboard. A moose is an L-shaped piece that covers three squares (see diagram). The idea is to cover the entire checkerboard, except for a single square, with mooses so that there is no overlap (each moose may be rotated as necessary). Use induction to prove that any $2^n \times 2^n$ checkerboard ($n \in \mathbf{N}$) can be moosed so that the one leftover square is a corner square.

A Moose:

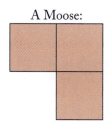

2.2 Strong Induction and the Well-Ordering Principle

If we are given an even number, we know intuitively that we can keep dividing out factors of 2 until the result is an odd number. In other words, every natural number can be written as a power of 2 times an odd number. For instance,

$$120 = 2^3 \cdot \mathbf{15}.$$

Naomi's Numerical Proof Preview: Example 1

Here is a formal version of the statement we would like to prove:

> Any natural number n can be written in the form $n = 2^b \cdot \boldsymbol{r}$,
> where $\boldsymbol{b}$ is a nonnegative integer and $\boldsymbol{r}$ is an odd integer. (1)

Imagine trying to prove this statement by induction. Let $P(n)$ be the statement that n can be written as a power of 2 times an odd integer. In the inductive step, we would assume that $P(k)$ is true (that k can be written as a power of 2 times an odd integer), and try to show that $P(k + 1)$ is true (that $k + 1$ can also be written as a power of 2 times an odd integer).

Unfortunately, there is little relationship between $P(k)$ and $P(k + 1)$. For example, suppose $k = 35$. Consider how the numbers $k = 35$ and $k + 1 = 36$ are written in the required form:

$$35 = 2^0 \cdot \mathbf{35}$$

$$36 = 2^2 \cdot \mathbf{9}$$

Knowing how to write 35 in the required form does not in any way help us see how to write 36 in the required form. It is unclear how we could get this inductive step to work—we will need a different proof technique.

While there is little relationship between $P(35)$ and $P(36)$, there is a strong relationship between $P(18)$ and $P(36)$. Knowing how 18 can be written as a power of 2 times an odd integer,

$$18 = 2^1 \cdot 9,$$

makes it easy to write 36 in the required form. Just multiply both sides of the equation by 2 to get

$$36 = 2^2 \cdot 9.$$

To use this idea as the basis of our proof, we will need a new form of induction. To prove that $P(36)$ is true, it is not enough to assume the standard inductive hypothesis that $P(35)$ is true. Instead, we would like to have a stronger inductive hypothesis that includes the statement that $P(18)$ is true. A powerful inductive hypothesis would be to assume that all of the statements $P(1), P(2), \ldots, P(35)$ are true.

In other words, when proving that $P(n)$ is true in our inductive step, we want license to assume that $P(k)$ is true for all natural numbers k smaller than n. This is precisely the inductive hypothesis that the *Principle of Strong Induction* allows us to use.

2.2.1   THE PRINCIPLE OF STRONG INDUCTION

Suppose that $P(n)$ is a statement about the natural number n.
If for every natural number n,

$$P(k) \textit{ is true for all natural numbers } k < n \implies P(n) \textit{ is true,} \qquad (2)$$

then $P(n)$ is true for all natural numbers n.

To use strong induction to prove that a statement $P(n)$ is true for all n, we must prove that statement (2) holds for every natural number n. To prove the conditional statement (2), we assume the **inductive hypothesis** that $P(k)$ **is true for all natural numbers** $k < n$, and use this assumption to prove that $P(n)$ **is true**.

Before we see an example of how to use the Principle of Strong Induction in a proof, let's consider the relationship between strong induction and our original Principle of Mathematical Induction (2.1.1).

While it may seem that the Principle of Strong Induction is stronger than the original Principle of Mathematical Induction (2.1.1), it turns out that the two principles are logically equivalent. The interested reader can find a proof of this fact in the Appendix (on the Student Companion Website).

Recall that the original Principle of Mathematical Induction (2.1.1) has a separate base case in which $P(1)$ is established. In contrast, the Principle of Strong Induction does not require a separate base case. This is because once you have proven that statement (2) holds for every natural number n, it follows that $P(1)$ is true. (See "Why strong induction is valid: An intuitive explanation," later.)

EXAMPLE 1

Prove that any natural number n can be written in the form $n = 2^b \cdot r$, where b is a nonnegative integer and r is an odd integer.

PROOF (By strong induction.) Let $P(n)$ denote the statement

There exist a nonnegative integer b and an odd integer r such that $n = 2^b \cdot r$. $\leftarrow P(n)$

Let n be a natural number. Assume the inductive hypothesis that $P(k)$ is true for all natural numbers $k < n$. In other words, we are assuming that every natural number $k < n$ can be written as a power of 2 times an odd integer.
[**We must show that $P(n)$ is true, i.e., that n may be written as a power of 2 times an odd integer.**]
Using the fact that n is either even or odd, we examine two cases separately:

Case 1 n is odd.
In this case,

$$n = 2^0 \cdot n$$

is a valid way to write n as a power of 2 times an odd integer. Thus, $P(n)$ is true.

Case 2 n is even.
Since n is even, there exists a natural number k such that

$$n = 2k. \tag{3}$$

It follows that $k < n$, so the inductive hypothesis tells us that $P(k)$ is true. This means that k can be written as a power of 2 times an odd integer; in other words, there exist a nonnegative integer b and an odd integer r such that

$$k = 2^b \cdot r.$$

Multiplying both sides by 2 yields

$$2k = 2 \cdot 2^b \cdot r.$$

Using equation (3), this becomes

$$n = 2^{(b+1)} \cdot r.$$

Since $b + 1$ is a nonnegative integer and r is an odd integer, we conclude that n has the required form. Thus, $P(n)$ is true.

By the Principle of Strong Induction (2.2.1), it now follows that for all natural numbers n, $P(n)$ is true.

EXAMPLE 2

Phil Lovett Your friend Phil Lovett is excited about a new sequence of numbers he's discovered. "It's easy," says Phil. "You start with 1 and 2. After that, the next term in the sequence is the previous term plus twice the one before that. I call them the Lovett numbers. Here's the formal definition of the Lovett numbers:

$$L_1 = 1$$
$$L_2 = 2 \qquad\qquad\qquad (4)$$
$$L_n = L_{n-1} + 2 \cdot L_{n-2}, \quad \text{for } n \geq 3.\text{"}$$

Convinced that Phil has made yet another fundamental contribution to mathematics, you decide to explore the properties of the Lovett numbers. You begin calculating:

$$L_3 = 4, \qquad L_4 = 8, \qquad L_5 = 16, \qquad L_6 = 32, \qquad \ldots$$

"Hold on a second, Phil," you exclaim. "This is no brilliant new discovery. These are just the powers of 2!"

Show that the Lovett numbers are just the powers of 2. That is, prove that for every positive integer n, $L_n = 2^{n-1}$.

PROOF (By strong induction.) Let $P(n)$ be the statement $L_n = 2^{n-1}$.

Let n be a natural number, and assume the inductive hypothesis that $P(k)$ is true for all natural numbers $k < n$.
[**To show:** $P(n)$ **is true.**]
We would like to use the recursive equation $L_n = L_{n-1} + 2 \cdot L_{n-2}$. However, this equation is valid only for $n \geq 3$, so we will need to treat the cases $n = 1$ and $n = 2$ separately.

Case 1 $n = 1$. By definition (4), $L_1 = 1$, so it is true that

$$L_1 = 2^{1-1}.$$

Case 2 $n = 2$. By definition (4), $L_2 = 2$, so it is true that

$$L_2 = 2^{2-1}.$$

Case 3 $n \geq 3$. In this case, definition (4) tells us that

$$L_n = L_{n-1} + 2 \cdot L_{n-2}. \qquad\qquad (5)$$

By the inductive hypothesis, $L_{n-1} = 2^{n-2}$, and $L_{n-2} = 2^{n-3}$. Substituting these into equation (5), we get

$$
\begin{aligned}
L_n &= 2^{n-2} + 2 \cdot 2^{n-3} \\
&= 2^{n-2} + 2^{n-2} \\
&= 2 \cdot 2^{n-2} \\
&= 2^{n-1}.
\end{aligned}
$$

Hence, $L_n = 2^{n-1}$. In other words, $P(n)$ is true.

We have thus shown that if $P(k)$ is true for all $k < n$, then $P(n)$ is true. By the Principle of Strong Induction (2.2.1), we conclude that for every natural number n, $P(n)$ is true. ■

EXAMPLE 3

Consider the sequence defined as follows:

$$
\begin{aligned}
a_1 &= 1 \\
a_2 &= 3 \\
a_3 &= 9 \\
a_n &= a_{n-1} + a_{n-2} + a_{n-3}, \quad \text{for } n > 3.
\end{aligned}
$$

Prove that for all natural numbers n, $a_n \leq 3^{n-1}$.

PROOF (By strong induction.) Let $P(n)$ be the statement $a_n \leq 3^{n-1}$.

Let $n \in \mathbf{N}$, and assume the inductive hypothesis that for all natural numbers $k < n$, $P(k)$ is true.

[**To show: $P(n)$ is true; i.e., $a_n \leq 3^{n-1}$.**]

We would like to use the recursive equation $a_n = a_{n-1} + a_{n-2} + a_{n-3}$. However, this equation is valid only for $n > 3$, so we will treat the cases $n = 1$, $n = 2$, and $n = 3$ separately.

Case 1 $n = 1$. We check that $a_1 \leq 3^{1-1}$. This is true, because $1 \leq 1$.

Case 2 $n = 2$. We check that $a_2 \leq 3^{2-1}$. This is true, because $3 \leq 3$.

Case 3 $n = 3$. We check that $a_3 \leq 3^{3-1}$. This is true, because $9 \leq 9$.

Case 4 $n > 3$. By definition of the sequence,

$$
a_n = a_{n-1} + a_{n-2} + a_{n-3}. \tag{6}
$$

We know by our inductive hypothesis that

$$a_{n-1} \leq 3^{n-2}, a_{n-2} \leq 3^{n-3}, \text{ and } a_{n-3} \leq 3^{n-4}.$$

Adding these three inequalities gives

$$a_{n-1} + a_{n-2} + a_{n-3} \leq 3^{n-2} + 3^{n-3} + 3^{n-4}.$$

Using equation (6), we can rewrite this as

$$a_n \leq 3^{n-2} + 3^{n-3} + 3^{n-4}. \tag{7}$$

It is easy to check that $3^2 + 3^1 + 3^0 < 3^3$. Multiply this inequality by the natural number 3^{n-4} to get:

$$3^{n-2} + 3^{n-3} + 3^{n-4} < 3^{n-1}. \tag{8}$$

Using the transitive property, we combine inequalities (7) and (8) to get

$$a_n \leq 3^{n-1}.$$

Therefore, by the Principle of Strong Induction (2.2.1), for all $n \in \mathbf{N}$, $a_n \leq 3^{n-1}$. ∎

Why strong induction is valid: An intuitive explanation

We have seen several examples in which we used strong induction to prove that a statement is true for all natural numbers. But why is strong induction a valid proof technique? In other words, suppose we know that for every natural number n, the conditional statement

$$P(k) \text{ is true for all natural numbers } k < n \Rightarrow P(n) \text{ is true} \tag{9}$$

holds. Why intuitively does it follow that $P(n)$ must be true for all natural numbers n?
Consider the following list of statements

$$P(1) \text{ is true.} \tag{10-1}$$

$$P(1) \text{ is true} \Rightarrow P(2) \text{ is true.} \tag{10-2}$$

$$P(1), P(2) \text{ are both true} \Rightarrow P(3) \text{ is true.} \tag{10-3}$$

$$P(1), P(2), P(3) \text{ are all true} \Rightarrow P(4) \text{ is true.} \tag{10-4}$$

$$P(1), P(2), P(3), P(4) \text{ are all true} \Rightarrow P(5) \text{ is true.} \tag{10-5}$$

$$\vdots$$

All of the statements in this list are specific instances of statement (9). For example, when $n = 4$, statement (9) reads

$P(k)$ is true for all natural numbers $k < 4 \Rightarrow P(4)$ is true,

which is exactly equivalent to statement (10-4).

Even statement (10-1) follows from the $n = 1$ instance of statement (9). When $n = 1$, statement (9) reads

$P(k)$ is true for all natural numbers $k < 1 \Rightarrow P(1)$ is true.

Since there are no natural numbers $k < 1$, the **hypothesis** of this statement is true (see Exercise 11); hence, the **conclusion** must also be true—that is, **$P(1)$ is true**, which is exactly statement (10-1).

The hypothesis of the Principle of Strong Induction (2.2.1) is that statement (9) holds for all $n \in \mathbf{N}$. From this hypothesis, it follows that all of the statements in list (10) hold. What can we conclude from the statements in list (10)? Well, we know that $P(1)$ is true by (10-1). It then follows from (10-2) that $P(2)$ is true. Now we have established that $P(1)$ and $P(2)$ are both true, so it follows from (10-3) that $P(3)$ is true. Continuing in this manner ad infinitum, we see that $P(n)$ must be true for all natural numbers n.

The Well-Ordering Principle

The smallest positive integer is 1. What is the smallest positive rational number? It's a trick question—as we proved in Section 1.3 (Example 1), there is no smallest positive rational number. No matter which positive rational number you choose, there are plenty of smaller positive rational numbers.

This contrast between the positive integers and the positive rational numbers points us toward a fundamental property of the ordering of the positive integers.

2.2.2	THE WELL-ORDERING PRINCIPLE

Every nonempty set of positive integers has a smallest element.

EXAMPLE 4

Let S be the set of positive integers greater than π. The smallest element of S is 4.

EXAMPLE 5

Let T be the set of natural numbers whose digits add up to 14. The smallest element of T is 59.

EXAMPLE 6

Let U be the set of positive integers not printed on this page. Then the smallest element of U is _____. (We'd better not print the answer here! We'll let you give the answer in Exercise 5.)

Whenever we consider a new property of natural numbers, the Well-Ordering Principle guarantees that if any natural number has the property, then there must be a *smallest* natural number that has the property. We will use this important fact about natural numbers in proofs throughout our study of number theory.

It turns out that the three order properties discussed so far in this chapter—the Principle of Mathematical Induction (2.1.1), the Principle of Strong Induction (2.2.1), and the Well-Ordering Principle (2.2.2)—are mathematically equivalent. That is, if one assumes any one of these three principles, it follows that the other two are also true. The proof of this equivalence is found in the Appendix (on the Student Companion Website).

The next two results, which you will prove in the exercises, are corollaries of the Well-Ordering Principle.

COROLLARY 2.2.3

Let S be a nonempty set of integers, and suppose there exists an integer m such that for every s ∈ S, s ≥ m. Then S has a smallest element.

PROOF See Exercise 12.

EXAMPLE 7

Let V be the set of all integers greater than -17.4. The smallest element of V is -17.

The following corollary states that every nonempty set of integers that is bounded above has a greatest element.

Let S be a nonempty set of integers, and suppose there exists an integer n such that for every $s \in S, s \le n$. Then S has a greatest element.

PROOF See Exercise 13.

EXAMPLE 8

Let W be the set of all negative integers. The greatest element of W is -1.

EXERCISES 2.2

Numerical Problems

1. Find the smallest positive integer in each of the following sets, or explain why the set is empty:

   a. Positive even integers

   b. Odd integers greater than 7

   c. The current ages of your siblings

   d. The current ages of your grandchildren

   e. The page numbers in this chapter

   f. $\{x \in \mathbf{N} \mid x > 2x\}$

2. Suppose a sequence A_n is defined by

$$A_1 = 1,$$
$$A_n = nA_{n-1}, \quad \text{for } n \ge 2.$$

   a. Find $A_2, A_3, A_4,$ and A_5.

   b. Based on your answer to part a, write an explicit (nonrecursive) formula expressing A_n as a function of n.

3. Suppose a sequence A_n is defined by

$$A_1 = 2,$$
$$A_n = A_{n-1} + 10, \quad \text{for } n \ge 2.$$

   a. Find the first five terms in the sequence.

   b. Based on your answer to part a, make a conjecture. That is, write an explicit (nonrecursive) formula expressing A_n as a function of n.

4. Let S be the set of natural numbers n that satisfy $n! > 20$. What is the smallest element of S?

5. Let U be the set defined in Example 6.

   a. What is the smallest element of U?

   b. In Example 6, we wrote a blank instead of printing the smallest element of U. Explain why there was no way for us to fill in that blank with the correct answer.

Reasoning and Proofs

6. If in the statement of the Well-Ordering Principle (2.2.2) we replaced "positive integers" by "negative integers," would the new statement be true?

7. Suppose a sequence G_n is defined by

$$G_1 - 1,$$
$$G_2 = 2,$$
$$G_n = 2G_{n-1} - G_{n-2}, \text{ for } n \geq 3.$$

   a. Find G_3, G_4, and G_5.

   b. Based on your answer to part a, make a conjecture. That is, write an explicit (nonrecursive) formula expressing G_n as a function of n.

   c. Prove the conjecture you made in part b.

8. Suppose a sequence J_n is defined by

$$J_1 = 1,$$
$$J_2 = 5,$$
$$J_n = 3J_{n-1} + 10J_{n-2}, \text{ for } n \geq 3.$$

   a. Find the values of J_1, J_2, J_3, and J_4.

   b. Based on your answer to part a, make a conjecture. That is, write an explicit (nonrecursive) formula expressing J_n as a function of n.

   c. Prove that your formula from part b is correct for all $n \in \mathbf{N}$.

9. Suppose a sequence K_n is defined by

$$K_1 = 5,$$
$$K_2 = 13,$$
$$K_n = 5K_{n-1} - 6K_{n-2}, \text{ for } n \geq 3.$$

Prove that for all $n \in \mathbf{N}$, $K_n = 2^n + 3^n$.

10. Suppose a sequence L_n is defined by

$$L_1 = 1,$$
$$L_2 = 5,$$
$$L_n = L_{n-1} + 2L_{n-2}, \quad \text{for } n \geq 3.$$

a. Find an explicit (nonrecursive) formula for L_n.

b. Prove that your formula from part a is correct for all $n \in \mathbf{N}$.

11. In this exercise, you will explain why for any statement $P(k)$, it is true that

$P(k)$ is true for all natural numbers $k < 1$.

This statement can be rewritten as

For all natural numbers $k < 1$, $P(k)$ is true. (11)

a. Consider the statement

For all dogs d in Chicago, d is a schnauzer.

Write the negation of this statement by filling in the blank:

"There exists a dog d in Chicago such that _____."

b. Now write the negation of statement (11) by filling in the blank:

"There exists a natural number $k < 1$ such that _____."

c. Explain why the statement that you wrote in part b is false, regardless of what $P(k)$ is.

d. What does part c allow you to conclude about statement (11)?

Advanced Reasoning and Proofs

12. Prove Corollary 2.2.3. [*Hint:* Consider the set $T = \{s - m + 1 \mid s \in S\}$, and apply the Well-Ordering Principle (2.2.2).]

13. Prove Corollary 2.2.4. [*Hint:* Consider the set $T = \{N - s + 1 \mid s \in S\}$, and apply the Well-Ordering Principle (2.2.2).]

14. Use the Well-Ordering Principle (2.2.2) to show that there is no integer n with $0 < n < 1$.

EXPLORATION A Combinatorial Game (Exercises 15–17)

You and your friend Amelia play a game with peanuts. The game begins with $n \geq 3$ peanuts divided into two heaps (a "heap" consists of a pile of one or more peanuts). For a player's turn, he or she chooses one of the heaps and eats the peanuts in the heap, and then takes the remaining heap and splits it into two heaps in any way the player wants (e.g., if the remaining heap contained 6 peanuts, the player could split it into heaps of sizes 1 and 5, sizes 2 and 4, or sizes 3 and 3). Play then passes to the other player.

A player loses the game if the player is presented with two heaps for his or her turn, and there is no way to follow the rule just described.

15. There is only one way that a player can have no way to follow the rule (and thus lose the game). What is it?

16. To determine a winning strategy for this game, let's look at the heaps that Amelia might leave you with after some number of moves. Decide whether you will necessarily lose in the following situations or whether you can win by making the proper move. If you can win, provide the winning move.

Amelia leaves you with two heaps of sizes:		Lose or win? Explain.
1	1	
1	2	
2	2	
1	3	
2	3	
3	3	
3	4	
4	4	

17. Amelia has played the game for years and asserts the following:

"If a game starts with $n \geq 3$ peanuts in two heaps, at least one of the two heaps contains an even number of peanuts, and if I move first, then I can always win."

Call this statement $P(n)$. Prove that $P(n)$ is true for all natural numbers $n \geq 3$, using the Principle of Strong Induction. There may be more than one way to prove this, as there may be more than one winning strategy. [*Hint:* In Exercise 16, which of the situations in which Amelia left you forced you to move a certain way? How might that be helpful to Amelia?]

EXPLORATION Induction and Graphs (Exercises 18–20)

In modeling connected networks, such as the Internet or a family tree, we are led to consider mathematical objects called *graphs*, which are collections of *vertices* and *edges*. Each edge connects two vertices. Here are three graphs:

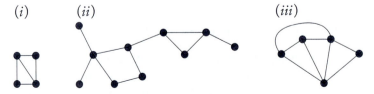

(*i*)   (*ii*)   (*iii*)

Graph (*i*) has 4 vertices (the large dots) and 5 edges.

These three graphs are special for two reasons. First, they are connected, meaning that it is possible to get from any vertex to any other vertex by traversing edges. Second, the graphs are drawn in the plane so that no edges intersect. We call such graphs *connected planar graphs*. A connected planar graph divides the plane into disjoint enclosed areas called *faces*. The area of the plane outside the graph is also considered to be a face. For example, graph (*ii*) has 3 faces: one triangle, one quadrilateral, and the outside face.

18. Let V be the number of vertices of a graph, E the number of edges of a graph, and F the number of faces of a graph. Count V, E, and F for each of the three graphs. Calculate $V - E + F$ in each case. What do you find?

19. Draw two new connected planar graphs, and verify that the pattern you found in Exercise 18 holds for these graphs.

20. The pattern you've found is called *Euler's formula*. Since graphs can come in all shapes and sizes, it might not be clear how one would prove this formula is true for all connected planar graphs. In this exercise, you will use induction to prove that Euler's formula is true for all connected planar graphs.

    a. The simplest graph consists of a single vertex with no edges:

    •

    Verify that Euler's formula is true for this graph.

    b. We can construct any connected planar graph by starting with a single vertex (the graph in part a) and repeatedly doing one of two things:

    Move 1: Starting at a vertex, draw a new edge that does not cross any existing edge, and then place a vertex at the end of your new edge.

    Move 2: Connect two existing vertices with a new edge that does not cross any existing edge.

    Show that in doing either move, the value of $V - E + F$ before the move is the same as the value of $V - E + F$ after the move.

    c. Why do these facts lead to a proof that Euler's formula holds true for all connected planar graphs?

2.3 The Fibonacci Sequence and the Golden Ratio

The Fibonacci numbers

In the year 1202, the Italian mathematician Leonardo of Pisa, also known as Fibonacci, introduced the following sequence in his celebrated book *Liber Abaci*:

$$1, 1, 2, 3, 5, 8, 13, 21, 34, 55, 89, 144, \ldots$$

What is the next number in the sequence? Each term in the sequence (after the first two) is the sum of the two preceding terms. For example, $21 = 8 + 13$. This property, along with the values of the first two terms, defines the sequence.

Fun Facts

Before Fibonacci, most people in Europe were still using cumbersome Roman numerals to represent numbers, rather than the Arabic system of numerals that we use today. Fibonacci was educated in North Africa, where Arabic numerals had been used for centuries. Fibonacci's explanation of Arabic numerals in *Liber Abaci* brought them into general use throughout Europe.

DEFINITION 2.3.1

The **Fibonacci sequence** $F_1, F_2, F_3, \ldots$ *is defined by*

$$F_1 = 1$$
$$F_2 = 1 \tag{1}$$
$$F_n = F_{n-1} + F_{n-2}, \quad for\ n \geq 3.$$

Check for Understanding

Date _____

Section 2.3

Pay to the order of *Understanding* _____

Use the definition of the Fibonacci sequence to find

1. F_6　　　2. F_7　　　3. F_8　　　4. F_{13}　　　5. F_{14}

Answer(s) _____　_____

The recursive formula (1) is easy to use if we want to list Fibonacci numbers starting with the beginning of the sequence. However, if we want to find the thousandth Fibonacci number, F_{1000}, then this recursive formula might not be very useful. To find F_{1000}, we would first have to find F_{998} and F_{999}, and to find those, we would first need to find the terms in the sequence that came before them, and so on, all the way back to F_1 and F_2. To find F_{1000} using the recursive formula, we would thus have to find *all* of the Fibonacci numbers from F_1 to F_{1000}.

There is also an explicit formula for the Fibonacci sequence known as *Binet's formula*. This explicit formula allows the nth Fibonacci number, F_n, to be found as a function of n, without the need to calculate any other numbers in the Fibonacci sequence.

$$F_n = \frac{1}{\sqrt{5}}\left[\left(\frac{1 + \sqrt{5}}{2}\right)^n - \left(\frac{1 - \sqrt{5}}{2}\right)^n\right] \qquad \leftarrow \textbf{Binet's formula} \qquad (2)$$

You may find it surprising that this formula always yields an integer, let alone the nth Fibonacci number! (In Exercise 10 of Section 2.4, you will prove that Binet's formula is correct.)

The explicit formula (2) is great for calculating any Fibonacci number without the need to calculate any others. However, the explicit formula (2) lacks the simplicity of the recursive formula (1) and thus may be harder to remember. The recursive definition also has the benefit that, since it relates each Fibonacci number to the immediately preceding ones, it is useful in inductive proofs, such as the proof we will give in Example 1.

The sum of the first five Fibonacci numbers is

$$1 + 1 + 2 + 3 + 5 = 12,$$

which is 1 less than the seventh Fibonacci number, $F_7 = 13$. The sum of the first six Fibonacci numbers is

$$1 + 1 + 2 + 3 + 5 + 8 = 20,$$

which is 1 less than the eighth Fibonacci number, $F_8 = 21$. At this point, we might conjecture that in general, the sum of the first n Fibonacci numbers is given by

$$F_1 + F_2 + \cdots + F_n = F_{n+2} - 1.$$

In the following example, we use induction to prove that this pattern always holds.

EXAMPLE 1

Prove that for every natural number n,

$$F_1 + \cdots + F_n = F_{n+2} - 1.$$

PROOF (By induction.) Let $P(n)$ be the statement

$$F_1 + \cdots + F_n = F_{n+2} - 1.$$

Base Case [To show: $F_1 = F_3 - 1$.]

$F_1 = 1$ and $F_3 = 2$. Since $1 = 2 - 1$, it is true that

$$F_1 = F_3 - 1.$$

Thus, $P(1)$ is true.

Inductive Step Let $k \in \mathbf{N}$. Assume the inductive hypothesis, $P(k)$:

$$F_1 + \cdots + F_k = F_{k+2} - 1. \tag{3}$$

[To show: $F_1 + \cdots + F_k + F_{k+1} = F_{(k+1)+2} - 1$,
i.e., $F_1 + \cdots + F_k + F_{k+1} = F_{k+3} - 1.$]

Add F_{k+1} to both sides of equation (3) to get

$$F_1 + \cdots + F_k + F_{k+1} = F_{k+2} - 1 + F_{k+1}$$
$$= F_{k+1} + F_{k+2} - 1.$$

From the definition of the Fibonacci sequence, $F_{k+3} = F_{k+1} + F_{k+2}$. Thus, we have

$$F_1 + \cdots + F_k + F_{k+1} = F_{k+3} - 1.$$

This completes the inductive step.

By the Principle of Mathematical Induction (2.1.1), we conclude that for every $n \in \mathbf{N}$,

$$F_1 + \cdots + F_n = F_{n+2} - 1. \qquad \blacksquare$$

The golden ratio

How quickly do the numbers in the Fibonacci sequence increase? One way to approach this question is to consider the ratios of each pair of consecutive Fibonacci numbers, as shown in Table 2.3.1.

Looking down the rightmost column of Table 2.3.1, what do you notice about the values? As n increases, it appears that successive values of the ratio $\dfrac{F_{n+1}}{F_n}$ converge to a limit. It can be shown (see Exercise 9) that they do converge to a limit, which we denote using the Greek letter φ (phi). We can thus write our observation about Table 2.3.1 as follows:

$$\varphi = \lim_{n \to \infty} \frac{F_{n+1}}{F_n}. \tag{4}$$

Let's try to find the exact value of φ. It follows from equation (4) that

$$\varphi = \lim_{n \to \infty} \frac{F_{n+2}}{F_{n+1}}.$$

TABLE 2.3.1 Ratios of consecutive Fibonacci numbers

n	$\dfrac{F_{n+1}}{F_n}$		fraction		decimal
1	$\dfrac{F_2}{F_1}$	$=$	$\dfrac{1}{1}$	$=$	1
2	$\dfrac{F_3}{F_2}$	$=$	$\dfrac{2}{1}$	$=$	2
3	$\dfrac{F_4}{F_3}$	$=$	$\dfrac{3}{2}$	$=$	1.5
4	$\dfrac{F_5}{F_4}$	$=$	$\dfrac{5}{3}$	$=$	1.66666 ...
5	$\dfrac{F_6}{F_5}$	$=$	$\dfrac{8}{5}$	$=$	1.6
6	$\dfrac{F_7}{F_6}$	$=$	$\dfrac{13}{8}$	$=$	1.625
7	$\dfrac{F_8}{F_7}$	$=$	$\dfrac{21}{13}$	$=$	1.61538 ...
8	$\dfrac{F_9}{F_8}$	$=$	$\dfrac{34}{21}$	$=$	1.61904 ...
9	$\dfrac{F_{10}}{F_9}$	$=$	$\dfrac{55}{34}$	$=$	1.61764 ...
10	$\dfrac{F_{11}}{F_{10}}$	$=$	$\dfrac{89}{55}$	$=$	1.61818 ...

By the definition of the Fibonacci sequence, $F_{n+2} = F_n + F_{n+1}$. Substituting this into the previous equation gives

$$\varphi = \lim_{n \to \infty} \frac{F_n + F_{n+1}}{F_{n+1}}$$

$$= \lim_{n \to \infty} \left(\frac{F_n}{F_{n+1}} + \frac{F_{n+1}}{F_{n+1}} \right).$$

$$= \lim_{n \to \infty} \frac{F_n}{F_{n+1}} + 1$$

By equation (4), $\lim_{n \to \infty} \dfrac{F_n}{F_{n+1}} = \dfrac{1}{\varphi}$. Substituting this into the previous equation gives

$$\varphi = \frac{1}{\varphi} + 1.$$

This leads to the quadratic equation

$$\varphi^2 - \varphi - 1 = 0, \tag{5}$$

which has the following solutions:

$$\varphi = \frac{1 + \sqrt{5}}{2} \quad \text{and} \quad \varphi = \frac{1 - \sqrt{5}}{2}.$$

Since every Fibonacci number is a positive integer, $\varphi = \lim\limits_{n \to \infty} \dfrac{F_{n+1}}{F_n}$ cannot be negative, so we eliminate the negative solution from consideration. We conclude that

$$\varphi = \frac{1 + \sqrt{5}}{2} = 1.6180339887\ldots. \tag{6}$$

The irrational number φ, defined by equation (6), is called the **golden ratio**. It is also known as the *golden section* or *golden cut* for reasons explored in Exercise 3.

Since the ratios of consecutive Fibonacci numbers approach φ, Fibonacci numbers increase at approximately the same rate as powers of φ. This is the idea behind the following lemma, which will be useful in Chapter 4.

LEMMA 2.3.2

For any natural number n,

$$\varphi^n < F_{n+2} < \varphi^{n+1}.$$

PROOF See Exercise 8.

Mathematically, the golden ratio is a thing of beauty. Mathematicians have been fascinated by the golden ratio and the Fibonacci sequence for centuries. In the exercises and in the next section, we will explore some of the amazing properties of these tantalizing mathematical objects.

Fibonacci numbers and plant phyllotaxis

The arrangement of leaves on a plant is known as *phyllotaxis*. Many plants undergo a growth pattern that causes the Fibonacci numbers to be clearly visible in the spiral patterns of their leaves, petals, buds, fruits, or florets. Examples include daisies, sunflowers, artichokes, pinecones, pineapples, cauliflower, and many others. These plants have spirals in both directions around the stalk or body, and the number of spirals in each direction are two consecutive Fibonacci numbers.

For example, Figure 1 shows the spirals of fruitlets around a pineapple. These fruitlets form rows with gradual slope in one direction around the pineapple (Figure 1a) and rows with steeper slope in the other direction around the pineapple (Figure 1b). Note that each row continues around the back of the pineapple. Counting the number of rows around the pineapple in each direction, we find that there are 8 of the gradual-sloped rows and 13 of the steeper-sloped rows. Thus, the pineapple exhibits two consecutive Fibonacci numbers: 8 and 13. In addition, the next Fibonacci number, 21, is also visible on some pineapples: it is sometimes possible to count the 21 steeper (almost vertical) columns of fruitlets shown in Figure 1c.

Figure 1   Illustration of the Fibonacci numbers 8, 13, and 21 in a pineapple.[2]

In other plants, you can observe other consecutive Fibonacci numbers. An artichoke, for example, has five gradual-sloped rows in one direction and eight steeper-sloped rows in the other direction (see Figure 2). Interestingly, five and eight are also the numbers of rows of leaves in each direction in the greenery at the top of a pineapple.

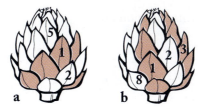

Figure 2   Illustration of the Fibonacci numbers 5 and 8 in an artichoke.

---

[2]On actual pineapples, it is easiest to count the rows by staying close to the middle of the pineapple (not too close to the top and not too close to the bottom). Also note that while many pineapples are relatively easy to count, some varieties of pineapple exhibit variable growth patterns that make them very difficult to count.

Numerical Exercises

1. Find these Fibonacci numbers: F_{10}, F_{13}, F_{15}.

2. Use the Fibonacci numbers $F_{21} = 10{,}946$ and $F_{23} = 28{,}657$ to find these Fibonacci numbers:

    a. F_{22} **b.** F_{24} **c.** F_{20}

3. The golden ratio is sometimes called the *golden section* or *golden cut* because it can be defined as a particular way of cutting (or sectioning) a line segment into two pieces. In Book VI, Definition 3 of the *Elements*, Euclid defines the golden ratio, which he calls the *extreme and mean ratio*, as follows:

    *A straight line is said to have been cut in **extreme and mean ratio** when, as the whole line is to the greater segment, so is the greater to the lesser.*

    In other words, the point B divides the line segment AC in the extreme and mean ratio if:

    $$\frac{AC}{AB} = \frac{AB}{BC}$$

    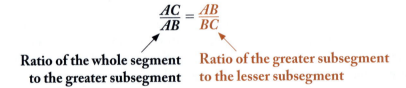

    Ratio of the whole segment to the greater subsegment   **Ratio of the greater subsegment to the lesser subsegment**

    Solve for the value of this common ratio.

Reasoning and Proofs

4. Prove that for all $n \in \mathbf{N}$, $F_1 + F_3 + F_5 + \cdots + F_{2n-1} = F_{2n}$.

5. Prove that for every $n \in \mathbf{N}$,

$$F_{3n-2} \text{ is odd,}$$

$$F_{3n-1} \text{ is odd,}$$

and F_{3n} is even.

6. In this exercise, you will discover a rule about the Fibonacci numbers and then prove that it holds.

   a. Calculate each of the following sums:

   $$F_1^2,$$

   $$F_1^2 + F_2^2,$$

   $$F_1^2 + F_2^2 + F_3^2,$$

   $$F_1^2 + F_2^2 + F_3^2 + F_4^2,$$

   $$F_1^2 + F_2^2 + F_3^2 + F_4^2 + F_5^2$$

   b. Use your answers from part a to conjecture a rule for $F_1^2 + F_2^2 + F_3^2 + \cdots + F_n^2$.

   c. Prove that your rule holds for every natural number n.

7. Suppose the sequence P_n is defined by

   $$P_1 = 2$$
   $$P_2 = 2$$
   $$P_n = P_{n-1} \cdot P_{n-2}, \quad \text{for } n \geq 3$$

   Prove that $P_n = 2^{F_n}$ for all $n \in \mathbf{N}$.

8. Prove Lemma 2.3.2 using strong induction. [*Hint:* It follows from equation (5) that $\varphi^2 = \varphi + 1$.]

9. Use Binet's formula (2), as well as methods for evaluating limits that you know from calculus, to show that the following limit exists and is equal to the golden ratio: $\lim_{n \to \infty} \dfrac{F_{n+1}}{F_n}$.

EXPLORATION Lucas Numbers (Exercises 10–12)

The sequence of *Lucas numbers*, L_n, is similar to the Fibonacci numbers but with a different starting point:

$$L_1 = 1$$
$$L_2 = 3$$
$$L_n = L_{n-1} + L_{n-2}, \quad \text{for } n \geq 3.$$

10. **a.** Make a table of the first 12 Lucas numbers.

   b. Use your results from part a to calculate (to several decimal places) the ratios of pairs of consecutive Lucas numbers. (Your table should be similar to Table 2.3.1, but for Lucas numbers instead of Fibonacci numbers.)

   c. Based on your results from part b, make a conjecture about the value of

   $$\lim_{n \to \infty} \frac{L_{n+1}}{L_n}.$$

11. **a.** Calculate each of the following sums:

   $$L_1, \ L_1 + L_2, \ L_1 + L_2 + L_3, \ L_1 + L_2 + L_3 + L_4.$$

   b. Make a conjecture about the relationship between the sum $L_1 + L_2 + \cdots + L_n$ and the number L_{n+2}.

   c. Prove that your relationship from part b holds for every natural number n.

12. Prove the following relationships between the Lucas numbers and the Fibonacci numbers.

   a. $L_n = F_{n-1} + F_{n+1}$ for all natural numbers $n > 1$.

   b. $5F_n = L_{n-1} + L_{n+1}$ for all natural numbers $n > 1$.

Advanced Reasoning and Proofs

13. Prove that for all $n \in \mathbf{N}$, $F_{n+1}^2 - F_n F_{n+2} = (-1)^n$.

14. Show that for all $m, n \in \mathbf{N}$ such that $m > 1$, $F_{m+n} = F_m F_{n+1} + F_{m-1} F_n$.

15. **a.** Use Binet's formula (2) to show that for any natural number n,

   $$\left| F_n - \frac{1}{\sqrt{5}} \left(\frac{1 + \sqrt{5}}{2} \right)^n \right| < \frac{1}{2}.$$

   b. Prove that the exact value of F_n can be obtained simpiy by rounding $\frac{1}{\sqrt{5}} \varphi^n$ to the nearest integer.

   c. Use a calculator and the method from part b to find F_{20}.

2.4 The Legend of the Golden Ratio

Along with its fellow irrational numbers π and e, the golden ratio,

$$\varphi = \frac{1 + \sqrt{5}}{2},$$

is one of the most well-known mathematical constants. While it is perhaps not as ubiquitous as π and e, the golden ratio has captured people's imagination since ancient times. There are good reasons to be excited about φ. Mathematically, it crops up in many different contexts, and it has a great many unusual properties that fans of mathematics are justifiably enthusiastic about.

But the popularity of the golden ratio has another side, which is based largely on rumors and unsubstantiated claims. Unfortunately, these misconceptions have been repeated so widely and for so many years that they often appear not only in the popular press but also in countless mathematics classrooms and even in mathematics textbooks. Before we delve deeper into this darker side of the golden ratio, however, we should remember why φ attracted people's attention in the first place: its beautiful mathematical properties.

Mathematical properties of the golden ratio

Let's explore a few of the reasons that mathematicians are so enamored with the number φ. In Section 2.3, we found that the quadratic equation

$$\varphi^2 - \varphi - 1 = 0 \tag{1}$$

has two solutions, one positive and one negative. The positive solution, φ, is the golden ratio. We call the negative solution $\overline{\varphi}$.

$$\varphi = \frac{1 + \sqrt{5}}{2} = 1.6180339\ldots.$$

$$\overline{\varphi} = \frac{1 - \sqrt{5}}{2} = -0.6180339\ldots.$$

You may notice that to the right of the decimal point, these numbers look the same:

$$\varphi - 1 = 0.6180339\ldots \qquad \leftarrow \textbf{The fractional part of } \varphi$$

$$-\overline{\varphi} = 0.6180339\ldots \qquad \leftarrow \textbf{The fractional part of } \overline{\varphi}$$

The similarity of the fractional part of these two numbers (the part to the right of the decimal point) is more than coincidence—they are equal. (See Exercise 1.)

Furthermore, the reciprocal of φ,

$$\frac{1}{\varphi} = 0.6180339\ldots,$$

is equal to the fractional part of φ. (See Exercise 2.)

Now consider the nested radical

$$\sqrt{1 + \sqrt{1 + \sqrt{1 + \sqrt{1 + \sqrt{1 + \cdots}}}}}. \tag{2}$$

As the number of terms in this radical expression increases, its value converges. The result is—you guessed it—the golden ratio! (See Exercises 5 and 11.)

When you study *continued fractions* in Chapter 14, you will see that the simplest continued fraction,

$$1 + \cfrac{1}{1 + \cfrac{1}{1 + \cfrac{1}{1 + \cfrac{1}{1 + \cfrac{1}{1 + \cdots}}}}}$$

is also equal to the golden ratio.

A rectangle whose side lengths are in the proportion φ to 1 is known as a *golden rectangle*. Applying the Euclidean algorithm (which you will learn in Chapter 4) to a golden rectangle produces a very pretty picture:

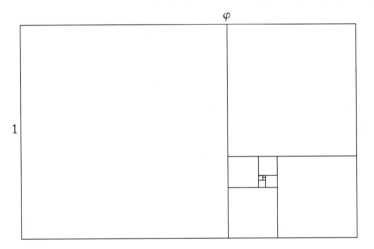

You may have seen this figure before in other mathematics books. Often, it is drawn with a spiral superimposed:

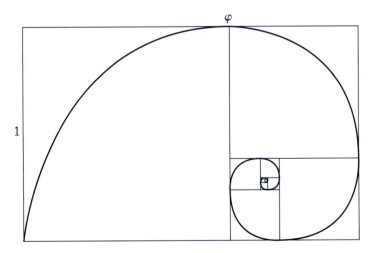

The favorite mystic symbol of the Pythagorean Brotherhood was the pentagram:

The golden ratio can be found all over the pentagram. More precisely, compare any segment in the pentagram with the segment of next shorter length. The ratio of the two lengths is the golden ratio! (See Exercise 4.)

Fun Facts

Like many symbols, the pentagram has been used by more than one cult. While the pentagram was the symbol of the Pythagoreans, the inverted pentagram is a satanic symbol. Thus, you may have seen the pentagram used in horror movies.

The exercises will reveal more evidence of the beauty of φ, exploring further examples of the golden ratio's appearance in geometric and algebraic patterns.

Misconceptions about φ's appearance in art, architecture, and nature

Since the Renaissance, a great deal has been written about φ. Much of this writing asserts that since the time of the ancient Greeks, the golden ratio has been featured in architecture, painting, and sculpture and that it is prevalent in nature as well. Simply put, most of these assertions are incorrect.

The golden ratio, also known as the golden section and (more poetically) as the *divine proportion*, rightfully deserves all of the excitement it has generated as a fascinating mathematical object. Unfortunately, the enthusiasm that many people have justifiably felt due to the mathematical beauty of φ has caused them to get caught up in the cult of φ, the belief that φ can be found practically everywhere in the world in both manmade and natural objects. Through the centuries, many authors have gotten caught up in the excitement of these stories about φ's prevalence in the physical world, and they have continued to propagate the misinformation that φ is one of the basic building blocks of nature.

The legend of φ has even found its way into popular fiction, most notably in Dan Brown's novel *The Da Vinci Code*, which has popularized some common myths about the golden ratio. The excitement about this novel even led some to make new claims, including the creation of the "The Diet Code," which claims that the ideal diet consists of eating certain fundamental food types in proportions based on the golden ratio.

During our own mathematics education, the authors of this book heard many stories about the golden ratio's repeated appearance in art and nature. We confess to having been caught up in the excitement for a long while before learning that these were not facts but fairy tales. This section addresses a few of the most commonly repeated false claims about φ, some of which you may have encountered during your own mathematical education. Learning the truth about these legends should not dampen your enthusiasm for φ in the slightest. Mathematically, φ is just as amazing as it ever was. The golden ratio's *mathematical* beauty, which we have glimpsed in this and the previous section, has never been called into question.

Myth 1: **The golden rectangle is the most aesthetically pleasing rectangle.** Many have claimed that the golden rectangle, a rectangle whose longer side is φ times as long as its shorter side, is more attractive than any other rectangle. When people are asked to choose which of a variety of rectangles is the most attractive, however, not all people choose the same rectangle. Furthermore, which one they choose is likely to depend on the order in which the rectangles are presented, as well as depending on the range of ratios of sides in the entire set of rectangles presented in the experiment.

It is often claimed that various works of art, architecture, and natural structures are based on the ratio φ. These claims are backed up by diagrams or measurements illustrating the golden ratio's presence in the structure, such as those in Figures 1 and 2. The problem with these claims is that in any given structure, a great many different distances can be measured. If enough measurements are taken, it is not surprising that some of them will have a ratio that is close to φ. If one were looking for two measurements that approximately yield a different ratio—say, π or $\sqrt{2}$—this would be no harder to find. The claims that we will discuss in myths 2–4 all depend on this.

Books perpetuating the legend of the golden ratio often include illustrations similar to the ones in Figures 1 and 2 to justify the presence of the golden ratio in works of art. These diagrams display a golden rectangle superimposed on a building, work of art, or animal form. Note that these diagrams use a number of fudge factors. Even in line drawings (such as those in Figures 1 and 2) that are put forth as evidence of the golden ratio, the golden rectangle is not a perfect fit. In addition, the choices of precisely which points in a work of art are used to produce the golden ratio (which points form the corners of a golden rectangle) are arbitrary. In the diagrams of animals and buildings, some portions of the objects do not quite reach the edge of the rectangle, and others slightly overlap the edge of the rectangle.

Myth 2: **The golden ratio has been used in architecture since antiquity.** It is often repeated, for example, that the dimensions of the Parthenon in Athens (built in the 5th century BCE), the Great Pyramid of Khufu (Cheops) in Egypt (built before 2500 BCE), and the United Nations building in New York (built in the 20th century), all embody the ratio φ. There is no evidence that the builders of any of these structures tried to incorporate φ into their designs. Pyramid enthusiasts compare all sorts of measurements until they find two that have roughly the ratio they are looking for. It is not surprising that some of these measurements yield good approximations to φ. The claim about the Great Pyramid appears to be based on a supposed ancient quote from Herodotus, which was actually fabricated by a 19th-century pyramidologist.

The dimensions of the Parthenon vary from source to source, in part because various authors are measuring from different points. A phi hunter is free to choose whichever of those measurements yield a ratio closest to φ. Figure 1a shows a golden rectangle superimposed on a line drawing of the Parthenon, purporting to show that the Parthenon was designed according to the golden ratio. Why are the steps at the base included in the height measurement but not in the width measurement? Why not measure from the edge of the pillars? In any event, this drawing of the Parthenon fits a golden rectangle much better than the actual Parthenon does.

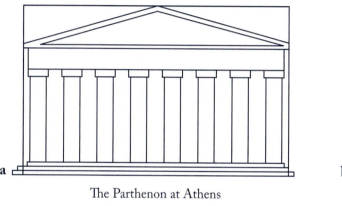
The Parthenon at Athens

Giotto's *Madonna Enthroned with Saints*

Figure 1 Line drawings such as these, which superimpose golden rectangles over famous works, are often used to substantiate claims that the golden ratio is found in art and architecture.

Myth 3: **Many famous works of art, including some by Leonardo da Vinci, were based on the golden ratio.** Leonardo da Vinci illustrated one of the earliest books about φ, *De Divina Proportione* (Divine Proportion), by Luca Pacioli. This has led many golden ratio enthusiasts to imagine that Leonardo based many of his great paintings on the ratio φ. But finding the golden ratio in these paintings would require fudging measurements and/or arbitrary selection of measurement points.

There are many different points in a painting from which distances can be measured. Out of all possible measurements, phi hunters will choose those whose ratio is as close as possible to the golden ratio. These choices are made arbitrarily, to get the desired results. Figure 1b shows a golden rectangle on a rendering of Giotto's (c. 1300) work *Madonna Enthroned with Saints*, which supposedly demonstrates the artist's use of the golden ratio. The exact placements of the corners of the rectangle are fairly arbitrary and were specifically chosen to make it a golden rectangle.

The historical claims of the golden ratio in art are without foundation. In the past century, however, some modern artists and composers, including Salvador Dali, have been inspired by the golden ratio and the legends that surround it. After hearing myths about the golden ratio's role in classical works of art, these 20th-century artists explicitly incorporated the golden ratio into their works.

Myth 4: **The golden ratio is present in the human body and in many animal forms.** Golden ratio buffs often claim that if you measure your total height and divide it by the height of your belly button, you will obtain the golden ratio. Similar claims are made about the ratio of the height of your hips above the ground to the height of your knee above the ground, and about the ratio of the distance from shoulder to fingertips to the distance from elbow to fingertips.

These claims are spurious, for many reasons. First, there are so many different body parts, it's not surprising that a phi hunter could find a few whose measurements have a ratio roughly equal to φ. Second, even without that problem, the results are still unconvincing. Suppose you take these body measurements for several different people. Each person will yield a different ratio. Some of the people's ratios may be close to φ, and some will not be so close. The ratios are not equal for different people. Finally, body parts are not points but physical features with nonzero size. When measuring the distance from elbow to fingertips, for example, the measurer gets to choose which point on the elbow to measure from and can choose which finger to measure to the tip of. A phi hunter will naturally choose precisely those points that yield the ratio closest to φ.

Many people are fooled by these claims that the golden ratio is an inherent feature of the human body. In fact, there is even a plastic surgeon who contends that beautiful faces exhibit φ in their measurements, and he surgically alters his patients' faces to better match a mask that he has constructed using the golden ratio!

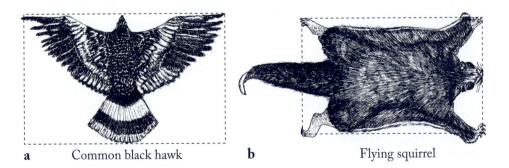

a   Common black hawk   b   Flying squirrel

Figure 2 Drawings that superimpose golden rectangles over various animals supposedly show that the golden ratio can be found in nature.

Claims of the golden ratio's presence in animals suffer from similar problems as the claims about its presence in humans. These claims are without merit, and are often backed up poorly even by line drawings, as illustrated in Figure 2.

There are a number of problems with these claims. First, even in the same species, different individual animals have a variety of shapes. Even if a few are close to the golden ratio, most will not be. Furthermore, animals' appendages (e.g., wings, legs, and fins) can be positioned in a wide range of positions, so golden ratio hunters can choose whichever position yields the ratio they desire. When the hawk pictured in Figure 2a spreads its wings farther or tucks them closer to its body, it does not fit nicely into a golden rectangle. Similarly, when the flying squirrel positions its limbs farther apart or closer together, it no longer fits into a golden rectangle. (Imagine the rectangle formed around the tips of your fingers and toes while you do jumping jacks—at one location, they will form a golden rectangle, but in all other locations, they will not.) The choice of limb position is arbitrary. You could just as easily find animal evidence for any other ratio (say 2 to 1, or π to 1). Since phi hunters are looking for the golden ratio, that is what they find.

Plant phyllotaxis: One place φ really does crop up in nature

As we saw in Section 2.3, the Fibonacci numbers appear in a number of plant forms, in the spiral patterns of their leaves, petals, buds, fruits, or florets. Each leaf, petal, bud, fruit, or floret begins as a tiny lump that forms around the apex of the plant. Biologists and mathematicians have determined that Fibonacci spirals emerge as a consequence of the fact that each new lump is formed at a particular constant angle around the apex from the previous lump. Strikingly, this angular interval is $\frac{360°}{\varphi}$, which is the angle that divides a circle into the golden section. This golden angle, which arises as a consequence of the dynamics of lumps that repel each other as they grow away from the apex, results in an extremely efficient packing of the leaves, petals, buds, fruits, or florets on the adult plant.

Further Reading

If you would like to learn more about the legend of the golden ratio and its history, here are some good places to start. The Markowsky article listed here is the most thorough, and it also gives references to a slew of other good sources.

- George Markowsky, "Misconceptions about the Golden Ratio," *College Mathematics Journal*, Vol. 23, No. 1, January 1992, pp. 2–19.

- Martin Gardner, "The Cult of the Golden Ratio," *Skeptical Inquirer*, Spring 1994, pp. 243–247.

(Continued)

Further Reading (*Continued*)

- Clement Falbo, "The Golden Ratio—A Contrary Viewpoint," *College Mathematics Journal*, Vol. 36, No. 2, March 2005, pp. 123–134.

- Keith Devlin, "Cracking the Da Vinci Code," *Discover*, June 2004, pp. 64–69.

- John H. Conway and Richard K. Guy, *The Book of Numbers*, 1996, pp. 113–126 (section on plant phyllotaxis).

Reader Beware: When you look for other sources of information about φ in the library or bookstore or on the Internet, please be aware that most books that have been written about the golden ratio simply propagate the legend, rather than setting the record straight, since they base their "facts" on the misconceptions in previous sources. Be a discerning reader, and you can avoid falling into the same trap.

EXERCISES 2.4

Numerical Exercises

1. Show that the fractional part of φ and the fractional part of $\overline{\varphi}$ are equal.

2. Show that the reciprocal of φ is equal to the fractional part of φ.

3. An isosceles triangle with angles $36° - 72° - 72°$ is sometimes referred to as a *golden triangle*, for a reason that you will discover in this exercise. Use similar triangles to find the ratio of length *AB* to length *AC* in this diagram of a golden triangle.

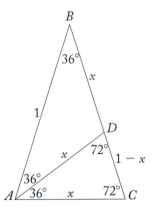

4. In this exercise, you will show that in the pentagram, the ratio of the length of each segment to the length of the next shortest segment is the golden ratio.

The diagram displays a pentagram inscribed in a regular pentagon. Show that each of the following ratios is equal to φ:

a. $\dfrac{AC}{AB}$ **b.** $\dfrac{AB}{BC}$ **c.** $\dfrac{BC}{BD}$

[*Hint:* You may want to use your results from Exercise 3.]

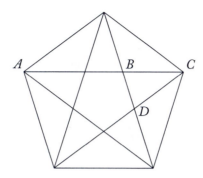

5. **a.** Evaluate each of the following radicals to several decimal places:

$$\sqrt{1}, \ \sqrt{1 + \sqrt{1}}, \ \sqrt{1 + \sqrt{1 + \sqrt{1}}}, \ \sqrt{1 + \sqrt{1 + \sqrt{1 + \sqrt{1}}}}, \ \sqrt{1 + \sqrt{1 + \sqrt{1 + \sqrt{1 + \sqrt{1}}}}}$$

   b. Continue for the next five terms in the pattern. What value does the sequence appear to be approaching?

   c. **Computer Exercise** Write a computer program to calculate the 1st, 10th, 100th, 1000th, 10,000th, and 100,000th terms in the pattern. For each one you calculate, calculate the difference between the term and φ, and express that difference as a percentage of φ.

Reasoning and Proofs

EXPLORATION Simplifying Powers of the Golden Ratio (Exercises 6–10)

6. As you will now explore, any power of the golden ratio, φ^n, can be expressed in the form $a\varphi + b$, where $a, b \in \mathbf{N}$. For instance, by rearranging equation (1), we can express φ^2 in the desired form:

$$\varphi^2 = \varphi + 1. \tag{3}$$

   a. Use equation (3) to express φ^3 in the form $a\varphi + b$, where $a, b \in \mathbf{N}$. Fill in the blanks with natural numbers:

$$\varphi^3 = \underline{\quad} \cdot \varphi + \underline{\quad}. \tag{4}$$

b. Perform similar operations on equation (4) to fill in the blanks with natural numbers:

$$\varphi^4 = \underline{\quad} \cdot \varphi + \underline{\quad}.$$

c. Continue your derivations to fill in the blanks in each equation with natural numbers:

$$\varphi^5 = \underline{\quad} \cdot \varphi + \underline{\quad}.$$

$$\varphi^6 = \underline{\quad} \cdot \varphi + \underline{\quad}.$$

$$\varphi^7 = \underline{\quad} \cdot \varphi + \underline{\quad}.$$

$$\varphi^8 = \underline{\quad} \cdot \varphi + \underline{\quad}.$$

d. Do you recognize the numbers you wrote in the blanks? They may look familiar. Once you recognize these numbers, fill in the boxes:

$$\varphi^n = F_{\boxed{}} \cdot \varphi + F_{\boxed{}}$$

7. a. Fill in the blanks with integers:

$$\varphi^{-1} = \underline{\quad} \cdot \varphi + \underline{\quad}.$$

b. Use similar reasoning as in Exercise 6 to fill in the blanks with integers:

$$\varphi^{-2} = \underline{\quad} \cdot \varphi + \underline{\quad}.$$

$$\varphi^{-3} = \underline{\quad} \cdot \varphi + \underline{\quad}.$$

$$\varphi^{-4} = \underline{\quad} \cdot \varphi + \underline{\quad}.$$

$$\varphi^{-5} = \underline{\quad} \cdot \varphi + \underline{\quad}.$$

$$\varphi^{-6} = \underline{\quad} \cdot \varphi + \underline{\quad}.$$

$$\varphi^{-7} = \underline{\quad} \cdot \varphi + \underline{\quad}$$

c. Do you recognize the numbers you wrote in the blanks? Using Fibonacci numbers, fill in the blanks:

$$\varphi^{-n} = \underline{\quad} \cdot \varphi + \underline{\quad}.$$

8. **a.** Use similar reasoning to that in Exercises 6 and 7 to find an expression for $\overline{\varphi}^n$.

b. Find each of the following (you may use your results from part a and Exercise 6 to help you calculate):

$$\varphi - \overline{\varphi}$$

$$\varphi^2 - (\overline{\varphi})^2$$

$$\varphi^3 - (\overline{\varphi})^3$$

$$\varphi^4 - (\overline{\varphi})^4$$

9. **a.** Prove the following identity: For every natural number $n > 2$,

$$\varphi^n = \varphi^{n-1} + \varphi^{n-2}.$$

b. Show that the identity in part a holds when φ is replaced by $\overline{\varphi}$.

c. Prove that the formula you wrote in Exercise 6d is correct.

[*Hint:* You may want to use the identity from part a.]

10. Prove Binet's formula for the Fibonacci sequence—that is, prove that for any natural number n,

$$F_n = \frac{1}{\sqrt{5}}\left[\left(\frac{1 + \sqrt{5}}{2}\right)^n - \left(\frac{1 - \sqrt{5}}{2}\right)^n\right].$$

[*Hint:* You may want to use the formulas you proved in Exercise 9 (but if you did not do Exercise 9, be sure to prove the formulas you use here).]

Advanced Reasoning and Proofs

11. Give an intuitive argument that the infinite radical expression (2),

$$\sqrt{1 + \sqrt{1 + \sqrt{1 + \sqrt{1 + \sqrt{1 + \cdots}}}}}$$

is equal to the golden ratio.

12. **Phil Lovett** You run into your friend Phil Lovett at a party, where he tells you,

"Dude, I don't know why they make such a big deal about Fibonacci numbers. They're really easy. Look, we know that $F_n = F_{n-1} + F_{n-2}$. We also know that $\varphi^n = \varphi^{n-1} + \varphi^{n-2}$. So I can use strong induction to prove that for every natural number n, $F_n = \varphi^n$. Here is my proof—like, I color-coded it and everything."

PHIL'S PROOF (By strong induction.) Let $P(n)$ be the statement $F_n = \varphi^n$.

Let $n \in \mathbf{N}$, and assume the inductive hypothesis that $P(k)$ is true for all $k < n$. We know that

$$\varphi^n = \varphi^{n-1} + \varphi^{n-2}. \tag{5}$$

By the inductive hypothesis, $\boldsymbol{F_{n-1}} = \boldsymbol{\varphi^{n-1}}$ and $\boldsymbol{F_{n-2}} = \boldsymbol{\varphi^{n-2}}$. Substituting these into equation (5) gives

$$\varphi^n = \boldsymbol{F_{n-1}} + \boldsymbol{F_{n-2}}.$$

By the definition of the Fibonacci numbers, the right side of this equation is equal to F_n, so the equation becomes

$$\varphi^n = F_n.$$

Thus, by strong induction, for every natural number n, $F_n = \varphi^n$. ●

Is Phil right that $F_n = \varphi^n$ for every n? If not, provide a counterexample, and explain what is wrong with Phil's proof.

13. This diagram shows a row of three small circles with diameter d inscribed in a semicircle with radius R. Show that $\dfrac{R}{d} = \varphi$, the golden ratio.

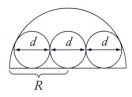

14. This diagram shows four large equilateral triangles with side length L and three smaller equilateral triangles with side length m, inscribed in a circle. Show that $\frac{L}{m} = \varphi$, the golden ratio.

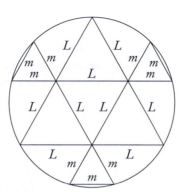

Chapter 3

Eratosthenes (276–194 BCE)

The astronomer Eratosthenes was born in the ancient Greek colony of Cyrene, which is now part of Libya. In 245 BCE, Ptolemy III, the ruler of Egypt, convinced Eratosthenes to move to Alexandria to become the tutor for his son and the director of the renowned library of Alexandria. In addition to his work in astronomy, Eratosthenes is well known for his achievements in mathematics, history, geography, poetry, and athletics. Eratosthenes was called by the nickname "Beta" (the second letter in the Greek alphabet), because he was considered second best in many fields, but not the very best in any field.

Eratosthenes wrote treatises on a broad range of topics and drew several maps of the world. His written works include *Good and Evil*, *Comedy*, *Geography*, *Chronology*, *Measurement of the Earth*, and *The Constellations*. Unfortunately, all of his writings and maps have been lost, so we know of him only through the writings of others. Eratosthenes' most famous contribution to number theory was his method for finding prime numbers, which has come to be called the sieve of Eratosthenes (see Section 3.3). Eratosthenes was also interested in geometry. He invented a mechanical device for creating a cube that has twice the volume of a given cube, and he had a column in Alexandria built and inscribed with a description of this device.

Eratosthenes' achievements in astronomy include estimating the circumference of the Earth, the distance to the moon, the distance to the sun, and the tilt of the Earth's axis (used to analyze the seasons). His measurements of the Earth's circumference and the tilt of its axis were quite accurate, while his measurements of the distance from the Earth to the moon and to the sun were both quite far off.

Eratosthenes also developed a calendar that included leap years. Bridging his interests in science and literature, Eratosthenes wrote a poem entitled "Hermes," in which he explained the basics of astronomy in verse. His writing in the humanities also included literary analyses of plays and essays on ethics.

It is believed that in his later years, Eratosthenes went blind and could no longer work. This led Eratosthenes to become depressed, and ultimately he starved himself to death.

A Bit of Eratosthenes' Math

Eratosthenes gave the first accurate estimate of the circumference of the Earth, using the following method. In the library of Alexandria, he read that at noon on the first day of summer, the sun shines directly into a particular well in the city of Syene. Eratosthenes reasoned that at that moment,

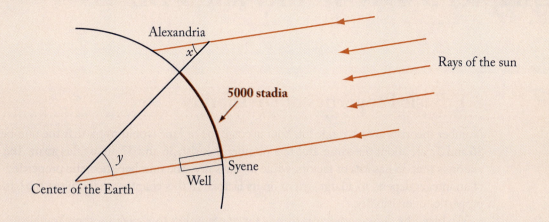

Rays of the sun

Alexandria

x

5000 stadia

y

Syene

Well

Center of the Earth

the line going from the center of the Earth to the sun passes directly through the well in Syene. He observed that in Alexandria at noon on that same day, objects would cast shadows. This meant that the sun was not directly overhead of Alexandria. He measured the angle between the sun and a vertical post in Alexandria, labeled x in the diagram, and found that it was $\frac{1}{50}$ of a circle. (The measure of angle x is $\frac{360°}{50} = 7.2°$.)

Since the sun is so far away from the Earth, Erastonenes figured it was reasonable to assume that the sun's rays that reach Alexandria are parallel to the sun's rays that reach Syene. So by Euclidean geometry, we know that angle x is congruent to angle y. Thus, the measure of angle y is also $\frac{360°}{50}$, and the arc of the Earth from Syene to Alexandria must be $\frac{1}{50}$ of the circumference of the Earth.

To complete his computation of the Earth's circumference, Eratosthenes sent a runner to pace off the distance between Alexandria and Syene. The runner found the distance to be 5000 stadia. (A *stadium* was a unit of distance equal to the length of one lap around a stadium, and the plural of *stadium* is *stadia*.) Using this information, Eratosthenes estimated that the circumference of the Earth is 50 · 5000 = 250,000 stadia. Many historians believe that one stadium was about $\frac{1}{10}$ of a mile. So according to Eratosthenes' estimate, the circumference of the Earth was about 25,000 miles. The actual circumference of the earth is 24,902 miles, so Eratosthenes' estimate seems remarkably accurate!

While the stadium was widely used in ancient Greece as a unit of measurement, the exact length of this unit was not fixed. As a unit of length, one stadium ranged from 0.0926 mile to 0.1796 mile, according to the region in which the stadium was located. Historians of science disagree on the exact length that Eratosthenes considered to be one stadium. In fact, some historians have asserted that Eratosthenes was only trying to find a rough value for the circumference of the Earth, that he used no specific value for the circumference of a stadium, and that he used 5000 stadia as a nice round estimate of the distance from Alexandria to Syene. In any event, Eratosthenes' measurement of the Earth's circumference is rightly remembered as one of the great scientific achievements of his day.

Chapter 3 Divisibility and Primes

3.1 Basic Properties of Divisibility

Number theorists are a strange lot. Your average Joe or Jane would think that the numbers 36 and 37 are about the same. But to a number theorist, 36 and 37 are worlds apart. The reason is that 36 has lots of factors, while 37 does not. The number-theoretic properties of an integer depend to a large extent on its factors. In this chapter, we'll explore the basic properties of divisibility.

We begin by defining what it means for one number to *divide* another. You have known since elementary school that 9 goes into 36 evenly. In mathematical terms, we say that 9 *divides* 36. We write this symbolically as

$$9 \mid 36 \qquad \leftarrow \text{``nine divides thirty-six''}$$

When we say that 9 divides 36, we mean that there is an integer q such that $36 = q \cdot 9$. (This is satisfied by $q = 4$.)

DEFINITION 3.1.1

*Let $a, d \in \mathbf{Z}$. We say that d **divides** a if there exists $q \in \mathbf{Z}$ such that $a = qd$. We express this in symbols as $d \mid a$ (which is read "d divides a").*

If d does not divide a, we can express this fact in symbols as $d \nmid a$. When we say that d divides a, we mean that $a = qd$ for some integer q. The requirement that q is an integer is crucial. For example, $4 \nmid 6$, even though $6 = \frac{3}{2} \cdot 4$.

EXAMPLE 1

Is it true that $13 \mid 52$?

SOLUTION We want to determine whether there exists an integer q such that $52 = q \cdot 13$. In fact, there is: $52 = 4 \cdot 13$, as every card player knows. Thus, $13 \mid 52$. ■

REMARK. The statement "13 divides 52" can be said in a number of different ways, all of which mean exactly the same thing:

- 13 **divides** 52.
- 13 is a **factor** of 52.
- 13 is a **divisor** of 52.
- 52 is a **multiple** of 13.
- 52 is **divisible** by 13.

EXAMPLE 2

Tell whether each of the following statements is true or false.

    a. $-14 \mid 28$ **b.** $15 \mid 25$ **c.** $27 \mid 9$ **d.** $13 \mid 0$

SOLUTION

    a. True. $-14 \mid 28$ because $28 = (-2) \cdot (-14)$.

    b. False. $15 \nmid 25$ since there is no integer q that satisfies the equation $25 = q \cdot 15$.

    c. False. $27 \nmid 9$ since there is no integer q that satisfies the equation $9 = q \cdot 27$. (Note, however, that $9 \mid 27$.)

    d. True. $13 \mid 0$ because $0 = 0 \cdot 13$.   ■

Any divisor of a natural number is smaller than or equal to the number, as we now prove.

LEMMA 3.1.2

Let d and a be natural numbers. If $d \mid a$, then $1 \le d \le a$.

PROOF Let $d, a \in \mathbf{N}$ such that $d \mid a$.

Since d is a natural number, we know that $1 \le d$.

[To show: $d \le a$]

By the definition of divides, there is an integer q such that

$$a = qd$$

Since a and d are positive, the integer q must also be positive. Thus, $1 \le q$. Multiplying both sides of this inequality by d, we obtain

$$d \le qd.$$

Hence,

$$d \le a.$$ ■

Linear combinations

Not only does 9 divide 36, but 9 also divides any multiple of 36. In general, if 9 divides a number x, then 9 also divides $2x$, $3x$, or any multiple of x. We state this as a general principle:

RULE 1. *Let $d, m, x \in \mathbf{Z}$. If $d \mid x$, then $d \mid mx$.*

If 9 divides a number x, and 9 also divides y, then 9 divides their sum, $x + y$. This is an instance of the following rule:

RULE 2. *Let $d, x, y \in \mathbf{Z}$. If $d \mid x$ and $d \mid y$, then $d \mid x + y$.*

We can combine Rule 1 and Rule 2 into a single statement, called the Linear Combination Lemma. A **linear combination** of x and y is an expression of the form $mx + ny$, such as $3x + 2y$, or $5x - 4y$. The Linear Combination Lemma will be used in many proofs throughout our study of number theory.

LEMMA 3.1.3	LINEAR COMBINATION LEMMA

Let $d, m, n, x, y \in \mathbf{Z}$. If $d \mid x$ and $d \mid y$, then $d \mid mx + ny$.

PROOF Let $d, m, n, x, y, \in \mathbf{Z}$ such that $d \mid x$ and $d \mid y$.

Since $d \mid x$, there exists some $k \in \mathbf{Z}$ such that

$$x = kd.$$

Multiplying both sides of this equation by m gives

$$mx = mkd. \tag{1}$$

Since $d \mid y$, there exists some $h \in \mathbf{Z}$ such that

$$y = hd.$$

Multiplying both sides of this equation by n gives

$$ny = nhd. \tag{2}$$

Adding equations (1) and (2) yields

$$mx + ny = mkd + nhd.$$
$$= (mk + nh)d.$$

Since $(mk + nh)$ is an integer, this equation shows that $mx + ny$ is equal to an integer times d. In other words,

$$d \mid mx + ny. \qquad \blacksquare$$

The Linear Combination Lemma implies Rules 1 and 2 (as you will show in Exercise 16). Whenever we use Rule 1 or Rule 2 in a proof, we refer to the Linear Combination Lemma (rather than referring to Rules 1 and 2).

EXERCISES 3.1

Numerical Problems

1. Tell whether each of the following statements is *true* or *false*. Explain your answer.

   a. $24 \mid 8$ b. $5 \mid 125$ c. $0 \mid 42$ d. $42 \mid 0$

   e. $0 \mid 0$ f. $8 \mid 1$ g. $1 \mid 8$ h. $-3 \mid 6$

   i. $20 \mid 3171$ j. $7 \mid -56$ k. $-10 \mid -5$ l. $1 \mid -1$

2. List a few integers b that satisfy the condition $20 \mid b$.

3. List all integers d that satisfy the condition $d \mid 20$.

4. List all integers d that satisfy the condition $d \mid -20$.

5. List every integer that divides both 28 and 24.

6. List every integer that divides both 27 and 100.

7. List every integer that divides all three of the numbers 12, 36, and 64.

8. A number is called *perfect* if it is equal to the sum of all of its positive divisors, excluding the number itself. For example, 6 is perfect—its only positive divisors (other than itself) are 1, 2, and 3, and the sum of these divisors is $1 + 2 + 3 = 6$.

   a. Is 18 a perfect number? Explain.

   b. Find a two-digit perfect number, and demonstrate that it equals the sum of its divisors.

Reasoning and Proofs

9. Find a counterexample to the following statement:

$$\text{For all } a, b, c \in \mathbf{Z}, \, ac \mid bc \text{ if and only if } a \mid b.$$

10. Let a be an integer.

   a. Prove that $a \mid a$.

   b. Prove that $1 \mid a$.

   c. Prove that $a \mid 0$.

11. Let a be an integer. Prove that $a \mid 1$ if and only if $a = 1$ or $a = -1$.

12. Let $a, b, c,$ and d be integers. Prove that if $a \mid b$ and $c \mid d$ then $ac \mid bd$.

13. Let $a, b,$ and c be integers. Prove that if $a \mid b$ and $b \mid c$ then $a \mid c$.

14. Let m be a nonzero integer, and let a and b be integers. Prove that $a \mid b$ if and only if $ma \mid mb$.

15. Let a and b be integers such that $a \mid b$ and $b \mid a$. What conclusion can you make about a and b? Prove that your conclusion is correct.

16. a. Prove that the Linear Combination Lemma (3.1.3) implies Rule 1.

   b. Prove that the Linear Combination Lemma (3.1.3) implies Rule 2.

17. a. Prove Rule 1 directly, without using the Linear Combination Lemma.

   b. Prove Rule 2 directly, without using the Linear Combination Lemma.

18. Let c be a positive integer such that $c < 5$. Prove that c cannot be expressed as a linear combination of 15 and 95.

19. Let $a, b,$ and c be integers. Suppose $a \mid (b + c)$ and $a \mid b$. Prove that $a \mid c$.

20. Let $a, b,$ and c be integers. Suppose $a \mid (b + c)$. Does it follow that $a \mid (b - c)$? Prove it or give a counterexample.

Conjecturer's Corner (Exercises 21–23)

21. Complete the following table, which lists all of the positive divisors of each number n.

n	List all of the divisors of n	Number of divisors of n
1	1	1
3	1, 3	2
4	1, 2, 4	3
5		
6		
7		
8		
9		
10		
11		
12		
13		
16		
17		
20		
25		

22. a. Use your answers from Exercise 21 to make a conjecture about which positive integers have exactly one positive divisor.

b. Prove your conjecture from part a.

23. Use your answers from Exercise 21 to make a conjecture about which positive integers have an odd number of positive divisors.

Advanced Reasoning and Proofs

24. Prove the following generalization of the Linear Combination Lemma.

Let $m_1, m_2, \ldots, m_r, x_1, x_2, \ldots, x_r \in \mathbf{Z}$.

If $d \mid x_i$ for $i = 1, 2, \ldots, r$, then $d \mid m_1 x_1 + m_2 x_2 + \cdots + m_r x_r$.

25. Let a and b be natural numbers. Suppose that $\dfrac{1}{a} + \dfrac{1}{b}$ is a natural number. Prove that either $a = b = 1$, or $a = b = 2$.

26. Let n be a nonnegative integer. Prove that $5 \mid (3^{3n+1} + 2^{n+1})$.

27. In each part below, prove or disprove the statement: For all integers $x, \ldots$

   a. $(x + 2) \mid (x^2 + 3x + 2)$ **b.** $(2 - x) \mid (2x - 4)$

   c. $x \mid x^2 + 12$ **d.** $2 \mid (x^2 + x)$

28. In Exercise 14 of Section 2.3, you proved that for all $m, n \in \mathbf{N}$ such that $m > 1$, $F_{m+n} = F_m F_{n+1} + F_{m-1} F_n$. Use this result to show that for all $a, b \in \mathbf{N}$, if $a \mid b$, then $F_a \mid F_b$.

3.2 Prime and Composite Numbers

Prime numbers have always been of fundamental interest in mathematics. A number $p > 1$ is called *prime* if it has no positive factors other than itself and 1.

DEFINITION 3.2.1

Let $p \in \mathbf{N}, p > 1$. p is **prime** *if its only positive factors are* 1 *and* p.

Numbers other than 1 that are not prime are called *composite* numbers.

DEFINITION 3.2.2

Let $n \in \mathbf{N}, n > 1$. n is **composite** *if it has at least one positive factor other than* 1 *and* n.

Note that the number 1 is neither prime nor composite.

EXAMPLE 1

Tell whether each number is prime or composite.

   a. 91   **b.** 17

SOLUTION

   a. The number 91 is composite. Since $7 \mid 91$, we know 91 has positive factors other than 1 and 91.

   b. The number 17 is prime, because 17 has no positive factors other than 1 and 17.

The definitions above are not the only way one can define prime and composite. The following conditions are equivalent to the definitions of prime and composite, as you will prove in the exercises.

Prime *Let $p \in \mathbf{N}, p > 1$. Then p is prime if and only if for every $a, b \in \mathbf{N}$, $p = ab$ implies $a = 1$ or $b = 1$.*

Composite *Let $n \in \mathbf{N}, n > 1$. Then n is composite if and only if there exist $a, b \in \mathbf{N}$ such that $n = ab, 1 < a < n$, and $1 < b < n$.*

PROOF See Exercise 6.

The only way for one prime number to divide another prime number is for the two numbers to be equal. We prove this in the following lemma.

LEMMA 3.2.4

Let p and q be prime numbers. If $p \mid q$, then $p = q$.

PROOF Let $p, q \in \mathbf{N}$ be prime such that $p \mid q$.

[To show: $p = q$.]

Since $p \mid q$ and p is a prime number, we know that p is a positive factor of q. However, since q is prime, by definition its only positive factors are 1 and q. Thus, either $p = 1$ or $p = q$. Since p is a prime, we know that $p > 1$. We conclude that $p = q$. ■

A burning question

Suppose for the moment that your house of natural numbers is on fire, and you only have a few seconds to salvage a couple of belongings on your way out the door. Rushing through the hallway, you manage to grab the plus sign. As you get to the door, you see all of the natural numbers sitting on the mail table. You can only take one with you—which one do you take? The logical choice would be the number 1. The number 1 is special, because starting with only the number 1 and the addition operation, we can generate all of the natural numbers.

Now imagine that you are in the same situation—your house of numbers is burning, and you're running out the door. This time, however, you go out the back way. You don't see the plus sign, but on your way through the living room, you manage to pick up the multiplication sign. As you approach the back door, there are some baskets nearby, each containing a set of numbers. One basket holds all the even numbers, one holds all the odd numbers, one holds all the prime numbers, and one holds all the composite numbers. You're able to take only one basket with you out of the burning house—which one do you take? In this case, the best choice is the prime numbers. The prime numbers are special, because using just the prime numbers and the multiplication operation, we can generate all of the natural numbers bigger than 1.

You probably learned years ago that every natural number greater than 1 can be factored as a product of primes and that there is only one way that it can be factored (other than rearranging the order of the prime factors). The importance of this result to number theory is indicated by its name: the Fundamental Theorem of Arithmetic. We will prove shortly that every natural number greater than 1 can be factored into primes. However, we do not yet have the tools to prove that this factorization is unique. We will not be able to prove that the factorization is unique until Chapter 6, when we finish our proof of the Fundamental Theorem of Arithmetic (6.1.1).

The existence of prime factorizations

Every number can be factored into primes. Before giving a formal statement of this basic property of the natural numbers, let's try to imagine how one might prove this property.

Naomi's Numerical Proof Preview: Theorem 3.2.5

How do we know that a number like 72, for example, has a factorization into primes? You learned the basic idea in elementary school: if the number is not already prime, then by definition, it can be factored as a product of two smaller numbers. For instance, we might write

$$72 = 8 \cdot 9.$$

If the factors are prime, then we are done. If the factors are composite, as 8 and 9 are, then we can continue factoring each of these composite numbers. Eventually, we will reach a point where all of the factors are prime,

$$72 = 2 \cdot 2 \cdot 2 \cdot 3 \cdot 3,$$

and we are done. If you get the feeling that this is an inductive argument, you are right. The proof will involve strong induction. In the case of the number 72, we can take advantage of the fact that the factors 8 and 9 are smaller than 72. Our inductive hypothesis will tell us that 8 and 9 can each be factored into primes. Given this, it is easy to deduce that 72 can be factored into primes.

Now that we have a good strategy for proving the existence of prime factorizations, we will give a formal statement and proof of this fact. When stating that every number n has a factorization into primes, a question arises: what if n is already prime? For example, does 11 have a prime factorization? We will say that 11 factors as the product of a single prime, 11. Initially, this may seem like a strange choice of language, but it allows us to assert that every number factors as the product of *one or more* primes.

THEOREM 3.2.5   THE EXISTENCE OF PRIME FACTORIZATIONS

Let $n > 1$ be a natural number. Then n can be written as a product of one or more prime numbers.

PROOF We prove this using strong induction.

Let $n \in \mathbf{N}$ such that $n > 1$. Assume the inductive hypothesis that for every natural number k such that $1 < k < n$, k can be written as a product of one or more prime numbers.

[**To show: n can be written as a product of one or more prime numbers.**]

Either n is prime or n is composite.

Case 1 n is prime.

Then n can be written as the product of the single prime n.

Case 2 n is composite.

By the equivalent condition for composite numbers (3.2.3), there exist $a, b \in \mathbf{N}$ such that

$$n = a \cdot b, \text{ where } 1 < a < n \text{ and } 1 < b < n.$$

Since a and b are each smaller than n, the inductive hypothesis guarantees that a and b may each be written as a product of primes. Thus, there exist primes $p_1, p_2, \ldots, p_e$ and $q_1, q_2, \ldots, q_f$ such that

$$a = p_1 \cdot p_2 \cdot \; \cdots \; \cdot p_e \text{ and } b = q_1 \cdot q_2 \cdot \; \cdots \; \cdot q_f.$$

Since $n = a \cdot b$, we can find a prime factorization for n simply by putting these two prime factorizations together:

$$n = p_1 \cdot p_2 \cdot \; \cdots \; \cdot p_e \cdot q_1 \cdot q_2 \cdot \; \cdots \; \cdot q_f.$$

Thus, n can be written as a product of primes.

By the Principle of Strong Induction, it follows that every number $n > 1$ can be written as a product of one or more primes.
●

EXERCISES 3.2

Numerical Problems

1. Based on the definitions of prime and composite (3.2.1 and 3.2.2), is 1 a prime number? Is 1 a composite number?

2. *Consecutive* integers are integers that differ by 1, such as 17 and 18.

   a. Find a pair of consecutive integers that are both prime numbers. How many such pairs are there?

   b. Find a pair of consecutive integers that are both composite numbers. How many such pairs are there?

3. *Twin primes* are pairs of prime numbers that differ by 2, such as the primes 5 and 7. Find four more pairs of twin primes.

4. For each statement, find an example of what is described, or explain why it cannot exist.

   a. An integer that is divisible by 2 and by 3, but not by 6

   b. An integer that is divisible by 3 and by 6, but not by 18

   c. An integer that is divisible by 5 and by -5, but not by 25

   d. Two integers whose product is divisible by 10, but neither integer is divisible by 10

   e. An integer n such that $n^3 + 1$ is prime

   f. An integer n such that $2^n - 1$ is not prime

Reasoning and Proofs

5. We would like to think about prime numbers in the integers, not just in the natural numbers. We can define a *prime* in $\mathbf{Z}$ as follows: An integer p is prime if and only if for every $a, b \in \mathbf{Z}, p = ab$ implies that either $a = \pm 1$ or $b = \pm 1$, but not both. (Here "$a = \pm 1$" is shorthand for the statement "$a = 1$ or $a = -1$.")

   a. List three numbers that are prime under this new definition but were not prime under the old definition (3.2.1).

   b. Prove the following statement, or find a counterexample.

      Let b be an integer. b is a prime by the new definition if and only if one of the following holds:

      (*i*) $b \in \mathbf{N}$ and b is prime according to Definition 3.2.1, or

      (*ii*) $-b \in \mathbf{N}$ and $-b$ is prime according to Definition 3.2.1.

6. a. Prove the equivalent condition for primality (3.2.3 Prime).

   b. Prove the equivalent condition for compositeness (3.2.3 Composite).

Conjecturer's Corner (Exercises 7–10)

Experiment with some numerical examples to make a conjecture about whether each assertion is true for all integers or only for some integers. If you determine that a particular assertion is true only for some integers, state additional requirements that make the assertion true for all integers meeting these requirements.

7. If $a \mid b$ and $a \mid (b + c)$, then $a \mid c$.

8. If $a \mid bc$, then $a \mid b$ or $a \mid c$.

9. If $a \mid c$ and $b \mid c$, then $ab \mid c$.

10. If $a \mid b^2$, then $a \mid b$.

For Exercises 11–13, recall our burning house and baskets of integers. Think in terms of the following categories: *even, odd, positive, negative, prime,* **and** *composite.*

11. Suppose that when you left your burning house, you grabbed the multiplication sign and the composite numbers. What numbers can you create with these? What numbers can't you make?

12. Suppose that when you left your burning house, you grabbed the addition sign and a pair of numbers. With each given pair of numbers, which integers can you create using the addition operation? Which numbers can't you create in this way?

   a. 2 and 5   **b.** 3 and 7   **c.** 6 and 9

13. Goldbach's conjecture states that every even integer greater than 4 can be written as the sum of two prime numbers. (For example: $24 = 11 + 13$, and $50 = 3 + 47$.) This very old conjecture is still unproven. Suppose that when you run out of the burning house, you grab the prime numbers and the addition sign, but you may only add three numbers at a time. Suppose that Goldbach's conjecture is true. Which numbers can you create by adding three prime numbers at a time? Which numbers can't you create in this way?

14. Prove that any prime that can be written in the form $3n + 1$ (for some integer n) can also be written in the form $6m + 1$ (for some integer m).

15. Prove that for all $n \in \mathbf{N}$, if $n > 1$, then $n^3 + 1$ is not prime.

16. Prove that 7 is the only prime that can be written in the form $n^3 - 1$ (where $n \in \mathbf{N}$).

17. Prove that if a natural number $n > 4$ is composite, then $n \mid (n - 1)!$.

18. Prove that for every natural number n, the number $8^n + 1$ is composite.

19. Prove that every integer greater than 11 can be written as the sum of two composite numbers.

20. Prove that any natural number that can be written in the form $3n + 2$ (for some integer n) has a prime factor that also has this form.

Advanced Reasoning and Proofs

21. Let a and n be natural numbers such that $n > 1$ and $a^n - 1$ is prime.

   a. Prove that $a = 2$.

   b. Prove that n must be prime. [*Hint:* Use your result from part a.]

22. At a party, a guest asked about the host's family. "I have three children," said the host. "The product of their ages is 72, and the sum of their ages happens to be our house number." The guest went to the door and checked the house number and, after thinking a moment, said that she needed more information. "Oh, very well," replied the host, "my oldest child is named Pat." The guest was then able to give the children's ages. What were they?

23. A fast food chain sells chicken nuggets in boxes of 6, 9, or 20. Some numbers of chicken nuggets are possible to buy exactly, but some numbers are not. For example, it is possible to buy exactly 55 chicken nuggets by purchasing one box of 6, one box of 9, and two boxes of 20. On the other hand, there is no way to buy exactly 14 chicken nuggets.

   a. Is it possible to buy exactly 53 chicken nuggets? If so, explain how.

   b. Is it possible to buy exactly 17 chicken nuggets? If so, explain how.

   c. Is it possible to buy exactly 34 chicken nuggets? If so, explain how.

   d. What is the largest number of chicken nuggets that it is impossible to purchase exactly?

   e. At a competing fast food chain, turkey nuggets are sold in boxes of 7, 10, and 15. What is the largest number of turkey nuggets that it is impossible to purchase exactly?

3.3 Patterns in the Primes

There are infinitely many primes

Suppose we were to make a list of the primes, starting with the smallest and continuing in increasing order:

$$2, 3, 5, 7, 11, 13, 17, 19, 23, 29, 31, 37, 41, 43, 47, 53, 59, 61, \ldots.$$

Does this list ever end? Do we ever run out of primes? More than 2000 years ago, Euclid proved that there are infinitely many primes. No matter how many primes we list, there will always be more prime numbers.

 Naomi's Numerical Proof Preview: Theorem 3.3.1
How can we prove that there are infinitely many primes? How can we convince ourselves, for example, that there is a prime number beyond 13? One way would be to test the subsequent natural numbers in order: 14, 15, and so on, until we find the prime 17. However, it is hard to see how to turn this trial-and-error method into a general argument.

Euclid had a good idea based on the following observation. Two consecutive integers cannot both be multiples of the same prime p. For example, 42 is a multiple of 7, so the next number, 43, cannot possibly be a multiple of 7. Thus, if we find a number B that is divisible by all of the primes 2, 3, 5, 7, 11, and 13, then $B + 1$ will be divisible by none of these primes.

The easiest way to create a number B that is divisible by all of the primes up to 13 is to let B equal their product:

$$B = 2 \cdot 3 \cdot 5 \cdot 7 \cdot 11 \cdot 13.$$

Then since B is clearly divisible by all primes less than or equal to 13, the number $B + 1$ will not be divisible by any of these primes. Hence, the prime factorization of $B + 1$ must consist entirely of primes greater than 13. In particular, there must be at least one prime greater than 13. This argument, unlike our trial-and-error method, can easily be turned into a general proof.

It is interesting to note that in this argument, $B + 1 = 2 \cdot 3 \cdot 5 \cdot 7 \cdot 11 \cdot 13 + 1$ is not a prime number:

$$2 \cdot 3 \cdot 5 \cdot 7 \cdot 11 \cdot 13 + 1 = 30031 = 59 \cdot 509.$$

Thus, multiplying the first few primes and adding 1 does not always result in a prime number. Indeed, in the argument above, we did not claim that $B + 1$ is prime; we showed only that the prime factors of $B + 1$ must be larger than 13 (which of course is true of 59 and 509).

THEOREM 3.3.1

There are infinitely many primes.

PROOF (By contradiction.)

Assumption: Suppose that there are only finitely many primes. That is, suppose that

$$p_1, p_2, p_3, \ldots, p_n \tag{1}$$

is a complete list of all the primes.

Let B be the product of all of the numbers in list (1):

$$B = p_1 \cdot p_2 \cdot p_3 \cdot \ \cdots \ \cdot p_n.$$

Consider the number $B + 1$. By the existence of prime factorizations (3.2.5), we know that $B + 1$ can be factored into prime numbers. Let q be any prime appearing in the prime factorization of $B + 1$. We claim that q is a new prime; that is,

Claim: The prime q is not equal to any of the primes in list (1): $p_1, p_2, p_3, \ldots, p_n$.

Proof of Claim (By contradiction.)

Suppose that q is equal to one of the primes in list (1); that is, $q = p_k$ for some $k = 1, 2, \ldots, n$. Since p_k is a prime factor of B,

$$q \mid B.$$

But q is a prime factor of $B + 1$, so

$$q \mid B + 1.$$

Thus, by the Linear Combination Lemma (3.1.3),

$$q \mid B + 1 - B.$$

Simplifying, we get

$$q \mid 1.$$

However, no prime number divides 1. ⇒⇐. Since we have reached a contradiction, our supposition, that q is equal to one of the primes in list (1), must be false. We have thus proven our Claim.

The Claim tells us that q is a prime number that is not in list (1). But list (1) was supposed to be a complete list containing all prime numbers. ⇒⇐. Since we have reached a contradiction, our initial Assumption, that there are only finitely many primes, must be false. We conclude that there are infinitely many primes. ■

Is it prime?

What do we have to check in order to be certain that a number n is prime? For example, suppose we want to determine whether $n = 853$ is prime. If we verify that none of the integers in the list

$$2, 3, 4, 5, 6, \ldots, 850, 851, 852$$

divides the number 853, then we will know that 853 is prime. In general, to show that a number n is prime, it is sufficient to check that none of the integers between 1 and n is a factor of n.

However, it is not necessary to check *all* of the natural numbers between 1 and n. In fact, it is sufficient to check that n is not divisible by any of the natural numbers that are less than or equal to $\sqrt{n}$. The reason is that if $n = ab$, then the factors a and b cannot both be greater than $\sqrt{n}$. (If it were true that both $a > \sqrt{n}$ and $b > \sqrt{n}$, then we would have $ab > n$, which contradicts the fact that $n = ab$.) Since $\sqrt{853} \approx 29.206$, in order to know that 853 is prime, it is sufficient to check that none of the integers in the list

$$2, 3, 4, 5, 6, \ldots, 27, 28, 29$$

divides the number 853. This is a huge improvement over checking all the numbers up to 852.

Yet this is still more work than is necessary. To verify that a number n is prime, we only need to make sure that it is not divisible by any of the primes up to $\sqrt{n}$. (See Exercise 10.) In particular, to know for sure that $n = 853$ is prime, we only need to check that it is not divisible by any of the prime numbers up to 29:

$$2, 3, 5, 7, 11, 13, 17, 19, 23, 29.$$

Checking divisibility by all of the primes up to $\sqrt{n}$ is vastly preferable to checking divisibility by all numbers less than n. Nonetheless, for extremely large numbers n, this method is still prohibitively slow. We will discuss much faster methods for primality testing in Chapter 12.

The sieve of Eratosthenes

We would like to have a good method for finding prime numbers. There is an ancient method based on the idea that if you start with a list of all of the positive integers between 1 and any number n, then eliminate all of the composite numbers up to n, everything that is left must be prime. The method is called the *sieve of Eratosthenes*. To find all of the prime numbers up to n, first write down all of the numbers from 1 to n. For example, to find all primes between 1 and 100, we first write down all the numbers from 1 to 100:

1	7	13	19	25	31	37	43	49	55	61	67	73	79	85	91	97
2	8	14	20	26	32	38	44	50	56	62	68	74	80	86	92	98
3	9	15	21	27	33	39	45	51	57	63	69	75	81	87	93	99
4	10	16	22	28	34	40	46	52	58	64	70	76	82	88	94	100
5	11	17	23	29	35	41	47	53	59	65	71	77	83	89	95	
6	12	18	24	30	36	42	48	54	60	66	72	78	84	90	96	

Since 1 is neither prime nor composite, we skip it and circle the next number on our list, 2. The circle indicates that 2 is prime. We then cross out every number on the list that is a multiple of 2:

1	7	13	19	25	31	37	43	49	55	61	67	73	79	85	91	97
②	8̶	1̶4̶	2̶0̶	2̶6̶	3̶2̶	3̶8̶	4̶4̶	5̶0̶	5̶6̶	6̶2̶	6̶8̶	7̶4̶	8̶0̶	8̶6̶	9̶2̶	9̶8̶
3	9	15	21	27	33	39	45	51	57	63	69	75	81	87	93	99
4̶	1̶0̶	1̶6̶	2̶2̶	2̶8̶	3̶4̶	4̶0̶	4̶6̶	5̶2̶	5̶8̶	6̶4̶	7̶0̶	7̶6̶	8̶2̶	8̶8̶	9̶4̶	1̶0̶0̶
5	11	17	23	29	35	41	47	53	59	65	71	77	83	89	95	
6̶	1̶2̶	1̶8̶	2̶4̶	3̶0̶	3̶6̶	4̶2̶	4̶8̶	5̶4̶	6̶0̶	6̶6̶	7̶2̶	7̶8̶	8̶4̶	9̶0̶	9̶6̶	

The next number on the list that is not crossed out, 3, must be prime because it is not divisible by 2 (the only prime smaller than 3). So, we circle 3, then cross out every multiple of 3 on the list:

1	7	13	19	25	31	37	43	49	55	61	67	73	79	85	91	97
②	8̶	1̶4̶	2̶0̶	2̶6̶	3̶2̶	3̶8̶	4̶4̶	5̶0̶	5̶6̶	6̶2̶	6̶8̶	7̶4̶	8̶0̶	8̶6̶	9̶2̶	9̶8̶
③	9̶	1̶5̶	2̶1̶	2̶7̶	3̶3̶	3̶9̶	4̶5̶	5̶1̶	5̶7̶	6̶3̶	6̶9̶	7̶5̶	8̶1̶	8̶7̶	9̶3̶	9̶9̶
4̶	1̶0̶	1̶6̶	2̶2̶	2̶8̶	3̶4̶	4̶0̶	4̶6̶	5̶2̶	5̶8̶	6̶4̶	7̶0̶	7̶6̶	8̶2̶	8̶8̶	9̶4̶	1̶0̶0̶
5	11	17	23	29	35	41	47	53	59	65	71	77	83	89	95	
6̶	1̶2̶	1̶8̶	2̶4̶	3̶0̶	3̶6̶	4̶2̶	4̶8̶	5̶4̶	6̶0̶	6̶6̶	7̶2̶	7̶8̶	8̶4̶	9̶0̶	9̶6̶	

The next number that has not been crossed out is 5. It must be prime, because it is not divisible by any prime smaller than itself. We circle 5 and cross out every multiple of 5. We then circle the next number that has not been crossed out, 7, and cross out every multiple of 7.

We could continue in the same manner, by circling 11 and crossing out its multiples, then circling 13 and crossing out its multiples, and so on. If we were to do so, however, we would not cross out any new numbers (other than the ones that have already been crossed out). The reason is that any composite number less than or equal to 100 must be divisible by a prime that is less than or equal to $\sqrt{100} = 10$. That is, every composite number up to 100 is divisible by one of the primes 2, 3, 5, or 7. But we have already crossed out all the multiples of these primes, so the remaining numbers on the list must be prime.

The last step of the sieve is to circle all numbers that remain, because they are all prime:

1	⑦	⑬	⑲	25	㉛	㊲	㊸	49	55	㊽	㊻	㊼	⑲	85	91	㊆

Figure 1 The completed sieve of Eratosthenes for the numbers up to 100.

We have now completed the sieve of Eratosthenes for the numbers up to 100. The circled numbers are all of the primes less than or equal to 100.

The Prime Number Theorem

As we showed in Theorem 3.3.1, there are infinitely many prime numbers. However, as we examine larger and larger numbers, primes appear less and less frequently (on average). To explore the density of primes among the natural numbers, we will consider the function $\pi(n)$, which is defined as follows.

DEFINITION 3.3.2

Let $n \in \mathbf{N}$. We define $\boldsymbol{\pi(n)}$ = the number of primes less than or equal to n.

EXAMPLE 1

$\pi(10) = 4$, because there are 4 primes that are less than or equal to 10: the primes 2, 3, 5, and 7.

Let's make a graph of $\pi(n)$ for some small values of n.

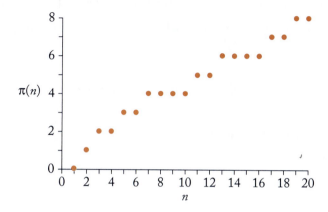

Notice that $\pi(n)$ is a step function. The value of $\pi(n)$ increases by 1 every time a new prime occurs. For example,

$$\pi(7) = \pi(8) = \pi(9) = \pi(10) = 4,$$

but $\pi(11) = 5$ since 11 is prime, and the number of primes less than or equal to 11 is one greater than the number of primes less than or equal to 10.

Let's see what happens when we zoom out and examine the graph of $\pi(n)$ for a greater range of values of n.

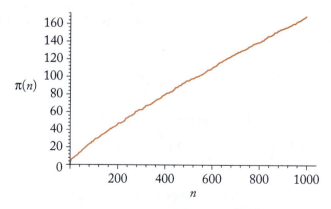

Notice that even though $\pi(n)$ is a discrete step function, when we zoom out and look at $\pi(n)$ for much larger values of n, the graph almost appears to be continuous. Perhaps we could find a continuous function that gives a close approximation to the discrete function $\pi(n)$ for large values of n.

In the 18th century, both Gauss and Legendre conjectured that for large values of n, the function $\pi(n)$ could be well approximated by the function $\dfrac{n}{\log n}$, where $\log n$ denotes the natural logarithm of n. Suppose you wish to know the number of primes less than one quadrillion, $n = 10^{15}$. To find the exact answer, you would need to count every single prime less than 10^{15}, which is a daunting task that would take a long time even using a computer. However, you could approximate the answer instantly using the formula $\dfrac{n}{\log n}$ on a pocket calculator and find that $\pi(10^{15})$ is approximately

$$\frac{10^{15}}{\log 10^{15}} \approx 28{,}952{,}965{,}460{,}217.$$

Let's compare this with the exact answer:

$$\pi(10^{15}) = 29{,}844{,}570{,}422{,}669.$$

The formula gives us a pretty good approximation. To determine precisely how good, we consider the ratio of the exact answer to the approximation:

$$\frac{\pi(10^{15})}{\left(\dfrac{10^{15}}{\log 10^{15}} \right)} \approx 1.0308.$$

The error of this approximation is only about 3%. Not too shabby.

More than a century after Gauss and Legendre made their conjecture, in 1896, two French mathematicians Hadamard and Vallée-Poussin independently proved that for large n,

$$\pi(n) \approx \frac{n}{\log n}.$$

More precisely, their result, known as the Prime Number Theorem, asserts that as n gets larger and larger, the ratio of $\pi(n)$ to $\dfrac{n}{\log n}$ approaches 1. This is stated in the following theorem.

THEOREM 3.3.3　THE PRIME NUMBER THEOREM

$$\lim_{n \to \infty} \frac{\pi(n)}{\left(\dfrac{n}{\log n} \right)} = 1.$$

A proof is too complex for us to include in this text.

The Prime Number Theorem tells how the primes are distributed on a grand scale, but it does not tell us precisely which numbers will be prime and which will not. For example, suppose you move to San Diego, and the phone company is going to randomly assign you a phone number in the 619 area code. What is the probability that your 10-digit phone number will be a prime number? The number of primes between 619 000 0001 and 620 000 0000 is equal to

$$\underbrace{\pi(6.20 \cdot 10^9)}_{\substack{\text{the number} \\ \text{of primes} \\ \leq 620\,000\,0000}} - \underbrace{\pi(6.19 \cdot 10^9)}_{\substack{\text{the number} \\ \text{of primes} \\ \leq 619\,000\,0000}}$$

We can approximate this using the Prime Number Theorem:

$$\pi(6.20 \cdot 10^9) - \pi(6.19 \cdot 10^9) \approx \frac{6.20 \cdot 10^9}{\log(6.20 \cdot 10^9)} - \frac{6.19 \cdot 10^9}{\log(6.19 \cdot 10^9)}$$

$$\approx 424{,}000.$$

Thus, the density of primes between the numbers 619 000 0001 and 620 000 0000 is

$$\frac{\pi(6.20 \times 10^9) - \pi(6.19 \times 10^9)}{6.20 \times 10^9 - 6.19 \times 10^9} \approx \frac{424{,}000}{10{,}000{,}000} \approx \frac{1}{24}.$$

The Prime Number Theorem has enabled us to calculate that a phone number in the 619 area code has about a 1 in 24 chance of being prime, but it does not tell us specifically which of the numbers are prime and which are not.

A formula for primes?

For centuries, mathematicians have dreamed of finding a simple formula that generates only primes—that is, a formula $f(n)$ into which one could put any $n \in \mathbf{N}$ and always obtain a prime number. Such a formula would generate an infinite sequence of prime numbers and no composite numbers. Fermat thought he had just such a formula when he conjectured that all numbers of the form

$$F(n) = 2^{2^n} + 1$$

are prime. The sequence of numbers produced by Fermat's function $F(n)$ are known as *Fermat numbers*. The first five Fermat numbers are

n	1	2	3	4	5
$F(n)$	5	17	257	65,537	4,294,967,297

Today, it is easy to check the primality of numbers of the size of $F(5) = 4,294,967,297$, but before computers, it was a trying task. Fermat's conjecture stood for a century until Euler showed that, in fact, $F(5)$ is composite. To date, no Fermat numbers beyond $F(4)$ are known to be prime.

Euler himself noticed that the function

$$f(n) = n^2 + n + 41$$

seemed to be particularly good at generating primes. Let's consider the first several values of $f(n)$.

n	1	2	3	4	5	6	7	8	9
$f(n) = n^2 + n + 41$	43	47	53	61	71	83	97	113	131

The first nine values of $f(n)$ are prime, and it does not stop there. Impressively, all of the first 39 values of the function are prime! Alas, the pattern breaks down eventually—the value of $f(40)$ is composite.

Given a function $g(n)$, we can also ask whether the sequence $g(n)$ generates infinitely many prime numbers (allowing composite values to be mixed in as well). For example, we can ask of the function $g(n) = n^2 + 1$ whether there are infinitely many primes among the values of $g(n)$.

n	1	2	3	4	5	6	7	8	9	10
$g(n) = n^2 + 1$	2	5	10	17	26	37	50	65	82	101

It is evident from the table that among the first few values of $n^2 + 1$, we find quite a few primes (shown in orange), but at present no one knows whether there are infinitely many primes of the form $n^2 + 1$.

If the function in question is linear, such as $g(n) = 5n + 2$, then the question of whether there are infinitely many primes of this form was answered by an important (and deep) theorem of Dirichlet. In the 19th century, Dirichlet proved that if a and b are natural numbers that have no common factors, then the sequence $D(n) = an + b$

(for $n \in \mathbf{N}$) contains infinitely many primes. For example, Dirichlet's theorem guarantees that the arithmetic progression

$$\{5n + 2 : n \in \mathbf{N}\} = \{7, 12, 17, 22, \ldots\}$$

contains infinitely many primes.

There are many sequences, such as the Fermat numbers $F(n)$ and the sequence $g(n) = n^2 + 1$, of which it is not known whether the number of primes in the sequence is infinite, or whether after some point the sequence will only generate composite numbers. Another sequence that is often used to generate very large primes, despite there being no guarantee that it will continue to produce primes, are the Mersenne numbers, which are numbers of the form

$$M(p) = 2^p - 1,$$

where p is prime. Primes of this form, such as $2^5 - 1 = 31$ and $2^7 - 1 = 127$, are known as *Mersenne primes*. Mersenne numbers are often used as candidates in the never-ending search for larger and larger primes.

At the time of the writing of this chapter, the largest known prime is a Mersenne prime,

$$2^{43,112,609} - 1.$$

This 12,978,189-digit prime was discovered in 2008 by Edson Smith of the University of California, Los Angeles (UCLA) as part of the Great Internet Mersenne Prime Search (GIMPS), in which tens of thousands of computers look for Mersenne primes when not in use by their owners. Smith, the Computing Manager for the UCLA Mathematics Department, had installed the GIMPS software to run during idle time on the machines in a large computer lab. Since this was the first known prime with 10 million or more digits, its discoverers were awarded a $100,000 prize from the Electronic Frontier Foundation.

The rapid progress of the search for larger and larger primes is illustrated in Figure 2.

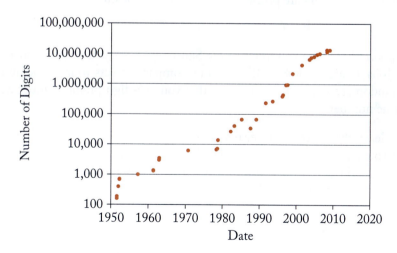

Figure 2 This graph plots the number of digits in the largest known prime for several years.

EXERCISES 3.3

Numerical Problems

1. a. Perform the sieve of Eratosthenes on the natural numbers up to 288:

1	7	13	19	25	31	37	43	49	55	61	67	73	79	85	91	97	103	109	115	121	127	133	139
2	8	14	20	26	32	38	44	50	56	62	68	74	80	86	92	98	104	110	116	122	128	134	140
3	9	15	21	27	33	39	45	51	57	63	69	75	81	87	93	99	105	111	117	123	129	135	141
4	10	16	22	28	34	40	46	52	58	64	70	76	82	88	94	100	106	112	118	124	130	136	142
5	11	17	23	29	35	41	47	53	59	65	71	77	83	89	95	101	107	113	119	125	131	137	143
6	12	18	24	30	36	42	48	54	60	66	72	78	84	90	96	102	108	114	120	126	132	138	144

145	151	157	163	169	175	181	187	193	199	205	211	217	223	229	235	241	247	253	259	265	271	277	283
146	152	158	164	170	176	182	188	194	200	206	212	218	224	230	236	242	248	254	260	266	272	278	284
147	153	159	165	171	177	183	189	195	201	207	213	219	225	231	237	243	249	255	261	267	273	279	285
148	154	160	166	172	178	184	190	196	202	208	214	220	226	232	238	244	250	256	262	268	274	280	286
149	155	161	167	173	179	185	191	197	203	209	215	221	227	233	239	245	251	257	263	269	275	281	287
150	156	162	168	174	180	186	192	198	204	210	216	222	228	234	240	246	252	258	264	270	276	282	288

b. Use your results from part a to find $\pi(200)$.

2. Use the primes you found in Exercise 1 to help you determine which of the following numbers are prime and which are composite:

a. 593   **b.** 617   **c.** 713

3. Suppose that when you move to the San Diego area, you actually move to La Jolla, in area code 858. If the phone company selects your phone number at random, estimate the probability that your 10-digit phone number will be a prime number.

4. You learn that your employer plans to relocate you to the town of Springfield, but you neglected to find out in which state, so you do not know the area code. Estimate the probability that your seven-digit phone number, chosen at random by the phone company, will be a prime number.

Reasoning and Proofs

5. This exercise concerns Euler's prime-generating function, $f(n) = n^2 + n + 41$. Without calculating the value of $f(40)$ or $f(41)$, show that $f(40)$ and $f(41)$ are both composite.

Phil Lovett   For Exercises 6–7: Your friend Phil Lovett tells you he has discovered a function $h(n)$ that generates primes for all natural numbers n. His function is $h(n) = n^2 - 79n + 1601$.

6. Without actually calculating any values of $h(n)$, give a value of n for which $h(n)$ will be composite. Explain your reasoning.

7. **Computer Exercise** Apply Phil's formula to all values of n between 1 and 10. Is $h(n)$ prime for all of those values? Find the first number n for which Phil's formula does not yield a prime number.

8. Euler's prime-generating function, $f(n) = n^2 + n + 41$, is a second-degree polynomial function with constant term 41. Show that if $g(n)$ is any polynomial function with integer coefficients and constant term $c > 1$, then there must be a natural number n for which $g(n)$ is composite.

9. **Phil Lovett** You tell your friend Phil Lovett about Dirichlet's theorem, which states that if $a, b \in \mathbf{N}$ have no common factors, then the sequence $D(n) = an + b$ (for $n \in \mathbf{N}$) contains infinitely many primes.

    Offended that you would hold any mathematician in higher regard than him, Phil cries, "I will not be outdone by some dead guy!" Phil proceeds to describe his own sequence: "Start with two natural numbers, a and b, that have a common factor [i.e., there is some natural number $d > 1$ such that $d \mid a$ and $d \mid b$]. Then the Lovett sequence is
    $$L(n) = an + b \text{ (for } n \in \mathbf{N}).$$
    Dude, my sequence contains way more primes than Dirichlet's!"

    a. Choose a pair of natural numbers a and b that share a common factor, $d > 1$. Write the first 10 terms of the sequence $L(n)$. Are any of them prime?

    b. Repeat part a with a different pair of values for a and b.

    c. Prove that no matter which values you use for a and b, the sequence $L(n)$ will contain no primes whatsoever.

10. a. Suppose that $n > 1$ is a natural number and n is not divisible by any natural number d such that $1 < d \le \sqrt{n}$. Prove that n is prime.

    b. Suppose that $n > 1$ is a natural number and n is not divisible by any prime number p such that $p \le 120$. Prove that n is prime.

Advanced Reasoning and Proofs

11. Let p_n denote the nth prime number; that is, $p_1 = 2$, $p_2 = 3$, $p_3 = 5$, $p_4 = 7$, and so on.

    a. Prove that for every $n \in \mathbf{N}$, $p_{n+1} \le p_1 \cdot p_2 \cdot \ \cdots \ \cdot p_n + 1$. [*Hint:* Use ideas from the proof that there are infinitely many primes (Theorem 3.3.1).]

b. Use induction and your result from part a to prove that for every $n \in \mathbf{N}$, $p_n < 2^{2^n}$.

c. Use your result from part b to conclude that for every $n \in \mathbf{N}$, $\pi(n) \geq \log_2(\log_2(2n))$.

12. **a.** Prove that there are infinitely many composite numbers of the form $5k + 2$, where $k \in \mathbf{N}$.

   b. Prove that there are infinitely many composite numbers of the form $3k + 1$, where $k \in \mathbf{N}$.

   c. Let a and b be natural numbers. Prove that there are infinitely many composite numbers of the form $ak + b$, where $k \in \mathbf{N}$.

13. **a.** Let n be a natural number. Prove that every factor of $n! + 1$ other than 1 is strictly greater than n.

   b. Use this fact to give another proof that there are infinitely many primes.

14. Let $n > 2$. Prove that there exists a prime p such that $n < p < n!$.

15. Find a million consecutive composite numbers. Explain your reasoning.

16. Prove that for every natural number n, there are at least n consecutive composite numbers.

17. Let $F(n)$ denote the nth Fermat number: $F(n) = 2^{2^n} + 1$.

   a. Show that no prime divides more than one Fermat number.
   [*Hint:* Assume that $p \mid F(m)$ and $p \mid F(n)$, with $m < n$. How are $F(m) - 1$ and $F(n) - 1$ related? Use what you know about binomial expansions to give a contradiction to this assumption.]

   b. Use the result you proved in part a to give another proof that there are infinitely many primes.

3.4 Common Divisors and Common Multiples

Given two integers, any integer that divides them both is called a *common divisor* of those two integers.

> **DEFINITION 3.4.1**
>
> *Let a and b be integers that are not both equal to* 0. *The* **greatest common divisor** *of a and b is the largest* $d \in \mathbf{N}$ *for which* $d \mid a$ *and* $d \mid b$. *We express this in symbols as*
>
> $$d = \gcd(a, b).$$

In this definition, we have assumed that such a number d always exists. This follows from Corollary 2.2.4 of the Well-Ordering Principle. (See Exercise 16.)

EXAMPLE 1

Find the greatest common divisor of 18 and 30.

SOLUTION

The divisors of 18 are 1, 2, 3, 6, 9, and 18.
The divisors of 30 are 1, 2, 3, 5, 6, 10, 15, and 30.
  The greatest common divisor of 18 and 30 is the largest number that appears in both of these lists: $\gcd(18, 30) = 6$. ●

You may be familiar with another method for finding greatest common divisors that involves factoring each number into primes. We demonstrate this method in the following example. Since the proof that this method works relies on the uniqueness of prime factorizations, we will have to wait until Section 6.2 to see a proof. Nonetheless, we will use this method for Example 2 as well as the numerical problems in the exercises.

EXAMPLE 2

Find each greatest common divisor.

    a. $\gcd(234, 540)$ **b.** $\gcd(19, -57)$ **c.** $\gcd(77, 250)$

SOLUTION To determine the gcd of each pair of numbers, we begin by factoring the numbers into primes. The greatest common divisor is the product of the highest power of each factor that divides both numbers in the pair. Note that this method assumes there is only one way to factor a number into primes, which we will not prove until Chapter 6. Nonetheless, we use the method in numerical examples to gain intuition about our definitions.

    a. $234 = 2^1 \cdot 3^2 \cdot 13^1$ $540 = 2^2 \cdot 3^3 \cdot 5^1$

        $\gcd(234, 540) = 2^1 \cdot 3^2 = 18$.

    b. 19 is prime $-57 = -3 \cdot 19$

        $\gcd(19, -57) = 19$.

    c. $77 = 7 \cdot 11$ $250 = 2 \cdot 5^3$

        Since 77 and 250 have no positive factors in common other than 1, $\gcd(77, 250) = 1$. ●

Numbers such as 77 and 250, which have no common factors greater than 1, are called *relatively prime*.

DEFINITION 3.4.2

Let $a, b \in \mathbf{Z}$. *We say a and b are* **relatively prime** *if* $\gcd(a, b) = 1$.

EXAMPLE 3

  a. The numbers 77 and 250 are relatively prime, because $\gcd(77, 250) = 1$.

  b. The numbers 234 and 540 are *not* relatively prime, because $\gcd(234, 540) = 18$.

Which numbers are relatively prime to 7? Since 7 is prime, any number that does not have 7 as a factor will have no common factors with 7 other than 1. In general, if a prime number p is not a factor of an integer n, then p and n are relatively prime.

LEMMA 3.4.3

Let p be prime, and let n be an integer. If p does not divide n, then $\gcd(p, n) = 1$.

PROOF See Exercise 7.

In part b of Example 3, we saw that 234 and 540 are not relatively prime since they share a common factor of 18. However, if we divide both numbers by their gcd, we get

$$\frac{264}{18} = 13 \qquad \text{and} \qquad \frac{540}{18} = 30.$$

The resulting integers, 13 and 30, are relatively prime. In general, if two integers are not relatively prime, we can divide them by their gcd to create a pair of relatively prime numbers. This fact is expressed in the following lemma.

LEMMA 3.4.4

Let a and b be integers that are not both equal to zero, and let $d = \gcd(a, b)$.
Let x and y be integers such that $a = xd$ *and* $b = yd$. *Then* $\gcd(x, y) = 1$.

Put it in Prose, Paul!

Start with any two integers, and factor out their gcd. The resulting numbers will be relatively prime.

PROOF Let $a, b \in \mathbf{Z}$ such that a and b are not both 0, and let $d = \gcd(a, b)$. Let $x, y \in \mathbf{Z}$ such that

$$a = xd \text{ and } b = yd. \tag{1}$$

Let $c = \gcd(x, y)$. **[To show: $c = 1$.]**

Since $c \mid x$ and $c \mid y$, there exist integers r and s such that

$$x = rc \qquad \text{and} \qquad y = sc.$$

Substituting these equations into the equations in (1), we get

$$a = rcd \qquad \text{and} \qquad b = scd.$$

Thus, $cd \mid a$ and $cd \mid b$, so cd is a common divisor of a and b. It follows that

$$cd \le d,$$

because d is the *greatest* common divisor of a and b. Dividing both sides of this inequality by d, we get

$$c \le 1.$$

Since $c \ge 1$, it follows that $c = 1$. ∎

Lemma 3.4.4 is the key to proving that every fraction can be written *in lowest terms*—that is, as a fraction whose numerator and denominator have no common factors. We state this familiar fact in the following corollary.

COROLLARY 3.4.5

Given any rational number r, there exist two relatively prime integers p and q, with $q \ne 0$, such that $r = \dfrac{p}{q}$.

PROOF See Exercise 13.

If two integers each divide a number m, then m is called a *common multiple* of the two integers.

DEFINITION 3.4.6

*Let a and b be nonzero integers. The **least common multiple** of a and b is the smallest $m \in \mathbf{N}$ for which $a \mid m$ and $b \mid m$. We express this in symbols as*

$$m = \operatorname{lcm}(a, b).$$

In this definition we have assumed that such a number m always exists. This follows from the Well-Ordering Principle (2.2.2). (See Exercise 15.)

EXAMPLE 4
· · · · · · · · · · · ·

Find each least common multiple.

    a. lcm(1960, 1100)   **b.** lcm(27, 85)   **c.** lcm(11, −132)

SOLUTION To determine the lcm of each pair of numbers, begin by factoring the numbers. The least common multiple is the product of the highest power of each factor that appears in either number.

    a. $1960 = 2^3 \cdot 5^1 \cdot 7^2$ $1100 = 2^2 \cdot 5^2 \cdot 11^1$

           $\mathrm{lcm}(1960, 1100) = 2^3 \cdot 5^2 \cdot 7^2 \cdot 11^1 = 107{,}800.$

    b. $27 = 3^3$ $85 = 5 \cdot 17$

           $\mathrm{lcm}(27, 85) = 3^3 \cdot 5 \cdot 17 = 2295.$

    c. $11 = 11$ $-132 = -3 \cdot 4 \cdot 11$

           $\mathrm{lcm}(11, -132) = 3 \cdot 4 \cdot 11 = 132.$

The method we used in Example 4 to find the least common multiple, like the method for finding the gcd from Example 2, will not be rigorously proven until Section 6.2. However, you should feel free to use this method for the numerical problems in the exercises.

The definitions of greatest common divisor and least common multiple can be extended to take 3, 4, or any number of arguments.

DEFINITION 3.4.7

Let $a_1, a_2, \ldots, a_n$ be integers that are not all 0. The **greatest common divisor** *of this set of numbers is the largest $d \in \mathbf{N}$ such that $d \mid a_k$ for all $k = 1, 2, \ldots, n$. We express this in symbols as*

$$d = \gcd(a_1, a_2, \ldots, a_n).$$

DEFINITION 3.4.8

Let $a_1, a_2, \ldots, a_n$ be nonzero integers. The **least common multiple** *of this set of numbers is the smallest $m \in \mathbf{N}$ such that $a_k \mid m$ for all $k = 1, 2, \ldots, n$. We express this in symbols as*

$$m = \mathrm{lcm}(a_1, a_2, \ldots, a_n).$$

EXAMPLE 5

    a. Find $\gcd(28, 35, 98)$.

    b. Find $\operatorname{lcm}(28, 35, 98)$.

SOLUTION First factor each number into primes.

$$28 = 2^2 \cdot 7 \qquad 35 = 5 \cdot 7 \qquad 98 = 2 \cdot 7^2$$

    a. $\gcd(28, 35, 98) = 7$.

    b. $\operatorname{lcm}(28, 35, 98) = 2^2 \cdot 5 \cdot 7^2 = 980$. ●

EXERCISES 3.4

Numerical Problems

1. Find each greatest common divisor.

    a. $\gcd(1331, 2431)$ **b.** $\gcd(-70, -203)$ **c.** $\gcd(-60, 207)$

    d. $\gcd(0, 99)$ **e.** $\gcd(3713, 3869)$ **f.** $\gcd(24, 34, 44)$

    g. $\gcd(15, 20, 48)$ **h.** $\gcd(1331, 2431, 2090)$ **i.** $\gcd(-60, 207, 99, 100)$

    j. $\gcd(3^5 \cdot 5^2, 2^3 \cdot 3 \cdot 7^2)$ **k.** $\gcd(2^2 \cdot 3^2 \cdot 5^2 \cdot 11, 5^3 \cdot 11 \cdot 17^5, 2^2 \cdot 3^2 \cdot 17)$

2. Find each least common multiple.

    a. $\operatorname{lcm}(24, 36)$ **b.** $\operatorname{lcm}(-50, -26)$ **c.** $\operatorname{lcm}(91, -49)$

    d. $\operatorname{lcm}(207, 414)$ **e.** $\operatorname{lcm}(96, 64)$ **f.** $\operatorname{lcm}(67, -67)$

    g. $\operatorname{lcm}(16, 24, 36)$ **h.** $\operatorname{lcm}(5, -17, 39)$ **i.** $\operatorname{lcm}(-22, -33, -44)$

    j. $\operatorname{lcm}(3^5 \cdot 5^2, 2^3 \cdot 3 \cdot 7^2)$ **k.** $\operatorname{lcm}(2^2 \cdot 3^2 \cdot 5^2 \cdot 11, 5^3 \cdot 11 \cdot 17^5, 2^2 \cdot 3^2 \cdot 17)$

Reasoning and Proofs

3. Find three integers a, b, and c so that any two of them share a common factor, but $\gcd(a, b, c) = 1$.

4. Find four integers a, b, c and d so that any three of them share a common factor, but $\gcd(a, b, c, d) = 1$.

5. Let a and n be natural numbers, and let $d = \gcd(a, n)$. Let $a' \in \mathbf{N}$ such that $a = a'd$. Does it follow that $\gcd(a', n) = 1$? If so, prove it. If not, give a counterexample.

6. Let $n > 1$ be an integer. Find

    a. $\gcd(n, 3n)$;

    b. $\gcd(n, n + 1)$;

    c. $\gcd(n - 1, n + 1)$, in the case that n is odd;

    d. $\gcd(n - 1, n + 1)$, in the case that n is even;

    e. $\gcd(n^2 - 1, n + 1)$.

7. Prove Lemma 3.4.3.

8. State and prove the converse of Lemma 3.4.3.

9. Let a and b be natural numbers. Prove that if $a \mid b$, then $\gcd(a, b) = a$.

10. Let $a, b, c \in \mathbf{Z}$ with $\gcd(a, b) = 1$ and $c \mid (a + b)$. Prove that $\gcd(a, c) = 1$ and $\gcd(b, c) = 1$.

11. Let a and b be nonzero integers. Prove that $\gcd(a, b) = \text{lcm}(a, b)$ if and only if $a = b$ or $a = -b$.

12. Let a and b be nonzero integers. Prove that $\gcd(a, a + b)$ divides b.

13. Prove Corollary 3.4.5.

14. Let $a, b,$ and c be nonzero integers. Prove that if $\gcd(a, b) = 1$ and $c \mid a$, then $\gcd(b, c) = 1$.

15. To justify our definition of least common multiple, we must prove the following:

    Let a and b be integers that are not both 0. Then the set
$S = \{c \in \mathbf{N} \mid c \text{ is a common multiple of } a \text{ and } b\}$ has a smallest element.

    Prove this assertion, using the Well-Ordering Principle (2.2.2).

16. To justify our definition of greatest common divisor, we must prove the following:

    Let a and b be integers that are not both 0. Then the set
$S = \{c \in \mathbf{N} \mid c \text{ is a common divisor of } a \text{ and } b\}$ has a largest element.

    Prove this assertion, using Corollary 2.2.4 of the Well-Ordering Principle.

Conjecturer's Corner (Exercises 17–22)

17. For each pair of numbers, find the greatest common divisor and the least common multiple. Then compute the product of the gcd and the lcm.

    a. 15 and 10   **b.** 19 and 10   **c.** 25 and 5   **d.** 30 and 70

18. Based on your answers to Exercise 17 (and by trying several more examples, if necessary), find an expression to put in the blank that you believe will make the statement true. Write the complete statement of your conjecture, with the blank filled in.

For all $a, b \in \mathbf{N}, \gcd(a,b) \cdot \mathrm{lcm}(a,b) = \underline{\hspace{2cm}}$.

For Exercises 19–22: By trying several examples, find an expression to put in the blank that you believe will make the statement true. Write the complete statement of your conjecture, with the blank filled in.

19. For all $a, b, k \in \mathbf{N}, \mathrm{lcm}(\gcd(k, a), \gcd(k, b)) = \gcd(k, \underline{\hspace{1.5cm}})$.

20. For all $a, b, c \in \mathbf{N}, \gcd(a, b, c) \cdot \mathrm{lcm}(ab, bc, ac) = \underline{\hspace{2cm}}$.

21. For all $a, b, c \in \mathbf{N}, \gcd(\gcd(a, b), c) = \underline{\hspace{2cm}}$.

22. For all $a, b, c \in \mathbf{N}, \mathrm{lcm}(\mathrm{lcm}(a, b), c) = \underline{\hspace{2cm}}$.

Advanced Reasoning and Proofs

23. Let k be a natural number, and let a and b be nonzero integers. Prove that

$$\mathrm{lcm}(ka, kb) = k \cdot \mathrm{lcm}(a, b).$$

24. Let $x, y \in \mathbf{Z}$ and let $a \in \mathbf{N}$. Prove that

$$\gcd(ax, ay) = a \cdot \gcd(x, y).$$

25. Prove that any two consecutive Fibonacci numbers are relatively prime. That is, show that for all $n \in \mathbf{N}, \gcd(F_n, F_{n+1}) = 1$.

26. **a.** Let a and n be integers, $n \neq 0$. Prove that $\gcd(a, n) = \gcd(n - a, n)$.
[*Hint:* Let $d = \gcd(a, n)$ and $e = \gcd(n - a, n)$. Prove that $e \leq d$ using the fact that every common divisor of a and n is less than or equal to $\gcd(a, n)$, and similarly prove that $d \leq e$.]

b. Let n be a natural number. Consider the following sequence of integers

$$1, 2, \ldots, n - 1.$$

Prove that an even number of integers in this list are relatively prime to n.

c. Consider the following sequence of fractions

$$\frac{1}{1000}, \frac{2}{1000}, \frac{3}{1000}, \ldots, \frac{999}{1000}.$$

Prove that an even number of these fractions are in lowest terms.

3.5 The Division Theorem

When you were in elementary school, you learned how to divide two numbers to produce a quotient and a remainder. For example, one divides 37 by 8 as follows:

$$\begin{array}{r} 4 \text{ REM } 5 \\ 8\overline{)\ 37} \\ -37 \\ \hline 5 \end{array}$$

$$37 = 4 \cdot 8 + 5$$

The division shown on the left is equivalent to the equation on the right. Both express the fact that when 37 is divided by 8, the quotient is **4** and the remainder is **5**.

A class divided

I n about 276 BCE, in the town of Cyrene, Greece, the distinguished scholar Eratosthenes was born. He is best known for his scientific and mathematical achievements, which include a surprisingly accurate measurement of the circumference of the Earth. In the middle of his life, Eratosthenes moved to Alexandria, an ancient Greek center of learning located in what is now Egypt. What's not so well known about Eratosthenes is that he was a high school football coach.

Every athletic coach at Alexandria High was also required to teach an academic subject. Eratosthenes taught driver's ed. He had wanted to teach mathematics, but that was the job of the school basketball coach, Archimedes. Eratosthenes consoled himself by grumbling about the many unpleasant aspects of his job. Teaching students to merge onto the main road was frustrating because the school's rickety old vehicles had such low horsepower. In addition, emissions standards in ancient Greece were not as strict as they are today, and the exhaust fumes coming from the rear of the vehicle ahead were often overpowering. Plus, it bothered Eratosthenes that no matter how much his students practiced, most of them never quite got the hang of parallel parking their chariots.

Eratosthenes accepted without hesitation when the school principal called one day and asked him to teach Archimedes' math class. Archimedes was out sick with a cold and a sore throat—he had run outside after his afternoon bath without dressing properly for the winter weather.

Eratosthenes had been told to cover the mathematics that students would need for the Advanced Placement test in alchemy. The topic for the day was division with remainder, which was an integral part of the Alch BC test.

"OK, team, here's the play: I want you to divide 37 by 8. In other words, I want you to express 37 as some number of 8's plus a remainder. Fill in the missing numbers:

$$37 = q \cdot 8 + r.$$

Let's go! Huddle up, and then get in formation. I want to see a score on the board!" Eratosthenes blew his whistle, and the classroom exploded with the sound of chalk against slate.

Eratosthenes was expecting the answer

$$37 = 4 \cdot 8 + 5.$$

He was stunned when he saw all the different answers that the mischievous students had come up with:

The students knew the answer Eratosthenes wanted, of course, but they took advantage of the poor substitute teacher. Eratosthenes rewarded the students for their creativity by requiring them to stay after school and run 5000 laps around the football stadium. The football players in the class got their own special reward—Eratosthenes made them run all the way to Syene for their next game against the Syene Suns.

In our myth, Eratosthenes learned the hard way that there are many ways to write 37 as some number of 8s plus a remainder. As his fiendish students demonstrated, there are many pairs of integers q and r that satisfy the equation

$$37 = q \cdot 8 + r \tag{1}$$

However, if we also add the requirement $0 \le r < 8$, then there will only be one pair of integers that satisfies equation (1): $q = 4$ and $r = 5$.

In general, given any integer a and any natural number b, we can divide a by b to get a quotient q and a remainder r:

$$\begin{array}{c} q \text{ REM } r \\ \overline{b)a} \end{array} \qquad \Leftrightarrow \qquad a = qb + r \qquad (2)$$

To **divide a by b** is to find integers q and r that satisfy the equation in (2) as well as the additional requirement that $0 \le r < b$.

The following theorem states that there is exactly one way to divide any integer by any natural number.

THEOREM 3.5.1   THE DIVISION THEOREM

Let $a \in \mathbf{Z}$ and $b \in \mathbf{N}$. Then there exist unique integers q and r such that $a = qb + r$ and $0 \le r < b$.

We will soon prove the Division Theorem. But first, let's consider some examples.

EXAMPLE 1

   a. Divide 329 by 67.

   b. Divide 23 by 42.

   c. Divide -120 by 50.

SOLUTION

   a. $329 = \mathbf{4} \cdot 67 + \mathbf{61}$

   b. $23 = \mathbf{0} \cdot 42 + \mathbf{23}$

   c. $-120 = \mathbf{-3} \cdot 50 + \mathbf{30}$

The Division Theorem can be extended to the case where b is an arbitrary nonzero integer (see Exercise 9). Before proving the Division Theorem, let's try to gain some geometric insight into the statement of the theorem.

A geometric view of the Division Theorem

To divide 37 by 8, make a number line with all of the multiples of 8. Locate 37 on the number line:

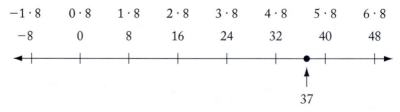

From this diagram, it is clear that $4 \cdot 8 = 32$ is the largest multiple of 8 that is less than or equal to 37. Thus, the quotient is **4**, and the remainder is **5**, the distance between 32 and 37 on the number line:

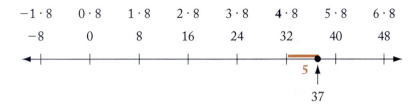

Given any integer a and any natural number b, we can use the same method to divide a by b. Make a number line with all of the multiples of b. Then find a on the number line, and look to the left for the nearest multiple of b that is less than or equal to a. This multiple is qb, where q is the quotient. The distance from qb to a is the remainder, r.

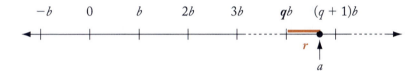

This diagram makes it clear that no matter where a is on the number line, when we look to the left of a for the nearest multiple of b, it is sure to be less than b units away. This is exactly what the Division Theorem (3.5.1) asserts.

Proof of the Division Theorem

Naomi's Numerical Proof Preview: The Division Theorem (3.5.1)

We will prove the Division Theorem shortly. Since the proof will be a little tricky, it will be instructive to recall the misfortune that befell Eratosthenes, our favorite substitute math teacher.

In the myth, when Eratosthenes wanted his students to divide 37 by 8, he asked them to find integers q and r such that

$$37 = q \cdot 8 + r.$$

The students came up with several values of q and r that satisfy this equation. Let's consider all of the values for the remainder r that make the equation true:

$$\ldots, -11, -3, 5, 13, 21, 29, 37, \ldots. \tag{3}$$

Looking at all of these numbers, how can we tell which one is the correct remainder? The correct remainder (the number **5**) is the smallest nonnegative number in the above list. Let T be the set of all nonnegative remainders—that is, the set of all nonnegative numbers in our list:

$$T = \{5, 13, 21, 29, 37, \ldots\}.$$

The Well-Ordering Principle (2.2.2) guarantees that T has a smallest element. In the following proof of the Division Theorem, we will show that this smallest element of T is, in fact, the remainder r that the Division Theorem is looking for.

PROOF OF **THE DIVISION THEOREM (3.5.1)**

This proof has two parts: the proof that q and r exist, and the proof that q and r are unique.

Proof of Existence

Let $a \in \mathbf{Z}$ and $b \in \mathbf{N}$.

[**To show:** There exist $q, r \in \mathbf{Z}$ such that $a = qb + r$ and $0 \leq r < b$.]

We wish to consider the set T of all nonnegative remainders. More precisely, we consider the set of all nonnegative integers that can be written as a minus an integer multiple of b:

$$T = \{a - xb \mid x \in \mathbf{Z} \text{ and } a - xb \geq 0\}. \tag{4}$$

First note that T is nonempty (see Exercise 13). Thus, by the Well-Ordering Principle (2.2.2), T has a smallest element, which we will call r. Since $r \in T$, r can be written as a minus an integer multiple of b. That is, there exists $q \in \mathbf{Z}$ such that $r = a - qb$. Equivalently,

$$a = qb + r.$$

This is the first part of the conclusion of the Division Theorem.

All that remains is to show that $0 \leq r < b$. Since $r \in T$, it follows that $0 \leq r$. To show that $r < b$, we will use proof by contradiction. The idea is that if $r \geq b$, then r cannot be the smallest element of T because $r - b$ would be a smaller element of T.

Claim: $r < b$

Proof of Claim (By contradiction.)

Assumption: Suppose $r \geq b$.
Then $r - b \geq 0$.
    Since $r = a - qb$, we have

$$r - b = a - qb - b.$$

Regrouping terms, we get

$$r - b = a - (q + 1)b.$$

In other words, $r - b$ is equal to a minus an integer multiple of b. Hence,

$$r - b \in T.$$

Thus, we have found an element of T that is smaller than r. This is a contradiction, as r is the smallest element of T. $\Rightarrow\Leftarrow$
Since we reached a contradiction, our Assumption must be false. This proves the Claim.

So, we have found $q, r \in \mathbf{Z}$ such that

$$a = qb + r \qquad \text{and} \qquad 0 \le r < b. \tag{5}$$

This completes the existence part of the proof.

Proof of Uniqueness

To show the uniqueness of q and r, we will assume that there exist q' and r' that satisfy the same properties as q and r.
   Let $q', r' \in \mathbf{Z}$ such that

$$a = q'b + r' \qquad \text{and} \qquad 0 \le r' < b. \tag{6}$$

[To show: $q = q'$ and $r = r'$.]
Without loss of generality, we may assume that $r' \le r$.
   Subtract the equation in (6) from the equation in (5) to obtain

$$0 = (q - q')b + (r - r'), \tag{7}$$

or, equivalently,

$$r - r' = -(q - q')b.$$

Thus, $r - r'$ is a multiple of b.
   Furthermore, since $r < b$ and $0 \le r' \le r$, it follows that

$$0 \le r - r' < b.$$

   At this point, we have established that the number $r - r'$ is a multiple of b that is nonnegative and less than b. However, all positive multiples of b are at least as large as b. (This follows from Lemma 3.1.2.) Thus, $r - r'$ cannot be positive. We conclude that $r - r' = 0$; that is,

$$r = r'.$$

   It remains only to show that $q = q'$. Since $r = r'$, equation (7) gives

$$0 = (q - q')b.$$

Since $b \neq 0$, we conclude that $q - q' = 0$. Thus,

$$q = q'.$$

This completes the uniqueness part of the proof. ∎

We can use the Division Theorem to prove the following lemma.

Let a, b, and x be integers such that a and b are nonzero. If $a \mid x$ and $b \mid x$, then $\mathrm{lcm}(a, b) \mid x$.

PROOF Let $a, b, x \in \mathbf{Z}$, with $a \neq 0$ and $b \neq 0$, such that $a \mid x$ and $b \mid x$. Let $M = \mathrm{lcm}(a, b)$.

[To show: $M \mid x$.]

First we divide x by M: the Division Theorem (3.5.1) tells us that there exist integers q and r such that

$$x = qM + r \qquad \text{and} \qquad 0 \leq r < M. \tag{8}$$

Solving this equation for r, we get

$$r = x - qM.$$

Since $a \mid x$ and $a \mid M$, we know that $a \mid r$ by the Linear Combination Lemma (3.1.3). Similarly, since $b \mid x$ and $b \mid M$, we know that $b \mid r$.

Thus, r is a nonnegative common multiple of a and b that is less than $\mathrm{lcm}(a, b)$. We conclude that $r = 0$. Substituting this into the equation in (8) gives

$$x = qM.$$

Hence, by definition, $M \mid x$. ∎

In the proof of Lemma 3.5.2, we used a common strategy for proving statements of the form $b \mid a$. This strategy was to use the Division Theorem to write $a = qb + r$ and then to show that r must equal 0.

EXERCISES 3.5

Numerical Problems

1. Perform each division indicated.

   a. Divide 85 by 19.   **b.** Divide 132 by 12.   **c.** Divide 573 by 191.

   d. Divide -45 by 7.   **e.** Divide 89 by 90.   **f.** Divide 0 by 17.

2. Explain why it is not possible to divide 17 by 0.

3. An integer is divided by 6. List all possible values for the remainder.

4. In our myth, Eratosthenes' students ignored the Division Theorem's requirement that the remainder r must be less than or equal to the divisor b. In this exercise, we will likewise ignore this restriction.

   a. Divide 1221 by 97 in three different ways. That is, find three different pairs of integers q and r such that $1221 = q \cdot 97 + r$.

   b. Out of all possible pairs of integers q, r that satisfy the equation $1221 = q \cdot 97 + r$, what is the smallest nonnegative value for r?

5. Define the set $S = \{3x + 5y \mid x, y \in \mathbf{Z} \text{ and } 3x + 5y > 0\}$. That is, S is the set of all positive numbers that can be expressed as an integral linear combination of 3 and 5.

   a. Show that the set S is not empty.

   b. According to the Well-Ordering Principle, S must have a smallest element. What is it?

   c. Which of the natural numbers from 1 to 10 are elements of S, and which are not?

   d. Make a conjecture that states precisely which natural numbers are elements of S.

6. Define the set $T = \{6x + 9y \mid x, y \in \mathbf{Z} \text{ and } 6x + 9y > 0\}$. That is, T is the set of all positive numbers that can be expressed as an integral linear combination of 6 and 9.

   a. Show that the set T is not empty.

   b. According to the Well-Ordering Principle, T must have a smallest element. What is it?

c. Which of the natural numbers from 1 to 15 are in T, and which are not?

d. Make a conjecture that states precisely which natural numbers are elements of T.

7. a. Find a counterexample to the following assertion:

Let $a, b, x \in \mathbf{Z}$ such that a and b are nonzero. If $a \mid x$ and $b \mid x$, then $ab \mid x$.

b. Verify that Lemma 3.5.2 is correct for the values of a, b, and x you gave in part a.

8. a. What is the hypothesis of the Division Theorem (3.5.1)?

b. What is the conclusion of the Division Theorem?

Reasoning and Proofs

9. In the Division Theorem (3.5.1), we require b to be a natural number. In this exercise, we explore what happens when we remove this condition.

a. Try to divide 27 by -5 using the Division Theorem. Remember, you want a remainder r so that $0 \le r < b$. Explain your difficulty in performing the division.

b. Alter the statement of the Division Theorem to make it true for all nonzero integers b. (You don't have to prove it.)

10. Use the Division Theorem to prove that every integer is either even or odd.[1]

11. State and prove the converse of Lemma 3.5.2.

12. a. Explain how the Division Theorem implies that every integer can be written in the form $4k$, $4k + 1$, $4k + 2$, or $4k + 3$, for some $k \in \mathbf{Z}$.

b. Prove that every odd integer can be written in the form $4k + 1$ or $4k + 3$, where $k \in \mathbf{Z}$.

c. Prove that every odd integer can be written in the form $6k + 1$, $6k + 3$, or $6k + 5$, where $k \in \mathbf{Z}$.

13. In the proof of the Division Theorem (3.5.1), we stated after equation (4) that the set T of all nonnegative remainders, $T = \{a - xb \mid x \in \mathbf{Z} \text{ and } a - xb \ge 0\}$, is nonempty. (Recall that a is an integer and b is a natural number.) Prove that T is nonempty.

---

[1] In the Appendix (on the Student Companion Website), we prove this using induction.

14. Let $a \in \mathbf{Z}$ and $b \in \mathbf{N}$.

    a. Prove that there are unique integers q and r such that $a = qb + r$ and $2b \le r < 3b$.

    b. Prove that there are unique integers q and r such that $a = qb + r$ and
    $$-\frac{b}{2} \le r < \frac{b}{2}.$$

    [*Hint:* Use the Division Theorem (3.5.1) as a starting point.]

15. Let n be an integer. Prove that one of the three integers n, $n + 2$, or $n + 4$ must be a multiple of 3. [*Hint:* Use the Division Theorem (3.5.1).]

16. Prove that given any integer n, the integer n^2 can be written as $3k$ or as $3k + 1$ for some integer k.

17. Prove that given any integer n, the integer n^4 can be written as $5k$ or as $5k + 1$ for some integer k.

EXPLORATION Patterns in the Sieve of Eratosthenes (Exercises 18–20)

18. In the completed sieve of Eratosthenes in Figure 1 of Section 3.3, you may have noticed that in four of the six rows, the entire row (other than the first number in the row) is crossed out. In this exercise, you will prove that this is so. There are six numbers in each column of the sieve. Thus, every number in row 2 can be written as $2 + 6k$, where k is the number of columns to the right of the number 2. Similarly, every number in row 3 can be written in the form $3 + 6k$. Use these facts (and similar ones) to prove each of the following:

    a. Every number in row 2 (other than 2 itself) is composite.

    b. Every number in row 3 (other than 3 itself) is composite.

    c. Every number in row 4 is composite.

    d. Every number in row 6 is composite.

    e. Every prime number greater than 3 must be of the form $6k + 1$ or $6k + 5$ for some integer k. [*Hint:* You may use the results from Exercise 12.]

19. You could also perform the sieve of Eratosthenes on the numbers up to 100 after listing the numbers in four rows, like this:

1	5	9	13	17	21	25	29	33	37	41	45	49	53	57	61	65	69	73	77	81	85	89	93	97
2	6	10	14	18	22	26	30	34	38	42	46	50	54	58	62	66	70	74	78	82	86	90	94	98
3	7	11	15	19	23	27	31	35	39	43	47	51	55	59	63	67	71	75	79	83	87	91	95	99
4	8	12	16	20	24	28	32	36	40	44	48	52	56	60	64	68	72	76	80	84	88	92	96	100

a. Perform the sieve of Eratosthenes on this list.

b. In which two rows of the list must all of the primes other than the number 2 appear? Why?

c. Which numbers in the list have the form $4k$, for some nonnegative integer k? Which numbers in the list have the form $4k + 1$? $4k + 2$? $4k + 3$?

d. Explain why every odd prime number is of the form $4k + 1$ or $4k + 3$ for some nonnegative integer k.

e. Prove that for any odd prime number p, it is the case that $8 \mid p^2 - 1$.

20. Some primes can be expressed as a sum of two squares, and some cannot. For example, the prime 29 can be expressed as the sum of two squares: $29 = 5^2 + 2^2$, but the prime 23 cannot be expressed as the sum of two squares.

a. Show that the prime 23 cannot be expressed as the sum of two squares; in other words, prove that there do not exist integers a and b such that $23 = a^2 + b^2$.

b. For every prime in the sieve you performed in Exercise 19, determine whether it can be expressed as the sum of two squares. Identify on your sieve every prime that can be so expressed.

c. What do you observe? Write a conjecture about which primes can be expressed as a sum of two squares and which primes cannot.

21. Consider the following numbers of the form $p^2 - 1$, where p is prime:

$$5^2 - 1 = 24; \qquad 7^2 - 1 = 48; \qquad 11^2 - 1 = 120.$$

Each of these numbers is divisible by 12. Prove or provide a counterexample to the following statement:

$$\text{If } p > 3 \text{ is prime, then } 12 \mid p^2 - 1.$$

EXPLORATION Converting Numbers into Other Bases (Exercises 22–26)

The arithmetic you've done your entire life has been with the base 10 number system. For example, we write the number **164** as shorthand for $1 \cdot 10^2 + 6 \cdot 10^1 + 4 \cdot 10^0$. Not all civilizations throughout history have used based 10. The Babylonians used a base 60 number system. In computer science, usage of base 2, base 8, and base 16 are all common.

To avoid ambiguity, we can use a subscript notation to indicate the base in which a number is written. For instance, the number 164 in base 10 is represented by

$$(164)_{10} = 1 \cdot 10^2 + 6 \cdot 10^1 + 4 \cdot 10^0,$$

whereas the number 164 in base 8 is represented by

$$(164)_8 = 1 \cdot 8^2 + 6 \cdot 8^1 + 4 \cdot 8^0$$
$$= (116)_{10}.$$

Note: You may assume that every number that is not written using the subscript notation is in base 10.

Digits in base b are always between 0 and $b - 1$. In base 10, for example, the possible digits are the numerals $0, 1, 2, \ldots, 9$. In base 8, the possible digits are the numerals $0, 1, 2, \ldots, 7$.

The following is a conversion algorithm that allows one to convert any natural number n into base b, where $b > 1$ is an integer.

CONVERSION ALGORITHM

Divide n by b using the Division Theorem (3.5.1):

$$n = q_0 b + r_0 \qquad 0 \leq r_0 < b$$

If the quotient $q_0 \neq 0$, divide q_0 by b:

$$q_0 = q_1 b + r_1 \qquad 0 \leq r_1 < b$$

If the quotient $q_1 \neq 0$, divide q_1 by b. Continue dividing b into each quotient q_i until you get a quotient that equals 0:

$$q_1 = q_2 b + r_2 \qquad 0 \leq r_2 < b$$
$$q_2 = q_3 b + r_3 \qquad 0 \leq r_3 < b$$
$$\vdots$$
$$q_{k-2} = q_{k-1} b + r_{k-1} \qquad 0 \leq r_{k-1} < b$$
$$q_{k-1} = 0 \cdot b + r_k \qquad 0 \leq r_k < b$$

CLAIM. In base b, the number n can be written

$$n = (r_k r_{k-1} \ldots r_2 r_1 r_0)_b;$$

that is,

$$n = r_k \cdot b^k + r_{k-1} \cdot b^{k-1} + \cdots + r_2 \cdot b^2 + r_1 \cdot b^1 + r_0 \cdot b^0.$$

22. Use the conversion algorithm just described to convert $(1000)_{10}$

    a. into base 3.   b.   into base 7.

23. Convert $(2345)_8$ into base 9.

24. For *hexadecimal* (base 16) numbers, the convention is to use the numerals 0–9 and the letters A–F as digits:

$$(A)_{16} = 10, \ (B)_{16} = 11, \ (C)_{16} = 12, \ (D)_{16} = 13, \ (E)_{16} = 14, \ \text{and} \ (F)_{16} = 15.$$

For example,

$$(FAB)_{16} = 15 \cdot 16^2 + 10 \cdot 16^1 + 11 \cdot 16^0$$
$$= (4011)_{10}.$$

a. Convert $(BEAD)_{16}$ into base 10.

b. Use the conversion algorithm to convert $(17837)_{10}$ into hexadecimal.

25. Use mathematical induction to prove the Claim that was stated at the end of the conversion algorithm. (That is, prove that the conversion algorithm is valid.)

26. Prove that there is a unique way to write any number n in base b. That is, prove that there is only one sequence of digits $r_k, r_{k-1}, \ldots, r_1, r_0$ (where $k \geq 0$) such that

$$n = r_k \cdot b^k + r_{k-1} \cdot b^{k-1} + \cdots + r_2 \cdot b^2 + r_1 \cdot b^1 + r_0 \cdot b^0,$$

$$0 \leq r_i < b \quad \text{for} \quad i = 0, 1, \ldots, k,$$

and

$$r_k \neq 0.$$

Advanced Reasoning and Proofs

27. Exercises 5 and 6 hint at an important fact: The greatest common divisor of two integers can be written as a linear combination of those two integers. In fact, it is the smallest positive linear combination. Let's write this more formally:

> THEOREM. *Let a and b be nonzero integers. Then the smallest positive linear combination of a and b is* $\gcd(a, b)$.

Using the following steps as an outline, write a proof of this theorem.

    I. Let a, b be nonzero integers. Let $S = \{ax + by \mid x, y \in \mathbf{Z} \text{ and } ax + by > 0\}$. Show that S is nonempty.

    II. Let d be the smallest element of S. (How do we know d exists?) Prove that $\gcd(a, b) \mid d$.

    III. Use the Division Theorem to show that $d \mid \gcd(a, b)$. Conclude that $d = \gcd(a, b)$.

28. Prove that any common divisor of two integers is also a divisor of their gcd. (That is, if $c \mid a$ and $c \mid b$, then $c \mid \gcd(a, b)$.)
[*Hint:* Use the theorem from Exercise 27.]

29. a. Prove the following:

> LEMMA. *Let a and b be integers. If both a and b have the form $4k + 1$ (where k is an integer), then ab also has the form $4k + 1$.*

b. The lemma from part a generalizes to products of n integers of the form $4k + 1$. State and prove the generalized lemma.

c. Prove that any natural number of the form $4k + 3$ has a prime factor of the form $4k + 3$.

30. The ideas behind the proof that there are infinitely many primes (3.3.1) can be adapted to prove the following theorem.

> THEOREM. *There are infinitely many primes of the form $4k + 3$, where k is a nonnegative integer.*

a. We have already proven that there are infinitely many primes, and we know that all primes other than 2 are odd integers. Why does that not automatically imply the Theorem?

b. Prove the Theorem, using the following steps as an outline.

   I. Use proof by contradiction. Assume that there are only finitely many primes of the form $4k + 3$ (where k is a nonnegative integer). That is, suppose that

$$p_1, p_2, p_3, \ldots, p_n \qquad (9)$$

   is a complete list of all the primes of the form $4k + 3$.

   II. Define

$$m = 4 \cdot p_1 \cdot p_2 \cdot p_3 \cdot \ \cdots \ \cdot p_n - 1.$$

   Use the result from part c of Exercise 29 to show that m has a prime factor q of the form $4k + 3$ (where k is a nonnegative integer).

   III. Prove that q is not equal to any of the primes in list (9). Finish the proof.

3.6 Applications of gcd and lcm

Gears

In the system of interlocking gears shown in Figure 1a, the two gears are aligned so that tooth number **0** on gear **A** is touching tooth number **0** on gear **B**. Both gears advance by one tooth with every tick of the clock. At $t = 0$, shown in Figure 1a, the gears are in position $(A = 0, B = 0)$. After one tick, at $t = 1$, the gears will be in position $(A = 1, B = 1)$. At $t = 2$, the gears will be in position $(A = 2, B = 2)$, and so on. At $t = 8$, shown in Figure 1b, gear **A** will return to its initial position, and the system will be in position $(A = 0, B = 8)$. After that, gear **A** will begin its cycle again, and at $t = 9$, the system of gears will be in position $(A = 1, B = 9)$.

a.

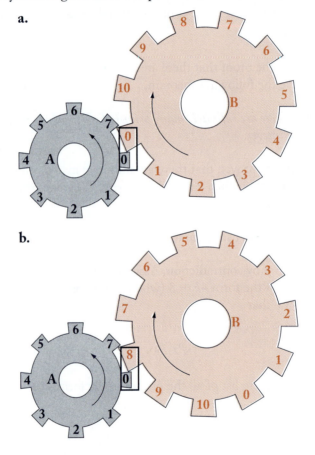

b.

Figure 1   A system of two interlocking gears, **A** and **B**.   **a.** At $t = 0$, the gears are in position $(A = 0, B = 0)$.   **b.** At $t = 8$, the gears are in position $(A = 0, B = 8)$.

Gear **A** will be in its original position after 0 seconds have elapsed, after 8 seconds have elapsed, and after 16 seconds have elapsed. In general, gear **A** will be in its original position when the elapsed time is any multiple of 8. Similarly, gear **B** will be in its original position when the elapsed time is any multiple of 11.

Thus, gear **A** and gear **B** will *both* be in their original positions (simultaneously) whenever the elapsed time is both a multiple of 8 and a multiple of 11. The elapsed time at which both gears first return to their original positions simultaneously is the least common multiple of 8 and 11, lcm(8, 11) = 88.

In general, a system comprising a gear with *a* teeth interlocked to a gear with *b* teeth will first return to its original position after lcm(*a*, *b*) ticks of the clock. After that, the system will again return to its initial position after another lcm(*a*, *b*) ticks of the clock.

Check for Understanding

Date _____

Section 3.6

Pay to the order of *Understanding* _____

A system is made up of two interlocking gears that turn at a rate of 1 tooth per second. When will the entire system first return to its original position, if

1. one gear has 15 teeth, and the other has 7 teeth?

2. one gear has 35 teeth, and the other has 15 teeth?

Answer(s) _____   _____

On a bicycle, the *chainring* is the gear that is attached to the pedals, and the *freewheel* is the gear that is attached to the rear wheel of the bicycle (see Figure 2). One full (360°) rotation of the pedals corresponds to one full rotation of the chainring. One full rotation of the freewheel corresponds to one full rotation of the rear tire of the bicycle. Although the two gears (the chainring and the freewheel) are not touching, the bicycle chain ensures that the teeth of the two gears move in lockstep, just as if the gears were interlocked. The effect of shifting gears on a bicycle is to change the number of teeth on the chainring and/or the freewheel.

Freewheel Chain   Chainring

Figure 2   The drive assembly of a one-speed bicycle.

EXAMPLE 1

A bicycle is put into a gear in which the chainring has 50 teeth and the freewheel has 18 teeth. After how many full rotations of the pedals will both the rear wheel and the pedals simultaneously return to their initial positions?

SOLUTION Let's assume that each gear advances by one tooth with each tick of the clock. Both gears will return to their original positions simultaneously after $\text{lcm}(50, 18) = 450$ ticks of the clock. Since the chainring has 50 teeth, this corresponds to $\dfrac{450}{50} = 9$ full rotations of the pedals. ●

Suppose we had a system of three interlocking gears, such as the one shown in Figure 3. Just as in the case of a two-gear system, the entire system of three gears will be in its original position whenever the elapsed time is a common multiple of a, b, and c. The first time the entire three-gear system will return to its original position is when the elapsed time is $\text{lcm}(a, b, c)$.

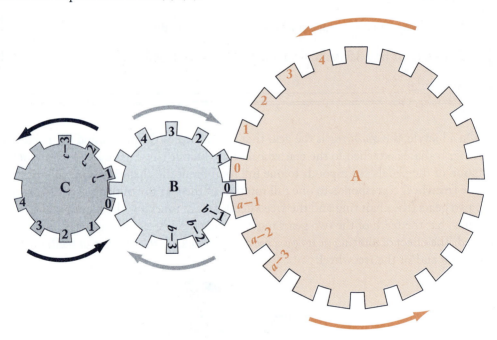

Figure 3 A system of three interlocking gears, A, B, and C. Gear A has a teeth, gear B has b teeth, and gear C has c teeth.

EXAMPLE 2

Gear A (with $a = 21$ teeth), gear B (with $b = 12$ teeth), and gear C (with $c = 10$ teeth) are all interlocking, as in Figure 3. After how many full rotations of gear A will all three gears simultaneously return to their initial positions?

SOLUTION The system will return to its initial position every $\mathrm{lcm}(21, 12, 10) = 420$ ticks of the clock. Since gear A has 21 teeth, this corresponds to $\dfrac{420}{21} = 20$ full rotations of gear A. ∎

Pitch perception and the missing fundamental

When you play a note on a musical instrument, or when you make a vowel sound or a voiced consonant while you are talking, listeners hear the sound as having a pitch. For example, when a violinist bows on the A string, the pitch that listeners hear corresponds to the frequency 440 Hertz (i.e., corresponds to a sound wave with a frequency of 440 cycles per second). When that note A is played, the sound wave generated has components not only at $f = 440$ Hz, but also at other frequencies: $2f = 880$ Hz, $3f = 1320$ Hz, $4f = 1760$ Hz, and other multiples of 440 Hz. The pitch that is heard by listeners corresponds to the *fundamental frequency*, $f = 440$ Hz, also known as the *first harmonic*. The other frequencies that are present in the sound wave ($2f$, $3f$, $4f$, etc.) are called the *upper harmonics*, or *overtones*.

If a clarinet plays a note at that same pitch, the sound wave it produces also has a component at the first harmonic (the fundamental frequency, 440 Hz) as well as components at the same upper harmonics as the violin sound. However, the relative magnitudes of the components at the various harmonics are different in the clarinet sound than they are in the violin sound. This difference in relative magnitudes is the main reason that the two instruments sound very different even though they are playing the same note.

It is possible to electronically reproduce a complex tone, such as the sound of a violin, by constructing the sound wave out of all of its frequency components (the fundamental frequency and the upper harmonics). We can even reproduce the sound while removing the component at the fundamental frequency, and the resulting sound will contain only the upper harmonics. For example, if we take a sound whose fundamental frequency is 200 Hz and modify it by removing its 200 Hz component, it will still have components at the overtones 400 Hz, 600 Hz, 800 Hz, 1000 Hz, and so on. Amazingly, even though the modified tone has no component at 200 Hz, listeners still hear the sound as if it were at 200 Hz! In other words, the pitch of the sound corresponds to the fundamental frequency, even though that frequency is no longer present in the sound. This auditory illusion is known as the *missing fundamental*.

Under a variety of circumstances, the missing fundamental frequency at which listeners perceive the sound is the greatest common divisor of the frequencies of the harmonics that are present.

EXAMPLE 3
. . . .

If pure tones at frequencies of 300 Hz and 450 Hz are played simultaneously, listeners can perceive a note at $\gcd(300, 450) = 150$ Hz, even though there is no energy in the sound at the frequency 150 Hz.

The missing fundamental illusion is exploited in the design of pipe organs. Because the frequency played by an organ pipe is inversely proportional to the length of the pipe, a very low note requires a very long organ pipe. But 32-foot-long organ pipes are bulky and costly. To make it sound as if a very low note is being played at the frequency f, many organs use two shorter pipes, one playing a note at frequency $2f$ and one playing a note at frequency $3f$. Listeners hear a sound at the missing fundamental frequency $f = \gcd(2f, 3f)$, without the need for a pipe that can produce such a low sound.

Although you may not realize it, you probably experience the missing fundamental illusion every day when you talk on the telephone. The inexpensive microphones used in many telephones do not pick up low frequencies well, and these telephones do not transmit low frequencies. Nonetheless, when you are listening on the phone, you have no trouble hearing even the lowest notes produced by someone with a deep voice, because your brain reconstructs the missing fundamental frequency from the upper harmonics that the phone transmits.

EXAMPLE 4

When Lou speaks his name into the telephone, Eva's phone at the other end of the line only transmits some of the frequencies of the sound into her ear. When Eva hears the sound "Lou," the sound wave on her end of the line has components at 345 Hz, 460 Hz, 575 Hz, and 690 Hz. Eva perceives Lou's voice at the greatest common divisor of the frequencies that are present, $\gcd(345, 460, 575, 690) = 115$ Hz, which is the missing fundamental frequency. Due to the phone's poor frequency response, no 115 Hz sound ever reaches Eva's ear, and she only receives some of the harmonics. Nonetheless, she perceives the sound at its actual fundamental frequency, 115 Hz.

EXERCISES 3.6

1. Assume that gears A and B are interlocked (similar to Figure 1). Gear A has a teeth and gear B has b teeth. Suppose you start with some initial position and then turn the gears. How many full turns of gear A are required until both gears simultaneously return to their initial positions, if the numbers of teeth are

   a. $a = 7$ and $b = 11$?   **b.** $a = 11$ and $b = 7$?   **c.** $a = 144$ and $b = 180$?

2. Three gears A, B, and C are interlocking, as in Figure 3. Gear A has 20 teeth, gear B has 16 teeth, and gear C has 25 teeth. The gears turn at a rate of 1 tooth per second.

   a. How long will it take until the first time gear A and gear B simultaneously return to their initial positions?

b. How long will it take until the first time gear A and gear C simultaneously return to their initial positions?

c. How long will it take until the first time all three gears simultaneously return to their original positions?

3. A two-speed bicycle has a chainring gear with 50 teeth and two freewheel gears.

   a. One freewheel gear has 15 teeth. When the bike is in this gear, how many full turns of the pedals are required until both the pedals and the rear wheel simultaneously return to their initial positions?

   b. The other freewheel gear has b teeth. When the bike is in this gear, how many full turns of the pedals are required before both the pedals and the rear wheel simultaneously return to their initial positions? [*Hint:* Your answer will be in terms of b.]

   c. Suppose that switching from the freewheel gear of part a to the freewheel gear of part b causes the bicycle to require twice as many full turns of the pedals for the system to first return to its initial position. What is the smallest number of teeth the second freewheel gear can have?

4. Harold is on the phone with Marian the librarian. He utters the words "ice cream" in a low voice at a frequency that is not carried over the phone due to its limited frequency range. She hears harmonics of his voice at the frequencies 375 Hz and 500 Hz. What is the (missing) fundamental frequency at which the sound of Harold's voice is perceived by Marian?

EXPLORATION Greatest Common Measure and the Missing Fundamental (Exercises 5–6)

5. The *greatest common measure* is a generalization of the greatest common divisor to real numbers. If two real numbers x and y are commensurable (see Section 1.4), then the *greatest common measure* of x and y is the largest real number f such that x is an integer times f and y is an integer times f.

   a. Find the greatest common measure of 3.9 and 5.2.

   b. Find the greatest common measure of 15.2 and 26.6.

   c. Write a formal definition for the greatest common measure of three real numbers x, y, and z.

   d. Find the greatest common measure of $15\sqrt{2}, 18\sqrt{2}$, and $33\sqrt{2}$.

6. Let f represent the fundamental frequency of a complex sound. The harmonics in the sound have frequencies that are integral multiples of f, whether or not f is an integer. If a missing fundamental is perceived at a noninteger frequency, then that fundamental frequency is the greatest common measure of the harmonic frequencies present.

   a. Suppose pure tones are played at 337.5 Hz and 562.5 Hz. At what frequency could a listener perceive the missing fundamental tone?

   b. Suppose pure tones are played at 666.9 Hz, 889.2 Hz, and 1333.8 Hz. At what frequency could a listener perceive the missing fundamental tone?

Chapter 4

Euclid (roughly 347–287 BCE)

Very little is known for certain about Euclid. It is generally thought that he was a contemporary of Archimedes, but the exact dates of Euclid's birth and death remain a mystery. Euclid is best known for writing the *Elements*, a textbook in which he explained much of the mathematics known at the time. Most of the material contained in the *Elements* was not actually due to Euclid, but he was the first person to develop and explain this material in a logical and rigorous way. Starting with a small number of definitions and basic assumptions (known as postulates or axioms), Euclid constructed a large body of mathematics, proving each new theorem using the postulates and the theorems that had come before. Known as the axiomatic method, this style of presentation has remained the predominant model for the development of mathematical knowledge up to the present day. Many mathematical historians have asserted that the *Elements* is the most widely read textbook of all time and that the Bible is the only book that has had more editions printed.

Most people today know the *Elements* as the foundation for their high school course on Euclidean geometry. However, only 9 of the 13 books of the *Elements* are about geometry. Books VII, VIII, and IX are devoted to number theory, beginning with the Euclidean Algorithm (which you will learn in this chapter). Book X concerns the related topic of line segments of rational and irrational lengths, including a discussion of Pythagorean triples (see "A Bit of Euclid's Math").

Around 306 BCE, Ptolemy I, the emperor of Egypt, established a university at Alexandria. Ptolemy selected Euclid to teach mathematics at the university because of Euclid's fame as the author of the *Elements*. It is believed that Euclid subsequently became known in Alexandria as a great teacher as well as a great author. According to legend, Ptolemy once asked Euclid whether there was an easier way to learn geometry than that presented in the *Elements*. Euclid explained that even a king has to work hard to learn math, telling Ptolemy, "There is no royal road to geometry." Another anecdote concerns a student who asked Euclid what was the use of studying geometry. Euclid responded, "Give him three obols, since he must needs make gain out of what he learns." The modern equivalent of Euclid's retort might be "Give that frosh a quarter, since he's only interested in math for the money."

A Bit of Euclid's Math

In Book X of the *Elements*, Euclid solved the problem of finding all right triangles whose sides are natural numbers. That is, Euclid showed how to find all triples of natural numbers x, y, and z satisfying the equation $x^2 + y^2 = z^2$. These are known as Pythagorean triples (see section 15.3). For example, $x = 12$, $y = 5$, $z = 13$ is a Pythagorean triple because $12^2 + 5^2 = 13^2$. Euclid was not the first to be interested in this problem—examples of Pythagorean triples have been found on Babylonian cuneiform tablets dating from as far back as 1500 BCE (about 1000 years before Pythagoras).

Euclid was also interested in how to construct geometric figures using only a straightedge (an unmarked ruler) and a compass. Such constructions are sometimes called *Euclidean constructions*. In the *Elements*, Euclid demonstrates how to do many of these types of construction problems. For example, given a rectangle, Euclid showed how to construct a square with the same area as that rectangle. Three construction problems dating from before 450 BCE were eventually shown to be impossible and have

since become known as the *three classical problems*:

1. **Squaring the circle**. Given a circle, construct a square whose area is equal to the area of that circle.

2. **Trisecting the angle**. Divide a given angle into three equal smaller angles.

3. **Doubling the cube**. Given a cube, construct a cube whose volume is twice the volume of the given cube.

It is believed that Euclid's construction in the *Elements* of a square whose area is equal to that of a given rectangle stimulated popular interest in the problem of squaring a circle. For over two millennia, mathematicians tried unsuccessfully to solve the three classical problems. Finally, in the 19th century, mathematicians used abstract algebra to prove that none of these constructions is possible. Because of the impossibility of these constructions, the phrase "squaring the circle" has come to mean "doing the impossible" in ordinary English usage.

Chapter 4　The Euclidean Algorithm
or Tales of a Master Baker

4.1　The Euclidean Algorithm

Euclid's story

Euclid, the great ancient Greek mathematician, lived in the 4th and 3rd centuries BCE. He is known for writing the *Elements*, a 13-volume mathematical work that became the foundation for two millennia of study in geometry and number theory. What's not so well known about Euclid is that he was not trained as a mathematician but rather as a baker. In fact, Euclid ran a very successful bakery in ancient Athens. Members of Greece's elite baking circles had long believed that pie was the true measure of a baker's skill. Euclid chose to specialize in cake, however, feeling that their constant obsession with pie was irrational.

We now take you to Athens in the 4th century BCE, where the master baker is hard at work.

Hi, I'm Euclid. Welcome to my bakery. I used to make only *square* cakes, because I got stuck with a whole bunch of square pans after a bet I made with a guy about squaring the circle. But a few days ago, a customer called and asked for a *rectangular* cake—a 175 × 65 lemon torte. At first I didn't think I could make it, because I own only square pans. Then I figured out a way.

I realized that I could make a rectangular cake by baking several square cakes, then fitting them together after they came out of the oven. I drew a picture of the 175 × 65 rectangular cake to see how I could break it up into squares. The largest square that fit into this rectangle was 65 × 65, so I put in as many of these as I could:

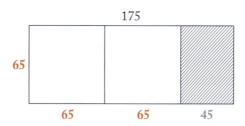

As you can see, after I put in two of these 65 × 65 squares, there was still a 65 × 45 rectangle left over. The largest square that would fit into this remaining piece was 45 × 45, which fit once:

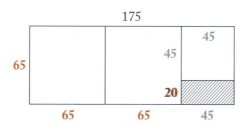

The leftover rectangle was now 45 × 20. The largest square that I could fit into this was 20 × 20, which fit twice:

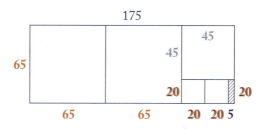

Now all that was left was a 20 × 5 rectangle. The largest square that fit into this piece was 5 × 5, and four of these fit:

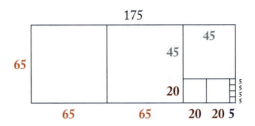

This time there was no rectangle remaining. Voilà! I had succeeded in dividing the entire rectangle into squares. So, I baked square cakes in these sizes and assembled them into a 175 × 65 rectangle using my diagram. The cake looked a bit like a jigsaw puzzle at first, but after I put the frosting on, it looked terrific!

Euclid's customers were thrilled that they could now order rectangular cakes, and the Euclidean Algorithm for dividing rectangles into squares became famous. The rest is history.

The Euclidean Algorithm

Let's examine the Euclidean Algorithm arithmetically. In our myth, Euclid wanted to bake a cake that measured 175 × 65. His first step was to see how many 65 × 65 squares fit into the 175 × 65 rectangle. This is equivalent to dividing 175 by 65.

Dividing 175 by **65**, we get

$$175 = 2 \cdot 65 + 45.$$

Next, divide **65** by the remainder, **45**:

$$65 = 1 \cdot 45 + 20$$

Then divide **45** by the remainder, **20**:

$$45 = 2 \cdot 20 + 5$$

Finally, divide **20** by the remainder, **5**:

$$20 = 4 \cdot 5 + 0$$

A remainder of 0 indicates that the entire rectangle has been successfully divided into squares. Notice that the lengths of the sides of the squares in our myth (**65**, **45**, **20**, and **5**) correspond to the divisors at each step of the Euclidean Algorithm.

4.1.1 THE EUCLIDEAN ALGORITHM

Given two natural numbers a and b, apply the Division Theorem (3.5.1) to divide a by b:

$$a = q_1 b + r_1 \qquad 0 \le r_1 < b$$

Provided that $r_1 \ne 0$, divide the divisor, b, by the remainder, r_1:

$$b = q_2 r_1 + r_2 \qquad 0 \le r_2 < r_1$$

Continue in this manner: at each step, divide the previous step's divisor by its remainder. When a remainder of 0 is obtained, stop.

The algorithm results in the following system of equations:

$$a = q_1 b + r_1 \qquad 0 \le r_1 < b$$
$$b = q_2 r_1 + r_2 \qquad 0 \le r_2 < r_1$$
$$r_1 = q_3 r_2 + r_3 \qquad 0 \le r_3 < r_2$$
$$r_2 = q_4 r_3 + r_4 \qquad 0 \le r_4 < r_3$$
$$\vdots$$
$$r_{n-2} = q_n r_{n-1} + r_n \qquad 0 \le r_n < r_{n-1}$$
$$r_{n-1} = q_{n+1} r_n + 0,$$

where all of the q_i and r_i are integers.

The inequalities in the right column tell us that

$$b > r_1 > r_2 > r_3 > \cdots .$$

Letting $r_0 = b$, we see that $r_0, r_1, r_2, r_3, \ldots$ is a decreasing sequence of nonnegative integers, so the Well-Ordering Principle (2.2.2) guarantees that we will eventually reach a remainder of 0 (see Exercise 10). Hence, starting with any pair of natural numbers a and b, the Euclidean Algorithm will halt after a finite number of steps.

EXAMPLE 1

Perform the Euclidean Algorithm on 138 and 61.

SOLUTION

$$138 = 2 \cdot 61 + 16$$

$$61 = 3 \cdot 16 + 13$$

$$16 = 1 \cdot 13 + 3$$

$$13 = 4 \cdot 3 + 1$$

$$3 = 3 \cdot 1 + 0$$

EXERCISES 4.1

Numerical Problems

1. Use the Euclidean Algorithm to divide this rectangle into squares. In your sketch, label the side lengths of each square.

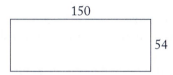

150

54

2. Perform the Euclidean Algorithm (4.1.1) on 175 and 77.

3. You might think that performing the Euclidean Algorithm on large numbers would take a longer time than performing it on small numbers. Test out this idea on the following pairs of numbers by performing the Euclidean Algorithm on each pair, noting how many steps it takes.

   a. 19 and 11   **b.** 124 and 36

   c. 13,753 and 55,029   **d.** 11,111 and 1,111,111

   (In fact, there is a relationship between the size of the numbers and the *maximum* number of steps necessary. You'll learn about this in the next section.)

4. Find two natural numbers, a and b, for which the Euclidean Algorithm performed on a and b takes only one step (only one division is required).

5. Find two natural numbers, a and b, for which the Euclidean Algorithm performed on a and b takes exactly two steps.

Reasoning and Proofs

6. Let k be a natural number. Perform the Euclidean Algorithm on $2k + 1$ and k.

7. Let k be a natural number. Perform the Euclidean Algorithm on $7k + 14$ and $3k + 6$. Exactly how many steps (how many divisions) were required?

8. **a.** Let $k > 3$ be an odd natural number. If the Euclidean Algorithm is performed on $2k$ and $k + 1$, exactly how many steps (how many divisions) are required? Explain your answer.

   b. Let $k > 3$ be an even natural number. If the Euclidean Algorithm is performed on $2k$ and $k + 1$, exactly how many steps (how many divisions) are required? Explain your answer.

9. **a.** The first step of the Euclidean Algorithm (4.1.1) is to find integers q_1 and r_1 such that $a = q_1 b + r_1$ and $0 \leq r_1 < b$. Show that $\gcd(a, b) \mid r_1$.

b. The second step of the Euclidean Algorithm is to find integers q_2 and r_2 such that $b = q_2 r_1 + r_2$ and $0 \leq r_2 < r_1$. Show that $\gcd(a, b) \mid r_2$.

c. Explain why $\gcd(a, b)$ must divide the remainder at every step of the Euclidean Algorithm.

10. In this exercise, you will prove that the Euclidean Algorithm (4.1.1) eventually reaches a remainder of 0. In other words, you will show that the Euclidean Algorithm halts in a finite number of steps.

   Beginning with any two natural numbers a and b, apply the Euclidean Algorithm (4.1.1). This yields the following sequence of equations:

$$a = q_1 b + r_1 \qquad 0 \leq r_1 < b$$
$$b = q_2 r_1 + r_2 \qquad 0 \leq r_2 < r_1$$
$$r_1 = q_3 r_2 + r_3 \qquad 0 \leq r_3 < r_2$$
$$r_2 = q_4 r_3 + r_4 \qquad 0 \leq r_4 < r_3$$
$$\vdots$$

a. If $b \mid a$, show that the Euclidean Algorithm will halt in a single step. Thus, for the rest of this exercise, we may assume that $b \nmid a$.

b. Consider the set R of positive remainders: $R = \{r_i \mid i \in \mathbf{N}, r_i > 0\}$. Prove that R has a smallest element.

c. The smallest element of R (the smallest positive remainder) must appear as the remainder at some step of the Euclidean Algorithm; call this step n. Thus, r_n is the smallest element of R. Prove that at step $n + 1$, the remainder must equal 0. (That is, prove that $r_{n+1} = 0$.) Conclude that the Euclidean Algorithm halts at step $n + 1$.

EXPLORATION The Euclidean Algorithm: Relating the Sizes of a and b to the Number of Steps (Exercises 11–13)

11. Suppose we are given two natural numbers a and b, where $a > b$, such that the Euclidean Algorithm performed on a and b takes exactly five steps:

$$a = q_1 b + r_1 \qquad 0 \leq r_1 < b$$
$$b = q_2 r_1 + r_2 \qquad 0 \leq r_2 < r_1$$
$$r_1 = q_3 r_2 + r_3 \qquad 0 \leq r_3 < r_2$$
$$r_2 = q_4 r_3 + r_4 \qquad 0 \leq r_4 < r_3$$
$$r_3 = q_5 r_4 + 0$$

a. Prove that $q_i \geq 1$ for all $i = 1, \ldots, 5$.

b. Since the Euclidean Algorithm required five steps, we know that $r_4 > 0$. Prove that $r_3 \geq 2$.

c. Prove that $r_2 \geq 3$.

d. Prove that $r_1 \geq 5$.

e. What is the smallest that b can be? State your answer by filling in the blank: $b \geq$ _____. Prove your answer.

f. What is the smallest that a can be? State your answer by filling in the blank: $a \geq$ _____. Prove your answer.

12. Suppose we are given two natural numbers a and b, where $a > b$, such that the Euclidean Algorithm performed on a and b takes exactly six steps. Reasoning as you did in Exercise 11, fill in the blanks with the largest numbers that make the inequalities true:

$$b \geq \text{_____} \qquad \text{and} \qquad a \geq \text{_____}.$$

Prove your answers.

13. Suppose we are given two natural numbers a and b, where $a > b$, such that the Euclidean Algorithm performed on a and b takes exactly n steps. Reasoning as in Exercises 11 and 12, make a conjecture by filling in the blanks with the largest numbers that make the inequalities true:

$$b \geq \text{_____} \qquad \text{and} \qquad a \geq \text{_____}.$$

EXPLORATION Euclid, the Game (Exercises 14–18)

The game of Euclid is played starting with an $a \times b$ rectangle, where $a, b \in \mathbf{N}$ and $a \geq b$. The two players alternate moves. After each move, the remaining $a \times b$ rectangle (where $a \geq b$) is smaller than the one that preceded the move. When it is your turn to move, you must remove some positive number of $b \times b$ squares, leaving a leftover rectangle. (It is then your opponent's turn to move.) In the event that b divides a evenly, you are permitted to remove the entire rectangle, leaving no leftover rectangle at all. In that case, you are said to *win* the game.

For example suppose you start with a 20×6 rectangle. You have three legal moves: removing one, two, or three squares of size 6×6. These moves leave, respectively, a 14×6 rectangle, an 8×6 rectangle, or a 2×6 rectangle.

If you were to choose the third option, leaving a 2×6 rectangle, then your opponent could win the game in her turn by removing the entire 2×6 rectangle. Starting with a 20×6 rectangle, a better move for you would be to remove two 6×6 squares, leaving an 8×6 rectangle for your opponent. Your opponent then only has one legal move: to remove one square, leaving a 2×6 rectangle for you. You then remove this entire rectangle and win.

As this analysis shows, if the game of Euclid is started with a 20 × 6 rectangle, the first player can play in a way that guarantees that he or she will win, no matter how good the second player is. We therefore say that a 20 × 6 rectangle is a *first-player win*. In general, an $a \times b$ rectangle is said to be a *first-player win* if the first player to move can be guaranteed to win the game by playing properly. Otherwise, we say that the rectangle is a *second-player win*.

14. For every $a \times b$ rectangle where $a, b \in \mathbf{N}$ and $1 \le b \le a \le 6$, determine whether an $a \times b$ rectangle is a first-player win or a second-player win.

15. Make a plot of your results from Exercise 14 in the Cartesian plane as follows: For every pair $a, b \in \mathbf{N}$ such that $1 \le b \le a \le 6$,

   • Place a blue dot at the point (a, b) if an $a \times b$ rectangle is a first-player win.

   • Place a red dot at the point (a, b) if an $a \times b$ rectangle is a second-player win.

16. Make a general conjecture about which $a \times b$ rectangles are first-player wins and which are second-player wins. (Your plot from Exercise 15 should be suggestive, but you may need more data to make a precise conjecture.)

17. It's your move starting with a 182 × 71 rectangle. Can you play in a way that guarantees you will win the game (is it a first-player win)? If so, what move should you make first?

*18. Prove that your conjecture from Exercise 16 is correct.

4.2 Finding the Greatest Common Divisor

A new way to find the gcd

Suppose you were asked to find the greatest common divisor of 504 and 308. To answer the question, you might start by factoring both numbers into primes:

$$504 = 2^3 \cdot 3^2 \cdot 7, \qquad 308 = 2^2 \cdot 7 \cdot 11.$$

From this, you can deduce[1] that $\gcd(504, 308) = 2^2 \cdot 7 = 28$.

   Taking prime factorizations seems to be the most natural way to find the gcd of two numbers. But suppose we wanted to solve

$$\gcd(974286723136045728122 0952, 2825240788613971993199718) = ? \qquad (1)$$

The prospect of factoring these numbers is daunting. Factoring 974286723136045728122 0952 by trial division could involve checking for factors

---

[1] As we mentioned in the previous chapter, the proof that this method works relies on the uniqueness of prime factorizations, which we will not prove until Chapter 6. Interestingly, the proof that prime factorizations are unique will fundamentally rely on the Euclidean Algorithm.

as large as $\sqrt{974286723136045728122952} \approx 3121356633158$. Even with the help of a computer, this would take a long time. One might wonder whether there is a simpler recipe for finding the greatest common divisor of two large numbers.

Euclid's insight

Perhaps we should take a break from factoring to learn more about the master baker of ancient Greece. From the moment he discovered how to make rectangular cakes, Euclid's business flourished—the volume of his sales doubled overnight. But all those cakes meant a lot of dirty dishes. And so it was that while cleaning his pans, Euclid was struck by a profound insight.

Hello—Euclid again. For the most part, cleaning up isn't so bad; it's a relaxing change of pace after a hard day's baking. All except for those puny little 1×1 pans. They're impossible to clean. I used to soak those tiny pans in detergent for hours, but now I've found a better solution. I simply refuse to bake any cakes that require 1×1 pans.

Fortunately, not all rectangular cakes require such small pans. When that customer ordered the 175×65 lemon torte the other day, the smallest pan I needed to use was **5 × 5**. I also realized that this number, **5**, is the greatest common divisor of 175 and 65. Coincidence? I think not! It always turns out this way: the side length of the smallest square pan is always the greatest common divisor of the cake dimensions. I can even prove it!

Euclidean Algorithm yields the gcd

: **Naomi's Numerical Proof Preview: Theorem 4.2.1**
In our myth, Euclid realized that the smallest square pan he needed always had sides of
length $\gcd(a, b)$. For example, let's recall how Euclid divided the 175×65 lemon torte:

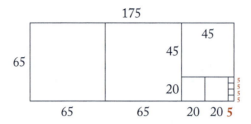

The smallest pan Euclid needed had sides of length **5**, which equals $\gcd(175, 65)$.

Arithmetically, the greatest common divisor turns out to be the last nonzero remainder in the Euclidean Algorithm:

$$175 = 2 \cdot 65 + 45$$
$$65 = 1 \cdot 45 + 20$$
$$45 = 2 \cdot 20 + 5 \leftarrow \textbf{The last nonzero remainder is 5.}$$
$$20 = 4 \cdot 5 + 0$$

Notice that at every step of the Euclidean Algorithm, the gcd of the dividend and the divisor equals the gcd of the next step's dividend and divisor:

$$\gcd(175, 65) = \gcd(65, 45) = \gcd(45, 20) = \gcd(20, 5) = \gcd(5, 0) = 5$$

This observation enables us to prove the following theorem.

THEOREM 4.2.1

Let a and b be natural numbers. Perform the Euclidean Algorithm (4.1.1) on a and b to obtain the remainders $r_0, r_1, r_2, \ldots, r_n$. Then $\gcd(a, b) = r_n$.

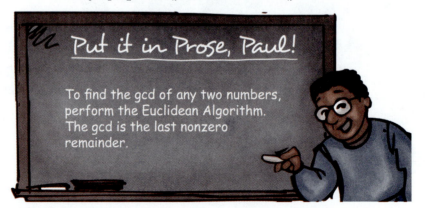

Put it in Prose, Paul!

To find the gcd of any two numbers, perform the Euclidean Algorithm. The gcd is the last nonzero remainder.

The main step in the proof of this theorem is the following lemma.

Let a, q, b, and r be integers such that a = qb + r. Then gcd(a, b) = gcd(b, r).

PROOF OF **LEMMA 4.2.2**

Let $a, q, b, r \in \mathbf{Z}$ such that

$$a = qb + r. \tag{2}$$

Let $d = \gcd(a, b)$, and let $e = \gcd(b, r)$.

[We must show that *d* = *e*. We will do this by showing that *e* ≤ *d* and *d* ≤ *e*.]

Since $e \mid b$ and $e \mid r$, we know by the Linear Combination Lemma (3.1.2) that

$$e \mid qb + r.$$

Substituting from equation (2), we see that $e \mid a$. Thus, e is a common divisor of a and b, so e must be less than or equal to the greatest common divisor of a and b; that is,

$$e \leq d.$$

From equation (2), we get

$$\boldsymbol{a - qb = r}. \tag{3}$$

Since $d \mid a$ and $d \mid b$, the Linear Combination Lemma (3.1.3) guarantees that

$$d \mid \boldsymbol{a - qb}.$$

Substituting from equation (3), we have

$$d \mid \boldsymbol{r}.$$

Thus, d is a common divisor of b and r. Hence, $d \leq \gcd(b, r)$; that is,

$$d \leq e.$$

We conclude that $d = e$. ●

Lemma 4.2.2 guarantees that the gcd remains the same from one step of the Euclidean Algorithm to the next. With this lemma under our belts, the proof of Theorem 4.2.1 will be a piece of cake.

Let $a, b \in \mathbf{N}$. Perform the Euclidean Algorithm on a and b. Then apply Lemma 4.2.2 to each equation.

Perform the Euclidean Algorithm on a and b		Apply Lemma 4.2.2 to each equation
$a = q_1 b + r_1$	$\Rightarrow$	$\gcd(a, b) = \gcd(b, r_1)$
$b = q_2 r_1 + r_2$	$\Rightarrow$	$\gcd(b, r_1) = \gcd(r_1, r_2)$
$r_1 = q_3 r_2 + r_3$	$\Rightarrow$	$\gcd(r_1, r_2) = \gcd(r_2, r_3)$
$r_2 = q_4 r_3 + r_4$	$\Rightarrow$	$\gcd(r_2, r_3) = \gcd(r_3, r_4)$
$\vdots$	$\vdots$	$\vdots$
$r_{n-2} = q_n r_{n-1} + r_n$	$\Rightarrow$	$\gcd(r_{n-2}, r_{n-1}) = \gcd(r_{n-1}, r_n)$
$r_{n-1} = q_{n+1} r_n + 0$	$\Rightarrow$	$\gcd(r_{n-1}, r_n) = \gcd(r_n, 0)$

Applying the transitive property of equality to the entire right column of equations, we conclude that $\gcd(a, b) = \gcd(r_n, 0)$, which equals r_n. This completes the proof of Theorem 4.2.1. ∎

EXAMPLE 1

Find $\gcd(153, 578)$.

SOLUTION Perform the Euclidean Algorithm on 578 and 153.

$$578 = 3 \cdot 153 + 119$$
$$153 = 1 \cdot 119 + 34$$
$$119 = 3 \cdot 34 + 17$$
$$34 = 2 \cdot 17 + 0$$

The final nonzero remainder is 17, so $\gcd(153, 578) = 17$. ∎

The Euclidean Algorithm is fast

Now we have two different methods for finding the greatest common divisor of two numbers: the Euclidean Algorithm and prime factorization. Prime factorization may work well for small numbers, but factoring large numbers can be quite a chore. For example, to solve equation (1) using prime factorization, we would need to factor both numbers. As we have discussed, factoring 9742867231360457281220952 could involve trying as many as $\sqrt{9742867231360457281220952} \approx 3121356633158$ potential factors.

Now let's see how many steps it would take to solve equation (1) using the Euclidean Algorithm. In Figure 1, we perform the Euclidean Algorithm on the two numbers, which yields their greatest common divisor:

$$\gcd(9742867231360457281220952, 282524078861397199319718) = \textbf{42}.$$

As we see from Figure 1, the Euclidean Algorithm takes only 49 steps to find the gcd of these two numbers! This is amazingly fast compared with the prime factorization method. The 25-digit numbers here may seem ridiculously huge, but in Chapter 9 we will see an application to cryptography in which we need to find the gcd of much larger numbers, some more than 100 digits long! How many steps do you think it would take to find the gcd of two 100-digit numbers? Hundreds? Thousands? Millions? Billions? Zillions?

Now look at Figure 1 from across the room. Approximately what shape do you see? The left and right edges of the numbers approximate a straight line. In other words, the number of digits in these numbers decreases *linearly*, which means that the numbers themselves decrease *exponentially*. Let's see if we can justify this empirical observation.

Recall the geometric version of the Euclidean Algorithm. To perform the Euclidean Algorithm on any two numbers a and b (with $a \geq b$), we start with an $a \times b$ rectangle. We then draw as many $b \times b$ squares as will fit:

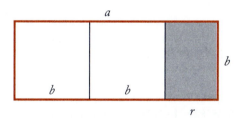

How big can the remaining rectangle (the shaded blue $b \times r$ rectangle) be, in proportion to the large orange rectangle? Since the large orange rectangle contains at least one $b \times b$ square, and since $r < b$, it follows that the area of the shaded blue rectangle is less than half the area of the large orange rectangle. In fact, at every stage of the Euclidean Algorithm, the area of the remaining rectangle will be less than half the area of the remaining rectangle from the previous step. Hence, after k steps of the Euclidean Algorithm, the remaining rectangle will have area less than $\left(\frac{1}{2}\right)^k$ times the area of the rectangle that we originally started with. We thus have the following theorem.

THEOREM 4.2.3

For any pair of natural numbers a and b, the Euclidean Algorithm takes at most $\log_2(ab)$ steps to find $\gcd(a, b)$.

1. $9742867231360457281220952 = 3 \cdot 2825240788613971993199718 + 1267144865518541301621798$
2. $2825240788613971993199718 = 2 \cdot 1267144865518541301621798 + 290951057576889389956122$
3. $1267144865518541301621798 = 4 \cdot 290951057576889389956122 + 103340635210983741797310$
4. $290951057576889389956122 = 2 \cdot 103340635210983741797310 + 84269787154921906361502$
5. $103340635210983741797310 = 1 \cdot 84269787154921906361502 + 19070848056061835435808$
6. $84269787154921906361502 = 4 \cdot 19070848056061835435808 + 7986394930674564618270$
7. $19070848056061835435808 = 2 \cdot 7986394930674564618270 + 3098058194712706199268$
8. $7986394930674564618270 = 2 \cdot 3098058194712706199268 + 1790278541249152219734$
9. $3098058194712706199268 = 1 \cdot 1790278541249152219734 + 1307779653463553979534$
10. $1790278541249152219734 = 1 \cdot 1307779653463553979534 + 482498887785598240200$
11. $1307779653463553979534 = 2 \cdot 482498887785598240200 + 342781877892357499134$
12. $482498887785598240200 = 1 \cdot 342781877892357499134 + 139717009893240741066$
13. $342781877892357499134 = 2 \cdot 139717009893240741066 + 63347858105876017002$
14. $139717009893240741066 = 2 \cdot 63347858105876017002 + 13021293681488707062$
15. $63347858105876017002 = 4 \cdot 13021293681488707062 + 11262683379921188754$
16. $13021293681488707062 = 1 \cdot 11262683379921188754 + 1758610301567518308$
17. $11262683379921188754 = 6 \cdot 1758610301567518308 + 711021570516078906$
18. $1758610301567518308 = 2 \cdot 711021570516078906 + 336567160535360496$
19. $711021570516078906 = 2 \cdot 336567160535360496 + 37887249445357914$
20. $336567160535360496 = 8 \cdot 37887249445357914 + 33469164972497184$
21. $37887249445357914 = 1 \cdot 33469164972497184 + 4418084472860730$
22. $33469164972497184 = 7 \cdot 4418084472860730 + 2542573662472074$
23. $4418084472860730 = 1 \cdot 2542573662472074 + 1875510810388656$
24. $2542573662472074 = 1 \cdot 1875510810388656 + 667062852083418$
25. $1875510810388656 = 2 \cdot 667062852083418 + 541385106221820$
26. $667062852083418 = 1 \cdot 541385106221820 + 125677745861598$
27. $541385106221820 = 4 \cdot 125677745861598 + 38674122775428$
28. $125677745861598 = 3 \cdot 38674122775428 + 9655377535314$
29. $38674122775428 = 4 \cdot 9655377535314 + 52612634172$
30. $9655377535314 = 183 \cdot 52612634172 + 27265481838$
31. $52612634172 = 1 \cdot 27265481838 + 25347152334$
32. $27265481838 = 1 \cdot 25347152334 + 1918329504$
33. $25347152334 = 13 \cdot 1918329504 + 408868782$
34. $1918329504 = 4 \cdot 408868782 + 282854376$
35. $408868782 = 1 \cdot 282854376 + 126014406$
36. $282854376 = 2 \cdot 126014406 + 30825564$
37. $126014406 = 4 \cdot 30825564 + 2712150$
38. $30825564 = 11 \cdot 2712150 + 991914$
39. $2712150 = 2 \cdot 991914 + 728322$
40. $991914 = 1 \cdot 728322 + 263592$
41. $728322 = 2 \cdot 263592 + 201138$
42. $263592 = 1 \cdot 201138 + 62454$
43. $201138 = 3 \cdot 62454 + 13776$
44. $62454 = 4 \cdot 13776 + 7350$
45. $13776 = 1 \cdot 7350 + 6426$
46. $7350 = 1 \cdot 6426 + 924$
47. $6426 = 6 \cdot 924 + 882$
48. $924 = 1 \cdot 882 + \mathbf{42}$
49. $882 = 21 \cdot 42 + 0$

Figure 1   The Euclidean Algorithm performed on two large numbers.

PROOF See Exercise 8.

Thus, for two 100-digit numbers a and b, the Euclidean Algorithm would take at most

$$\log_2(10^{100} \cdot 10^{100}) = \log_2(10^{200})$$

$$\approx 664 \text{ steps.}$$

While this many steps would be a chore to do by hand, on a computer they could be done in a fraction of a second.

The Euclidean Algorithm and the Fibonacci numbers

Naomi's Numerical Proof Preview: Theorem 4.2.4

With the help of our old friends the Fibonacci numbers, we can get an even better bound for the number of steps required to perform the Euclidean Algorithm on two numbers a and b.

Suppose we are given two natural numbers a and b such that the Euclidean Algorithm performed on a and b takes exactly five steps:

$$a = q_1 b + r_1 \qquad 0 \le r_1 < b$$
$$b = q_2 r_1 + r_2 \qquad 0 \le r_2 < r_1$$
$$r_1 = q_3 r_2 + r_3 \qquad 0 \le r_3 < r_2$$
$$r_2 = q_4 r_3 + r_4 \qquad 0 \le r_4 < r_3$$
$$r_3 = q_5 r_4 + 0$$

How big must b be? To answer this question, let's start with the bottom equation and work our way up. First note that since the Euclidean Algorithm takes five steps, $r_4 \ne 0$. In other words,

$$r_4 \ge 1.$$

Since $r_4 < r_3$, it follows that

$$r_3 \ge 2.$$

How big must r_2 be? Recall that $r_2 = q_4 r_3 + r_4$. Thus, $r_2 \ge r_3 + r_4$ (because $q_4 \ge 1$; see Exercise 6). Hence, $r_2 \ge 2 + 1$. That is,

$$r_2 \ge 3.$$

By similar reasoning, $r_1 \geq r_2 + r_3$. Hence, $r_1 \geq 3 + 2$. That is,

$$r_1 \geq 5.$$

Continuing in the same fashion, we find that $b \geq r_1 + r_2$. Hence, $b \geq 5 + 3$. That is,

$$b \geq 8.$$

Do you see a pattern in these inequalities? You may notice that the successive remainders, r_4, r_3, r_2, r_1, and b are bounded by successive Fibonacci numbers: $1, 2, 3, 5$, and 8. This follows from the fact that each remainder is at least the sum of the previous two remainders.

We have just shown that for the Euclidean Algorithm to require five steps, b must be at least 8, which is the 6th Fibonacci number. In other words:

If the Euclidean Algorithm on a and b takes five steps, then $b \geq F_6$.

An equivalent statement is the contrapositive:

If $b < F_6$, then the Euclidean Algorithm on a and b takes at most 4 steps.

In general, starting with any natural numbers a and b, if $b < F_{n+2}$, then the Euclidean Algorithm will take at most n steps. We will now prove this general statement, which uses the size of b to provide an excellent upper bound on the number of steps the Euclidean Algorithm takes.

THEOREM 4.2.4

Let $a, b, n \in \mathbf{N}$. If $b < F_{n+2}$, then the Euclidean Algorithm performed on a and b takes at most n steps.

In the statement of Theorem 4.2.4, the bound $b < F_{n+2}$ is as strong as it can possibly be, in the following sense. If $b = F_{n+2}$, then the Euclidean Algorithm can take more than n steps. In fact, as you will show in Exercise 10, if $a = F_{n+3}$ and $b = F_{n+2}$ (for any $n \geq 0$), then the Euclidean Algorithm takes exactly $n + 1$ steps. Thus, in the statement of Theorem 4.2.4, if we were to replace the strict inequality $b < F_{n+2}$ by the slightly weaker inequality $b \leq F_{n+2}$, then the statement would no longer be true. In this sense, Theorem 4.2.4 gives us the best possible estimate of the number of steps required by the Euclidean Algorithm.

PROOF OF THEOREM 4.2.4

We prove this by strong induction.

Let $P(n)$ denote the statement

> For any $a, b \in \mathbf{N}$ such that $b < F_{n+2}$,
> the Euclidean Algorithm on a and b takes at most n steps. $\qquad \leftarrow P(n)$

Let n be a natural number, and suppose that $P(k)$ is true for all natural numbers $k < n$.

[To show: $P(n)$ is true.]

Let $a, b \in \mathbf{N}$ such that $b < F_{n+2}$.

If $n = 1$, then $b < F_3$ (i.e., $b < 2$). Thus, $b = 1$, and hence the Euclidean Algorithm takes only one step.

If $n = 2$, then $b < F_4$ (i.e., $b < 3$). Hence, $b = 1$ or $b = 2$, and thus the Euclidean Algorithm takes at most two steps.

With $n = 1$ and $n = 2$ out of the way, we may now assume that $n \geq 3$.

Let $a, b \in \mathbf{N}$ with $b < F_{n+2}$. Dividing a by b gives

$$a = q_1 b + r_1 \qquad 0 \leq r_1 < b.$$

We may assume $r_1 > 0$ (if $r_1 = 0$, then the Euclidean Algorithm starting with a and b takes only 1 step, so the number of steps required is $\leq n$). Thus, we may divide b by r_1:

$$b = q_2 r_1 + r_2 \qquad 0 \leq r_2 < r_1 \qquad (4)$$

Now we may assume $r_2 > 0$ (if $r_2 = 0$, then the Euclidean Algorithm starting with a and b takes only 2 steps, so the number of steps required is $\leq n$).

Claim $r_1 < F_{n+1}$ or $r_2 < F_n$.

Proof of Claim (By contradiction.)

Assumption: Suppose $\boldsymbol{r_1 \geq F_{n+1}}$ and $\boldsymbol{r_2 \geq F_n}$.

Starting with equation (4), we have

$$b = q_2 r_1 + r_2$$
$$\geq r_1 + r_2$$
$$\geq F_{n+1} + F_n \qquad \leftarrow \textbf{by the Assumption}$$
$$= F_{n+2}.$$

Thus, $b \geq F_{n+2}$, which contradicts the hypothesis that $b < F_{n+2}$. $\Rightarrow\Leftarrow$

This completes the proof of the Claim.

This Claim allows us to separate into two cases:

Case 1 $r_1 < F_{n+1}$.

By the inductive hypothesis, we know that $P(n-1)$ is true, so the Euclidean Algorithm on b and r_1 takes at most $n-1$ steps. Since the Euclidean Algorithm on a and b has one more step than the Euclidean Algorithm on b and r_1, we conclude that the Euclidean Algorithm on a and b takes at most n steps.

Case 2 $r_2 < F_n$.

By the inductive hypothesis, we know that $P(n-2)$ is true, so the Euclidean Algorithm on r_1 and r_2 takes at most $n-2$ steps. Since the Euclidean Algorithm on a and b has two more steps than the Euclidean Algorithm on r_1 and r_2, we conclude that the Euclidean Algorithm on a and b takes at most n steps.

Thus, $P(n)$ is true.

By the Principle of Strong Induction, we conclude that for all $n \in \mathbf{N}$, $P(n)$ is true. ∎

EXAMPLE 2

Suppose you wish to perform the Euclidean Algorithm on two numbers a and b, where $b \leq 100$. Use Theorem 4.2.4 to give an upper bound on the number of steps the Euclidean Algorithm will take.

SOLUTION We know that $b \leq 100$. The first Fibonacci number greater than 100 is $F_{12} = 144$. Thus, $b < F_{12}$, so Theorem 4.2.4 guarantees that at most 10 steps will be needed to perform the Euclidean Algorithm on a and b. ∎

The Euclidean Algorithm is good as gold!

Naomi's Numerical Proof Preview: Theorem 4.2.5

In Example 2, we were given that $b \leq 100$, and in order to estimate the number of steps that the Euclidean Algorithm would take, we first needed to determine the first Fibonacci number larger than 100. Although we could determine this by trial and error, a better method is suggested by Lemma 2.3.2, which tells us that the Fibonacci numbers can be approximated by powers of the golden ratio, φ. Specifically, Lemma 2.3.2 tells us that

$$\varphi^n < F_{n+2}. \tag{5}$$

Thus, if we want to find a Fibonacci number greater than 100, it is sufficient to find a natural number n such that

$$100 \leq \varphi^n. \tag{6}$$

To do this, we harness the power of logarithms. Taking the log base φ of both sides, we see that this inequality is equivalent to

$$\log_\varphi 100 \leq n.$$

Since $\log_\varphi 100 \approx 9.57$, we can round up to the nearest integer and find that $n = \mathbf{10}$ satisfies this inequality. It follows from (5) and (6) that $F_{12} > 100$, and we've succeeded in finding a Fibonacci number greater than 100. Thus, $b < F_{12}$, and we can then use Theorem 4.2.4 to determine that the Euclidean Algorithm takes at most $\mathbf{10}$ steps.

A similar argument can be used to show that when the Euclidean Algorithm is performed on *any* pair of natural numbers a and b, the number of steps required is at most $\mathbf{\log_\varphi b}$, **rounded up to the nearest integer**. We state this in the following theorem.

THEOREM 4.2.5

Let $a, b \in \mathbf{N}$ with $b > 1$. Then the number of steps required when the Euclidean Algorithm is performed on a and b is at most

$$\lceil \log_\varphi b \rceil. \qquad \leftarrow \textbf{The symbol } \lceil x \rceil \textbf{, read "ceiling of } x \textbf{," is}$$
$$\textbf{the result of rounding } x \textbf{ up to the}$$
$$\textbf{nearest integer.}$$

Before proving Theorem 4.2.5, let's look at an example.

EXAMPLE 3

According to Theorem 4.2.5, how many steps are required to perform the Euclidean Algorithm on the following values of a and b?

$a = 616294798705487364509840003$ and $b = 317831490619591482513697 3$

SOLUTION We compute

$$\log_\varphi b = 117.2423\ldots .$$

Thus, the Euclidean Algorithm on a and b takes at most

$$\lceil 117.2423\ldots \rceil = 118 \text{ steps.}$$

(In actual fact, the Euclidean Algorithm on a and b takes only 53 steps, as you will demonstrate in Exercise 5.)

Let $a, b \in \mathbf{N}$ with $b > 1$. Let $n = \lceil \log_\varphi b \rceil$.
Since $\log_\varphi b \leq n$,

$$b \leq \varphi^n.$$

Recall from Lemma 2.3.2 that $\varphi^n < F_{n+2}$. Thus,

$$b < F_{n+2}.$$

Now applying Theorem 4.2.4, we conclude that the Euclidean Algorithm performed on a and b takes at most n steps. ∎

Suppose we wish to perform the Euclidean Algorithm on two 100-digit numbers, a and b. According to Theorem 4.2.5, this would require at most

$$\lceil \log_\varphi 10^{100} \rceil = \lceil 478.49 \ldots \rceil$$

$$= 479 \text{ steps.}$$

On a computer, 479 steps of the Euclidean Algorithm could be done in an instant.
Now say we chose numbers that were about 10 times as large. To find the gcd of two whopping 101-digit numbers, would the Euclidean Algorithm require 10 times as many steps? No! In fact, this time we find that we need at most

$$\lceil \log_\varphi 10^{101} \rceil = \lceil 483.27 \ldots \rceil$$

$$= 484 \text{ steps.}$$

Thus, going from two 100-digit numbers to two 101-digit numbers adds at most 5 additional steps to our computation.
In general, we have the identity

$$\log_\varphi(10b) = \log_\varphi 10 + \log_\varphi b$$

$$< 4.79 + \log_\varphi b.$$

This shows that in general, whenever we multiply each of two initial numbers a and b by a factor of about 10, we expect that this will add about 5 steps to our computation of their gcd. We might have thought that *multiplying* the initial numbers by 10 would cause the maximum number of steps required to be *multiplied* by some constant. However, because the maximum number of steps required is proportional to the logarithm of the initial number b, *multiplying* the initial numbers by a constant only *adds* a constant to the number of steps required.

A computational note

At various points throughout this book, we encounter algorithms related to our theorems of number theory. When we meet such an algorithm, we would like to know how efficient the algorithm is. In other words, we would like to know approximately how long the algorithm takes to run on a computer as a function of the sizes of the input numbers.

Computer scientists draw a sharp distinction between algorithms that run in *polynomial time* and those that do not. A **polynomial-time** algorithm is one whose running time can be bounded by a polynomial function of the length of the input. For example, if d represents the length of the input, an algorithm requiring d^3 steps is considered a polynomial-time algorithm. In contrast, many algorithms that are not polynomial time require *exponential* time. For instance, an algorithm requiring 2^d steps is considered an *exponential-time algorithm*.

Polynomial functions (such as d^3) grow much, much, much more slowly than exponential functions (such as 2^d) as d gets large. If $d = 1000$, for instance, there is no comparison between 1000^3, which is a mere billion, and 2^{1000}, which is a humongous number with more than 300 digits. For this reason, computer scientists consider an algorithm feasible if it is polynomial-time, and infeasible otherwise.

This being a number theory text, the algorithms we encounter will take as their input one or more natural numbers. If the input is a d-digit natural number, n, then the length of the input is d (the number of digits in n), which is approximately $\log n$. Thus, for our algorithm to be considered polynomial-time, the running time must be a polynomial in log n. For instance, it follows from Theorem 4.2.5 that the Euclidean Algorithm is a polynomial-time algorithm.

Contrast this with the algorithm for factoring a d-digit number, n, using trial division by all numbers up to $\sqrt{n}$. The number of steps required for this algorithm can be as many as $\sqrt{n}$, which is approximately $\sqrt{10^d} = \left(\sqrt{10}\right)^d$. Thus, the running time of this algorithm is an exponential function in the input size d. We conclude that this algorithm for factoring a number is an exponential-time algorithm. It is not a feasible algorithm for factoring large numbers. Although better factoring algorithms exist, there is no known polynomial-time algorithm for factoring.

Throughout our study of number theory, as we meet algorithms, it is worth considering whether these algorithms can be performed in polynomial time. We shall see mixed results. Some number-theoretic operations, such as gcd's, turn out to be quickly computable if one uses the right algorithm. Others, such as factoring, pose more of a challenge and are believed to be computationally infeasible no matter what algorithm is used.

EXERCISES 4.2

Numerical Problems

1. Use the Euclidean Algorithm to find each greatest common divisor.

   a. $\gcd(17, 39)$ b. $\gcd(108, 15)$ c. $\gcd(826, 624)$

   d. $\gcd(468, 864)$ e. $\gcd(42823, 6409)$ f. $\gcd(11111, 1111111)$

2. Use the Euclidean Algorithm to find the greatest common divisor of the following pairs of consecutive Fibonacci numbers. Describe any patterns you find in the computation.

   a. $\gcd(2, 3)$ b. $\gcd(3, 5)$ c. $\gcd(5, 8)$

   d. $\gcd(8, 13)$ e. $\gcd(13, 21)$ f. $\gcd(21, 34)$

Conjecturer's Corner (Exercises 3 and 4)

3. Find the greatest common divisor of the following pairs of Fibonacci numbers. If your answer is a Fibonacci number, express it in the form F_n.

   a. $\gcd(F_4, F_8)$ b. $\gcd(F_5, F_{10})$ c. $\gcd(F_3, F_6)$

   d. $\gcd(F_6, F_9)$ e. $\gcd(F_8, F_{12})$

4. Based on the patterns that you notice from Exercise 3, make a conjecture that completes the following statement:

   $\gcd(F_m, F_n) =$ _____.

Reasoning and Proofs

5. **Computer Exercise** Write a program to perform the Euclidean Algorithm on any two natural numbers a and b. Apply your program to the values of a and b from Example 3. Number each step to verify that it requires exactly 53 steps.

6. Suppose we are given two natural numbers a and b, where $a > b$. When the Euclidean Algorithm is performed on a and b, can any of the quotients q_i equal 0? Explain your reasoning.

7. Let a and b be natural numbers with $1{,}000{,}000 > a > b$.

   a. What bound does Theorem 4.2.3 give for the number of steps the Euclidean Algorithm will take when performed on a and b?

   b. What bound does Theorem 4.2.4 give for the number of steps the Euclidean Algorithm will take when performed on a and b? Compare your answer with the bound from part a.

8. Prove Theorem 4.2.3. (Do not use Theorem 4.2.4 or 4.2.5 in your proof.)

9. Let $a, b, k \in \mathbf{N}$. Use the Euclidean Algorithm to prove that $\gcd(ka, kb) = k \cdot \gcd(a, b)$.

10. Show that for any $n \geq 0$, the Euclidean Algorithm starting with the consecutive Fibonacci numbers F_{n+3} and F_{n+2} takes exactly $n + 1$ steps.

11. **Phil Lovett** In this exercise, you will compare the amount of time required to compute the gcd of very large numbers with the time required to factor very large numbers using trial division.[2]

    a. Your friend Phil Lovett stops by one day with two whopping 1000-digit numbers, a and b. According to Theorem 4.2.5, the Euclidean Algorithm will be guaranteed to terminate in how many steps?

    b. Your computer takes about 1/10,000 of a second to do one step of the Euclidean Algorithm. (Actually, modern computers can do divisions much faster than this.) Approximately how long will it take your computer to find $\gcd(a, b)$?

    c. "Dude," Phil explains, "I want to find $\gcd(a, b)$, so I've decided to factor a and b. I know factoring can take a long time on an ordinary PC, but I just got a new Rasta-4000 that runs at 1,000,000 GHz, and can do a quadrillion (10^{15}) trial divisions every second. That should be able to factor a and b really fast!"

    Using trial division (up to the square root of a), approximately how long will it take Phil's computer to factor the number a? How does this compare with scientists' estimate of the present age of the universe (approximately 14 billion years)?

---

[2] There are faster factoring methods, but they still require an enormous amount of time for numbers of this size.

12. *Use your answers to Exercise 11 to answer the following questions.* Assuming the Rasta-4000 consumes only 10^{-9} joules per second (1 nanowatt of power), how much mass would be required to power the computer to the completion of the factorization? How does this compare with scientists' estimate of the mass of the observable universe? (Assume that there are $3 \cdot 10^{52}$ kg of matter in the observable universe and that each kilogram of matter can be converted to 10^{17} joules of energy.)

Advanced Reasoning and Proofs

13. Prove that for all natural numbers a and b, $\gcd(F_a, F_b) = F_{\gcd(a,b)}$.

    [*Hint:* Use the Euclidean Algorithm and the fact that $F_{n+m} = F_m F_{n+1} + F_{m-1} F_n$ (which was proved in Section 2.3, Exercise 14).]

14. Prove that $\gcd(a^m - 1, a^n - 1) = a^{\gcd(m,n)} - 1$.

4.3 A Greeker Argument that $\sqrt{2}$ Is Irrational

The Euclidean Algorithm with fractions

How could we divide a $\frac{9}{2} \times \frac{5}{3}$ rectangle into squares? Although the Euclidean Algorithm is only defined for rectangles with integral sides, we can still try using the same method—at each step, we fit in the largest square possible. In this way, we can apply the Euclidean Algorithm to a rectangle whose side lengths are arbitrary real numbers. Bear in mind that although the Euclidean Algorithm eventually halts when applied to any rectangle with integer dimensions, we are not guaranteed that the Euclidean Algorithm will halt for every rectangle whose dimensions are real numbers.

EXAMPLE 1

In Figure 1, we use the Euclidean Algorithm to divide a $\frac{9}{2} \times \frac{5}{3}$ rectangle into squares.

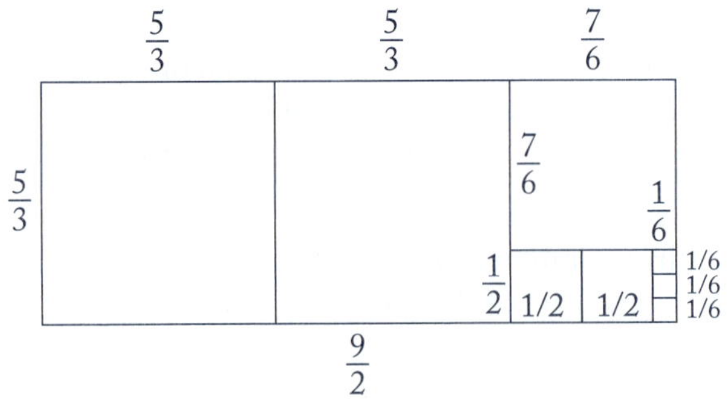

Figure 1 The Euclidean Algorithm applied to a $\frac{9}{2} \times \frac{5}{3}$ rectangle.

The Euclidean Algorithm successfully divides the rectangle in Example 1 even though this rectangle has noninteger sides. This may not be surprising, because the rectangle is similar to a rectangle that has integer sides. When we scale up the $\frac{9}{2} \times \frac{5}{3}$ rectangle of the example by a factor of 6, we obtain a 27×10 rectangle. In Figure 2, we examine what happens when the Euclidean Algorithm is applied to this larger rectangle.

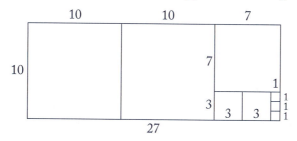

Figure 2   The Euclidean Algorithm applied to a 27×10 rectangle.

The result is identical to Figure 1, except that all of the lengths are multiplied by the scale factor, 6, as shown in Figure 2. In general, the Euclidean Algorithm treats similar rectangles identically up to a scale factor, as noted in the following observation.

If two rectangles are similar with scale factor r, the Euclidean Algorithm produces the same picture for both, except that all lengths are scaled by a factor of r.

In Section 4.2, we saw that the Euclidean Algorithm halts for all rectangles whose sides have integral length. But given any rectangle whose sides are rational numbers, it is easy to find a similar rectangle whose sides are positive integers: an $\frac{a}{b} \times \frac{c}{d}$ rectangle is similar to an $ad \times bc$ rectangle. Thus:

REMARK. *The Euclidean Algorithm will halt for any rectangle with rational sides.*

The Euclidean Algorithm with irrational numbers

We are now ready to consider a $\sqrt{2} \times 1$ rectangle. What happens when we try to divide this rectangle into squares using the Euclidean Algorithm?

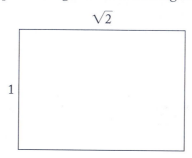

The largest square that fits into this rectangle has dimensions 1×1. Since $\sqrt{2} \approx 1.41$, only one 1×1 square fits, and the remaining rectangle **R** has dimensions $1 \times (\sqrt{2} - 1)$:

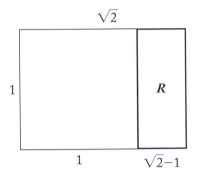

How many $(\sqrt{2} - 1) \times (\sqrt{2} - 1)$ squares fit into **R**? Since $\sqrt{2} - 1 \approx 0.41$, we know it goes into 1 twice, and so two squares fit:

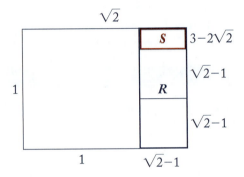

Figure 3 The first two steps of the Euclidean Algorithm performed on a $\sqrt{2} \times 1$ rectangle.

The length of the smaller side of the remaining rectangle **S** is

$$1 - 2\left(\sqrt{2} - 1\right) = 3 - 2\sqrt{2},$$

as indicated in the diagram. Our work is still not done. The next step would be to see how many $(3 - 2\sqrt{2}) \times (3 - 2\sqrt{2})$ squares fit into **S**. Then we would continue, hoping that at some point the process would stop, leaving our rectangle successfully divided into squares. The curious reader is encouraged to try a few more steps of this process.

But instead of continuing mechanically along, let us step back and make a remarkable observation.

REMARKABLE OBSERVATION 4.3.2

*Rectangle **R** and rectangle **S** are similar.*

PROOF [To show: $\dfrac{\text{width of } R}{\text{length of } R} = \dfrac{\text{width of } S}{\text{length of } S}$, i.e., $\dfrac{\sqrt{2}-1}{1} = \dfrac{3-2\sqrt{2}}{\sqrt{2}-1}$.]

$$\frac{\text{width of } R}{\text{length of } R} = \frac{\sqrt{2}-1}{1}$$

$$= \frac{\sqrt{2}-1}{1} \cdot \frac{\sqrt{2}-1}{\sqrt{2}-1}$$

$$= \frac{2 - \sqrt{2} - \sqrt{2} + 1}{\sqrt{2}-1}$$

$$= \frac{3 - 2\sqrt{2}}{\sqrt{2}-1} = \frac{\text{width of } S}{\text{length of } S}.$$ ∎

What does this observation have to do with the Euclidean Algorithm? We saw earlier that

*Applying one step of the Euclidean Algorithm to **R** divides **R** into two squares and a remainder rectangle, **S**, that is similar to **R**.*

We know that the Euclidean Algorithm will treat the similar rectangles **R** and **S** identically up to a scale factor (Observation 4.3.1). It follows that

*Applying one step of the Euclidean Algorithm to **S** divides **S** into two squares and a remainder rectangle, **T**, that is similar to **S**:*

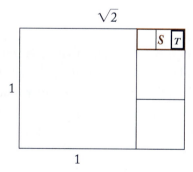

Figure 4   The third step of the Euclidean Algorithm performed on a $\sqrt{2} \times 1$ rectangle.

Continuing the Euclidean Algorithm from here, we find that at each step, the rectangle is divided into two squares and a remainder rectangle that is similar to the rectangle that began the step.

So, how many steps will it take before the Euclidean Algorithm halts? No finite number of steps will be enough! At each step, we are no closer to being done than at the previous step. The remainder rectangles become smaller and smaller, but at no stage can the remainder rectangle be evenly divided into squares—we always get two squares and an even smaller rectangle.

Thus, we've proven the following theorem:

THEOREM 4.3.3

The Euclidean Algorithm applied to a $\sqrt{2} \times 1$ rectangle never halts.

Interestingly, this result gives us another way to prove an old theorem.

THEOREM 4.3.4 (A.K.A. THEOREM 1.4.1)

$\sqrt{2}$ *is irrational.*

PROOF (By contradiction.)

Assumption: Suppose that $\sqrt{2} \in \mathbf{Q}$.

This means that there exist $a, b \in \mathbf{N}$ such that $\sqrt{2} = \frac{a}{b}$. Then a $\sqrt{2} \times 1$ rectangle is similar to an $a \times b$ rectangle. Since a and b are natural numbers, the Euclidean Algorithm (4.1.1) halts when applied to an $a \times b$ rectangle. Because the Euclidean Algorithm treats similar rectangles identically (Observation 4.3.1), the Euclidean Algorithm also halts for a $\sqrt{2} \times 1$ rectangle. This contradicts Theorem 4.3.3. $\Rightarrow\Leftarrow$.

Hence, $\sqrt{2}$ is irrational. ■

Let's compare the two proofs we have seen that $\sqrt{2}$ is irrational. Recall that the first proof we gave (the proof of Theorem 1.4.1) involved manipulating the equation $\sqrt{2} = \frac{a}{b}$ algebraically and carefully considering the parity (odd vs. even) of a and b. In contrast, the proof of Theorem 4.3.4 is pure geometry. At the heart of the proof lies the observation that two rectangles are similar. This second proof is much more in the spirit of the Greeks, who viewed number theory in terms of geometry. For another geometric proof that $\sqrt{2}$ is irrational, see Exercise 9.

EXERCISES 4.3

Numerical Problems

1. Use the Euclidean Algorithm to divide a $\frac{19}{14} \times \frac{1}{2}$ rectangle into squares.

For Exercises 2–3: In the text, we defined greatest common divisor only for integers, but we can also define the gcd of a pair of rational numbers. Given rational numbers a and b, perform the Euclidean Algorithm geometrically on a and b. The gcd of a and b is the side length of the smallest square that is needed.

2. Find the gcd of each of the following pairs of rational numbers. Describe any patterns you observe.

   a. $a = \dfrac{1}{3}, b = \dfrac{1}{5}$ b. $a = \dfrac{1}{8}, b = \dfrac{1}{12}$

   c. $a = \dfrac{1}{10}, b = \dfrac{1}{6}$ d. $a = \dfrac{1}{6}, b = \dfrac{1}{9}$

3. Find the gcd of each of the following pairs of rational numbers. Describe any patterns you observe in the numerators of your answers.

   a. $a = \dfrac{1}{10}, b = \dfrac{1}{4}$ b. $a = \dfrac{3}{10}, b = \dfrac{15}{4}$

   c. $a = \dfrac{17}{10}, b = \dfrac{5}{4}$ d. $a = \dfrac{21}{10}, b = \dfrac{35}{4}$

Reasoning and Proofs

4. Apply the Euclidean Algorithm to a $\sqrt{3} \times 2$ rectangle. Continue until you get similar rectangles. Show algebraically that the rectangles are similar.

5. Apply the Euclidean Algorithm to a $\sqrt{5} \times 1$ rectangle. Continue until you get similar rectangles. Show algebraically that the rectangles are similar.

6. Use your results from Exercise 5 to show that $\sqrt{5}$ is irrational.

7. Does the Euclidean Algorithm halt for a $5\sqrt{2} \times 17\sqrt{2}$ rectangle? Prove your answer.

8. Suppose you wish to perform the Euclidean Algorithm on a $\sqrt{3} \times 1$ rectangle. Instead of drawing a diagram, you could do the Euclidean Algorithm algebraically. Here are the first three steps:

$$\sqrt{3} = 1 \cdot 1 + (\sqrt{3} - 1) \qquad\qquad q_1 = 1, \ r_1 = \sqrt{3} - 1,$$
$$1 = 1 \cdot (\sqrt{3} - 1) + (2 - \sqrt{3}) \qquad\qquad q_2 = 1, \ r_2 = 2 - \sqrt{3},$$
$$(\sqrt{3} - 1) = 2 \cdot (2 - \sqrt{3}) + (3\sqrt{3} - 5) \qquad\qquad q_3 = 2, \ r_3 = 3\sqrt{3} - 5$$
$$\vdots \qquad\qquad\qquad\qquad \vdots$$

   a. Verify that the remainders are smaller at each step.

   b. Write the next two steps of the Euclidean Algorithm.

   c. Explain why this process will continue forever (will never halt).

Advanced Reasoning and Proofs

EXPLORATION Another Greeker Argument that $\sqrt{2}$ Is Irrational
(Exercises 9–14)

In this Exploration, you will write a geometric proof that $\sqrt{2}$ is irrational. You will do this by proving that the hypotenuse of an isosceles right triangle is incommensurable with its leg.

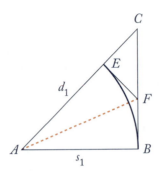

Suppose triangle ABC is an isosceles right triangle with leg $AB = s_1$ and hypotenuse $AC = d_1$, as shown in the diagram. To prove that s_1 and d_1 are incommensurable, we argue by contradiction. We assume that s_1 and d_1 are commensurable. Thus, there is a distance δ such that s_1 and d_1 are both integral multiples of δ. In Exercises 9–13, you will show how to construct a smaller isosceles right triangle such that its leg and hypotenuse are also multiples of δ.

Intuitively, this can be done by cutting out triangle ABC and folding leg AB over to coincide with the hypotenuse AC. More formally, proceed as follows. Draw a circle with center A and radius AB. Let E be the point where this circle intersects AC. Next, draw the tangent to the circle at E, intersecting side BC at F. Note that EF is perpendicular to AC.

9. Prove that $EC = EF$.

Now we have seen that $EC = EF$ and that EC is perpendicular to EF. Thus, CEF is an isosceles right triangle.

10. Prove that right triangle AEF is congruent to right triangle ABF.

11. Let $s_2 = EF$. Prove that s_2 is a multiple of δ.

12. Let $d_2 = CF$. Prove that d_2 is a multiple of δ. [*Hint:* Start by writing d_2 in terms of s_1 and s_2.]

13. Prove that $s_2 < s_1$.

At this point, we have shown that given any isosceles right triangle whose leg and hypotenuse are both multiples of δ, we can find a smaller isosceles right triangle whose leg and hypotenuse are again both multiples of δ. Iterating this construction gives us a sequence of triangles with leg lengths $s_1 > s_2 > s_3 > \cdots$. We also know that each s_n is a multiple of δ. Thus, we may write $s_n = a_n \delta$ where a_n is a natural number.

14. Apply the Well-Ordering Principle to the set $\{a_1, a_2, a_3, \ldots\}$ to obtain a contradiction. What can you conclude?

Chapter 5

Diophantus (*fl.* 250 CE)

Almost nothing is known about Diophantus's life except for the "facts" included in the following word problem:

> *God granted him to be a boy for the sixth part of his life, and adding a twelfth part to this, He clothed his cheeks with down; He lit him the light of wedlock after a seventh part, and five years after his marriage He granted him a son. Alas! Late-born wretched child; after attaining the measure of half his father's life, chill Fate took him. After consoling his grief by the science of numbers for four years, he ended his life.*

This puzzle was contained in a problem book entitled *Greek Anthology* that was written in the fifth or sixth century. From this problem, it can be deduced (see Exercise 1) that Diophantus married when he was 33, had a son who lived to be 42, and died at the age of 84. However, it is not clear whether these "facts" are true or whether they just made a good word problem. Historians are not sure when Diophantus lived—while most believe that he lived in Alexandria around the year 250, it has also been suggested that he might have lived one or two centuries earlier.

Diophantus is primarily known for his treatise *Arithmetica*, a book of algebra problems and their solutions. The introduction to *Arithmetica* states that the work is divided into 13 books. However, only 10 of the books have

Title page of the 1621 *edition of Diophantus's Arithmetica.*

been found, and 4 of these were not discovered until 1973! Diophantus is often considered to be the founder of algebra because he introduced the use of symbolic notation, making it possible to solve more complex problems. Prior to the publication of *Arithmetica*, mathematicians did not use symbols to solve algebra problems, instead relying exclusively on words. Prior to *Arithmetica*, exponents had to be expressed in terms of geometry. Since dimensions beyond three are not easy to visualize, this geometrical viewpoint prevented the ancient Greeks from considering powers other than squares and cubes. In *Arithmetica*, Diophantus introduced a system to symbolically represent arithmetic operations, an unknown variable, and powers of this unknown variable up to degree 6.

Arithmetica is not a theoretical treatise giving postulates about algebraic functions or operations. Rather, it is a collection of 150 problems whose solutions were presented (though not always in complete detail) using Diophantus's new symbolic notation. One difficulty that Diophantus faced was that he always used the same Greek letter to represent unknowns. Hence, he was unable to represent more than one variable at a time. Also, Diophantus considered only positive rational solutions to problems. For example, he states that the equation $4 = 4x + 20$ is "absurd" because it has no positive solution, and he states that the equation $3x + 18 = 5x^2$ is "not rational" because it has no rational solution.

In addition to introducing symbolic notation, *Arithmetica* contains many results in number theory. One such result involves expressing a particular perfect square as a sum of two squares. In contrast with Euclid (see "A Bit of Euclid's Math" at the beginning of Chapter 4), Diophantus does not solve this problem in general but only finds particular integer solutions. Many centuries later, while reading this section of *Arithmetica*, Fermat proposed what has become known as Fermat's Last Theorem (15.2.1), which states that there do not exist nonzero integers x, y, z, and $n > 2$ such that $x^n + y^n = z^n$.

In the margin of his copy of *Arithmetica*, Fermat wrote in Latin

> *Cubum autem in duos cubos, aut quadrato-quadratum in duos quadrato-quadratos, et generaliter nullam in infinitum ultra quadratum potestatem in duos eiusdem nominis fas est dividere cuius rei demonstrationem mirabilem sane detexi. Hanc marginis exiguitas non caperet.*

Translated into English, this means:

> *It is impossible for a cube to be the sum of two cubes, a fourth power to be the sum of two fourth powers, or in general for any number that is a power greater than the second to be the sum of two like powers. I have discovered a truly remarkable proof of this proposition that this margin is too small to contain.*

Fermat left no written proof of his claim, and for more than 300 years people tried unsuccessfully to prove it. Quite recently, in 1995, the theorem was finally proven! To find out more about Fermat's Last Theorem and its fascinating history, see Sections 15.2−15.5.

A bit of Diophantus's math

While the problems in *Arithmetica* were stated in general form, Diophantus also specified particular numerical values or conditions in order to make sure that each problem had only one solution. For example, Book I, Problem 4 asks for "two numbers in a given ratio, their difference also being given." The problem then specifies, "Given ratio 5:1, given difference 20." In other words, we are looking for two numbers whose difference is 20 such that one number is 5 times the other. Diophantus's solution was as follows:

"Numbers $5x, x$. Therefore, $x = 5$, and the numbers are 25, 5."

A modern approach would be to write the equation $5x - x = 20$ and then use elementary algebra to deduce that $x = 5$ and conclude that the two numbers are 5 and 25. Since we know that Diophantus was looking only for positive solutions, we would not also consider the equation $x - 5x = 20$.

Chapter 5 Linear Diophantine Equations
or General Potato Theory

5.1 The Equation $aX + bY = 1$

Diophantus and the potato

The ancient Greek mathematician Diophantus of Alexandria wrote *Arithmetica*, a celebrated treatise on algebra and number theory that introduced the use of symbols for algebra. The importance of Diophantus's work has been widely recognized—over the centuries, great mathematicians such as Fibonacci and Fermat have used *Arithmetica* as a basis for much of their work. What's not so well known about Diophantus is that he was a potato farmer. One of the old sage's greatest insights came to him while he was preparing for the year's harvest.

With the harvest fast approaching, Diophantus sent his son out to the tool shed to fetch the potato scale. To his dismay, the son noticed that Diophantus had neglected to clean the scale after the previous year's harvest. There was an old potato stuck to one pan of the scale.

Diophantus and his son both tried to remove the potato from the scale, but it would not budge. "It's no use, Sonny. This 1-lb potato is good and stuck. We might as well give up on getting this year's harvest to market."

Diophantus's son quickly spoke up. "I've got an idea, Dad. If we put a 1-lb brick on the pan opposite the potato, the scale will balance. Then we can use it just like we normally would."

"Well Sonny, your plan would work if we had any 1-lb bricks, but we don't. We've got plenty of 5-lb bricks and 7-lb bricks, but I doubt we can balance the scale with those."

Diophantus shook his head and stared at the scale, until his son blurted out, "Hold on, Pop. I think I've figured out a way to balance the scale using 5-lb bricks and 7-lb bricks! Just add three 5-lb bricks to the side opposite the potato, and—"

"Be quiet, Sonny. I'm trying to think. While I'm working on the scale, why don't you keep yourself busy with those arithmetic problems you seem to enjoy so much."

Diophantus fiddled with the scale for some time, stacking 5-lb and 7-lb bricks higher and higher on both sides of the scale without success. Feeling he was close to solving his problem, Diophantus reached down to get a few more bricks. That's when it hit him. A precariously placed 5-lb brick had fallen from the very top of the stack, and it struck Diophantus, suddenly, that he should have come up with a plan instead of haphazardly adding bricks to the scale. In dismay, he cried out, "Ow, my head!"

Diophantus's son looked up from his work. "Hey, Dad, I've got great news! I finished th—"

"Son, I just can't concentrate with you yabbering on like that."

"Are you still working on that scale, Pop? Just put two 7-lb bricks on the pan with the potato and three 5-lb bricks on the side opposite the potato. Now, about my good news: I finished those arithmetic problems I've been working on. I wrote them all down in this book—do you want to see?"

"No, Son, that's OK. Just put my name on it and send it off to the publisher."

"Don't you think I should copy it over, Dad? I'm worried that the margins may be a little too narrow."

"This is no time to worry about margins, Son. We've got the potato harvest coming up."

"But, Dad, I haven't thought of a title yet."

"Just call it *Arithmetic I*. No, better yet, call it *Arithmetic A*."

Linear Diophantine equations

To balance his scale in our myth, Diophantus had to figure out how to balance a 1-lb potato using only 5-lb bricks and 7-lb bricks. We can formulate Diophantus's problem as the following equation:

$$7X + 5Y = 1 \qquad (1)$$

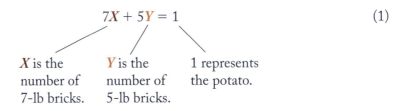

X is the number of 7-lb bricks.

Y is the number of 5-lb bricks.

1 represents the potato.

Now let's express algebraically the solution that balanced the scale:

$$5 \cdot 3 = 7 \cdot 2 + 1$$

We can rearrange this equation to match the form of equation (1):

$$7 \cdot (-2) + 5 \cdot 3 = 1.$$

Thus, we have found that equation (1) has the following solution:

$$X = -2, Y = 3.$$

In our myth, Diophantus was not interested in *all* of the real solutions to the linear equation $7X + 5Y = 1$. For example, $X = \frac{1}{2}$, $Y = -\frac{1}{2}$ is a solution to the equation, but it would not have helped Diophantus with his scale. Since X and Y represent numbers of bricks, Diophantus only cared about solutions in which X and Y are both integers. Equations for which only integral solutions are sought are called **Diophantine equations.** Linear equations, such as $7X + 5Y = 1$, are called **linear Diophantine equations** if only integral solutions are desired. For a geometric interpretation of linear Diophantine equations, see "Graphing Solutions to a Linear Diophantine Equation".

As we have seen, the linear Diophantine equation $7X + 5Y = 1$ has solutions in the integers. More generally, we can start with any two integers a and b (in place of 7 and 5) and consider the linear Diophantine equation

$$aX + bY = 1.$$

For which constants a and b does this equation have integral solutions?

Graphing Solutions to a Linear Diophantine Equation

When you first studied graphs of equations, you learned that $7X + 5Y = 1$ represents a line in the coordinate plane. Each point on the line represents a real solution to the equation.

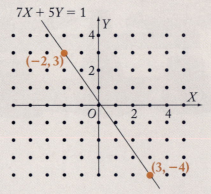

When $7X + 5Y = 1$ is considered as a Diophantine equation, however, only integral solutions are desired. Thus, we are looking for points with integer coordinates that lie on the line $7X + 5Y = 1$, such as $(-2, 3)$ and $(3, -4)$.

Points in the plane that have integer coordinates (the dots in the grid above) are called *lattice points*. Finding solutions to a linear Diophantine equation amounts to finding lattice points that lie on the line determined by the equation.

For example, does the Diophantine equation

$$10X + 15Y = 1 \tag{1}$$

have any solutions? Try substituting different integer values for X and Y into the left side of equation (1). Can you explain why the left side will never equal 1?

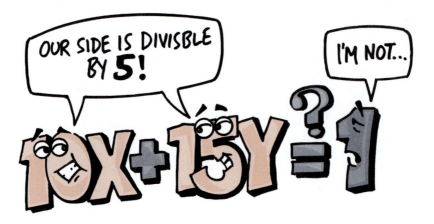

What is it about the coefficients 10 and 15 that prevents the Diophantine equation $10X + 15Y = 1$ from having a solution? As the reader may have guessed, this equation has no solution because 10 and 15 have a common factor, 5. In general, the equation $aX + bY = 1$ does not have integral solutions if a and b share a common factor that is larger than 1. This is not hard to prove:

THEOREM 5.1.1

Let a and b be integers. If $\gcd(a, b) > 1$, *then the equation* $aX + bY = 1$ *has no integral solutions.*

PROOF Let $a, b \in \mathbf{Z}$ such that $\gcd(a, b) > 1$. Let $d = \gcd(a, b)$.

[We must show that the equation $aX + bY = 1$ has no integral solutions. We will prove this by contradiction.]

Assumption: Suppose that there exist $X, Y \in \mathbf{Z}$ such that

$$aX + bY = 1. \tag{2}$$

Since $d \mid a$ and $d \mid b$, the Linear Combination Lemma (3.1.3) guarantees that

$$d \mid aX + bY.$$

It now follows from equation (2) that $d \mid 1$. Thus, $d = 1$, which contradicts the hypothesis that $\gcd(a, b) > 1$. $\Rightarrow\Leftarrow$

Hence, there are no integers X and Y that satisfy $aX + bY = 1$. ■

EXAMPLE 1

The Diophantine equation $35X + 21Y = 1$ has no solutions, because 35 and 21 are not relatively prime.

Theorem 5.1.1 tells us about Diophantine equations such as $35X + 21Y = 1$, in which the coefficients share a common factor. But what about Diophantine equations in which the coefficients are relatively prime? We saw earlier that the equation $7X + 5Y = 1$ has a solution. You might guess that whenever the coefficients a and b are relatively prime, the Diophantine equation $aX + bY = 1$ has a solution. It turns out that this is the case, as the following theorem asserts.

THEOREM 5.1.2   SOLVING LINEAR DIOPHANTINE EQUATIONS

Let a and b be integers. The Diophantine equation $aX + bY = 1$ *has a solution if and only if* $\gcd(a, b) = 1$.

PROOF Let $a, b \in \mathbf{Z}$.

[$\Rightarrow$] Suppose the Diophantine equation $aX + bY = 1$ has a solution. By Theorem 5.1.1, it follows that $\gcd(a, b) = 1$. (This is actually the contrapositive of Theorem 5.1.1; see Exercise 8.)

[$\Leftarrow$] Now assume that $\gcd(a, b) = 1$.

[To show: The Diophantine equation $aX + bY = 1$ has a solution.]

Consider the set S of all linear combinations of a and b:

$$S = \{ax + by \mid x, y \in \mathbf{Z}\}.$$

It follows from the Well-Ordering Principle (see Exercise 10) that S has a smallest positive element, d.

Claim: $d \mid a$.

Proof of Claim Divide a by d: by the Division Theorem (3.5.1), there exist integers q and r such that

$$a = qd + r \tag{3}$$

and

$$0 \le r < d. \tag{4}$$

Solving equation (3) for r gives

$$r = a - qd. \tag{5}$$

Since $d \in S$, there exist integers x and y such that

$$d = ax + by. \tag{6}$$

Substituting this into equation (5) yields

$$r = a - q(ax + by)$$
$$= a(1 - qx) + b(-qy).$$

Thus, $r \in S$. Since d is the smallest positive element of S, it follows from inequality (4) that $r = 0$. Thus, equation (3) becomes $a = qd$, and hence $d \mid a$. This completes the proof of the Claim.

The Claim tells us that $d \mid a$. By similar reasoning, $d \mid b$. Thus, d is a common divisor of a and b. Since $\gcd(a, b) = 1$, we conclude that $d = 1$. Hence, by equation (6), there exist integers x and y such that

$$ax + by = 1.$$

Theorem 5.1.2 tells us that every linear Diophantine equation whose coefficients are relatively prime, such as $19X + 7Y = 1$, has a solution. However, the theorem does not tell us how to actually find a solution. We might find solutions of $19X + 7Y = 1$ using trial and error, but this seems an inappropriate method for solving more difficult equations such as $343X + 218Y = 1$. In the next section, we will discover a general method for solving linear Diophantine equations whose coefficients are relatively prime.

EXERCISES 5.1

Numerical Problems

1. The riddle of Diophantus's life, which began our biography of Diophantus at the beginning of this chapter, can be summarized in modern language as follows:

    For the first sixth of his life, Diophantus was a boy. After another twelfth of his life, Diophantus grew a beard. One-seventh of his life after this, Diophantus married. Five years after his marriage, Diophantus's son was born. Diophantus's son died at a relatively young age. Four years after his son died, Diophantus himself died. The total number of years the son lived is one-half of the total number of years Diophantus lived.

    a. Write one or more equations that express the information in the riddle algebraically.

    b. Solve your equation(s) from part a to determine the following:

    - The total number of years Diophantus lived

    - The number of years Diophantus spent as a boy

    - The age at which Diophantus grew a beard

    - The age at which Diophantus got married

    - Diophantus's age when his son was born

    - Diophantus's age when his son died

2. For each of the following Diophantine equations, determine whether Theorem 5.1.1 applies. If so, what does the theorem allow you to conclude? Explain.

    a. $18X + 30Y = 1$ b. $27X + 20Y = 1$ c. $57X + 95Y = 1$

3. $(3, -2)$ is one solution to the Diophantine equation $5X + 7Y = 1$. Find at least two other solutions.

4. $(3, -2)$ is one solution to the Diophantine equation $5X + 7Y = 1$. Use this to find a solution to each of the following Diophantine equations.

    a. $5X + 7Y = 11$ b. $5X + 7Y = -1$ c. $5X + 7Y = 2$

    d. $5X + 7Y = -2$ e. $5X + 7Y = -3$ f. $5X + 7Y = -13$

5. **a.** Find three lattice tpoints on the line $2X + 3Y = 1$.

    b. Find three lattice points on the line $6X + 9Y = 3$.

6. **a.** Find three lattice points on the line $5X + 7Y = 0$.

    b. Find three lattice points on the line $10X + 14Y = 0$.

7. Find the equation of the line containing the lattice points $(-2, 5)$ and $(4, -9)$. Write your equation with integer coefficients. Does your line pass through any other lattice points? If so, find three more.

8. Write the contrapositive of Theorem 5.1.1.

Reasoning and Proofs

9. Give an equation for a line that contains no lattice points. Explain how you know it contains no lattice points.

10. Consider the set S from the proof of Theorem 5.1.2.

    a. In the proof of Theorem 5.1.2, we needed to conclude that the set S has a smallest positive element. To what set T should you apply the Well-Ordering Principle (2.2.2) in order to reach this conclusion?

    b. In order to apply the Well-Ordering Principle, we need to know that T is nonempty. Prove that T is nonempty. (First recall that gcd $(0, 0)$ is not defined, and use this to show that a and b are not both zero.)

11. Does the Diophantine equation $21X + 51Y + 12Z = 1$ have a solution? Prove that your answer is correct.

Advanced Reasoning and Proofs

For Exercises 12–17: Many interesting Diophantine equations are not linear. For each of the following nonlinear Diophantinc cquations, either find a solution or prove that no solution exists. (To learn more about nonlinear Diophantine equations, see Chapter 15.)

12. $X^2 + Y^2 = 13$ 13. $X^2 - Y^2 = 13$ 14. $X^2 + Y^2 = 11$

15. $X^2 - 2Y^2 = 1$ 16. $X^2 - 2Y^2 = 0$ 17. $X^2 - 5Y = 2$

EXPLORATION Pythagorean Triples (Exercises 18–22)

As you know, the Pythagorean Theorem states that if $x, y,$ and z are the lengths of the two legs and hypotenuse of a right triangle, respectively, then $x^2 + y^2 = z^2$. It is especially pleasing when the lengths $x, y,$ and z are integers, and we call such a triple of integers a *Pythagorean triple*.

18. Give three examples of Pythagorean triples.

19. Show that if (x, y, z) is a Pythagorean triple and a is any natural number, then (ax, ay, az) is also a Pythagorean triple.

Plato is usually known for philosophy, but the ancient discipline of philosophy encompassed all knowledge, including astronomy and mathematics. A formula ascribed to Plato for generating Pythagorean triples is the following:

$$x = 2n \qquad y = n^2 - 1 \qquad z = n^2 + 1$$

for any integer $n \geq 2$.

20. Show that every (x, y, z) generated by Plato's formula is a Pythagorean triple.

21. Use Plato's formula to find a Pythagorean triple that you did not already list in Exercise 18.

22. Not all Pythagorean triples can be generated by Plato's formula.[1] Give an example of a Pythagorean triple that cannot be generated by the formula, and explain how you know that it cannot.

5.2 Using the Euclidean Algorithm to Find a Solution

Theorem 5.1.2 guarantees that every linear Diophantine equation with relatively prime coefficients has a solution. But if we want to actually find a solution, then the theorem is not much help. To figure out how to solve linear Diophantine equations, let's return to the most celebrated potato farmer of the ancient world, Diophantus of Alexandria.

Heavier bricks

You may remember that when Diophantus's scale had a 1-lb potato stuck to it, his son saved the day by balancing the scale using 5-lb and 7-lb bricks. With the scale in working order, Diophantus was able to bring that year's bountiful crop to market. The following year, Diophantus once again sent his son out to fetch the scale. The potato harvest was only days away, and again they had a problem.

Diophantus's son did not like what he found in the tool shed. "Dad, remember that 1-lb potato that almost put us out of business last year? Well, it's still stuck to the scale! I thought you were going to clean the scale after last year's harvest."

"Well, Sonny, I was fixin' to clean it, but then I got caught up in building this new doghouse for Phido. I'm so sorry, Son."

"That's OK, Pop. I'll go get three 5-lb bricks and two 7-lb bricks, and we can do the same thing we did last year. Everything will be all right."

"No, it won't, Sonny. We can't do what we did last year because we don't have any 5-lb bricks. I used them all to build the doghouse.

---

[1]In Section 15.3, we discuss a method for generating all Pythagorean triples.

Then Diophantus had an idea. "Look on the bright side, Son. We don't have any 5-lb bricks, but we've still got plenty of 7-lb bricks. And we've got a ton of these old 19-lb bricks. There must be something we can do to balance the scale."

Can you balance this scale using only 19-lb and 7-lb bricks?

Sonny's solution

Sonny remembered that in the previous year, they balanced the scale using three 5-lb bricks and two 7-lb bricks:

This year, there were no 5-lb bricks, but Sonny realized that each 5-lb brick could be replaced by one 19-lb brick and two 7-lb bricks.

Since this scale balances, . . .   *. . . so does this one:*

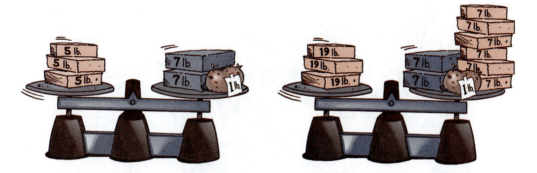

The second scale in the diagram shows Sonny's solution: the scale can be balanced using three 19-lb bricks and eight 7-lb bricks.

Let's examine Sonny's solution algebraically. Sonny observed that a 5-lb brick on the left side of the scale can be replaced by putting one 19-lb brick on the left side of the scale and two 7-lb bricks on the right side of the scale. This can be represented by the following equation:

$$\boxed{5} = \boxed{19} - 2 \cdot \boxed{7} \qquad \textbf{Sonny's Spud-Saving Substitution} \qquad (1)$$

The solution that Diophantus and Sonny came up with for the first year's harvest (a solution to $7X + 5Y = 1$) was

$$(-2) \cdot \boxed{7} + 3 \cdot \boxed{5} = 1.$$

Using Sonny's Spud-Saving Substitution (1), this becomes

$$(-2) \cdot \boxed{7} + 3 \cdot (\boxed{19} - 2 \cdot \boxed{7}) = 1 \qquad \leftarrow \textbf{Substitute from (1).}$$
$$(-2) \cdot \boxed{7} + 3 \cdot \boxed{19} - 6 \cdot \boxed{7} = 1 \qquad \leftarrow \textbf{Distribute the 3.}$$
$$3 \cdot \boxed{19} + (-8) \cdot \boxed{7} = 1 \qquad \leftarrow \textbf{Combine like terms.}$$

Sonny's Spud-Saving Substitution (1) has solved the problem of the second year's harvest by leading us from a solution of the Diophantine equation $7X + 5Y = 1$ to a solution of the Diophantine equation $19X + 7Y = 1$. This solution is $(X, Y) = (3, -8)$.

Another way to view Sonny's Spud-Saving Substitution (1) is to rewrite it in the form

$$\boxed{19} = 2 \cdot \boxed{7} + \boxed{5}.$$

This equation is simply the division of 19 by 7. Thus, dividing 19 by 7 enables us to reduce the Diophantine equation $19X + 7Y = 1$ to the simpler equation $7X + 5Y = 1$.

In general, whenever we divide two integers a and b to get a remainder r, we can reduce the problem of solving $aX + bY = 1$ to the simpler problem of solving $bX + rY = 1$. We prove this in the following proposition.

PROPOSITION 5.2.1

Let a, q, b, and r be integers such that $a = qb + r$.
If $bX + rY = 1$ has a solution in the integers, then so does $aX + bY = 1$.

PROOF By hypothesis, there exist $x, y \in \mathbf{Z}$ such that

$$bx + ry = 1.$$

Rewrite $a = qb + r$ in the form $r = a - qb$, and substitute this into the previous equation:

$$bx + (a - qb)y = 1.$$

Regrouping terms gives

$$ay + b(x - qy) = 1.$$

So we have produced the solution $(X, Y) = (y, x - qy)$ to the Diophantine equation $aX + bY = 1$. ◼

Solving $aX + bY = 1$

If one division can reduce a Diophantine equation to a simpler one, then several divisions in a row will keep reducing a Diophantine equation, making it simpler and simpler. Performing several divisions in a row sounds like a job for the Euclidean Algorithm. In fact, the Euclidean Algorithm is exactly what we need to solve linear Diophantine equations.

For example, suppose we want a solution to the Diophantine equation

$$223X + 58Y = 1. \tag{2}$$

Let's write out all the steps of the Euclidean Algorithm (4.1.1) performed on 223 and 58, shown in the first column below. To the right of each step, in the second column, we rewrite each equation, solving for the remainder, so that the equation resembles Sonny's Spud-Saving Substitution (1).

Perform the Euclidean Algorithm	**Solve each equation for the remainder**	**What Proposition 5.2.1 tells us**
$\boxed{233} = 3 \cdot \boxed{58} + \boxed{49}$ (A)$\Rightarrow$	$\boxed{49} = \boxed{223} - 3 \cdot \boxed{58}$	I say: we can solve $223X + 58Y = 1$ if we can solve $58X + 49Y = 1$.
$\boxed{58} = 1 \cdot \boxed{49} + \boxed{9}$ (B)$\Rightarrow$	$\boxed{9} = \boxed{58} - 1 \cdot \boxed{49}$	I say: we can solve $58X + 49Y = 1$ if we can solve $49X + 9Y = 1$.
$\boxed{49} = 5 \cdot \boxed{9} + \boxed{4}$ (C)$\Rightarrow$	$\boxed{4} = \boxed{49} - 5 \cdot \boxed{9}$	I say: we can solve $49X + 9Y = 1$ if we can solve $9X + 4Y = 1$.
$\boxed{9} = 2 \cdot \boxed{4} + \boxed{1}$ (D)$\Rightarrow$	$\boxed{1} = \boxed{9} - 2 \cdot \boxed{4}$	I say: *we can solve* $9X + 4Y = 1$.
$\boxed{4} = 4 \cdot \boxed{1} + 0$		

The first three equations in the second column (labeled (A), (B), and (C)) simplify the problem, but the last equation in the second column (labeled (D)) does more. It gives us an explicit solution to $9X + 4Y = 1$, namely, $X = 1$ and $Y = -2$.

Our strategy is really panning out: solving this problem is turning out to be a piece of cake. Read from bottom to top, equations Ⓒ, Ⓑ, and Ⓐ tell us how to convert this solution into a solution to equation (2), the Diophantine equation $223X + 58Y = 1$. This is done by systematically substituting, as follows:

$$\boxed{9} - 2 \cdot \boxed{4} = 1 \quad \leftarrow \text{Rewrite equation } Ⓓ.$$

$$\boxed{9} - 2 \cdot (\boxed{49} - 5 \cdot \boxed{9}) = 1 \quad \leftarrow \text{Use } Ⓒ \text{ to substitute } \boxed{49} - 5 \cdot \boxed{9} \text{ for } \boxed{4}.$$

$$\boxed{9} - 2 \cdot \boxed{49} + 10 \cdot \boxed{9} = 1 \quad \leftarrow \text{Distribute}$$

$$- 2 \cdot \boxed{49} + 11 \cdot \boxed{9} = 1 \quad \leftarrow \text{Simplify}$$

$$-2 \cdot \boxed{49} + 11 \cdot (\boxed{58} - 1 \cdot \boxed{49}) = 1 \quad \leftarrow \text{Use } Ⓑ \text{ to substitute } \boxed{58} - 1 \cdot \boxed{49} \text{ for } \boxed{9}.$$

$$-2 \cdot \boxed{49} + 11 \cdot \boxed{58} - 11 \cdot \boxed{49} = 1 \quad \leftarrow \text{Distribute}.$$

$$11 \cdot \boxed{58} - 13 \cdot \boxed{49} = 1 \quad \leftarrow \text{Simplify}.$$

$$11 \cdot \boxed{58} - 13 \cdot (\boxed{223} - 3 \cdot \boxed{58}) = 1 \quad \leftarrow \text{Use } Ⓐ \text{ to substitute } \boxed{223} - 3 \cdot \boxed{58} \text{ for } \boxed{49}.$$

$$11 \cdot \boxed{58} - 13 \cdot \boxed{223} + 39 \cdot \boxed{58} = 1 \quad \leftarrow \text{Distribute}.$$

$$-13 \cdot \boxed{223} + 50 \cdot \boxed{58} = 1 \quad \leftarrow \text{Simplify}.$$

We're done! The last equation gives us a solution to the Diophantine equation $223X + 58Y = 1$. The solution is $X = -13, Y = 50$.

With the help of our old friend the Euclidean Algorithm, we have systematically solved what at first appeared to be a difficult Diophantine equation. In fact, this method will solve any equation of the form $aX + bY = 1$, provided that a and b are relatively prime. Here is a summary of the method:

METHOD 5.2.2   **EUCLID'S ROYAL ROAD TO SOLVING LINEAR DIOPHANTINE EQUATIONS**

When confronted with an equation of the form $aX + bY = 1$, with $\gcd(a, b) = 1$, follow these steps:

Step 1   Perform the Euclidean Algorithm on a and b.

Step 2   Rewrite each equation by solving for the remainder.

Step 3   Starting with the final equation from step 2, systematically substitute larger boxed numbers by working your way back through the equations from step 2.

This method worked quite well for solving the Diophantine equation $223X + 58Y = 1$, and later we will see more examples of Euclid's Royal Road (Method 5.2.2) in action.

But before we do, let's take a moment to understand why this method will solve *any* equation of the form $aX + bY = 1$ in which a and b are relatively prime. First note that in step 1, when we perform the Euclidean Algorithm on a and b, we know (by Theorem 4.2.1) that the last nonzero remainder will be $\gcd(a, b)$, which equals **1**. The appearance of the number **1** at the end of the Euclidean Algorithm is the key, because when we perform step 2, this gives us a **1** on one side of the final equation. Then in step 3, every equation has a **1** on one side, and we end up with an equation that has the number **1** on one side and a linear combination of a and b on the other. The Diophantine equation is now solved.

In fact, we can use the ideas of this section to give another proof that if $\gcd(a, b) = 1$, then the Diophantine equation $aX + bY = 1$ always has a solution (the hard direction of Theorem 5.1.2, Solving Linear Diophantine Equations). See Exercise 15 for such a proof.

EXAMPLE 1

Find a solution to the Diophantine equation $113X + 42Y = 1$.

SOLUTION Use Method 5.2.2 (Euclid's Royal Road) to solve this linear Diophantine equation. First perform the Euclidean Algorithm on 113 and 42. Then rewrite each equation by solving it for the remainder.

Perform the Euclidean Algorithm

$$113 = 2 \cdot 42 + 29$$
$$42 = 1 \cdot 29 + 13$$
$$29 = 2 \cdot 13 + 3$$
$$13 = 4 \cdot 3 + 1$$
$$3 = 3 \cdot 1 + 0$$

$\Rightarrow$

Solve each equation for the remainder

$$29 = 113 - 2 \cdot 42$$
$$13 = 42 - 1 \cdot 29$$
$$3 = 29 - 2 \cdot 13$$
$$1 = 13 - 4 \cdot 3$$

Rewrite the last equation from the second column above. Then systematically substitute in larger constants, using the other equations from the second column from bottom to top.

$$13 - 4 \cdot 3 = 1 \qquad \leftarrow \textbf{Rewrite last equation in second column.}$$
$$13 - 4 \cdot (29 - 2 \cdot 13) = 1 \qquad \leftarrow \textbf{Substitute } 29 - 2 \cdot 13 \textbf{ for } 3.$$
$$-4 \cdot 29 + 9 \cdot 13 = 1 \qquad \leftarrow \textbf{Distribute and simplify.}$$
$$-4 \cdot 29 + 9 \cdot (42 - 1 \cdot 29) = 1 \qquad \leftarrow \textbf{Substitute } 42 - 1 \cdot 29 \textbf{ for } 13.$$
$$9 \cdot 42 - 13 \cdot 29 = 1 \qquad \leftarrow \textbf{Distribute and simplify.}$$
$$9 \cdot 42 - 13 \cdot (113 - 2 \cdot 42) = 1 \qquad \leftarrow \textbf{Substitute } 113 - 2 \cdot 42 \textbf{ for } 29.$$
$$-13 \cdot 113 + 35 \cdot 42 = 1 \qquad \leftarrow \textbf{Distribute and simplify.} \qquad (3)$$

So, $(X, Y) = (-13, 35)$ is a solution to the Diophantine equation $113X + 42Y = 1$. $\blacksquare$

If one (or both) of the coefficients in a linear Diophantine equation is negative, then the method of solution is almost identical, as the following example demonstrates.

EXAMPLE 2

Find a solution to the Diophantine equation $113X - 42Y = 1$.

SOLUTION To solve this Diophantine equation, we begin as if we were solving $113X + 42Y = 1$. This process was already done in the solution to Example 1 and ended with equation (3),

$$-13 \cdot 113 + 35 \cdot 42 = 1$$

By inspection of this equation, we conclude that $(X, Y) = (-13, -35)$ is a solution to the Diophantine equation $113X - 42Y = 1$.

EXERCISES 5.2

Numerical Problems

1. Consider the Diophantine equation $11X + 4Y = 1$.

    a. Find a solution using trial and error.

    b. Draw a diagram of the two pans of a balanced scale with a 1-lb potato and bricks corresponding to the solution you gave in part a. Include the appropriate number of each type of brick, and label each brick (and the potato) with its weight.

    c. Apply the Division Theorem (3.5.1) to divide 81 by 11. Then put the resulting equation in the same form as Sonny's Spud-Saving Substitution (1) by solving for the remainder.

    d. Use your equation from part c and your solution from part a to find a solution to the Diophantine equation $81X + 11Y = 1$. Interpret your solution in terms of a balanced scale with potato and bricks.

In Exercises 2–10, find a solution to each Diophantine equation or explain why no solution exists.

2. $47X + 49Y = 1$ 3. $23X + 97Y = 1$ 4. $355X + 113Y = 1$

5. $500X + 423Y = 1$ 6. $223X - 75Y = 1$ 7. $-47X + 93Y = 1$

8. $113X - 99Y = 1$ 9. $35X + 49Y = 1$ 10. $21X - 77Y = 1$

Reasoning and Proofs

11. Suppose that (x_0, y_0) is a solution to the Diophantine equation $47X + 6Y = 1$. Using the fact that $6 = 100 - 2 \cdot 47$, find a solution to the Diophantine equation $100X + 47Y = 1$ in terms of x_0 and y_0.

12. a. Let a be an integer. Prove that there exist integers X and Y such that $aX + (a + 1)Y = 1$.

    b. Find a solution (X, Y) to the equation from part a.

13. a. For which integers a does the Diophantine equation $aX + (a + 2)Y = 1$ have a solution?

    b. Let a be an integer such that the Diophantine equation $aX + (a + 2)Y = 1$ has a solution. Find a solution (X, Y) in terms of a.

Advanced Reasoning and Proofs

14. In this exercise, we will consider linear Diophantine equations whose coefficients are Fibonacci numbers.

   a. Find a solution to the Diophantine equation $13X + 8Y = 1$.

   b. Find a solution to the Diophantine equation $21X + 13Y = 1$.

   c. Based on your answers to parts a and b, make a conjecture by filling in the blanks:

   A solution to the Diophantine equation $F_{n+1}X + F_nY = 1$ is given by $X = $ _____ , $Y = $ _____ .

   d. Prove your conjecture from part c.

15. In this exercise, you will give another proof of the harder direction of Theorem 5.1.2 (Solving Linear Diophantine Equations) based the ideas of this section (in particular, Proposition 5.2.1). That is, you will prove the following theorem:

   THEOREM. *For all nonnegative integers a and b, if* $\gcd(a, b) = 1$, *then the Diophantine equation* $aX + bY = 1$ *has a solution.*

   To do this, you will use strong induction on $m = \min(a, b)$. (That is, m is equal to the smaller of a and b.) Base your proof on the outline below.

   I. Suppose $m = 0$. Show that the theorem holds in this case. [*Hint:* Recall that $\gcd(0, 0)$ is not defined.]

   II. Now assume that $m > 0$. Explain why you can assume that $a \geq b$. Apply the Division Theorem (3.5.1) to divide a by b, and get $a = qb + r$.

   III. Explain why $\min(b, r) < \min(a, b)$. This inequality enables you to use the induction hypothesis, which guarantees that a certain Diophantine equation has a solution.

   IV. Use Proposition 5.2.1 to complete the proof.

5.3 The Diophantine Equation $aX + bY = n$

A pile of potatoes

In the previous two sections, we learned that the Diophantine equation $aX + bY = 1$ has a solution if and only if the coefficients a and b are relatively prime. We also learned how to find a solution (when a solution exists) using the Euclidean Algorithm.

But suppose we have a Diophantine equation of the form $aX + bY = n$, such as $7X + 5Y = 3$. Can we find a method for solving this kind of equation? We already know how to solve $7X + 5Y = 1$, so let's start there. Recall that the solution we found was

$$(-2) \cdot \boxed{7} + 3 \cdot \boxed{5} = 1.$$

Let's see what happens when we multiply both sides of this equation by 3:

$$(-6) \cdot \boxed{7} + 9 \cdot \boxed{5} = 3.$$

With little effort, a solution to the Diophantine equation $7X + 5Y = 3$ has appeared: $X = -6, Y = 9$.

In general, whenever $\gcd(a, b) = 1$, we know (by Theorem 5.1.2) that the Diophantine equation $aX + bY = 1$ has a solution, and we can multiply this solution by n to obtain a solution to $aX + bY = n$. This is stated in the following theorem.

THEOREM 5.3.1

Suppose $a, b \in \mathbf{Z}$ and $\gcd(a, b) = 1$. Then for any $n \in \mathbf{Z}$, the Diophantine equation $aX + bY = n$ has a solution.

PROOF Since a and b are relatively prime, Theorem 5.1.2 (Solving Linear Diophantine Equations) tells us that there exist $r, s \in \mathbf{Z}$ such that

$$ar + bs = 1.$$

Multiplying by n, we get

$$arn + bsn = n.$$

This gives the solution $(X, Y) = (rn, sn)$ to the Diophantine equation $aX + bY = n$. ●

Theorem 5.3.1 says that given two relatively prime integers a and b, we can express *any* integer as a linear combination of a and b. But what if we start with two integers that are not relatively prime, such as 12 and 15? Which integers can be expressed as a linear combination of 12 and 15? In other words, for which values of n does the Diophantine equation $12X + 15Y = n$ have a solution?

For instance, let's consider the equation

$$12X + 15Y = 19. \tag{1}$$

For every pair of integer values X and Y, the left side of this equation will be divisible by 3. However, the right side, 19, is not divisible by 3. This implies that the equation has no integral solutions.

Now consider the equation

$$12X + 15Y = 33. \tag{2}$$

Are there any integral solutions? Looking at the equation, we notice that every term is divisible by 3. The equation thus reduces to

$$4X + 5Y = 11. \tag{3}$$

Now we have an equation in which the coefficients, 4 and 5, are relatively prime. We know by Theorem 5.3.1 that such equations always have integral solutions. Since equation (3) has solutions, the equivalent equation (2) has the exact same solutions.

Thus, equation (2) has solutions, whereas equation (1) does not. These two cases suggest that the Diophantine equation $12X + 15Y = n$ has a solution if and only if n is divisible by 3, the greatest common divisor of 12 and 15. This answers the question of which numbers can be expressed as a linear combination of 12 and 15—namely, all multiples of 3. This result is an instance of the following theorem.

THEOREM 5.3.2

Suppose a, b, $n \in \mathbf{Z}$ such that a and b are not both equal to zero. The Diophantine equation $aX + bY = n$ has a solution if and only if $\gcd(a, b) \mid n$.

PROOF Let $d = \gcd(a, b)$.

[$\Rightarrow$] Suppose the Diophantine equation $aX + bY = n$ has a solution. Then there exist $x, y \in \mathbf{Z}$ such that $ax + by = n$. We know that $d \mid a$ and $d \mid b$, so by the Linear Combination Lemma (3.1.3), $d \mid ax + by$. Hence, $d \mid n$, as was to be shown.

[$\Leftarrow$] Suppose $d \mid n$. Then a, b, and n are all divisible by d, so we may factor d from each of them; that is, we may write

$$a = da', b = db', n = dn',$$

where a', b', and n' are integers. Since a' and b' are obtained from a and b by factoring out $d = \gcd(a, b)$, it follows (by Lemma 3.4.4) that a' and b' are relatively prime. Thus, by Theorem 5.3.1, there are integers x and y such that

$$a'x + b'y = n'.$$

Multiplying by d, this equation becomes

$$ax + by = n.$$

We conclude that the Diophantine equation $aX + bY = n$ has a solution, as desired. ■

EXAMPLE 1

Are there any solutions to the Diophantine equation $24X + 14Y = 6$? If so, find a solution.

SOLUTION First observe that $\gcd(24, 14) = 2$. Since $2 \mid 6$, Theorem 5.3.2 tells us that there are integral solutions to the equation

$$24X + 14Y = 6. \tag{4}$$

To find a solution, we begin by dividing this equation by 2, yielding the equivalent equation

$$12X + 7Y = 3. \tag{5}$$

To solve this equation, we first solve the equation

$$12X + 7Y = 1. \tag{6}$$

which we can do by using Method 5.2.2 (Euclid's Royal Road). We find that $X = 3$, $Y = -5$ is a solution to equation (6). Multiplying by 3, we find that $X = 9$, $Y = -15$ is a solution to equation (5), and hence is a solution to our original equation (4). ■

An immediate yet important consequence of Theorem 5.3.2 is that $\gcd(a, b)$ is itself a linear combination of a and b. This is expressed in the following corollary, which you will prove in Exercise 14. This corollary will be extremely useful in proving facts about the gcd.

COROLLARY 5.3.3   GCD AS A LINEAR COMBINATION

Let a and b be nonzero integers, and let $d = \gcd(a, b)$. Then there exist $x, y \in \mathbf{Z}$ such that $ax + by = d$.

EXERCISES 5.3

Numerical Problems

1. Find $\gcd(91, 21)$. Then, without solving, determine whether each of the following Diophantine equations has a solution.

   a. $91X + 21Y = 3$ **b.** $91X + 21Y = 7$ **c.** $91X + 21Y = 15$

   d. $91X + 21Y = -28$ **e.** $91X + 21Y = 0$ **f.** $91X + 21Y = 30$

In Exercises 2–10, find a solution to each Diophantine equation or explain why no solution exists.

2. $17X + 85Y = 1$

3. $22X + 111Y = 1$

4. $15X + 23Y = 4$

5. $105X + 91Y = 70$

6. $47X + 49Y = 100$

7. $24X + 36Y = 18$

8. $25X - 10Y = 35$

9. $-33X + 55Y = -121$

10. $23X + 97Y = -57$

11. Express the greatest common divisor of 246 and 540 as a linear combination of 246 and 540.

Reasoning and Proofs

12. Which integers can be expressed as a linear combination of 20 and 88?

13. Which integers can be expressed as a linear combination of 21 and 88?

14. Prove Corollary 5.3.3.

15. **a.** Solve the linear Diophantine equation $7X + 24Y = 2$.

    b. You have a 7-liter bucket and a 24-liter bucket. Using these, you can add or remove exactly 7 or 24 liters of water at a time into or out of a large tank. If the large tank begins empty, explain how you can use the buckets to end with exactly 2 liters of water in the large tank. How does this relate to part a?

    c. You have a 73-liter bucket and a 55-liter bucket. How can you use these to put exactly 8 liters of water into a large tank that begins empty?

    d. You have two hourglasses: one can time a 15-minute period, and the other can time a 41-minute period. How can you use these two hourglasses together to boil an egg for precisely 3 minutes?

16. Let a and b be nonzero integers. Prove that every common divisor of a and b divides $\gcd(a, b)$.

17. Let a, b, c be nonzero integers. Prove that $\gcd(a, b, c) = \gcd(\gcd(a, b), c)$. [*Hint:* Use the result of Exercise 16.]

EXPLORATION Linear Diophantine Equations in More than Two Variables (Exercises 18–22)

18. Find a solution to the Diophantine equation $6X + 10Y + 45Z = 1$. [*Hint:* First express $\gcd(6, 10)$ as a linear combination of 6 and 10. Then express 1 as a linear combination of 45 and $\gcd(6, 10)$.]

19. Find a solution to the Diophantine equation $28X + 63Y + 90Z = 1$. [*Hint:* First express $\gcd(28, 63)$ as a linear combination of 28 and 63. Then express 1 as a linear combination of 90 and $\gcd(28, 63)$.]

20. Let a, b, and c be nonzero integers. Prove that the Diophantine equation $aX + bY + cZ = 1$ has a solution if and only if $\gcd(a, b, c) = 1$. [*Hint:* Use the result of Exercise 17.]

21. Let a, b, and c be nonzero integers, and let $n \in \mathbf{Z}$. Prove that the Diophantine equation $aX + bY + cZ = n$ has a solution if and only if $\gcd(a, b, c) \mid n$.

*22. Let $a_1, a_2, \ldots, a_r$ be nonzero integers, and let $n \in \mathbf{Z}$. Prove that the Diophantine equation $a_1 X_1 + a_2 X_2 + \cdots + a_r X_r = n$ has a solution if and only if $\gcd(a_1, a_2, \ldots, a_r) \mid n$.

Advanced Reasoning and Proofs

23. Which numbers can be expressed as a linear combination of 33, 66, and 90? Prove that your answer is correct.

24. The following is a problem from sixth-century China: Suppose that a cock costs 5 coins, a hen costs 3 coins, and three chicks together cost 1 coin. You want to buy a total of exactly 100 birds by spending exactly 100 coins. What combination of cocks, hens, and chicks can you buy for exactly 100 coins?

25. Let $a, b, s, t \in \mathbf{N}$. Consider the following two arithmetic sequences:

$$s, \ s + a, \ s + 2a, \ldots \qquad \text{and} \qquad t, \ t + b, \ t + 2b, \ldots .$$

Prove that these two sequences have at least one term in common if and only if $\gcd(a, b) \mid (t - s)$.

5.4 Finding All Solutions to a Linear Diophantine Equation

Striking a new balance

At this point, we know all about finding a solution to the Diophantine equation $aX + bY = n$. We know that solutions exist if and only if $\gcd(a, b) \mid n$. Given such an equation, our method for solving it produces one solution. Perhaps there are other solutions. If so, how can we find them?

In our myth in Section 5.1, Diophantus found one solution to the equation

$$7X + 5Y = 1 \tag{1}$$

(namely, $X = -2$ and $Y = 3$), which enabled him to balance his potato scale. Can you find a different solution to this Diophantine equation?

The solution $(X, Y) = (-2, 3)$ corresponds to two 7-lb bricks on the side with the potato and three 5-lb bricks on the side opposite the potato. What happens if we add five 7-lb bricks to one side of the scale and seven 5-lb bricks to the other side?

This adds the same amount of weight, 35 lbs, to both sides of the scale, so the scale will still balance:

Since this scale balances, ... so does this one:

With little effort, we've found a new solution to the Diophantine equation (1). The diagram of the scale on the right represents the solution $(X, Y) = (-7, 10)$.

Let's examine this algebraically. To get from the original solution $(-2, 3)$ to the new solution $(-7, 10)$, the value of X decreases by 5, so the value of **7X** decreases by 35. Likewise, the value of Y increases by 7, so the value of **5Y** increases by 35. Thus, the total value of **7X + 5Y** does not change—it still equals 1.

We can easily find other solutions by continuing the process of subtracting 5 from X and adding 7 to Y. The next solutions obtained in this way are

$$(-12, 17), \ (-17, 24), \ (-22, 31), \ldots.$$

If we start with the original solution $(-2, 3)$ and repeat this process k times, we get the solution $(-2 - 5k, 3 + 7k)$.

This process gives us infinitely many solutions to the Diophantine equation (1). In fact, as we will prove shortly, *every* solution to this Diophantine equation has the form $(-2 - 5k, 3 + 7k)$, where k is an integer. In other words, the entire solution set is given by

$$\{(-2 - 5k, 3 + 7k) \mid k \in \mathbf{Z}\}. \tag{2}$$

Check for Understanding

In the solution set given by (2):

1. If $k = 10$, what solution do we get to $7X + 5Y = 1$?

2. If $k = -2$, what solution do we get to $7X + 5Y = 1$?

Given any one solution to any linear Diophantine equation, we can use the above method to obtain *all* solutions to that equation, as the following theorem states.

THEOREM 5.4.1

Let $a, b \in \mathbf{Z}$ be relatively prime, and let (x, y) be a solution to the Diophantine equation $aX + bY = n$. Then the solution set of this Diophantine equation is exactly

$$S = \{(x - kb, y + ka) \mid k \in \mathbf{Z}\}.$$

To prove Theorem 5.4.1, we will need the help of a lemma. Suppose n is a natural number, and you know that $4 \mid 9n$. What can you say about n? Because 4 and 9 are relatively prime, it seems reasonable to conclude that $4 \mid n$. The following lemma generalizes this observation.

LEMMA 5.4.2   EUCLID'S LEMMA

Let $d, m,$ and n be integers. If $d \mid mn$ and $\gcd(d, m) = 1$, then $d \mid n$.

> Put it in Prose, Paul!
>
> If the number d divides a product and is relatively prime to one of the factors, then d must divide the other factor.

PROOF OF **EUCLID'S LEMMA (5.4.2)**

Let $d, m, n \in \mathbf{Z}$ such that $d \mid mn$ and $\gcd(d, m) = 1$.

Since m and d are relatively prime, Theorem 5.1.2 (Solving Linear Diophantine Equations) tells us that the Diophantine equation $mX + dY = 1$ has a solution. Thus, there exist $s, t \in \mathbf{Z}$ such that

$$ms + dt = 1$$

Multiplying by n, we get

$$nms + ndt = n. \tag{3}$$

Since $d \mid mn$ (by hypothesis) and $d \mid d$, it follows from the Linear Combination Lemma (3.1.3) that

$$d \mid nms + ndt.$$

Thus, by equation (3), we know that $d \mid n$. ■

We are now ready to prove Theorem 5.4.1.

PROOF of **THEOREM 5.4.1**

Let $a, b, x, y, n \in \mathbf{Z}$ such that $\gcd(a, b) = 1$ and $ax + by = n$. There are two things we must show. First (i), we will show that every element of

$$S = \{(x - kb, y + ka) \mid k \in \mathbf{Z}\}$$

satisfies the equation

$$aX + bY = n. \tag{4}$$

Second (ii), we will show that any solution to this Diophantine equation belongs to S.

(i)   Let $k \in \mathbf{Z}$.

[To show: $(x - kb, y + ka)$ is a solution to equation (4).]

We simply substitute $(x - kb)$ for X and $(y + ka)$ for Y in the left side of equation (4), and show that this equals the right side of equation (4).

$$a(x - kb) + b(y + ka) = ax - akb + by + bka$$
$$= ax + by$$
$$= n \quad \leftarrow \text{by hypothesis.}$$

Thus, for every $k \in \mathbf{Z}$, $(x - kb, y + ka)$ is a solution to equation (4).

(ii)   Let (x', y') be a solution to equation (4).

[To show: There exists $k \in \mathbf{Z}$ such that $x' = x - kb$ and $y' = y + ka$.]

Since (x', y') and (x, y) are both solutions to equation (4), we have:

$$ax' + by' = n$$

and

$$ax + by = n$$

Subtracting these equations yields

$$a(x' - x) + b(y' - y) = 0,$$

or equivalently,

$$-a(x' - x) = b(y' - y). \tag{5}$$

From this equation, it is clear that $a \mid b(y' - y)$. We know that a and b are relatively prime by hypothesis, so Euclid's Lemma (5.4.2) allows us to conclude that

$$a \mid y' - y.$$

This means there exists $k \in \mathbf{Z}$ such that

$$y' - y = ka;$$

that is,

$$y' = y + ka. \tag{6}$$

Now we have shown that y' has the desired form. To show that x' has the desired form, substitute equation (6) into equation (5) and simplify:

$$-a(x' - x) = b(ka)$$

Now divide both sides by $-a$ and solve for x':

$$(x' - x) = -bk$$

$$x' = x - kb.$$

Thus, we have succeeded in proving that an arbitrary solution (x', y') has the form $(x - kb, y + ka)$. ∎

EXAMPLE 1

Find the entire solution set for the Diophantine equation $113X + 42Y = 1$.

SOLUTION In Example 1 of Section 5.2, we found a solution to the Diophantine equation $113X + 42Y = 1$. The particular solution we found was $(-13, 35)$. By Theorem 5.4.1, the solution set is given by

$$\{(-13 - 42k, 35 + 113k) \mid k \in \mathbf{Z}\}.$$

EXAMPLE 2

Find the entire solution set for the Diophantine equation $4X + 10Y = 102$.

SOLUTION First we divide the equation above by the common factor of 2 to get the equivalent equation

$$2X + 5Y = 51. \tag{7}$$

To solve this equation, we first solve the equation

$$2X + 5Y = 1 \tag{8}$$

which we can do by using Method 5.2.2. We find that $(-2, 1)$ is a particular solution to equation (8). Multiplying by 51, we find that $(-102, 51)$ is a particular solution to equation (7). Now by Theorem 5.4.1,

$$S = \{(-102 - 5k, 51 + 2k) \mid k \in \mathbf{Z}\} \tag{9}$$

is the entire solution set to equation (7). Hence, S is also the solution set to our original equation, $4X + 10Y = 102$.

Note that in equation (9), by setting $k = -25$ we obtain the particularly pleasant solution $(23, 1)$ to the Diophantine equation $4X + 10Y = 102$. Starting from this particular solution, we can use Theorem 5.4.1 to express the solution set in a way that is slightly simpler than equation (9):

$$S = \{(23 - 5k, 1 + 2k) \mid k \in \mathbf{Z}\}. \tag{10}$$

■

EXERCISES 5.4

Numerical Exercises

In Exercises 1–16:

    a. Use the methods of Sections 5.2 and 5.3 to find a solution to the Diophantine equation, or explain why no solution exists.

    b. If you found a solution in part a, use it to find the entire solution set.

  1. $19X + 11Y = 1$ **2.** $33X + 27Y = 1$ **3.** $1203X + 2411Y = 1$

  4. $99X + 991Y = 1$ **5.** $1155X + 29393Y = 1$ **6.** $986X + 485Y = 1$

  7. $242X + 429Y = 1$ **8.** $169X + 48Y = 1$ **9.** $169X - 48Y = 1$

10. $55X + 34Y = 7$ **11.** $39X + 66Y = 24$ **12.** $13X + 91Y = 10$

13. $81X + 99Y = -9000$ **14.** $56X + 84Y = -30$ **15.** $49X + 106Y = 50$

16. $-49X + 106Y = 50$

Reasoning and Proofs

17. Prove that the set defined by equation (9) and the set defined by equation (10) are exactly the same set. [*Hint:* One way to do this is to prove that every element of the first set is an element of the second set, and vice versa.]

18. Let $a, b, c \in \mathbf{Z}$ such that $a \mid bc$. Use Euclid's Lemma (5.4.2) to prove that $a \mid \gcd(a, b) \cdot \gcd(a, c)$.

19. Phil Lovett You run into your friend Phil Lovett at a party, and he tells you,

"Dude, Diophantine equations are no sweat. If a Diophantine equation has one solution, it's got infinitely many, and I can find them all. That Royal Road thing works even if the gcd of the coefficients is bigger than 1. So I can use the Royal Road to find one solution, then use that $x - kb, y + ka$ trick to get the rest.

"Like, suppose you want to find all the solutions to the Diophantine equation $9X + 6Y = 3$. You know there are solutions, since 3 is the gcd of 9 and 6. The Royal Road gives $(1, -1)$ as a particular solution. Now by Theorem 5.4.1, the solution set to the Diophantine equation is given by

$$S = \{(1 - 6k, -1 + 9k) \mid k \in \mathbf{Z}\}.$$

It's wicked easy."

a. Show that Phil's answer is wrong. That is, prove that S is not the entire solution set of the Diophantine equation $9X + 6Y = 3$.

b. What is wrong with Phil's reasoning?

20. Let $a, b, k \in \mathbf{Z}$ such that $a \mid k$, $b \mid k$, and $\gcd(a, b) = 1$. Prove that $ab \mid k$.

21. Let $a, b, c, d \in \mathbf{Z}$. Prove that if $\gcd(a, b) = 1$, $d \mid a$, and $d \mid bc$, then $d \mid c$.

22. Let $a, b, c \in \mathbf{Z}$. Prove that if $\gcd(a, b) = \gcd(a, c) = 1$, then $\gcd(a, bc) = 1$.

23. Let $a, b, c \in \mathbf{Z}$. Prove that if $\gcd(a, b) = 1$, then $\gcd(ac, b) = \gcd(b, c)$.

EXPLORATION Positive Solutions to Linear Diophantine Equations
(Exercises 24–30)

In this Exploration, we will consider just the *positive* solutions to linear Diophantine equations. We begin with some numerical problems.

For the Diophantine equations given in Exercises 24−27:

    a. Find a solution with X, Y both positive integers, or explain why no such solution exists.

    b. Find all solutions with X, Y both positive integers.

24. $15X + 17Y = 81$

25. $15X + 27Y = 9$

26. $5X + 6Y = 50$ [*Hint:* you can check your answer by looking at a U.S. flag.]

27. $23X + 20Y = 99$

You will now prove some general results about positive solutions to linear Diophantine equations.

28. Let $a, b, c \in \mathbf{N}$ such that $\gcd(a, b) = 1$ and $c > ab$. Prove that $aX + bY = c$ has a solution in which X and Y are both positive integers.

29. Let $a, b, c, n \in \mathbf{N}$ such that $\gcd(a, b) = 1$ and $c > abn$. Prove that $aX + bY = c$ has at least n positive solutions in which X and Y are both positive integers.

30. Let $a, b, c \in \mathbf{N}$ such that $\gcd(a, b) = 1$. Prove that there are infinitely many pairs of positive integers (X, Y) such that $aX - bY = c$.

Advanced Reasoning and Proofs

31. At a popular fast food restaurant, you can buy chicken nuggets in boxes of 5 or 7 pieces.

   a. You want to purchase exactly 34 chicken nuggets. How many boxes of 5 and how many boxes of 7 should you buy in order to total exactly 34 nuggets?

   b. Explain why it is impossible to buy exactly 18 chicken nuggets.

   c. What is the largest number of chicken nuggets that it is impossible to buy?

   d. Prove that it is possible to buy any number of chicken nuggets larger than your answer to part c.

32. In the country of Erehwon, football is played differently than in the United States. In Erehwonian football, a touchdown is worth s points, and a field goal is worth t points. It is impossible to get to a score of 58, and there are precisely 34 other scores that no team can possibly have. What are the values of s and t?

Chapter 6

Marin Mersenne (1588–1648)

The French Franciscan monk Marin Mersenne belonged to a religious order devoted to prayer, study, and scholarship. The intellectual pursuits that interested Mersenne the most were mathematics, physics, philosophy, and music.

Mersenne contributed to the development of mathematics by helping researchers stay abreast of one another's work. He invited mathematicians, scientists, and other intellectuals to come to his convent each week to discuss their recent discoveries and learn about the discoveries of others. The mathematicians and scientists who met each week in Mersenne's cell were the core group that later formed the French Academy of Sciences. Not only did Mersenne help the mathematics community in Paris by hosting these meetings, but he also helped mathematicians throughout Europe by writing letters informing them of the discoveries of the Parisian mathematicians and inquiring about their own mathematical research. Since scientific journals did not exist at the time, it was only through Mersenne's letters that many results in mathematics were communicated. Although Mersenne did not make any important mathematical discoveries of his own, his mathematical questions, conjectures, meetings, and letters played a significant role in the progress of mathematical research.

Mersenne wrote a number of books about mathematics and translated Galileo's lectures

and writings into French. His translations are credited with bringing the work of Galileo to the attention of scholars outside Italy. He also attempted to defend Galileo against the criticism of the church. In addition to Mersenne's work in mathematics, he wrote treatises on mechanics, mathematical physics, and acoustics.

In mechanics, there is a formula called Mersenne's Law that relates the pitch produced by an instrument's string to the string's length, tension, and mass per unit length. This relationship explains why low-pitched strings on a piano must be made thicker (have more mass per unit length) than high-pitched strings.

A bit of Mersenne's math

Mersenne was interested in finding a formula for all primes. He considered numbers of the form $m = 2^p - 1$, where p is a prime, and tried to determine which of these numbers are prime. Although Euclid was the first to consider primes of this form, these primes m have become known as *Mersenne primes* because Mersenne's letters induced many mathematicians to study them. (See Section 3.3 for more about Mersenne primes.)

Mersenne claimed that for the primes $p = 2, 3, 5, 7, 13, 17, 19, 31, 67, 127$, and 257, the number $m = 2^p - 1$ is prime, and that for all other primes $p < 257$, the number $m = 2^p - 1$ is composite. However, Mersenne was not able to test this claim because the numbers involved were too big. It turns out that Mersenne was right except for five values of p. For $p = 67$ and 257, Mersenne claimed that $m = 2^p - 1$ was prime but m is actually composite, and for $p = 61, 89$, and 107 Mersenne claimed that $m = 2^p - 1$ was composite but m is actually prime. To this day, mathematicians continue to search for larger and larger Mersenne primes, yet there are still only 47 such primes known.

Mersenne primes are related to another type of numbers called *perfect numbers*.

A natural number n is said to be *perfect* if it is equal to the sum of all of its positive divisors that are less than n. For example, $6 = 1 + 2 + 3$, and $28 = 1 + 2 + 4 + 7 + 14$, so 6 and 28 are both perfect numbers. Euclid showed that if $2^p - 1$ is prime, then $n = 2^{p-1}(2^p - 1)$ is perfect. Conversely, Euler showed that every even perfect number has the form $n = 2^{p-1}(2^p - 1)$, where $2^p - 1$ is prime. Thus, if we had a complete list of the Mersenne primes, that would give us a complete list of the even perfect numbers (and vice versa). The question of whether there are any *odd* perfect numbers baffled the ancient Greeks, and it remains unsolved to this day.

Chapter 6 The Fundamental Theorem of Arithmetic
or Monopolizing the Internet

6.1 The Fundamental Theorem

Mersenne in his prime

Born in 1588, French monk Marin Mersenne was instrumental in disseminating mathematical and scientific knowledge throughout Europe. What's not so well known about Mersenne is that he was a corporate lawyer. He built a successful career as an attorney in 17th-century France's booming software industry. As a young man, he was fortunate to land a position with multinational software giant Millisoft.

Millisoft wanted to expand its influence from software to all facets of the computer and communications industries. The founder and CEO of Millisoft, the legendary Gil Bates, had a plan to get in on the ground floor of the nascent but fast-growing Internet.

Today's Internet domains, such as .com and .edu, had not yet been invented. In the 17th century, the Internet was numerical. For example, you could visit Millisoft's help page, www.127.num, after one of Millisoft's programs invariably caused your computer to crash.

Bates had a plan to buy up as many Internet addresses as he could. Not all Internet addresses were available for purchase, however. The government was selling only addresses with prime numbers, reserving the composite numbers for later use. When Mersenne joined the company, Bates had already purchased all of the prime addresses from www.2.num up to www.524287.num, and he showed no sign of stopping.

It happened that young Mersenne had studied Internet law in school, where he had come across a peculiar statute called the Writ of Habeas Productus Internetibus. This law states that if a party owns both of the Internet addresses www.x.num and www.y.num, then the same party automatically owns the product address www.xy.num. For example, if Millisoft owned the Internet addresses www.3.num and www.5.num, the law would give Millisoft automatic possession of the product address, www.15.num.

Bates was thrilled to learn of this law, because it gave him automatic ownership of a lot more Internet addresses. This saved him the unpleasantness of having to ruthlessly buy out every company that crossed his path. For bringing the Writ to his attention, Bates rewarded Mersenne by promoting him to Head of Millisoft's Legal Division.

Writ of Habeas
Productus Internetibus:

Supposimus personus possessus
addressus internetibus

www.X.num and www.y.num

Automaticalus, personus also possessus
productus addressibus

www.xy.num

Being fluent in Latin, Mersenne realized that this law would help Millisoft's profits continue to multiply.

Government officials had been trying for years to break up Millisoft's monopoly on the industry, and they hated the idea of Millisoft suddenly gaining control of so many Internet addresses. The Justice Department filed a lawsuit asking the Supreme Court to strike down the Writ of Habeas Productus Internetibus on the grounds that it was inconsistent and therefore unconstitutional. To prove their case, the government lawyers needed to find a natural number, N, that could be factored into primes in two different ways. Then the Writ would be inconsistent: If one party owned the primes in one factorization of N, and another party owned the primes in a different factorization of N, then the Writ would give ownership of the address www.N.num to both parties.

The attorney general sent a memo to all departments requesting help in finding a number that factors into primes in more than one way. The only response came from the Justice Department's Chief Deputy Assistant Undersecretary of Software Engineering, Phil Lovett. Assistant Undersecretary Lovett was confident that any number small enough to fit on a calculator would have only one prime factorization. However, he figured that by taking advantage of the government's awesome computing power, he could test lots of really big numbers, and surely one of them would have two different prime factorizations.

At the pretrial hearing, Lovett testified before the High Court wearing sunglasses and a wrinkled Hawaiian shirt. "Yo, Your Honors! My software has found a natural number that can be factored into primes in two different ways."

Exhibit A:

$$17^{11} \cdot 19^2 \cdot 67 \cdot 101^{101} \cdot 251 = 13^{79} \cdot 47^{76} \cdot 59 \cdot 677 \cdot 1181$$

"I'm sure it's true, dudes—I multiplied both products out to, like, 15 decimal places, and got the same answer."

Exhibit B:

$$5.68398094458232 \cdot 10^{222}$$

Mersenne, in charge of Millisoft's defense, punched each of the products from Exhibit A into Millisoft's multiplication software. To his surprise, both products produced the same result on his computer screen: $5.68398094458232 \cdot 10^{222}$. Could Lovett really be correct? Mersenne thought not. He had a feeling that there must be something wrong with Lovett's software. Mersenne asked Bates to order Millisoft's entire debugging staff to find the mistake in Lovett's factoring program.

"Debugging staff?" Bates responded. "We don't debug our software. We just come out with new versions."

After the hearing, Mersenne was not hopeful—he still needed a defense, and the trial was only days away. After many hours scouring the dusty tomes in the law library, he found a centuries-old precedent in Book 9 of Euclid's *Elements of Law*.

"Aha!" Mersenne exclaimed." 17 divides the product on the left side of Exhibit A. If Lovett's equation is correct, then 17 must also divide the product on the right side of Exhibit A, $13^{79} \cdot 47^{76} \cdot 59 \cdot 677 \cdot 1181$. So 17 would have to divide one of the primes 13, 47, 59, 677, or 1181, which is just plain wrong. Now I have the perfect defense!"

In court the next day, Mersenne's eloquent argument proved beyond a reasonable doubt that Lovett's example was flawed, and the Court dismissed the case against Millisoft. The ruling enabled Millisoft to secure its place in the industry for centuries to come.

In Lovett's next annual review, the Justice Department praised his efforts on Exhibit A as "close enough for government work," and he was promoted to Chief Deputy Associate Undersecretary.

Naomi's Numerical Proof Preview: Fundamental Theorem of Arithmetic (6.1.1)

In our myth, Mersenne wanted to prove that Lovett's Exhibit A,

$$17^{11} \cdot 19^2 \cdot 67 \cdot 101^{101} \cdot 251 = 13^{79} \cdot 47^{76} \cdot 59 \cdot 677 \cdot 1181$$

could not be correct. He used indirect reasoning. Assuming this equation is correct, 17 must divide the right side of the equation:

$$17 \mid 13^{79} \cdot 47^{76} \cdot 59 \cdot 677 \cdot 1181. \tag{1}$$

Mersenne was able to convince the court that this implied that 17 must divide one of the factors:

$$17 \mid \mathbf{13} \quad \text{or} \quad 17 \mid \mathbf{47} \quad \text{or} \quad 17 \mid \mathbf{59} \quad \text{or} \quad 17 \mid \mathbf{677} \quad \text{or} \quad 17 \mid \mathbf{1181}. \tag{2}$$

But 17 cannot divide any of these numbers, since they are all prime. This contradiction implies that Lovett's Exhibit A must be incorrect.

We will use these ideas to prove that no number can have more than one prime factorization. This statement, together with the Existence of Prime Factorizations (3.2.5), is known as the Fundamental Theorem of Arithmetic (FTA).

Let $n \in \mathbf{N}$, $n > 1$. Then n may be written as a product of one or more prime numbers. Furthermore, the prime factorization of n is unique up to order.

When we say that every number has a unique factorization, we clearly do not wish to count $2 \cdot 3 \cdot 5$ and $5 \cdot 2 \cdot 3$ as different factorizations. We express this idea in the statement of the Fundamental Theorem of Arithmetic by saying that prime factorizations are unique *up to order*. That is, two factorizations are to be considered the same if the only difference between them is the order of the factors.[1]

A basic property of primes

We will prove the Fundamental Theorem of Arithmetic shortly. But first, let's explore Mersenne's argument from our story to understand why it held up in court. The main factor in Mersenne's defense was his realization that (1) implies (2). He needed to know that if 17 divides a product, then 17 must divide one of the factors in the product:

$$\text{If } 17 \mid \boldsymbol{ab}, \text{ then } 17 \mid \boldsymbol{a} \text{ or } 17 \mid \boldsymbol{b}.$$

This is a special case of the following proposition. Our proof of this proposition relies on the theory of linear Diophantine equations, with Euclid's Lemma (5.4.2) playing the key role.

| 6.1.2 | FUNDAMENTAL PROPERTY OF PRIMES |

Let p be a prime number, and let a, b $\in \mathbf{Z}$. If p $\mid$ ab, then p $\mid$ a or p $\mid$ b.

> **Put it in Prose, Paul!**
>
> If a prime p divides a product, then p must divide one of the factors.

---

[1]The phrase "up to . . ." is used similarly in many other contexts; for example, 40 and -40 are equal up to sign, $\dfrac{7\pi}{2}$ and $-\dfrac{\pi}{2}$ are equal up to a multiple of 2π, the functions $12x^2$ and x^2 are equal up to a constant factor, and so forth.

PROOF Let p be a prime, and let a and b be integers such that $p \mid ab$.

[To show: $p \mid a$ or $p \mid b$.]

We know that either $p \mid a$ or $p \nmid a$.

If $p \mid a$, then we are done. So we may assume that $p \nmid a$. Now we need to show that $p \mid b$. Since p is prime and $p \nmid a$, it follows that $\gcd(p, a) = 1$ (by Lemma 3.4.3). Since p divides the product ab and p is relatively prime to the factor a, we may use Euclid's Lemma (5.4.2) to conclude that p divides the other factor, i.e., $p \mid b$. ∎

The Fundamental Property of Primes (6.1.2) describes a property that is true for all prime numbers. Does this property hold for composite numbers? No, it does not. For example, $6 \mid 4 \cdot 3$, but $6 \nmid 4$ and $6 \nmid 3$.

The following corollary generalizes the Fundamental Property of Primes (6.1.2) to an arbitrary number of factors:

COROLLARY 6.1.3

Let p be a prime number. Let $a_1, a_2, \ldots, a_r \in \mathbf{Z}$.

If $p \mid a_1 \cdot a_2 \cdot \; \cdots \; \cdot a_r$, then $p \mid a_i$ for some $i = 1, 2, \ldots, r$.

PROOF You will prove this in Exercise 7.

Proof of the FTA

Now that we have the Fundamental Property of Primes, we are finally ready to prove that prime factorizations are unique.

PROOF of **THE FUNDAMENTAL THEOREM OF ARITHMETIC (6.1.1)**

Since we have already proved the existence part of the FTA, that every number has a prime factorization (Theorem 3.2.5), all that remains to prove is the uniqueness part of the FTA: we must show that every number $n > 1$ has at most one prime factorization up to order. We will prove this using strong induction.

Let $n \in \mathbf{N}$, $n > 1$, and assume the inductive hypothesis that every natural number k for which $1 < k < n$ has at most one prime factorization up to order.

[To show: n has at most one prime factorization up to order.]

Suppose that n can be written as the product of primes in two different ways; that is, there exist primes $p_1, \ldots, p_e$ and $q_1, \ldots, q_f$ such that $n = p_1 \cdot p_2 \cdot \; \cdots \; \cdot p_e = q_1 \cdot q_2 \cdot \; \cdots \; \cdot q_f$. We must show that these two prime factorizations of n are the same up to order.

Starting with

$$p_1 \cdot p_2 \cdot \; \cdots \; \cdot p_e = q_1 \cdot q_2 \cdot \; \cdots \; \cdot q_f, \tag{3}$$

we observe that the number $q_1 \cdot q_2 \cdot \; \cdots \; \cdot q_f$ is equal to p_1 multiplied by an integer. Hence,

$$p_1 \mid q_1 \cdot q_2 \cdot \; \cdots \; \cdot q_f.$$

By Corollary 6.1.3, p_1 divides one of the primes $q_1, q_2, \ldots, q_f$. Reordering the factors if necessary (which is explicitly permitted in the statement of the FTA), we may assume without loss of generality that $p_1 \mid q_1$. Since p_1 and q_1 are both prime, it follows that $p_1 = q_1$ (by Lemma 3.2.4). Thus, in equation (3), we may cancel a factor of p_1 from both sides of the equation to obtain

$$\frac{n}{p_1} = p_2 \cdot \ \cdots \ \cdot p_e = q_2 \cdot \ \cdots \ \cdot q_f \tag{4}$$

If $n = p_1$, then $n = p_1 = q_1$, so the two factorizations in equation (3) are identical. Hence, we may assume that $n \neq p_1$. Thus, $1 < \dfrac{n}{p_1} < n$, so we may invoke the inductive hypothesis, which tells us that the two factorizations in equation (4) are the same up to order. But the factorizations in equation (3) are precisely the factorizations of equation (4) with an additional factor of p_1. We conclude that the two factorizations in equation (3) are themselves the same up to order. Thus, we have shown that n has at most one prime factorization up to order.

By the principle of strong induction, it follows that every natural number $n > 1$ has at most one prime factorization up to order, as was to be shown. ∎

The Fundamental Theorem of Arithmetic (6.1.1) tells us that every number $n > 1$ is the product of one or more prime numbers. For example,

$$1500 = 2 \cdot 2 \cdot 3 \cdot 5 \cdot 5 \cdot 5.$$

This prime factorization can be written more compactly:

$$1500 = 2^2 \cdot 3^1 \cdot 5^3.$$

This way of writing the prime factorization is often more convenient. In general, we may write any number $n > 1$ in the form

$$p_1^{a_1} \cdot p_2^{a_2} \cdot \ \cdots \ \cdot p_r^{a_r} \tag{5}$$

where $p_1, p_2, \ldots, p_r$ are distinct primes, and $a_1, a_2, \ldots, a_r$ are natural numbers.

Enjoying the view

Having proved the Fundamental Theorem of Arithmetic, we have earned the right to take a brief rest from our climb up number theory mountain and look back at the important steps we took en route to the FTA. The main ingredient in the proof of the FTA is the Fundamental Property of Primes (6.1.2), which is a close cousin of Euclid's Lemma (5.4.2). To prove this lemma, we needed to know about solutions to linear Diophantine equations (Theorem 5.1.2). We learned how to find solutions to these using the Euclidean Algorithm (4.1.1), which involves repeated application

of the Division Theorem (3.5.1). Finally, the proof of the Division Theorem relied on the Well-Ordering Principle (2.2.2), which we took as an axiom.

EXERCISES 6.1

Numerical Problems

1. Factor each of the following numbers. Write the factorization in the form $p_1^{a_1} \cdot p_2^{a_2} \cdot \ \cdots \ \cdot p_r^{a_r}$.

   a. 2160   b. 343   c. 1024

   d. 1000   e. 36   f. 2125

2. Write out the prime factorization of each number in the form $p_1^{a_1} \cdot p_2^{a_2} \cdot \ \cdots \ \cdot p_r^{a_r}$.

   a. 5!   b. 10!   c. 20!

Reasoning and Proofs

3. Suppose that $p = 2^k - 1$ is prime. Prove that either $k = 2$ or k is odd.

4. Let n be an integer, $n > 1$. Prove that n is a perfect square if and only if n's prime factorization, $n = p_1^{a_1} \cdot p_2^{a_2} \cdot \ \cdots \ \cdot p_r^{a_r}$, has only even exponents.

5. Let p be a prime number. Suppose you were to write out the prime factorization of $p!$. How many factors of p would it contain?

6. Let p be a prime number. Suppose you were to write out the prime factorization of $(p^2)!$. How many factors of p would it contain?

7. Prove Proposition 6.1.3.

8. Let $n \in \mathbf{N}$, $n > 1$. Show that n is prime if and only if $n \nmid (n-1)!$

Advanced Reasoning and Proofs

9. Note that the number $10! = 3628800$ ends in two zeros.

   a. Suppose you were to write out all of the digits of $100!$. How many zeros would be at the end of this number?

   b. Suppose you were to write out all of the digits of $1000!$. How many zeros would be at the end of this number?

10. The set of six consecutive integers 2, 3, 4, 5, 6, and 7 can be arranged into two disjoint subsets, the product of whose elements is almost equal: $3 \cdot 4 \cdot 6 = 72$, and $2 \cdot 5 \cdot 7 = 70$. Are there any sets of six consecutive integers that can be partitioned into two subsets (not necessarily of equal size) to make two products that are *exactly* equal?

11. Suppose that q is prime and that a and n are natural numbers. Prove that if $q \mid a^n$, then $q \mid a$.

12. Let p be prime and let n be a natural number. Suppose that $d \in \mathbf{N}$ is a factor of p^n. Prove that $d = p^m$ for some natural number m such that $0 \le m \le n$.

13. Let a and b be integers such that $25 \mid ab$, but $25 \nmid a$. Prove that $5 \mid b$.

14. Let p and q be primes, and suppose $pq = ab$, where $a, b \in \mathbf{Z}$ such that $a, b > 1$. Show that either $a = p$ and $b = q$, or $a = q$ and $b = p$.

15. Let a and b be integers.

   a. Suppose that the prime 5 occurs exactly three times in the prime factorization of a, and that 5 occurs exactly twice in the prime factorization of b. What can you say about the number of times that 5 occurs in the prime factorization of $a + b$?

   b. Suppose that the prime p occurs exactly m times in the prime factorization of a, and that p occurs exactly n times in the prime factorization of b. Also assume that $m > n$. What can you say about the number of times that p occurs in the prime factorization of $a + b$? Justify your answer.

   c. Now suppose that 5 occurs exactly twice in the prime factorization of a, and exactly twice in the prime factorization of b. What can you say about the number of times that 5 occurs in the prime factorization of $a + b$?

d. Assuming that the prime p occurs exactly n times in the prime factorization of a and exactly n times in the prime factorization of b, what can you say about the number of times that p occurs in the prime factorization of $a + b$? Justify your answer.

16. Prove that every natural number n can be expressed uniquely in the form $2^b r$, where b is a nonnegative integer and r is an odd natural number. (*Note:* The existence but not the uniqueness part of this statement was proved in Example 1 of Section 2.2 using strong induction.)

17. A natural number n is called *squarefree* if it is not divisible by any perfect square other than 1.

    a. Let n be a natural number. Prove that n is squarefree if and only if n can be written as the product of distinct primes.

    b. Express the number 216,000 as the product of a squarefree number and a perfect square.

    c. Prove that any natural number can be expressed as the product of a squarefree number and a perfect square.

18. Let $r \in \mathbf{Z}$ and let p be prime. Suppose that $p \mid r^n$ for some natural number n. Prove that $p^n \mid r^n$.

19. Let $a, b \in \mathbf{N}$ such that $\gcd(a, b) = 1$. Prove that $\gcd(a^2, b^2) = 1$.

20. Let $a, b, n \in \mathbf{N}$ such that $\gcd(a, b) = 1$. Prove that $\gcd(a^n, b^n) = 1$.

21. Let $p > 3$ and $q > 3$ be primes. Prove that $p^2 - q^2$ is divisible by 24.

22. Find all primes p such that $17p + 1$ is a perfect square.

23. Let $p > 3$ and suppose that both p and $p + 2$ are prime. Show that their sum is divisible by 12.

24. Find all primes p and q such that $q = p + 2$ and both $p + 2$ and $pq - 2$ are prime.

6.2 Consequences of the Fundamental Theorem

Finding all the divisors of a number

How many factors does a million have? One way to answer this question is to try each of the numbers less than a million to see whether it is a factor of a million. Fortunately, the Fundamental Theorem of Arithmetic (FTA) gives a better way to approach this problem, as well as many other problems involving divisors.

Naomi's Numerical Proof Preview: Proposition 6.2.1

Suppose that d is a natural number that divides $1{,}000{,}000 = 10^6$. Then there exists $e \in \mathbf{Z}$ such that $10^6 = d \cdot e$. Since $10^6 = 2^6 \cdot 5^6$, we have

$$2^6 \cdot 5^6 = d \cdot e.$$

It follows from the FTA (6.1.1) that the prime factorization of d can contain only the primes 2 and 5, with each of these primes appearing no more than 6 times. Thus, d has the form

$$d = 2^i \cdot 5^j, \quad \text{where } i, j = 0, 1, 2, 3, 4, 5, \text{ or } 6.$$

This identifies all the divisors of 10^6. For example, taking $i = 5, j = 2$ produces $d = 800$, which is a divisor of 10^6.

We now know how to figure out how many factors 10^6 has: since there are 7 choices for the value of i (namely 0, 1, 2, 3, 4, 5, and 6), and 7 choices for j, and each of these choices gives us a different value for d, there are a total of exactly $7 \cdot 7 = 49$ positive factors of $1{,}000{,}000$.

In general, we can use these ideas to identify and count all the factors of a given number n, provided that we know the prime factorization of n.

PROPOSITION 6.2.1

Let $n \in \mathbf{N}, n > 1$, and let $p_1^{a_1} \cdot \cdots \cdot p_r^{a_r}$ be the prime factorization of n. If d is a natural number, then $d \mid n$ if and only if $d = p_1^{b_1} \cdot \cdots \cdot p_r^{b_r}$, where $0 \le b_i \le a_i$ for all $i = 1, \ldots, r$.

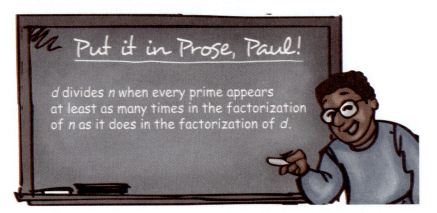

Put it in Prose, Paul!

d divides n when every prime appears at least as many times in the factorization of n as it does in the factorization of d.

PROOF Let $n, d \in \mathbf{N}$, and let $p_1^{a_1} \cdot \cdots \cdot p_r^{a_r}$ be the prime factorization of n.

[$\Rightarrow$] Assume that $d \mid n$.

[**To show:** $d = p_1^{b_1} \cdot \cdots \cdot p_r^{b_r}$, where $0 \le b_i \le a_i$ for all $i = 1, \ldots, r$.]

Since $d \mid n$, there exists $e \in \mathbf{Z}$ such that

$$n = d \cdot e \tag{1}$$

The existence part of the FTA (6.1.1) tells us that d and e can each be factored into a product of primes:

$$d = v_1 \cdot v_2 \cdot \; \cdots \; \cdot v_s \qquad \text{and} \qquad e = w_1 \cdot w_2 \cdot \; \cdots \; \cdot w_t.$$

Thus, equation (1) may be rewritten as

$$p_1^{a_1} \cdot \; \cdots \; \cdot p_r^{a_r} = (v_1 \cdot v_2 \cdot \; \cdots \; \cdot v_s) \cdot (w_1 \cdot w_2 \cdot \; \cdots \; \cdot w_t).$$

By the uniqueness part of FTA (6.1.1), the two sides of this equation must be the same prime factorizations up to order. Hence, the factorization $v_1 \cdot v_2 \cdot \; \cdots \; \cdot v_s$ contains no primes other than $p_1, p_2, \ldots, p_r$, and each prime p_i appears at most a_i times in this factorization. In other words,

$$d = v_1 \cdot v_2 \cdot \; \cdots \; \cdot v_s = p_1^{b_1} \cdot \; \cdots \; \cdot p_r^{b_r},$$

with $0 \le b_i \le a_i$ for all $i = 1, \ldots, r$. We have thus shown that d has the required form.

[$\Leftarrow$] Now assume that $d = p_1^{b_1} \cdot \; \cdots \; \cdot p_r^{b_r}$, where $0 \le b_i \le a_i$ for all $i = 1, \ldots, r$.

[To show: $d \mid n$.]

Since

$$p_1^{a_1} \cdot \; \cdots \; \cdot p_r^{a_r} = \left(p_1^{b_1} \cdot \; \cdots \; \cdot p_r^{b_r}\right)\left(p_1^{a_1 - b_1} \cdot \; \cdots \; \cdot p_r^{a_r - b_r}\right),$$

we know that

$$p_1^{b_1} \cdot \; \cdots \; \cdot p_r^{b_r} \mid p_1^{a_1} \cdot \; \cdots \; \cdot p_r^{a_r},$$

which says exactly that $d \mid n$. ∎

EXAMPLE 1

List the positive divisors of 135.

SOLUTION The prime factorization of 135 is $3^3 \cdot 5$. Thus, by Proposition 6.2.1, the divisors of 135 are all the numbers of the form

$$d = 3^a \cdot 5^b,$$

where $a = 0, 1, 2$, or 3, and $b = 0$ or 1. Hence, 135 has a total of 8 divisors:

$$3^0 \cdot 5^0 = 1, \qquad 3^1 \cdot 5^0 = 3, \qquad 3^2 \cdot 5^0 = 9, \qquad 3^3 \cdot 5^0 = 27,$$
$$3^0 \cdot 5^1 = 5, \qquad 3^1 \cdot 5^1 = 15, \qquad 3^2 \cdot 5^1 = 45, \qquad 3^3 \cdot 5^1 = 135.$$

∎

In Example 1, we found that the number $3^3 \cdot 5^1$ has a total of $(3 + 1) \cdot (1 + 1) = 8$ positive factors. This is a specific example of the following corollary.

Let $n \in \mathbf{N}$, $n > 1$, and let $p_1^{a_1} \cdot p_2^{a_2} \cdot \ \cdots \ \cdot p_r^{a_r}$ be the prime factorization of n. Then the number of positive factors of n is

$$(a_1 + 1) \cdot (a_2 + 1) \cdot \ \cdots \ \cdot (a_r + 1).$$

PROOF By Proposition 6.2.1, the divisors of n are all numbers d of the form

$$d = p_1^{b_1} \cdot p_2^{b_2} \cdot \ \cdots \ \cdot p_r^{b_r}, \text{ where } 0 \le b_i \le a_i \text{ for all } i = 1, 2, \ldots, r.$$

Since b_i may be any of the numbers $0, 1, 2, \ldots, a_i$, there are $a_i + 1$ possible values for b_i. Thus, the total number of ways to choose the exponents $b_1, b_2, \ldots, b_r$ is $(a_1 + 1) \cdot (a_2 + 1) \cdot \ \cdots \ \cdot (a_r + 1)$.

To complete the proof, we must show that different choices of the exponents b_i yield different values of d. To do this, we will assume that two choices of exponents result in the same value of d and show that the exponents must be identical.

Suppose there exist integers $b_1, \ldots, b_r$ and $c_1, \ldots, c_r$ such that

$$d = p_1^{b_1} \cdot p_2^{b_2} \cdot \ \cdots \ \cdot p_r^{b_r} = p_1^{c_1} \cdot p_2^{c_2} \cdot \ \cdots \ \cdot p_r^{c_r},$$

Then by the FTA (6.1.1), it follows that $b_i = c_i$ for all $i = 1, \ldots, r$. ●

Prime factorizations and the gcd

Naomi's Numerical Proof Preview: Proposition 6.2.3

Using the Euclidean Algorithm, we can quickly find the greatest common divisor of any two numbers m and n without having to find their prime factorizations. However, if we happen to know the prime factorizations of m and n, then it is even easier to find $\gcd(m, n)$. For example, say $m = 3^6 \cdot 7 \cdot 13^2 \cdot 29$ and $n = 2^3 \cdot 3^4 \cdot 7^3 \cdot 11^3 \cdot 13^3$.

$$\gcd(3^6 \cdot 7 \cdot 13^2 \cdot 29, \ 2^3 \cdot 3^4 \cdot 7^3 \cdot 11^3 \cdot 13^3) = 3^4 \cdot 7 \cdot 13^2.$$

As this equation demonstrates, If m has a factor of p^a and n has a factor of p^b, then m and n have a common factor of p^c, where c is the smaller of the two exponents a and b. We introduced this method of finding the gcd in Chapter 3, without justifying it. Now that we have proven the FTA (6.1.1), we are in a position to prove that this method always yields the gcd.

To give a general statement of this method, it will be useful to have our two numbers m and n expressed in terms of the *same* list of primes $p_1, \ldots, p_r$. This is always possible if we allow ourselves to include factors with exponents of 0 when we write prime factorizations. For example, for the values of m and n from our gcd calculation above, both m and n may be written in the form $2^{a_1} \cdot 3^{a_2} \cdot 7^{a_3} \cdot 11^{a_4} \cdot 13^{a_5} \cdot 29^{a_6}$, as follows:

$$m = 2^0 \cdot 3^6 \cdot 7^1 \cdot 11^0 \cdot 13^2 \cdot 29^1 \quad \text{and} \quad n = 2^3 \cdot 3^4 \cdot 7^3 \cdot 11^3 \cdot 13^3 \cdot 29^0.$$

We use the expression **min(a, b)** to denote the smaller of a and b. Similarly, **max(a, b)** denotes the larger of a and b.

PROPOSITION 6.2.3

Let $m, n \in \mathbf{N}$ have the prime factorizations

$$m = p_1^{a_1} \cdot \,\cdots\, \cdot p_r^{a_r} \qquad \text{and} \qquad n = p_1^{b_1} \cdot \,\cdots\, \cdot p_r^{b_r},$$

with each integer $a_i, b_i \geq 0$. Then

$$\gcd(m, n) = p_1^{c_1} \cdot \,\cdots\, \cdot p_r^{c_r},$$

where $c_i = \min(a_i, b_i)$ for all $i = 1, \ldots, r$.

PROOF By Proposition 6.2.1, a natural number d will be a common divisor of m and n provided that the prime factorization of d is given by

$$d = p_1^{e_1} \cdot \,\cdots\, \cdot p_r^{e_r},$$

where for all $i = 1, \ldots, r$, the exponent e_i is a nonnegative integer such that

$$e_i \leq a_i \qquad \text{and} \qquad e_i \leq b_i.$$

These two inequalities are equivalent to the single inequality

$$e_i \leq \min(a_i, b_i).$$

The largest possible value of e_i is therefore $\min(a_i, b_i)$. Thus, the largest of all the common divisors of m and n is $d = p_1^{c_1} \cdot \,\cdots\, \cdot p_r^{c_r}$, where $c_i = \min(a_i, b_i)$. ∎

Prime factorizations and the lcm

Like the gcd, the lcm of two numbers m and n can easily be found if we know their prime factorizations. For example, consider the same values of *m* and *n* from our earlier calculations.

$$\text{lcm}(3^6 \cdot 7 \cdot 13^2 \cdot 29,\, 2^3 \cdot 3^4 \cdot 7^3 \cdot 11^3 \cdot 13^3) = 2^3 \cdot 3^6 \cdot 7^3 \cdot 11^3 \cdot 13^3 \cdot 29.$$

As this equation demonstrates, if m has a factor of p^a and n has a factor of p^b, then any common multiple of m and n is divisible by p^f, where f is the larger of a and b. This idea gives us a method to find the lcm of any two numbers, given their prime factorizations:

PROPOSITION 6.2.4

Let $m, n \in \mathbf{N}$ have the prime factorizations

$$m = p_1^{a_1} \cdot \,\cdots\, \cdot p_r^{a_r} \quad \text{and} \quad p_1^{b_1} \cdot \,\cdots\, \cdot p_r^{b_r},$$

with each $a_i, b_i \geq 0$. Then

$$\mathrm{lcm}(m, n) = p_1^{f_1} \cdot \cdots \cdot p_r^{f_r}$$

where $f_i = \max(a_i, b_i)$ for all $i = 1, \ldots, r$.

PROOF See Exercise 20.

What happens when we multiply the gcd of two natural numbers by the lcm of the same two numbers? For example, $\gcd(6, 8) = 2$ and $\mathrm{lcm}(6, 8) = 24$. In this case, the product of the gcd and the lcm is $2 \cdot 24 = 48$, which is the same as the product of the original numbers, 6 and 8. Is it true for any pair of natural numbers that the product of the gcd and the lcm is the same as the product of the original numbers?

For the values

$$m = 3^6 \cdot 7 \cdot 13^2 \cdot 29 \qquad \text{and} \qquad n = 2^3 \cdot 3^4 \cdot 7^3 \cdot 11^3 \cdot 13^3,$$

we get

$$\gcd(m, n) \cdot \mathrm{lcm}(m, n) = (3^4 \cdot 7 \cdot 13^2) \cdot (2^3 \cdot 3^6 \cdot 7^3 \cdot 11^3 \cdot 13^3 \cdot 29). \qquad (2)$$

Let's compare this with the product of the original numbers:

$$mn = 3^6 \cdot 7 \cdot 13^2 \cdot 29 \cdot 2^3 \cdot 3^4 \cdot 7^3 \cdot 11^3 \cdot 13^3. \qquad (3)$$

Is the product in (2) equal to the product in (3)? We can answer this question without even doing a single calculation! Simply by rearranging the factors, we can see that the right sides of equations (2) and (3) are identical.

We now prove that this fundamental relationship between the gcd and the lcm is true in general.

PROPOSITION 6.2.5

Let $m, n \in \mathbf{N}$. Then $\gcd(m, n) \cdot \mathrm{lcm}(m, n) = mn$.

PROOF Suppose that m and n have the prime factorizations

$$m = p_1^{a_1} \cdot \cdots \cdot p_r^{a_r} \quad \text{and} \quad n = p_1^{b_1} \cdot \cdots \cdot p_r^{b_r}$$

with each $a_i, b_i \geq 0$. Then by Propostion 6.2.3 and Proposition 6.2.4, we have

$$\gcd(m, n) = p_1^{c_1} \cdot \cdots \cdot p_r^{c_r} \quad \text{and} \quad \mathrm{lcm}(m, n) = p_1^{f_1} \cdot \cdots \cdot p_r^{f_r},$$

where $c_i = \min(a_i, b_i)$ and $f_i = \max(a_i, b_i)$. Thus,

$$\gcd(m, n) \cdot \mathrm{lcm}(m, n) = p_1^{c_1 + f_1} \cdot \cdots \cdot p_r^{c_r + f_r}. \qquad (4)$$

The quantity $c_i + f_i$ is the smaller of the two numbers a_i and b_i added to the larger of these two numbers. The result is just the sum of the two numbers, $a_i + b_i$. Thus,

$$c_i + f_i = a_i + b_i$$

for any $i = 1, \ldots, r$. Substituting into equation (4), we get

$$\gcd(m, n) \cdot \operatorname{lcm}(m, n) = p_1^{a_1 + b_1} \cdot \ \cdots \ \cdot p_r^{a_r + b_r}.$$

The product on the right side of this equation is equal to mn. Thus, $\gcd(m, n) \cdot \operatorname{lcm}(m, n) = mn$, as was to be shown.

COROLLARY 6.2.6

Let $m, n \in \mathbf{N}$. If $\gcd(m, n) = 1$, then $\operatorname{lcm}(m, n) = mn$.

PROOF The corollary follows immediately from Proposition 6.2.5.

Periodical cicadas and the lcm

The largest that the lcm of any two natural numbers can possibly be is the product of the two numbers. It follows from Corollary 6.2.6 that for distinct primes p and q, this largest possible value is achieved:

$$\operatorname{lcm}(p, q) = pq.$$

This simple mathematical fact may help to explain the life cycles of certain types of cicadas.

Cicadas are among the many insects that we often see in the summertime. However, in addition to the most common types of cicadas, which appear every year, certain parts of the United States are occasionally invaded by overwhelming numbers of *periodical cicadas*.[2] These cicadas appear all at once and can reach densities exceeding one million cicadas per acre.

Each species of periodical cicadas is divided into *broods* according to how often they appear: one brood appears every 7 years, another brood appears every 13 years, and a third brood appears every 17 years. The cicadas in a given brood live underground for most of their life, then all the cicadas in the brood emerge at exactly the same time. Once they emerge, they live only for a few more weeks, during which time they mate, lay their eggs, and then die.

Scientists believe that the life cycles of periodical cicadas reduce the danger from predators. Since all of the cicadas in a given brood appear at the same time, they are so numerous that even if many of them are eaten by birds or other small animals, the species itself is never threatened.

---

[2] Although it might seem more correct to call these *periodic* cicadas, they are more commonly known as *periodical* cicadas.

Curiously, the life spans of all of the different broods of periodical cicadas are prime numbers. Scientists do not completely understand why, but some possible explanations have been offered. Many of these explanations are based on the fact that having prime life spans makes it very rare that two different broods will emerge in the same year. For example, the 13-year brood and the 17-year brood will emerge simultaneously only once every $\text{lcm}(13, 17) = 13 \cdot 17 = 221$ years, a long time indeed! And it is *extremely* rare that all three broods emerge simultaneously (see Exercise 9).

One theory is that this decreases the probability of interbreeding between different broods. Thus, there will not be much gene mixing, and the lengths of the life cycles of the separate broods will be preserved. Another theory holds that it is advantageous for the broods to emerge individually because then the large numbers of cicadas in a given brood do not have to compete with those in another brood for limited quantities of food.

A separate theory is based on the fact that the life cycles of many parasites and poisonous fungi are 2, 3, 4, or 6 years. Since these smaller numbers have no common divisors with 7, 13, and 17, it is relatively rare that the parasites and fungi threaten a given brood of periodical cicadas (see Exercise 10). The same advantage would be achieved if a brood of periodical cicadas were to have a non-prime life cycle that is relatively prime to 2, 3, 4, and 6 years. However, the smallest composite number that is relatively prime to both 2 and 3 is 25, so to achieve this advantage with a non-prime life cycle, a brood would have to have an extraordinarily long life cycle—all known broods of cicadas have life cycles less than 25 years.

Irrational numbers and the FTA

The Fundamental Theorem of Arithmetic can be used to prove that certain numbers are irrational. We begin by considering the square roots of positive integers. In Chapter 1, we proved that $\sqrt{2}$ is irrational. Are all of the square roots $\sqrt{2}, \sqrt{3}, \sqrt{4}, \sqrt{5}, \ldots$ irrational? Certainly we cannot make that claim, since $\sqrt{4} = 2$ is a rational number. Similarly, the square root of any other perfect square will be an integer, and thus a rational number. However, the square roots of all other natural numbers are irrational, as we now prove.

THEOREM 6.2.7

Let n be a natural number that is not a perfect square. Then $\sqrt{n}$ is irrational.

To prove this theorem, it will be useful to have the following lemma, whose proof relies on the FTA.

LEMMA 6.2.8

Let r and s be natural numbers. If $s^2 \mid r^2$ then $s \mid r$.

PROOF OF LEMMA 6.2.8

Let $r, s \in \mathbf{N}$ such that $s^2 \mid r^2$.

[To show: $s \mid r$.]

By the FTA (6.1.1), r and s have prime factorizations:

$$r = p_1^{a_1} \cdot \; \cdots \; \cdot p_k^{a_k} \qquad \text{and} \qquad s = p_1^{b_1} \cdot \; \cdots \; \cdot p_k^{b_k}$$

where for every $i = 1, \ldots, k$, the number p_i is prime, and a_i and b_i are nonnegative integers. Squaring these equations yields the prime factorizations of r^2 and s^2:

$$r^2 = p_1^{2a_1} \cdot \; \cdots \; \cdot p_k^{2a_k} \qquad \text{and} \qquad s^2 = p_1^{2b_1} \cdot \; \cdots \; \cdot p_k^{2b_k}$$

Since $s^2 \mid r^2$, it follows from Proposition 6.2.1 that

$$0 \le 2b_i \le 2a_i \quad \text{for all } i = 1, \ldots, k.$$

Hence,

$$0 \le b_i \le a_i.$$

Now we may apply Proposition 6.2.1 again to conclude that $s \mid r$, as was to be shown. ●

We are now ready to prove Theorem 6.2.7.

PROOF of **THEOREM 6.2.7**

Let $n \in \mathbf{N}$ such that n is not a perfect square.

[We must show that $\sqrt{n}$ is irrational. We will prove this by contradiction.]

Assumption: Suppose that $\sqrt{n}$ is rational.
   Then there exist $r, s \in \mathbf{N}$, with $s \ne 0$, such that

$$\sqrt{n} = \frac{r}{s}. \tag{5}$$

Squaring both sides and clearing denominators yields

$$s^2 \cdot n = r^2.$$

Hence, $s^2 \mid r^2$, and so by Lemma 6.2.8, $s \mid r$. This means that $\frac{r}{s}$ is an integer. Thus, $\sqrt{n}$ is an integer by equation (5). This contradicts the hypothesis that n is not a perfect square. $\Rightarrow\Leftarrow$
   So, our Assumption must be false, and we conclude that $\sqrt{n}$ is irrational. ●

We can also use the FTA to prove that certain logarithms are irrational.

$\log_2 3$ *is irrational.*

PROOF (By contradiction.)

Assumption: Suppose $\log_2 3$ is rational.

Then there exist $r, s \in \mathbf{Z}$, with $s \neq 0$, such that

$$\log_2 3 = \frac{r}{s}.$$

Since $\log_2 3$ is positive, we may assume r and s are natural numbers. Using the definition of logarithm, this equation is equivalent to

$$3 = 2^{r/s}.$$

Hence,

$$3^s = 2^r.$$

This equation says that a particular number can be expressed both as a power of 3 and as a power of 2, which contradicts the uniqueness part of the FTA (6.1.1). ⇒⇐

We conclude that $\log_2 3$ is irrational.

Fun Facts

As asserted in the proof of Proposition 6.2.9, the FTA implies that a power of 2 cannot equal a power of 3: the equality $3^s = 2^r$ is impossible for natural numbers r and s.

This fact has an interesting corollary: It is impossible to tune a piano! More precisely, it is impossible to tune a piano so that all of the octaves and fifths are in their acoustically correct frequency ratios (2:1 for octaves and 3:2 for fifths). For more details, see Section 14.3.

EXERCISES 6.2

Numerical Problems

1. How many positive divisors does each number have?

    a. 30   **b.** 16   **c.** 63   **d.** 210

    e. 100   **f.** 101^5 **g.** $5^{12} \cdot 7^9$ **h.** 10^{100}

2. List all of the positive divisors of each of the following numbers:

    a. 36   **b.** 93   **c.** 756

3. Let $m = 5^3 \cdot 7^{12} \cdot 13 \cdot 17^2$ and $n = 2^9 \cdot 3^2 \cdot 5^4 \cdot 7^6 \cdot 13^3 \cdot 29$.

   a. Find $\gcd(m, n)$.   b. Find $\text{lcm}(m, n)$.

4. Recall that $2^{10} = 1024$.

   a. Find $\gcd(1024, 1000000)$.   b. Find $\text{lcm}(1024, 1000000)$.

5. Find three numbers that each have exactly 2 positive divisors.

6. Find three numbers that each have exactly 3 positive divisors.

7. Find three numbers that each have exactly 4 positive divisors.

8. a. Find all pairs of natural numbers (a, b) with $a < b$ such that $\text{lcm}(a, b) = 20$.

   b. Find all pairs of natural numbers (a, b) with $a < b$ such that $\text{lcm}(a, b) = 30$.

9. In this exercise, you will explore the assertion that by having prime life spans, the various broods of periodical cicadas minimize the frequency of all broods emerging simultaneously.

   a. How often will the 7-year brood and the 13-year brood of periodical cicadas emerge simultaneously?

   b. How often will the 7-year brood and the 17-year brood of cicadas emerge simultaneously?

   c. How often will all three of the 7-year brood, the 13-year brood, and the 17-year brood emerge simultaneously?

   d. Suppose that the country of Septania has three broods of periodical cicadas: a 7-year brood, a 14-year brood, and a 21-year brood. After a year in which all three broods emerge simultaneously in Septania, how many years will pass before all three broods again emerge simultaneously?

10. Suppose that the parasitic fungus *mythohexenniaspora* emerges every 6 years, sometimes in the same year as an n-year brood of periodical cicadas. For each given value of the cicada brood's life span, n, determine in what fraction of the years that the brood emerges it will be plagued by the fungus.

    a. $n = 7$ years   b. $n = 8$ years   c. $n = 9$ years

    d. $n = 10$ years   e. $n = 11$ years   f. $n = 12$ years

    g. Explain why it is advantageous for a cicada brood to have a life span that is relatively prime to the period of the fungus.

Reasoning and Proofs

11. Two natural numbers, a and b, satisfy $\gcd(a, b) = 5$ and $\text{lcm}(a, b) = 540$. Find all possible pairs of natural numbers satisfying these conditions.

12. Two natural numbers, a and b, satisfy $\gcd(a, b) = 2$ and $\text{lcm}(a, b) = 600$. Find all possible pairs of natural numbers satisfying these conditions.

13. Prove that $\log_2 10$ is irrational.

14. Prove that $\log_5 100$ is irrational.

15. Prove that $\sqrt[3]{5}$ is irrational, using the ideas from this section.

16. a. Of all natural numbers less than 100, which has the greatest number of factors?

    b. Of all natural numbers less than 1000, which has the greatest number of factors?

    c. Of all natural numbers less than 1,000,000, which has the greatest number of factors?

17. Find natural numbers a, b, and c such that $\gcd(a, b) = 30$ and $\gcd(b, c) = 24$.

18. Find natural numbers a, b, and c such that $\gcd(a, b) = 30$, $\gcd(b, c) = 24$, and $\gcd(a, c) = 54$.

19. Can you find natural numbers a, b, and c such that $\gcd(a, b) = 30$, $\gcd(b, c) = 24$, and $\gcd(a, c) = 84$? Prove that your answer is correct.

20. Prove Proposition 6.2.4. [*Hint:* Consider a common multiple of m and n, and apply Proposition 6.2.1.]

21. Let $a, b, n \in \mathbf{N}$ such that $a^n \mid b^n$. Prove that $a \mid b$.

22. Let a, b, and c be natural numbers. Suppose that $a \mid c$, $b \mid c$, and $\gcd(a, b) = 1$. Prove that $ab \mid c$.

23. Prove that if $\gcd(a, b) = 1$, then $\gcd(a, bc) = \gcd(a, c)$ for any integer c.

24. The nth harmonic number, H_n, is defined by $H_n = 1 + \dfrac{1}{2} + \dfrac{1}{3} + \cdots + \dfrac{1}{n}$. Show that for all $n > 1$, H_n is not an integer. [*Hint:* Let $L = \text{lcm}(1, 2, \ldots, n)$. Show that $L \cdot H_n$ is an odd integer.]

Advanced Reasoning and Proofs

25. Prove that if m and n are natural numbers, then $\sqrt[m]{n}$ is irrational provided that n is not a perfect mth power. (That is, provided that $\sqrt[m]{n}$ is not an integer.)

26. Prove that if p is a prime number, then $\log_p n$ is irrational provided that n is not a power of p. (That is, provided that $\log_p n$ is not an integer.)

27. Let $c_0, \ldots, c_{n-1} \in \mathbf{Z}$, and suppose x is a real number such that
$$x^n + c_{n-1}x^{n-1} + \cdots + c_1 x + c_0 = 0.$$
Prove that either x is an integer or x is irrational.

28. What is the smallest natural number that is divisible by exactly 28 natural numbers?

29. Let $n \in \mathbf{N}$ be a perfect square. Suppose that $n = ab$, where $a, b \in \mathbf{N}$ such that $\gcd(a, b) = 1$. Prove that a and b are both perfect squares.

30. Generalize Proposition 6.2.3 to three numbers. That is, prove the following:

PROPOSITION. *Let $m, n, k \in \mathbf{N}$ have prime factorizations*

$$m = p_1^{a_1} \cdot \cdots \cdot p_r^{a_r}, \qquad n = p_1^{b_1} \cdot \cdots \cdot p_r^{b_r}, \qquad and \qquad k = p_1^{c_1} \cdot \cdots \cdot p_r^{c_r},$$

with $a_i, b_i, c_i \geq 0$ for every $i = 1, \ldots, r$. Then $\gcd(m, n, k) = p_1^{d_1} \cdot \cdots \cdot p_r^{d_r}$, where $d_i = \min(a_i, b_i, c_i)$.

31. Generalize Proposition 6.2.4 to three numbers. That is, prove the following:

PROPOSITION. *Let $m, n, k \in \mathbf{N}$ have prime factorizations*

$$m = p_1^{a_1} \cdot \cdots \cdot p_r^{a_r}, \qquad n = p_1^{b_1} \cdot \cdots \cdot p_r^{b_r}, \qquad and \qquad k = p_1^{c_1} \cdot \cdots \cdot p_r^{c_r},$$

with $a_i, b_i, c_i \geq 0$ for every $i = 1, \ldots, r$. Then $\mathrm{lcm}(m, n, k) = p_1^{d_1} \cdot \cdots \cdot p_r^{d_r}$, where $d_i = \max(a_i, b_i, c_i)$.

In Exercises 32–35, fill in the blank to make the statement true. Then prove the statement.

32. For all $a, b, k \in \mathbf{N}$, $\mathrm{lcm}(\gcd(k, a), \gcd(k, b)) = \gcd(k, \underline{\hspace{1.5cm}})$.

33. For all $a, b, c \in \mathbf{N}$, $\gcd(a, b, c) \cdot \mathrm{lcm}(ab, bc, ac) = \underline{\hspace{1.5cm}}$.

34. For all $a, b, c \in \mathbf{N}$, $\gcd(\gcd(a, b), c) = \underline{\hspace{1.5cm}}$.

35. For all $a, b, c \in \mathbf{N}$, $\mathrm{lcm}(\mathrm{lcm}(a, b), c) = \underline{\hspace{1.5cm}}$.

Chapter 7

Carl Friedrich Gauss

(1777–1855)

Considered to be one of the greatest mathematicians of all time, Carl Friedrich Gauss made important discoveries in a wide range of fields, including pure and applied mathematics, statistics, physics, and astronomy. Due in large part to Gauss's work, Germany became the mathematical leader of the Western world. In recognition of Gauss's importance in the history of Germany, his picture appeared on the 10 Deutschmark bill, shown below.

From a very young age, Gauss demonstrated an unusual aptitude for mathematics as well as an impressive memory. Gauss mastered the rules of arithmetic at the age of three. When Gauss was 14, the Duke of Brunswick began paying him a stipend to encourage him in his studies. At the age of 18, Gauss went to the University of Göttingen to study mathematics and classical languages. As a student, he began to keep a diary of his mathematical and scientific discoveries. The diary, which is only 19 pages long, contains a total of 146 discoveries that Gauss made between the ages of 18 and 37. The first entry in his diary was a proof that a regular 17-gon

The 10-Deutschmark bill featured a picture of Gauss, as well as a graph of the Gaussian distribution and even its equation! (The Deutschmark was the official currency of Germany before the introduction of the euro in 2002.)

could be constructed using only a compass and straightedge (see "A Bit of Gauss's Math"). When Gauss's professor, Abraham Kaestner, would not believe his proof, Gauss described Kaestner to his friends as "the leading mathematician among poets and leading poet among mathematicians." Gauss stayed at Göttingen for only three years and then returned home without a degree. He eventually completed his degree in absentia at the University of Helmstedt.

When Gauss was only 24 years old, he published an extraordinary book entitled *Disquisitiones Arithmeticae*, in which he explained all of the major results in number theory that were known at the time together with many important new results of his own. Mathematicians consider the publication of the *Disquisitiones* as the birth of modern number theory. In particular, the concept of congruence modulo n (the topic of this chapter) and the corresponding notation $a \equiv b \pmod{n}$ were developed by Gauss and explained in the *Disquisitiones*. This concept revolutionized the way that mathematicians thought about number theory. As a result of the *Disquisitiones*, Gauss immediately gained fame throughout the European mathematical community.

In spite of his great success in mathematics, Gauss's personal life was not easy. In 1805, he married his childhood sweetheart, Johanna Osthoff. When she moved into his apartment, she described the accommodations as "shabby, filthy rooms, a smoky and drafty kitchen, ancient and phlegmatic landlord and landlady." Nonetheless, Gauss and Johanna lived happily together for four years. After the death of Gauss's father in 1808, his mother moved in with them and their two children. In 1809, Johanna died after the birth of their third child, and this new baby died several months later. Gauss was left with

two young children and a blind mother to care for. Gauss remarried and had three more children within the next few years. However, his second wife, Minna, became bedridden after the birth of her third child and eventually died in 1831. In spite of all the hardships he faced, though, Gauss's work continued unabated.

Gauss was supported financially by the Duke of Brunswick from 1791 until the duke died in 1806. In 1807, Gauss became a professor of astronomy and director of the observatory at the University of Göttingen, and he remained in this position for most of his life. The position suited him well, since Gauss had a strong interest in both astronomy and physics. While on the faculty at Göttingen, Gauss began doing research on magnetism and the magnetic force of the Earth. In recognition of Gauss's work on magnetism, the word *Gauss* now refers to a unit of magnetic field strength. (The strength of the Earth's magnetic field, which causes compasses to point to magnetic north, is approximately 0.5 Gauss.) Together with his colleague Wilhelm Weber, Gauss built the first electromagnetic telegraph, which enabled them to communicate over a distance of 5000 feet via a pair of wires.

In total, Gauss's collected works amount to more than 300 papers. However, most of this work was not published during his lifetime, and some of his unpublished results were independently rediscovered and published by his contemporaries. This led to conflicts over who discovered a particular result first. For example, Gauss did not publish his method of least squares (see "A Bit of Gauss's Math"), and Legendre discovered the method himself ten years later. When Gauss claimed priority, Legendre accused Gauss of plagiarism.

Though there was tension over priority between Gauss and some of his contemporaries, nobody doubted his brilliance or the phenomenal importance of his work. In fact, people were so impressed with his intelligence and productivity that when he died, his brain was weighed and measured to determine whether it was unusually large or heavy. Although his brain was found to be of ordinary size and weight, it was preserved by the Department of Physiology at Göttingen, where it remains to this day.

After Gauss's death, a friend wrote, "As he was in his youth, so he remained through his old age until his dying day, the unaffectedly simple Gauss. A small study, a little work table with a green cover, a standing-desk painted white, a narrow sofa and, after his seventieth year, an armchair, a shaded lamp, an unheated bedroom, plain food, a dressing gown and velvet cap, these were becomingly all his needs."

A bit of Gauss's math

The first entry in Gauss's diary, from when he was 18 years old, concerned straightedge and compass constructions of polygons. The ancient Greeks were able to construct regular polygons with 3, 4, 5, 6, 8, 10, 12, 15, and 16 sides. However, it was not known whether the regular 17-gon could be constructed using only a compass and straightedge. In 1796, Gauss proved in his diary that the 17-gon could indeed be constructed with only these tools. In fact, he proved that for any number n that is a Fermat prime or a product of distinct Fermat primes, a regular n-gon can be constructed. (A Fermat prime is a prime that has the form $2^{2^k} + 1$; see Section 3.3 for a discussion of Fermat numbers.) As a result of this discovery, Gauss decided to pursue mathematics as a career.

In his Ph.D. dissertation, Gauss proved the *Fundamental Theorem of Algebra*, which states that every nonconstant polynomial with real or complex coefficients has at least one root. Several earlier mathematicians had stated this theorem but were unable to prove it. Over the course of his lifetime, Gauss actually gave four distinct proofs of this important result.

Gauss's research was not restricted to number theory or even to pure mathematics. In his quest to apply mathematics to the study of the natural world, Gauss developed the method of *least squares* as a way of finding the line that best approximates a collection of data points. This led him to observe that the errors of an experiment tend to occur in a particular bell-shaped probability distribution, which is now known as a *Gaussian distribution* or *normal distribution*. (A graph of this distribution and its equation appear on the 10-Deutschmark note pictured earlier.) As you may have learned in statistics, a Gaussian distribution is completely determined by two parameters: its mean and standard deviation (denoted by μ and σ on the 10-Deutschmark bill).

Gauss's first result in astronomy was his calculation of the orbit of the recently discovered asteroid, Ceres, based on data from only 9° of the asteroid's orbit. Gauss developed general methods to estimate the orbit of a planet or comet based on only three observations and to predict how the gravitational pull of the sun and other planets would affect a particular orbit. Using his method of least squares, Gauss was able to calculate orbits significantly faster and more accurately than his predecessors.

Chapter 7 Modular Arithmetic
or Interplanetary Math

7.1 Congruence Modulo *n*

Gauss's mathematical journey

Carl Friedrich Gauss made important contributions to many fields of mathematics and science, including number theory, statistics, optics, and electromagnetism. Much of his work was motivated by his lifelong interest in astronomy. What's not so well known about Gauss is that he was abducted by aliens.

Gauss's adventure began one evening in his backyard when he was observing the planets through his telescope. While taking his nightly measurements of the position of Venus, he was startled by a cold, bony finger tapping on his shoulder.

"Would you like to visit my planet?" asked a strange blue creature.

It turned out that Gauss had no choice in the matter. Powerful sedatives were already at work in his bloodstream, and the great mathematician fell into a long, deep sleep. Gauss awoke to the ugliest sunrise he had ever seen.

"Hi, my name is Milo. Welcome to my planet." After brief introductions, the blue alien gave Gauss a full tour of Venus. Gauss was surprised to find that life on Venus was similar in many respects to life in Germany—the locals worked hard at high-paying technical jobs, the beer was good, and all of the hovercraft ran on schedule.

Due to Gauss's personal magnetism, the Venusians were drawn to him and welcomed him into their society. Before long, however, Gauss became homesick. He felt stifled by the hot, hazy climate. The pace of life was hectic—it seemed like every day there was a year's worth of work to get done. It was not a healthy atmosphere for Gauss. He just couldn't take the pressure.

To make matters worse, the cost of living on Venus was sky-high; Gauss's air conditioning bills alone were astronomical. Living in this wealthy, technologically advanced society made Gauss feel like he was from the third world.

Despite these difficulties, Gauss remained on Venus because he was fascinated by the Venusians' unusual counting system. Once they got past five, they would start over again at zero:

0, 1, 2, 3, 4, 5, 0, 1, 2, 3, 4, 5, 0, 1, 2, ...

Gauss soon realized that on Venus, there were only six numbers. If Gauss pointed to six objects, the Venusians would count them and say there were 0. If Gauss pointed to seven objects, the Venusians would count them as 1, and so on.

Gauss kept a diary containing his notes about Venusian numbers and the Venusian economy. His diary was published back on Earth and earned high marks from the German financial press. Gauss's valuable analysis quickly gained currency, and before long, copies of Gauss's notes were circulating throughout all of Germany.

A Venusian who has 17 snozberries would say she has **5** snozberries. This is because 17 leaves a remainder of **5** when divided by 6. To convert from one of our numbers to the corresponding Venusian number, we just take the remainder after division by 6.

A Venusian who has 29 snozberries would also say she has **5** snozberries, because 29 leaves a remainder of **5** when divided by 6. The Earth numbers 17 and 29 are both represented by the same Venusian number, because 17 and 29 leave the same remainder when divided by 6. Such numbers are said to be *congruent modulo* 6.

There is another way to look at the idea of congruence modulo 6. Notice that the numbers 17 and 29 are separated by a multiple of 6. That is, the difference $29 - 17 = 12$ is divisible by 6. Using this observation, we are now ready to define congruence modulo 6 and, in fact, congruence modulo any natural number n.

DEFINITION 7.1.1

Let $a, b \in \mathbf{Z}$ and let $n \in \mathbf{N}$. We say that a is **congruent** *to b* **modulo** *n if $n \mid (a - b)$. In symbols, we write this as $a \equiv b \pmod{n}$.*

In the congruence $a \equiv b \pmod{n}$, the number n is called the **modulus**.[1]

EXAMPLE 1

    a. Are 34 and 54 congruent modulo 10?

    b. Are 48 and 25 congruent modulo 11?

    c. Are 15 and 3 congruent modulo 12?

SOLUTION

    a. Since $34 - 54 = -20$ is a multiple of 10, we know that $10 \mid (34 - 54)$. Thus, $34 \equiv 54 \pmod{10}$.

    b. We want to determine whether 11 divides evenly into $(48 - 25) = 23$. Since $11 \nmid 23$, we see that $48 \not\equiv 25 \pmod{11}$.

    c. Since $15 - 3 = 12$, we know that $12 \mid (15 - 3)$. Thus, $15 \equiv 3 \pmod{12}$. ●

Counting modulo 12 is cyclical just like the time of day is cyclical: Starting with 12 o'clock, we count the hours 1 o'clock, 2 o'clock, and so forth. After 11 o'clock, the cycle repeats. Every 12 hours, we return to where we started. We can visualize the numbers modulo 12 using a clock, such as the one pictured here. As we saw in part c of Example 1 above, $15 \equiv 3 \pmod{12}$, which corresponds to the fact that 15 and 3 represent the same location on the clock face.

---

[1] When we use the words *modulus* and *modulo*, we are actually speaking Latin. The words *modulus* and *modulo* are two different cases of the same noun. *Modulus* is in the nominative case (used as the subject of a sentence), whereas *modulo* is in the ablative case of the same noun (often used for the *means* by which an action is accomplished.) Thus, when we say that two numbers are congruent *modulo n*, we are saying that the numbers are congruent *by means of the modulus n*.

EXAMPLE 2

List a few numbers that are congruent to 5 modulo 7.

SOLUTION By definition, an integer a will satisfy

$$a \equiv 5 \pmod 7 \tag{1}$$

if and only if

$$7 \mid (a - 5). \tag{2}$$

By definition of divides, this happens exactly when there exists $k \in \mathbf{Z}$ such that $a - 5 = 7k$. In other words, this happens if and only if

$$\text{there exists } k \in \mathbf{Z} \text{ such that } a = 5 + 7k. \tag{3}$$

Thus, a number a will be congruent to 5 modulo 7 exactly when a has the form $5 + 7k$ for some integer k. The numbers of this form are

$$\ldots, -23, -16, -9, -2, 5, 12, 19, 26, \ldots. \qquad \blacksquare$$

In Example 2, we found three different ways, (1), (2), and (3), to express the congruence $a \equiv 5 \pmod 7$. In general, the relationship $a \equiv b \pmod n$ can be expressed in several ways. Four of these statements are written below in the Equivalent Conditions for Congruence (7.1.2). This theorem allows us to use these statements interchangeably.

THEOREM 7.1.2 EQUIVALENT CONDITIONS FOR CONGRUENCE

Let $a, b \in \mathbf{Z}, n \in \mathbf{N}$. The following statements are equivalent:

$$a \equiv b \pmod n \quad \Leftrightarrow \quad n \mid (a - b) \quad \Leftrightarrow \quad \exists k \in \mathbf{Z} \textit{ such that } a = b + kn$$

$\Leftrightarrow$ *when a and b are divided by n, they leave the same remainder.*

PROOF You will prove this theorem in Exercise 11.

Reducing a number modulo *n*

If you ever visit Venus as a tourist, you must be careful to respect the local modulo 6 customs. If you find yourself in possession of 26 snozberries, you should refer to them as 2 snozberries, because 2 is the remainder when 26 is divided by 6. It's fine to walk around quietly thinking about your 26 snozberries, but you're well advised never to speak the number 26. Nobody will understand you, and besides, it's just not polite.

Proper etiquette demands that you *reduce* your numbers modulo 6. That is, you take the remainder after dividing by 6.

DEFINITION 7.1.3

Let $a, b \in \mathbf{Z}$, $n \in \mathbf{N}$. *To* **reduce** *a* **modulo** *n means to find the remainder when a is divided by n. This remainder is called the* **reduction** *of a* **modulo** *n or the* **value** *of a* **modulo** *n.*

EXAMPLE 3

   a. Reduce 41 modulo 15.

   b. Reduce $6 + 5$ modulo 7.

SOLUTION

   a. Divide 41 by 15:

$$41 = 2 \cdot 15 + 11$$

     11 is the reduction of 41 modulo 15.

   b. We know from standard integer arithmetic that $6 + 5 = 11$.

     Divide 11 by 7:

$$11 = 1 \cdot 7 + 4$$

     4 is the reduction of $6 + 5$ modulo 7.

In part a of Example 3, you were asked to reduce 41 modulo 15, and you obtained the answer 11. This number 11 has two special properties:

   (*i*)   11 is congruent to the original number 41 modulo 15, and

   (*ii*)   11 is in the range $0, 1, 2, \ldots, 14$.

In fact, 11 is the only integer with both of these properties. The following two lemmas state that this is true in general.

LEMMA 7.1.4

Given an integer a and a natural number n, if r is the reduction of a modulo n then

   (*i*) $a \equiv r \pmod{n}$;

   (*ii*)   *r is one of the integers* $0, 1, 2, \ldots, n - 1$.

PROOF To see why property (*i*) is true, note that since r is the remainder when a is divided by n, we have

$$a = qn + r,$$

for some $q \in \mathbf{Z}$. By the Equivalent Conditions for Congruence (7.1.2), it follows that

$$a \equiv r \pmod{n}.$$

Property (*ii*) follows from the Division Theorem (3.5.1) because r is the remainder when a is divided by n, and hence, $0 \le r < n$. ∎

LEMMA 7.1.5

Let $n \in \mathbf{N}$. Given any $a \in \mathbf{Z}$, a is congruent modulo n to exactly one of the numbers $0, 1, 2, \dots, n - 1$.

PROOF Let $n \in \mathbf{N}$ and $a \in \mathbf{Z}$.

First, we need to prove that a is congruent modulo n to at least one of the numbers $0, 1, 2, \dots, n - 1$. This is exactly the content of Lemma 7.1.4. Let r denote the reduction of a modulo n. Then $a \equiv r \pmod{n}$, and r is in the desired range $0, 1, 2, \dots, n - 1$.

Next, we must show that a cannot be congruent modulo n to more than one number in the list

$$0, 1, 2, \dots, n - 1.$$

To do this, suppose that

$$a \equiv r \pmod{n} \quad \text{and} \quad a \equiv r' \pmod{n}$$

with r and r' each in the range $0, 1, 2, \dots, n - 1$.

[To show: $r = r'$.]

By the Equivalent Conditions for Congruence (7.1.2), there exist integers q and q' such that

$$a = qb + r \quad \text{and} \quad a = q'b + r'.$$

By the uniqueness assertion of the Division Theorem (3.5.1), it follows that $r = r'$. ∎

Congruence is an equivalence relation

Congruence modulo n is a special type of relation, known as an *equivalence relation*.

THEOREM 7.1.6

Let $n \in \mathbf{N}$. Congruence modulo n is an equivalence relation. That is, the following three properties are satisfied:

(*i*)   *Reflexive Property*

   For all $a \in \mathbf{Z}$, $a \equiv a \pmod{n}$.

(*ii*) Symmetric Property

   For all $a, b \in \mathbf{Z}$, if $a \equiv b \pmod{n}$, then $b \equiv a \pmod{n}$.

(*iii*) Transitive Property

   For all $a, b, c \in \mathbf{Z}$, if $a \equiv b \pmod{n}$ and $b \equiv c \pmod{n}$, then $a \equiv c \pmod{n}$.

PROOF

(*i*) Proof of the Reflexive Property
   Let $n \in \mathbf{N}, a \in \mathbf{Z}$.

   Since $n \mid 0$,

$$n \mid a - a.$$

   Thus, by the definition of congruence modulo n,

$$a \equiv a \pmod{n}.$$

(*ii*) Symmetric Property

(*iii*) Transitive Property

In Exercises 12–13, you will prove that congruence modulo n satisfies these properties.

   You will explore other examples of equivalence relations, as well as general properties of equivalence relations, in the exercises.

Congruence classes

In Example 2, we asked which integers are congruent to 5 modulo 7. We found that the numbers with this property are in the set

$$\{\ldots, -23, -16, -9, -2, 5, 12, 19, 26, \ldots\}.$$

This set is called the *congruence class* of 5 modulo 7.

DEFINITION 7.1.7

*Let $n \in \mathbf{N}$, and let $b \in \mathbf{Z}$. Then the **congruence class** of b modulo n is the set of all integers congruent to b modulo n:*

$$\{x \in \mathbf{Z} \mid x \equiv b \pmod{n}\}. \tag{4}$$

The congruence class of b modulo n may also be defined as

$$\{b + kn \mid k \in \mathbf{Z}\}.$$

This set is equal to the set (4) by the Equivalent Conditions for Congruence (7.1.2). We may also think of this congruence class as the set of all integers that are b more than a multiple of n.

EXAMPLE 4
. . . .

The congruence class of 2 modulo 3, the set of integers congruent to 2 modulo 3, is

$$\{2 + 3k \mid k \in \mathbf{Z}\} = \{\ldots, -4, -1, 2, 5, 8, 11, \ldots\}.$$

This is the set of all numbers that are 2 more than a multiple of 3.

The congruence class of 2 modulo 3 contains all numbers of the form $2 + 3k$. What do the other congruence classes modulo 3 look like? The congruence classes of 0, 1, and 2 are listed here:

Congruence class of 0 modulo 3 = $\{0 + 3k \mid k \in \mathbf{Z}\} = \{\ldots, -6, -3, 0, 3, 6, 9, \ldots\}$

Congruence class of 1 modulo 3 = $\{1 + 3k \mid k \in \mathbf{Z}\} = \{\ldots, -5, -2, 1, 4, 7, 10, \ldots\}$

Congruence class of 2 modulo 3 = $\{2 + 3k \mid k \in \mathbf{Z}\} = \{\ldots, -4, -1, 2, 5, 8, 11, \ldots\}$

Lemma 7.1.5 guarantees that every integer is congruent modulo 3 to exactly one of the numbers 0, 1, or 2. Thus, every integer appears in exactly one of the three sets listed above. This can be expressed by saying that these three congruence classes *partition* the integers.

Naomi's Numerical Proof Preview: Theorem 7.1.8

Above, we listed the congruence classes of 0, 1, and 2 modulo 3. Now let's examine the congruence class of a larger number modulo 3. What does the congruence class of 5 modulo 3 look like? By definition, this class is the set of numbers congruent to 5 modulo 3:

$$\{5 + 3k \mid k \in \mathbf{Z}\} = \{\ldots, -4, -1, 2, 5, 8, 11, \ldots\}.$$

Notice that this set, the congruence class of 5 modulo 3, is exactly the same as the congruence class of 2 modulo 3. This is not hard to explain: 5 and 2 are congruent modulo 3, so if a number x is congruent to 2 modulo 3, then x will also be congruent to 5 modulo 3 (by the transitive property from Theorem 7.1.6).

THEOREM 7.1.8

Let $n \in \mathbf{N}$, and let $a, b \in \mathbf{Z}$. Then the congruence class of a modulo n is equal to the congruence class of b modulo n if and only if $a \equiv b \pmod{n}$.

PROOF Let $n \in \mathbf{N}$, and let $a, b \in \mathbf{Z}$.
Let A and B denote the congruence classes of a and b modulo n:

$$A = \{x \in \mathbf{Z} \mid x \equiv a \;(\text{mod } n)\}, \qquad B = \{x \in \mathbf{Z} \mid x \equiv b \;(\text{mod } n)\}$$

[$\Rightarrow$] Suppose that $A = B$. [**To show: $a \equiv b$ (mod n).**]

Since $a \equiv a$ (mod n) by the reflexive property (from Theorem 7.1.6), we know that $a \in A$. But the sets A and B are equal by hypothesis, so $a \in B$. Thus, $a \equiv b$ (mod n).

[$\Leftarrow$] Now suppose that $a \equiv b$ (mod n). [**To show: $A = B$.**]

Let $x \in A$. Then $x \equiv a$ (mod n). Since $x \equiv a$ (mod n) and $a \equiv b$ (mod n), the transitive property (from Theorem 7.1.6) gives $x \equiv b$ (mod n). Hence, $x \in B$.

    We have now shown that every element of A is an element of B. A similar argument shows that every element of B is an element of A (see Exercise 17). Thus, the sets A and B are equal.   ■

    The property of congruence classes that was proven in Theorem 7.1.8 is a special case of a more general fact about equivalence relations, as you will show in Exercise 22. Theorem 7.18 has the following corollary.

COROLLARY 7.1.9

Let $n \in \mathbf{N}$. There are exactly n congruence classes modulo n:

$$\text{the class of } 0, \text{ the class of } 1, \ldots, \text{ the class of } n - 1. \tag{5}$$

PROOF We will first show that every congruence class modulo n is equal to one of the congruence classes in the list (5). Let $a \in \mathbf{Z}$. By Lemma 7.1.5, a is congruent modulo n to exactly one of the numbers in the list:

$$0, 1, 2, \ldots, n - 1.$$

Thus, by Theorem 7.1.8, the congruence class of a is equal to exactly one of the congruence classes in list (5).

    Now we will show that the congruence classes in list (5) are all distinct. To do this, consider two classes in list (5), the class of a and the class of b, where $0 \le a \le n - 1$ and $0 \le b \le n - 1$, and assume that the class of a equals the class of b.

[**To show: $a = b$.**]

Since the class of a equals the class of b, Theorem 7.1.8 tells us that $a \equiv b$ (mod n). So by Lemma 7.1.5, $a = b$. Thus, the classes in list (5) are all distinct.   ■

EXAMPLE 5

Congruence modulo 2 divides the integers into two congruence classes:

$$\text{Congruence class of 0 modulo 2} = \{\ldots, -6, -4, -2, 0, 2, 4, 6, \ldots\}$$
$$\text{Congruence class of 1 modulo 2} = \{\ldots, -5, -3, -1, 1, 3, 5, \ldots\}$$

The first congruence class contains all of the even integers, and the second contains all of the odd integers.

EXAMPLE 6

Congruence modulo 10 separates the integers into 10 congruence classes:

$$\text{Congruence class of 0 modulo 10} = \{\ldots, -20, -10, 0, 10, 20, \ldots\}$$
$$\text{Congruence class of 1 modulo 10} = \{\ldots, -19, -9, 1, 11, 21, \ldots\}$$
$$\text{Congruence class of 2 modulo 10} = \{\ldots, -18, -8, 2, 12, 22 \ldots\}$$

$$\vdots$$

$$\text{Congruence class of 9 modulo 10} = \{\ldots, -11, -1, 9, 19, 29, \ldots\}.$$

One more useful fact about mods

We end this section with one final observation about congruences. If two numbers are congruent modulo 10, then they must also be congruent modulo 5. In general, if two numbers are congruent modulo n, then they must also be congruent modulo any divisor of n.

LEMMA 7.1.10

Let $a, b \in \mathbf{Z}$ and $m, n \in \mathbf{N}$ such that $m \mid n$. If $a \equiv b \pmod{n}$, then $a \equiv b \pmod{m}$.

PROOF You will prove this in Exercise 14.

EXERCISES 7.1

Numerical Problems

1. Show that the following congruence relationships hold:

   a. $17 \equiv -163 \pmod{45}$ **b.** $617 \equiv 1 \pmod{88}$

   c. $-17 \equiv -71 \pmod{9}$ **d.** $192 \equiv 0 \pmod{8}$

2. Determine which of the following congruence statements are true and which are false:

   a. $34 \equiv 144 \pmod{10}$

   b. $-1 \equiv 99 \pmod{8}$

   c. $12 \equiv 144 \pmod{7}$

   d. $-20 \equiv 20 \pmod{40}$

3. Perform each reduction.

   a. Reduce 36 modulo 13.

   b. Reduce 15 modulo 3.

   c. Reduce $5 \cdot 12$ modulo 7.

4. Congruence modulo 10 divides the integers into 10 congruence classes. Which negative integers are in the congruence class of 4 modulo 10?

5. Consider the congruence classes modulo 4. How many congruence classes are there? Which integers does each congruence class contain?

6. Consider the congruence classes modulo 1. How many congruence classes are there? Which integers does each congruence class contain?

7. List several integers x, some positive and some negative, for which the given congruence relation holds:

   a. $x \equiv 15 \pmod{20}$

   b. $x \equiv 0 \pmod{9}$

   c. $37 \equiv x \pmod{35}$

8. If possible, find a pair of integers x and y that satisfy the given condition. If it is not possible, explain why not.

   a. x and y are congruent modulo 3 but not modulo 6.

   b. x and y are congruent modulo 6 but not modulo 3.

   c. x and y are congruent modulo 5 but not modulo 10.

   d. x and y are congruent modulo 10 but not modulo 5.

9. Find all integers $m > 1$ for which the given congruence relation holds:

   a. $5 \equiv 75 \pmod{m}$

   b. $-1 \equiv 100 \pmod{m}$

   c. $-5 \equiv 15 \pmod{m}$

   d. $-46 \equiv 3 \pmod{m}$

Reasoning and Proofs

10. a. Let r be the reduction of an odd integer $\pmod 5$. List all possible values for r.

    b. Let r be the reduction of an odd integer $\pmod 6$. List all possible values for r.

11. Prove that the statements in the Equivalent Conditions for Congruence (7.1.2) are all equivalent.

In Exercises 12 and 13, you will finish proving Theorem 7.1.6.

12. Prove that congruence modulo n satisfies the symmetric property.

13. Prove that congruence modulo n satisfies the transitive property.

14. Prove Lemma 7.1.10.

15. Let $a, b \in \mathbf{Z}$ and $n \in \mathbf{N}$ such that $a \equiv b \pmod{n}$. Let k be a natural number. Prove that $ka \equiv kb \pmod{kn}$.

16. Prove that if $x \equiv a \pmod{n}$, then either $x \equiv a \pmod{2n}$ or $x \equiv a + n \pmod{2n}$.

17. Fill in the missing detail in the proof of Theorem 7.1.8 by proving that every element of B is an element of A. If you are careful, you will notice that the argument is slightly different than the argument that every element of A is an element of B. How is the symmetric property used in the proof?

18. Let $k \in \mathbf{Z}$ and $n \in \mathbf{N}$. Consider the set

$$S = \{k, k + 1, \ldots, k + n - 1\}.$$

Prove that every integer is congruent modulo n to some element of S.

EXPLORATION Equivalence Relations (Exercises 19–28)

Congruence modulo n and the congruence classes modulo n are examples of the more general concepts of *equivalence relations* and *equivalence classes*.

DEFINITION. *A relation,* $\sim$, *is an* **equivalence relation** *on the set S if the following three properties are satisfied:*

(i) *Reflexive Property*

   For all $a \in S$, $a \sim a$.

(ii) *Symmetric Property*

   For all $a, b \in S$, if $a \sim b$ then $b \sim a$.

(iii) *Transitive Property*

   For all $a, b, c \in S$, if $a \sim b$ and $b \sim c$, then $a \sim c$.

DEFINITION. *If $\sim$ is an equivalence relation on S and $a \in S$, then the set $\{b \in S \mid b \sim a\}$ is called the* **equivalence class** *of a.*

19. Determine which of the following are equivalence relations. If the relation is not an equivalence relation, explain which properties it fails to satisfy.

   a. The set $S = \mathbf{R}$, the real numbers, where the relation $x \sim y$ is defined by $x = y$.

b. The set $S = \mathbf{R}$, the real numbers, where the relation $x \sim y$ is defined by $x \leq y$.

c. The set S of all living people, where the relation $x \sim y$ holds whenever x has the same birth date as y.

d. The set $S = \mathbf{Z}$, the integers, where the relation $x \sim y$ is defined by $x \mid y$.

e. The set $S = \mathbf{R}$, the real numbers, where the relation $x \sim y$ is defined by $x^2 = y^2$.

f. The set $S = \mathbf{R}$, the real numbers, where the relation $x \sim y$ holds whenever $x - y$ is an integer.

g. The set $S = \mathbf{R}$, the real numbers, where the relation $x \sim y$ is defined by $|x| = |y| - 1$.

h. The set S of students in your number theory class, where the relation $x \sim y$ holds whenever x has the same number of siblings as y.

i. The set S of lines in the plane where the relation $l_1 \sim l_2$ holds whenever l_1 and l_2 are perpendicular.

j. The set S of all students at your institution, where the relation $x \sim y$ holds whenever x has ever had lunch with y.

k. The set $S = \mathbf{Z}$, the integers, where the relation $x \sim y$ is defined by $5 \mid x + y$.

l. The set $S =$ the nonzero real numbers, where the relation $x \sim y$ is defined by $xy > 0$.

20. For each relation in Exercise 19 that is an equivalence relation, describe the equivalence classes associated with that equivalence relation.

21. Give your own example of each of the following:

    a. A set S with a relation $\sim$ that is reflexive and symmetric, but not transitive.

    b. A set S with a relation $\sim$ that is reflexive and transitive, but not symmetric.

    c. A set S with a relation $\sim$ that is transitive and symmetric, but not reflexive.

22. Let $\sim$ be an equivalence relation on a set S. Show that for all $a, b \in S$, the equivalence class of a equals the equivalence class of b if and only if $a \sim b$. [*Hint:* Argue as in the proof of Theorem 7.1.8.]

23. Let $\sim$ be an equivalence relation on a set S. Show that the equivalence classes of $\sim$ partition the set S. That is, show that each element of S is in exactly one of the equivalence classes.

24. Let S be a set that is partitioned into nonempty subsets $P_1, P_2, \ldots, P_n$. (This means that every element of S is in exactly one of these subsets). Let the relation $\sim$ on the set S be defined by

$a \sim b$ whenever there is some P_i that contains both a and b.

Prove that $\sim$ is an equivalence relation.

25. In Exercise 24, we saw that one way to define an equivalence relation on a set S is to partition S into subsets. For example, if $S = \{1, 2, 3, 4, 5\}$, then the partition

$$P_1 = \{1, 3\}, P_2 = \{2, 4\}, P_3 = \{5\}$$

defines an equivalence relation, $\sim$, on S.

a. List all nine ordered pairs (a, b) for which $a \sim b$.

b. Define a new equivalence relation, $\approx$, on S, by partitioning S in a different way. List all ordered pairs (a, b) for which $a \approx b$.

26. List two different equivalence relations on the set $\{1, 2, 3, 4\}$.

27. List all possible equivalence relations on the set $\{1, 2, 3\}$.

28. Let $\sim$ and $\approx$ be any two equivalence relations on a set S.

a. Define a new relation, $\blacklozenge$, on S by

$x \blacklozenge y$ whenever both $x \sim y$ and $x \approx y$.

Must $\blacklozenge$ be an equivalence relation on S? Either prove that it must, or give a counterexample.

b. Define a new relation, $\spadesuit$, on S by

$x \spadesuit y$ whenever $x \sim y$ or $x \approx y$.

Must $\spadesuit$ be an equivalence relation on S? Either prove that it must, or give a counterexample.

7.2 Arithmetic with Congruences

As you've known since high school algebra, you can add two equations together to produce a new equation. Is the same thing true of two congruences?

For example, here are two valid congruences modulo 6:

$$40 \equiv 4 \ (\text{mod } 6)$$

$$21 \equiv 3 \ (\text{mod } 6)$$

If we add these two congruences, is the result true?

$$61 \stackrel{?}{\equiv} 7 \ (\text{mod } 6)$$

As the reader may verify, this is a valid congruence: 61 is in fact congruent to 7 modulo 6. Does this work in general?

The answer is yes. As the following theorem asserts, two congruences can be added or multiplied, provided that they have the same modulus.

THEOREM 7.2.1 CONGRUENCES ADD AND MULTIPLY

Let a, b, r, s $\in$ **Z** *and let n* $\in$ **N.**

If a $\equiv$ *b* (mod *n*) *and r* $\equiv$ *s* (mod *n*), *then*:

(*i*) $a + r \equiv b + s \ (\text{mod } n)$

(*ii*) $ar \equiv bs \ (\text{mod } n)$

PROOF

Proof of (*i*) You will do this part of the proof in Exercise 11.

Proof of (*ii*) By hypothesis, $a \equiv b$ (mod *n*) and $r \equiv s$ (mod *n*). By the definition of congruence modulo *n*,

$$n \mid a - b \quad \text{and} \quad n \mid r - s.$$

[To show: $ar \equiv bs \pmod{n}$, i.e., $n \mid ar - bs$.]

The trick behind this proof is to start with $ar - bs$ and add **0** in a clever way:

$$ar - bs = ar - as + as - bs.$$

Regrouping terms,

$$ar - bs = a(r - s) + s(a - b). \tag{1}$$

Since $n \mid r - s$ and $n \mid a - b$, the Linear Combination Lemma (3.1.3) allows us to conclude that n divides the right side of equation (1):

$$n \mid a(r - s) + s(a - b).$$

Since n divides the right side of equation (1), it also divides the left side:

$$n \mid ar - bs.$$

Thus, $ar \equiv bs \pmod{n}$. ∎

Suppose you know that $a \equiv b \pmod{n}$. Then by part (ii) of Theorem 7.2.1, you can multiply this congruence by itself to obtain the congruence

$$a^2 \equiv b^2 \pmod{n}.$$

Multiplying by $a \equiv b \pmod{n}$ again gives

$$a^3 \equiv b^3 \pmod{n}.$$

COROLLARY 7.2.2

Let $n, k \in \mathbf{N}$. If $a \equiv b \pmod{n}$ then $a^k \equiv b^k \pmod{n}$.

PROOF You will prove this in Exercise 12.

EXAMPLE 1

Reduce each expression modulo 10.

    a. $90987 + 7269 + 2341014 + 758776$

    b. $90987 \cdot 7269 \cdot 2341014 \cdot 758776$

SOLUTION

    a. First reduce each term modulo 10:

$90987 \equiv 7 \pmod{10}$, $7269 \equiv 9 \pmod{10}$, $2341014 \equiv 4 \pmod{10}$, $758776 \equiv 6 \pmod{10}$

By Theorem 7.2.1, these four congruences may be added:

$$90987 + 7269 + 2341019 + 758776 \equiv 7 + 9 + 4 + 6 \pmod{10}.$$

To evaluate $7 + 9 + 4 + 6$ modulo 10, the easiest way is to add successive terms, reducing as we go along:

$$
\begin{aligned}
90987 + 7269 + 2341014 + 758776 &\equiv 7 + 9 + 4 + 6 &&\leftarrow \textbf{by Theorem 7.2.1} \\
&\equiv 16 + 4 + 6 \\
&\equiv 6 + 4 + 6 &&\leftarrow \textbf{by Theorem 7.2.1} \\
&\equiv 10 + 6 \\
&\equiv 0 + 6 &&\leftarrow \textbf{by Theorem 7.2.1} \\
&\equiv 6 \pmod{10}.
\end{aligned}
$$

b. Reduce each factor modulo 10, then multiply successive factors, reducing as we go along:

$$
\begin{aligned}
90987 \cdot 7269 \cdot 2341014 \cdot 758776 &\equiv 7 \cdot 9 \cdot 4 \cdot 6 &&\leftarrow \textbf{by Theorem 7.2.1} \\
&\equiv 63 \cdot 4 \cdot 6 \\
&\equiv 3 \cdot 4 \cdot 6 &&\leftarrow \textbf{by Theorem 7.2.1} \\
&\equiv 12 \cdot 6 \\
&\equiv 2 \cdot 6 &&\leftarrow \textbf{by Theorem 7.2.1} \\
&\equiv 12 \\
&\equiv 2 \pmod{10}.
\end{aligned}
$$

Divisibility tests

Is 6435 divisible by 9? There is a simple trick to figure this out without actually performing the division. Simply add the digits:

$$6 + 4 + 3 + 5 = 18.$$

Since 18 is divisible by 9, you conclude that the original number, 6435, is also divisible by 9. You may already be familiar with this test, but do you know why it works? In this section, we use modular arithmetic to explain why this test works, showing that the test follows from a method for reducing any number modulo 9. We then derive methods for reducing modulo other numbers, such as 4 and 11, and formulate the corresponding divisibility tests.

Our familiar decimal (base 10) system expresses every number as a linear combination of powers of 10. For instance, the base 10 number **2935107** is a compact way to express the following sum:

$$\mathbf{2935107} = \mathbf{2} \cdot 10^6 + \mathbf{9} \cdot 10^5 + \mathbf{3} \cdot 10^4 + \mathbf{5} \cdot 10^3 + \mathbf{1} \cdot 10^2 + \mathbf{0} \cdot 10^1 + \mathbf{7} \cdot 10^0. \quad (2)$$

Analogous to equation (2), any n-digit natural number, d, whose digits are denoted $d_{n-1}d_{n-2}\ldots d_2 d_1 d_0$, can be written as follows:

$$d_{n-1}d_{n-2}\ldots d_2 d_1 d_0 = d_{n-1} \cdot 10^{n-1} + d_{n-2} \cdot 10^{n-2} + \cdots + d_2 \cdot 10^2 + d_1 \cdot 10^1 + d_0 \cdot 10^0$$

$$= \sum_{k=0}^{n-1} d_k \cdot 10^k.$$

Suppose we want to test whether a natural number, such as $d = \mathbf{2935107}$, is divisible by 9. In other words, we want to know whether $\mathbf{2935107} \equiv 0 \pmod 9$.

EXAMPLE 2

Reduce $d = \mathbf{2935107}$ modulo 9, and determine whether d is a multiple of 9.

SOLUTION Equation (2) expresses d as a linear combination of powers of 10. Thus, before we can reduce d modulo 9, it behooves us to evaluate the powers of 10 modulo 9. Since

$$\mathbf{10} \equiv \mathbf{1} \pmod 9,$$

it follows from Theorem 7.2.1 (Congruences Add and Multiply) that

$$10^2 = \mathbf{10} \cdot \mathbf{10}$$

$$\equiv \mathbf{1} \cdot \mathbf{1} \pmod 9$$

$$= \mathbf{1} \pmod 9.$$

Similarly, by Theorem 7.2.1,

$$10^3 = 10^2 \cdot \mathbf{10}$$

$$\equiv \mathbf{1} \cdot \mathbf{1} \pmod 9$$

$$\equiv 1 \pmod 9.$$

By similar reasoning, for any nonnegative integer n,

$$10^n \equiv 1 \pmod 9. \quad (3)$$

(You will prove this congruence rigorously in Exercise 13.) We can use this to reduce d modulo 9:

$$d = 2935107$$
$$= 2 \cdot 10^6 + 9 \cdot 10^5 + 3 \cdot 10^4 + 5 \cdot 10^3 + 1 \cdot 10^2 + 0 \cdot 10^1 + 7 \cdot 10^0$$
$$\equiv 2 \cdot 1 + 9 \cdot 1 + 3 \cdot 1 + 5 \cdot 1 + 1 \cdot 1 + 0 \cdot 1 + 7 \cdot 1 \pmod 9 \quad \leftarrow \textbf{by Theorem 7.2.1}$$
$$\equiv 2 + 9 + 3 + 5 + 1 + 0 + 7 \pmod 9$$
$$\equiv 27 \pmod 9.$$

At this point, we could just reduce 27 modulo 9 by inspection. For completeness, however, let's use the same method to reduce 27 modulo 9:

$$27 = 2 \cdot 10^1 + 7 \cdot 10^0$$
$$\equiv 2 \cdot 1 + 7 \cdot 1 \pmod 9$$
$$\equiv 2 + 7 \pmod 9$$
$$\equiv 9 \pmod 9$$
$$\equiv 0 \pmod 9.$$

Thus, $d \equiv 0 \pmod 9$. Hence, d is a multiple of 9. ∎

The method from Example 2 can be used to reduce any natural number, d, modulo 9. We can sum up the method from Example 2 as follows:

METHOD 7.2.3 REDUCING MODULO 9

To reduce the natural number $d = d_{n-1} \, d_{n-2} \ldots d_2 \, d_1 \, d_0$ modulo 9, sum up the digits:

$$d \equiv d_{n-1} + d_{n-2} + \cdots + d_2 + d_1 + d_0 \pmod 9.$$

Repeat the process, if necessary, until the final result is a single-digit number. If the result is a number in the range $0, 1, 2, \ldots, 8$, then this is the reduction of d modulo 9. If the result is 9, then $d \equiv 0 \pmod 9$.

Since d is divisible by 9 if and only if $d \equiv 0 \pmod 9$, Method 7.2.3 immediately leads to the standard test for divisibility by 9:

COROLLARY 7.2.4 TEST FOR DIVISIBILITY BY 9

A natural number is divisible by 9 if and only if the sum of its digits is divisible by 9.

As you may already know, there is a very similar test for divisibility by 3. You will state and justify this method in Exercise 20.

Looking back at Example 2, we see that the key behind Method 7.2.3 was to know the reductions of the powers of 10 (mod 9), which are given by equation (3). Since the powers of 10 modulo 9 are very simple (they are all congruent to 1), the method is particularly easy to use. To develop similar methods for reducing with other moduli, we need to evaluate the powers of 10 in these other moduli.

For instance, to discover a method for reducing modulo 11, consider the reductions of the first several powers of 10 (mod 11):

n	0	1	2	3	4	5	6	7
10^n **(mod 11)**	1	10	1	10	1	10	1	10

The general pattern (as shown in Exercise 13) is that for any nonnegative integer n,

$$10^n \equiv 1 \ (\text{mod } 11), \ \text{if } n \text{ is even}$$

$$10^n \equiv 10 \equiv -1 \ (\text{mod } 11), \ \text{if } n \text{ is odd.}$$

(4)

These congruences are quite useful for reducing numbers modulo 11, as we will now see.

EXAMPLE 3
· · · · · · · · · · · · · · · ·

Reduce $d = \mathbf{1829764}$ modulo 11.

SOLUTION Writing the digits in reverse order, we express d as a linear combination of the powers of 10. Then we use the congruences in (4) to reduce the powers of 10 modulo 11 and apply Theorem 7.2.1 to substitute either 1 or -1 for each power of 10.

$$\mathbf{1829764} = \mathbf{4} \cdot 10^0 + \mathbf{6} \cdot 10^1 + \mathbf{7} \cdot 10^2 + \mathbf{9} \cdot 10^3 + \mathbf{2} \cdot 10^4 + \mathbf{8} \cdot 10^5 + \mathbf{1} \cdot 10^6$$

$$\equiv \mathbf{4} \cdot 1 + \mathbf{6} \cdot (-1) + \mathbf{7} \cdot 1 + \mathbf{9} \cdot (-1) + \mathbf{2} \cdot 1 + \mathbf{8} \cdot (-1) + \mathbf{1} \cdot 1 \ (\text{mod } 11)$$

$$\equiv \mathbf{4} - \mathbf{6} + \mathbf{7} - \mathbf{9} + \mathbf{2} - \mathbf{8} + \mathbf{1} \ (\text{mod } 11)$$

$$\equiv -9 \ (\text{mod } 11)$$

$$\equiv 2 \ (\text{mod } 11).$$

Thus, 2 is the reduction of d (mod 11).

We can sum up the method from Example 3 as follows:

METHOD 7.2.5   REDUCING MODULO 11

To reduce the natural number $d = d_{n-1} \, d_{n-2} \cdots d_2 \, d_1 \, d_0$ modulo 11, write all of the digits in reverse order, and alternately add and subtract digits:

$$d \equiv d_0 - d_1 + d_2 - d_3 + d_4 - \cdots d_{n-1} \ (\text{mod } 11).$$

Reduce the result modulo 11 to get the final answer.

The purpose of writing the digits in reverse order is to make it easy to remember which digits should be added and which should be subtracted. If our only purpose is to determine whether a number is divisible by 11, however, we do not need to reverse the order, because it is okay if we get the signs completely reversed. The reason is that (as shown in Exercise 15)

$$d_0 - d_1 + d_2 - d_3 + d_4 - \cdots d_{n-1} \equiv 0 \ (\text{mod } 11)$$

if and only if (5)

$$- d_0 + d_1 - d_2 + d_3 - d_4 + \cdots d_{n-1} \equiv 0 \ (\text{mod } 11).$$

This leads to the following corollary.

COROLLARY 7.2.6 | TEST FOR DIVISIBILITY BY 11

Take a natural number d, and alternately add and subtract subsequent digits. The result is divisible by 11 if and only if d is divisible by 11.

To find a rule for reducing modulo 4, let's reduce the first several powers of 10 modulo 4:

n	0	1	2	3	4	5	6	7
$10^n \ (\text{mod } 4)$	1	2	0	0	0	0	0	0

It is evident from the table that

$$\text{for any integer } n \geq 2, \ 10^n \equiv 0 \ (\text{mod } 4). \tag{6}$$

(You will prove this congruence rigorously in Exercise 16.) This leads to the following method for reducing any natural number modulo 4, which you will justify in Exercise 17.

METHOD 7.2.7 | REDUCING MODULO 4

To reduce a natural number $d = d_{n-1} d_{n-2} \cdots d_2 d_1 d_0$ modulo 4, consider the two-digit number, e, formed by the last two digits of d:

$$e = d_1 d_0.$$

Then

$$d \equiv e \ (\text{mod } 4).$$

Reduce the two-digit number e modulo 4 to get the final answer.

The natural number d is divisible by 4 if and only if the two-digit number formed by the last two digits of d is divisible by 4.

EXAMPLE 4

Reduce $d = $ **37140871839394201257056** modulo 4.

SOLUTION The last two digits of d are **56**. Since $56 \equiv 0 \pmod 4$, we know that $d \equiv 0 \pmod 4$. In other words, d is divisible by 4. ∎

There is another version of the above method for reducing a number modulo 4, which you will explore in Exercise 18.

EXERCISES 7.2

Numerical Problems

1. Evaluate each of the two given numbers modulo 7. Then use your answers to reduce the sum of the two numbers modulo 7.

   a. 53 and 17   b. 29 and 103

   c. 99 and -32 d. 7,000,019 and 28,000,000,012

2. For each part of Exercise 1, reduce the product of the two given numbers modulo 7.

3. Use the methods of this section to reduce each number modulo 9.

   a. 681654   b. 688045

   c. 840971832313256   d. 10187589573425

   e. 681654 + 688045 + 840971832313256 + 10187589573425

4. For each part of Exercise 3, reduce the number modulo 4 (instead of modulo 9).

5. For each part of Exercise 3, reduce the number modulo 11 (instead of modulo 9).

6. Reduce each expression.

   a. $108 + 2534 + 3976 + 321539 \pmod 4$

   b. $1486 + 2365 + 68773 + 186474 \pmod 9$

   c. $108 \cdot 2534 \cdot 3976 \cdot 321539 \pmod 4$

   d. $1486 \cdot 2365 \cdot 63704 \cdot 186474 \pmod 9$

7. Find all values of x (in the range $0, 1, \ldots, 9$) that make the congruence true.

    a. $x + 8 \equiv 0 \pmod{10}$ **b.** $7 + 3x \equiv 0 \pmod{10}$

    c. $4x + 6 \equiv 0 \pmod{10}$ **d.** $1 + 8x \equiv 0 \pmod{10}$

8. List all possible values of the missing digit, e, that will make each divisibility statement true.

    a. $9 \mid 4052830e9582$ **b.** $4 \mid 984018742e$ **c.** $11 \mid 3274e3287$

9. Show that $61! \equiv 63! \pmod{71}$. [*Hint:* First reduce $62 \cdot 63$ modulo 71.]

Reasoning and Proofs

10. Prove that for any $x \in \mathbf{Z}$ and $m \in \mathbf{N}$, $-x \equiv m - x \pmod{m}$.

11. Prove part (i) of Theorem 7.2.1.

12. Prove Corollary 7.2.2. [*Hint:* Let $n \in \mathbf{N}$. Then show that the statement is true for all k by induction on k.]

13. Use induction to prove that for any nonnegative integer n, $10^n \equiv 1 \pmod{9}$.

14. Prove that for any nonnegative integer n, the congruences in (4) are true.

15. Prove that statement (5) is true for any integers $d_0, d_1, \ldots, d_{n-1}$.

16. Prove statement (6).

17. Explain why Method 7.2.7 is correct.

18. Here is an alternative to Method 7.2.7 for reducing modulo 4.

> **ALTERNATE METHOD FOR REDUCING MODULO 4**
>
> To reduce a natural number $d = d_{n-1} d_{n-2} \cdots d_2 d_1 d_0$ modulo 4, compute
> $$d \equiv 2 \cdot d_1 + d_0 \pmod{4},$$
> and reduce the result modulo 4 to get the final answer.

    a. Use this alternate method to reduce the number $d = 9024583672037894573$ modulo 4.

    b. Explain why this alternate method is correct.

19. Explain why each of the following divisibility tests is correct:

    a. A number is divisible by 10 if and only if its final digit is 0.

    b. A number is divisible by 5 if and only if its final digit is 0 or 5.

20. State, and justify, a method for reducing a number modulo 3 and a corresponding test for divisibility by 3. [*Hint:* It is similar to the method for reducing modulo 9 (method 7.2.3).]

21. State, and justify, a method for reducing a number modulo 8 and a corresponding test for divisibility by 8. [*Hint:* Your method can be similar to Method 7.2.7 or the method from Exercise 18.]

22. Consider the sum $1 + 2 + \cdots + (m - 1) \pmod{m}$.

    a. Evaluate this sum for each of the following values of m:

    $$m = 2, \; m = 3, \; m = 4, \; m = 5, \; m = 6, \; m = 7, \; m = 8.$$

    b. Make a conjecture giving the value of $1 + 2 + \cdots + (m - 1) \pmod{m}$ for any natural number m.

    c. Prove that your answer to part b is correct.

23. A common error in banking is to interchange two of the digits in an amount. Prove that the difference between the correct amount and the amount with the two digits interchanged is always divisible by 9.

24. Your friend does the following trick. She asks you to perform these steps in order:

    a. Write down a positive integer, which can have any number of digits.

    b. Write down a second positive integer that has exactly the same digits as the first, but scrambled into any order you wish. (For example, if the first number is 5258, then the second number could be 8255.)

    c. Subtract the smaller integer from the larger integer.

    d. Circle any one of the digits in your number—"But not zero," your friend says cryptically. "Zero is a place holder, not really a digit."

    e. Read aloud all of the other digits in the number (other than the digit you circled), in any order.

    Your friend, who has been across the room the entire time, then reveals the value of the digit you circled.

    How did she do it? And what was the reason for her cryptic remark—why are you not allowed to circle a zero? [*Hint:* See Exercise 23.]

25. A *palindrome* is a word or phrase that is spelled the same backward and forward, such as

    "A man, a plan, a canal—Panama!"

We can refer to natural numbers as palindromes as well, if the digits read forward are the same as read backward, such as 14641 or 999999. Are all palindrome numbers divisible by 11? If so, prove it. If not, describe explicitly which ones are and which ones are not.

26. **Phil Lovett**   To solve the congruence $2x + 6 \equiv 4 \pmod 8$, your friend Phil Lovett writes down the following steps:

$$2x + 6 \equiv 4 \pmod 8 \tag{A}$$
$$\Leftrightarrow \quad x + 3 \equiv 2 \pmod 8 \tag{B}$$
$$\Leftrightarrow \quad x \equiv -1 \pmod 8 \tag{C}$$

Phil reasons that congruence (B) is equivalent to congruence (A), and that congruence (C) is equivalent to congruence (B). From congruence (C), Phil concludes that for the congruence $2x + 6 \equiv 4 \pmod 8$, the entire solution set is $\{x \mid x \equiv -1 \pmod 8\}$. Phil's answer is incorrect.

a. What is the correct solution set to the congruence $2x + 6 \equiv 4 \pmod 8$?

b. Which step in Phil's reasoning is incorrect? Explain what is wrong with it.

c. Find a modulus, n, for which the congruence $2x + 6 \equiv 4 \pmod n$ actually does have the solution set $\{x \mid x \equiv -1 \pmod n\}$.

27. This exercise concerns the reduction of the Fibonacci sequence in various moduli.

a. Suppose you know the value of the following Fibonacci numbers modulo 3:

$$F_{78} \equiv 2 \pmod 3 \quad \text{and} \quad F_{79} \equiv 1 \pmod 3.$$

Reduce $F_{80} \pmod 3$. Justify your answer.

b. Find the first several numbers in the Fibonacci sequence, $F_1, F_2, F_3, \ldots$, reduced modulo 3. Does the resulting sequence eventually become periodic? If so, continue writing terms until the sequence repeats. [*Hint:* Starting with F_1 and F_2, proceed by adding successive terms, reducing as you go along.]

c. Find the first several numbers in the Fibonacci sequence reduced modulo 7. Does the resulting sequence eventually become periodic? If so, continue writing terms until the sequence repeats.

d. Find the first several numbers in the Fibonacci sequence reduced modulo 6. Does the resulting sequence eventually become periodic? If so, continue writing terms until the sequence repeats.

e. Imagine writing the Fibonacci sequence modulo 12 out to the 150th term (but do not actually do so). Explain how you know that the sequence will start repeating before this term.

f. Let $m \in \mathbf{N}$. Explain how you know that the Fibonacci sequence reduced modulo m is eventually periodic.

∗g. Let $m \in \mathbf{N}$. Prove that the Fibonacci sequence reduced modulo m is (purely) periodic. That is, prove that there exists a natural number p (the period) such that for all $k \in \mathbf{N}$, $F_k = F_{k+p} \pmod{m}$.

Advanced Reasoning and Proofs

28. Suppose that n soccer teams wish to have a tournament in which each team plays every other team exactly once. Such a tournament is called a *round-robin* tournament. Assume that no team can play more than one match on a given day.

   a. Suppose that the number of teams, n, is even. Show that it is possible to schedule a round-robin tournament among n teams in $n - 1$ days. [*Hint*: Begin by scheduling some of the matches according to the following rule. On day k, for teams i and j where $1 \le i < j \le n - 1$, make sure to schedule a match between teams i and j if $i + j \equiv k \pmod{n - 1}$.]

   b. Now suppose n is odd. Show that it is possible to schedule a round-robin tournament among n teams in n days. [*Hint*: Introduce a phantom $(n + 1)$st team, and apply the result of part a.]

29. Given a natural number n and an m-digit natural number $d = d_{m-1}\, d_{m-2} \cdots d_2 d_1 d_0$,

   a. Prove that $2^n \mid d$ if and only if 2^n divides $d_{n-1} \cdots d_2\, d_1\, d_0$ (the number that is equal to the last n digits of d).

   b. Use part a to determine the largest power of 2 that divides the number $d = 398769876760987022345534098709803945870393509814 62976$.

   c. Find a method similar to part a for determining whether 5^n divides d. Use your method to determine the largest power of 5 that divides the number $d = 65744765876876093284706982340928460982409428568929375$.

30. State, and justify, a method for reducing any natural number modulo 7 and a corresponding test for divisibility by 7.

31. State, and justify, a method for reducing any natural number modulo 13 and a corresponding test for divisibility by 13.

32. Let $a = 111^{111}$.

   a. Reduce a modulo 17.

   b. Reduce a^a modulo 17.

33. Prove that $2^{644} - 1$ is divisible by 645. [*Hint*: First reduce 2^{14} modulo 645.]

7.3 Check-Digit Schemes

Identification numbers are used in almost every aspect of modern life. To name just a few examples, numbers are used to uniquely identify people, credit card accounts, retail products, books, receipts, and airline tickets. Most retail and business transactions involve the transmission of identification numbers by people or by machines. An error can occur any time one of these numbers is transmitted, such as a number being written, typed, spoken, or heard incorrectly by a person, a bar code being scanned incorrectly by a machine, or a digit being altered during electronic transmission over a noisy channel.

By far the most common transmission error for identification numbers is the *single-digit error*, in which one digit of the number is incorrect. The next most common error is the *transposition error*, in which two digits (usually adjacent to each other) are switched. Single-digit errors and transposition errors, illustrated in the table below, account for 90% of all identification number transmission errors (about 79% of all errors are single-digit errors, 10% are adjacent transposition errors, and 1% are non-adjacent transposition errors). The table gives three examples of how the number 65432 could be incorrectly transmitted.

Correct number	Single-digit error	Transposition errors	
		Adjacent digits	Nonadjacent digits
65432	65482	65342	62435

Many organizations, aware that transmission errors are inevitable, have designed their identification numbers so that a machine can usually tell whether a number contains a mistake, even without knowing what the correct number is supposed to be.[2] This is commonly accomplished by including a *check digit* in the identification number.

In this section, we will use subscripts to indicate the position of a digit in an identification number. For example, in the five-digit identification number $a = 65432$, the first digit is $a_1 = 6$, the second digit is $a_2 = 5$, and the rest are $a_3 = 4$, $a_4 = 3$, and $a_5 = 2$.

U.S. Postal money orders

The United States Postal Service (USPS) uses 11-digit serial numbers on its money orders. The first ten digits identify the document, and the last digit is the check digit.

---

[2] Indeed, it is even possible to encode information in such a way that on encountering most transmission errors, a machine can not only determine that a mistake has been made but also fix the mistake. Such encoding schemes are known as *error-correcting codes*.

For example, the money order in Figure 1 has serial number $a = 16094004377$. The money order is identified by the first 10 digits, 1609400437. The eleventh digit, **7**, is the check digit.

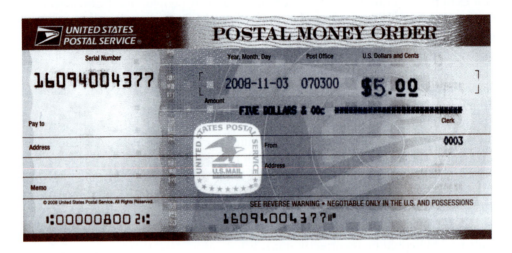

Figure 1 A U.S. Postal Service money order. In the serial number, 16094004377, the check digit is the final digit, **7**.

In a USPS money order with serial number a, the check digit a_{11} is calculated by taking the sum of the first 10 digits, $a_1 + a_2 + a_3 + \cdots + a_{10}$, and reducing the result modulo 9.

EXAMPLE 1

The first ten digits of the serial number on a USPS money order are 7306125986. Their sum is

$$a_1 + a_2 + \cdots + a_{10} = 7 + 3 + 0 + 6 + 1 + 2 + 5 + 9 + 8 + 6 = 47.$$

Since the reduction of 47 (mod 9) is **2**, the check digit must be $a_{11} = $ **2**, and the full 11-digit serial number is 73061259862.

Because addition is commutative, transposing any two of the first ten digits in a USPS money order number does not affect the check digit. If there were a transposition error in the first 10 digits of an 11-digit USPS money order number, the resulting 11-digit number would still be a valid USPS money order number according to the check-digit scheme. Thus, the USPS money order check-digit scheme fails to detect most transposition errors. However, the scheme *does* detect most single-digit errors. The only single-digit errors the scheme does not catch are cases where in one of the first ten digits, a **0** is replaced with a **9** or vice versa.

Universal Product Codes

Virtually every product in every store in the United States is marked with a Universal Product Code (UPC). When you are buying groceries and the bar code on a product is scanned at a cash register, the scanner transmits the UPC (the identification number) to the cash register, which uses the number to determine the price of the item. The UPC pictured, 3-02993-92180-**6**, is an example of a 12-digit UPC. (In this text, we will only consider 12-digit UPCs.)

3   02993 92180   6

Figure 2   A 12-digit Universal Product Code (UPC). In the identification number, 3-02993-92180-**6**, the check digit is the final digit, **6**.

A 12-digit UPC is divided into four parts. The first digit indicates the type of product: 0 for general groceries, 2 for meat and produce, 3 for drugs and health products, 4 for nonfood items, 5 for coupons, and 6 or 7 for other items. (The first-digit values 1, 8, and 9 are reserved for future use.) In the UPC pictured, the first digit is 3. The second part of the UPC, 02993, identifies the manufacturer. The third part, 92180, identifies the product. The fourth part of the UPC, the single digit 6, is the check digit.

In a UPC a, the check digit a_{12} is chosen to satisfy the following congruence:

$$3a_1 + a_2 + 3a_3 + a_4 + 3a_5 + a_6 + 3a_7 + a_8 + 3a_9 + a_{10} + 3a_{11} + a_{12} \equiv 0 \ (\mathrm{mod}\ 10). \quad (1)$$

EXAMPLE 2

On the package for a whiteboard eraser, the third digit of the UPC is illegible. What should be the value of the third digit, m, in the UPC **0-7m641-81505-6**?

SOLUTION   We need to solve the following congruence for m:

$$3 \cdot 0 + 7 + 3m + 6 + 3 \cdot 4 + 1 + 3 \cdot 8 + 1 + 3 \cdot 5 + 0 + 3 \cdot 5 + 6 \equiv 0 \ (\mathrm{mod}\ 10).$$

Combining the constant terms, we get

$$3m + 87 \equiv 0 \ (\mathrm{mod}\ 10).$$

By Theorem 7.2.1, since $87 \equiv 7 \ (\mathrm{mod}\ 10)$ we can substitute:

$$3m + 7 \equiv 0 \ (\mathrm{mod}\ 10)$$
$$3m \equiv -7 \ (\mathrm{mod}\ 10)$$
$$3m \equiv 3 \ (\mathrm{mod}\ 10).$$

It does not take long to verify that out of all of the possible values $0, 1, 2, \ldots, 9$ for the digit m, the only one that satisfies this last congruence is

$$m = 1.$$

Thus, the missing digit is **1**, and the complete UPC is 0-7**1**641-81505-6.   ●

The UPC check-digit scheme is an improvement over the U.S. Postal money order check-digit scheme because it detects all single-digit errors (see Exercise 10). However, the UPC scheme does not detect all adjacent transposition errors. For example, suppose the UPC

0-747<u>20</u>-08957-7 is incorrectly received as **0-74<u>27</u>0-08957-7**.

Because

$$7 + 3 \cdot 2 \equiv 2 + 3 \cdot 7 \pmod{10},$$

Theorem 7.2.1 (Congruences Add and Multiply) tells us that

$$3 \cdot 0 + 7 + 3 \cdot 4 + 7 + 3 \cdot 2 + \cdots \equiv 3 \cdot 0 + 7 + 3 \cdot 4 + 2 + 3 \cdot 7 + \cdots \pmod{10},$$

and thus, the same check digit is correct for both numbers.

We can generalize this result. Suppose that there is an adjacent transposition error in a UPC, in which adjacent digits d and e are transposed. Without loss of generality, assume that d is in an even position (d is not multiplied by 3 in (1)) and that e is in an odd position (e is multiplied by 3 in (1)). The check-digit scheme will fail to catch the error only if

$$d + 3 \cdot e \equiv e + 3 \cdot d \pmod{10}$$

or, equivalently,

$$2d \equiv 2e \pmod{10}.$$

By the Equivalent Conditions for Congruence (7.1.2), this congruence is equivalent to the statement

$$\exists k \in \mathbf{Z} \text{ s.t. } 2d = 2e + 10k.$$

Simplifying the equality, this statement is equivalent to

$$\exists k \in \mathbf{Z} \text{ s.t. } d = e + 5k.$$

Since d and e are distinct single digits in the range $0, 1, \ldots, 9$, this will only be true if

$$|\, d - e \,| = 5.$$

We conclude that the UPC check-digit scheme detects all adjacent transposition errors except for those in which the values of the transposed digits differ by 5.   ●

International Standard Book Numbers

Every book that is published is assigned a unique ten-digit International Standard Book Number (ISBN). An ISBN has four parts. For example, the ISBN for a popular number theory textbook is **0-470424-13-3**.

The first part of this ISBN, the number **0**, indicates the language group or country group in which the book was published. The first part of the ISBN of any book published in the English-speaking world is either a 0 or a 1. The second part of the ISBN, **470424**, identifies the book's publisher, and the third part of the ISBN, **13**, is a number that the publisher assigns to identify the particular book. The last part of the ISBN is the check digit, **3**.

In an ISBN a, the check digit a_{10} is chosen to satisfy the following congruence:

$$a_1 + 2a_2 + 3a_3 + 4a_4 + 5a_5 + 6a_6 + 7a_7 + 8a_8 + 9a_9 + 10a_{10} \equiv 0 \pmod{11}. \quad (2)$$

or, in summation notation,

$$\sum_{k=1}^{10} ka_k \equiv 0 \pmod{11}. \qquad (3)$$

It is sometimes necessary for the check digit, a_{10}, to take on the value 10. In these cases, the check digit is the letter X, which represents the value 10.

The ISBN check-digit scheme is the best of the three that we have seen. It can detect all single-digit errors, as you will prove in Exercise 9. In addition, it can detect any transposition error, whether adjacent or not, as we now prove.

CLAIM 7.3.1

The ISBN check-digit scheme detects any transposition error.

PROOF (By contradiction.)

Assumption: Suppose that there were a transposition error that the scheme does not detect. In other words, suppose that there were two ten-digit numbers, a and b, that were identical except that their mth and nth digits were transposed, where $m \neq n$. That is,

$$a_m = b_n, \ a_n = b_m,$$

$$\text{and} \quad a_k = b_k \text{ for all } k \neq m, n.$$

Further suppose that both a and b satisfy congruence (3) which defines the ISBN check-digit scheme.

From congruence (3), we get

$$\sum_{k=1}^{10} ka_k \equiv \sum_{k=1}^{10} kb_k \,(\text{mod } 11).$$

Every term other than the mth and nth terms will be the same on both sides of this congruence. Subtracting the equal terms from both sides of the congruence, we get

$$ma_m + na_n \equiv mb_m + nb_n \,(\text{mod } 11).$$

Since $a_m = b_n$ and $a_n = b_m$, this becomes

$$ma_m + na_n \equiv ma_n + na_m \,(\text{mod } 11),$$

from which we get

$$m(a_m - a_n) + n(a_n - a_m) \equiv 0 \,(\text{mod } 11).$$

Regrouping terms yields

$$(m - n) \cdot (a_m - a_n) \equiv 0 \,(\text{mod } 11).$$

By the Equivalent Conditions for Congruence (7.1.2), this is equivalent to the statement

$$11 \mid (m - n) \cdot (a_m - a_n).$$

Since 11 is prime, the Fundamental Property of Primes (6.1.2) allows us to conclude that

$$11 \mid (m - n) \quad \text{or} \quad 11 \mid (a_m - a_n). \tag{4}$$

But m and n are distinct integers in the range $1, 2, \ldots, 10$, and a_m and a_n are distinct integers in the range $0, 1, \ldots, 9$, so neither of the conditions in (4) can be true. $\Rightarrow\Leftarrow$

Since we have reached a contradiction, our Assumption must be false. We conclude that the ISBN check-digit scheme detects any transposition error. ∎

Fun Facts

In 2007, the ISBN check-digit scheme was updated to have 13 digits (the 13 digits underneath the bar code) rather than 10. Mathematically, the new scheme is a step backward—it uses a method similar to the one for UPC codes, which does not detect all transposition errors.

EXERCISES 7.3

Numerical Problems

1. What is the value of the check digit, c, in each of the following USPS money order numbers?

   a. 1121344312c b. 8383001267c c. 2736548199c

2. Each of the following numbers is the serial number of a USPS money order. One of the digits (indicated by a letter) is unreadable. What is the value of the unreadable digit? (If there is more than one possible value for the unreadable digit, then state all possible values.)

   a. m8812499557   b. 274493x2410   c. 319701014d5

3. What is the value of the check digit, c, in each UPC below?

   a. 5-11001-43245-c b. 0-33210-56473-c c. 3-12980-66115-c

4. Each of the following numbers is a UPC for some product. One of the digits (indicated by a letter) is unreadable. What is the value of the unreadable digit? (If there is more than one possible value for the unreadable digit, then state all possible values.)

   a. 5-119m3-45627-0   b. 0-11871-x3333-6   c. 6-22131-1234d-6

5. What is the value of the check digit, c, in each ISBN below?

   a. 0-7102-0679-c b. 0-8014-9363-c c. 0-8135-1490-c

6. Each of the following numbers is the ISBN of some book. One of the digits (indicated by a letter) is unreadable. What is the value of the unreadable digit? (If there is more than one possible value for the unreadable digit, then state all possible values.)

   a. 0-a65-01572-7   b. 0-395-b9088-0   c. 2-890c1-065-2

Reasoning and Proofs

7. Although the USPS money order check-digit scheme fails to detect *most* transposition errors, there are a few transposition errors that it does detect.

   a. Create a valid 11-digit USPS money order number.

   b. Modify the number you created in part a by making a transposition error that the check-digit scheme does *not* detect. (That is, the modified number should be a valid number according to the check-digit scheme.)

   c. Modify the number you created in part a by making a transposition error that the check-digit scheme *does* detect. (That is, the modified number should *not* be a valid number according to the check-digit scheme.)

8. This exercise concerns transposition errors in UPCs.

   a. Create a valid 12-digit UPC.

   b. Modify the number you created in part a by making an adjacent transposition error that the check-digit scheme does *not* detect. (That is, the modified number should be a valid number according to the check-digit scheme.)

   c. Modify the number you created in part a by making a nonadjacent transposition error that the check-digit scheme does *not* detect. (That is, the modified number should be a valid number according to the check-digit scheme.)

   d. Modify the number you created in part a by making a transposition error that the check-digit scheme *does* detect. (That is, the modified number should *not* be a valid number according to the check-digit scheme.)

Advanced Reasoning and Proofs

9. Prove that the ISBN check-digit scheme detects all single-digit errors.

10. Prove that the UPC check-digit scheme detects all single-digit errors.

7.4 The Chinese Remainder Theorem

Food for thought

During his years on Venus, Gauss became a big fan of one of the most popular sports in the galaxy: pizza eating. One Thursday evening in June, every carbon-based life form in the solar system was awaiting news from Jupiter's smelliest snack shack, Callisto's Pizzeria. Callisto's was not the greatest spot on Jupiter, but it was the home pizzeria of Juno the Jolly Gas Giant, who was trying to beat the current record in pizza eating. Juno was trying to unseat Phat Phoebe, one of the titans of pizza eating from the outskirts of Saturn. Phat Phoebe, who held the current record of a whopping 12 slices, had been eating rings around the competition for years. As was the custom, Juno was trying to finish as many slices as he could gobble of a gargantuan galactic 15-slice pizza.

Gauss couldn't wait to find out whether Juno had broken the worlds' record, but his local paper, the *Daily Planet*, had not yet arrived at the newsstand. "How improbable," Gauss muttered. "I mean, the paper is normally distributed by the moment I arrive at the newsstand." Forced to deviate from his standard routine, Gauss decided to buy the newspapers from Mars and Neptune. Gauss knew that the Martians used the modulo 3 number system and that the Neptunians used modulo 5.

From the news reports, Gauss gathered that the number of slices Juno had eaten was congruent to 1 **(mod 3)** and congruent to 3 **(mod 5)**. Could Gauss use this information to determine how many slices Juno had actually eaten?

After some thought, Gauss made a table:

Number of slices, s	0	1	2	3	4	5	6	7	8	9	10	11	12	13	14	15
Mars (mod 3)	0	1	2	0	1	1	0	1	1	0	1	2	0	1	2	0
Neptune (mod 5)	0	1	2	3	4	0	1	2	3	4	0	1	2	3	4	0

"By Jove!" Gauss exclaimed, looking at his table. "It's clear as a bell. Juno set a new worlds' record!"

In our myth, Gauss learned from the newspapers that the number of slices, s, that Juno had eaten satisfied both of the following congruences:

$$s \equiv 1 \ (\text{mod } 3)$$

$$s \equiv 3 \ (\text{mod } 5)$$

From his table, Gauss could see that the number $s = 13$ does the trick. In fact, 13 is the only nonnegative integer less than 15 that simultaneously satisfies both of these congruences, so Gauss concluded that Juno must have eaten exactly 13 slices. (There are other, larger, integer solutions to these congruences, but we assume that Juno didn't eat more than one whole pizza.)

Naomi's Numerical Proof Preview:
The Chinese Remainder Theorem (7.4.1)

Let's consider another system of congruences.

$$s \equiv 31 \pmod{49} \tag{1}$$

$$s \equiv 6 \pmod{20}$$

If we made a table to solve this system of congruences, as Gauss did in our myth, our table would have $20 \cdot 49 = 980$ columns. This would be slow and tedious. Perhaps there is a better way to find the solution.

Recalling the Equivalent Conditions for Congruence (7.1.2), we can rewrite the congruences in (1) as follows:

$$s \equiv 31 \pmod{49} \quad \Leftrightarrow \quad \text{there exists an integer } X \text{ such that } s = \mathbf{31 + 49X} \tag{2}$$

$$s \equiv 6 \pmod{20} \quad \Leftrightarrow \quad \text{there exists an integer } Y \text{ such that } s = \mathbf{6 + 20Y}$$

Combining the two equations on the right, we get

$$\mathbf{31 + 49X = 6 + 20Y}.$$

Simplifying yields

$$-49X + 20Y = 25. \tag{3}$$

We have succeeded in reducing the original system of congruences to the problem of finding an integral solution to equation (3). The hungry reader may look at equation (3) and see potatoes. Equation (3) is a linear Diophantine equation, which we know how to solve! Since the coefficients, -49 and 20, are relatively prime, we know from our study of linear Diophantine equations in Chapter 5 that this equation has integral solutions. Using the methods of Chapter 5, we find the solution

$$(X, Y) = (-225, -550).$$

From these values of X and Y, we can find s:

$$s = 6 + 20Y$$
$$= 6 + 20(-550)$$
$$= -10994.$$

Thus, we have successfully found a solution, $s = -10994$, to our system of congruences (1). (The reader is invited to check that $s = -10994$ does indeed satisfy this system of congruences.)

Let's call the one solution we have found $s_0 = -10994$. Now suppose we wanted to find other solutions to the system of congruences (1). Since $\text{lcm}(49, 20) = 980$ is a multiple of both 49 and 20, adding 980 does not change the value of s_0 modulo 49 or modulo 20. Hence, $s_0 + 980$ is another solution to (1). Similarly s_0 plus any multiple of 980 will also be a solution to (1). In other words, any s satisfying

$$s \equiv -10994 \pmod{980} \tag{4}$$

will be a solution to our system of congruences. In fact, as we will see (Theorem 7.4.1), the single congruence (4) gives the entire solution set to our original system of congruences. That is, the system of two congruences (1) is actually equivalent to the single congruence (4).

The congruence (4) can be simplified. Since -10994 reduces to 766 modulo 980, the congruence can be rewritten as

$$s \equiv 766 \pmod{980}.$$

Let's look back at how we solved the system of congruences (1). We solved it by reducing our problem to the linear Diophantine equation (3). More generally, as we will see in the proof of Theorem 7.4.1, solving the system of congruences

$$s \equiv k \pmod{a}$$
$$s \equiv l \pmod{b} \tag{5}$$

reduces to finding a solution to the Diophantine equation

$$aX + bY = l - k. \tag{6}$$

If a and b are relatively prime, this is always possible. These observations are summarized in the following theorem, which is known as the Chinese Remainder Theorem.

THEOREM 7.4.1　THE CHINESE REMAINDER THEOREM

Let $a, b \in \mathbf{N}$ *such that* $\gcd(a, b) = 1$, *and let* $k, l \in \mathbf{Z}$. *Then there exists* $s \in \mathbf{Z}$ *such that*

$$s \equiv k \pmod{a}$$

$$\text{and} \quad s \equiv l \pmod{b}. \tag{7}$$

Furthermore, the solution is unique modulo ab. That is, if s_0 *is any solution to this system of congruences, then an integer s is a solution to this system if and only if*

$$s \equiv s_0 \pmod{ab}.$$

PROOF Since a and b are relatively prime, the Diophantine equation $aX + bY = l - k$ has a solution, by Theorem 5.3.1. Thus, there exist integers x and y such that

$$ax + by = l - k.$$

Rearranging terms gives

$$k + ax = l - by.$$

Let $s = k + ax = l - by$.

By the Equivalent Conditions for Congruence (7.1.2), we know

$$s \equiv k \pmod{a} \quad \text{and} \quad s \equiv l \pmod{b}.$$

We conclude that the system (7) has the solution s.

In Exercise 23, you will complete the proof of the theorem by showing that this solution is unique modulo ab.

If the moduli a and b are not relatively prime, the system of congruences (5) does not always have a solution. (For example, see Exercise 10.) The system of congruences (5) will have a solution when the Diophantine equation (6) has a solution. By Theorem 5.3.2, this happens provided that $\gcd(a, b)$ divides $l - k$. This observation is captured by the following more general version of the Chinese Remainder Theorem.

Let $a, b \in \mathbf{N}$ and $k, l \in \mathbf{Z}$. Then there exists $s \in \mathbf{Z}$ such that

$$s \equiv k \,(\mathrm{mod}\ a)$$

and $\quad s \equiv l \,(\mathrm{mod}\ b).$ $\qquad\qquad$ (8)

if and only if

$$\gcd(a, b) \mid l - k.$$

Furthermore, the solution is unique modulo $m = \mathrm{lcm}(a, b)$. That is, if s_0 is any solution to this system of congruences (8), then an integer s is a solution to this system if and only if

$$s \equiv s_0 \,(\mathrm{mod}\ m).$$

PROOF You will prove this theorem in Exercise 24.

EXAMPLE 1

Using the Chinese Remainder Theorem, answer the questions below about the following system of congruences:

$$s \equiv 13 \,(\mathrm{mod}\ 104)$$

$$s \equiv 49 \,(\mathrm{mod}\ 60) \qquad\qquad (9)$$

a. Find a solution to the system.

b. Find the entire solution set of the system.

c. What is the smallest nonnegative integer solution to the system?

SOLUTION

a. Using the Equivalent Conditions for Congruence (7.1.2), we rewrite the congruences in (9):

$s \equiv 13 \,(\mathrm{mod}\ 104) \quad \Leftrightarrow \quad$ there exists an integer X such that $s = \mathbf{13 + 104}X$

$s \equiv 49 \,(\mathrm{mod}\ 60) \quad \Leftrightarrow \quad$ there exists an integer Y such that $s = \mathbf{49 + 60}Y$

Combining the two equations on the right, we get the linear Diophantine equation:

$$\mathbf{13 + 104}X = \mathbf{49 + 60}Y,$$

which simplifies to

$$104X - 60Y = 36.$$

Dividing both sides of this equation by the common factor, 4, yields the Diophantine equation

$$26X - 15Y = 9.$$

Using the methods of Chapter 5, we find the solution

$$(X, Y) = (-36, -63).$$

From these values of X and Y, we can find s:

$$s = 13 + 104X$$
$$= 13 + 104(-36)$$
$$= -3731.$$

Thus, $s = -3731$ is a solution to the system of congruences (9).

b. Now we compute the lcm of the two moduli in the system of congruences:

$$\text{lcm}(104, 60) = 3120.$$

By the General Version of the Chinese Remainder Theorem (7.4.2), the entire solution set is

$$\{s \in \mathbf{Z} \mid s \equiv -3731 \ (\text{mod } 3120)\}.$$

c. The smallest nonnegative integer in this solution set can be found by reducing -3731 modulo 3120:

$$s = 2509. \qquad \bullet$$

The Chinese Remainder Theorem (7.4.1 and 7.4.2) enables us to convert a system of two congruences into a single congruence with a larger modulus. We can also convert a larger system (with more than two congruences) into a single congruence. The following theorem is a generalization of Theorem 7.4.1 to this multiple-congruences case. (The multiple-congruences extension of Theorem 7.4.2 is stated in Exercise 39.)

In the following theorem, we will assume that the moduli $a_1, a_2, \ldots, a_r$ are *pairwise relatively prime*. This means that any pair of them are relatively prime; i.e., for each $i \neq j$, $\gcd(a_i, a_j) = 1$.

Let $a_1, a_2, \ldots, a_r \in \mathbf{N}$ be pairwise relatively prime, and let $k_1, k_2, \ldots, k_r \in \mathbf{Z}$. Then there exists $s \in \mathbf{Z}$ such that

$$s \equiv k_1 \ (\mathrm{mod}\ a_1),$$
$$s \equiv k_2 \ (\mathrm{mod}\ a_2),$$
$$\vdots \tag{10}$$
$$and \qquad s \equiv k_r \ (\mathrm{mod}\ a_r).$$

Furthermore, the solution is unique modulo the product $m = a_1 a_2 \cdots a_r$. That is, if s_0 is any solution to this system of congruences (10), then an integer s is a solution to this system if and only if

$$s \equiv s_0 \ (\mathrm{mod}\ m).$$

PROOF You will prove this theorem in Exercise 38.

The Chinese Remainder Theorem and congruences in composite moduli

Suppose you wish to reduce the integer $x = 1{,}000{,}000{,}001$ modulo 36. Dividing x by 36 in order to find the remainder could be tedious. However, we know how to quickly reduce x modulo 9 and modulo 4. Modulo 9, any number is congruent to the sum of its digits (Method 7.2.3). Modulo 4, any number is congruent to its last two digits (Method 7.2.7). Thus, we obtain:

$$x \equiv 2 \ (\mathrm{mod}\ 9)$$
$$x \equiv 1 \ (\mathrm{mod}\ 4) \tag{11}$$

The Chinese Remainder Theorem (7.4.1) enables us to rewrite this system of congruences as a single congruence modulo $36 = 9 \cdot 4$. Solving the system (11) to find this congruence yields

$$x \equiv 29 \ (\mathrm{mod}\ 36).$$

Thus, we have solved our problem: 1,000,000,0001 reduces to 29 modulo 36.

In general, if a and b are relatively prime integers, then the Chinese Remainder Theorem (7.4.1) tells us that a system of two congruences, one modulo a and the other modulo b, are equivalent to a single congruence modulo ab. Since congruences are much easier to work with in small moduli, we might prefer to work with separate congruences modulo a and b, rather than a single congruence modulo ab.

The following observation expresses the fact that two numbers are congruent modulo ab if and only if they are congruent both modulo a and modulo b. This observation follows directly from the Chinese Remainder Theorem (7.4.1).

Let $a, b \in \mathbf{N}$ such that $\gcd(a, b) = 1$, and let $x, y \in \mathbf{Z}$. Then

$$x \equiv y \;(\mathrm{mod}\; ab) \qquad \text{if and only if} \qquad \begin{cases} x \equiv y \;(\mathrm{mod}\; a) \\ \quad and \\ x \equiv y \;(\mathrm{mod}\; b). \end{cases} \qquad (12)$$

As we have just seen, it is sometimes more convenient to work with the two simpler congruences on the right side of (12) rather than the one toughie on the left. Here's another example.

EXAMPLE 2

Find a solution x to the congruence

$$x^3 \equiv 2 \;(\mathrm{mod}\; 55). \qquad (13)$$

SOLUTION By Observation 7.4.4, an integer x satisfies the congruence (13) if and only if both

$$x^3 \equiv 2 \;(\mathrm{mod}\; 5) \qquad (14)$$

and

$$x^3 \equiv 2 \;(\mathrm{mod}\; 11). \qquad (15)$$

These two congruences are not difficult to solve by trial and error. One finds that congruence (14) has solution

$$x \equiv 3 \;(\mathrm{mod}\; 5), \qquad (16)$$

and congruence (15) has solution

$$x \equiv 7 \;(\mathrm{mod}\; 11). \qquad (17)$$

Using the Chinese Remainder Theorem (7.4.1), we can convert the system of congruences (16) and (17) into an equivalent single congruence:

$$x \equiv 18 \;(\mathrm{mod}\; 55). \qquad (18)$$

Thus, any integer that satisfies congruence (18) is a solution to the initial congruence (13). ●

The next observation, which follows from Theorem 7.4.3, extends Observation 7.4.4 to more than two congruences.

Let $a_1, a_2, \ldots, a_r \in \mathbf{N}$ be pairwise relatively prime, and let $x, y \in \mathbf{Z}$. Then

$$x \equiv y \pmod{a_1 a_2 \cdots a_r} \quad \text{if and only if} \quad \begin{cases} x \equiv y \pmod{a_1}, \\ x \equiv y \pmod{a_2}, \\ \quad \vdots \\ \text{and} \quad x \equiv y \pmod{a_r}. \end{cases}$$

EXERCISES 7.4

Numerical Problems

For each system of congruences in Exercises 1–9,

    a. Find a solution to the system, if possible. If no solution exists, explain how you know.

    b. Find the entire solution set of the system.

    c. What is the smallest nonnegative integer solution to the system?

1. $s \equiv 4 \pmod 5$ and $s \equiv 1 \pmod 2$

2. $s \equiv 1 \pmod{16}$ and $s \equiv 3 \pmod 6$

3. $s \equiv 9 \pmod{25}$ and $s \equiv 5 \pmod 7$

4. $s \equiv 10 \pmod{13}$ and $s \equiv 13 \pmod{15}$

5. $s \equiv 2 \pmod{21}$ and $s \equiv 1 \pmod{91}$

6. $s \equiv 2 \pmod{54}$ and $s \equiv 20 \pmod{45}$

7. $x \equiv 37 \pmod{182}$ and $x \equiv 25 \pmod{202}$

8. $x \equiv 17 \pmod{105}$ and $x \equiv 72 \pmod{161}$

9. $x \equiv 21 \pmod{512}$ and $x \equiv 81 \pmod{603}$

10. Show that each system of congruences has no solution.

    a. $t \equiv 2 \pmod 8$ and $t \equiv 1 \pmod{10}$

    b. $t \equiv 78 \pmod{169}$ and $t \equiv 118 \pmod{325}$

11. Let $x = 1{,}040{,}302$.

    a. Reduce x modulo 2.

    b. Reduce x modulo 9.

    c. Use your results from parts a and b to find the reduction of $x \pmod{18}$.

12. Let $x = 3{,}100{,}997{,}996$.

    a. Reduce x modulo 9.

    b. Reduce x modulo 10.

    c. Use your results from parts a and b to find the reduction of $x \pmod{90}$.

Reasoning and Proofs

13. a. Let $s, t \in \mathbf{Z}$. Suppose $s \equiv t \pmod 7$ and $s \equiv t \pmod{11}$. Does it follow that $s \equiv t \pmod{77}$? If so, justify your answer. If not, give a counterexample.

    b. Let $s, t \in \mathbf{Z}$. Suppose $s \equiv t \pmod 6$ and $s \equiv t \pmod{15}$. Does it follow that $s \equiv t \pmod{90}$? If so, justify your answer. If not, give a counterexample.

14. Let $s, t \in \mathbf{Z}$. Suppose $s \equiv t \pmod{20}$ and $s \equiv t \pmod{35}$.

    a. Does it follow that $s \equiv t \pmod{700}$? If so, justify your answer. If not, give a counterexample.

    b. Does it follow that $s \equiv t \pmod{140}$? If so, justify your answer. If not, give a counterexample.

    c. Does it follow that $s \equiv t \pmod{70}$? If so, justify your answer. If not, give a counterexample.

In Exercises 15–20, you are given a congruence of the form $ax \equiv c \pmod n$. Find all solutions x to this congruence such that $0 \leq x \leq n - 1$.
[*Hint:* Finding integers x such that $ax \equiv b \pmod n$ is equivalent to finding solutions to the Diophantine equation $ax + ny = c$.]

15. $4x \equiv 16 \pmod{18}$ 16. $7x \equiv 11 \pmod{15}$ 17. $3x \equiv 2 \pmod 9$

18. $21x \equiv 10 \pmod{52}$ 19. $9x \equiv 8 \pmod{12}$ 20. $22x \equiv 4 \pmod{30}$

21. Let $a, c \in \mathbf{Z}$ and $n \in \mathbf{N}$. Prove that $ax \equiv c \pmod n$ has a solution if and only if c is a multiple of $d = \gcd(a, n)$.

22. Let $a, c \in \mathbf{Z}$ and $n \in \mathbf{N}$, and let $d = \gcd(a, n)$. Suppose that $ax \equiv c \pmod{n}$ has a solution. Prove that $ax \equiv c \pmod{n}$ has exactly d solutions such that $0 \le x \le n - 1$.

23. Prove that the solution given by the Chinese Remainder Theorem (7.4.1) is unique modulo ab. That is, if s_0 is any solution to the system of congruences (7), then an integer s is a solution to (7) if and only if $s \equiv s_0 \pmod{ab}$.

24. Prove the General Version of the Chinese Remainder Theorem (7.4.2).

25. **a.** Find a solution to this system of congruences: $s \equiv 1 \pmod{13}$ and $s \equiv 0 \pmod{32}$

   b. Find a solution to this system of congruences: $s \equiv 0 \pmod{13}$ and $s \equiv 1 \pmod{32}$

   c. Let c and d be integers. Give a formula (in terms of c and d) that expresses a solution to the following system of congruences:

   $$s \equiv c \pmod{13} \text{ and } s \equiv d \pmod{32}.$$

   [*Hint:* Use your answers to parts a and b.]

26. Your number theory professor, whose handwriting is notoriously illegible, gives you the following handwritten quiz:

   > Solve the following system of congruences modulo m, where $m = 22 \cdot 25 \cdot 29$.
   >
   > $x \equiv 1 \pmod{29}$
   > $x \equiv 3 \pmod{25}$
   > $x \equiv 4 \pmod{22}$.

   From your professor's scrawl, you deduce that

   $$x \equiv 1 \text{ or } 7 \pmod{29},$$
   $$x \equiv 3 \text{ or } 8 \pmod{25},$$
   $$\text{and} \quad x \equiv 4 \text{ or } 9 \pmod{22}.$$

   How many different possibilities are there for the value of x modulo m? Explain.

For Exercises 27–28, consider the system of congruences

$$s \equiv k \pmod{a}$$
$$s \equiv l \pmod{b}. \tag{19}$$

27. Suppose $\gcd(a, b) = 1$.

   a. Prove that for any integers k and l, the system of congruences (19) has a unique solution in the range $0 \le s < ab$.

   b. Prove that for every integer s in the range $0 \le s < ab$, there exist integers k and l such that s is a solution to the system of congruences (19).

28. Suppose $\gcd(a, b) = d$, where $d > 1$.

    a. Prove that for any integers k and l, if $d \mid (k - l)$, then the system of congruences (19) has a unique solution in the range $0 \le s < \text{lcm}(a, b)$.

    b. Prove that for every integer s in the range $0 \le s < \text{lcm}(a, b)$, there exist integers k and l such that s is a solution to the system of congruences (19).

EXPLORATION Solving Systems of Three or More Congruences
(Exercises 29–34)

29. Consider the following system of congruences:

$$s \equiv 3 \;(\text{mod } 5)$$

$$s \equiv 1 \;(\text{mod } 6)$$

$$s \equiv 8 \;(\text{mod } 13).$$

    a. The Chinese Remainder Theorem tells us that the system of the first two congruences is equivalent to a single congruence modulo $30 = \text{lcm}(5, 6)$. Find this congruence.

    b. Using the Chinese Remainder Theorem again, the system comprising the congruence you found in part a and the third congruence above is equivalent to a single congruence modulo $390 = \text{lcm}(30, 13)$. Find this congruence. Congratulations! You have now found a single congruence that expresses the entire solution set of the original system of three congruences.

In Exercises 30–33, find the entire solution set to the given system of congruences. Express your answer as a single congruence.

30. $r \equiv 1 \;(\text{mod } 21), r \equiv 0 \;(\text{mod } 22), r \equiv 0 \;(\text{mod } 23)$

31. $r \equiv 3 \;(\text{mod } 8), r \equiv 9 \;(\text{mod } 15), r \equiv 3 \;(\text{mod } 12)$

32. $r \equiv 0 \;(\text{mod } 20), r \equiv 14 \;(\text{mod } 24), r \equiv 0 \;(\text{mod } 25)$

33. $r \equiv 1 \;(\text{mod } 2), r \equiv 2 \;(\text{mod } 3), r \equiv 4 \;(\text{mod } 7), r \equiv 6 \;(\text{mod } 11).$

34. **a.** Find the smallest natural number r that is evenly divisible by each of the numbers 2, 3, 4, 5, 6, 7, 8, 9, 10, 11, and 12, and that leaves a remainder of 1 when divided by 13.

    b. Find the smallest natural number s that is evenly divisible by 13 and that leaves a remainder of 1 when divided by each of the numbers 2, 3, 4, 5, 6, 7, 8, 9, 10, 11, and 12.

35. You are at a carnival, on a spinning ride whose lever has gotten stuck in the ON position. The ride has rotated 824,137 degrees clockwise since it began spinning.

    a. Reduce 824,137 modulo 9.

    b. Reduce 824,137 modulo 4.

    c. Reduce 824,137 modulo 10.

    d. Use your results from parts a–c to find the direction the ride is currently pointing, in degrees clockwise from its initial position (that is, to reduce 824,127 modulo 360).

36. According to a well-known story, a Chinese general was fond of using modular arithmetic to count his troops. One day, he ordered the troops to line up in rows of 2, and found that there was 1 soldier left over at the end. Then he ordered the troops to line up in rows of 3, and again there was 1 soldier left over. Next he ordered the troops to line up in rows of 5, then rows of 7, and in each case there was exactly 1 soldier left over. Finally, when the troops lined up in rows of 11, there were no soldiers left over, which pleased the general. What can you say about the number of soldiers in the company?

Advanced Reasoning and Proofs

37. Let $a_1, \ldots, a_r \in \mathbf{N}$ be pairwise relatively prime, and let $b = a_1 a_2 \cdots a_r$. Let $f(x)$ be a polynomial with integer coefficients. Prove that x is a solution to the congruence $f(x) \equiv 0 \pmod{b}$ if and only if x is a solution to $f(x) \equiv 0 \pmod{a_i}$ for each $i = 1, 2, \ldots, r$.

38. Prove Theorem 7.4.3 (the Multiple-Congruences Extension of the Chinese Remainder Theorem).

39. Prove the following theorem, which is an extension of Theorem 7.4.2 to the case of multiple congruences.

THEOREM. *Let* $a_1, a_2, \ldots, a_r \in \mathbf{N}$ *and* $k_1, k_2, \ldots, k_r \in \mathbf{Z}$ *such that* $\gcd(a_i, a_j) \mid k_j - k_i$ *for each* $i \neq j$. *Then there exists* $s \in \mathbf{Z}$ *such that*

$$s \equiv k_1 \pmod{a_1},$$
$$s \equiv k_2 \pmod{a_2}, \qquad\qquad (20)$$
$$\vdots$$

and $\quad s \equiv k_r \pmod{a_r}$.

Furthermore, this solution is unique modulo $m = \mathrm{lcm}(a_1, a_2, \ldots, a_r)$. *That is, if* s_0 *is any solution to the system of congruences* (20), *then an integer* s *is a solution to this system if and only if*

$$s \equiv s_0 \pmod{m}.$$

7.5 The Gregorian Calendar

On what day of the week were you born? You could ask your parents, or you could impress them by figuring it out yourself. In this section, we will learn how to quickly determine the day of the week (Sunday, Monday, etc.) for any date—past, present, or future—using the Gregorian calendar (see "Fun Facts" for the history of the Gregorian calendar). The method we present is relatively easy to memorize and do without a calculator, making it an ideal way to impress people at parties. Our method is a variant of the "Doomsday Rule," invented by John Conway.[3]

Fun Facts

The calendar used around the world today is called the Gregorian calendar, named for Pope Gregory XIII. The previous calendar, the Julian Calendar, was named for Julius Caesar who adopted it in the 1st century BCE. Three-fourths of the years in the Julian calendar had 365 days, but every fourth year was a leap year, a year of length 366 days with its extra day at the end of February. Thus, the Julian year had an average length of 365.25 days. This is slightly longer than the actual length of the solar year (the period of Earth's revolution around the Sun), which is approximately 365.2424. As a result, over the centuries the Julian calendar got farther and farther off from the solar year until by 1582, there was a discrepancy of 10 days.

In 1582, Pope Gregory XIII fixed the problem by instituting the Gregorian calendar. On its institution, the Gregorian calendar skipped 10 days, so that the day after October 4, 1582, was October 15, 1582. (Many at the time were opposed to the Gregorian calendar, because they thought that the landlords were proposing it as a way to charge them a full month's rent for a month with only 21 days.) The Gregorian calendar also changed the rules for leap years, making it so that not every year ending in 00 is a leap year. This gave the average year a length that is much closer to the length of the solar year than the Julian calendar's 365.25 days. (See Exercise 17.)

Italy, France, and Spain adopted the Gregorian calendar in 1582, but the rest of the world only gradually adopted the new reforms. Britain and the American colonies did not officially adopt the Gregorian calendar until 1752, when they skipped 11 days (September 3–13, 1752, did not exist in Britain and its colonies). Many countries outside of the Western world did not switch to the Gregorian calendar until the 20th century.

Finding the Doomsday for any year

For any given year, we focus on a particular day of the week, which we call the *Doomsday* for that year. The Doomsday for a given year is defined as the day of the week of the

---

[3]The original Doomsday Rule is explained in *Winning Ways for Your Mathematical Plays* by Elwyn R. Berlekamp, John H. Conway, and Richard K. Guy.

last day of February in that year. For example, the Doomsday for 2007 is Wednesday, because the last day of February in 2007 was Wednesday, February 28.

Finding the day of the week is inherently a question modulo 7, so before we continue, let's assign a number modulo 7 to each day of the week. Along with a number corresponding to each day of the week, the following table includes a mnemonic to help remember which day of the week is associated with each number. (The mnemonic is not necessary for understanding the system, but it may make it easier to do in one's head.)

TABLE 7.5.1 Converting days of the week into numbers

Each day of the week corresponds to a different number modulo 7. The third column contains a mnemonic that may help you remember which number corresponds to each day of the week.

Day	Number	Mnemonic
Sunday	0, 7	None-day or Se'en-day
Monday	1	One-day
Tuesday	2	Twos-day
Wednesday	3	Trebles-day
Thursday	4	Fours-day
Friday	5	Five-day
Saturday	6	Sixer-day

The Doomsday for the year 2000 was Tuesday (the last day of February in 2000 was Tuesday, February 29). By Table 7.5.1, we can express this as follows:

$$D_{2000} \equiv 2 \ (\text{mod } 7). \tag{1}$$

This is easy to remember, because "**two** thousand" starts with the same sound as "**two**s-day." The Doomsday for the year 1900 was Wednesday:

$$D_{1900} \equiv 3 \ (\text{mod } 7).$$

(At the end of this section we will discuss how to find Doomsdays for other centuries.) Knowing the Doomsday for the years 1900 and 2000, we can use the following method to determine the Doomsday for any year in the 20th or 21st century. We illustrate by explaining the method for any year in the 21st century.

Let Y represent the last two digits of the year (e.g., $Y = 23$ represents the year 2023), and let L represent the **number of leap years** that occur after 2000, up to (and including) year Y. After the 00 year of any century, every fourth year is a leap year, so L is the quotient when Y is divided by 4. The **number of nonleap years** that occur after 2000, up to (and including) year Y, is $Y - L$. The Doomsday for a leap year comes **366** days after the Doomsday for the previous year, but the Doomsday for a non-leap year comes **365** days after the Doomsday for the previous year. Thus, the formula for D_Y, the Doomsday in year Y of the 21st century, is

$$D_Y \equiv D_{2000} + \mathbf{365} \cdot (\mathbf{Y - L}) + \mathbf{366} \cdot L \ (\text{mod } 7),$$

which simplifies to

$$D_Y \equiv D_{2000} + 365Y + L \ (\text{mod } 7).$$

Since $365 \equiv 1 \ (\text{mod } 7)$, it follows that

$$D_Y \equiv D_{2000} + Y + L \ (\text{mod } 7) \qquad (2)$$

by Theorem 7.2.1 (Congruences Add and Multiply).

EXAMPLE 1

Find the Doomsday for 2068.

SOLUTION The number of leap years after 2000, up to and including 2068, is $L = 17$ (the quotient when 68 is divided by 4). Using formula (2),

$$
\begin{aligned}
D_{2068} &\equiv D_{2000} + 68 + 17 \\
&\equiv 2 + 5 + 3 \\
&\equiv \mathbf{3} \ (\text{mod } 7).
\end{aligned}
$$

Checking Table 7.5.1 for the number **3**, we conclude that the Doomsday for 2068 is **Wednesday**. ●

Here is a summary of what we have learned so far.

METHOD 7.5.1	HOW TO FIND THE DOOMSDAY FOR ANY YEAR

The Doomsday for any year is the day of the week of the last day of February in that year. To find the Doomsday, D, for any given year, use the following formula:

$$D \equiv D_{00} + Y + L \ (\text{mod } 7), \qquad (3)$$

where D_{00} is the Doomsday for the 00 year of the given century, Y is the last two digits of the given year, and L is the quotient when Y is divided by 4. (L is the number of leap years after the 00 year.)

EXAMPLE 2

Find the Doomsday for 1912.

SOLUTION Recall that the Doomsday for 1900 is Wednesday, so by Table 7.5.1, $D_{1900} = 3$. Now use formula (3):

$$D_{1912} \equiv D_{1900} + Y + L \;(\text{mod } 7)$$
$$\equiv 3 + 12 + 3 \;(\text{mod } 7)$$
$$\equiv 3 + 5 + 3 \;(\text{mod } 7)$$
$$\equiv \mathbf{4} \;(\text{mod } 7).$$

Thus (using Table 7.5.1), the Doomsday for 1912 is **Thursday**. ●

Finding the day of the week for any date

If we know the Doomsday for a given year, then we can determine the day of the week of any date in that year. The method relies on remembering key dates that fall on the same day of the week as the Doomsday in every year. To begin with, the last day of February (28th or 29th in a nonleap year or leap year, respectively) and the "last" day of January (31st or 32nd in a nonleap year or leap year, respectively) both fall on the same day of the week as the Doomsday. (Note that to make it easier to handle dates in January, we think of January 32nd as a date, but January 32nd really refers to February 1st.) For March, just think of the last day of February (the Doomsday) as the 0th day of March. In the even months after February, the dates 4/4, 6/6, 8/8, 10/10, and 12/12 all fall on the same day of the week as the Doomsday. For the odd months after March, the dates 5/9, 9/5, 7/11, and 11/7 all fall on the same day of the week as the Doomsday. In addition, you can optionally remember that three holidays, 7/4 (Independence Day), 10/31 (Halloween), and 12/26 (Boxing Day, the day after Christmas) always fall on the same day of the week as the Doomsday. These are all shown in Table 7.5.2.

TABLE 7.5.2   Dates that always occur on the same day of the week as the Doomsday

Months	Mnemonic	Dates				
First two months	"Last" day of month in nonleap year/leap year	Jan. 31/32	Feb. 28/29 (= Mar. 0)			
Even months	4/4, 6/6, 8/8, 10/10, 12/12	April 4	June 6	Aug. 8	Oct. 10	Dec. 12
Odd months	5/9, 9/5, 7/11, 11/7	May 9	July 11	Sept. 5	Nov. 7	
Holidays	Independence Day, Halloween, Boxing Day	July 4	Oct. 31	Dec. 26		

The only step that remains is to get good at subtracting or adding from these reference dates to find the difference between any date and the Doomsday modulo 7.

EXAMPLE 3
....

Find the day of the week for each date:

    a. November 25, 1966

    b. January 22, 2012

SOLUTION

    a. First find the Doomsday for 1966 using Method 7.5.1:

$$D \equiv D_{1900} + Y + L \ (\text{mod } 7)$$
$$\equiv 3 + 66 + 16 \ (\text{mod } 7)$$
$$\equiv 3 + 3 + 2 \ (\text{mod } 7)$$
$$\equiv \mathbf{1} \ (\text{mod } 7)$$

Choose a key date near November 25 that occurs on the same day of the week as the Doomsday. We choose November 7. November 25 is **18** days **after** November 7, so we find

$$\mathbf{1 + 18} \equiv 1 + 4 \ (\text{mod } 7)$$
$$\equiv \mathbf{5} \ (\text{mod } 7),$$

which corresponds to **Friday**, as shown in Table 7.5.1.

    b. By formula (3), the Doomsday for 2012 is

$$D \equiv D_{2000} + Y + L \ (\text{mod } 7)$$
$$\equiv 2 + 12 + 3 \ (\text{mod } 7)$$
$$\equiv \mathbf{3} \ (\text{mod } 7).$$

Choose a key date near January 22 that occurs on the same day of the week as Doomsday. We choose January 32. (Since 2012 is a leap year, the "last" day of January is January 32.) January 22 is **10** days **before** January 32, so the day of the week is

$$\mathbf{3 - 10} \equiv 3 - 3 \equiv \mathbf{0} \ (\text{mod } 7).$$

Thus, January 22, 2012, falls on a **Sunday**.   ●

Using the Doomsday method for other centuries

In the Julian calendar, which preceded our Gregorian calendar, every year that was divisible by 4 was a leap year. In the Gregorian calendar, however, only one-fourth of the years that end in 00 are leap years. In the Gregorian calendar, years that are divisible by 400 (such as the year 2000) are leap years, but other multiples of 100 (such as the year 1900) are not. In other words, any year that is divisible by 4 is a leap year, unless it is a multiple of 100 that is not divisible by 400. For example, 1900 is divisible by 100 but not divisible by 400, so 1900 was not a leap year. On the other hand, 2000 is divisible by 400, so 2000 was a leap year.

Check for Understanding

Determine whether each of the following is a leap year in the Gregorian calendar:

1. 2010   2. 1968   3. 1836   4. 1800   5. 1600   6. 2100

Knowing that the Doomsday for 2000 is Tuesday, $D_{2000} \equiv 2 \pmod 7$, we can determine the Doomsday for the 00 year of any century. The year 2100 is not a leap year, since 2100 is divisible by 100 but not by 400. Thus, the number of leap years occurring after 2000, up to (and including) 2100, is the same as the number of leap years occurring after 2000, up to 2096. That number is $L = 96/4 = 24$. From formula (2), the Doomsday for 2100 is

$$D_{2100} \equiv D_{2000} + 100 + 24 \pmod 7$$
$$\equiv D_{2000} + 2 + 3 \pmod 7$$
$$\equiv D_{2000} - 2 \pmod 7.$$

Thus, the Doomsday for 2100 is 2 days earlier than the Doomsday for 2000. Since the Doomsday for 2000 is Tuesday ($\equiv 2$), the Doomsday for 2100 is Sunday ($\equiv 0$).

By similar reasoning, for any century whose 00 year is not a leap year (e.g., 2100), the Doomsday of its 00 year will be 2 days earlier than the Doomsday of the previous century's 00 year. On the other hand, if the 00 year of a century is a leap year (e.g., 2000), then the Doomsday of its 00 year will be 1 day earlier than the Doomsday of the previous century's 00 year (as you will demonstrate in Exercise 12).

EXAMPLE 4

a. Since 2000 was a leap year, its Doomsday is one day earlier than the Doomsday for 1900. The Doomsday for 2000 is Tuesday ($\equiv 2$), so the Doomsday for 1900 must be Wednesday ($\equiv 3$).

b. Since 1900 was not a leap year, its Doomsday is two days earlier than the Doomsday for 1800. The Doomsday for 1900 is Wednesday ($\equiv 3$), so the Doomsday for 1800 must be Friday ($\equiv 5$).

Continuing with the method demonstrated in Example 4, we can find the Doomsday of the 00 year of any century. Some of them are shown in Table 7.5.3.

TABLE 7.5.3 The Doomsday for the 00 year of each century (1600–2100)

Year	1600	1700	1800	1900	2000	2100
Doomsday	Tuesday ($\equiv 2$)	Sunday ($\equiv 0$)	Friday ($\equiv 5$)	Wednesday ($\equiv 3$)	Tuesday ($\equiv 2$)	Sunday ($\equiv 0$)

EXAMPLE 5

Find the day of the week of March 15, 1859.

SOLUTION The Doomsday for 1859 is

$$D \equiv D_{1800} + Y + L \, (\text{mod } 7)$$
$$\equiv 5 + 59 + 14 \, (\text{mod } 7)$$
$$\equiv 5 + 3 + 0 \, (\text{mod } 7)$$
$$\equiv 1 \, (\text{mod } 7).$$

Thus, the Doomsday for 1859 is **Monday**. Choose a key date near March 15 that occurs on the same day of the week as the Doomsday. We choose the last day of February (February 28) as our key date. March 15 is precisely 15 days ($\equiv 1$ modulo 7) after the last day of February, so the day of the week is

$$1 + 1 \equiv 2 \, (\text{mod } 7).$$

Thus, March 15, 1859 was a **Tuesday**.

EXERCISES 7.5

Numerical Problems

In Exercises 1–7, find the day of the week of each important date in history.

1. October 22, 1685   Repeal of the Edict of Nantes, France

2. July 4, 1776   Declaration of Independence, United States

3. June 20, 1837   Victoria becomes queen of England

4. August 6, 1945   Atomic bomb dropped on Hiroshima, Japan

5. July 20, 1969   First man on the moon

6. November 9, 1989   Fall of the Berlin Wall, Germany

7. September 11, 2001   Terrorist attacks on the World Trade Center and Pentagon

8. Calculate on what day of the week you were born.

9. Calculate the day of the week of your birthday this year.

10. Find the day of the week of your 47th birthday.

11. Impress a friend by figuring out the day of the week on which your friend was born.

Reasoning and Proofs

12. Show that the Doomsday for 2000 is 1 day earlier than the Doomsday for 1900. In general, show that if the 00 year of a century is a leap year (e.g., 2000), then the Doomsday of its 00 year will be 1 day earlier than the Doomsday of the previous century's 00 year.

13. Explain why the calendar for the year 1999 was identical to the calendar for the year 1915.

14. Given any two years between 1801 and 1899 that are 28 years apart, the calendar will be identical for both years. Explain why.

15. What is the first year in the future for which the calendar will be identical to this year's calendar? Explain your answer.

16. What is the first year after 2010 in which Halloween will fall on the same day of the week as in 2010, but for which the calendar is not identical to the calendar in 2010? Explain your answer.

17. The true length of the solar year (how long it takes Earth to orbit the Sun) is approximately 365.2424 days.

    a. Find the average length of a year in the Gregorian calendar.

    b. The Gregorian calendar was aligned with the solar year in 1582. Use your answer to part a to estimate in what year there will be a 1-day discrepancy between the Gregorian calendar and the solar year.

    c. To correct for this discrepancy, some have suggested that years divisible by 4000 should not be leap years. If that change were instituted, what would be the new average year length?

18. **Doing it in your head** In formula (3), the values of Y and L can be large, making the calculations harder to do in one's head. To keep the numbers smaller (to make the Doomsday method easier to do in one's head), we could replace $Y + L$ by $D + R + F$, where D is the number of dozens in the year number (D is the quotient when Y is divided by 12), R is the remainder after

this (R is the remainder when Y is divided by 12), and F is the number of fours in this remainder (F is the quotient when R is divided by 4).

a. Use both methods to find the day of the week of the same date in history. Did both methods give the same result?

b. Prove that both methods give the same result, by showing that
$Y + L \equiv D + R + F \pmod 7$.

7.6 The Mayan Calendar

Mayan civilization existed in Central America for thousands of years, from approximately 1000 BCE to 1521 CE. The Maya's lands covered much of what is now Mexico, Guatemala, El Salvador, Honduras, and Belize. At its height, between 250 and 900 CE, the Mayan civilization was one of the greatest civilizations of antiquity. The Maya had the most complex writing system in the new world, elaborate systems for farming and irrigation, and large city-states that flourished for many hundreds of years. Mathematically, the Maya had a *vigesimal* (base 20) number system, and invented zero as a place saver in their written numbers. The Maya also had an impressive and elaborate calendar, which they used to record important events in the lives of their kings on stone pillars.

The Mayan calendar system used two different methods to specify dates: the *calendar round* and the *long count*. The calendar round was used to identify a unique date within a 52-year period. The long count was used to uniquely identify dates over a longer time span than 52 years (see "Fun Facts"). In this section, we will focus on the Mayan calendar round and its relation to modular arithmetic.

Fun Facts

The Maya used the *long count* to specify dates over very long time spans. They believed that time is cyclic, based on a great cycle of 1,872,000 days (more than 5000 years). The Maya used the long count to locate their own time in relation to the start of this great cycle. They believed that at the end of a great cycle the world would be destroyed, only to be re-created for the next great cycle. According to the Mayan long count, the hour of our destruction is at hand—the current great cycle will end on December 23, 2012.

The calendar round

The calendar round was based on three simultaneous cycles: a 13-day cycle called the *Trecena*, a 20-day cycle called the *Veintena*, and a 365-day cycle called the *Haab*. Every day in the calendar round's cycle was specified by a unique combination of three numbers: a day number that identified which of the 13 days in the Trecena it was, a day name that specified which of the 20 days in the Veintena it was, and a *uinal* (month) name and day number that together specified which of the 365 days in the Haab it was.

The calendar round can be thought of as a system of three interlocking gears, illustrated in Figure 1. The smallest gear has 13 teeth (and therefore 13 possible positions), each representing a different day in the Trecena. The next largest gear has 20 teeth (20 possible positions), each representing a different day in the Veintena, and the largest gear has 365 teeth (365 possible positions), each representing a different day in the Haab. Each new day, all three wheels are moved forward by one notch. We will use the term *era* to refer to the amount of time required for the calendar round to cycle through every possible combination of settings of the three wheels.

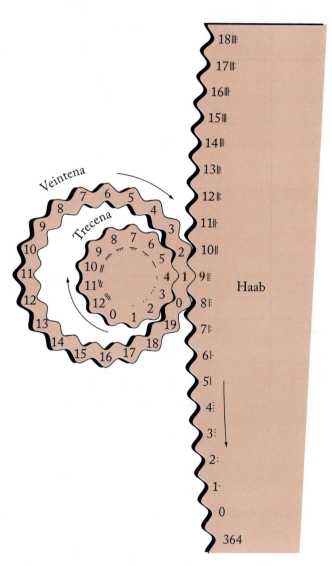

Figure 1 The Mayan calender round was based on three cycles, represented here by three gears: the 13-day Trecena, the 20-day Veintena, and the 365-day Haab. The location where the three gears interlock specifies the current day in each cycle (shown in **brown**): $T = 4$, $V = 1$, $H = 9$.

Throughout this section, we will use T to denote a day's position in the Trecena (modulo 13), V to denote the day's position in the Veintena (modulo 20), and H to denote the day's position in the Haab (modulo 365). We will use E to denote the day's position in the entire era. Assume that on the first day of the era (when $E = 0$),

$$T \equiv 0 \ (\text{mod } 13), V \equiv 0 \ (\text{mod } 20), \text{ and } H \equiv 0 \ (\text{mod } 365).$$

The following table illustrates how each day in an era is specified by the day number in each of the three cycles.

Day in era E	Day in Trecena $T\,(\text{mod } 13)$	Day in Veintena $V\,(\text{mod } 20)$	Day in Haab $H\,(\text{mod } 365)$
0	0	0	0
1	1	1	1
2	2	2	2
⋮	⋮	⋮	⋮
12	12	12	12
13	0	13	13
14	1	14	14
⋮	⋮	⋮	⋮
19	6	19	19
20	7	0	20
21	8	1	21
⋮	⋮	⋮	⋮
364	0	4	364
365	1	5	0
366	2	6	1
⋮	⋮	⋮	⋮

EXAMPLE 1

Find the day number in each of the three cycles for day 4322 in the era.

SOLUTION The day in the era is $E = 4322$.

T is the reduction of E (mod 13), V is the reduction of E (mod 20), and H is the reduction of E (mod 365):

$$T \equiv E \equiv 6 \ (\text{mod } 13)$$
$$V \equiv E \equiv 2 \ (\text{mod } 20)$$
$$H \equiv E \equiv 307 \ (\text{mod } 365).$$

The Tzolkin

The Maya combined the 13-day Trecena cycle and the 20-day Veintena cycle into a single cycle called the *Tzolkin*. The Maya specified any day in the Tzolkin by giving the day number in the Trecena and the day name in the Veintena. What is the length of the Tzolkin? That is, if we start at day 0 of the Trecena and day 0 of the Veintena, how many days will it take before both cycles return to day 0? Suppose it has been d days since both cycles were at 0. The Trecena will be back at 0 if d is a multiple of 13, and the Veintena will be back at 0 if d is a multiple of 20. Thus, both cycles will be at 0 simultaneously if d is a common multiple of 13 and 20 days. The first time this happens is when $d = \text{lcm}(13, 20) = 260$, so the length of the Tzolkin is 260 days.

The system consisting of the Trecena and the Veintena is analogous to two inter-locking gears, one with 13 teeth and one with 20 teeth. This system of gears will return to its starting point every $\text{lcm}(13, 20) = 260$ ticks of the clock. (See Section 3.6 for more details on systems of gears.)

Suppose we know which day today is in the Trecena and in the Veintena. How can we determine which day today is in the Tzolkin? As the following example illustrates, answering this question is easy if we can solve a system of congruences, a skill we mastered in our discussion of the Chinese Remainder Theorem in Section 7.4.

EXAMPLE 2

Suppose today is day 6 in the Trecena and day 11 in the Veintena. Which day of the Tzolkin is it?

SOLUTION Let Z represent the day number in the Tzolkin.

$$Z \equiv 6 \ (\text{mod } 13) \quad \text{and} \quad Z \equiv 11 \ (\text{mod } 20) \tag{1}$$

Thus, there exist integers X and Y such that

$$Z = 6 + 13X \quad \text{and} \quad Z = 11 + 20Y. \tag{2}$$

Combining these two equations gives

$$6 + 13X = 11 + 20Y,$$

which reduces to the linear Diophantine equation

$$13X - 20Y = 5. \tag{3}$$

Since $\gcd(13, 20) = 1$, we know a solution exists (by Theorem 5.3.1).

We can solve this equation (3) using the methods of Chapter 5. We obtain the solution

$$(X, Y) = (-15, -10).$$

Substituting $X = -15$ into the first equation in (2) yields

$$Z = 6 + 13(-15)$$
$$= -189.$$

Since the answer $Z = -189$ is not between 0 and 259, we reduce it modulo 260 to get

$$Z \equiv 71 \ (\text{mod } 260).$$

The reader can verify that 71 is a solution to the original system of congruences (1).

Thus, it must be day 71 in the Tzolkin. Furthermore, by the General Version of the Chinese Remainder Theorem (7.4.2), this solution is unique modulo $260 = \text{lcm}(13, 20)$.

Finding the date in the era

The Maya divided the three wheels of the calendar round system into two sub-systems: the 260-day Tzolkin and the 365-day Haab. Recall that to specify a day in the era, the Maya gave the day in the Trecena, the day in the Veintena, and the day in the Haab. This is equivalent to giving the day in the Tzolkin and the day in the Haab. Thus, the length of the calendar round era is $\text{lcm}(260, 365) = 18{,}980$ days, which is 52 years.

EXAMPLE 3

Suppose today is day 6 in the Trecena, day 11 in the Veintena, and day 226 in the Haab. What is the day number in the era?

SOLUTION First we find the day number in the Tzolkin. We already did this in the previous example and found that day 6 in the Trecena and day 11 in the Veintena corresponds to day 71 in the Tzolkin.

Now we know that it is day 71 in the Tzolkin and day 226 in the Haab, and we must find E, the day in the era. We have

$$E \equiv 71 \ (\text{mod } 260) \quad \text{and} \quad E \equiv 226 \ (\text{mod } 365).$$

By the Equivalent Conditions for Congruence (7.1.2), there exist integers X and Y such that

$$E = 71 + 260X \quad \text{and} \quad E = 226 + 365Y. \tag{4}$$

Combining these two equations gives

$$71 + 260X = 226 + 365Y,$$

which reduces to the linear Diophantine equation

$$260X - 365Y = 155.$$

Since $\gcd(260, 365) = 5$ and $5 \mid 155$, we know a solution exists (by Theorem 5.3.2). Dividing both sides of the equation by the common factor 5 gives

$$52X - 73Y = 31.$$

Using the methods of Chapter 5, we can find the solution:

$$(X, Y) = (-217, -155).$$

Substituting into either equation in (4) gives

$$E = -56{,}349.$$

We reduce this modulo the length of the era, which is 18,980:

$$E \equiv 591 \ (\text{mod } 18{,}980).$$

Thus, today is day 591 in the era, which is in the middle of the second year of the 52-year era.

The reader is invited to check that this solution satisfies the conditions stated in the example:

$$E \equiv 6 \ (\text{mod } 13), \ E \equiv 11 \ (\text{mod } 20), \quad \text{and} \quad E \equiv 226 \ (\text{mod } 365).$$

EXERCISES 7.6

Numerical Problems

1. Find the day number in each of the three cycles (Trecena, Veintena, and Haab) for the given day of the era.

   a. 10,000   b. 7409   c. 303

2. Find the day number in the Tzolkin, given the day number in the Trecena and the Veintena.

   a. Day 5 in the Trecena and day 3 in the Veintena

   b. Day 0 in the Trecena and day 11 in the Veintena

   c. Day 10 in the Trecena and day 0 in the Veintena

3. Find the day number in the era, given the number in the Trecena, the Veintena, and the Haab.

   a. Day 9 in the Trecena, day 11 in the Veintena, and day 161 in the Haab

   b. Day 0 in the Trecena, day 5 in the Veintena, and day 0 in the Haab

   c. Day 1 in the Trecena, day 1 in the Veintena, and day 196 in the Haab

4. According to the Mayan long count, the world will be destroyed on December 23, 2012. Suppose this is correct. Use the Doomsday method (from Section 7.5) to determine on what day of the week the world will end.

Reasoning and Proofs

5. **A trip back in time** You travel back in time to the ancient Mayan civilization. The king of the city-state plans to honor you by offering you up as a ritual sacrifice. It is day 11,342 of the current era, and the king decrees that your sacrifice will take place when the date is day 4 in the Trecena, day 19 in the Veintena, and day 10 in the Haab. How long do you have to live? Explain.

EXPLORATION A New Calendar Unearthed (Exercises 6–9)

You are an archeologist who finds the remains of a previously undiscovered civilization. You find that their calendar, like the Mayan calendar, has three wheels. The first wheel of this newly discovered calendar counts 14 days, the second wheel counts 31 days, and the third wheel counts 731 days.

6. How many days are in their "era"? That is, if the wheels all start at 0, how long does it take all three wheels to return to 0?

7. If it is day 15,000 in their era, which day appears on each wheel?

8. Find the day in their era that corresponds to each of the following:

   a. Day 1 on the first wheel, day 0 on the second wheel, and day 0 on the third wheel

   b. Day 0 on the first wheel, day 1 on the second wheel, and day 0 on the third wheel

   c. Day 0 on the first wheel, day 0 on the second wheel, and day 1 on the third wheel

9. **a.** Find a formula for the day in their era that corresponds to day *a* on the first wheel, day *b* on the second wheel, and day *c* on the third wheel. [*Hint:* Use your answers to Exercise 8.]

b. Use your formula from part a to determine which day in their era corresponds to day 3 on the first wheel, day 19 on the second wheel, and day 100 on the third wheel.

Chapter 8

Alan Turing (1912–1954)

The brilliant mathematician Alan Turing played a fundamental role in founding the field of computer science. Turing spent his early childhood living in foster homes in England while his parents were in India. Alan's father was in the Indian Civil Service, and he and his wife did not return to England until his retirement, when Alan was 14.

From a very young age, Turing showed unusual ability and interest in math and chemistry. Nonetheless, he did not do particularly well in school, even in his math and science classes, because he preferred to develop his own approaches to problems rather than learn the techniques presented by his teachers. While in high school, Turing read Einstein's papers about relativity on his own and taught himself quantum mechanics.

As an undergraduate at Cambridge University, Turing developed an interest in the foundations of mathematics. At age 24, he wrote a paper introducing a mathematical model of a computer, which he called a *universal machine*. These theoretical machines, now known as *Turing machines*, played a central role in the development of theoretical computer science and mathematical logic.

In 1937, Turing entered graduate school at Princeton, where he completed his Ph.D. in mathematical logic and number

theory in just two years. Shortly after Turing returned from Princeton to England, war broke out with Germany, and the British Government Code and Cypher School (GCCS) invited him to join its code-breaking operations. The Germans used a device called the *Enigma machine* to encrypt their radio messages. The system of encryption that the Enigma machine used was very complex and could be changed from one day to the next, making messages extremely hard to decrypt. In 1939, the British learned of a Polish device known as *Bomba Kryptologiczna*, meaning cryptologic bomb, which was used to decrypt Enigma messages that used simple key systems. To address German enhancements to the Enigma machine, in 1940 Turing designed his own machine, known as the *Bombe*, that improved on the Polish *Bomba*. Turing's *Bombe* could decrypt any Engima message of the German Air Force if a small part of the original message could be correctly guessed. The German Navy had its own version of the Engima encryption system that was even more complex than that of the air force, and many people considered it to be unbreakable. Finally, in 1941, using statistical methods, Turing and his colleagues at the GCCS succeeded in deciphering messages of the German Navy as well.

From 1942 until the end of World War II, Turing focused on developing a speech secrecy system so that Roosevelt and Churchill could speak securely on the telephone. During the same period, as the Germans continued to make their encryption systems more and more complicated, Turing acted as a consultant to the British code-breaking operations. The success that the Allies had in deciphering German messages throughout the war played a significant role in the Allied victory, and that success was in large part due to the work of Alan Turing.

While the British government applauded Turing's work, the authorities were not so happy about his personal life. Turing was gay during an era when homosexuality was illegal in Britain. He reported a blackmail threat to the police in 1952, and in answer to their questions Turing told the police details of a gay relationship that he had had. He was then tried on homosexuality charges and found guilty. Turing did not deny the charges but instead asserted that there was nothing wrong with homosexuality. The judge gave him the choice of one year in prison or one year of estrogen injections. Turing chose the injections.

After his conviction, Turing attempted to continue his work, but his security clearance was revoked, and he lost his job with the Code unit of the Government Communications Headquarters. Furthermore, the police were suspicious of any foreign visitors that Turing had, as well as any trips that he made abroad. Turing was no longer free to do his work or to live his life the way he was used to. At the age of 41, Turing died of an apparent suicide.

A bit of Turing's math

Mathematicians in the early 20th century, including David Hilbert, proposed the development of formal axiomatic systems in the hopes that all mathematical truths could be proved using such a system.[1] Given a fixed axiom system (set of axioms), one would like to find a procedure to decide whether or not any given mathematical statement is provable using that system. Hilbert asked whether every axiom system has such a decision procedure. Answering Hilbert's question required a formal definition of what is meant by "procedure." In order to give such a definition, Turing introduced a theoretical machine (now called a *Turing machine*) that follows a finite set of rules to read, write, and delete symbols on a virtual paper tape. It has become widely accepted that the Turing machine exactly captures the intuitive notion of what can be computed: any mathematical function that can be performed by any computing device can be computed using a Turing machine.[2] Thus, a procedure may be reasonably defined as anything that can be carried out by a Turing machine. Turing proved that there is no Turing machine that can determine whether an arbitrary mathematical statement is provable. The answer to Hilbert's question, therefore, is no—there is no procedure that can decide whether any mathematical statement is provable!

Long before the technology to build an electronic computer existed, the Turing machine provided a blueprint for how a computer might work. At the end of World War II, the National Physical

---

[1] The limitations of this approach later became apparent as a result of Kurt Gödel's Incompleteness Theorem as well as the work of Turing mentioned here.

[2] This assertion is known as the Church-Turing Thesis.

Laboratory of Britain asked Turing to design a computer. In 1946, he produced a design for a computer that he called the Automatic Computing Engine (ACE), and in 1947, he designed the first programming language. Though the ACE was never actually built, Turing's ideas were fundamental in constructing the first real computers and in developing the field of computer science in general.

Turing was also interested in the philosophical question of whether computers could have intelligence. He suggested an "imitation game" in which a subject would type questions to a computer and to a human. The computer and the human would each reply with typed responses. Turing argued that if the subject were unable to tell which responses came from the computer and which came from the human, then the computer must be as intelligent as the human. This "imitation game" has come to be known as the *Turing test* for intelligence.

Chapter 8 Modular Number Systems

8.1 The Number System Z_n: An Informal View

A new number system

Let's revisit the modulo 6 world, in which there are only 6 numbers: 0, 1, 2, 3, 4, and 5. (This is the world that Gauss found on Venus in our math myth in Chapter 7.) When we add and multiply these numbers, we do our arithmetic modulo 6; that is, after performing the addition or multiplication, we reduce the answer modulo 6. For example,

$$5 + 4 \equiv 3 \ (\text{mod } 6) \qquad \text{and} \qquad 5 \cdot 4 \equiv 2 \ (\text{mod } 6). \tag{1}$$

It is not hard to write down complete tables for addition modulo 6 and multiplication modulo 6 for the set $\{0, 1, 2, 3, 4, 5\}$:

Addition modulo 6

+	0	1	2	3	4	5
0	0	1	2	3	4	5
1	1	2	3	4	5	0
2	2	3	4	5	0	1
3	3	4	5	0	1	2
4	4	5	0	1	2	3
5	5	0	1	2	3	4

Multiplication modulo 6

·	0	1	2	3	4	5
0	0	0	0	0	0	0
1	0	1	2	3	4	5
2	0	2	4	0	2	4
3	0	3	0	3	0	3
4	0	4	2	0	4	2
5	0	5	4	3	2	1

The set of numbers $\{0, 1, 2, 3, 4, 5\}$, together with these operations (addition and multiplication modulo 6), is called $\mathbf{Z_6}$.

Arithmetic properties of Z_6

Even though arithmetic in this modulo 6 number system may seem somewhat peculiar to the average Earthling, we might wonder which familiar arithmetic properties hold in $\mathbf{Z_6}$. For example, is addition modulo 6 *commutative*? Yes, surely adding $5 + 4$ and reducing modulo 6 gives the same answer as adding $4 + 5$ and reducing modulo 6. This is because $5 + 4$ and $4 + 5$ are equal as integers, and therefore they reduce to the same number modulo 6. Similarly, multiplication modulo 6 is commutative.

Another basic property of many familiar number systems is the existence of *additive inverses*. In $\mathbf{Z}$, for example, the additive inverse of 2 is -2, because

2 and -2 add up to 0. In fact, every integer has an additive inverse. Is the same thing true in our modulo 6 number system? For instance, does 2 have an additive inverse in $\mathbf{Z}_6$? In other words, we want to know whether there is a member y of our number system such that

$$2 + y \equiv 0 \ (\text{mod } 6).$$

The number $y = 4$ does the trick: $2 + 4 \equiv 0 \ (\text{mod } 6)$. Thus, 4 is the additive inverse of 2 in $\mathbf{Z}_6$. Similarly, the reader can verify that every element of $\mathbf{Z}_6$ has an additive inverse in $\mathbf{Z}_6$.

Check for Understanding

1. Find the additive inverse of 1 in Z_6.

2. Find the additive inverse of 3 in Z_6.

It seems that $\mathbf{Z}_6$ possesses many of the same properties as the number systems we are already familiar with, such as $\mathbf{Z}, \mathbf{Q}, \mathbf{R}$, and $\mathbf{C}$. In this section, we work informally with $\mathbf{Z}_6$ to explore some of these properties. Of course, there's nothing special about the modulus 6. We could similarly define a modulo 10 number system, or a modulo 31 number system; and in the exercises, you will work informally with such systems. Once we have gained an intuitive feel for these number systems, we will be ready in Section 8.2 to formally define the modulo n number system, $\mathbf{Z}_n$, for any $n \in \mathbf{N}$ and prove that it satisfies a host of familiar arithmetic properties.

Patterns in the tables

One interesting difference between $\mathbf{Z}_6$ and the more familiar number systems $\mathbf{Z}, \mathbf{Q}, \mathbf{R}$, and $\mathbf{C}$ is that whereas the more familiar number systems are all infinite, $\mathbf{Z}_6$ has only a finite number of elements. While the idea of a number system with finitely many elements may seem somewhat unusual at first, this finiteness is quite a pleasing feature of modular number systems. Unlike the addition and multiplication tables we learned in elementary school up to $9 + 9$ or $12 \cdot 12$, the addition and multiplication tables for $\mathbf{Z}_6$ are complete: they tell how to add and multiply any pair of numbers in $\mathbf{Z}_6$. Thus, the operations in the number system $\mathbf{Z}_6$ are completely specified by the tables.

Even a brief glance at the tables reveals many patterns and symmetries. Indeed, many of the important properties of $\mathbf{Z}_n$ can be seen directly by examining its addition and multiplication tables. Before continuing, you're invited to find as many patterns as possible in the $\mathbf{Z}_6$ tables shown above.

One pattern shared by both the addition and multiplication tables is symmetry about the diagonal running from top-left to bottom-right. For instance, consider the multiplication table for Z_6:

Multiplication modulo 6

·	0	1	2	3	4	5
0	0	0	0	0	0	0
1	0	1	2	3	4	5
2	0	2	4	0	2	4
3	0	3	0	3	0	3
4	0	4	2	0	4	2
5	0	5	4	3	2	1

If the entire table is reflected across the diagonal line shown, none of the entries change. Why is this symmetry present? Is there an algebraic property of our number system that explains this symmetry? To answer this, it helps to focus in on a single entry—say, the entry at the $5 \cdot 2$ position, which is a **4**. If we reflect the position of this 4 about the diagonal, we find another **4**, which occupies the $2 \cdot 5$ position. Now it becomes clear why the symmetry is present: commutativity of multiplication dictates that $5 \cdot 2$ and $2 \cdot 5$ are equal, and hence the entries at these positions in the table will be the same.

You will explore other patterns in the tables, and their relation to algebraic properties of Z_n, in the exercises.

EXERCISES 8.1

Numerical Problems

Exercises 1–5 concern the number system Z_n, which is the set $\{0, 1, \ldots, n - 1\}$ together with the operations of addition and multiplication modulo n.

1. Construct addition and multiplication tables for Z_2.

2. Construct addition and multiplication tables for Z_3.

3. Construct addition and multiplication tables for Z_4.

4. Construct addition and multiplication tables for Z_5.

5. Construct addition and multiplication tables for Z_{10}.

6. Use the tables you made in Exercises 1–5 to answer the following questions.

   a. For each element x of Z_2, find the additive inverse of x, if it has one.

   b. For each element x of Z_3, find the additive inverse of x, if it has one.

   c. For each element x of Z_4, find the additive inverse of x, if it has one.

d. For each element x of $\mathbf{Z}_5$, find the additive inverse of x, if it has one.

e. For each element x of $\mathbf{Z}_{10}$, find the additive inverse of x, if it has one.

7. Use your tables from Exercise 2 to find all solutions to the following equations in $\mathbf{Z}_3$. (If there are no solutions, say so.)

    a. $x + 2 = 0$ **b.** $2x = 1$ **c.** $2x + 2 = 1$

    d. $2x = 0$ **e.** $x^2 = 1$ **f.** $x^2 = 2$

8. Use your tables from Exercise 3 to find all solutions to the following equations in $\mathbf{Z}_4$. (If there are no solutions, say so.)

    a. $x + 3 = 2$ **b.** $2x = 1$ **c.** $3x = 1$

    d. $3x + 2 = 1$ **e.** $2x + 1 = 3$ **f.** $x^2 = 0$

9. Use your tables from Exercise 4 to find all solutions to the following equations in $\mathbf{Z}_5$. (If there are no solutions, say so.)

    a. $x + 3 = 0$ **b.** $2x = 1$ **c.** $4x = 1$

    d. $3x + 3 = 1$ **e.** $x^2 = 3$ **f.** $x^2 = 4$

10. Use your tables from Exercise 5 to find all solutions to the following equations in $\mathbf{Z}_{10}$. (If there are no solutions, say so.)

    a. $x + 7 = 3$ **b.** $2x = 4$ **c.** $5x = 2$

    d. $3x + 5 = 4$ **e.** $x^2 = 6$ **f.** $x^2 = 0$

Reasoning and Proofs

11. Like the multiplication table, the addition table for $\mathbf{Z}_6$ is symmetric about the diagonal running from top-left to bottom-right. What property of the number system explains this symmetry?

12. It is possible to tell by looking at the $\mathbf{Z}_6$ addition table that every element has an additive inverse. Explain how.

13. A *multiplicative inverse* of an element $x \in \mathbf{Z}_6$ is an element $y \in \mathbf{Z}_6$ such that $xy = 1$. For each element x of $\mathbf{Z}_6$, find a multiplicative inverse, if one exists.

8.2 The Number System $\mathbf{Z}_n$: Definition and Basic Properties

Elements of Z_n are congruence classes

In the last section, we introduced $\mathbf{Z}_6$ as the number system consisting of the numbers $0, 1, 2, 3, 4$, and 5, under the operations of addition modulo 6 and multiplication modulo 6.

Before making a formal definition of $\mathbf{Z}_6$ (or defining $\mathbf{Z}_n$ in general), we will discuss a few notational issues that will enable us to make a definition that is easy to work with.

When working in a modular number system such as $\mathbf{Z}_6$, we will write a bar over each number:

$$\mathbf{Z}_6 = \{\bar{0}, \bar{1}, \bar{2}, \bar{3}, \bar{4}, \bar{5}\}.$$

This notation allows us to distinguish the elements of $\mathbf{Z}_6$ from elements of $\mathbf{Z}$. This is particularly helpful in writing equations, such as

$$\bar{5} + \bar{4} = \bar{3} \quad \text{and} \quad \bar{5} \cdot \bar{4} = \bar{2}, \tag{1}$$

which hold in $\mathbf{Z}_6$. On the other hand, the equations $5 + 4 = 3$ and $5 \cdot 4 = 2$ are false, because without the bars, these numbers refer to ordinary integers.

The numbers $\bar{0}, \bar{1}, \bar{2}, \bar{3}, \bar{4}$, and $\bar{5}$ may be thought of as the individual elements that make up the number system $\mathbf{Z}_6$. But there is another way to think of these barred numbers. When the inhabitants of a modulo 6 world (such as Venus) use the number $\bar{2}$, we Earthlings know that they may be thinking about any number that is congruent to 2 modulo 6. Thus, it is reasonable to think of $\bar{2}$ as the set of all integers that are congruent to 2 modulo 6:

$$\bar{2} = \{\ldots, -10, -4, 2, 8, 14, \ldots\} = \{2 + 6k \mid k \in \mathbf{Z}\}. \tag{2}$$

The reader will recognize this set as the congruence class of 2 modulo 6. At first it may seem odd to have $\bar{2}$, a single element of $\mathbf{Z}_6$, representing an infinite set of integers. However, this idea turns out to be so useful and powerful[3] that we will take equation (2) as our *definition* of $\bar{2}$ (in Definition 8.2.1). The other elements $\bar{a}$ in $\mathbf{Z}_6$ are defined similarly:

$$\bar{a} = \{\ldots, a - 12, a - 6, a, a + 6, a + 12, \ldots\} = \{a + 6k \mid k \in \mathbf{Z}\}; \tag{3}$$

that is, the number $\bar{a}$ in $\mathbf{Z}_6$ is the congruence class of a modulo 6. With this definition, the set $\mathbf{Z}_6 = \{\bar{0}, \bar{1}, \bar{2}, \bar{3}, \bar{4}, \bar{5}\}$, consists of six elements, each of which is a congruence class modulo 6.

Note that equation (3) makes sense even for numbers outside the range $0, 1, \ldots, 5$. Thus, for example, we have

$$\bar{8} = \{\ldots, -4, 2, 8, 14, 20, \ldots\} = \{8 + 6k \mid k \in \mathbf{Z}\}.$$

This set of numbers is easily seen to be the same as $\bar{2}$. This is because $8 \equiv 2 \pmod 6$, so by Theorem 7.1.8, the congruence class of 8 modulo 6 is equal to the congruence class of 2 modulo 6. Thus, $\bar{8}$ is just another name for the set $\bar{2}$:

$$\bar{8} = \bar{2}.$$

---

[3]The idea of doing arithmetic with (or finding other structure in) congruence classes is ubiquitous in higher mathematics.

Is $\bar{8}$ an element of $\mathbf{Z}_6$? At first might it seem that the answer is no, since the elements of $\mathbf{Z}_6$ are $\bar{0}, \bar{1}, \bar{2}, \bar{3}, \bar{4}$, and $\bar{5}$. However, $\bar{8} = \bar{2}$, so indeed $\bar{8}$ is an element of $\mathbf{Z}_6$. Similarly, $\overline{23}$ is an element of $\mathbf{Z}_6$, since $\overline{23} = \bar{5}$. Thus, we see that every element of $\mathbf{Z}_6$ goes by many different names.

More generally, if a is any integer, then $\bar{a}$ is an element of $\mathbf{Z}_6$. It follows directly from Corollary 7.1.9 that the congruence class $\bar{a}$ is equal to one of the classes $\bar{0}, \bar{1}, \bar{2}, \bar{3}, \bar{4}$, or $\bar{5}$. Furthermore, Theorem 7.1.8 tells us exactly when two barred numbers are equal:

$$\bar{a} = \bar{b} \text{ in } \mathbf{Z}_6 \text{ if and only if } a \equiv b \pmod{6}.$$

> ## Check for Understanding
>
> 1. In $\mathbf{Z}_6$, the element $\bar{4}$ represents a certain congruence class. List a few elements of this class.
>
> 2. In $\mathbf{Z}_6$, the number $\overline{74}$ goes by another more familiar name. What is it?

Arithmetic with congruence classes

Thus far, we've introduced $\mathbf{Z}_6$ as the set $\{\bar{0}, \bar{1}, \bar{2}, \bar{3}, \bar{4}, \bar{5}\}$ of all congruence classes modulo 6. We still need to formally define the operations of addition and multiplication on this set. Informally, we already know how to do this. To add or multiply two barred numbers, we just add or multiply the numbers under the bars, and then reduce the result modulo 6. For example,

$$\bar{3} + \bar{5} = \bar{2}. \tag{4}$$

because $3 + 5 = 8$, which reduces to 2 modulo 6.

Rather than defining addition and multiplication this way, however, we will take advantage of the fact that every element of $\mathbf{Z}_6$ goes by many different names. Since $\bar{2} = \bar{8}$, we can write equation (4) equivalently as

$$\bar{3} + \bar{5} = \bar{8}. \tag{5}$$

As a candidate for a definition of addition, equation (5) is easier to work with than equation (4), because equation (5) simply requires us to add the numbers under the bars and does not require us to reduce the result.

In general, we will define addition of two elements in $\mathbf{Z}_6$ simply by adding the numbers under the bars:

$$\bar{a} + \bar{b} = \overline{a + b}.$$

This equation is remarkable since it defines addition in $\mathbf{Z}_6$ without any mention of reducing modulo 6.

Similarly, we will define multiplication in $\mathbf{Z}_6$ by

$$\bar{a} \cdot \bar{b} = \overline{ab}.$$

For example, to find $\bar{2} \cdot \bar{5}$, we may simply multiply the numbers under the bars to get $\overline{10}$, which is the correct answer. Since $\overline{10} = \bar{4}$, one may also report the answer as $\bar{4}$, a form which may be preferable (especially if one wants to communicate with a Venusian).

Definition of $\mathbf{Z}_n$

Having discussed $\mathbf{Z}_6$ and its operations, we are now ready to define the number system $\mathbf{Z}_n$ for any natural number n.

DEFINITION 8.2.1

Let $n \in \mathbf{N}$. We define $\mathbf{Z}_n$, the integers modulo n, by

$$\mathbf{Z}_n = \{\bar{0}, \bar{1}, \bar{2}, \ldots, \overline{n-1}\}.$$

where $\bar{a}$ is defined, for any $a \in \mathbf{Z}$, to be the congruence class of a modulo n:

$$\bar{a} = \{a + kn \mid k \in \mathbf{Z}\}.$$

The operations of **addition and multiplication in $\mathbf{Z}_n$** *are defined by*

$$\bar{a} + \bar{b} = \overline{a+b}, \qquad and$$

$$\bar{a} \cdot \bar{b} = \overline{ab}.$$

REMARK. The meaning of a number like $\bar{2}$ changes depending on which modular number system we are working in. Working in $\mathbf{Z}_6$, for example, $\bar{2} = \{\ldots, -10, -4, 2, 8, 14, \ldots\}$ is the congruence class of 2 modulo 6. In $\mathbf{Z}_7$, though, $\bar{2} = \{\ldots, -12, -5, 2, 9, 16, \ldots\}$ is the congruence class of 2 modulo 7. This ambiguity will not be a problem, since the value of n, which determines which modular system we are working in, will be carefully specified before we use any barred numbers.

It might appear from our definition that $\mathbf{Z}_n$ consists of only *some* of the congruence classes modulo n—namely,

the class of 0, the class of 1, ..., the class of $n-1$.

However, this is actually a list of *all* the congruence classes modulo n, by Corollary 7.1.9. Thus, for any $a \in \mathbf{Z}$, $\bar{a}$ is an element of $\mathbf{Z}_n$, even if a is not in the range $0, 1, 2, \ldots, n - 1$. We could also define $\mathbf{Z}_n$ as

$$\mathbf{Z}_n = \{\bar{a} \mid a \in \mathbf{Z}\}.$$

It easily follows from this alternate definition that $\mathbf{Z}_n$ is **closed** under the operations $+$ and $\cdot$ defined in 8.2.1. That is, given any two elements of $\mathbf{Z}_n$, their sum is an element of $\mathbf{Z}_n$, and so is their product.

LEMMA 8.2.2

Let $n \in \mathbf{N}$. Then $\mathbf{Z}_n$ is closed under the operations $+$ and $\cdot$.

PROOF We first show that $\mathbf{Z}_n$ is closed under addition.

Let $\bar{a}, \bar{b} \in \mathbf{Z}_n$.

[To show: $\bar{a} + \bar{b} \in \mathbf{Z}_n$.]

By definition, $\bar{a} + \bar{b} = \overline{a + b}$. Since a and b are integers, the ordinary sum $a + b$ is an integer. Thus, $\overline{a + b} \in \mathbf{Z}_n$, and hence $\bar{a} + \bar{b} \in \mathbf{Z}_n$. We conclude that $\mathbf{Z}_n$ is closed under addition.

A similar argument shows that $\mathbf{Z}_n$ is closed under multiplication. ●

In our discussion of $\mathbf{Z}_6$, we noted that each element of $\mathbf{Z}_6$ goes by many different names. For example, we noted that $\bar{8} = \bar{2}$ in $\mathbf{Z}_6$. This is true because $8 \equiv 2 \pmod{6}$, so the congruence class of 8 modulo 6 is the same as the congruence class of 2 modulo 6 (by Theorem 7.1.8). In general, two elements $\bar{a}, \bar{b} \in \mathbf{Z}_n$ are equal if and only if a and b are congruent modulo n. We state this in the following lemma, which is a restatement of Theorem 7.1.8.

LEMMA 8.2.3

Let $n \in \mathbf{N}$. For any $a, b \in \mathbf{Z}$, $\bar{a} = \bar{b}$ in $\mathbf{Z}_n$ if and only if $a \equiv b \pmod{n}$.

PROOF This lemma follows immediately from Theorem 7.1.8.

This lemma allows us to expand our Equivalent Conditions for Congruence (7.1.2):

THEOREM 8.2.4   EQUIVALENT CONDITIONS FOR CONGRUENCE

Let $a, b \in \mathbf{Z}, n \in \mathbf{N}$.

The following statements are all equivalent.

$$\bar{a} = \bar{b} \text{ in } \mathbf{Z}_n \iff a \equiv b \,(\mathrm{mod}\ n) \iff n \mid a - b \iff \exists k \in \mathbf{Z} \text{ such that } a = b + kn$$

$$\iff \text{ when } a \text{ and } b \text{ are divided by } n \text{ they leave the same remainder.}$$

EXAMPLE 1

Working in $\mathbf{Z}_5 = \{\bar{0}, \bar{1}, \bar{2}, \bar{3}, \bar{4}\}$, use the definition of $\mathbf{Z}_n$ to find the sum $\bar{3} + \bar{4}$.

SOLUTION   By the definition of addition in $\mathbf{Z}_5$,

$$\bar{3} + \bar{4} = \overline{3 + 4} = \bar{7}.$$

At this point, we may be satisfied that we have found the correct answer. However, it would be preferable to express our answer as one of the numbers $\bar{0}, \bar{1}, \bar{2}, \bar{3},$ or $\bar{4}$. Since $7 \equiv 2 \,(\mathrm{mod}\ 5)$, we have $\bar{7} = \bar{2}$ in $\mathbf{Z}_5$. Thus, we conclude that in $\mathbf{Z}_5, \bar{3} + \bar{4} = \bar{2}$.   ●

Note that in this example, we added $3 + 4$ and reduced modulo 5. Thus, our definition of addition in $\mathbf{Z}_n$ captures our intuitive idea (from Section 8.1) that to add numbers in $\mathbf{Z}_n$, we just add the numbers as integers and then reduce modulo n.

The operations on $\mathbf{Z}_n$ are well defined

Before we can further explore our newly defined number system $\mathbf{Z}_n$, there is a subtle but important feature of Definition 8.2.1: we must check that addition and multiplication in $\mathbf{Z}_n$ are *well defined*. To understand what this means, we will consider an example of an operation defined on a more familiar set: $\mathbf{Q}$, the set of rational numbers.

Phil Lovett   Your friend Phil Lovett is excited about his discovery of a cool new operation, denoted #, on the set of rational numbers. His definition of the "Lovett product" of any two rational numbers $\frac{a}{c}$ and $\frac{b}{d}$ is

$$\frac{a}{c} \,\#\, \frac{b}{d} = \frac{a + b}{c + d}$$

Phil says: Just add the numerators and add the denominators.

Intrigued by your friend's enthusiasm, you decide to compute the Lovett product of your favorite two rational numbers, 0.4 and 0.5. Since $0.4 = \dfrac{4}{10}$ and $0.5 = \dfrac{5}{10}$, you compute

$$\frac{4}{10} \mathbin{\#} \frac{5}{10} = \frac{9}{20}. \tag{6}$$

The next day, you decide to think some more about Phil's operation. Having forgotten your original computation, you decide to compute the Lovett product $0.4 \mathbin{\#} 0.5$ again. This time you figure that since $0.4 = \dfrac{2}{5}$ and $0.5 = \dfrac{1}{2}$ the Lovett product will be

$$\frac{2}{5} \mathbin{\#} \frac{1}{2} = \frac{3}{7},$$

quite a different answer than the $\dfrac{9}{20}$ you got in equation (6). Apparently, the value of the Lovett product $0.4 \mathbin{\#} 0.5$ can change, depending on how you choose to represent 0.4 and 0.5 as fractions. You have discovered a major problem with your friend's proposed operation: it is not well defined. For $\#$ to be a valid operation on $\mathbf{Q}$, the value of $x \mathbin{\#} y$ must depend only on *what the numbers x and y are*, and must not depend on *how x and y are represented* as fractions.

Our number system $\mathbf{Z}_n$ shares with $\mathbf{Q}$ the property that each of its elements may be expressed in many different ways. For example, in $\mathbf{Z}_6$, $\overline{-1} = \overline{5} = \overline{11} = \overline{17}$. Because of this, we must check to see that addition and multiplication, as given in Definition 8.2.1, are well defined. Before doing so, let's look at an example.

Naomi's Numerical Proof Preview: Proposition 8.2.5

Suppose that in $\mathbf{Z}_6$, you wish to use Definition 8.2.1 to compute the sum $\overline{5} + \overline{2}$. The definition tells us that this sum is:

$$\overline{5} + \overline{2} = \overline{7}.$$

Now let's choose a different way of representing the numbers $\overline{5}$ and $\overline{2}$. Say we choose $\overline{17}$ (which equals $\overline{5}$) and $\overline{8}$ (which equals $\overline{2}$) and compute the sum again using Definition 8.2.1. The definition tells us that the sum $\overline{17} + \overline{8}$ is:

$$\overline{17} + \overline{8} = \overline{25}.$$

Note that since $\overline{25} = \overline{7}$, we did in fact get the same result even though we represented each of the numbers being added ($\overline{5}$ and $\overline{2}$) in a different way.

In general, if $\overline{a} = \overline{x}$ and $\overline{b} = \overline{y}$, then the value that Definition 8.2.1 gives for the sum $\overline{a} + \overline{b}$ is equal to the value that Definition 8.2.1 gives for the sum $\overline{x} + \overline{y}$. That is, $\overline{a + b} = \overline{x + y}$. We prove this in the following proposition.

The operations addition and multiplication in $\mathbf{Z}_n$, as given by the formulas in Definition 8.2.1, are well defined. That is,

(i)   For any $a, b, x, y \in \mathbf{Z}$, if $\overline{a} = \overline{x}$ and $\overline{b} = \overline{y}$, then $\overline{a + b} = \overline{x + y}$.

(ii)   For any $a, b, x, y \in \mathbf{Z}$, if $\overline{a} = \overline{x}$ and $\overline{b} = \overline{y}$, then $\overline{ab} = \overline{xy}$.

PROOF

Proof of (i)   Suppose that $\overline{a} = \overline{x}$ and $\overline{b} = \overline{y}$.

Then

$$a \equiv x \;(\text{mod } n) \qquad \text{and} \qquad b \equiv y \;(\text{mod } n)$$

by Theorem 8.2.4 (Equivalent Conditions for Congruence). Adding these two congruences (using Theorem 7.2.1, Congruences Add and Multiply), we obtain

$$a + b \equiv x + y \;(\text{mod } n).$$

It follows that

$$\overline{a + b} = \overline{x + y},$$

again by Theorem 8.2.4.

Proof of (ii)   The proof of (ii) is similar to the proof of (i).   ●

If this proof seemed short and sweet, that's because we had already done the real work when we proved Theorem 7.2.1 (Congruences Add and Multiply). Indeed, Proposition 8.2.5 is simply a restatement of Theorem 7.2.1 in the language of barred numbers.

$\mathbf{Z}_n$ is a ring

In Section 8.1, we noticed that $\mathbf{Z}_n$ possesses familiar algebraic properties, such as commutativity of addition and multiplication, and existence of additive inverses. The following theorem gives us a more complete list of the algebraic properties enjoyed by $\mathbf{Z}_n$. Mathematicians use the word **ring** to refer to any number system that possesses all of the properties listed in the following theorem.[4] Thus, all eight assertions in the following theorem can be encapsulated in a single sentence: $\mathbf{Z}_n$ is a ring.

---

[4]Often the definition of *ring* does not include our last two properties of multiplication (commutativity of multiplication and the existence of a multiplicative identity element). We have actually defined what most mathematicians would call a *commutative ring with unity*.

THEOREM 8.2.6 $\mathbf{Z}_n$ IS A RING

Let $n \in \mathbf{N}$. Then $\mathbf{Z}_n$ satisfies the following arithmetical properties.

Properties of Addition

- Associativity. *For every $\bar{a}, \bar{b},$ and $\bar{c}$ in $\mathbf{Z}_n$, $(\bar{a} + \bar{b}) + \bar{c} = \bar{a} + (\bar{b} + \bar{c})$.*

- Commutativity. *For every $\bar{a}$ and $\bar{b}$ in $\mathbf{Z}_n$, $\bar{a} + \bar{b} = \bar{b} + \bar{a}$.*

- Identity. *The element $\bar{0}$ in $\mathbf{Z}_n$ is an **additive identity** element:*

  For every $\bar{a}$ in $\mathbf{Z}_n$, $\bar{a} + \bar{0} = \bar{a}$.

- Additive Inverses. *For every $\bar{a}$ in $\mathbf{Z}_n$, there exists an **additive inverse**, $-\bar{a}$ in $\mathbf{Z}_n$, such that $\bar{a} + (-\bar{a}) = \bar{0}$.*

Properties of Multiplication

- Associativity. *For every $\bar{a}, \bar{b},$ and $\bar{c}$ in $\mathbf{Z}_n$, $(\bar{a} \cdot \bar{b}) \cdot \bar{c} = \bar{a} \cdot (\bar{b} \cdot \bar{c})$.*

- Commutativity. *For every $\bar{a}$ and $\bar{b}$ in $\mathbf{Z}_n$, $\bar{a} \cdot \bar{b} = \bar{b} \cdot \bar{a}$.*

- Identity. *The element $\bar{1}$ in $\mathbf{Z}_n$ is a **multiplicative identity** element:*

  For every $\bar{a}$ in $\mathbf{Z}_n$, $\bar{a} \cdot \bar{1} = \bar{a}$.

Properties Relating Addition and Multiplication

- Distributivity. *For every $\bar{a}, \bar{b},$ and $\bar{c}$ in $\mathbf{Z}_n$, $\bar{a} \cdot (\bar{b} + \bar{c}) = (\bar{a} \cdot \bar{b}) + (\bar{a} \cdot \bar{c})$.*

The reader is no stranger to rings; the familiar number systems $\mathbf{Z}, \mathbf{Q}, \mathbf{R},$ and $\mathbf{C}$ with their usual operations of addition and multiplication are all examples of rings. The study of rings in general belongs to the subject of abstract algebra. For a more thorough discussion of rings and their properties, see the Appendix (on the Student Companion Website). For our purposes here, knowing that $\mathbf{Z}_n$ is a ring means that we can use all of the properties listed in the theorem above when working with this number system. Thus, when we're dealing with $\mathbf{Z}_n$, it's pretty much arithmetic as usual.[5]

PROOF OF **THEOREM 8.2.6**

Let $n \in \mathbf{N}$.

We first prove associativity of addition. Let $\bar{a}, \bar{b}, \bar{c} \in \mathbf{Z}_n$. We must show that $(\bar{a} + \bar{b}) + \bar{c} = \bar{a} + (\bar{b} + \bar{c})$. To do this, we begin manipulating the left side:

---

[5] The astute reader may have noticed that the definition of a ring requires the existence of additive inverses but does not require the existence of multiplicative inverses. We will consider the subject of multiplicative inverses in Section 8.3.

$$(\bar{a} + \bar{b}) + \bar{c} = \overline{a + b} + \bar{c} \quad \leftarrow \textbf{by definition of addition in } \mathbf{Z}_n$$
$$= \overline{(a + b) + c} \quad \leftarrow \textbf{again using the definition of addition in } \mathbf{Z}_n$$
$$= \overline{a + (b + c)} \quad \leftarrow \textbf{by associativity of addition in the integers}$$
$$= \bar{a} + \overline{b + c} \quad \leftarrow \textbf{by definition of addition in } \mathbf{Z}_n$$
$$= \bar{a} + (\bar{b} + \bar{c}), \quad \leftarrow \textbf{by definition of addition in } \mathbf{Z}_n$$

and associativity of addition is established.

Note that all one needs for this proof, aside from the definition of addition in $\mathbf{Z}_n$, is associativity of addition in the integers. This idea may be expressed by saying that associativity of addition in $\mathbf{Z}_n$ is *inherited* from the integers. The properties of commutativity of addition, associativity and commutativity of multiplication, and distributivity of multiplication over addition in $\mathbf{Z}_n$ are similarly inherited from $\mathbf{Z}$. Their proofs are left to you in Exercise 11.

The element $\bar{0} \in \mathbf{Z}_n$ serves as an additive identity because for every $\bar{a} \in \mathbf{Z}_n$,

$$\bar{a} + \bar{0} = \overline{a + 0}$$
$$= \bar{a}.$$

The proof that $\bar{1} \in \mathbf{Z}_n$ is a multiplicative identity is similar, and it is left to you in Exercise 11.

It only remains to establish that every element $\bar{a} \in \mathbf{Z}_n$ has an additive inverse.

Claim: The element $\overline{-a} \in \mathbf{Z}_n$ is an additive inverse of $\bar{a}$.

Proof of Claim

$$\bar{a} + \overline{-a} = \overline{a + (-a)} \quad \leftarrow \textbf{by definition of addition in } \mathbf{Z}_n$$
$$= \bar{0} \quad \leftarrow \textbf{since } a + (-a) = 0 \textbf{ in } \mathbf{Z}.$$

Hence, $\overline{-a}$ is an additive inverse of $\bar{a}$, and the Claim is established.

Thus, $\mathbf{Z}_n$ is a ring. ●

In this proof, we showed that the element $\overline{-a}$ is an additive inverse of $\bar{a}$ in $\mathbf{Z}_n$ (i.e., $-\bar{a} = \overline{-a}$). As you will prove in Exercise 14, every element of $\mathbf{Z}_n$ has a *unique* additive inverse. Thus, instead of referring to *an* additive inverse of $\bar{a}$, we may refer to *the* additive inverse of $\bar{a}$.

Now that we have addition and additive inverses in $\mathbf{Z}_n$, it is easy to define *subtraction* in $\mathbf{Z}_n$, as follows. For any $\bar{a}, \bar{b} \in \mathbf{Z}_n$, we define $\bar{b} - \bar{a} = \bar{b} + (-\bar{a})$.

EXAMPLE 2

Find the additive inverse of $\bar{7}$ in $\mathbf{Z}_9$.

SOLUTION As we saw in the proof that $\mathbf{Z}_n$ is a ring, the additive inverse of $\overline{7}$ is $\overline{-7}$ because

$$\overline{7} + \overline{-7} = \overline{7 + (-7)} = \overline{0}.$$

Since $\overline{-7} = \overline{2}$ in $\mathbf{Z}_9$, we may also say that $\overline{2}$ is the additive inverse of $\overline{7}$. This can be checked directly as well:

$$\overline{7} + \overline{2} = \overline{7 + 2} = \overline{9} = \overline{0},$$

and hence $\overline{7}$ and $\overline{2}$ are additive inverses in $\mathbf{Z}_9$.

In this example, we saw that $\overline{7}$ and $\overline{2}$ are additive inverses in $\mathbf{Z}_9$. In general, if a and b are integers such that $a + b = n$, then $\overline{a}$ and $\overline{b}$ are additive inverses in $\mathbf{Z}_n$ (see Exercise 10).

EXERCISES 8.2

Numerical Problems

1. Consider the element $\overline{4}$ in $\mathbf{Z}_{11}$. In addition to representing a single element in $\mathbf{Z}_{11}$, $\overline{4}$ also represents a set of integers. Describe this set and list a few of its elements.

2. By Definition 8.2.1, $\mathbf{Z}_4$ consists of four elements: $\overline{0}, \overline{1}, \overline{2}, \overline{3}$. From this it may seem that $\overline{11}$ is not an element of $\mathbf{Z}_4$. Explain why, in spite of this, $\overline{11}$ is indeed an element of $\mathbf{Z}_4$.

3. Find the additive inverse of the given element of $\mathbf{Z}_{20}$. Express your answer in the form $\overline{a}$, where a is a number in the range $0, 1, \ldots, 19$.

    a. $\overline{8}$ b. $\overline{5}$ c. $\overline{2}$

4. Find the additive inverse of the given element of $\mathbf{Z}_{100}$. Express your answer in the form $\overline{a}$, where a is a number in the range $0, 1, \ldots, 99$.

    a. $\overline{8}$ b. $\overline{5}$ c. $\overline{2}$

5. Find all of the perfect squares in $\mathbf{Z}_{13}$.

6. Find all of the perfect squares in $\mathbf{Z}_{17}$.

7. Find all solutions to the equation $x(x + 1) = 0$ in $\mathbf{Z}_6$.

8. Rewrite each of the following statements in the other four equivalent forms using the Equivalent Conditions for Congruence (8.2.4).

    a. $\overline{5} = \overline{1}$ in $\mathbf{Z}_4$

    b. $3 \mid (26 - 2)$

c. There exists $k \in \mathbf{Z}$ such that $7311 = 11 + 100k$.

d. $91 \equiv 0 \pmod{13}$

Reasoning and Proofs

9. Explain why the natural numbers are not a ring.

10. Show that if $n \in \mathbf{N}$ and $a, b \in \mathbf{Z}$ such that $a + b = n$, then $\bar{a}$ and $\bar{b}$ are additive inverses in $\mathbf{Z}_n$.

11. Complete the proof of Theorem 8.2.5 that $\mathbf{Z}_n$ is a ring by proving the following:

   a. Commutativity of addition

   b. Associativity and commutativity of multiplication

   c. Distributivity of multiplication over addition

   d. $\bar{1}$ is an identity element for multiplication

12. Let $n \in \mathbf{N}$. Prove that in $\mathbf{Z}_n$, $\bar{0} \cdot \bar{x} = \bar{0}$.

13. Let $n \in \mathbf{N}$, and let $\bar{a}, \bar{x}, \bar{y} \in \mathbf{Z}_n$. Prove that if $\bar{a} + \bar{x} = \bar{a} + \bar{y}$, then $\bar{x} = \bar{y}$.

14. In this exercise, you will prove that the additive inverse of any element of $\mathbf{Z}_n$ is unique. (In fact, this is true not only in $\mathbf{Z}_n$ but in any ring, as we prove in the Appendix on the Student Companion Website.) Let $n \in \mathbf{N}$, and let $\bar{a} \in \mathbf{Z}_n$. Suppose $\bar{x}, \bar{y} \in \mathbf{Z}_n$ are both additive inverses of $\bar{a}$ (i.e., suppose $\bar{a} + \bar{x} = \bar{0}$ and $\bar{a} + \bar{y} = \bar{0}$), and show that $\bar{x} = \bar{y}$. (*Note:* You may wish to do Exercise 13 and use that result in your proof.)

15. Let $n \in \mathbf{N}$. Prove that in $\mathbf{Z}_n$, $(\bar{a} + \bar{b})^2 = \bar{a}^2 + \bar{2}\,\bar{a}\,\bar{b} + \bar{b}^2$.

16. Prove that $(\bar{a} + \bar{b})^3 = \bar{a}^3 + \bar{b}^3$ in $\mathbf{Z}_3$.

17. Recall that $\max(a, b)$ is defined to be the greater of a and b. Let n be a natural number. For any $\bar{a}, \bar{b} \in \mathbf{Z}_n$, define $\bar{a} * \bar{b}$ as follows: $\bar{a} * \bar{b} = \overline{\max(a, b)}$.
   Explain why the operation $*$ is not well defined.

18. Let n be a natural number. Define $\bar{a}$ in $\mathbf{Z}_n$ to be **even** if a is an even integer.

   a. Find integers a and b such that $\bar{a} = \bar{b}$ in $\mathbf{Z}_9$, where a is an even integer but b is not. Conclude that the concept of being even is not well defined in $\mathbf{Z}_9$.

   b. Let $n \in \mathbf{N}$ be even. Prove that the concept of being even is well defined in $\mathbf{Z}_n$.

19. Let n be a natural number. For $\bar{a}, \bar{b} \in \mathbf{Z}_n$, define the operation $\wedge$ as follows: $\bar{a} \wedge \bar{b} = \overline{a^b}$
   Is this operation well defined? Explain.

8.3 Multiplicative Inverses in Z_n

A closer look at the multiplication tables

We began our study of $\mathbf{Z}_6$ in Section 8.1 by examining the addition and multiplication tables for this number system:

Addition in $\mathbf{Z}_6$

+	0	1	2	3	4	5
0	0	1	2	3	4	5
1	1	2	3	4	5	0
2	2	3	4	5	0	1
3	3	4	5	0	1	2
4	4	5	0	1	2	3
5	5	0	1	2	3	4

Multiplication in $\mathbf{Z}_6$

·	0	1	2	3	4	5
0	0	0	0	0	0	0
1	0	1	2	3	4	5
2	0	2	4	0	2	4
3	0	3	0	3	0	3
4	0	4	2	0	4	2
5	0	5	4	3	2	1

Both of these tables contain many patterns, but the addition table in particular displays a great deal of regularity. Going across any row or down any column, we simply count up by 1, keeping in mind that if our counting reaches 6, we will start our counting back at 0. Similar regularity is found in the addition table of $\mathbf{Z}_n$ for any n. Having examined the $\mathbf{Z}_6$ addition table, we will not find many surprises in the $\mathbf{Z}_7$ addition table or even in the $\mathbf{Z}_{30}$ addition table.

The multiplication tables for $\mathbf{Z}_n$ are more subtle and deserve a closer look. Let's start by comparing the multiplication tables for $\mathbf{Z}_{10}$ and $\mathbf{Z}_{11}$. (The reader should note that since all the numbers in the tables refer to elements of $\mathbf{Z}_n$, technically each number should be barred. However, putting bars on all the numbers would make the table hard to read, so we will permit ourselves to omit the bars, *in tables only*, as long as we remember that the bars really ought to be there.)

Table 8.3.1   Multiplication in $\mathbf{Z}_{10}$

·	0	1	2	3	4	5	6	7	8	9
0	0	0	0	0	0	0	0	0	0	0
1	0	1	2	3	4	5	6	7	8	9
2	0	2	4	6	8	0	2	4	6	8
3	0	3	6	9	2	5	8	1	4	7
4	0	4	8	2	6	0	4	8	2	6
5	0	5	0	5	0	5	0	5	0	5
6	0	6	2	8	4	0	6	2	8	4
7	0	7	4	1	8	5	2	9	6	3
8	0	8	6	4	2	0	8	6	4	2
9	0	9	8	7	6	5	4	3	2	1

Table 8.3.2 Multiplication in Z_{11}

·	0	1	2	3	4	5	6	7	8	9	10
0	0	0	0	0	0	0	0	0	0	0	0
1	0	1	2	3	4	5	6	7	8	9	10
2	0	2	4	6	8	10	1	3	5	7	9
3	0	3	6	9	1	4	7	10	2	5	8
4	0	4	8	1	5	9	2	6	10	3	7
5	0	5	10	4	9	4	8	2	7	1	6
6	0	6	1	7	2	8	3	9	4	10	5
7	0	7	3	10	6	2	9	5	1	8	4
8	0	8	5	2	10	7	4	1	9	6	3
9	0	9	7	5	3	1	10	8	6	4	2
10	0	10	9	8	7	6	5	4	3	2	1

These two tables have very different flavors. Perhaps the most striking difference is that every row and column of the Z_{11} multiplication table (except for the $\overline{0}$ row and $\overline{0}$ column) contains each element of Z_{11} exactly once.[6] For example, the row of $\overline{4}$ contains the numbers

$$\overline{0}, \overline{4}, \overline{8}, \overline{1}, \overline{5}, \overline{9}, \overline{2}, \overline{6}, \overline{10}, \overline{3}, \overline{7}.$$

(It is natural to number the rows starting with 0, so that this row is called the **row of $\overline{4}$**, or the **fourth row**, even though it appears fifth in the table.) Though they are jumbled up, all 11 elements of Z_{11} appear exactly once in this row.

For Z_{10}, the story is quite different. A glance at the multiplication table for Z_{10} reveals that many rows of the table have repeated elements and do not contain all of the elements of Z_{10}. For example, in the fourth row of the Z_{10} table, the numbers

$$\overline{0}, \overline{4}, \overline{8}, \overline{2}, \text{ and } \overline{6}$$

all appear twice, while the numbers

$$\overline{1}, \overline{3}, \overline{5}, \overline{7}, \text{ and } \overline{9}$$

are not present. The row of $\overline{5}$ is even worse in this regard: it contains only the numbers $\overline{0}$ and $\overline{5}$, each repeated five times. Nonetheless, some of the rows of the Z_{10} multiplication table do contain every element exactly once. For example the row of $\overline{7}$, consisting of the numbers

$$\overline{0}, \overline{7}, \overline{4}, \overline{1}, \overline{8}, \overline{5}, \overline{2}, \overline{9}, \overline{6}, \overline{3},$$

---

[6] You sudoku addicts know what we're talking about!

has this property. Do you see any other rows with this property? There are a total of four rows in the $\mathbf{Z}_{10}$ table that contain every number exactly once: the rows of $\overline{1}, \overline{3}, \overline{7},$ and $\overline{9}$.

In the $\mathbf{Z}_{10}$ table, we have noticed that there is a big difference between the rows that contain every element of $\mathbf{Z}_{10}$, namely the rows of

$$\overline{1}, \overline{3}, \overline{7}, \text{ and } \overline{9}, \tag{1}$$

and the remaining rows, which are the rows of

$$\overline{0}, \overline{2}, \overline{4}, \overline{5}, \overline{6}, \text{ and } \overline{8}. \tag{2}$$

What distinguishes the **orange** numbers (the first list) from the **brown** numbers (the second list)? Numbers in the first list (1) are relatively prime to the modulus 10, while each of the numbers in the second list (2) shares a common factor with 10. This might lead one to conjecture that in the $\mathbf{Z}_n$ multiplication table, the row of $\overline{a}$ will contain every number exactly once if and only if the number a is relatively prime to n. We will prove this later (Theorem 8.3.5).

Multiplicative inverses

In general, two numbers are said to be *multiplicative inverses* if their product is 1, the multiplicative identity element. A familiar property of the real numbers is that every nonzero element of $\mathbf{R}$ has a multiplicative inverse in $\mathbf{R}$. For example, the multiplicative inverse of 7 is $\frac{1}{7}$. Similarly, every nonzero rational number has a multiplicative inverse in $\mathbf{Q}$. In contrast, in the integers, only 1 and -1 have multiplicative inverses in $\mathbf{Z}$.

In our new $\mathbf{Z}_n$ number systems, $\overline{x}$ is called a **multiplicative inverse** of $\overline{a}$ if $\overline{a} \cdot \overline{x} = \overline{1}$. In $\mathbf{Z}_{10}$, for example, $\overline{3} \cdot \overline{7} = \overline{1}$, so $\overline{7}$ is a multiplicative inverse of $\overline{3}$.

Which elements of $\mathbf{Z}_n$ have multiplicative inverses? The multiplication tables can help us answer this question. Every time we see a $\overline{1}$ in the multiplication table—say, in the $\overline{x} \cdot \overline{y}$ position—we know that $\overline{x}$ and $\overline{y}$ are multiplicative inverses. For example, the $\mathbf{Z}_{11}$ multiplication table tells us that $\overline{5} \cdot \overline{9} = \overline{1}$; hence, $\overline{5}$ and $\overline{9}$ are multiplicative inverses.

Does $\overline{8}$ have a multiplicative inverse in $\mathbf{Z}_{11}$? The question is whether there is number x such that $\overline{8} \cdot \overline{x} = \overline{1}$. All of the products $\overline{8} \cdot \overline{x}$ appear in the row of $\overline{8}$, so we simply scan across this row looking for the number $\overline{1}$.

Multiplication in $\mathbf{Z}_{11}$

·	0	1	2	3	4	5	6	7	8	9	10
⋮											
8	0	8	5	2	10	7	4	1	9	6	3
⋮											

We find it in the 7th column, as $\overline{8} \cdot \overline{7} = \overline{1}$. Thus, $\overline{7}$ is the multiplicative inverse of $\overline{8}$.

In Exercise 9, you will prove that if an element $\bar{b} \in \mathbf{Z}_n$ has a multiplicative inverse, then this inverse is unique. Thus, if $\bar{b} \cdot \bar{x} = \bar{1}$, instead of referring to $\bar{x}$ as *a* multiplicative inverse of $\bar{b}$, we may refer to it as *the* multiplicative inverse of $\bar{b}$. We often express this using the notation $\bar{x} = \bar{b}^{-1}$.

EXAMPLE 1

a. Determine the multiplicative inverse of $\bar{2}$ in $\mathbf{Z}_{11}$.

b. Using the multiplication table, determine which elements of $\mathbf{Z}_{11}$ have multiplicative inverses, and find their inverses.

c. Using the multiplication table, determine which elements of $\mathbf{Z}_{10}$ have multiplicative inverses, and find their inverses.

SOLUTION

Multiplication in $\mathbf{Z}_{11}$

·	0	1	2	3	4	5	6	7	8	9	10
0	0	0	0	0	0	0	0	0	0	0	0
1	0	1	2	3	4	5	6	7	8	9	10
2	0	2	4	6	8	10	1	3	5	7	9
3	0	3	6	9	1	4	7	10	2	5	8
4	0	4	8	1	5	9	2	6	10	3	7
5	0	5	10	4	9	4	8	2	7	1	6
6	0	6	1	7	2	8	3	9	4	10	5
7	0	7	3	10	6	2	9	5	1	8	4
8	0	8	5	2	10	7	4	1	9	6	3
9	0	9	7	5	3	1	10	8	6	4	2
10	0	10	9	8	7	6	5	4	3	2	1

a. Scanning the row of $\bar{2}$ in the $\mathbf{Z}_{11}$ multiplication table, we find a 1 in the 6th column. Thus, $\bar{2} \cdot \bar{6} = \bar{1}$, which means that $\bar{6}$ is the multiplicative inverse of $\bar{2}$.

b. In the $\mathbf{Z}_{11}$ multiplication table, we find that

$$\bar{1} \cdot \bar{1} = \bar{1}, \quad \bar{2} \cdot \bar{6} = \bar{1}, \quad \bar{3} \cdot \bar{4} = \bar{1}, \quad \bar{5} \cdot \bar{9} = \bar{1}, \quad \bar{7} \cdot \bar{8} = \bar{1}, \quad \overline{10} \cdot \overline{10} = \bar{1}.$$

Thus, $\bar{1}$ is its own inverse, $\bar{2}$ and $\bar{6}$ are inverses, $\bar{3}$ and $\bar{4}$ are inverses, $\bar{5}$ and $\bar{9}$ are inverses, $\bar{7}$ and $\bar{8}$ are inverses, and $\overline{10}$ is its own inverse. We see that every nonzero element of $\mathbf{Z}_{11}$ has a multiplicative inverse! More on this shortly.

Multiplication in Z_{10}

·	0	1	2	3	4	5	6	7	8	9
0	0	0	0	0	0	0	0	0	0	0
1	0	1	2	3	4	5	6	7	8	9
2	0	2	4	6	8	0	2	4	6	8
3	0	3	6	9	2	5	8	1	4	7
4	0	4	8	2	6	0	4	8	2	6
5	0	5	0	5	0	5	0	5	0	5
6	0	6	2	8	4	0	6	2	8	4
7	0	7	4	1	8	5	2	9	6	3
8	0	8	6	4	2	0	8	6	4	2
9	0	9	8	7	6	5	4	3	2	1

c. Examining the Z_{10} multiplication table, we find that

$$\bar{1} \cdot \bar{1} = \bar{1}, \quad \bar{3} \cdot \bar{7} = \bar{1}, \quad \text{and} \quad \bar{9} \cdot \bar{9} = \bar{1},$$

while no other products yield $\bar{1}$. We conclude that $\bar{1}, \bar{3}, \bar{7},$ and $\bar{9}$ have multiplicative inverses: $\bar{3}$ and $\bar{7}$ are inverses of each other, $\bar{1}$ is its own inverse, and $\bar{9}$ is its own inverse. The other elements, $\bar{0}, \bar{2}, \bar{4}, \bar{5}, \bar{6},$ and $\bar{8}$, do not have multiplicative inverses. ●

In Example 1, we found that the elements of Z_{10} that have multiplicative inverses are $\bar{1}, \bar{3}, \bar{7},$ and $\bar{9}$. We saw these same numbers previously in list (1) when we examined which rows of the Z_{10} multiplication table contain every element exactly once, and we noted that 1, 3, 7, and 9 are precisely those numbers between 0 and 9 that are relatively prime to 10. We will soon prove (in Theorem 8.3.1) that this holds in general: a number $\bar{a}$ has a multiplicative inverse in Z_n if and only if a and n are relatively prime.

Finding multiplicative inverses without a table

Naomi's Numerical Proof Preview: Theorem 8.3.1

Suppose we are working in the ring Z_{178}, and we wish to find the multiplicative inverse of $\overline{37}$. If we were lucky enough to have the multiplication table for this ring on hand, we could solve our problem by looking across the $\overline{37}$ row of our table until we found a $\bar{1}$. But even if we already had the table, this might not be much fun, so let's figure out a better way. Algebraically, we wish to find a number $\bar{x}$ in Z_{178} that satisfies

$$\overline{37} \cdot \bar{x} = \bar{1}.$$

Using the definition of multiplication in $\mathbf{Z}_n$, this is the same as

$$\overline{37x} = \overline{1}.$$

By the Equivalent Conditions for Congruence (8.2.4), this will hold if and only if there is a number $y \in \mathbf{Z}$ such that

$$37x = 1 + 178y. \tag{3}$$

At this point, we're on familiar ground—this is a linear Diophantine equation. Since the coefficients 37 and 178 are relatively prime, Theorem 5.1.2 (Solving Linear Diophantine Equations) guarantees that there will be a solution. Using the Euclidean Algorithm to solve this equation (see Method 5.2.2), we arrive at the solution

$$x = 77, \ y = 16.$$

We conclude that in $\mathbf{Z}_{178}$,

$$\overline{x} = \overline{77} \text{ is the multiplicative inverse of } \overline{37}. \tag{4}$$

We have just seen that finding the inverse of $\overline{37}$ in $\mathbf{Z}_{178}$ is equivalent to solving the linear Diophantine equation $37x = 1 + 178y$. In general, finding the inverse of $\overline{a}$ in $\mathbf{Z}_n$ is equivalent to solving the Diophantine equation $ax = 1 + ny$. This is the key idea in the proof of the following theorem.

THEOREM 8.3.1

Let $n \in \mathbf{N}$, and let $\overline{a} \in \mathbf{Z}_n$. Then $\overline{a}$ has a multiplicative inverse in $\mathbf{Z}_n$ if and only if $\gcd(a, n) = 1$.

PROOF

[$\Rightarrow$] Suppose $\overline{a}$ has a multiplicative inverse in $\mathbf{Z}_n$. **[To show: $\gcd(a, n) = 1$.]**

Then there exists $\overline{x} \in \mathbf{Z}_n$ such that $\overline{a} \cdot \overline{x} = \overline{1}$. Hence $\overline{ax} = \overline{1}$, and so by the Equivalent Conditions for Congruence (8.2.4), there exists $y \in \mathbf{Z}$ such that

$$ax = 1 + ny.$$

We can write this as

$$ax + n(-y) = 1,$$

and hence the linear Diophantine equation

$$aX + nY = 1$$

has a solution. By Theorem 5.1.2 (Solving Linear Diophantine Equations), it follows that the coefficients a and n in this equation are relatively prime. Thus, $\gcd(a, n) = 1$.

[⇐] Now suppose that $\gcd(a, n) = 1$. **[To show: $\bar{a}$ has a multiplicative inverse in $\mathbf{Z}_n$.]** Again by Theorem 5.1.2, there exist integers x and y such that

$$ax + ny = 1.$$

It follows that

$$ax \equiv 1 \;(\text{mod } n).$$

Hence, $\overline{ax} = \bar{1}$, and so $\bar{a} \cdot \bar{x} = \bar{1}$. Thus, $\bar{x}$ is the multiplicative inverse of $\bar{a}$ in $\mathbf{Z}_n$. ◼

EXAMPLE 2

Find each multiplicative inverse in $\mathbf{Z}_{55}$, if one exists.

  a. $\overline{24}^{-1}$ **b.** $\overline{22}^{-1}$

SOLUTION

   a. Since $\gcd(24, 55) = 1$, we know (by Theorem 8.3.1) that $\overline{24}$ has a multiplicative inverse in $\mathbf{Z}_{55}$. Call this inverse $\bar{x}$. We then have

$$\overline{24} \cdot \bar{x} = \bar{1},$$

Or, equivalently, there exists $y \in \mathbf{Z}$ such that

$$24x = 1 + 55y.$$

This linear Diophantine equation can be solved using the Euclidean Algorithm (see Method 5.2.2), which yields the solution

$$x = -16, y = -7.$$

Therefore, $\overline{-16} = \overline{39}$ is the multiplicative inverse of $\overline{24}$ in $\mathbf{Z}_{55}$.

   b. Since $\gcd(22, 55) = 11$, Theorem 8.3.1 tells us that there is no multiplicative inverse of $\overline{22}$ in $\mathbf{Z}_{55}$. ◼

Cancellation property

One of the most useful rules you learned in high school algebra is the law of cancellation: for any real numbers a, x, and y, if

$$ax = ay, \tag{5}$$

then provided that $a \neq 0$, we can cancel the a from both sides to obtain

$$x = y.$$

One way to justify this rule is to say that we divide both sides of equation (5) by a. We can also say that we multiply both sides of equation (5) by a^{-1}, the multiplicative inverse of a.

In the next theorem, we show that in $\mathbf{Z}_n$, we can cancel any factor $\bar{a}$ provided that $\bar{a}$ has a multiplicative inverse—that is, provided that $\gcd(a, n) = 1$.

THEOREM 8.3.2 MODULAR CANCELLATION LAW

Let $n \in \mathbf{N}$, and let $\bar{a}, \bar{x}, \bar{y} \in \mathbf{Z}_n$ such that $\gcd(a, n) = 1$. If $\bar{a} \cdot \bar{x} = \bar{a} \cdot \bar{y}$, then $\bar{x} = \bar{y}$.

Note that in this theorem, the condition that $\gcd(a, n) = 1$ is necessary. For example, if we let $a = 3$ and $n = 6$, then $\gcd(a, n) \neq 1$. In this case, we see that in $\mathbf{Z}_6$, $\bar{a} \cdot \bar{5} = \bar{a} \cdot \bar{3}$, but $\bar{5} \neq \bar{3}$.

PROOF OF **THEOREM 8.3.2**

Since $\gcd(a, n) = 1$, we know that $\bar{a}$ has a multiplicative inverse in $\mathbf{Z}_n$ (by Theorem 8.3.1), which we will denote by $\bar{a}^{-1}$. Hence, we may take the equation

$$\bar{a} \cdot \bar{x} = \bar{a} \cdot \bar{y}$$

and multiply both sides by $\bar{a}^{-1}$, which yields

$$\bar{a}^{-1} \cdot (\bar{a} \cdot \bar{x}) = \bar{a}^{-1} \cdot (\bar{a} \cdot \bar{y}).$$

Now by associativity,

$$(\bar{a}^{-1} \cdot \bar{a}) \cdot \bar{x} = (\bar{a}^{-1} \cdot \bar{a}) \cdot \bar{y},$$

which simplifies to

$$\bar{1} \cdot \bar{x} = \bar{1} \cdot \bar{y},$$

and hence,

$$\bar{x} = \bar{y}. \qquad \blacksquare$$

The Modular Cancellation Law can also be expressed in terms of congruences:

THEOREM 8.3.3 MODULAR CANCELLATION LAW (Restatement)

Let $n \in \mathbf{N}$, and let $a, x, y \in \mathbf{Z}$ such that $\gcd(a, n) = 1$. If $ax \equiv ay \pmod{n}$, then $x \equiv y \pmod{n}$.

PROOF We leave the proof as Exercise 8.

Z_p is a field

In Example 1 part b, the multiplication table helped us determine that every nonzero element of Z_{11} has a multiplicative inverse. This fact can also be proved using Theorem 8.3.1. According to this theorem, an element $\bar{a}$ of Z_{11} has a multiplicative inverse if and only if a is relatively prime to 11. But since 11 is prime, all of the numbers $1, 2, 3, \ldots, 10$ are relatively prime to 11. Hence, all of the elements $\bar{1}, \bar{2}, \bar{3}, \ldots, \overline{10}$ of Z_{11} have multiplicative inverses.

We can use the same reasoning to prove that if p is any prime number, then every nonzero element of Z_p has a multiplicative inverse.

COROLLARY 8.3.4

Assume $p \in \mathbf{N}$ is prime. Then every nonzero element of $\mathbf{Z}_p$ has a multiplicative inverse in $\mathbf{Z}_p$.

PROOF Let p be a prime number.

[To show: Every nonzero element of Z_p has a multiplicative inverse.]

Let $\bar{a} \in \mathbf{Z}_p$ with $\bar{a} \neq \bar{0}$. It follows from the Equivalent Conditions for Congruence (8.2.4) that $p \nmid a$. Hence $\gcd(a, p) = 1$ (by Lemma 3.4.3). We may now apply Theorem 8.3.1 and conclude that $\bar{a}$ has a multiplicative inverse in $\mathbf{Z}_p$. ∎

When we introduced the concept of a ring in the previous section, you may have noticed that the definition of ring does not include the existence of multiplicative inverses. A ring in which every nonzero element has a multiplicative inverse is called a **field**.

Thus, Corollary 8.3.4 can be restated as follows:

$$\text{If } p \text{ is prime, then } \mathbf{Z}_p \text{ is a field.}$$

The converse to Corollary 8.3.4, which you will prove in Exercise 10, is also true for any natural number $n > 1$:

$$\text{If } \mathbf{Z}_n \text{ is a field, then } n \text{ is prime.}$$

This can be stated another way (the contrapositive):

$$\text{If } n \text{ is not prime, then } \mathbf{Z}_n \text{ is not a field.}$$

When does $xy = 0$?

If x and y are real numbers and you know that their product $xy = 0$, what can you conclude about x and y? Certainly, one of them must equal 0. For any real numbers x and y,

$$\text{If } xy = 0, \text{ then } x = 0 \text{ or } y = 0. \tag{6}$$

You used this fact often in high school algebra when solving quadratic equations.

Does this work in the modular number system $\mathbf{Z}_n$? That is, if the product of two elements of $\mathbf{Z}_n$ equals $\bar{0}$, does it follow that one of the factors must be $\bar{0}$? To answer this, let's take another look at the $\mathbf{Z}_{10}$ multiplication table (Table 8.3.1). We see plenty of $\bar{0}$'s appearing in the middle of the table. For example, $\bar{4} \cdot \bar{5} = \bar{0}$, but neither $\bar{4}$ nor $\bar{5}$ is equal to $\bar{0}$. Thus, the familiar property (6), which holds for real numbers, fails in the number system $\mathbf{Z}_{10}$.

When we consider $\mathbf{Z}_{11}$, the situation is completely different. If we look at the $\mathbf{Z}_{11}$ multiplication table (Table 8.3.2), the only place we find $\bar{0}$'s are at the edge of the table. In $\mathbf{Z}_{11}$, the only way to get $\bar{0}$ as a product is to have $\bar{0}$ as one of the factors. As you may have guessed, this holds in $\mathbf{Z}_p$ for any prime number p.

THEOREM 8.3.5

Let p be a prime number, and let $\bar{x}, \bar{y} \in \mathbf{Z}_p$. If $\bar{x} \cdot \bar{y} = \bar{0}$, then $\bar{x} = \bar{0}$ or $\bar{y} = \bar{0}$.

PROOF Suppose that

$$\bar{x} \cdot \bar{y} = \bar{0}. \tag{7}$$

If $\bar{x} = \bar{0}$, then we are done. So we may suppose that $\bar{x} \neq \bar{0}$. Then $\bar{x}$ has a multiplicative inverse in $\mathbf{Z}_p$ by Corollary 8.3.4. Now we multiply both sides of equation (7) by $\bar{x}^{-1}$:

$$\bar{x}^{-1}(\bar{x} \cdot \bar{y}) = \bar{x}^{-1} \cdot \bar{0}.$$

It follows that

$$\bar{y} = \bar{0},$$

as desired. ■

Good rows

At the beginning of this section, we noticed that some rows in the multiplication table of $\mathbf{Z}_n$ have the pleasant property that they contain each element of $\mathbf{Z}_n$ exactly once. In $\mathbf{Z}_{10}$, the rows with this property are the rows of $\bar{1}, \bar{3}, \bar{7},$ and $\bar{9}$. We noted that these are exactly the elements $\bar{a}$ of $\mathbf{Z}_{10}$ for which a is relatively prime to the modulus, 10. Theorem 8.3.1 gives us another way to say this: $\bar{1}, \bar{3}, \bar{7},$ and $\bar{9}$ are the elements of $\mathbf{Z}_{10}$ that have multiplicative inverses.

Indeed, multiplicative inverses play a key role in proving the following theorem, which confirms our guess about which rows contain all the elements of $\mathbf{Z}_n$ exactly once.

Let $n \in \mathbf{N}$, and let $\bar{a} \in \mathbf{Z}_n$. The row of $\bar{a}$ in the $\mathbf{Z}_n$ multiplication table—namely, the list of numbers

$$\bar{a} \cdot \bar{0}, \quad \bar{a} \cdot \bar{1}, \quad \bar{a} \cdot \bar{2}, \ldots, \bar{a} \cdot \overline{n-1}, \tag{8}$$

contains every element of $\mathbf{Z}_n$ exactly once if and only if $\gcd(a, n) = 1$.

PROOF

[$\Rightarrow$] Suppose every element of $\mathbf{Z}_n$ appears exactly once in list (8). In particular, $\bar{1}$ appears somewhere in the list; that is, there is a number $\bar{x} \in \mathbf{Z}_n$ such that $\bar{a} \cdot \bar{x} = \bar{1}$. This means that $\bar{a}$ has a multiplicative inverse, so by Theorem 8.3.1, we conclude that $\gcd(a, n) = 1$.

[$\Leftarrow$] Now suppose that $\gcd(a, n) = 1$. We must show that every element of $\mathbf{Z}_n$ appears exactly once in list (8). To do this, we must show two things:

(i)   Every element of $\mathbf{Z}_n$ appears in the list, and

(ii)   No element of $\mathbf{Z}_n$ appears twice.

Proof of (i): Let $\bar{b} \in \mathbf{Z}_n$. We wish to show that $\bar{b}$ appears in the list

$$\bar{a} \cdot \bar{0}, \quad \bar{a} \cdot \bar{1}, \quad \bar{a} \cdot \bar{2}, \ldots, \bar{a} \cdot \overline{n-1}.$$

In other words, we wish to find an $\bar{x} \in \mathbf{Z}_n$ satisfying $\bar{a} \cdot \bar{x} = \bar{b}$.

To find such an $\bar{x}$, we take advantage of the fact that $\bar{a}$ has a multiplicative inverse. We set $\bar{x} = \bar{a}^{-1} \cdot \bar{b}$ and check that this $\bar{x}$ has the required property:

$$\bar{a} \cdot \bar{x} = \bar{a} \cdot (\bar{a}^{-1} \cdot \bar{b})$$

$$= (\bar{a} \cdot \bar{a}^{-1}) \cdot \bar{b} \quad \leftarrow \textbf{by associativity of multiplication in } \mathbf{Z}_n$$

$$= \bar{1} \cdot \bar{b} \quad \leftarrow \textbf{since } \bar{a} \textbf{ and } \bar{a}^{-1} \textbf{ are multiplicative inverses}$$

$$= \bar{b}.$$

Proof of (ii): We know by (i) that every element of $\mathbf{Z}_n$ appears in list (8). Since $\mathbf{Z}_n$ has n distinct elements, this list must contain n distinct elements. However, the list is precisely n elements long. Thus, no element of $\mathbf{Z}_n$ can appear twice in the list.

Note that although this theorem makes a statement about the *rows* of the $\mathbf{Z}_n$ multiplication table, the same statement may also be made about the *columns* of the table. (See Exercise 15.)

Finding a number in a row

Imagine the multiplication table for $\mathbf{Z}_{178}$, a very large table. Theorem 8.3.6 tells us which rows of the table contain every element of $\mathbf{Z}_{178}$ exactly once. For example, since $\gcd(38, 178) \neq 1$, the 38th row does not contain every element. (In Exercise 12, you will explore the question of precisely which elements of $\mathbf{Z}_{178}$ appear in the 38th row.)

On the other hand, the row of $\overline{37}$ *does* contain each of the numbers $\overline{0}, \overline{1}, \overline{2}, \ldots, \overline{177}$ exactly once. Thus, we are guaranteed by Theorem 8.3.6 that the number $\overline{50}$, for example, must appear somewhere in the row of $\overline{37}$. But suppose we wanted to know exactly where in this row the number $\overline{50}$ occurs. Here's a picture:

$\cdot$	0	1	2	3	4	5	6	$\cdots$	x	$\cdots$	176	177
$\vdots$												
37	0	37	74	111	148	7	44	$\cdots$	**50**	$\cdots$	104	141
$\vdots$												

To find where $\overline{50}$ appears in this row, we must solve the equation

$$\overline{37} \cdot \overline{x} = \overline{50}.$$

To solve, we multiply both sides by the multiplicative inverse of $\overline{37}$ to obtain

$$\overline{x} = \overline{37}^{-1} \cdot \overline{50}.$$

As we computed earlier (4), the multiplicative inverse of $\overline{37}$ is $\overline{77}$, and hence we get

$$\overline{x} = \overline{77} \cdot \overline{50}.$$

Taking this product and then reducing modulo 178 yields

$$\overline{x} = \overline{3850} = \overline{112}.$$

We have successfully found the location of $\overline{50}$ in the $\overline{37}$ row of the table: $\overline{50}$ appears in the **112** column. One can check this answer quite easily by multiplying and then reducing:

$$\overline{37} \cdot \overline{112} = \overline{4414} = \overline{50}.$$

Wilson's Theorem

What happens if we multiply all the nonzero elements of $\mathbf{Z}_5$ together? That is, what is the product

$$\overline{1} \cdot \overline{2} \cdot \overline{3} \cdot \overline{4}$$

in $\mathbf{Z}_5$? The product is $\overline{24} = \overline{4}$. This can also be written as $\overline{-1}$.

Let's try the same thing in $\mathbf{Z}_7$. This time the product of the nonzero elements

$$\bar{1} \cdot \bar{2} \cdot \bar{3} \cdot \bar{4} \cdot \bar{5} \cdot \bar{6}$$

yields $\overline{720} = \bar{6}$. Again, this can be written as $\overline{-1}$.

Naomi's Numerical Proof Preview: Wilson's Theorem (8.3.7)

Now let's consider in $\mathbf{Z}_{11}$ the product of all the nonzero elements

$$\bar{1} \cdot \bar{2} \cdot \bar{3} \cdot \bar{4} \cdot \bar{5} \cdot \bar{6} \cdot \bar{7} \cdot \bar{8} \cdot \bar{9} \cdot \overline{10}. \qquad (9)$$

If we evaluate this product, will we find an answer of $\overline{-1}$ once again? Instead of whipping out our calculators at this point, let's try to be a bit more clever. One thing we know about $\mathbf{Z}_{11}$ is that every nonzero element of $\mathbf{Z}_{11}$ has a multiplicative inverse. In fact, in Example 1(b), we found that $\bar{1}$ is its own inverse and $\overline{10}$ is its own inverse, while the remaining elements pair up: $\bar{2}$ and $\bar{6}$ are inverses, $\bar{3}$ and $\bar{4}$ are inverses, $\bar{5}$ and $\bar{9}$ are inverses, and $\bar{7}$ and $\bar{8}$ are inverses. Thus, we can reorder the product (9) and find

$$\bar{1} \cdot \bar{2} \cdot \bar{3} \cdot \bar{4} \cdot \bar{5} \cdot \bar{6} \cdot \bar{7} \cdot \bar{8} \cdot \bar{9} \cdot \overline{10} = \bar{1} \cdot \overline{10} \cdot (\bar{2} \cdot \bar{6}) \cdot (\bar{3} \cdot \bar{4}) \cdot (\bar{5} \cdot \bar{9}) \cdot (\bar{7} \cdot \bar{8})$$

$$= \bar{1} \cdot \overline{10} \cdot \bar{1} \cdot \bar{1} \cdot \bar{1} \cdot \bar{1}.$$

Observe that almost all of the terms from our original product have canceled out, and we are left with an answer of $\overline{10}$. This equals $\overline{-1}$.

In fact, we can use a similar argument to establish that if p is any prime, then the product of all the nonzero elements of $\mathbf{Z}_p$,

$$\bar{1} \cdot \bar{2} \cdot \bar{3} \cdot \ \cdots \ \cdot (\overline{p-1}),$$

is equal to $\overline{-1}$.

THEOREM 8.3.7   WILSON'S THEOREM

Let p be a prime number. Then in $\mathbf{Z}_p$,

$$\bar{1} \cdot \bar{2} \cdot \bar{3} \cdot \ \cdots \ \cdot (\overline{p-1}) = \overline{-1}.$$

Note that the product on the left side of this equation equals $\overline{(p-1)!}$. Thus, Wilson's theorem may also be stated as follows:

If p is prime, then $(p-1)! \equiv -1 \ (\mathrm{mod} \ p)$.

The converse of Wilson's Theorem is also true. (See Exercise 18.)

Our proof of Wilson's Theorem, which we give shortly, will follow the argument of the Numerical Proof Preview above. In the Proof Preview, we observed that every element of $\mathbf{Z}_{11}$ pairs up with its inverse except for the numbers $\bar{1}$ and $\overline{10}$, which are their own inverses.

LEMMA 8.3.8

Let p be prime. If $\bar{x} \in \mathbf{Z}_p$ is its own multiplicative inverse, then $\bar{x} = \bar{1}$ or $\bar{x} = \overline{p-1}$.

PROOF Let p be prime, and suppose that $\bar{x} \in \mathbf{Z}_p$ is its own inverse. Then $\bar{x} \cdot \bar{x} = \bar{1}$. Thus,

$$\bar{x}^2 = \bar{1},$$

so

$$\bar{x}^2 - \bar{1} = \bar{0}.$$

Using the properties of Theorem 8.2.6, it follows (see Exercise 17) that

$$(\bar{x} + \bar{1}) \cdot (\bar{x} - \bar{1}) = \bar{0}.$$

So by Theorem 8.3.5,

$$\bar{x} + \bar{1} = \bar{0} \qquad \text{or} \qquad \bar{x} - \bar{1} = \bar{0},$$

and we conclude that

$$\bar{x} = \overline{-1} \qquad \text{or} \qquad \bar{x} = \bar{1}.$$

Since $\overline{-1} = \overline{p-1}$, we conclude that $\bar{x} = \bar{1}$ or $\bar{x} = \overline{p-1}$, as desired. ●

We are now in a position to prove Wilson's Theorem.

PROOF OF **WILSON'S THEOREM (8.3.7)**

Let p be a prime number. In $\mathbf{Z}_p$, consider the elements in the list

$$\bar{2}, \bar{3}, \ldots, \overline{p-2}. \tag{10}$$

By Lemma 8.3.8, none of these numbers is its own inverse. Thus, the inverse of each of the numbers in the list (10) is another number in the list. Furthermore, these inverses are unique (see Exercise 9). Hence, in the product

$$\bar{1} \cdot \bar{2} \cdot \bar{3} \cdot \ldots \cdot (\overline{p-2}) \cdot (\overline{p-1}),$$

each of the factors $\bar{2}, \bar{3}, \ldots, \overline{p-2}$ can be cancelled with its multiplicative inverse, and we get

$$\bar{1} \cdot \bar{2} \cdot \bar{3} \cdot \ \ldots \ \cdot (\overline{p-2}) \cdot (\overline{p-1}) = \bar{1} \cdot (\overline{p-1})$$
$$= \overline{p-1}$$
$$= \overline{-1}.$$

EXERCISES 8.3

Numerical Problems

1. Find each multiplicative inverse, if it exists.

   a. The inverse of $\overline{53}$ in $\mathbf{Z}_{111}$.

   b. The inverse of $\overline{299}$ in $\mathbf{Z}_{901}$.

   c. The inverse of $\overline{51}$ in $\mathbf{Z}_{187}$.

   d. The inverse of $\overline{41}$ in $\mathbf{Z}_{300}$.

2. Find each multiplicative inverse in $\mathbf{Z}_{13}$:

   a. $\bar{7}^{-1}$

   b. $\bar{3}^{-1}$

   c. $\bar{2}^{-1}$

3. Use your answers to Exercise 2 to solve these equations in $\mathbf{Z}_{13}$:

   a. $\bar{7} \cdot \bar{x} = \overline{10}$

   b. $\bar{3} \cdot \bar{x} - \bar{1} = \bar{0}$

   c. $\bar{2} \cdot \bar{x} + \bar{4} = \bar{1}$

4. Find each multiplicative inverse in $\mathbf{Z}_{19}$:

   a. $\bar{5}^{-1}$

   b. $\bar{3}^{-1}$

5. Use your answers to Exercise 4 to solve these equations in $\mathbf{Z}_{19}$:

   a. $\bar{5} \cdot \bar{x} = \bar{4}$

   b. $\bar{3} \cdot \bar{x} + \bar{2} = \overline{15}$

6. Use your solution to Exercise 1b to solve the equation $\overline{299} \cdot \bar{y} + \overline{20} = \overline{17}$ in $\mathbf{Z}_{901}$.

7. Find $\overline{(n-1)!}$ in $\mathbf{Z}_n$ for each of the following values of n.

   a. $n = 7$ b. $n = 101$ c. $n = 15$ d. $n = 91$

Reasoning and Proofs

8. Prove Theorem 8.3.3, the restatement of the Modular Cancellation Law.

9. Let $n \in \mathbf{N}$, and let $\bar{a} \in \mathbf{Z}_n$. Prove that if $\bar{a}$ has a multiplicative inverse, then this inverse is unique. In other words, suppose $\bar{x}, \bar{y} \in \mathbf{Z}_n$ are both multiplicative inverses of $\bar{a}$ (i.e., suppose $\bar{a}\,\bar{x} = \bar{1}$ and $\bar{a}\,\bar{y} = \bar{1}$), and show that $\bar{x} = \bar{y}$.

10. Prove the converse of Corollary 8.3.4:

   Let $n > 1$ be a natural number. If $\mathbf{Z}_n$ is a field, then n is prime.

11. Let $a, b, c \in \mathbf{Z}$ and let $n \in \mathbf{N}$. Let $d = \gcd(c, n)$. Prove that if $ca \equiv cb \pmod{n}$, then $a \equiv b \left(\bmod \dfrac{n}{d}\right)$.

12. **a.** Which elements of $\mathbf{Z}_{15}$ appear in the 10th row of the multiplication table? How about the 11th row?

   b. Which elements of $\mathbf{Z}_{18}$ appear in the 12th row of the multiplication table? How about the 15th row?

   c. Make a conjecture about which elements of $\mathbf{Z}_{178}$ appear in the 38th row of the multiplication table.

   d. Make a conjecture about which elements of $\mathbf{Z}_{178}$ appear in the 102nd row of the multiplication table.

13. **a.** Let n be a natural number, and let $\bar{a}$ be an element of $\mathbf{Z}_n$. Make a conjecture about precisely which elements $\bar{b}$ of $\mathbf{Z}_n$ appear in the row of $\bar{a}$ in the multiplication table.

   b. Prove your conjecture.

   c. How many distinct elements appear in the row of $\bar{a}$ in the multiplication table for $\mathbf{Z}_n$?

14. Let $n \in \mathbf{N}$, and let $\bar{a}, \bar{b} \in \mathbf{Z}_n$. Let $d = \gcd(a, n)$. Prove that if $\bar{b}$ appears in the $\bar{a}$ row of the $\mathbf{Z}_n$ multiplication table, then $\bar{b}$ appears exactly d times in the $\bar{a}$ row of the table.

15. What property of the ring $\mathbf{Z}_n$ allows you to conclude that the statement of Theorem 8.3.6 would still be true if the word *row* were replaced by the word *column*?

16. Prove that the UPC check digit scheme, described in Section 7.3, will detect all single-digit errors.

17. Let $n \in \mathbf{N}$, and let $\bar{x}, \bar{y} \in \mathbf{Z}_n$. Prove that

$$\bar{x}^2 - \bar{y}^2 = (\bar{x} + \bar{y}) \cdot (\bar{x} - \bar{y}).$$

Be sure not to skip any steps and to carefully justify each step. [*Hint:* You may want to use the definition of subtraction in $\mathbf{Z}_n$: $\bar{b} - \bar{a} = \bar{b} + (-\bar{a})$. The properties of Theorem 8.2.6 may also be useful.]

18. Prove the converse of Wilson's theorem:

Let $n > 2$ be a natural number. If in $\mathbf{Z}_p$,

$$\bar{1} \cdot \bar{2} \cdot \bar{3} \cdot \ \ldots \ \cdot (\overline{p - 1}) = \overline{-1},$$

then p is prime.

8.4 Elementary Cryptography

There are many reasons why people need to communicate privately. For instance, the U.S. secretary of state may wish to communicate with the ambassador to China without foreign governments knowing the content of the message. When you buy music or books over the Internet, you don't want your credit card number to end up in the wrong hands. And, of course, secret communication is fundamental to military operations. In fact, during World War II, a major factor in the victory of the Allies was their success at reading secret messages sent by the Germans and Japanese. The development of systems for transforming text to conceal its meaning is called *cryptography*.

Turing's travels

Alan Turing was a British mathematician who played a major role in breaking German military ciphers during World War II. He also made significant discoveries in theoretical computer science and pioneered the development of the digital computer. What's not so well known about Turing is that he was a chocoholic. After two years of suffering through bland, lumpy puddings in the student cafeterias at Cambridge University, he gave in to his passion for chocolate. Turing spent his junior year abroad in Paris, where he studied the chocolate sciences at France's most prestigious candy research institute, the Sorbonbon.

Turing's halting French was a problem, because the Parisian students would not accept a language other than their own. Fortunately, Turing met two other English students, Alice and Bob, in his quantum fudge theory class. While Turing liked his two new friends, they were something of an enigma to him. For some reason, Alice and Bob always seemed to be communicating mysteriously with one another about the Sorbonbon's greatest treasure, the Truffle de Triomphe. This historic piece of chocolate was commissioned by Napoleon in 1806 to commemorate the many victories of the great French chefs over all the other chefs in Europe. The priceless truffle was not completed until 1836, 15 years after Napoleon's death.

As graduation approached, Alice and Bob were sitting outside at a café speaking in low tones, planning their latest method of secret communication. Alice smiled. "Oh, I get it—we're just going to shift every letter in the alphabet by three letters. So we change **A** to **D**, **B** to **E**, and so on. At the end of the alphabet we wrap around, so that we change **X** to **A**, **Y** to **B**, and **Z** to **C**."

Bob nodded. "Right. Your name **ALICE** encrypts to **DOLFH**, while my name **BOB** becomes **ERE**. If you want, you can do it with this table."

Replace this letter:	A	B	C	D	E	···	V	W	X	Y	Z
by this letter:	D	E	F	G	H	···	Y	Z	A	B	C

Alice took a sip of her hot cocoa. "Hey, we could do this with numbers instead of letters. This process of shifting by three letters and wrapping around is just like adding 3 modulo 26. Here, look at the third row of the addition table modulo 26."

+	0	1	2	3	4	$\cdots$	21	22	23	24	25	$\leftarrow x$
3	3	4	5	6	7	$\cdots$	24	25	0	1	2	$\leftarrow x + 3$

"So shifting the alphabet is the same as doing modular addition," agreed Bob.

Alice frowned. "I'm not sure this shifting method is secure enough to protect our secret plans, though. Maybe instead of modular addition we could use more complicated operations, like modular multiplication or even modular expon—"

"Shhh," interrupted Bob. "Here comes the waiter."

"Your Caesar salad, sir."

When the coast was clear, Alice continued. "These are all grand ideas. But I'm still worried that someone really clever, like our friend Al Turing, could break these codes in a second. Just give that guy a strip of paper and something to write with, and he's like a computing machine. They say that if anything can be computed, Turing can do it!"

Bob nodded. "Good point. We need a system that would even test Turing."

Overview of cryptography

Suppose two parties wish to communicate in such a way that their messages will be unintelligible to eavesdroppers. The two parties, often called Alice and Bob, must agree in advance on a method to **encrypt** their messages—to transform the messages into a form that conceals their meaning. For example, in our story Alice and Bob considered encrypting messages by shifting three letters down the alphabet. Using this system, to send the message

PARTY ON DUDE,

Alice would encrypt it as

SDUWB RQ GXGH

The initial message, PARTY ON DUDE, is called the **plaintext**, and the encrypted message, SDUWB RQ GXGH, is the **ciphertext**. Of course, when Bob receives the ciphertext, he needs to **decrypt** it, or transform the ciphertext back into the original plaintext.

Check for Understanding

Suppose Alice and Bob decide to shift by 4 letters instead of 3. Using the 4-letter shift:

1. Encrypt the message HOT WHEAT BRAN.

2. Decrypt the message AMPH EFSYX QEXL.

Mathematically, we can think of the encryption process as a function that takes a plaintext message as input and returns the ciphertext as output. Similarly, decryption can be thought of as a function that takes ciphertext to plaintext. The encryption function and the decryption function are inverse functions: if Alice encrypts a plaintext message, and Bob then decrypts the resulting ciphertext, he will recover Alice's original message exactly.

Encryption using modular addition

In our story, Alice noticed that the encryption method of shifting the alphabet by three letters can be viewed as adding 3 modulo 26. To decrypt, Bob shifts the alphabet backward by three letters, which may be seen as **subtracting 3 modulo 26**. In modulo 26 arithmetic, subtracting 3 is the same as adding the additive inverse of 3, which is 23. Thus, if Bob prefers addition to subtraction, he can also decrypt messages by **adding 23 modulo 26**.

In Alice and Bob's modular addition (alphabet shift) encryption method, every time a given letter appears in the plaintext, it always encrypts to the same letter in the ciphertext. (For example, the letter A is always encrypted as D.) Encryption methods that have this property are called *simple substitution ciphers*. You may have seen such ciphers in the puzzle section of the newspaper. Simple substitution ciphers are not limited to alphabet shifts—they can involve an arbitrary rearrangement of the letters of the alphabet. Even so, these ciphers can be solved quickly by an experienced puzzle solver or a simple computer program.

Fun Facts

More than two thousand years ago, Julius Caesar used the method of shifting the alphabet 3 places to encrypt military messages. This type of encryption by shifting the alphabet (i.e., addition modulo 26) is known as a *Caesar cipher*.

Part of the reason these ciphers are easy to break is that there are only 26 letters in the alphabet, so one only needs to determine 26 correspondences to break the code.

One way to get around this is to group the plaintext message into blocks of more than one letter and encrypt the message one block at a time. Let's see how we can use this idea to improve Alice and Bob's modular addition encryption system.

The word *NUMBER* could be divided into two-letter blocks: **NU MB ER**. Since **N** is the **14**th letter in the alphabet and **U** is the **21**st, the block **NU** corresponds to the number **1421**. Similarly, the block **MB** becomes **1302**, and **ER** becomes **518**. We have now converted the entire message into numbers:

$$NU \ \ MB \ \ ER = 1421 \ \ 1302 \ \ 518$$

Notice that we have not yet encrypted the message—we have merely converted it into numbers. The list of numbers 1421, 1302, 518 is just another form of the plaintext message.[7]

If we want to encrypt messages that are more than one word long, we will want to have a character, –, that we can use to separate words. This character can be thought of as a space, and it will be represented by the number 27. This character also has another use: if the last block in a plaintext message does not have enough letters, this character can be used to fill out the rest of the block.

You may find the following table useful when encrypting and decrypting.

Table 8.4.1 Correspondences for encoding letters of the alphabet into numbers

A	B	C	D	E	F	G	H	I	J	K	L	M	N	O	P	Q	R	S	T	U	V	W	X	Y	Z	–
1	2	3	4	5	6	7	8	9	10	11	12	13	14	15	16	17	18	19	20	21	22	23	24	25	26	27

Suppose Alice and Bob wish to exchange secret messages using modular addition. Since the numbers representing two-letter blocks may be as large as 2727 (which corresponds to the two-letter block "– –"), Alice and Bob must choose a modulus at least this large. For example, Alice and Bob might decide to encrypt each block by adding 1519 modulo 2730.

EXAMPLE 1

Encrypt the message NUMBER using two-letter blocks and encryption by adding 1519 modulo 2730.

SOLUTION

$$
\begin{aligned}
NU &= 1421 & 1421 + 1519 &\equiv 210 \pmod{2730} \\
MB &= 1302 & 1302 + 1519 &\equiv 91 \pmod{2730} \\
ER &= 518 & 518 + 1519 &\equiv 2037 \pmod{2730}
\end{aligned}
$$

Thus, the entire message NUMBER encrypts to the ciphertext 210 91 2037. ●

---

[7]One might say that the text message has been *encoded* but not encrypted.

After Bob receives the ciphertext 210 91 2037 from Alice, he needs to subtract 1519 from each number modulo 2730. Bob can accomplish this subtraction easily by adding the inverse of 1519 modulo 2730. Recall that this additive inverse of 1519 may be found by subtracting:

$$2730 - 1519 = 1211. \qquad \leftarrow \textbf{Modulo 2730, the additive inverse of 1519 is 1211.}$$

Thus, to decrypt, Bob simply adds 1211 to each block modulo 2730.

EXAMPLE 2

The following ciphertext was encrypted using two-letter blocks and adding 1519 modulo 2730 (the encryption method from Example 1). Using the system described above, decrypt the following ciphertext:

$$793 \quad 2034 \quad 614.$$

SOLUTION

$$793 + 1211 \equiv 2008 \ (\text{mod } 2730)$$
$$2034 + 1211 \equiv 515 \ (\text{mod } 2730)$$
$$614 + 1211 \equiv 1825 \ (\text{mod } 2730)$$

Thus, we recover the plaintext 2008 515 1825. Converted into text (using Table 8.4.1), this becomes TH EO RY. ■

EXERCISES 8.4

Numerical Problems

1. Encrypt the following statement using two-letter blocks and encryption by adding 1343 modulo 2727. (Here we use the "–" character to indicate spaces between words and to complete the final two-letter block.)

   I–LOVE–MATH–

2. The following statement was encrypted by shifting each letter in the alphabet by 3. Spaces between words were not encrypted. Decrypt the message.

   PHH WPH DWW KHX VXD OSO DFH DWH LJK WRF ORF N.

3. The following statement was encrypted by shifting each letter in the alphabet by some amount. Spaces between words were not encrypted. Decrypt the message.

   ZGYNQ DFTQA DKUEG EQRGX RADQZ OUBTQ DUZSY QEEMS QE.

4. Encrypt the following quote of Will Rogers using two-letter blocks and encryption by adding 1500 modulo 2800. (Here we use the "–" character to indicate spaces between words.)

   EVERYBODY–IS–IGNORANT–ONLY–ON–DIFFERENT–SUBJECTS

5. To encrypt a message three letters at a time, you need to use a modulus at least as large as 272727. Use a modulus of 279999 and encryption by adding 123456 to encrypt the following quote in three-letter blocks. (Here we use the "–" character to indicate spaces between words and to complete the final three-letter block.)

   THE–FUTURE–WILL–BE–BETTER–TOMORROW– –

6. The following message was encrypted using the scheme in Exercise 5 (three-letter blocks, modulus of 279999, and addition of 123456). Decrypt the message.

   214083   73984 234861   76180   203576 115762 115762

   23984 164965 264183 215483   74978 243883 264976

   113662 113758 244661 166174 175361 135259 206183

7. To encrypt a message five letters at a time, you need to use a modulus that is at least 2727272727. Use a modulus of 2727272727 and encryption by adding 1122334455 to encrypt the following quote of Mark Twain in five-letter blocks. (Here we use the "–" character to indicate spaces between words.)

   FICTION–IS–OBLIGED–TO–STICK–TO–POSSIBILITIES–TRUTH–ISNT

8. A famous actor said, "I don't want to achieve immortality through my work. . . . I want to achieve it through not dying." His name is encrypted below using the same method as Exercise 7 (five-letter blocks, modulus of 2727272727, and addition of 1122334455). Decrypt this actor's name:

   710212153 1234454969

8.5 Encryption Using Modular Multiplication

In the last section, we saw how to encrypt messages using modular addition. One problem with modular addition is that the encryption function does not mix things up very well. Consecutive numbers in the plaintext map to consecutive numbers in the ciphertext. For example, using the modular addition method of the examples in Section 8.4 (adding 1519 modulo 2730), the plaintext TH = 2008 encrypts as 793, and the plaintext TI = 2009 encrypts as 794. This kind of regularity makes the system easier to crack.

Earlier in this chapter, we observed that the rows of the $\mathbf{Z}_n$ multiplication table are generally much more scrambled than the rows of the $\mathbf{Z}_n$ addition table. Unlike encryption using modular addition, encryption using modular multiplication does not encrypt consecutive numbers in the plaintext to consecutive ciphertexts.

Let's look at an example of encryption using modular multiplication. Suppose that Alice and Bob agree to encrypt messages in two-letter blocks using multiplication by 11 modulo 2800.

EXAMPLE 1

Encrypt the message FAT CAT in two-letter blocks using multiplication by 11 modulo 2800.

SOLUTION First separate the message into two-letter blocks, using the "–" character to represent spaces between words and to complete the final two-letter block. Use Table 8.4.1 to convert each two-letter block into a number. Then multiply each number by 11 and reduce modulo 2800.

Plaintext	Ciphertext
FA = 601	$601 \cdot 11 \equiv 1011 \pmod{2800}$
T– = 2027	$2027 \cdot 11 \equiv 2697 \pmod{2800}$
CA = 301	$301 \cdot 11 \equiv 511 \pmod{2800}$
T– = 2027	$2027 \cdot 11 \equiv 2697 \pmod{2800}$

Thus, the message FAT CAT encrypts to the ciphertext 1011 2697 511 2697.

Phil Lovett When your friend Phil Lovett sees how much fun you're having in number theory class, he wants to encrypt messages, too. He decides to encrypt in two-letter blocks using multiplication by 385 modulo 2800. Proud that he is lead vocalist of the rock group Dogbreath, Phil tries to encrypt the message I SING:

Plaintext	Ciphertext
I– = 927	$927 \cdot 385 \equiv 1295 \pmod{2800}$
SI = 1909	$1909 \cdot 385 \equiv 1365 \pmod{2800}$
NG = 1407	$1407 \cdot 385 \equiv 1295 \pmod{2800}$

Thus, Phil encrypts I SING to the ciphertext 1295 1365 1295.

Do you notice a problem with Phil's encryption? The first and third plaintext blocks, I– and NG, encrypt to the same number, 1295. The intended recipient of

Phil's message cannot possibly know how to decrypt the ciphertext 1295, which could represent either the plaintext I– or the plaintext NG. Phil's procedure is therefore not a valid method for encryption: his encryption function has no inverse function. (Phil's encryption function is not *one-to-one*; see Exercise 9.)

If Phil's encryption procedure of multiplication by 385 modulo 2800 does not work, how do we know that our method from Example 1, multiplying by 11 modulo 2800, is any better? We need to make sure that no two plaintext numbers encrypt to the same ciphertext.

Suppose x and y are two plaintext numbers. Then x encrypts to the reduction of $11 \cdot x \pmod{2800}$, and y encrypts to the reduction of $11 \cdot y \pmod{2800}$. Is it possible for x and y to encrypt to the same ciphertext? This would imply that

$$11x \equiv 11y \pmod{2800}.$$

Since $\gcd(11, 2800) = 1$, we can cancel the 11's (by Modular Cancellation 8.3.3), and get

$$x \equiv y \pmod{2800}.$$

Thus, the only way that x and y can encrypt to the same ciphertext is if x and y are equal to begin with. We can conclude that multiplication by 11 modulo 2800 is a valid encryption method.

The key to this argument is the fact that 11 and 2800 are relatively prime. Similar reasoning can be used to show that whenever a and n are relatively prime, multiplication by a modulo n is a valid encryption method. Note that we cannot use a similar argument to justify Phil Lovett's lousy encryption procedure of multiplying by 385 modulo 2800, because 385 and 2800 are not relatively prime.

Encrypting and decrypting with modular multiplication tables

The encryption method of Example 1, multiplying by 11 modulo 2800, can be thought of in terms of modular multiplication tables. Here is the row of $\overline{11}$ in the $\mathbf{Z}_{2800}$ multiplication table:

$\cdot$	0	1	2	3	$\cdots$	254	255	256	$\cdots$	2798	2799	← Plaintext
$\vdots$												
11	0	11	22	33	$\cdots$	2794	5	16	$\cdots$	2778	2789	← Ciphertext
$\vdots$												

For example, the plaintext message 254 encrypts to the ciphertext 2794.

The table provides another way to see why the encryption method of multiplication by 11 modulo 2800 does not encrypt two different plaintext numbers to the same ciphertext. Since $\gcd(11, 2800) = 1$, we know by Theorem 8.3.6 that there are no repeated elements in the $\overline{11}$ row; in other words, no ciphertext occurs twice. Each ciphertext corresponds to a single plaintext message, so multiplication by 11 modulo 2800 is a valid encryption scheme.

EXAMPLE 2

Decrypt the ciphertext 1011 1520 98, which was encrypted using the method just discussed (multiplication by 11 modulo 2800). Use the $\mathbf{Z}_{2800}$ multiplication table provided.

·	⋯	517	518	517	⋯	599	600	601	⋯	1920	1921	1922	⋯
⋮													
11	⋯	87	98	109	⋯	989	1000	1011	⋯	1520	1531	1542	⋯
⋮													

SOLUTION Find each ciphertext number in the $\overline{11}$ row of the table; then look in the top row to find the corresponding plaintext.

Ciphertext	Plaintext
$1011 \equiv 11 \cdot 601 \pmod{2800}$	$601 = $ FA
$1520 \equiv 11 \cdot 1920 \pmod{2800}$	$1920 = $ ST
$98 \equiv 11 \cdot 518 \pmod{2800}$	$518 = $ ER

Thus, the ciphertext 1011 1520 98 decrypts to the plaintext message FASTER. ●

In Example 2, we used the method of *reverse table lookup* to decrypt. As a general method for decryption, this is not very efficient. Not only do we have to make the entire 2800-column table, but we then have to look through the 2800 columns of the table every time we wish to decrypt a two-letter block. The message from Example 2 is clear. We need a faster method for decrypting messages. To get some new ideas for decrypting, let's return our attention to Alan Turing and his fellow English exchange students, Alice and Bob.

Truffle trouble

It was the day before graduation. Everyone at the Sorbonbon was eagerly awaiting the commencement ceremony, where the world-famous Truffle de Triomphe would be available for the annual public viewing and sniffing. Preparations were well under way. The truffle had been removed from its secure underground vault and transferred to a refrigerated gazebo built specially for the occasion. Security was tight since everyone knew that the students at Paris's rival institution of higher chocolate learning, the Institut des Hautes Études Chocolatiques, would do anything to get their hands on the prized truffle.

Returning from a party late that night, Turing was surprised to see police cars in front of his dorm with their lights flashing. He ran upstairs and discovered a group of police officers gathered around the desk in Alice's room.

"I say, what's going on?" Turing demanded.

The officer in charge, who was sitting behind Alice's desk, stood up and brushed the croissant crumbs off of his wrinkled uniform. "Bonjour, dude," said the officer. "I am Deputy Inspector Philippe L'Ovette."

Pointing to a piece of paper on Alice's desk, Inspector L'Ovette continued, "We found this encrypted note in Bob's room. Then we rushed over here, where we discovered a calculator with these encryption directions beside it. During the past five hours, we've managed to decrypt most of the message. Et voilà! We now have all the evidence we need to arrest Alice and Bob."

Turing looked down and saw a note with numbers in Alice's handwriting. Under most of the numbers, the police had written the decrypted text in orange.

Inspector L'Ovette glared at Turing. "It would appear that your compatriots are planning to steal the Truffle de Triomphe."

Turing was indignant. "I refuse to believe that Alice and Bob could be involved in such a crime. How can you arrest them when you haven't even finished decrypting Alice's message?"

"There is no time to finish the decryption! We'd need at least another hour to decrypt the remaining numbers using our reverse table lookup method, and midnight is only 10 minutes away. I am sorry, mon ami, but we've got to arrest Alice and Bob immediately."

Turing studied the message again, wondering what it said. Could his friends really be planning such a crime? He thought for a moment, then grabbed a pencil and hastily jotted down a few equations. Soon he was furiously pushing buttons on Alice's calculator. A minute later, Turing looked up. "Wait! I know what the rest of the message says. Alice and Bob are innocent!"

But the officers were already out the door. Turing rushed after the police, catching up to them just before they reached the gazebo. "Look here, chaps! Look at this." Turing thrust the decrypted note into the Inspector's hand.

"*Mon Dieu!* So Alice and Bob are not thieves after all." Inspector L'Ovette patted Turing on the back. "Do not worry. We French know all about these things. Let us leave your friends to carry out their little plan in peace."

In our myth, the plodding police had decrypted the beginning of Alice's message to Bob using the method of reverse table lookup (the same method we used in Example 2). Their method was so slow that they did not have time to finish decrypting the message before the alleged crime was going to take place. Turing saved his friends from arrest by coming up with a faster method for decryption, a more direct decryption method that did not use the multiplication table.

Decrypting without using the table

Alice and Bob's plaintext message was encrypted using multiplication by 11 modulo 2800. To decrypt the ciphertext, Turing needed to undo the effect of multiplication by 11 modulo 2800. Effectively, he needed to divide by 11. To accomplish this, he multiplied the ciphertext by the multiplicative inverse of 11 modulo 2800.

Using the same methods that we learned in Section 8.3, Turing found that

the multiplicative inverse of 11 modulo 2800 is **2291**.

Turing was then able to decrypt the message quickly by multiplying each ciphertext number by **2291** modulo 2800.

EXAMPLE 3

The following ciphertext was encrypted using multiplication by 11 modulo 2800. Use the method just described to decrypt the following ciphertext:

$$465 \ 1897 \ 1765 \ 2397.$$

SOLUTION

$$465 \cdot \mathbf{2291} \equiv 1315 \ (\text{mod } 2800)$$
$$1897 \cdot \mathbf{2291} \equiv 427 \ (\text{mod } 2800)$$
$$1765 \cdot \mathbf{2291} \equiv 415 \ (\text{mod } 2800)$$
$$2397 \cdot \mathbf{2291} \equiv 727 \ (\text{mod } 2800)$$

Thus, we've recovered the plaintext: 1315 427 415 727.
Converted into text, this becomes: M O D− DO G−. ●

The method we used in this example is much faster than the reverse table lookup method of Example 2. In Example 2, we had to look through as many as 2800 columns of the table to decrypt each block. To use this new method, however, we only needed to do a single modular multiplication to decrypt each block. Of course, we first needed to find the multiplicative inverse of 11 modulo 2800. But this inverse can be found very quickly—it only takes a handful of steps using the Euclidean Algorithm.

EXERCISES 8.5

Numerical Problems

1. Encrypt the following message using multiplication by 11 modulo 2800.

I DIG ARCHAEOLOGY

2. Decrypt the following message, which was encrypted using multiplication by 11 modulo 2800.

   1643 1854 2720 1711 1855 409 197 2565 1876 2555 2598 2597

3. Decrypt the following message, which was encrypted using multiplication by 11 modulo 2800.

   2155 2632 2577 2597 365 621 1597

4. Suppose you want to encrypt messages using multiplication by a modulo 2727. Which positive integer values of a would make this a valid encryption method?

5. You and your friend have decided to encrypt messages using two-letter blocks and multiplication by 5 modulo 2727.

   a. Encrypt this message for your friend:

   DENY EVERYTHING

   b. Find the multiplicative inverse of 5 modulo 2727.

   c. Use the inverse you found in part b to decrypt this message from your friend:

   1859 2625 2647 1621 727 1894 2181 1151 1540

6. You and your friend have decided to encrypt messages using two-letter blocks and multiplication by 21 modulo 2750.

   a. Encrypt this message for your friend:

   R U PHIL LOVETT

   b. Find the multiplicative inverse of 21 modulo 2750.

   c. Your friend's response reveals the true origin of the name Phil Lovett. Use the inverse you found in part b to decrypt this response from your friend:

   2215 1889 1721 367 2041 702 2015 2167 70

7. Decrypt the rest of Alice's message to Bob from the myth. What did Deputy Inspector Phillippe L'Ovette mean by "We French know all about these things"?

Reasoning and Proofs

8. Prove that if $\gcd(a, n) = 1$, then multiplication by a modulo n is a valid encryption method.

9. A function f is said to be *one-to-one* (or *injective*) if for every a and b in the domain of f, $f(a) = f(b)$ implies that $a = b$. Explain why encryption functions must be one-to-one.

10. **a.** **Computer Exercise** Encrypt the following quote in four-letter blocks, using multiplication by 37 modulo 27989898.

COMPUTERS ARE USELESS—THEY CAN ONLY GIVE YOU ANSWERS

    b. Find the inverse of $\overline{37}$ in $\mathbf{Z}_{27989898}$.

    c. The following ciphertext message was encrypted using the method you used in part a (four-letter blocks, multiplication by 37 modulo 27989898). Decrypt the following message to reveal the name of the famous person to whom the quote in part a is attributed.

4589986 5251573 27471309 5292939

Chapter 9

Pierre de Fermat (1601–1665)

Pierre de Fermat laid the foundation for modern number theory. A lawyer by profession, Fermat did his brilliant mathematics in his spare time, earning himself the appellation "the prince of amateurs."

Fermat's mother Claire de Long was a member of the French nobility, and his father, a wealthy merchant, encouraged him to pursue a career in law. After completing his law degree at age 30, Fermat married his mother's cousin, Louise de Long. He bought himself two positions in the parliament of Toulouse: the office of Councilor and that of Commissioner of Requests. Purchasing positions in parliament, which was not unusual for someone of Pierre Fermat's economic class, enabled him to add the title of "de" to his name.

In the mid-1600s, the Black Plague killed countless people throughout Europe. Fermat contracted the plague in 1652, and one of his friends even announced that Fermat was dead! Fortunately Fermat recovered, though many of his colleagues in parliament were not so lucky. As one after another of his superiors died, Fermat was promoted to fill their positions. Ultimately, he attained the position of liaison between the local people and the king and also served as a judge on many serious cases. On one occasion, Fermat sentenced a priest who had abused his position to be burned at the stake.

Although Fermat had little formal training in mathematics and was not living near any center of mathematical activity, he had a passion for mathematics and devoted all of his free time to it. His discoveries had a fundamental impact on the development of number theory, probability, calculus, and analytic geometry. After reading Diophantus's *Arithmetica*, Fermat developed a deep and lasting interest in number theory. In contrast with Diophantus, who was interested in rational solutions to equations, Fermat focused exclusively on integer solutions. This change in focus, together with Fermat's many results on divisibility and primes, opened many new directions in number theory. As Europeans had made little progress in number theory from Diophantus's time to Fermat's time, Fermat is considered to be the founder of modern number theory. In addition to the theorems he proved, Fermat pioneered the *method of infinite descent* (equivalent to the Well-Ordering Principle), a technique that has been used to prove a number of important results in number theory (see Theorem 15.3.2 for an example of a proof using this method).

Until the middle of the 17th century, probability had not been developed into a mathematical theory—the subject was merely a hodge-podge of tricks and the intuitions of professional gamblers. In 1654, the Chevalier de Méré, a French gambler and nobleman, asked the mathematician Blaise Pascal a question concerning a popular dice game. Pascal's efforts to answer de Méré's question ignited a correspondence between Pascal and Fermat in which they laid the foundation for the theory of probability.

In addition to his pivotal work in both number theory and probability, Fermat's ideas were fundamental in the development of analytic geometry and calculus. Though he did not have the concepts of limit or derivative that were developed later by Newton and Leibniz, Fermat used the idea of infinitesimals to describe methods for finding the tangent line to a curve of the form $y = x^n$ and for finding the area under such a curve. In his treatise *Method of Finding Maxima and Minima,* he presented techniques for finding the maxima and minima of a polynomial, similar to the techniques taught in calculus classes today. Although the invention of calculus is attributed to Newton and Leibniz, Newton wrote a note acknowledging that his ideas for calculus were based on Fermat's method of drawing tangents.

In spite of Fermat's profound effect on the development of mathematics, he published few papers and had little contact with other mathematicians. Most of his results were contained in his letters, notes, or manuscripts that he circulated among his friends, and these generally included only sketches of proofs or no proofs at all. When he did publish his work, Fermat insisted that his papers appear anonymously. Often he would write a letter to a friend stating that he knew how to prove a particular result and challenge the recipient to try to prove it as well. He also wrote statements of some of his theorems on scraps of paper

or as notes in the margins of his copy of *Arithmetica* without including proofs.

In 1636, Fermat wrote a manuscript entitled *Introduction to Plane and Surface Loci,* which explained his ideas for analytic geometry. When Marin Mersenne (whom we met in Chapter 6) was shown a copy of the manuscript, he initiated a correspondence with Fermat. After that, Fermat began to write news of his latest discoveries to Mersenne, who then shared this news with the French and Italian mathematical communities.

In his later years, Fermat focused almost entirely on number theory, an unpopular subject among mathematicians at the time. As a result, Fermat became increasingly isolated mathematically. After Fermat's death, his son Clément-Samuel carefully transcribed Fermat's notes from the margins of *Arithmetica* and published a book entitled *Diophantus' Arithmetica Containing Observations by P. de Fermat.* Nonetheless, few mathematicians were aware of Fermat's work until the revival of number theory around the beginning of the 19th century.

A Bit of Fermat's Math

Fermat's most important contributions belong to the field of number theory. In this chapter, we will learn about Fermat's Little Theorem (9.1.1), which has played a significant role in many proofs in number theory. After proving his Little Theorem, Fermat made the conjecture that any number that can be written in the form $2^{2^n} + 1$ must be prime (see Section 3.3); this conjecture turned out to be false.

Another number-theoretic result of Fermat's was stimulated by Diophantus's *Arithmetica.* Diophantus had asserted without proof that every number of the form $4n + 3$ is neither a square nor the sum of two squares and that every prime number of the form $4n + 1$ is the sum of two squares. Fermat proved both assertions and further showed that every prime number of the form $4n + 1$ can be written *uniquely* as the sum of two squares. We will prove these results in Chapter 13; see Fermat's Two Squares Theorem (Section 13.6). Fermat also gave a formula for the number of ways that any perfect square can be expressed as the sum of two squares.

Over the years, mathematicians have given proofs of all of the remaining unproven assertions that Fermat made. The most notorious of these claims, which became known as Fermat's Last Theorem, remained unproven for almost 350 years. (See Sections 15.2–15.5 to learn more about Fermat's Last Theorem and its history.) This assertion, as well as Fermat's many others, have provided tremendous stimulation to the development of number theory and of mathematics in general.

Chapter 9 Exponents Modulo *n*

9.1 Fermat's Little Theorem

The Grapes of Math

Frenchman Pierre de Fermat made major contributions to the field of number theory. What's not so well known about Fermat is that he smelled. In fact, he smelled much better than anyone else in 17th-century France. Some historians have even contended that Fermat had one of the greatest mathematical noses of all time. Due to his highly discriminating sense of smell, Fermat was appointed Chief of Nasal Operations of a small developing country vineyard in southwestern France.

It was Fermat's job to select grapes that exuded the delicate aroma for which the Château Petite Odeur was famous. Less exclusive vineyards in the region removed foul odors from their wine by sending it off to a de-scenting facility, where duck fat was added to mask the smell. But due to Fermat's olfactory prowess, Petite Odeur did not need to send its wine out for descenting. Fermat's method of in-vineyard descent was to weed out the malodorous grapes in the field before they were pressed.

The most important part of Fermat's job each year was sniffing out the select grapes used to make the vineyard's Limited Edition Chardon-nez Grand Cru. This extraordinary wine was carefully packed into special pentagonal cases, with 5 bottles to a case.

Every bottle of the Limited Edition Grand Cru was individually stamped to certify its authenticity. Each stamp contained a 4-digit serial number made up of 1's and 2's. For example, one bottle was stamped 1121, while another was stamped 2122.

Since the serial numbers were 4 digits long, and each digit had 2 possible values, a total of $2^4 = 16$ bottles of the precious Grand Cru were produced each year. When the time came to pack the 16 bottles into pentagonal cases (with 5 bottles per case), Fermat packed three cases full, with one bottle left over.

$$2^4 \equiv 1 \ (\text{mod } 5) \qquad \leftarrow \textbf{exactly 1 bottle left over.}$$

Following long-standing tradition, Fermat carefully placed this leftover bottle in the display case at the château.

The following year, the owner of the vineyard, a sophisticated gentleman named Pascal, came to Fermat with a problem. Pascal's two great passions in life were wine and gambling. Unfortunately, the gambling was as bad for his bank account as the wine was for his liver.

Pascal explained his problem. "You've got to help me, Fermat. I'm in big trouble—I've gambled away so much money that I can't afford this year's liver transplant."

Fermat offered, "Maybe we could produce more bottles of the Grand Cru. That should increase our profits considerably."

"Great idea. I'll drink to that!" exclaimed Pascal, raising his glass. "But if we are going to increase production, we'll need a new numbering system. The great thing about our current numbering system is that after packing as many full cases as possible, there is exactly 1 bottle of wine left over to put on display. It may be impossible to come up with a new system for which this is true."

"I'm sure I can figure out a way," boasted Fermat.

"You wanna bet?" asked Pascal.

Fermat accepted Pascal's wager and tried to concoct a new serial number system. To win the bet, Fermat needed to have a single bottle of Grand Cru remaining after packing the rest into cases of 5 bottles each. In other words, the total number of bottles had to be congruent to 1 modulo 5.

The problem plagued Fermat for weeks, but finally he had a powerful insight, and he summoned Pascal to tell him the news. "The solution is quite simple, *mon cher* Pascal. As before, we will use 4-digit-long serial numbers, but they will now be composed of the numerals 1, 2, and 3. This results in a total of $3^4 = 81$ serial numbers. You see, 80 of those bottles will fit perfectly into sixteen cases of 5 bottles each, with exactly 1 bottle remaining to put on display."

$$3^4 \equiv 1 \ (\text{mod } 5) \qquad \leftarrow \textbf{exactly 1 bottle left over}.$$

"*C'est fantastique!*" exclaimed Pascal. "I have never been so happy to lose a bet! Once we sell those sixteen cases, I will be able to afford my liver transplant and have just enough money left to pay off all of the losing bets I made this year."

"Including, of course, the bet that you just lost to me?" asked Fermat.

"Oh. I hadn't considered that," Pascal responded soberly, taking another swig from his carafe. "Now I can't possibly pay all my debts. I'm ruined! Unless you can figure out a way for us to boost production even further. . . ."

"Not to worry, Monsieur," replied Fermat. "If we use 4-digit-long serial numbers composed of the numerals 1, 2, 3, and 4, then your problems will again be solved. We will have $4^4 = 256$ bottles of the Grand Cru. After packing these into cases of 5 bottles each, there will be exactly one bottle left over once again."

$$4^4 \equiv 1 \ (\text{mod } 5) \qquad \leftarrow \textbf{exactly 1 bottle left over}.$$

Pascal cried tears of joy. "To your health, Fermat!"

To win his bet with Pascal in our myth, Fermat proposed increasing production from the original 2^4 bottles to 3^4 bottles and finally to 4^4 bottles. Pascal was pleased because all of these numbers are congruent to 1 modulo 5.

Let's explore exponents in $\mathbf{Z}_5$ in more detail. We can begin by finding a^b modulo 5 for various values of a and b. For example, here is a table of the powers of 2 modulo 5.

2^n (mod 5)

$2^1 \equiv 2$
$2^2 \equiv 4$
$2^3 \equiv 3$
$2^4 \equiv 1$
$2^5 \equiv 2$
$2^6 \equiv 4$
$2^7 \equiv ?$
$\vdots$

How would we compute 2^7 modulo 5 to finish the seventh row of this table? One approach would be to first calculate $2^7 = 128$ and then reduce 128 modulo 5. An easier method is to start with the previous row of the table:

$$2^6 \equiv 4 \ (\text{mod } 5).$$

Multiplying both sides by 2 (using Theorem 7.2.1, Congruences Add and Multiply), we get

$$2 \cdot 2^6 \equiv 2 \cdot 4 \ (\text{mod } 5)$$
$$\equiv 3 \ (\text{mod } 5).$$

Thus, $2^7 \equiv 3 \ (\text{mod } 5)$.

Check for Understanding

Make a table of the powers of 2 modulo 9, similar to the one above (but reducing modulo 9 instead of modulo 5), from 2^0 through 2^{12}.

Does your table have a pattern?

Exponents in $\mathbf{Z}_n$

If $\bar{a} \in \mathbf{Z}_n$ and k is a natural number, we can define $\bar{a}^k$ to be the product of a with itself k times:

$$\bar{a}^k = \underbrace{\bar{a} \cdot \bar{a} \cdot \bar{a} \cdot \ \cdots \ \cdot \bar{a}}_{k \text{ times}}.$$

We also allow an exponent of 0, defining $\bar{a}^0 = \bar{1}$. We can even define negative exponents, provided that $\bar{a}$ has a multiplicative inverse: For any $k \in \mathbf{N}$, we define $\bar{a}^{-k}$ to be the product of $\bar{a}^{-1}$ with itself k times. With these definitions, all of the usual laws of exponents hold.

Earlier we computed the powers of 2 modulo 5. In other words, we computed $\bar{2}^1, \bar{2}^2, \bar{2}^3, \ldots$ in $\mathbf{Z}_5$. The following exponent table shows the powers (up to the 5th power) of all of the nonzero elements of $\mathbf{Z}_5$. (As was the case in Chapter 8, putting bars above all of the numbers in a table would make it harder to read. We will again omit bars, *inside tables only*. Although we don't always show the bars, every number in the table is really a barred number.)

Exponents in $\mathbf{Z}_5$

$\bar{a}$	1	2	3	4
$\bar{a}^2$	1	4	4	1
$\bar{a}^3$	1	3	2	4
$\bar{a}^4$	1	1	1	1
$\bar{a}^5$	1	2	3	4

Check for Understanding

1. Use this table of exponents to evaluate $\bar{3}^4$ in $\mathbf{Z}_5$.

2. Use this table of exponents to evaluate $\bar{4}^5$ in $\mathbf{Z}_5$.

A pattern perceptible

What patterns do you notice in the $\mathbf{Z}_5$ exponent table above? One pattern you may see is that the $\bar{a}^4$ row consists entirely of $\bar{1}$'s. That is, in $\mathbf{Z}_5$,

$$\bar{1}^4 = \bar{1},$$

$$\bar{2}^4 = \bar{1},$$

$$\bar{3}^4 = \bar{1},$$

$$\bar{4}^4 = \bar{1}.$$

In our myth, these congruences helped Fermat win his bet with Pascal.

Now let's have a look at the exponent table for $\mathbf{Z}_7$.

Exponents in $\mathbf{Z}_7$

$\bar{a}$	$\bar{1}$	$\bar{2}$	$\bar{3}$	$\bar{4}$	$\bar{5}$	$\bar{6}$
$\bar{a}^2$	1	4	2	2	4	1
$\bar{a}^3$	1	1	6	1	6	6
$\bar{a}^4$	1	2	4	4	2	1
$\bar{a}^5$	1	4	5	2	3	6
$\bar{a}^6$	1	1	1	1	1	1
$\bar{a}^7$	1	2	3	4	5	6

Do you see a pattern in this table that is similar to the pattern we found in the $\mathbf{Z}_5$ exponent table? This time, it is the $\bar{a}^6$ row that consists entirely of $\bar{1}$'s. We can state this formally:

$$\text{For every nonzero } \bar{a} \in \mathbf{Z}_7, \ \bar{a}^6 = \bar{1}. \tag{1}$$

You may wonder whether this observation can be generalized to other moduli. To explore this question, let's examine the exponent tables for $\mathbf{Z}_{10}$ and $\mathbf{Z}_{11}$.

Exponents in $\mathbf{Z}_{10}$

$\bar{a}$	$\bar{1}$	$\bar{2}$	$\bar{3}$	$\bar{4}$	$\bar{5}$	$\bar{6}$	$\bar{7}$	$\bar{8}$	$\bar{9}$
$\bar{a}^2$	1	4	9	6	5	6	9	4	1
$\bar{a}^3$	1	8	7	4	5	6	3	2	9
$\bar{a}^4$	1	6	1	6	5	6	1	6	1
$\bar{a}^5$	1	2	3	4	5	6	7	8	9
$\bar{a}^6$	1	4	9	6	5	6	9	4	1
$\bar{a}^7$	1	8	7	4	5	6	3	2	9
$\bar{a}^8$	1	6	1	6	5	6	1	6	1
$\bar{a}^9$	1	2	3	4	5	6	7	8	9
$\bar{a}^{10}$	1	4	9	6	5	6	9	4	1

Exponents in $\mathbf{Z}_{11}$

$\bar{a}$	$\bar{1}$	$\bar{2}$	$\bar{3}$	$\bar{4}$	$\bar{5}$	$\bar{6}$	$\bar{7}$	$\bar{8}$	$\bar{9}$	$\overline{10}$
$\bar{a}^2$	1	4	9	5	3	3	5	9	4	1
$\bar{a}^3$	1	8	5	9	4	7	2	6	3	10
$\bar{a}^4$	1	5	4	3	9	9	3	4	5	1
$\bar{a}^5$	1	10	1	1	1	10	10	10	1	10
$\bar{a}^6$	1	9	3	4	5	5	4	3	9	1
$\bar{a}^7$	1	7	9	5	3	8	6	2	4	10
$\bar{a}^8$	1	3	5	9	4	4	9	5	3	1
$\bar{a}^9$	1	6	4	3	9	2	8	7	5	10
$\bar{a}^{10}$	1	1	1	1	1	1	1	1	1	1
$\bar{a}^{11}$	1	2	3	4	5	6	7	8	9	10

In the $\bar{a}^{10}$ row of the table of exponents modulo 11, the only element that appears is $\bar{1}$. Thus, pattern (1) that we saw for $\mathbf{Z}_7$ does generalize to $\mathbf{Z}_{11}$:

$$\text{For every nonzero } \bar{a} \in \mathbf{Z}_{11}, \ \bar{a}^{10} = \bar{1}. \tag{2}$$

However, notice that in the $\mathbf{Z}_{10}$ exponents table, the same pattern does not hold: there is no row that contains only $\bar{1}$.

For which other moduli does the pattern observed in (1) and (2) hold? We have already observed that it holds in $\mathbf{Z}_5, \mathbf{Z}_7$, and $\mathbf{Z}_{11}$; the pattern appears to hold whenever the modulus is prime. This important result is known as Fermat's Little Theorem.

THEOREM 9.1.1 FERMAT'S LITTLE THEOREM

Let $p \in \mathbf{N}$ be prime, and let $\bar{a} \in \mathbf{Z}_p$ such that $\bar{a} \neq \bar{0}$. Then

$$\bar{a}^{p-1} = \bar{1}.$$

Fermat's Little Theorem can be restated in terms of congruences as follows.

THEOREM 9.1.2 RESTATEMENT OF FERMAT'S LITTLE THEOREM

Let $p \in \mathbf{N}$ be prime, and let $a \in \mathbf{Z}$ such that $p \nmid a$. Then

$$a^{p-1} \equiv 1 \ (\text{mod } p).$$

The two statements of Fermat's Little Theorem, 9.1.1 and 9.1.2, are equivalent (see Exercise 19). Thus, we get to choose which of the two statements we wish to prove. We will prove Theorem 9.1.1 in a moment; we could prove Theorem 9.1.2 using an almost identical proof.

Fermat's Little Theorem can also be stated as a congruence that is valid for all integers a, even those divisible by p:

COROLLARY 9.1.3

Let $p \in \mathbf{N}$ be prime, and let $a \in \mathbf{Z}$. Then $a^p \equiv a \pmod{p}$.

PROOF You will prove this in Exercise 15.

> **Check for Understanding**
>
> Use Fermat's Little Theorem to:
>
> 1. Reduce 38^{102} modulo 103
> 2. Reduce 8^{17} modulo 17

Proof of Fermat's Little Theorem

The proof of Fermat's Little Theorem (9.1.1) will be easier to understand if we first explore a special case of the proof.

Naomi's Numerical Proof Preview: Fermat's Little Theorem (9.1.1)
Let's consider how we might show that

$$\overline{3}^{\,4} = \overline{1} \text{ in } \mathbf{Z}_5.$$

Of course, we could just compute 3^4 and reduce modulo 5. But this will not help us prove the general case. Instead, let's find a method that will generalize better.

The trick behind this method is to go back and reexamine the *multiplication* table for $\mathbf{Z}_5$. Consider the $\overline{3}$ row of the $\mathbf{Z}_5$ multiplication table (note that we omit the $\overline{0}$ row and the $\overline{0}$ column of the table):

$\cdot$	$\overline{1}$	$\overline{2}$	$\overline{3}$	$\overline{4}$
$\overline{1}$				
$\overline{2}$				
$\overline{3}$	$\overline{3}$	$\overline{1}$	$\overline{4}$	$\overline{2}$
$\overline{4}$				

Every element inside this modulo 5 multiplication table represents a product. For example, the $\overline{1}$ in the third row of the table represents the product $\overline{3} \cdot \overline{2} = \overline{1}$. Let's write out all four products in the row of $\overline{3}$:

$$\overline{3} \cdot \overline{1} = \overline{3}$$
$$\overline{3} \cdot \overline{2} = \overline{1}$$
$$\overline{3} \cdot \overline{3} = \overline{4}$$
$$\overline{3} \cdot \overline{4} = \overline{2}.$$

Multiplying all four of these equations together gives

$$(\overline{3} \cdot \overline{1}) \cdot (\overline{3} \cdot \overline{2}) \cdot (\overline{3} \cdot \overline{3}) \cdot (\overline{3} \cdot \overline{4}) = \overline{3} \cdot \overline{1} \cdot \overline{4} \cdot \overline{2}.$$

Since multiplication in $\mathbf{Z}_5$ is commutative, we can rearrange the terms in the products:

$$\overline{3} \cdot \overline{3} \cdot \overline{3} \cdot \overline{3} \cdot \overline{1} \cdot \overline{2} \cdot \overline{3} \cdot \overline{4} = \overline{1} \cdot \overline{2} \cdot \overline{3} \cdot \overline{4},$$

or

$$\overline{3}^4 \cdot \overline{1} \cdot \overline{2} \cdot \overline{3} \cdot \overline{4} = \overline{1} \cdot \overline{2} \cdot \overline{3} \cdot \overline{4}.$$

Since the modulus 5 is prime (and therefore relatively prime to all nonzero elements of $\mathbf{Z}_5$), we can apply the Modular Cancellation Law (8.3.2) and cancel $\overline{1} \cdot \overline{2} \cdot \overline{3} \cdot \overline{4}$ from both sides of the equation. The result is the equation

$$\overline{3}^4 = \overline{1},$$

which is precisely what we wanted to show!

While this may seem like a roundabout way to compute $\overline{3}^4$, close inspection reveals that this method is completely general. In fact, this example forms the basis for the following proof of Fermat's Little Theorem.

PROOF OF **FERMAT'S LITTLE THEOREM (9.1.1)**

Let $p \in \mathbf{N}$ be prime, and let $\overline{a} \in \mathbf{Z}_p$ such that $\overline{a} \neq \overline{0}$.

Consider the $\overline{a}$ row of the $\mathbf{Z}_p$ multiplication table:

Multiplication in $\mathbf{Z}_p$

$\cdot$	$\overline{1}$	$\overline{2}$	$\overline{3}$	$\cdots$	$\overline{p-1}$
$\overline{1}$					
$\overline{2}$					
$\vdots$					
$\overline{a}$	$\overline{a} \cdot \overline{1}$	$\overline{a} \cdot \overline{2}$	$\overline{a} \cdot \overline{3}$	$\cdots$	$\overline{a} \cdot \overline{p-1}$
$\vdots$					

$\leftarrow$ **This row contains every nonzero element of $\mathbf{Z}_p$ exactly once.**

Since $\bar{a} \neq \bar{0}$, we know that $p \nmid a$; hence, $\gcd(a, p) = 1$. Thus (by Theorem 8.3.6), every nonzero element of $\mathbf{Z}_p$ appears exactly once in the row of $\bar{a}$. The product of all of the elements of the row will therefore be equal to the product of all of the nonzero elements of $\mathbf{Z}_p$:

$$(\bar{a} \cdot \bar{1}) \cdot (\bar{a} \cdot \bar{2}) \cdot \ \cdots \ \cdot (\bar{a} \cdot \overline{p-1}) = \bar{1} \cdot \bar{2} \cdot \ \cdots \ \cdot \overline{p-1}.$$

By rearranging the terms in the products on the left side, we get

$$\bar{a} \cdot \bar{a} \cdot \ \cdots \ \cdot \bar{a} \cdot \bar{1} \cdot \bar{2} \cdot \ \cdots \ \cdot \overline{p-1} = \bar{1} \cdot \bar{2} \cdot \ \cdots \ \cdot \overline{p-1}$$

or

$$\bar{a}^{p-1} \cdot \bar{1} \cdot \bar{2} \cdot \ \cdots \ \cdot \overline{p-1} = \bar{1} \cdot \bar{2} \cdot \ \cdots \ \cdot \overline{p-1}.$$

Since the modulus p is prime (and therefore relatively prime to all nonzero elements of $\mathbf{Z}_p$), we can apply the Modular Cancellation Law (8.3.2) and cancel $\bar{1} \cdot \bar{2} \cdot \ \cdots \ \cdot \overline{p-1}$ from both sides of the equation. This results in

$$\bar{a}^{p-1} = \bar{1}. \qquad \blacksquare$$

Fermat's Little Theorem can sometimes be used to quickly reduce expressions with exponents in a prime modulus.

EXAMPLE 1

Reduce 74^{62} modulo 11.

SOLUTION First we reduce the base, 74, modulo 11:

$$74 \equiv 8 \ (\text{mod } 11).$$

Thus,

$$74^{62} \equiv 8^{62} \ (\text{mod } 11) \qquad \leftarrow \textbf{by Corollary 7.2.2}.$$

Our problem is now simpler: we must reduce 8^{62} modulo 11.
By Fermat's Little Theorem (Restatement, 9.1.2),

$$8^{10} \equiv 1 \ (\text{mod } 11). \qquad (3)$$

Hence,

$$8^{62} \equiv 8^{10} \cdot 8^{10} \cdot 8^{10} \cdot 8^{10} \cdot 8^{10} \cdot 8^{10} \cdot 8^2 \pmod{11}$$

$$\equiv 1 \cdot 1 \cdot 1 \cdot 1 \cdot 1 \cdot 1 \cdot 8^2 \pmod{11} \qquad \leftarrow \textbf{by Theorem 7.2.1}$$
$$\textbf{(Congruences Add and Multiply)}$$

$$\equiv 8^2 \pmod{11}$$

$$\equiv 9 \pmod{11}.$$

We can think about this solution in another way: In order to reduce 8^{62} modulo 11, we need to divide the exponent, **62**, by 10:

$$8^{62} \equiv 8^{10 \cdot 6 + 2} \pmod{11}$$

$$\equiv (8^{10})^6 \cdot 8^2 \pmod{11}$$

$$\equiv 1^6 \cdot 8^2 \pmod{11} \qquad \leftarrow \textbf{by (3) and Corollary 7.2.2}$$

$$\equiv 8^2 \pmod{11}.$$
∎

Notice that to reduce 8^{62} modulo 11, we ended up reducing the exponent, **62**, modulo 10, resulting in the much smaller exponent **2**. In general, when we use Fermat's Little Theorem to reduce a^d modulo p, we end up reducing the exponent, d, modulo $p - 1$.

EXERCISES 9.1

Numerical Problems

1. Reduce 43^{38} modulo 37.

2. Reduce 87^{60} modulo 31.

In Exercises 3–5, evaluate each expression in Z_{13}:

3. $\overline{8}^{98}$　　　　4. $\overline{16}^{99}$　　　　5. $\overline{26}^{1000}$

Reasoning and Proofs

6. Fermat's Little Theorem gives us a method for finding multiplicative inverses modulo any prime p.

   a. Let p be prime and let $\bar{a} \in \mathbf{Z}_p$ be nonzero. Prove that $\bar{a}^{p-2}$ is the multiplicative inverse of $\bar{a}$ in $\mathbf{Z}_p$.

   b. Use part a and the table of exponents in $\mathbf{Z}_{11}$ to find the multiplicative inverse of $\overline{7}$ in $\mathbf{Z}_{11}$.

7. For this exercise, you will work in the number system $\mathbf{Z}_7$.

   a. You know by Fermat's Little Theorem that $\overline{3}^6 = \overline{1}$. Find three other natural numbers m that satisfy $\overline{3}^m = \overline{1}$. What do these exponents m have in common?

   b. Find three natural numbers m that satisfy $\overline{3}^m = \overline{3}$. What do these exponents m have in common?

   c. Find three natural numbers m that satisfy $\overline{3}^m = \overline{2}$. What do these exponents m have in common?

8. For this exercise, you will work in the number system $\mathbf{Z}_{13}$.

   a. You know by Fermat's Little Theorem that $\overline{2}^{12} = \overline{1}$. Find three other natural numbers m that satisfy $\overline{2}^m = \overline{1}$. What do these exponents m have in common?

   b. Find three natural numbers m that satisfy $\overline{2}^m = \overline{2}$. What do these exponents m have in common?

   c. Find three natural numbers m that satisfy $\overline{2}^m = \overline{4}$. What do these exponents m have in common?

9. Generalize Exercises 7 and 8 by filling in the blank to make the following statement true. Then prove the statement.

   Let p be prime, let $m, n \in \mathbf{N}$, and let $a \in \mathbf{Z}$ such that $p \nmid a$.
   If $m \equiv n \pmod{\underline{\hspace{0.6cm}}}$, then $a^m \equiv a^n \pmod{p}$.

10. The number 10^{100} is known as a *googol*. Reduce a googol modulo 7.

11. a. You can't use the method from Example 1 to reduce 4^{129} modulo 119. Explain why not.

    b. Reduce 4^{129} modulo 7.

    c. Reduce 4^{129} modulo 17.

    d. Use your answers from parts b and c to evaluate $4^{129} \pmod{119}$. [*Hint:* Use the Chinese Remainder Theorem (see Observation 7.4.4).]

12. a. Prove that if p is prime, then $1^{p-1} + 2^{p-1} + \cdots + (p-1)^{p-1} \equiv -1 \pmod{p}$.

    b. The converse of the statement in part a is:

    Let $n \in \mathbf{N}, n > 1$. If $1^{n-1} + 2^{n-1} + \cdots + (n-1)^{n-1} \equiv -1 \pmod{n}$, then n is prime. (4)

    This statement (4) is known as *Guiga's conjecture*, and it is unknown whether it is true. It *is* known that if any counterexample exists to Guiga's conjecture, it must have at least thousands of digits! Show that the composite numbers $n = 6$ and $n = 8$ are not counterexamples to Guiga's conjecture. That is, show that $n = 6$ and $n = 8$ do not satisfy the congruence in (4).

13. Let a and b be positive integers, and let p be prime. Prove that if $a^p \equiv b^p \pmod{p}$, then $a \equiv b \pmod{p}$.

14. **Phil Lovett** Your friend Phil Lovett was looking over your shoulder when you were reading Example 1, reducing $74^{62} \pmod{11}$.

"That's easy!" exclaimed Phil. "Reducing 74 modulo 11 gets you **8**, and reducing 62 modulo 11 gets you **7**. Then 8^7 is congruent to 2 modulo 11. Dude, the answer they give in your book is a mistake." Explain what is wrong with Phil's method.

15. Prove Corollary 9.1.3.

16. Suppose $\gcd(a, 35) = 1$.

   a. Show that $a^{12} - 1$ is divisible by 5. [*Hint:* Use Fermat's Little Theorem.]

   b. Show that $a^{12} - 1$ is divisible by 7.

   c. Show that $a^{12} - 1$ is divisible by 35. [*Hint:* Use the Chinese Remainder Theorem (see Observation 7.4.4).]

17. For all natural numbers n, show the following:

   a. $n^7 - n$ is divisible by 7.

   b. $n^7 - n$ is divisible by 2.

   c. $n^7 - n$ is divisible by 3.

   d. $n^7 - n$ is divisible by 42.

18. Let $p > 2$ be prime. Evaluate the following sum modulo p. Explain your reasoning.

$$1^p + 2^p + \cdots + (p - 1)^p$$

19. Explain why Fermat's Little Theorem (9.1.1) is equivalent to the Restatement of Fermat's Little Theorem (9.1.2).

Advanced Reasoning and Proofs

20. Let p be prime, and let a and b be integers. Prove that if $a \equiv b \pmod{p}$, then $a^p \equiv b^p \pmod{p^2}$. [*Hint:* Consider the expansion of $(b + kp)^p$.]

21. Let a be a positive integer, and let p and q be distinct odd primes such that $p - 1 \mid q - 1$. Suppose that $\gcd(a, pq) = 1$. Prove that $a^{q-1} \equiv 1 \pmod{pq}$.

22. Let p and q be distinct primes. Prove that $p^{q-1} + q^{p-1} \equiv 1 \pmod{pq}$.

9.2 Reduced Residues and the Euler φ-Function

Fermat's Little Theorem (9.1.1) is a powerful statement about exponents in prime mods. Our goal in the next two sections is to generalize Fermat's Little Theorem to cases in which the modulus is not prime. To do so, we need to become familiar with some new concepts.

Reduced residues

When we looked at the $\mathbf{Z}_n$ multiplication tables in Section 8.3, we noticed that not all rows are created equal. In particular, we proved that if $\gcd(a, n) = 1$, then we know that $\bar{a}$ has a multiplicative inverse in $\mathbf{Z}_n$ (Theorem 8.3.1) and that the row of $\bar{a}$ contains every element of $\mathbf{Z}_n$ (Theorem 8.3.6). Because of these important properties, we give such elements of $\mathbf{Z}_n$ a special name.

DEFINITION 9.2.1

Let $n \in \mathbf{N}$ and let $a \in \mathbf{Z}$. We call $\bar{a} \in \mathbf{Z}_n$ a **reduced residue** *if* $\gcd(a, n) = 1$.

EXAMPLE 1

a. In $\mathbf{Z}_{10}$, the reduced residues are $\{\bar{1}, \bar{3}, \bar{7}, \bar{9}\}$.

b. In $\mathbf{Z}_{12}$, the reduced residues are $\{\bar{1}, \bar{5}, \bar{7}, \overline{11}\}$.

c. In $\mathbf{Z}_7$, the reduced residues are $\{\bar{1}, \bar{2}, \bar{3}, \bar{4}, \bar{5}, \bar{6}\}$.

Before making use of reduced residues, we must make sure that the concept of a reduced residue is well defined. There are many ways to refer to each element of $\mathbf{Z}_n$. For example, the number system $\mathbf{Z}_{10}$ has only 10 elements, but there are infinitely many ways to refer to each element. For instance, $\bar{3}, \overline{13}, \overline{23}$, and $\overline{33}$ are different names for the exact same element:

$$\text{In } \mathbf{Z}_{10}, \quad \bar{3} = \overline{13} = \overline{23} = \overline{33}.$$

Using the definition (9.2.1), we can determine that $\bar{3}$ is a reduced residue because $\gcd(3, 10) = 1$. We can determine that $\overline{13}$ is a reduced residue because $\gcd(13, 10) = 1$. We can similarly verify that $\overline{23}$ and $\overline{33}$ are reduced residues.

What if we could find an integer b such that $\bar{b} = \bar{3}$, but $\gcd(b, 10) \neq 1$. In that case, $\bar{b}$ would not be a reduced residue, which contradicts the fact that $\bar{3}$ is a reduced residue. If we could find such a b, then *reduced residue* would not be a well-defined concept in $\mathbf{Z}_{10}$.

We will now prove that in fact, *reduced residue* is a well-defined concept for all $\mathbf{Z}_n$. To do so, we suppose that a and b are integers such that $\bar{a} = \bar{b}$ in $\mathbf{Z}_n$, and show that if $\bar{a}$ satisfies the definition of reduced residue, then so does $\bar{b}$.

Let $n \in \mathbf{N}$ and $a, b \in \mathbf{Z}$ such that in $\mathbf{Z}_n$, $\bar{a} = \bar{b}$. If $\gcd(a, n) = 1$, then $\gcd(b, n) = 1$.

PROOF (By contradiction.) Let $n \in \mathbf{N}$ and $a, b \in \mathbf{Z}$ such that $\bar{a} = \bar{b}$ in $\mathbf{Z}_n$ and $\gcd(a, n) = 1$.

Assumption: Suppose $\gcd(b, n) > 1$.
Let $d = \gcd(b, n)$. By our Assumption, $d > 1$.
   By Theorem 8.2.4 (Equivalent Conditions for Congruence), there exists an integer k such that

$$a = b + kn.$$

Since $d \mid b$ and $d \mid n$, it follows from the Linear Combination Lemma (3.1.3) that $d \mid a$. Thus, a and n have a common divisor, d, which is greater than 1. This contradicts our hypothesis that $\gcd(a, n) = 1$. $\Rightarrow\Leftarrow$
   Since we have reached a contradiction, our Assumption must be false. We conclude that $\gcd(b, n) = 1$. ■

EXAMPLE 2

In $\mathbf{Z}_{10}$, $\bar{7}$ is a reduced residue. Because $\overline{30597} = \bar{7}$, we know (by Lemma 9.2.2) that 30597 is a reduced residue.

Reduced residue multiplication tables

Consider the $\mathbf{Z}_{10}$ multiplication table:

Multiplication in $\mathbf{Z}_{10}$

·	$\bar{0}$	$\bar{1}$	$\bar{2}$	$\bar{3}$	$\bar{4}$	$\bar{5}$	$\bar{6}$	$\bar{7}$	$\bar{8}$	$\bar{9}$
$\bar{0}$	0	0	0	0	0	0	0	0	0	0
$\bar{1}$	0	1	2	3	4	5	6	7	8	9
$\bar{2}$	0	2	4	6	8	0	2	4	6	8
$\bar{3}$	0	3	6	9	2	5	8	1	4	7
$\bar{4}$	0	4	8	2	6	0	4	8	2	6
$\bar{5}$	0	5	0	5	0	5	0	5	0	5
$\bar{6}$	0	6	2	8	4	0	6	2	8	4
$\bar{7}$	0	7	4	1	8	5	2	9	6	3
$\bar{8}$	0	8	6	4	2	0	8	6	4	2
$\bar{9}$	0	9	8	7	6	5	4	3	2	1

In Section 8.3, we observed that some of the rows of the multiplication table contain every element of Z_{10} exactly once, while other rows have repeated elements. By Theorem 8.3.6, the rows that contain every element of Z_{10} are precisely the rows of the reduced residues: $\overline{1}, \overline{3}, \overline{7},$ and $\overline{9}$. Similarly, the columns that contain every element of Z_{10} are the columns of the reduced residues: $\overline{1}, \overline{3}, \overline{7},$ and $\overline{9}$.

Let's see what happens when we cross out all of the "bad" rows and columns (the rows and columns of the numbers that are not reduced residues).

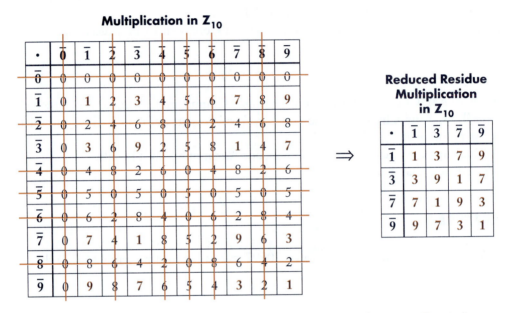

Multiplication in Z_{10}

Reduced Residue Multiplication in Z_{10}

Strikingly, the products in the table that were not crossed out (shown in **brown**) are all reduced residues! In fact, this observation holds for any modulus. In other words, the product of two reduced residues is itself a reduced residue, as we now show.

LEMMA 9.2.3   REDUCED RESIDUES ARE CLOSED UNDER MULTIPLICATION

Let $n \in \mathbf{N}$, and let $\overline{a}, \overline{b} \in \mathbf{Z}_n$ such that $\overline{a}$ and $\overline{b}$ are both reduced residues. Then $\overline{a} \cdot \overline{b}$ is also a reduced residue.

PROOF You will prove this lemma in Exercise 12.

Look again at the modulo 10 reduced residue multiplication table. You may notice that all of the rows of the table share a common property. Each row of the table contains every reduced residue exactly once. The following corollary states that this property holds for any modulus.

COROLLARY 9.2.4

Let $n \in \mathbf{N}$, and let $\overline{a}$ be a reduced residue in $\mathbf{Z}_n$. Then the row of $\overline{a}$ in the reduced residue multiplication table for $\mathbf{Z}_n$ contains every reduced residue exactly once.

PROOF Let $\bar{a}$ be a reduced residue in $\mathbf{Z}_n$. Then $\gcd(a, n) = 1$.

The $\bar{a}$ row of the reduced residue multiplication table for $\mathbf{Z}_n$ can be obtained by deleting some of the elements from the $\bar{a}$ row of the entire modulo n multiplication table. No element of $\mathbf{Z}_n$ appears more than once in the $\bar{a}$ row of the entire modulo n multiplication table (by Theorem 8.3.6), and thus no element of $\mathbf{Z}_n$ can appear more than once in the $\bar{a}$ row of the reduced residue multiplication table.

Since reduced residues are closed under multiplication (Lemma 9.2.3), every element in the $\bar{a}$ row of the reduced residue multiplication table must be a reduced residue. No reduced residue can appear more than once in the row, and the number of elements in the row is equal to the number of reduced residues. It follows that the $\bar{a}$ row of the reduced residue multiplication table for $\mathbf{Z}_n$ must contain every reduced residue exactly once. ■

The Euler φ-function

The number of reduced residues modulo n has a special name, $\varphi(n)$, read "phi of n." For example, $\varphi(10) = 4$, because $\mathbf{Z}_{10}$ has **4** reduced residues: $\bar{1}, \bar{3}, \bar{7},$ and $\bar{9}$. Counting the number of reduced residues modulo n is the same as counting the number of nonnegative integers less than n that are relatively prime to n.

DEFINITION 9.2.5

The **Euler φ-function** *is defined for any natural number n as follows:*

$$\varphi(n) = \text{the number of integers a that satisfy } 0 \le a < n \text{ and } \gcd(a, n) = 1.$$

EXAMPLE 3

    a. Find $\varphi(14)$.

    b. Find $\varphi(20)$.

    c. Find $\varphi(1)$.

SOLUTION

    a. We must determine how many nonnegative integers less than 14 are relatively prime to 14. The numbers 1, 3, 5, 9, 11, and 13 satisfy this condition. Hence, $\varphi(14) = 6$.

        Another way to think about this is that $\varphi(14) = 6$ counts the number of reduced residues in $\mathbf{Z}_{14}$. These reduced residues are $\bar{1}, \bar{3}, \bar{5}, \bar{9}, \overline{11},$ and $\overline{13}$.

    b. The numbers 1, 3, 7, 9, 11, 13, 17, and 19 are relatively prime to 20. Thus, $\varphi(20) = 8$.

    c. $\varphi(1) = 1$ because there is exactly 1 integer, the number $a = 0$, satisfying the conditions in the definition. ●

EXAMPLE 4
....

$\varphi(29) = 28$. Since 29 is prime, every nonzero element of $\mathbf{Z}_{29}$ is a reduced residue.

This example is a particular case of the following lemma.

LEMMA 9.2.6

Let $p \in \mathbf{N}$ be prime. Then $\varphi(p) = p - 1$.

PROOF You will prove this lemma in Exercise 13.

Naomi's Numerical Proof Preview: Lemma 9.2.7

Suppose we wish to find $\varphi(35)$. The prime factorization of 35 is $35 = 5 \cdot 7$. Thus, every number that is not relatively prime to 35 must be a multiple of 5 or 7. Let's list all of the integers from 0 to 34 and cross out all the **multiples of 5** and all the **multiples of 7**. The remaining numbers are relatively prime to 35:

~~0~~	6	12	18	24	~~30~~
1	7	13	19	~~25~~	31
2	8	~~14~~	~~20~~	26	32
3	9	~~15~~	~~21~~	27	33
4	~~10~~	16	22	~~28~~	34
~~5~~	11	17	23	29	

Now let's count the number of elements that were crossed off. We crossed off all **5 multiples of 7** and all **7 multiples of 5**. However, 0 is a multiple of both 5 and 7, so we crossed it off twice. Thus, the total number of elements we crossed off is $(5 + 7 - 1)$ elements.

We began with 35 elements, so the number of remaining elements (the number of reduced residues) is

$$\varphi(35) = 35 - (5 + 7 - 1) \qquad (1)$$

At this point, we have found that $\varphi(35) = 24$. In the interest of trying to generalize our result, however, let's manipulate the form of equation (1):

$$\varphi(35) = 35 - 5 - 7 + 1$$

$$\varphi(5 \cdot 7) = 5 \cdot 7 - 5 - 7 + 1$$

$$\varphi(5 \cdot 7) = (5 - 1)(7 - 1). \qquad (2)$$

Equation (2) can be generalized to a formula for the number of reduced residues in any modulus that is the product of two distinct primes.

Let p and q be distinct primes. Then

$$\varphi(pq) = (p-1)(q-1).$$

PROOF Consider the set S of all nonnegative integers less than pq:

$$S = \{0, 1, 2, 3, \ldots, pq - 1\}.$$

By definition, $\varphi(pq)$ equals the number of elements of S that are relatively prime to pq.

Every element of S that is not relatively prime to pq must be a **multiple of p** or a **multiple of q**. In the set S, the **multiples of q** are

$$0 \cdot q, \ 1 \cdot q, \ 2 \cdot q, \ldots, (p-1) \cdot q. \tag{3}$$

Notice that S contains exactly **p multiples of q**.

The **multiples of p** in the set S are

$$0 \cdot p, \ 1 \cdot p, \ 2 \cdot p, \ldots, (q-1) \cdot p. \tag{4}$$

Notice that S contains exactly **q multiples of p**.

As you will prove in Exercise 14, there is only **1** number, the number zero, that is contained in both lists (3) and (4).

To summarize, the set S contains

p multiples of q, **q** multiples of p, and **1** common multiple of p and q.

Therefore, the total number of multiples of p or q in S is $(p + q - 1)$. Since there are a total of pq elements of S, the number of elements of S that are relatively prime to pq is given by

$$\varphi(pq) = pq - (p + q - 1)$$
$$= pq - p - q + 1$$
$$= (p-1)(q-1). \qquad \blacksquare$$

EXAMPLE 5

The number of reduced residues in $\mathbf{Z}_{77}$ is

$$\varphi(77) = \varphi(7 \cdot 11)$$
$$= 6 \cdot 10 \qquad \leftarrow \textbf{by Lemma 9.2.7}$$
$$= 60.$$

Using Lemma 9.2.6, we can rewrite the result of Lemma 9.2.7: For distinct primes p and q,

$$\varphi(pq) = \varphi(p)\,\varphi(q).$$

This is a special case of the following lemma.

LEMMA 9.2.8

Let m and n be relatively prime natural numbers. Then $\varphi(mn) = \varphi(m)\,\varphi(n)$.

PROOF In Exercises 21 and 22, we sketch two different proofs of this lemma.

EXAMPLE 6

$$
\begin{aligned}
\varphi(105) &= \varphi(3 \cdot 5 \cdot 7) \\
&= \varphi(3 \cdot 5) \cdot \varphi(7) &&\leftarrow \textbf{by Lemma 9.2.8} \\
&= \varphi(3) \cdot \varphi(5) \cdot \varphi(7) &&\leftarrow \textbf{by Lemma 9.2.8} \\
&= 2 \cdot 4 \cdot 6 &&\leftarrow \textbf{by Lemma 9.2.6} \\
&= 48.
\end{aligned}
$$

Given any natural number n, we can find $\varphi(n)$ using the following formula.

THEOREM 9.2.9 GENERAL FORMULA FOR THE EULER φ-FUNCTION

Let $p_1, p_2, \ldots, p_k$ be distinct prime numbers, and let $a_1, a_2, \ldots, a_k$ be nonnegative integers. Then

$$\varphi(p_1^{a_1} p_2^{a_2} \cdots p_k^{a_k}) = (p_1 - 1)p_1^{a_1-1} \cdot (p_2 - 1)p_2^{a_2-1} \cdot \; \cdots \; \cdot (p_k - 1)p_k^{a_k-1}.$$

PROOF In the exercises, we sketch two different proofs of this theorem, one in Exercise 23 and one in Exercises 24–26.

EXAMPLE 7

Find $\varphi(97200)$.

SOLUTION The prime factorization of 97200 is

$$97200 = 2^4 \cdot 3^5 \cdot 5^2$$

By the General Formula for the Euler φ-Function (9.2.9),

$$\varphi(2^4 \cdot 3^5 \cdot 5^2) = (2-1)2^3 \cdot (3-1)3^4 \cdot (5-1)5^1.$$

Thus,

$$\varphi(97200) = 25920. \qquad \blacksquare$$

Using Theorem 9.2.9, we can easily find φ of any number, *provided we can factor the number*. If we do not know a number's prime factorization, however, then we have no efficient method for finding φ of that number.

EXERCISES 9.2

Numerical Problems

1. **a.** Find $\varphi(5)$ by listing all the reduced residues in $\mathbf{Z}_5$.

   b. Find $\varphi(6)$ by listing all the reduced residues in $\mathbf{Z}_6$.

   c. Find $\varphi(30)$ by listing all the reduced residues in $\mathbf{Z}_{30}$.

2. Make a table of the values of $\varphi(n)$ for all $1 \le n \le 40$.

3. Use the formula for $\varphi(n)$ given in Theorem 9.2.9 to calculate each of the following:

   a. $\varphi(656)$

   b. $\varphi(2905)$

   c. $\varphi(27720)$

   d. $\varphi(10^{100})$ (Recall that the number 10^{100} is known as a *googol*.)

4. For the following values of a and m, reduce $a^{\varphi(m)}$ modulo m.

   a. $a = 2, m = 9$ **b.** $a = 3, m = 9$

   c. $a = 12, m = 25$ **d.** $a = 15, m = 25$

   e. $a = 2, m = 10$ **f.** $a = 3, m = 10$

5. Compute the following sums.

   a. $\displaystyle\sum_{\substack{d \mid 9 \\ d \in \mathbf{N}}} \varphi(d)$ ← **This denotes the sum of $\varphi(d)$ over all natural numbers d that divide 9.**

b. $\displaystyle\sum_{\substack{d \,\mid\, 12 \\ d \,\in\, \mathbf{N}}} \varphi(d)$

c. $\displaystyle\sum_{\substack{d \,\mid\, 20 \\ d \,\in\, \mathbf{N}}} \varphi(d)$

6. Compute $\varphi(2^m)$ for the following values of m.

    a. $m = 2$ **b.** $m = 3$

    c. $m = 4$ **d.** $m = 5$

7. Let p be prime and let a be a nonnegative integer.

    a. Use Theorem 9.2.9 to find a formula for $\varphi(p^a)$.

    b. Use your formula from part a to find $\varphi(81)$.

Reasoning and Proofs

8. **Phil Lovett** Envious that you are learning so many new concepts in your number theory class, your friend Phil Lovett makes his own "cool" new definition of the "Lovett residue." In $\mathbf{Z}_n$, $\overline{a}$ is a *Lovett residue* if $3 \mid a$. For example, $\overline{9}$ is a Lovett residue in $\mathbf{Z}_{11}$, and $\overline{15}$ is a Lovett residue in $\mathbf{Z}_{12}$. Is the concept of a Lovett residue well defined? Prove your answer.

9. Prove that for all natural numbers $n > 2$, $\varphi(n)$ is even.

10. Find all natural numbers n such that $\varphi(n) = 12$.

11. Prove that there does not exist an integer n such that $\varphi(n) = 98$.

12. Prove Lemma 9.2.3. [*Hint:* Take a look at Euclid's Lemma (5.4.2).]

13. Prove Lemma 9.2.6.

14. Prove the following assertion, which we used in the proof of Lemma 9.2.7:

    If p and q are distinct primes, then the sets

$$\{0 \cdot q,\ 1 \cdot q,\ 2 \cdot q, \ldots, (p-1) \cdot q\} \text{ and } \{0 \cdot p,\ 1 \cdot p,\ 2 \cdot p, \ldots, (q-1) \cdot p\}$$

    have only one element in common.

15. For all parts of this problem, do not use Theorem 9.2.9. Let p be prime. We will consider the integers modulo p^2.

a. Complete the following table:

p	p^2	List of reduced residues in Z_{p^2}	$\varphi(p^2)$
2	4		
3	9		
5	25		

Find a formula for $\varphi(p^2)$.

b. Without using Theorem 9.2.9, prove that the formula you found in part a is correct. [*Hint:* To calculate the number of reduced residues modulo p^2, start with the p^2 integers from 0 to $p^2 - 1$ and remove any that are not relatively prime to p^2. How many remain?]

c. Using similar reasoning (and without using Theorem 9.2.9), find a formula that gives $\varphi(p^k)$ for any $k \in \mathbf{N}$, and prove that it is correct.

16. Suppose that both p and $2p - 1$ are odd primes. Let $n = 2(2p - 1)$. Prove that $\varphi(n) = \varphi(n + 2)$.

17. Prove that there are infinitely many natural numbers n such that $\varphi(n)$ is a perfect square.

18. Let n be a positive integer.

a. Prove that if n is odd, then $\varphi(2n) = \varphi(n)$.

b. Prove that if n is even, then $\varphi(2n) = 2\varphi(n)$.

Advanced Reasoning and Proofs

19. Let $m, n > 1$ be integers. Prove that $\varphi(mn) = \dfrac{\gcd(m, n)\varphi(m)\varphi(n)}{\varphi(\gcd(m, n))}$.

20. Let n be a composite natural number such that $\varphi(n) \mid (n - 1)$. Prove that

a. n is a product of distinct primes;

b. n is the product of at least three distinct primes.

21. In this exercise, you will prove Lemma 9.2.8, which asserts that if m and n are relatively prime integers, then $\varphi(mn) = \varphi(m)\varphi(n)$. We know $\varphi(mn)$ is the number of integers a in the range $0, 1, 2, \ldots, mn - 1$ that are relatively prime to mn.

a. Let $a \in \mathbf{Z}$. Prove that a is relatively prime to mn if and only if a is relatively prime to both m and n.

b. For any b in the range $0, 1, 2, \ldots, m - 1$, define s_b as the number of integers a in the range $0, 1, 2, \ldots, mn - 1$ such that $a \equiv b \pmod{m}$ and $\gcd(a, mn) = 1$. Prove that if $\gcd(b, m) = 1$, then $s_b = \varphi(n)$. [*Hint:* Use the Chinese Remainder Theorem.]

c. With notation as in part b, prove that if $\gcd(b, m) > 1$, then $s_b = 0$.

d. Explain why $\varphi(mn) = \displaystyle\sum_{b=0}^{m-1} s_b$.

e. Using your results from the previous parts of this exercise, prove that $\varphi(mn) = \varphi(m)\varphi(n)$.

22. In this exercise, you will prove Lemma 9.2.8 using the language of functions.

    Let $m, n \in \mathbf{N}$ be relatively prime. Define a function f from $\mathbf{Z}_{mn}$ to $\mathbf{Z}_m \times \mathbf{Z}_n$ as follows: For any $a \in \mathbf{Z}$,

    $$f(\bar{a}) = (\bar{a}, \bar{a}).$$

    (Here $\bar{a}, \bar{a},$ and $\bar{a}$ represent elements of $\mathbf{Z}_{mn}, \mathbf{Z}_m,$ and $\mathbf{Z}_n$, respectively.)

    a. Prove that f is well defined. That is, prove that if $\bar{a} = \bar{x}$ in $\mathbf{Z}_{mn}$, then $f(\bar{a}) = f(\bar{x})$.

    b. Prove that f is one-to-one (i.e., injective).

    c. Prove that f is onto (i.e., surjective).

    d. If $f(\bar{x}) = (\bar{y}, \bar{z})$, prove that $\bar{x}$ is a reduced residue in $\mathbf{Z}_{mn}$ if and only if $\bar{y}$ is a reduced residue in $\mathbf{Z}_m$ and $\bar{z}$ is a reduced residue in $\mathbf{Z}_n$.

    e. You have now shown that f defines a one-to one correspondence between reduced residues in $\mathbf{Z}_{mn}$ and ordered pairs $(\bar{y}, \bar{z})$, where $\bar{y} \in \mathbf{Z}_n$ and $\bar{z} \in \mathbf{Z}_m$ are reduced residues. Conclude that $\varphi(mn) = \varphi(m)\varphi(n)$.

23. Prove Theorem 9.2.9 using the result of Exercise 15c and Lemma 9.2.8.

EXPLORATION Proof of Theorem 9.2.9 (Exercises 24−26)

In this series of exercises, we will prove the General Formula for the Euler φ-Function (9.2.9).

24. In this exercise, let m and n be natural numbers, and suppose every prime that divides n also divides m. We will prove that $\varphi(nm) = n\varphi(m)$.

    a. We begin by writing the integers from 0 to $nm - 1$ in the following grid:

0	1	2	$\cdots$	$m - 1$
m	$m + 1$	$m + 2$	$\cdots$	$2m - 1$
$2m$	$2m + 1$	$2m + 2$	$\cdots$	$3m - 1$
$\vdots$	$\vdots$	$\vdots$	$\vdots$	$\vdots$
$(n - 1)m$	$(n - 1)m + 1$	$(n - 1)m + 2$	$\cdots$	$nm - 1$

We will calculate $\varphi(nm)$ by starting with these integers and crossing out all of those that are *not* relatively prime to nm. Explain why this is the same as crossing off only those integers in the table that are not relatively prime to m.

b. Prove that for every column, either the entire column is crossed out or no entries in the column are crossed out.

c. How many of the integers in the first row are *not* crossed out?

d. Use parts b and c to determine how many integers in the entire table are not crossed out. Explain why this completes the proof that $\varphi(nm) = n\varphi(m)$.

25. We will prove a formula similar to that in Exercise 24 using the same kind of logic. Assume that p is a prime that does not divide m. In this exercise, we will prove that $\varphi(pm) = (p-1)\varphi(m)$. We begin by writing the integers from 0 to $pm - 1$ in the following grid:

0	1	2	$\cdots$	$m - 1$
m	$m + 1$	$m + 2$	$\cdots$	$2m - 1$
$2m$	$2m + 1$	$2m + 2$	$\cdots$	$3m - 1$
$\vdots$	$\vdots$	$\vdots$	$\vdots$	$\vdots$
$(p-1)m$	$(p-1)m + 1$	$(p-1)m + 2$	$\cdots$	$pm - 1$

We will calculate $\varphi(pm)$ by starting with these integers and crossing off all of those that are not relatively prime to pm. This is the same as crossing off those integers in the table that are not relatively prime to m as well as those that are not relatively prime to p.

a. How many integers in the table are not relatively prime to m? [*Hint:* Look back at how you did Exercise 24.]

b. How many integers in the table are not relatively prime to p?

c. How many integers in the table are *both* not relatively prime to m and not relatively prime to p? (This counts how many numbers were crossed out twice—once in part a and once in part b.)

d. Use parts a, b, and c to prove the formula $\varphi(pm) = (p-1)\varphi(m)$.

26. Use the results of Exercises 24 and 25 to give a proof of Theorem 9.2.9.

9.3 Euler's Theorem

Fermat's Little Theorem applies only when we are working in a prime modulus. Our goal in this section is to find a generalization of Fermat's Little Theorem (9.1.1) to cases in which the modulus is not prime. Let's take another look at the tables of exponents modulo 7 (a prime modulus) and modulo 10 (a composite modulus).

Exponents in Z_7

$\bar{a}$	$\bar{1}$	$\bar{2}$	$\bar{3}$	$\bar{4}$	$\bar{5}$	$\bar{6}$
$\bar{a}^2$	1	4	2	2	4	1
$\bar{a}^3$	1	1	6	1	6	6
$\bar{a}^4$	1	2	4	4	2	1
$\bar{a}^5$	1	4	5	2	3	6
$\bar{a}^6$	1	1	1	1	1	1
$\bar{a}^7$	1	2	3	4	5	6

Exponents in Z_{10}

$\bar{a}$	$\bar{1}$	$\bar{2}$	$\bar{3}$	$\bar{4}$	$\bar{5}$	$\bar{6}$	$\bar{7}$	$\bar{8}$	$\bar{9}$
$\bar{a}^2$	1	4	9	6	5	6	9	4	1
$\bar{a}^3$	1	8	7	4	5	6	3	2	9
$\bar{a}^4$	1	6	1	6	5	6	1	6	1
$\bar{a}^5$	1	2	3	4	5	6	7	8	9
$\bar{a}^6$	1	4	9	6	5	6	9	4	1
$\bar{a}^7$	1	8	7	4	5	6	3	2	9
$\bar{a}^8$	1	6	1	6	5	6	1	6	1
$\bar{a}^9$	1	2	3	4	5	6	7	8	9
$\bar{a}^{10}$	1	4	9	6	5	6	9	4	1

The statement of Fermat's Little Theorem corresponds to the pattern of $\bar{1}$'s in the 6th row of the table of exponents modulo 7. Unfortunately, there does not appear to be any row of $\bar{1}$'s in the table of exponents modulo 10.

Now let's cross out the columns of all elements of $\mathbf{Z}_{10}$ that are not reduced residues.

Exponent Table for Z_{10}

$\bar{a}$	$\bar{1}$	$\bar{2}$	$\bar{3}$	$\bar{4}$	$\bar{5}$	$\bar{6}$	$\bar{7}$	$\bar{8}$	$\bar{9}$
$\bar{a}^2$	1	4	9	6	5	6	9	4	1
$\bar{a}^3$	1	8	7	4	5	6	3	2	9
$\bar{a}^4$	1	6	1	6	5	6	1	6	1
$\bar{a}^5$	1	2	3	4	5	6	7	8	9
$\bar{a}^6$	1	4	9	6	5	6	9	4	1
$\bar{a}^7$	1	8	7	4	5	6	3	2	9
$\bar{a}^8$	1	6	1	6	5	6	1	6	1
$\bar{a}^9$	1	2	3	4	5	6	7	8	9
$\bar{a}^{10}$	1	4	9	6	5	6	9	4	1

$\Rightarrow$

Exponent Table for Reduced Residues in Z_{10}

$\bar{a}$	$\bar{1}$	$\bar{3}$	$\bar{7}$	$\bar{9}$
$\bar{a}^2$	1	9	9	1
$\bar{a}^3$	1	7	3	9
$\bar{a}^4$	1	1	1	1
$\bar{a}^5$	1	3	7	9
$\bar{a}^6$	1	9	9	1
$\bar{a}^7$	1	7	3	9
$\bar{a}^8$	1	1	1	1
$\bar{a}^9$	1	3	7	9
$\bar{a}^{10}$	1	9	9	1

We now have a row that contains only $\bar{1}$'s, just as in Fermat's Little Theorem! The first row of $\bar{1}$'s is the $\bar{a}^4$ row.

Naomi's Numerical Proof Preview: Euler's Theorem (9.3.1)

Why does the $\overline{a}^4$ row of the reduced residue exponent table turn out to consist entirely of $\overline{1}$'s? To answer this question, let's explore how we might show that

$$\overline{7}^4 = \overline{1} \text{ in } \mathbf{Z}_{10}.$$

This argument will be very similar to the Numerical Proof Preview for Fermat's Little Theorem (9.1.1).

Consider the $\overline{7}$ row of the $\mathbf{Z}_{10}$ reduced residue multiplication table:

Reduced Residue Multiplication Modulo 10

·	$\overline{1}$	$\overline{3}$	$\overline{7}$	$\overline{9}$
$\overline{1}$				
$\overline{3}$				
$\overline{7}$	$\overline{7}$	$\overline{1}$	$\overline{9}$	$\overline{3}$
$\overline{9}$				

We now write out all four products in the row of $\overline{7}$:

$$\overline{7} \cdot \overline{1} = \overline{7}$$
$$\overline{7} \cdot \overline{3} = \overline{1}$$
$$\overline{7} \cdot \overline{7} = \overline{9}$$
$$\overline{7} \cdot \overline{9} = \overline{3}.$$

Multiplying all four of these equations together gives

$$(\overline{7} \cdot \overline{1}) \cdot (\overline{7} \cdot \overline{3}) \cdot (\overline{7} \cdot \overline{7}) \cdot (\overline{7} \cdot \overline{9}) = \overline{7} \cdot \overline{1} \cdot \overline{9} \cdot \overline{3}.$$

Since multiplication in $\mathbf{Z}_{10}$ is commutative, we can rearrange the terms in the products:

$$\overline{7} \cdot \overline{7} \cdot \overline{7} \cdot \overline{7} \cdot \overline{1} \cdot \overline{3} \cdot \overline{7} \cdot \overline{9} = \overline{1} \cdot \overline{3} \cdot \overline{7} \cdot \overline{9} \qquad (1)$$

or

$$\overline{7}^4 \cdot \overline{1} \cdot \overline{3} \cdot \overline{7} \cdot \overline{9} = \overline{1} \cdot \overline{3} \cdot \overline{7} \cdot \overline{9}.$$

Recall that $\overline{1}, \overline{3}, \overline{7},$ and $\overline{9}$ are reduced residues in $\mathbf{Z}_{10}$ precisely because the integers $1, 3, 7,$ and 9 are each relatively prime to 10. Thus, we can apply the Modular Cancellation Law (8.3.2) and cancel $\overline{1} \cdot \overline{3} \cdot \overline{7} \cdot \overline{9}$ from both sides of this equation. We obtain

$$\overline{7}^4 = \overline{1},$$

which is precisely what we wanted to show.

One can use a similar argument to show that any reduced residue in $\mathbf{Z}_{10}$ raised to the 4th power equals $\overline{1}$. But why the **4**th power instead of some other power?

To solve this mystery, let's look at equation (1). In this equation, we see one factor of $\overline{7}$ for each column in the reduced residue table. The number of factors of $\overline{7}$ is thus the number of reduced residues modulo 10, which is $\varphi(10) = \mathbf{4}$. Mystery solved!

In general, we can use this method to show that any reduced residue in $\mathbf{Z}_n$ raised to the power $\varphi(n)$ equals $\overline{1}$. This important fact was discovered by Swiss mathematician Leonhard Euler[1] and is known as Euler's Theorem. (See Chapter 13 for a historical summary of Euler's life and work.)

THEOREM 9.3.1   EULER'S THEOREM

Let $n \in \mathbf{N}$, and let $\overline{a}$ be a reduced residue in $\mathbf{Z}_n$. Then

$$\overline{a}^{\,\varphi(n)} = \overline{1}.$$

Euler's Theorem can also be stated in terms of congruence modulo n.

THEOREM 9.3.2   EULER'S THEOREM (Restatement)

Let $n \in \mathbf{N}$, $a \in \mathbf{Z}$ such that $\gcd(a, n) = 1$. Then

$$a^{\varphi(n)} \equiv 1 \ (\mathrm{mod}\ n). \tag{2}$$

The two statements of Euler's Theorem (9.3.1 and 9.3.2) are equivalent (see Exercise 13).

What does Euler's theorem say in the case that n happens to be a prime number? If n is prime, then $\varphi(n) = n - 1$. So equation (2) becomes

$$a^{n-1} \equiv 1 \ (\mathrm{mod}\ n).$$

This is exactly the conclusion of Fermat's Little Theorem (Restatement, 9.1.2). Thus, we see that Euler's Theorem is a direct generalization of Fermat's Little Theorem (see Exercise 14).

Proof of Euler's Theorem

The proof of Euler's Theorem (9.3.1) is quite similar to the proof of Fermat's Little Theorem (9.1.1). We will base our proof of Euler's Theorem on the Numerical Proof Preview.

PROOF of EULER'S THEOREM (9.3.1)

Let $n \in \mathbf{N}$, and let $f = \varphi(n)$. Let $\overline{a}$ be a reduced residue in $\mathbf{Z}_n$. By the definition of reduced residue (9.2.1), the integer a is relatively prime to n. Let the list

$$\overline{r_1}, \ \overline{r_2}, \ \ldots, \ \overline{r_f}$$

represent all of the reduced residues in $\mathbf{Z}_n$, in order from $\overline{r_1} = \overline{1}$ to $\overline{r_f} = \overline{n-1}$.

---

[1]The name *Euler* is pronounced "oiler."

Consider the $\bar{a}$ row of the $\mathbf{Z}_n$ reduced residue multiplication table:

Reduced Residue
Multiplication Modulo *n*

$\cdot$	$\overline{r}_1$	$\overline{r}_2$	$\overline{r}_3$	$\cdots$	$\overline{r}_f$
$\overline{r}_1$					
$\overline{r}_2$					
$\vdots$					
$\overline{a}$	$\overline{a}\cdot\overline{r}_1$	$\overline{a}\cdot\overline{r}_2$	$\overline{a}\cdot\overline{r}_3$	$\cdots$	$\overline{a}\cdot\overline{r}_f$
$\vdots$					
$\overline{r}_f$					

← **This row contains every reduced residue exactly once.**

Every reduced residue in $\mathbf{Z}_n$ appears exactly once in the row of $\bar{a}$ (by Corollary 9.2.4). The product of all of the elements of the row will therefore be equal to the product of all of the reduced residues in $\mathbf{Z}_n$:

$$(\bar{a}\cdot\overline{r}_1)\cdot(\bar{a}\cdot\overline{r}_2)\cdot\ \cdots\ \cdot(\bar{a}\cdot\overline{r}_f) = \overline{r}_1\cdot\overline{r}_2\cdot\ \cdots\ \cdot\overline{r}_f.$$

By rearranging the terms in the products on the left side, we get

$$\bar{a}\cdot\bar{a}\cdot\ \cdots\ \cdot\bar{a}\cdot\overline{r}_1\cdot\overline{r}_2\cdot\ \cdots\ \cdot\overline{r}_f = \overline{r}_1\cdot\overline{r}_2\cdot\ \cdots\ \cdot\overline{r}_f$$

or

$$\bar{a}^f\cdot\overline{r}_1\cdot\overline{r}_2\cdot\ \cdots\ \cdot\overline{r}_f = \overline{r}_1\cdot\overline{r}_2\cdot\ \cdots\ \cdot\overline{r}_f.$$

Since the elements $\overline{r}_1, \overline{r}_2, \ldots, \overline{r}_f$ are reduced residues in $\mathbf{Z}_n$, the integers $r_1, r_2, \ldots, r_f$ are each relatively prime to n. Thus, we can apply the Modular Cancellation Law (8.3.2) and cancel $\overline{r}_1\cdot\overline{r}_2\cdot\ \cdots\ \cdot\overline{r}_f$ from both sides of this equation. This results in

$$\bar{a}^f = \bar{1}. \qquad\blacksquare$$

Euler's Theorem can sometimes be used to quickly reduce terms with exponents in modular number systems.

EXAMPLE 1

Reduce 52^{62} modulo 21.

SOLUTION First we reduce the base, 52, modulo 21:

$$52 \equiv 10 \ (\text{mod } 21).$$

Thus,

$$52^{62} \equiv 10^{62} \pmod{21} \qquad \leftarrow \textbf{by Corollary 7.2.2.}$$

We are now left with a simpler problem: reducing 10^{62} modulo 21.

We factor the modulus, 21, in order to find $\varphi(21)$. Since 21 factors into primes as $21 = 3 \cdot 7$,

$$\varphi(21) = (3 - 1)\,(7 - 1) \qquad \leftarrow \textbf{by Lemma 9.2.7}$$
$$= 12.$$

Since 10 is relatively prime to 21, we can apply Euler's Theorem (Restatement, 9.3.2) to get

$$10^{12} \equiv 1 \pmod{21}. \tag{3}$$

We now use this to reduce 10^{62} modulo 21.

$$10^{62} \equiv 10^{12 \cdot 5 + 2} \pmod{21}$$
$$\equiv (10^{12})^5 \cdot 10^2 \pmod{21}$$
$$\equiv 1^5 \cdot 10^2 \pmod{21} \qquad \leftarrow \textbf{by Corollary 7.2.2}$$
$$\equiv 16 \pmod{21}. \qquad\qquad\qquad\qquad\blacksquare$$

In this example, in order to reduce 10^{62} modulo 21, we ended up reducing the exponent, **62**, modulo $\varphi(21) = 12$, resulting in the much smaller exponent **2**. In general, when we use Euler's Theorem to reduce a^d modulo n, we end up reducing the exponent, d, modulo $\varphi(n)$.

Repeated squaring: an efficient method for modular exponentiation

As we have seen, Fermat's Little Theorem (9.1.1, 9.1.2) and Euler's Theorem (9.3.1, 9.3.2) can be useful when reducing expressions that have large exponents. Even with the help of these powerful theorems, however, reducing exponential expressions may still require some effort. For example, suppose we want to evaluate

$$9^{105} \pmod{137}.$$

Fermat's Theorem tells us that $9^{136} \equiv 1 \pmod{137}$, but that does not help us to reduce 9^{105}.

We will now see a general computational method for raising any number to any exponent in any modulus. This method, which we call *repeated squaring*, is based on the following observation. If you start with a number x, then square it, then square the result, then continue to square successive results,

$$x, x^2, x^4, x^8, x^{16}, x^{32}, x^{64}, x^{128}, \ldots, \tag{4}$$

the exponents get large *very* quickly. For instance, after squaring just 50 times—a snap for a computer—we've computed the value of $x^{1125899906842624}$, an enormous power of x.

To compute x^n for an arbitrary exponent n, we start by repeatedly squaring x, computing the sequence (4) until we have raised x to every power of 2 smaller than n. We can then express x^n as the product of numbers selected from the list (4). Now let's see this method in action.

EXAMPLE 2

Reduce 9^{105} modulo 137.

SOLUTION 64 is the largest power of 2 that is less than or equal to the exponent, 105. So, we begin by finding the reduction (mod 137) of 9 raised to every power of 2, up to 9^{64}. We do this by squaring and reducing successive results:

$$9^1 \equiv 9 \ (\text{mod } 137)$$
$$9^2 \equiv 81 \ (\text{mod } 137)$$
$$9^4 \equiv 81^2 \equiv 6561$$
$$\equiv 122 \ (\text{mod } 137)$$
$$9^8 \equiv 122^2 \equiv 14884$$
$$\equiv 88 \ (\text{mod } 137)$$
$$9^{16} \equiv 88^2 \equiv 7744$$
$$\equiv 72 \ (\text{mod } 137)$$
$$9^{32} \equiv 72^2 \equiv 5184$$
$$\equiv 115 \ (\text{mod } 137)$$
$$9^{64} \equiv 115^2 \equiv 13225$$
$$\equiv 73 \ (\text{mod } 137).$$

The original exponent, 105, can be expressed as a sum of powers of 2:

$$105 = 64 + 32 + 8 + 1. \tag{5}$$

Now we can use these results to find 9^{105}:

$$9^{105} \equiv 9^{64+32+8+1}$$
$$\equiv 9^{64} \cdot 9^{32} \cdot 9^8 \cdot 9^1$$
$$\equiv 73 \cdot 115 \cdot 88 \cdot 9$$
$$\equiv 93 \ (\text{mod } 137).$$

In equation (5) of this example, we expressed the exponent as a sum of powers of 2, with each power of 2 appearing at most once. The method we used in the example is completely general, because any natural number can be expressed as a sum of powers of 2 with each power appearing in the sum at most once. Expressing a number in this form is equivalent to writing the number in *binary*, or base 2.

For a brief review of binary, observe that by equation (5), the number 105 is represented in binary as 1101001, as we now demonstrate:

Place	64 2^6	32 2^5	16 2^4	8 2^3	4 2^2	2 2^1	1 2^0
Binary representation	1	1	0	1	0	0	1

EXAMPLE 3

Reduce 17^{43} modulo 210.

SOLUTION Use the method of repeated squaring.

$$17^1 \equiv 17 \ (\text{mod } 210)$$

$$17^2 \equiv 79 \ (\text{mod } 210)$$

$$17^4 \equiv 79^2 \equiv 6241$$

$$\equiv 151 \ (\text{mod } 210)$$

$$17^8 \equiv 151^2 \equiv 22801$$

$$\equiv 121 \ (\text{mod } 210)$$

$$17^{16} \equiv 121^2 \equiv 14641$$

$$\equiv 151 \ (\text{mod } 210)$$

$$17^{32} \equiv 151^2 \equiv 22801$$

$$\equiv 121 \ (\text{mod } 210).$$

We now use these results to find 17^{43}. We express 43 as a sum of powers of 2, with each power appearing in the sum at most once (as noted earlier, this is equivalent to expressing 43 as a binary number):

$$43 = 32 + 8 + 2 + 1.$$

The rest is basic modular arithmetic:

$$17^{43} \equiv 17^{32+8+2+1}$$
$$\equiv 17^{32} \cdot 17^8 \cdot 17^2 \cdot 17^1$$
$$\equiv 121 \cdot 121 \cdot 79 \cdot 17$$
$$\equiv 143 \ (\text{mod } 210).$$

Repeated squaring is lightning fast

Examples 2 and 3 suggest that the method of repeated squaring is an efficient way to compute exponents modulo n. We would like to make a precise statement about how efficient this algorithm is.

Let's revisit our computation of 9^{105} modulo 137 from Example 2. Computationally, the hardest part of this calculation was doing several modular multiplications. Thus, counting how many multiplications we performed gives us a good idea of the overall efficiency of our method. How many multiplications did we need to perform to arrive at the value of 9^{105} (mod 137)? We started with 9 and squared a total of **6** times. To find the end result, we then needed to multiply 4 quantities together, which required another **3** modular multiplications. Thus, our calculation involved a total of **9** multiplications. Notice that to calculate 9^{105} modulo 137 in the more naive way of multiplying 9 by itself 105 times, we would need to perform a total of **104** multiplications. In this example, therefore, the method of repeated squaring is considerably faster.

In general, suppose that we want to compute a^b modulo n. If we do this using the method of repeated squaring, how many modular multiplications will be necessary? Well, with one multiplication, we find a^2; with a second multiplication, we find a^4; and so on. After k multiplications, we have computed the number a^{2^k}. We can stop when the number 2^k becomes as large as b. In other words, we can safely stop after k squarings if

$$2^k \geq b.$$

Taking logarithms base 2, we see that this happens exactly when

$$k \geq \log_2 b.$$

Thus, after completing a total of $\log_2 b$ squarings, we may stop squaring.

The second phase of the algorithm involves taking some of the results we obtained in the squaring phase and multiplying these together. Since we only had to square a total of $\log_2 b$ times, multiplying some of these numbers together requires at most $\log_2 b$ additional multiplications.

We conclude that each of the two phases of our algorithm takes at most $\log_2 b$ multiplications. Thus, we have the following observation:

OBSERVATION 9.3.3

To perform the modular exponentiation a^b (mod n), the method of repeated squaring requires at most $2 \cdot \log_2 b$ modular multiplications.

Suppose you have gargantuan 1000-digit numbers a, b, and n, and you wish to reduce a^b modulo n. Will your computer be able to do this in a reasonable amount of time using the method of repeated squaring, or should you give up all hope of ever finding the answer? Since the number b is less than 10^{1000}, Observation 9.3.3 tells us that the number of multiplications your computer will need to perform is at most

$$2 \cdot \log_2 (10^{1000}) = 2 \cdot 1000 \cdot \log_2 10 \approx 6635.$$

Even though your numbers a, b, and n are much larger than the number of atoms in the observable universe, you can find the value of $a^b \pmod{n}$ in only a few thousand steps. On today's computers, this can be done before you can say the words "modular exponentiation." As noted in Chapter 4, where we met the Euclidean algorithm, the logarithm function turns huge numbers into much smaller numbers. Since the number of steps required by repeated squaring is proportional to the logarithm of the input numbers, this algorithm is fast indeed. In particular, our repeated squaring method is a *polynomial-time* algorithm (a term we first discussed in Section 4.2).

Reducing a^b modulo n, where a, b, and n are extremely large numbers, may seem like a pointless exercise. This could not be further from the truth! As we will see later in this chapter, modular exponentiation is the basis of the RSA cryptosystem, one of the most useful and widely used encryption systems today. The fact that modular exponentiation can be done efficiently is crucial for secure communication in the modern world.

EXERCISES 9.3

Numerical Problems

In Exercises 1–5, perform each reduction using Fermat's Little Theorem or Euler's Theorem.

1. Reduce 7^{21} modulo 33.

2. Reduce 17^8 modulo 20.

3. Reduce 5^{14} modulo 13.

4. Reduce 31^{122} modulo 56.

5. Reduce 217^{73} modulo 108.

6. Write each number as a sum of distinct powers of 2.

   a. 35   **b.** 89   **c.** 172   **d.** 255

In Exercises 7–9, evaluate the expression using the method of repeated squaring.

7. Evaluate 3^{-10} in $\mathbf{Z}_{17}$.   8. Evaluate 5^{-21} in $\mathbf{Z}_{67}$.   9. Reduce 11^{15} modulo 101.

In Exercises 10–11, evaluate each expression in $\mathbf{Z}_{35}$.

10. $\overline{11}^{\,130}$ **11.** $\overline{47}^{\,485}$

12. Reduce 100^{100} modulo 31.

Reasoning and Proofs

13. Explain why Euler's Theorem (9.3.1) and the Restatement of Euler's Theorem (9.3.2) are equivalent.

14. Explain why Fermat's Little Theorem is just a special case of Euler's Theorem. Be sure to address the difference between the hypotheses of the two theorems.

15. Find the last digit of 123^{456}.

16. Euler's Theorem gives us a method for finding multiplicative inverses in $\mathbf{Z}_n$.

    a. Let $n \in \mathbf{N}$ and let $\bar{a}$ be a reduced residue in $\mathbf{Z}_n$. Prove that $\bar{a}^{(\varphi(n)-1)}$ is the multiplicative inverse of $\bar{a}$ in $\mathbf{Z}_n$.

    b. Use part a to find the multiplicative inverse of $\overline{13}$ in $\mathbf{Z}_{20}$.

    c. Use part a to find the multiplicative inverse of $\overline{5}$ in $\mathbf{Z}_{36}$.

    d. Use part a to find the multiplicative inverse of $\overline{10}$ in $\mathbf{Z}_{63}$.

17. Use the results of Exercise 16 to solve the following equations for $\bar{x}$:

    a. $\overline{13}\,\bar{x} + \overline{1} = \overline{10}$ in $\mathbf{Z}_{20}$ **b.** $\overline{5}\,\bar{x} + \overline{22} = \overline{9}$ in $\mathbf{Z}_{36}$ **c.** $\overline{10}\,\bar{x} = \overline{33}$ in $\mathbf{Z}_{63}$

Advanced Reasoning and Proofs

18. The number $10^{10^{100}}$ (ten raised to the power of a googol) is known as a *googolplex*. Reduce a googolplex modulo 7.

19. Reduce $7^{5^{4^3}}$ modulo 12.

20. Let p and q be distinct primes and $a \in \mathbf{N}$ with $\gcd(a, pq) = 1$. Prove that $\bar{a}^{(p-1)(q-1)} = \overline{1}$ in $\mathbf{Z}_{pq}$.

21. Let $a = 3^{27}$. Let B be the expansion of a in base 8. What is the last digit (the units digit) of B? What is the second-to-last digit (the "eights" digit) of B?

22. Let $n \in \mathbf{N}$, and let $f = \varphi(n)$.

    a. Let $\bar{a} \in \mathbf{Z}_n$ be a reduced residue. Consider the sequence of powers of $\bar{a}$:

$$\bar{a}^0,\ \bar{a}^1,\ \bar{a}^2,\ \bar{a}^3,\ldots \qquad\qquad (7)$$

    Prove that this sequence (7) is periodic with period f. That is, show that for all $j \geq 0$,

$$\bar{a}^{\,j+f} = \bar{a}^{\,j}.$$

b. Now suppose that $\bar{a} \in \mathbf{Z}_n$, but do not assume that $\bar{a}$ is a reduced residue. Give a few examples to show that sequence (7) need not be periodic.

c. Though sequence (7) is not necessarily periodic for every $\bar{a} \in \mathbf{Z}_n$, it turns out that if we start with $\bar{a}^f$, the power sequence

$$\bar{a}^f, \; \bar{a}^{f+1}, \; \bar{a}^{f+2}, \ldots$$

is periodic with period f. Prove this fact. In other words, prove that for all $j \geq f$, $a^{j+f} = a^j$. [*Hint:* It is enough to show that $\bar{a}^f = \bar{a}^{2f}$. To do this, factor n into primes, $n = p_1^{e_1} \cdot \; \cdots \; \cdot p_r^{e_r}$, and show that $a^f \equiv a^{2f} \pmod{p_i^{e_i}}$ for all i. It will help to distinguish the cases where p_i divides a and where p_i does not divide a.]

9.4 Exponentiation Ciphers with a Prime Modulus

In Section 8.4, we learned to encrypt messages using modular addition: To encrypt a message, we added a constant modulo n. Then in Section 8.5, we learned to encrypt using modular multiplication: To encrypt a message, we multiplied by a constant modulo n. You may be wondering whether it is possible to encrypt messages using modular exponentiation. That is, can we encrypt a message by raising to a constant power modulo n? Let's try it and see if it works.

Encryption by exponentiation modulo 29

We will begin to explore exponentiation ciphers using a small, prime modulus: 29. For encrypting messages, it will be useful to have Table 9.4.1, which contains the results of raising every nonzero element of $\mathbf{Z}_{29}$ to every power between 1 and 29. Above each of the numbers in the first row of this table, we have written the corresponding letters from Table 8.4.1. We can use this correspondence to convert a plaintext message into numbers, one letter at a time.

EXAMPLE 1

Encrypt the message ALMOST one character at a time by raising to the power 11 modulo 29.

SOLUTION Using the top row of Table 9.4.1, we can convert each character of the plaintext message to a number. We then refer to the $\bar{x}^{11}$ row of Table 9.4.1 to raise

TABLE 9.4.1 Z_{29} Exponent Table

	A	B	C	D	E	F	G	H	I	J	K	L	M	N	O	P	Q	R	S	T	U	V	W	X	Y	Z		−
$\bar{x}$	1	2	3	4	5	6	7	8	9	10	11	12	13	14	15	16	17	18	19	20	21	22	23	24	25	26	27	28
$\bar{x}^2$	1	4	9	16	25	7	20	6	23	13	5	28	24	22	22	24	28	5	13	23	6	20	7	25	16	9	4	1
$\bar{x}^3$	1	8	27	6	9	13	24	19	4	14	26	17	22	18	11	7	12	3	15	25	10	5	16	20	23	2	21	28
$\bar{x}^4$	1	16	23	24	16	20	23	7	7	24	25	1	25	20	20	25	1	25	24	7	7	23	20	16	24	23	16	1
$\bar{x}^5$	1	3	11	9	22	4	16	27	5	8	14	12	6	19	10	23	17	15	21	24	2	13	25	7	20	18	26	28
$\bar{x}^6$	1	6	4	7	23	24	25	13	16	22	9	28	20	5	5	20	28	9	22	16	13	25	24	23	7	4	6	1
$\bar{x}^7$	1	12	12	28	28	28	1	17	28	17	12	17	28	12	17	1	12	17	12	1	12	28	1	1	1	17	17	28
$\bar{x}^8$	1	24	7	25	24	23	7	20	20	25	16	1	16	23	23	16	1	16	25	20	20	7	23	24	25	7	24	1
$\bar{x}^9$	1	19	21	13	4	22	20	15	6	18	2	12	5	3	26	24	17	27	11	23	14	9	7	25	16	8	10	28
$\bar{x}^{10}$	1	9	5	23	20	16	24	4	25	6	22	28	7	13	13	7	28	22	6	25	4	24	16	20	23	5	9	1
$\bar{x}^{11}$	1	18	15	5	13	9	23	3	22	2	10	17	4	8	21	25	12	19	27	7	26	6	20	16	24	14	11	28
$\bar{x}^{12}$	1	7	16	20	7	25	16	24	24	20	23	1	23	25	25	23	1	23	20	24	24	16	25	7	20	16	7	1
$\bar{x}^{13}$	1	14	19	22	6	5	25	18	13	26	21	12	9	2	27	20	17	8	3	16	11	4	24	23	7	10	15	28
$\bar{x}^{14}$	1	28	28	1	1	1	1	28	1	28	28	28	1	28	28	1	28	28	28	1	28	1	1	1	1	28	28	1
$\bar{x}^{15}$	1	27	26	4	5	6	7	21	9	19	18	17	13	15	14	16	12	11	10	20	8	22	23	24	25	3	2	28
$\bar{x}^{16}$	1	25	20	16	25	7	20	23	23	16	24	1	24	7	7	24	1	24	16	23	23	20	7	25	16	20	25	1
$\bar{x}^{17}$	1	21	2	6	9	13	24	10	4	15	3	12	22	11	18	7	17	26	14	25	19	5	16	20	23	27	8	28
$\bar{x}^{18}$	1	13	6	24	16	20	23	22	7	5	4	28	25	9	9	25	28	4	5	7	22	23	20	16	24	6	13	1
$\bar{x}^{19}$	1	26	18	9	22	4	16	2	5	21	15	17	6	10	19	23	12	14	8	24	27	13	25	7	20	11	3	28
$\bar{x}^{20}$	1	23	25	7	23	24	25	16	16	7	20	1	20	24	24	20	1	20	7	16	16	25	24	23	7	25	23	1
$\bar{x}^{21}$	1	17	17	28	28	28	1	12	28	12	17	12	28	17	12	1	17	12	17	1	17	28	1	1	1	12	12	28
$\bar{x}^{22}$	1	5	22	25	24	23	7	9	20	4	13	28	16	6	6	16	28	13	4	20	9	7	23	24	25	22	5	1
$\bar{x}^{23}$	1	10	8	13	4	22	20	14	6	11	27	17	5	26	3	24	12	2	18	23	15	9	7	25	16	21	19	28
$\bar{x}^{24}$	1	20	24	23	20	16	24	25	25	23	7	1	7	16	16	7	1	7	23	25	25	24	16	20	23	24	20	1
$\bar{x}^{25}$	1	11	14	5	13	9	23	26	22	27	19	12	4	21	8	25	17	10	2	7	3	6	20	16	24	15	18	28
$\bar{x}^{26}$	1	22	13	20	7	25	16	5	24	9	6	28	23	4	4	23	28	6	9	24	5	16	25	7	20	13	22	1
$\bar{x}^{27}$	1	15	10	22	6	5	25	11	13	3	8	17	9	27	2	20	12	21	26	16	18	4	24	23	7	19	14	28
$\bar{x}^{28}$	1	1	1	1	1	1	1	1	1	1	1	1	1	1	1	1	1	1	1	1	1	1	1	1	1	1	1	1
$\bar{x}^{29}$	1	2	3	4	5	6	7	8	9	10	11	12	13	14	15	16	17	18	19	20	21	22	23	24	25	26	27	28

that number to the 11th power in $\mathbf{Z}_{29}$. The relevant rows of Table 9.4.1 are duplicated here.

	A	B	C	D	E	F	G	H	I	J	K	L	M	N	O	P	Q	R	S	T	U	V	W	X	Y	Z	–
$\bar{x}$	1	2	3	4	5	6	7	8	9	10	11	12	13	14	15	16	17	18	19	20	21	22	23	24	25	26	27	28
$\vdots$													$\ldots$														
$\bar{x}^{11}$	1	18	15	5	13	9	23	3	22	2	10	17	4	8	21	25	12	19	27	7	26	6	20	16	24	14	11	28

Plaintext	Ciphertext
A = 1	$1^{11} \equiv 1 \pmod{29}$
L = 12	$12^{11} \equiv 17 \pmod{29}$
M = 13	$13^{11} \equiv 4 \pmod{29}$
O = 15	$15^{11} \equiv 21 \pmod{29}$
S = 19	$19^{11} \equiv 27 \pmod{29}$
T = 20	$20^{11} \equiv 7 \pmod{29}$

Thus, the entire message ALMOST encrypts to the ciphertext **1 17 4 21 27 7**. ●

Now put yourself in the shoes of the recipient of the ciphertext message **1 17 4 21 27 7**. How are you going to decrypt the message? The most naive way to do it is to use the table from Example 1 and apply the method of reverse table lookup. For example, to decrypt the second number, **17**, you would find where **17** appears in the bottom row, and then read the corresponding number **12 = L** from the top row.

Check for Understanding

1. Encrypt the message FUN MATH one character at a time by raising to the power 11 modulo 29.

2. The ciphertext 5 22 23 22 7 was obtained one character at a time by raising to the power 11 modulo 29. Decrypt the message.

Decryption using Fermat's Little Theorem

Decrypting a message using reverse table lookup works well enough in $\mathbf{Z}_{29}$, but in real life, we must use a modulus much larger than 29 if we want any hope of having a secure system. When the modulus is large, the method of reverse table lookup is extremely inefficient. We need a better method for decryption.

In Section 8.4, we learned how to encrypt using modular addition: to encrypt a message, we added a number e to the plaintext (mod n). Decryption was accomplished by adding the additive inverse of e to the ciphertext (mod n). In Section 8.5, we studied encryption using modular multiplication: to encrypt a message, we multiplied the plaintext by a number e (mod n). To decrypt in this case, we multiplied the ciphertext by the multiplicative inverse of e (mod n).

When we encrypted using either modular addition or modular multiplication, there was a straightforward way of undoing, or *inverting*, the encryption operation. In this section, we are encrypting using modular exponentiation: to encrypt, we raise the plaintext to a power e modulo a prime p. In this case, is there a straightforward way to invert the encryption process? Can we find an inverse to the operation of raising to the power e, analogous to the additive inverse and multiplicative inverse of Sections 8.4 and 8.5?

Naomi's Numerical Proof Preview: Method 9.4.1

Before considering the general case, let's try to find the answer for a specific example. In Example 1, we encrypted the fifth letter of the message, S $=$ 19, by raising it to the power 11 modulo 29:

$$19^{11} \equiv 27 \text{ (mod 29)}.$$

Perhaps we can raise the ciphertext 27 to some power (mod 29) to recover the plaintext 19. Inspection of Table 9.4.1 reveals that the exponent 23 has the desired decrypting effect:

$$27^{23} \equiv 19 \text{ (mod 29)}.$$

In fact, the same exponent, 23, will also successfully decrypt any other letter in the message! Why does this work? In other words, when working modulo 29, why does raising a number to the power 11 and then raising the result to the power 23 always return the original number? Let's examine this in more detail. Suppose we have any plaintext message m, which is a natural number less than 29. By Fermat's Little Theorem (Restatement, 9.1.2), $m^{28} \equiv 1$ (mod 29). Hence,

$$(m^{11})^{23} \equiv m^{11 \cdot 23}$$
$$\equiv m^{253}$$
$$\equiv m^{28 \cdot 9 + 1}$$
$$\equiv (m^{28})^9 \cdot m^1$$
$$\equiv 1^9 \cdot m^1$$
$$\equiv m \text{ (mod 29)}.$$

We conclude that raising to the power **23** is the inverse of raising to the power **11** (mod 29). Our argument rests on two crucial facts. First, Fermat's Little Theorem told us that

$$m^{28} \equiv 1 \ (\text{mod } 29).$$

Second, $11 \cdot 23 = 28 \cdot 9 + 1$. That is, the product of the encrypting exponent **11** and the decrypting exponent **23** is 1 more than a multiple of **28**, or

$$11 \cdot 23 \equiv 1 \ (\text{mod } 28).$$

In other words, **11** and **23** are multiplicative inverses modulo **28**.

We now know enough to state the method of encryption and decryption in general.

METHOD 9.4.1 | **EXPONENTIAL ENCRYPTION AND DECRYPTION WITH A PRIME MODULUS**

SETUP:

Let p be prime.

Choose an encryption exponent, $e \in \mathbf{N}$, such that $\gcd(e, p - 1) = 1$.

The decryption exponent, $d \in \mathbf{N}$, is the multiplicative inverse of e modulo $p - 1$:

$$e \cdot d \equiv 1 \ (\text{mod } p - 1), \tag{1}$$

with $d < p - 1$.

TO ENCRYPT:

Let $m \in \mathbf{N}$ be the plaintext message we wish to encrypt, $m < p$.

The ciphertext, c, is obtained by raising the message to the encryption exponent modulo p:

$$c \equiv m^e \ (\text{mod } p). \tag{2}$$

TO DECRYPT:

To recover the plaintext, raise the ciphertext to the decryption exponent d modulo p:

$$recovered \ message \equiv c^d \ (\text{mod } p).$$

Verifying that decryption undoes encryption

We now show that in exponential encryption and decryption with a prime modulus (Method 9.4.1), the *recovered message* is equal to the original plaintext message, m.

$$
\begin{aligned}
recovered \ message &\equiv c^d \\
&\equiv (m^e)^d \qquad \leftarrow \textbf{by (2) and Corollary 7.2.2} \\
&\equiv m^{e \cdot d} \ (\text{mod } p). \tag{3}
\end{aligned}
$$

By (1) and Theorem 8.2.4 (Equivalent Conditions for Congruence), there exists an integer k such that

$$e \cdot d = (p - 1)k + 1.$$

Continuing from (3) yields

$$\begin{aligned}
\text{recovered message} &\equiv m^{e \cdot d} \\
&\equiv m^{(p-1)k+1} \\
&\equiv (m^{p-1})^k \cdot m^1 \pmod{p}.
\end{aligned}$$

By Fermat's Little Theorem (Restatement, 9.1.2), $m^{p-1} \equiv 1 \pmod{p}$. Thus, $(m^{p-1})^k \equiv 1^k \pmod{p}$ by Corollary 7.2.2, which gives

$$\begin{aligned}
\text{recovered message} &\equiv 1^k \cdot m \qquad\qquad\qquad (4)\\
&\equiv m \pmod{p}.
\end{aligned}$$

This completes our demonstration that the encryption scheme of Method 9.4.1 is valid, in the sense that the decryption method is in fact the inverse of the encryption method.

Examples of encryption and decryption

When we learned encryption using modular addition (in Section 8.4) and encryption using modular multiplication (in Section 8.5), we grouped the text message in blocks of two or more letters to convert the message into numbers. Encryption by modular exponentiation would not be very good if we could only encrypt one character at a time.

Now that we understand the concepts behind exponential encryption in prime moduli, we can use a modulus much larger than 29, which will enable us to encrypt in blocks of two or more letters. (For a refresher on how to convert a two-letter block to a number, see Section 8.4.)

EXAMPLE 2

Encrypt the message SO BIG in two-letter blocks by raising to the exponent 9 modulo 2819.

SOLUTION The modulus 2819 is prime. We first verify that the encryption exponent, **9**, is relatively prime to $2819 - 1$, so that we know our message can be decrypted later:

$$\gcd(9, 2818) = 1.$$

We separate the message into two-letter blocks, and use the top row of Table 9.4.1 to convert each two-letter block into a number:

$$SO = 1915, \quad -B = 2702, \quad IG = 907.$$

To encrypt the message, we raise each of these plaintext numbers to the power **9** modulo 2819, using the method of repeated squaring from Section 9.3.

x	1915	2702	907
x^2	$1915^2 \equiv \mathbf{2525} \pmod{2819}$	$2702^2 \equiv \mathbf{2413} \pmod{2819}$	$907^2 \equiv \mathbf{2320} \pmod{2819}$
x^4	$2525^2 \equiv \mathbf{1866} \pmod{2819}$	$2413^2 \equiv \mathbf{1334} \pmod{2819}$	$2320^2 \equiv \mathbf{929} \pmod{2819}$
x^8	$1866^2 \equiv \mathbf{491} \pmod{2819}$	$1334^2 \equiv \mathbf{767} \pmod{2819}$	$929^2 \equiv \mathbf{427} \pmod{2819}$

Now $x^9 \equiv x^8 \cdot x \pmod{2819}$, so

$$1915^9 \equiv 491 \cdot 1915 \equiv \mathbf{1538} \pmod{2819}$$
$$2702^9 \equiv 767 \cdot 2702 \equiv \mathbf{469} \pmod{2819}$$
$$907^9 \equiv 427 \cdot 907 \equiv \mathbf{1086} \pmod{2819}.$$

Thus, the ciphertext is: **1538 469 1086**. ●

EXAMPLE 3

Your friends Alistair and Bobbie have been clandestinely exchanging coded messages, and you've been dying to know what they say. In response to your pleas, Alistair eventually tells you, "We encrypt in two-letter blocks by raising to the power **1957** modulo 3181. But I refuse to reveal our secret method for decryption under any circumstances!" Later that day, you intercept a message from Bobbie's backpack. The message reads

2284 1688 510.

Even though Alistair did not give you instructions for decrypting, can you still figure out what the message says?

SOLUTION To decrypt the message using Method 9.4.1, we need to know the decryption exponent, d. The encryption exponent is $e = \mathbf{1957}$. The decryption exponent, d, is the multiplicative inverse of **1957** modulo $3181 - 1$:

$$1957d \equiv 1 \pmod{3180}, \quad \text{with } d < 3180.$$

Since $\gcd(\mathbf{1957}, 3180) = 1$, we know this multiplicative inverse exists, and (as discussed in Section 8.3) we can use the Euclidean Algorithm to find it:

$$d = \mathbf{13}.$$

To decrypt the message, we raise each of the ciphertext numbers to the power **13** modulo 3181. Using the method of repeated squaring from Section 9.3, we find

$$2284^{13} \equiv 1905 \ (\text{mod } 3181)$$

$$1688^{13} \equiv 318 \ (\text{mod } 3181)$$

$$510^{13} \equiv 520 \ (\text{mod } 3181)$$

Thus, the plaintext is: **1905 318 520**. Converting this into letters using the top row of Table 9.4.1 reveals the decrypted message: SECRET.

As this example demonstrates, exponentiation ciphers using a prime modulus have the following limitation: If someone tells you how messages are encrypted, it is easy to figure out how to decrypt. One only needs to find a multiplicative inverse using the Euclidean Algorithm, which is fast and easy, especially if you have a computer handy. This property of exponentiation ciphers using a prime modulus is undesirable for many applications. Sometimes you'd like to reveal the method of encryption (so that anyone can send you messages), without revealing how to decrypt (so that other people cannot read encrypted messages). In the next section, we will see that this can be achieved with a very small modification to the encryption systems of this section.

EXERCISES 9.4

Numerical Problems

1. Encrypt the message SAY IT one character at a time by raising to the power **9** modulo 29.

2. A message was encrypted one character at a time by raising to the power **17** modulo 29. The ciphertext is

    15 19 24 24 12 9.

    a. Find the decryption exponent, d.

    b. Use your answer to part a to decrypt the message. (You may use Table 9.4.1.)

3. To use modular exponentiation on two-letter blocks, you need a prime modulus that is at least 2727. For each prime modulus p given, list three valid encryption exponents e (that is, values of e that satisfy the conditions of Method 9.4.1).

    a. $p = 2819$ **b.** $p = 2729$

4. You want to send your friend a message that is encrypted in two-letter blocks, using an exponent of 15 and a prime modulus of 2819.

    a. Confirm that this is a valid encryption method (that it satisfies the "Setup" hypotheses of Method 9.4.1).

b. Use this method to encrypt the following message: LEAVE AT DAWN

c. Find the decryption exponent, d, that will undo this cipher.

5. Your friend sent you a message that was encrypted in two-letter blocks using an exponent of 2317 and a prime modulus of 2897.

    a. Confirm that this is a valid encryption method (i.e., that it satisfies the "Setup" hypotheses of Method 9.4.1).

    b. Find the decryption exponent, d, that will undo this cipher.

    c. Decrypt this message from your friend, which was encrypted in two-letter blocks using an exponent of 2317 and prime modulus 2897:

$$71 \quad 2450 \quad 404 \quad 324 \quad 633 \quad 1804.$$

6. A message was encrypted in two-letter blocks by raising to the power 673 modulo 2861. The ciphertext is: 776 1411 2033. Decrypt the message.

Reasoning and Proofs

7. **Phil Lovett** Convinced that he is the master of encryption, your friend Phil Lovett tries to encrypt a message one character at a time by raising to the power 18 modulo 29. Proud of his curly locks, Phil wants to encrypt the message "long hair." Lacking the patience to encrypt such a long message, however, Lovett decides to merely encrypt the shorter message, LONG.

    a. Using Table 9.4.1, use Phil's method to encrypt the word LONG.

    b. Explain why it would be impossible to decrypt the ciphertext you found in part a.

    c. Is this a mistake in Method 9.4.1, or yet another case in which Phil got overexcited and forgot to check an important hypothesis? Explain.

8. In the demonstration that exponentiation with a prime modulus (Method 9.4.1) is a valid encryption scheme, to obtain congruence (4) we used the fact that $m^{p-1} \equiv 1 \pmod{p}$, which we justified using Fermat's Little Theorem. Fermat's Little Theorem requires that the base, m, must be relatively prime to the modulus, p. Explain how we know that in this case, $\gcd(m, p) = 1$.

9. **Phil Lovett** Your friend Phil Lovett is so sure of his encryption prowess that his prediction for your outcome in the upcoming cryptography competition you both entered is succinct: DEFEAT. To practice for the competition, he suggests that you both race to encrypt the six-letter word DEFEAT in two-letter blocks by raising to the power 33 modulo 2861. He beats you by a few seconds, but when you look at your answer, you aren't as impressed by Phil's alleged cryptography prowess. Why not?

9.5 The RSA Encryption Algorithm

Public key cryptography

In all of the encryption schemes we have studied so far, if a person knows how to encrypt using the scheme, then it is not too difficult for him or her to figure out how to decrypt. This could be a problem. For example, suppose you run a company that sells surfboards over the Internet. You need to give your customers a way to encrypt their credit card numbers so that they can place orders securely. But if someone with questionable motives is able to decrypt messages, that person could intercept your incoming messages and steal your customers' credit card numbers. For people to buy things securely on the Internet, it is essential to have an encryption scheme in which the public knows how to encrypt messages but not how to decrypt them.

The goal is to have a cryptosystem in which one key, known as the *public key*, is used to encrypt messages, but a different key, known as the *private key*, is used to decrypt messages. This type of cryptosystem is known as a *public key* cryptosystem. The public key is distributed to everyone who has a need to encrypt (such as surfboard customers), while the private key is kept secret. We wish to arrange the system so that even if one knows the public key, it is computationally infeasible to compute the private key. The reader may be surprised that such a cryptosystem is even possible!

The system we seek resembles a *trapdoor*, through which one can pass easily in one direction but cannot pass in the other direction without great difficulty. How could we possibly create a mathematical function that is easy to do but hard to undo? As we will see shortly, number theory provides a surprisingly elegant and effective solution.

Though relatively new (see "The history of public key cryptography and RSA" at the end of this section), public key cryptosystems are crucial to a number of technologies central to modern life, including secure transactions over the Internet, financial transfers among banks, digital signatures, and keeping confidential email private.

The RSA algorithm

The RSA encryption algorithm is a public key cryptosystem named for the mathematicians who discovered it: Rivest, Shamir, and Adelman. Their idea was to create a mathematical trapdoor based on the following observation: it is easy to multiply large numbers together but very difficult to go in the other direction—that is, to factor large numbers. For example, you could multiply 53549 by 89561 using pencil and paper and arrive at the answer, 4795901989, in a few minutes. But suppose instead that you were given the number 4795901989 and asked to factor it. Certainly it would take you *much* longer than a few minutes to factor this number by hand.

The situation is similar with larger numbers on a computer. Multiplying two 300-digit numbers is a cinch for a computer, requiring only microseconds to arrive at their 600-digit product. But factoring a given 600-digit number could take countless millennia, even using the most sophisticated algorithms on the latest supercomputers (see "Factoring Huge Numbers").

Factoring Huge Numbers

Until 2007, RSA Laboratories, a private computer security company, offered cash prizes for successfully factoring large numbers. While the challenge was active, the largest challenge number to be factored was the 193-digit number

3107418240490043721350750035888567930037346022842727545720161948 823206440518081504556346829671723286782437916272838033415471073108 50191954852900733772482227835257423864540146917366024776523466099.

The research team of F. Bahr, M. Boehm, J. Franke, and T. Kleinjung earned a $20,000 prize for factoring this number. Completed on November 2, 2005, their solution took over five months of calendar time—30 years of processor time—on a cluster of 2.2GHz Opteron processors. RSA Laboratories had also proposed several more difficult factoring challenges, with prizes up to $200,000 for the following 617-digit number:

25195908475657893494027183240048398571429282126204032027777137 83604366202070759555626401852588078440691829064124951508218929 85591497176184502808489120072844992687392807287776735971418347 27026189637501497182469116507761337985909570009733045974880842 84017974291006424586918171951187461215151726546322822168699875 49182422433637259085141865462043576798423387184774447920739934 23658482382428119816381501067481045166037730605620161967625613 38441436038339044149526344321901146575444541784240209246165157 23350778707749817125772467962926386356373289912154831438167899 88504044536402352738195137863656439121201039712282212072035 7

As computers get faster and algorithms become more efficient, factoring large numbers requires less and less time. In a *Scientific American* article in 1977, Rivest was quoted as saying that it would take 40 quadrillion years of computing time to factor a particular 129-digit number. However, in 1994, the number was factored using only eight months of computer time, working with many computers in parallel. Based on today's technology, it is predicted that factoring a 230-digit number in one year would require 215,000 Pentium-class machines. Using historical data about when different numbers were factored, Richard Brent at the Oxford University Computing Laboratory gave the following formula as a best guess for the year Y in which a number with D digits would be factored:

$$Y = 13.24D^{1/3} + 1928.6.$$

According to this formula, the RSA challenge with 617 digits would be factored in 2041!

So it seems that we've found our mathematical trapdoor. But how can this be used to create a cryptosystem in which it's possible to tell someone how to encrypt without revealing how to decrypt?

The RSA algorithm is quite similar to exponentiation ciphers with a prime modulus, which we studied in Section 9.4. The only difference is that instead of using a prime modulus, RSA uses a composite modulus. Although this seems like a minor change, it makes a world of difference in the power of the encryption scheme.

In exponential encryption with a prime modulus (Method 9.4.1), the decryption method relied on Fermat's Little Theorem (9.1.1). In the RSA encryption algorithm, the modulus is no longer prime, so the decryption method relies on Euler's Theorem (9.3.1) instead. The modulus used in RSA is a product of two large prime numbers.

9.5.1 THE RSA ENCRYPTION ALGORITHM

SETUP:

Choose large prime numbers p and q, and let $n = pq$.
By Lemma 9.2.6,

$$\varphi(n) = (p - 1)(q - 1).$$

Choose an encryption exponent, $e \in \mathbf{N}$, such that $\gcd(e, \varphi(n)) = 1$.
The decryption exponent, $d \in \mathbf{N}$, is the multiplicative inverse of e modulo $\varphi(n)$:

$$e \cdot d \equiv 1 \ (\mathrm{mod} \ \varphi(n)), \tag{1}$$

with $d < \varphi(n)$.

Private (known only by person who is decrypting)	**Public** (known by person who is encrypting)
Large prime numbers p and q $\varphi(n) = (p - 1)(q - 1)$ Decryption exponent d	The modulus n (where $n = pq$) Encryption exponent e

TO ENCRYPT:

Let $m \in \mathbf{N}$ be the plaintext message we wish to encrypt, $m < n$.
The ciphertext, c, is obtained by raising the message to the encryption exponent modulo n:

$$c \equiv m^e \ (\mathrm{mod} \ n). \tag{2}$$

TO DECRYPT:

To recover the plaintext, raise the ciphertext to the decryption exponent d modulo n:

$$recovered\ message \equiv c^d \ (\mathrm{mod} \ n).$$

Verifying that decryption undoes encryption

We now show that in the RSA Encryption Algorithm (9.5.1), the *recovered message* is equal to the original plaintext message, m.

$$recovered\ message \equiv c^d$$
$$\equiv (m^e)^d \qquad \leftarrow \textbf{by (2) and Corollary 7.2.2}$$
$$\equiv m^{ed} \pmod{n}. \qquad\qquad\qquad\qquad (3)$$

By Congruence (1) and Theorem 8.2.4 (Equivalent Conditions for Congruence), there exists an integer k such that

$$ed \equiv k\varphi(n) + 1.$$

Continuing from (3) yields

$$recovered\ message \equiv m^{ed}$$
$$\equiv m^{k\varphi(n)+1}$$
$$\equiv (m^{\varphi(n)})^k \cdot m^1 \pmod{n}.$$

Euler's Theorem (Restatement, 9.3.2) tells us that $m^{\varphi(n)} \equiv 1 \pmod{n}$. Thus,

$$recovered\ message \equiv 1^k \cdot m \qquad \leftarrow \textbf{by Corollary 7.2.2} \qquad (4)$$
$$\equiv m \pmod{n}.$$

We have thus demonstrated that decryption recovers the original plaintext message.

Note that to get to congruence (4), we used Euler's Theorem, which requires that the base, m, must be relatively prime to the modulus, n. In practice, however, it is not necessary to verify that $\gcd(m, n) = 1$ because if the primes p and q are large, then it is almost certain that this will be true (see Exercise 14).

Examples of encryption and decryption with RSA

EXAMPLE 1

Encrypt the message FERMAT in two-letter blocks using the RSA algorithm, with modulus $n = 3233$ and encryption exponent $e = 7$.

SOLUTION We separate the message into two-letter blocks and use Table 8.4.1 (or the top row of Table 9.4.1) to convert each two-letter block into a number:

$$FE\ =\ 605$$
$$RM\ =\ 1813$$
$$AT\ =\ 120.$$

To encrypt the message, we raise each of these plaintext numbers to the power **7** modulo 3233, using the method of repeated squaring from Section 9.3. We find

$$605^7 \equiv \mathbf{1663} \pmod{3233}$$

$$1813^7 \equiv \mathbf{2410} \pmod{3233}$$

$$120^7 \equiv \mathbf{55} \pmod{3233}.$$

Thus, the ciphertext is **1663 2410 55**.

EXAMPLE 2

You intercept a message that was encrypted in two-letter blocks with RSA using the values $n = 3901$ and $e = \mathbf{343}$. The ciphertext reads as follows: **1930 208 1910**. Decrypt the message.

SOLUTION You would like to use the decrypting instructions from the RSA Encryption Algorithm (9.5.1). To do this, you need to raise each of the numbers in the ciphertext to the power d and reduce modulo 3901. However, you do not know the decryption exponent d. Recall that d is the multiplicative inverse of **343** modulo $\varphi(\mathbf{3901})$. Thus, before you can find d, you must calculate the value of $\varphi(\mathbf{3901})$. The best way to do this is to first factor 3901 as $83 \cdot 47$ (note that this step would be prohibitively time-consuming if the number to be factored were large). We then have

$$\varphi(\mathbf{3901}) = (83 - 1)(47 - 1) = \mathbf{3772},$$

by Lemma 9.2.5.

We now know that the decryption exponent d is the multiplicative inverse of **343** modulo **3772**. Using the Euclidean Algorithm to find this inverse (as discussed in Section 8.3), we get

$$d = \mathbf{11}.$$

Now that you know the decryption key, you can decrypt the message using the instructions from the RSA Encryption Algorithm (9.5.1). Raising each of the ciphertext numbers to the power **11** modulo 3901 (which can be accomplished using the method of repeated squaring), we find

$$1930^{11} \equiv \mathbf{521} \pmod{3901}$$

$$208^{11} \equiv \mathbf{1205} \pmod{3901}$$

$$1910^{11} \equiv \mathbf{1827} \pmod{3901}.$$

Thus, the plaintext is **521 1205 1827**. Converting this into letters using Table 8.4.1 (or the top row of Table 9.4.1) reveals the decrypted message: EULER.

The security of RSA depends on the difficulty of factoring

In Example 2, you were able to "crack" the RSA system. That is, you were able to recover the plaintext even though you were not given the decryption key d. The reason for your success is that you were able to factor the modulus $n = 3901$, which is a relatively small number. If n had been a lot larger, the system would not have been so easy to crack.

Suppose that you have set up an RSA cryptosystem so that people can send you private email messages. The primes p and q that you have chosen are 300 digits each, making your n a 600-digit number. You freely give out your values of n and e, and pretty soon the encrypted messages start pouring in. Now imagine that your nemesis, Evil Eve the Eavesdropper, has been intercepting your incoming messages and wants to read them. What would Eve need to do in order to decrypt your secrets?

Like everyone else, Eve knows the values of n and e. To decrypt, she must raise the encrypted messages to the d power modulo n. To do this, Eve needs to find d, the multiplicative inverse of e modulo $\varphi(n)$. Of course, to do this she needs to know the value of $\varphi(n)$. But $\varphi(n) = (p - 1)(q - 1)$, so to find $\varphi(n)$, Eve needs to find p and q. In other words, Eve needs to factor n, a 600-digit number! In practice, this is impossible, even for Evil Eve.

Thus, the security of RSA rests on the difficulty of factoring huge numbers, which is widely believed to be impossible in practice. This is why people believe in the security of the RSA algorithm. It's why individuals, corporations, and governments trust RSA with their most valued secrets.

Wanted: large prime numbers

There is one more computational issue that you may have been wondering about: How do we find the large prime numbers, p and q, needed for RSA? Suppose we want to find a 300-digit prime number. A natural way to do this is to generate 300-digit random numbers[2] until we get one that is prime. How many numbers will we have to check until we are lucky enough to find a prime? According to the Prime Number Theorem (3.3.3), the probability that a random 300-digit number is prime is approximately $\dfrac{1}{\log(10^{300})}$, or about 1 chance in 691. Thus, by the time we have tested a few thousand numbers, it is quite likely that we will have found a prime.

For this method to succeed, we need a quick way to determine whether a given number is prime or composite. You might think that the easiest way to do this would be to factor the number, which (as discussed earlier) would be a Herculean task, even with the help of a supercomputer. It may surprise you to learn that factoring is not the fastest way to determine whether a large number is prime. Determining whether a particular large number is prime can be accomplished quickly with the aid of a computer; this is known as *primality testing* (see Chapter 12).

---

[2]See Section 10.4 for a discussion of methods for generating random numbers.

The history of public key cryptography and RSA

With the invention of the computer in the middle of the 20th century, both code making and code breaking became more complex. Encryption systems could now employ enormous arithmetical calculations; likewise, someone trying to break a code could use a computer to try millions of possibilities. As computers became more widespread, not only did the government and the military use encryption to protect sensitive information, but businesses of all sorts began to use encryption for money transfers and trade negotiations. To make it easier for companies to send encrypted messages to one another, the National Bureau of Standards suggested that there should be a standard method of encryption that all companies could use. IBM developed the Data Encryption Standard (DES) in the early 1970s, and it quickly caught on. To use DES, the sender and receiver first agree on a secret number, called the *key*. (Unlike the public keys used in RSA, this key must be kept private by the two parties.) The sender then divides the message into blocks of 64 digits, which are each encrypted by an algorithm making use of this key. Although the algorithm of DES is public, an encrypted message can only be decrypted by someone who knows the key that was used to encrypt it. To break the code, a person would have to correctly guess the key or use a computer to try every possibility until it found the right number. If the key is large enough, this is practically impossible. In 1976, when DES officially became the national standard, the National Security Agency (NSA) restricted keys to be less than 2^{56}—that is, at most 56 bits (binary digits). This restriction meant that only the tremendously powerful computers at the NSA had the speed to search through all possible keys within a reasonable amount of time. So businesses could use DES to communicate with one another without having to worry about other companies or individuals spying on them, yet the NSA could break any DES code if it were deemed necessary for national security.

Using DES requires both the sender and the recipient to know a shared secret key. The most obvious way to share this information is for one party to communicate the key directly to the other. Of course, this communication runs the risk of being intercepted. The key could even be handed over in person, but this could be hard to arrange. This difficulty, known as the *key distribution problem*, seemed unsolvable until 1976, when Whitfield Diffie, Martin Hellman, and Ralph Merkle came up with an ingenious solution based on modular exponentiation. We explain their solution, known as *Diffie-Hellman key exchange* (or Diffie-Hellman-Merkle key exchange), in Section 10.4. Diffie-Hellman key exchange is a great way for two parties to securely agree on a key. The one difficulty with this method is that it requires the parties to send several messages back and forth before they can send each other encrypted messages.

The combination of Diffie-Hellman key exchange and DES is effective for many applications. However, it has some drawbacks. Communication would be faster if there were no need to agree on a key. In addition, as we discussed at the beginning of this section, with DES everyone who knows how to encrypt messages also knows how to decrypt them. This is undesirable for many applications.

While working on the solution to the key distribution problem, Diffie thought up a totally new approach, public key cryptography, which would avoid the need for a key exchange. Until then, all cryptographic systems were based on the idea of a *symmetric* key, that is, a key which is used for both encryption and decryption. Diffie's idea was that there could be *asymmetric* keys, so that the sender would use one key to encrypt the message and the recipient would use a different key to decrypt the message. With an asymmetric system, a party could post a public key that anyone could use to encrypt a message to them, but only the party would know the private key needed to decrypt the message. In 1975, Diffie and Hellman published a paper explaining their idea for public key cryptography. However, they were still missing one crucial piece: Their method required an encryption function that could be computed easily using the public key, but whose inverse was practically impossible to compute without the private key.

Three computer scientists at MIT, Ron Rivest, Adi Shamir, and Leonard Adleman, read Diffie and Hellman's paper, and within a year they had figured out a function that could only be reversed by someone who knew the private key. As we have already seen, their method of public key cryptography, the RSA Encryption Algorithm (9.5.1), is based on the difficulty of factoring very large numbers. Secure key exchange is not necessary for RSA, and for large enough values of the private key, RSA is virtually unbreakable. For these reasons, RSA has become the norm for encryption and will probably continue to be the norm for many years to come. In 1982, Ron Rivest, Adi Shamir, and Leonard Adleman obtained a patent on the RSA algorithm and formed a company known as RSA Data Security, Inc., devoted to protecting online identities and digital assets. One measure of the current success of the RSA algorithm is that in 1996, RSA Data Security, Inc., was sold for $200 million.

This would be the end of our story, were it not for recent revelations that the RSA algorithm has an earlier history centered around the British security agency known as Government Communications Headquarters (GCHQ). In 1997, GCHQ revealed that they had secretly discovered RSA a few years before it was discovered by Rivest, Shamir, and Adleman in 1976.

In 1969, a cryptographer at GCHQ, John Ellis, developed the idea of public key cryptography (which was known within GCHQ as *nonsecret encryption*, or NSE). Ellis was missing the same crucial piece as Diffie: He was unable to find a function to implement his method. Number theorist Clifford Cocks joined GCHQ in 1973, and within a half an hour of hearing about NSE, Cocks had figured out a function that would work—he had developed the same algorithm that would later become known as RSA. When cryptographer Malcolm Williamson joined GCHQ in 1974, he thought Cocks's algorithm was too good to be true. In the process of unsuccessfully trying to find a flaw in it, Williamson discovered the key exchange method that Diffie, Hellman, and Merkle would later discover independently.

In 1974, both RSA and Diffie-Hellman key exchange were known to GCHQ, but they were unknown outside the intelligence community. This made them all the more valuable to British intelligence. In the early 1980s, after these methods were well known to the public, GCHQ informed their American counterparts at the NSA of their earlier discovery of the RSA algorithm, and considered informing the public as well. However, just at that time, the book *Spycatcher* was released, in which Peter Wright revealed his experiences as a British intelligence officer, detailing irregularities and illegalities that he had observed within the agency. The British attorney general tried unsuccessfully to prevent the publication of the book, which eventually became a best seller. As a result of the publicity surrounding *Spycatcher*, GCHQ decided not to release any classified information about their work on RSA.

Finally, in 1997, Cocks wanted to present an unclassified paper on his recent research about the RSA algorithm, and GCHQ was worried that someone might ask about his role in the original development of RSA. After considering its options, GCHQ gave Cocks permission to begin his talk by revealing the history of public key cryptography at GCHQ. News of this historical revelation spread quickly, and it soon appeared in the *New York Times*.

Even knowing about the work of Ellis, Cocks, and Williamson at GCHQ, we still cannot be certain of the full history of public key cryptography. The NSA has asserted that it, too, had the idea of RSA in the 1960s and was using it by the mid-1970s. However, details of the NSA's role in its development have not been made public. It is even possible that there are other secret histories in which other governments developed their own versions of RSA before Rivest, Shamir, and Adleman discovered it in 1976.

Quantum factoring

A completely new model for computation, called *quantum computation*, may eventually provide a fast way to factor large numbers and hence break RSA.

Today's computers represent information in terms of bits (0's and 1's), and do computations by performing logical operations (e.g., NOT, AND, OR) on these bits. The idea of quantum computing is to create a computer that acts at its logical level according to the principles of quantum mechanics, a branch of physics that was created in the early 20th century to describe the behavior of matter at very small scales. Quantum mechanics has some interesting and unintuitive consequences. For example, rather than being in a fixed state, such as 0 or 1, a quantum system may be in linear combination of states, called a *superposition* of states. This allows us to represent information in a new way. Whereas a classical (nonquantum) computer processes bits (each of which is equal to either 0 or 1), a quantum computer processes *quantum bits*, or *qubits*, each of which is a linear combination of 0 and 1. The rules of quantum mechanics tell us which types of operations we may perform on quantum bits.[3] Using this new model of information processing, one may design algorithms for quantum computers. For a given problem, such as factoring integers, one may ask whether there is a quantum algorithm that solves the problem more efficiently than the best classical algorithms.

---

[3]The allowable operations on qubits are certain linear transformations called unitary transformations.

In 1994, Peter Shor gave a fast (polynomial-time) algorithm for factoring large integers on a quantum computer. Thus, in theory, quantum computers can be used to break RSA. However, actually building a quantum computer is no easy task! Doing so requires the ability to control extremely small physical systems in extremely complicated ways. Physicists are working on this problem and have succeeding in building quantum computers capable of manipulating a very small number of qubits. But it is still uncertain whether we will eventually be able to build quantum computers that can quickly factor huge numbers, such as the ones currently used for RSA.

EXERCISES 9.5

Numerical Problems

1. Suppose $p = 41, q = 67$ for an RSA encryption scheme.

   a. Find n.

   b. Find $\varphi(n)$.

   c. Which of the following are valid encryption exponents: $7, 9, 15, 35, 49, 91$?

2. Suppose $p = 31, q = 97$ for an RSA encryption scheme.

   a. Find n.

   b. Find $\varphi(n)$.

   c. Which of the following are valid encryption exponents: $7, 9, 10, 77, 143$?

3. Suppose $p = 13, q = 29$ for an RSA encryption scheme. For the encryption exponent $e = 55$, what is the decryption exponent, d?

4. Suppose $p = 17, q = 23$ for an RSA encryption scheme. For the encryption exponent $e = 7$, what is the decryption exponent, d?

5. Using RSA encryption with $p = 43, q = 67$, and $e = 5$, encrypt the following message in two-letter blocks:

   DELIVER IT

6. Using RSA encryption with $p = 47, q = 61$, and $e = 7$, encrypt the following message in two-letter blocks:

   TOFU

7. Using RSA encryption with $p = 47, q = 61$, and $e = 7$, encrypt the following message in two-letter blocks:

   FAKE

8. The following ciphertext was encrypted in two-letter blocks using RSA with modulus $n = 4747$ and encryption exponent $e = 3067$.

$$1563 \quad 4362 \quad 2416 \quad 1730 \quad 3660 \quad 192$$

   a. Find the decryption exponent, d.

   b. Decrypt the message.

9. The following ciphertext was encrypted in two-letter blocks using RSA with modulus $n = 3127$ and encryption exponent $e = 1371$.

$$845 \quad 1065$$

   a. Find the decryption exponent, d.

   b. Decrypt the message.

10. **Computer Exercise** Consider the RSA encryption scheme described in Exercise 5.

   a. Find the decryption exponent, d.

   b. Decrypt the following message, which was encrypted using this system:

$$2755 \quad 920 \quad 623 \quad 28 \quad 410 \quad 2874$$

11. **Computer Exercise** Consider the RSA encryption scheme described in Exercise 7.

   a. Find the decryption exponent, d.

   b. Decrypt the following message, which was encrypted using this system:

$$1404 \quad 2553 \quad 1103 \quad 1408 \quad 92$$

12. **Computer Exercise** The following ciphertext was encrypted in a 15-letter block using RSA with modulus

$$n = 5{,}967{,}296{,}933{,}762{,}411{,}848{,}589{,}059{,}030{,}457$$

and encryption exponent $e = 449$.

Ciphertext: 2,292,807,516,720,144,034,331,713,382,356

   a. Find the decryption exponent, d.

   b. Decrypt the message.

Reasoning and Proofs

13. You intercept a ciphertext message, 600 1450 2023, that was encrypted using RSA with modulus $n = 3233$ and encryption exponent $e = 1783$.

    a. Decrypt the message.

    b. RSA is supposed to be an ultrasecure encryption method, yet you just decrypted a message for which you knew only the encryption key. How could the person setting up the RSA encryption system make it much harder to crack?

14. In our demonstration that decryption is the inverse of encryption in the RSA algorithm, to obtain congruence (4) we used the fact that $m^{\varphi(n)} \equiv 1 \pmod{n}$, which we justified using Euler's Theorem (9.3.1). Euler's Theorem requires that the base, m, must be relatively prime to the modulus, n. For a randomly chosen message $m < n$, find the probability that $\gcd(m, n) > 1$.

15. Note that the RSA algorithm involves exponential encryption with a composite modulus. In particular, this composite modulus is the product of two primes. Would this method of encryption be valid if the modulus were the product of three distinct primes? If so, explain what would change in the algorithm. If not, explain why the method would fail.

Chapter 10

Joseph Louis Lagrange
(1736–1813)

Often thought of as a Frenchman, mathematician Joseph Louis Lagrange was actually Italian. Born Giuseppe Lodovico Lagrangia, he did not move to France until after the age of 50. Although Lagrange's parents were both Italian, his great-grandfather was a French cavalry captain. Lagrange's great-grandfather moved to Turin while it was part of Savoy, and he remained there after it became part of Sardinia-Piedmont, which is now known as Italy. Lagrange identified with his French ancestors, and even as a child he chose to sign his name Lagrange rather than Lagrangia. His father held the prestigious and well-paid position of Treasurer of the Office of Public Works and Fortifications. However, the family did not have much money, because Lagrange's father had lost a lot of money in bad investments.

Lagrange became interested in mathematics and physics as a teenager, writing his first mathematics paper at the age of 18. In this paper, he generalized the product rule for derivatives by finding a formula for the nth derivative of the product of two functions. However, Lagrange later discovered that his result had already appeared in a letter from Leibniz to Johann Bernoulli almost 50 years earlier. Terribly afraid that he would be accused of plagiarism, Lagrange worked

intensively to redeem himself by coming up with some truly new results. When he was 19, Lagrange sent a letter to Leonhard Euler[1] explaining his current research (the foundations of the field of the calculus of variations). Euler wrote back, "I cannot admire you as much as you deserve!" As a result of this impressive work, young Lagrange was hired as a professor of geometry at the Royal Artillery School in Turin. Within a year, he was offered a prestigious chair in mathematics at the Berlin Academy, where Euler was director of mathematics. Known for his shyness and modesty, Lagrange turned down the offer, professing that he had nothing to offer the academy beyond what Euler could already offer.

While in Turin, Lagrange wrote papers on the calculus of variations, probability theory, and the foundations of dynamics and fluid mechanics. He also completed important research on the propagation of sound and the theory of vibrating strings. Lagrange impressively won prizes five times in the competitions on applications of mathematics to astronomy that were sponsored by the French Academy of Sciences.

---

[1] See Chapter 13 for a historical summary of Euler's life and work.

[2] Lagrange did not drink alcohol and was a vegetarian.

In 1766, Euler left Berlin to accept a position at St. Petersburg. The king of Prussia, Frederick the Great, appointed Lagrange to replace Euler (who was blind in one eye) as director of Mathematics at the Berlin Academy. Frederick the Great had originally offered the position to d'Alembert, who turned it down, explaining that Lagrange would be "much more useful to the academy than I could be. . . . Mr. de la Grange is young, and I am almost old; his ardor is rising, mine is on the decline; he is getting up, and I am going to bed." Frederick the Great, who had found Euler difficult, responded by writing, "I am thankful to you for having replaced a half-blind mathematician by a mathematician with both eyes, which will especially please the anatomical members of my academy." Lagrange was well paid and had no teaching responsibilities at the Berlin Academy, which allowed him to concentrate entirely on his research.

Lagrange married shortly after moving to Berlin. In a letter, he wrote, "My wife, who is one of my cousins, and who even lived for a long time with my family, is a very good housewife and has no pretensions at all." They had no children, and after 16 years of marriage, his wife died, leaving Lagrange quite depressed. In 1787, he left Berlin to accept a research position at the French Academy of Sciences in Paris. However, for several years he was overcome by depression and produced little mathematics.

As a result of the French Revolution, the Academy of Sciences was abolished in 1793 because it was seen as a symbol of the aristocracy. Shortly afterward, the revolutionaries called for the arrest and confiscation of the property of anyone who had been born in enemy territory. This law applied to Lagrange, but he was spared because the chemists Lavoisier and Morveau intervened on his behalf. Ironically, Lavoisier was himself later arrested and executed by guillotine for insisting that the Academy of Sciences should be saved.

The Academy of Sciences was replaced by the École Normale Supérieure, which trained citizens to be teachers, and the École Polytechnique, a school of engineering. Lagrange was first asked to give lectures at the École Normale and was later invited to give lectures on analysis at the École Polytechnique. When he began teaching at the École Normale Supérieure, it had been 40 years since he had last taught a class. He was not very interested in teaching, and he attained a reputation at the École Polytechnique for teaching at too high a level. Nonetheless, the published lecture notes for several of his classes became classic texts, including a book on college algebra, two books on calculus, and the first book on the theory of functions of a real variable. In fact, his book on college algebra was translated into English and became a widely used text in America entitled *Lectures on Elementary Mathematics*. While at the École Polytechnique, Lagrange was a member of the committee to standardize weights and measures, where he helped to develop the metric system.

Lagrange and Euler are considered by many to be the greatest mathematicians of the 18th century. In addition to the outstanding theorems Lagrange proved, the unusual clarity, elegance, and beauty of his writing style were impressive. In recognition of Lagrange's contributions to astronomy and mathematics, Napoleon appointed him a senator in 1799 and later conferred on him the title of Count of the Empire. On his death, Lagrange had the high honor of being buried in the Panthéon in Paris.

A bit of Lagrange's math

Lagrange's contributions spanned many fields, including analysis, number theory, and classical and celestial mechanics. Most people consider his most important contribution to be the foundation of the calculus of variations, a generalization of calculus that deals with functions of functions (as opposed to ordinary calculus, which deals with functions of numbers).

In 1958, France produced this postage stamp in honor of Lagrange.

In his work on the development of calculus, Lagrange introduced the notation

$$f'(x), f''(x), f'''(x), \ldots, f^n(x)$$

for derivatives and introduced the term *derived function*, which later turned into the word *derivative*. The mean value theorem was another of his contributions to calculus. Lagrange also developed the theory of differential equations and wrote a comprehensive four-volume set on the mathematics of mechanics.

Lagrange's work on the theory of equations was also influential. Lagrange developed a method to approximate roots of polynomials with continued fractions, and he related the solvability of equations to permutations of the roots of the equations. He attempted to determine whether there is an algebraic formula to solve polynomial equations of the fifth degree. His work in this area laid the foundation not only for the proof that such an algebraic formula is impossible but also for the general theory of groups.

Lagrange also proved many outstanding theorems in number theory. He proved that a natural number p is prime if and only if $(p - 1)! \equiv -1 \pmod{p}$, which was originally stated without proof by John Wilson. Although Lagrange supplied the proof, this theorem has come to be known as Wilson's Theorem (8.3.7). Lagrange had various results on representing primes in specific forms, and he developed a method to find all integer solutions to the quadratic Diophantine equation $X^2 - aY^2 = 1$. (When a is not a perfect square, this is known as *Pell's equation*, the subject of Section 15.1.) One of Lagrange's most widely known results in number theory is that every positive integer can be expressed as the sum of at most four perfect squares. This important result has come to be known as Lagrange's Four Square Theorem. Another of Lagrange's major contributions to number theory was the first proof of the Primitive Root Theorem, which we will study in Section 10.3.

Chapter 10 Primitive Roots

10.1 The Order of an Element of Z_n

Lagrange's day in court

In addition to proving important results in number theory, Joseph Louis Lagrange, a mathematician during the French Revolution, made major contributions to applied mathematics and was one of the founders of the metric system. What's not so well known about Lagrange is that he was the kingpin of the Italian math mafia. He made his money by smuggling theorems across the border and selling them on the streets of Paris. Lagrange trafficked primarily in pure math, which was irresistible to France's growing multitude of math addicts. In the criminal underworld of Paris, everyone feared Lagrange—nobody dared to cross Lemma Lips Louie.

Government agents had been trying to nail Lagrange for years, but somehow he always slipped through their grasp. When they finally brought him into court, Lagrange was indicted on four felony counts: deriving under the influence, obstruction of logic, additive identity theft, and conspiracy to solve equations in the fifth degree. The base gangster raised funds from the shadiest elements of his crime ring to take his case to a high-powered defense lawyer. This slippery attorney, "Mendacious Manny" Morveau, negotiated a settlement in which the government agreed to drop the felony charges. In exchange, Lagrange pled guilty to the single misdemeanor charge of failure to restate a hypothesis, which carried a maximum sentence of 24 hours in jail.

As Lagrange waited in the courtroom to be sentenced on that Sunday afternoon, he was pleased that he would be released on a Monday. Lagrange was looking forward to celebrating his release by watching *Monday Night Football* with his friend Lavoisier.

The judge banged her gavel and declared, "Joseph Louis Lagrange, I hereby sentence you to one night in prison."

"Awesome!" blurted Lagrange. "I'll be released on Monday night. I'm ready for some football!"

Just as the judge was about to bang her gavel again to close the session, a tinny melody recognizable as the *Marseillaise* was heard throughout the courtroom. Lagrange reached for his pocket. "Hold on a minute, Your Honor—that's my cell phone. Do you mind if I put this in speakerphone mode?"

Everyone in the courtroom heard a staticky voice from Lagrange's cell phone. "Hi, Lagrange. It's me, Lavoisier. Hope I'm not interrupting anything important. But I can't plan your release party all by myself! I mean, who should we invite to watch the game? What food should we serve? Help me out here!"

Lagrange spoke loudly into the phone. "Don't lose your head, Lavoisier. Just pick up a couple bags of those Camembert nachos and a few six-packs of champagne. And let's invite all my rowdy friends—Euler, Bernoulli, Legendre. . . . Oh, and that clean-cut young Leblanc fellow. He's quite charming—"

"Order in the court!" roared the judge. "Bailiff, confiscate the defendant's cell phone at once! Lagrange, for your lack of respect, I triple your sentence to three days in jail!"

"What! Three days? That's awful," whined Lagrange. "Today is Sunday, so a 3-day sentence means I'll be released on Wednesday. We can't watch *Monday Night Football* on a Wednesday!"

"Silence!" ordered the judge. "For your continued insolence, I triple your sentence again. Nine days."

With a 9-day sentence, Lagrange would be released on a Tuesday. This did not please him at all, because he wanted to be released on a Monday. Lagrange wondered whether any further triplings of his sentence would result in a Monday release. He had already determined that a sentence of 3 or 9 days would not result in a Monday release, but what about a sentence of 27 days, or some higher power of 3?

Lagrange reasoned that the day of his release was determined by the number of days in his sentence modulo 7, so he grabbed a legal pad and jotted down the powers of 3 reduced modulo 7:

LÉGAL

$3^0 = 1 \equiv 1 \pmod 7 \rightarrow$ MONDAY RELEASE $\rightarrow$

$3^1 = 3 \equiv 3 \pmod 7 \rightarrow$ WEDNESDAY RELEASE

$3^2 = 9 \equiv 2 \pmod 7 \rightarrow$ TUESDAY RELEASE

$3^3 = 27 \equiv 6 \pmod 7 \rightarrow$ SATURDAY RELEASE

$3^4 = 81 \equiv 4 \pmod 7 \rightarrow$ THURSDAY RELEASE

$3^5 = 243 \equiv 5 \pmod 7 \rightarrow$ FRIDAY RELEASE

$3^6 = 729 \equiv 1 \pmod 7 \rightarrow$ MONDAY RELEASE $\rightarrow$

$\checkmark$ CHIPS, DIP, CHAMPAGNE

Looking at his list, Lagrange realized the answer. His original sentence of 1 day would need to be tripled a total of **6** times in order for Lagrange to be released on a Monday. Since the judge had already tripled his sentence two times, he just needed to get her to triple it four more times.

Lagrange cleared his throat. "Your Honor, I would like to convince you to triple my sentence four more times. I'm ready with four rude and witty insults that I could blurt out to make fun of the court. But I'd like to save us all the embarrassment of that and simply request that you triple my sentence four times."

The judge stood up, incensed. "How dare you tell me how to do my job! You are the most presumptuous, repugnant, arrogant, and disrespectful defendant I have ever had in my courtroom. For your outrageous behavior, I triple your sentence four more times. You must serve 729 days in jail. That's almost 2 years in prison—I hope it teaches you a lesson!" The judge banged her gavel and said angrily, "Case closed!"

Lagrange jumped for joy. "Woo-hoo! Out on a Monday. Thanks, Judge!"

The order of an integer modulo n

In our myth, Lagrange needed his initial 1-day sentence to be tripled a total of **6** times to achieve his highly coveted Monday release. This is because

$$3^6 \equiv 1 \;(\mathrm{mod}\;7),$$

and

no smaller positive power of 3 gives 1 modulo 7.

We express this by saying that **6** is the *order* of 3 modulo 7. We may also say that **6** is the order of the element $\bar{3}$ in $\mathbf{Z}_7$.

*Let n be an integer, and let $\bar{a} \in \mathbf{Z}_n$ be a reduced residue. The **order** of $\bar{a}$ is the smallest positive integer r such that $\bar{a}^r = \bar{1}$. We will use the notation $\mathrm{ord}(\bar{a})$ to denote the order of an element $\bar{a} \in \mathbf{Z}_n$.*

EXAMPLE 1

Determine the order of $\bar{2}$ in $\mathbf{Z}_7$.

SOLUTION We have

$$\bar{2}^1 = \bar{2}$$
$$\bar{2}^2 = \bar{4}$$
$$\bar{2}^3 = \bar{1}.$$

Since $\bar{2}^3$ is the smallest positive power of $\bar{2}$ that equals $\bar{1}$, we conclude that $\mathrm{ord}(\bar{2}) = \mathbf{3}$.

Definition 10.1.1 concerns the order of a reduced residue in $\mathbf{Z}_n$. We may also refer to the order of an *integer* modulo n. For instance, in Example 1 we found that in $\mathbf{Z}_7$, the order of $\bar{2}$ is **3**. We may also state this by saying that the order of 2 modulo 7 is **3**. In general, the definition of the order of an integer modulo n can be stated as follows.

DEFINITION 10.1.2

*Suppose $n \in \mathbf{N}$, and $a \in \mathbf{Z}$ such that $\gcd(a, n) = 1$. Then the **order** of a modulo n is the smallest positive integer r such that $a^r \equiv 1 \;(\mathrm{mod}\;n)$. We use the notation $\mathrm{ord}_n(a)$ to denote the order of a modulo n.*

EXAMPLE 2

Find the order of 8 modulo 15.

SOLUTION We wish to find the smallest power of 8 that is congruent to 1 modulo 15.

$$8^1 \equiv 8 \pmod{15}$$
$$8^2 \equiv 4 \pmod{15}$$
$$8^3 \equiv 2 \pmod{15}$$
$$8^4 \equiv 1 \pmod{15}.$$

Thus, $\mathrm{ord}_{15}(8) = 4$. ∎

The examples we have seen illustrate that the order of a reduced residue in $\mathbf{Z}_n$ can be found by taking powers of that element until we obtain an answer of $\bar{1}$. But how can we be sure that we will eventually obtain an answer of $\bar{1}$? Is it possible that we could compute higher and higher powers of a reduced residue, $\bar{a}$, and never get $\bar{1}$ as an answer? Fortunately, Euler's Theorem (9.3.1) helps us here by telling us that $\bar{a}^{\varphi(n)} = \bar{1}$. Hence, there is indeed some positive power of $\bar{a}$ that equals $\bar{1}$. The Well-Ordering Principle (2.2.2) then guarantees that there is a smallest positive integer r for which $\bar{a}^r = \bar{1}$. This shows us that for any reduced residue $\bar{a}$ in $\mathbf{Z}_n$, the order of $\bar{a}$ is defined.

In Example 2, we found the order of 8 modulo 15 by computing the powers of 8 until we got an answer of 1. If we wanted to find $\mathrm{ord}_n(a)$ for a very large modulus n, this method could take an extremely long time. In fact, there is no known efficient method for finding $\mathrm{ord}_n(a)$ for large numbers n. Indeed, if there were an efficient method, then one could factor n quickly, which is believed to be impossible (see Exercises 26–31).

EXAMPLE 3

Find the order of each of the nonzero elements of $\mathbf{Z}_{11}$.

SOLUTION This is easy to do if we refer to the exponent table for $\mathbf{Z}_{11}$, which we completed in Chapter 9.

Exponents in $\mathbf{Z}_{11}$

	$\bar{1}$	$\bar{2}$	$\bar{3}$	$\bar{4}$	$\bar{5}$	$\bar{6}$	$\bar{7}$	$\bar{8}$	$\bar{9}$	$\overline{10}$
$\bar{a}^1$	1	2	3	4	5	6	7	8	9	10
$\bar{a}^2$	1	4	9	5	3	3	5	9	4	1
$\bar{a}^3$	1	8	5	9	4	7	2	6	3	10
$\bar{a}^4$	1	5	4	3	9	9	3	4	5	1
$\bar{a}^5$	1	10	1	1	1	10	10	10	1	10
$\bar{a}^6$	1	9	3	4	5	5	4	3	9	1
$\bar{a}^7$	1	7	9	5	3	8	6	2	4	10
$\bar{a}^8$	1	3	5	9	4	4	9	5	3	1
$\bar{a}^9$	1	6	4	3	9	2	8	7	5	10
$\bar{a}^{10}$	1	1	1	1	1	1	1	1	1	1
$\bar{a}^{11}$	1	2	3	4	5	6	7	8	9	10

We can determine the order of $\overline{4}$, for example, by looking down the $\overline{4}$ column until we find a $\overline{1}$. The first time a $\overline{1}$ appears in this column is in the $\overline{a}^5$ row. Thus, the order of $\overline{4}$ is equal to 5.

Using the table in this way, one sees that

$$\mathrm{ord}(\overline{1}) = 1,$$

$$\mathrm{ord}(\overline{10}) = 2,$$

$$\mathrm{ord}(\overline{3}) = \mathrm{ord}(\overline{4}) = \mathrm{ord}(\overline{5}) = \mathrm{ord}(\overline{9}) = 5,$$

$$\mathrm{ord}(\overline{2}) = \mathrm{ord}(\overline{6}) = \mathrm{ord}(\overline{7}) = \mathrm{ord}(\overline{8}) = 10.$$

In this example, we determined that in $\mathbf{Z}_{11}$, the order of the element $\overline{4}$ is 5. By definition, this means that $\overline{4}^5 = \overline{1}$, and no smaller positive power of $\overline{4}$ equals $\overline{1}$. Are there any *larger* powers of $\overline{4}$ that equal $\overline{1}$? Reading down the $\overline{4}$ column of the $\mathbf{Z}_{11}$ exponent table from the example, we see that the powers of $\overline{4}$ follow the repeating pattern

$$\overline{4}, \overline{5}, \overline{9}, \overline{3}, \overline{1}, \overline{4}, \overline{5}, \overline{9}, \overline{3}, \overline{1}, \overline{4}, \dots. \tag{1}$$

Thus, we observe that

$$\overline{4}^5 = \overline{1}, \ \overline{4}^{10} = \overline{1}, \ \overline{4}^{15} = \overline{1}, \ \overline{4}^{20} = \overline{1}, \dots$$

and that $\overline{4}$ raised to any other positive power does not equal $\overline{1}$. In other words,

$$\overline{4}^e = \overline{1} \text{ if and only if } 5 \mid e.$$

Furthermore, since the repeating pattern of the powers of the element $\overline{4}$ has period 5, we see that two powers $\overline{4}^j$ and $\overline{4}^k$ are equal if and only if the exponents j and k are congruent modulo 5. The following proposition and its corollary generalize these observations.

PROPOSITION 10.1.3

Let n be a natural number, and let $\overline{a} \in \mathbf{Z}_n$ be a reduced residue. Let r be the order of $\overline{a}$.

(*i*)   *For any integer e,*

$$\overline{a}^e = \overline{1} \text{ if and only if } r \mid e.$$

(*ii*) *For any integers j and k,*

$$\overline{a}^j = \overline{a}^k \text{ if and only if } j \equiv k \pmod{r}.$$

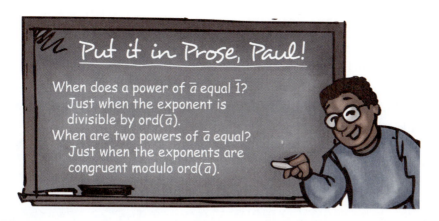

Put it in Prose, Paul!

When does a power of $\bar{a}$ equal $\bar{1}$?
Just when the exponent is divisible by $\text{ord}(\bar{a})$.
When are two powers of $\bar{a}$ equal?
Just when the exponents are congruent modulo $\text{ord}(\bar{a})$.

PROOF

Proof of (*i*) Let $n \in \mathbf{N}$, let $\bar{a} \in \mathbf{Z}_n$ be a reduced residue, and let $r = \text{ord}(\bar{a})$. Let $e \in \mathbf{Z}$.

[$\Leftarrow$] Suppose $r \mid e$. [**To show: $\bar{a}^e = \bar{1}$.**]

Since $r \mid e$, there exists $d \in \mathbf{Z}$ such that $e = rd$.
Hence,

$$\bar{a}^e = \bar{a}^{rd} = (\bar{a}^r)^d = \bar{1}^d = \bar{1},$$

as desired.

[$\Rightarrow$] Now suppose that $\bar{a}^e = \bar{1}$. [**To show: $r \mid e$.**]

We divide e by r: by the Division Theorem (3.5.1), there exist $q, s \in \mathbf{Z}$ such that

$$e = qr + s, \text{ where } 0 \le s < r. \tag{2}$$

We then have

$$\bar{a}^e = \bar{a}^{qr+s}$$
$$= \bar{a}^{qr}\bar{a}^s$$
$$= (\bar{a}^r)^q\bar{a}^s$$
$$= \bar{1}^q\bar{a}^s$$
$$= \bar{a}^s.$$

That is,

$$\bar{a}^e = \bar{a}^s.$$

But $\bar{a}^e = \bar{1}$, so we have $\bar{a}^s = 1$. However, since s is less than r, which is the order of $\bar{a}$, $\bar{a}^s = \bar{1}$ implies $s = 0$. Thus, equation (2) becomes $e = qr$. We conclude that $r \mid e$.

Proof of (*ii*) You will prove this part in Exercise 8.

Naomi's Numerical Proof Preview: Corollary 10.1.4

In Example 3, we found that every element of $\mathbf{Z}_{11}$ has order 1, 2, 5, or 10. These orders are all divisors of 10. How can we explain this pattern? Fermat's Little Theorem (9.1.1) tells us that $\overline{a}^{10} = \overline{1}$ for any nonzero element $\overline{a} \in \mathbf{Z}_{11}$. It now follows from Proposition 10.1.3 that the order of $\overline{a}$ must divide 10. This proves our observation that the order of every element of $\mathbf{Z}_{11}$ is a factor of 10.

For any prime p, a similar argument shows that in $\mathbf{Z}_p$, the order of every nonzero element must divide $p - 1$.

COROLLARY 10.1.4

Let p be prime, and let $\overline{a} \in \mathbf{Z}_p$ with $\overline{a} \neq \overline{0}$. Then $\mathrm{ord}(\overline{a}) \mid p - 1$.

PROOF By Fermat's Little Theorem, we have

$$\overline{a}^{p-1} = \overline{1}.$$

It follows from Proposition 10.1.3(i) that $\mathrm{ord}(\overline{a}) \mid p - 1$.

The following lemma gives a relation between the order of a reduced residue $\overline{a}$ in $\mathbf{Z}_n$ and the order of any power of $\overline{a}$.

LEMMA 10.1.5

Let $n \in \mathbf{N}$, and let $\overline{a}$ be a reduced residue in $\mathbf{Z}_n$. Then for any $j \in \mathbf{N}$,

$$\mathrm{ord}(\overline{a}^j) \mid \mathrm{ord}(\overline{a}).$$

PROOF Let s be the order of $\overline{a}$. Then $\overline{a}^s = \overline{1}$, so

$$(\overline{a}^j)^s = (\overline{a}^s)^j = \overline{1}^j = \overline{1}.$$

Thus,

$$(\overline{a}^j)^s = \overline{1}.$$

It follows directly from Proposition 10.1.3(i) that the order of $\overline{a}^j$ divides s. This is what we needed to show.

We can make a statement more precise than Lemma 10.1.5:

Let $n \in \mathbf{N}$, and let $\bar{a}$ be a reduced residue in $\mathbf{Z}_n$. Let $s = \mathrm{ord}(\bar{a})$. For any $j \in \mathbf{N}$,

$$\mathrm{ord}(\bar{a}^{\,j}) = \frac{s}{\gcd(j, s)}.$$

PROOF You will prove this in Exercise 13.

Perfect shuffles

To shuffle a deck of cards, one typically cuts the deck roughly in half and then interweaves the cards from the two halves. Some very skilled magicians can perform a *perfect riffle shuffle*, which involves

1. cutting the deck into two exactly equal halves, and
2. interweaving the cards in a perfectly alternating manner.

When the shuffler begins to interweave the two piles, he or she has the choice of which pile to start with. For our discussion of shuffling, we will always make this choice so that the card originally on the top of the deck stays on the top, and the card originally on the bottom of the deck stays on the bottom. (To explore what happens in the other case, see Exercises 16–19.) For example, if the deck consists of 10 cards labeled $0, 1, 2, \ldots, 9$ in order, performing a perfect riffle shuffle has the following result:

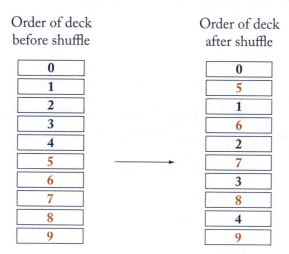

Figure 1 A perfect riffle shuffle of a 10-card deck.

Let's examine where each card is after the shuffle is completed. Card 0 stayed in position **0**, card 1 moved to position **2**, card 2 moved to position **4**, and so forth. Let's make a table.

Original position of card	New position of card
0	0
1	2
2	4
3	6
4	8
5	1
6	3
7	5
8	7
9	9

Observe that the new position is just the original position multiplied by 2 modulo 9. Thus, the card originally in position j ends up in a position congruent to $2j$ (mod 9).

What happens when we do this procedure repeatedly? Performing several perfect riffle shuffles in a row will return the deck to its original order—an ideal situation for a magic trick! But how many shuffles will this take?

Since one perfect riffle shuffle multiplies the position by 2 modulo 9, performing r such shuffles multiplies the position by 2^r (mod 9). Thus, a card originally in position j ends up in a position congruent to $2^r j$ (mod 9) after r perfect riffle shuffles.

Now it is easy to see when the deck returns to its original order. This happens when

$$2^r \equiv 1 \ (\text{mod } 9).$$

We want to know the smallest positive integer r that makes this congruence true; that is, we want the order of 2 modulo 9. Since

$$\text{ord}_9(2) = 6,$$

we conclude that after 6 perfect riffle shuffles, our 10-card deck will return to its original order.

It is not hard to generalize to decks of arbitrary size. To cut a deck evenly, the number of cards must be even. Let $2n$ denote the number of cards in a deck. Reasoning as before, we see that a single perfect riffle shuffle multiplies the position of each card by 2 modulo $2n - 1$. Hence, we have the following result:

$$\textit{Number of shuffles necessary to return a} \atop \textit{deck with 2n cards to its original order} = \text{ord}_{2n-1}(2).$$

For a regular deck of 52 playing cards, the number of perfect shuffles necessary is

$$\text{ord}_{51}(2) = 8,$$

because $2^8 = 256$ is the smallest positive power of 2 that is congruent to 1 (mod 51). Very few people in the world are capable of returning a deck of cards to its original order by performing 8 perfect shuffles in a row. One of them is Stanford mathematician Persi Diaconis. (The authors have witnessed him performing this amazing feat!)

EXERCISES 10.1

Numerical Problems

1. What does Corollary 10.1.4 tell us about the order of any element of $\mathbf{Z}_{29}$?

2. Working in $\mathbf{Z}_{10}$, find each of the following orders.

   a. $\text{ord}(\bar{1})$ **b.** $\text{ord}(\bar{7})$ **c.** $\text{ord}(\bar{9})$

3. Find each of the following orders.

   a. $\text{ord}_{15}(4)$ **b.** $\text{ord}_{15}(7)$ **c.** $\text{ord}_{20}(7)$

4. How many perfect shuffles are required to return each of these decks to its original order?

   a. A deck of 10 cards **b.** A deck of 20 cards **c.** A deck of 30 cards

Reasoning and Proofs

5. Let $n \in \mathbf{N}$. Prove that in $\mathbf{Z}_n$, the order of any reduced residue divides $\varphi(n)$.

6. Let $n \in \mathbf{N}$. Prove that in $\mathbf{Z}_n$, the order of any reduced residue equals the order of its multiplicative inverse.

7. Suppose you have an element $\bar{a} \in \mathbf{Z}_n$ such that powers of $\bar{a}$ give all of the reduced residues in $\mathbf{Z}_n$. Determine the order of $\bar{a}$. Explain your answer.

8. Prove Proposition 10.1.3 part (ii). (You may use part (i) of the proposition in your proof.)

9. Poker chips can be shuffled similarly to cards, but shuffling chips is easier to master. With a few days' practice, a novice can learn to shuffle poker chips, performing perfect shuffles with small stacks of chips. If one begins with a stack consisting of n red chips atop n blue chips, then after one shuffle, the chips alternate red and blue.

   a. Starting with a stack of 4 red chips on top of 4 blue chips, how many perfect shuffles are required before the stacks will again be one stack of 4 red chips atop 4 blue chips?

   b. Starting with a stack of 8 red chips on top of 8 blue chips, how many perfect shuffles are required before the stacks will again be one stack of 8 red chips atop 8 blue chips?

10. Let $d, n \in \mathbf{N}$ such that $d \mid n$, and let $a \in \mathbf{Z}$ such that $\gcd(a, n) = 1$. Prove that $\text{ord}_d(a) \mid \text{ord}_n(a)$.

11. Let $n \in \mathbf{N}$, and let $\bar{a} \in \mathbf{Z}_n$ be a reduced residue. Let $r = \text{ord}(\bar{a})$. Prove that if $r = st$ for positive integers s and t, then $\text{ord}(\bar{a}^t) = s$.

12. Let $n \in \mathbf{N}$, and let $\bar{a} \in \mathbf{Z}_n$ be a reduced residue. Let $r = \text{ord}(\bar{a})$. Prove that if a natural number j is relatively prime to r, then $\text{ord}(\bar{a}^j) = r$.

13. Prove Lemma 10.1.6.

14. For this exercise, use the result of Exercise 11 and the fact that $\text{ord}_{35}(3) = 12$.

    a. Find $\text{ord}_{35}(9)$.

    b. Find $\text{ord}_{35}(27)$.

15. For this exercise, use the result of Exercises 11 and 12 and the fact that in $\mathbf{Z}_{37}$, $\text{ord}(\bar{2}) = 36$.

    a. Find $\text{ord}(\bar{8})$ in $\mathbf{Z}_{37}$.

    b. Find $\text{ord}(\overline{32})$ in $\mathbf{Z}_{37}$.

EXPLORATION Inside Shuffles (Exercises 16–19)

In our discussion of perfect shuffles, we assumed that each shuffle is performed in such a way that the card originally on the top of the deck stays on the top, and the card originally on the bottom of the deck stays on the bottom. This is called a perfect *outside shuffle*, because the two cards on the outside of the deck stay on the outside of the deck. It is also possible to perform a perfect *inside shuffle*, in which the card originally on the bottom of the deck becomes the card next to the bottom card, and the card originally on top of the deck becomes the card next to the top card. Here is what a perfect inside shuffle looks like for a deck of 8 cards, labeled 1–8.

Order of deck before shuffle	Order of deck after shuffle
1	5
2	1
3	6
4	2
5	7
6	3
7	8
8	4

16. Compare the inside shuffle of an 8-card deck illustrated above to the outside shuffle of a 10-card deck that is illustrated in Figure 1. Fill in the blanks:

> An inside shuffle of an 8-card deck is identical to an outside shuffle of a 10-card deck in which the _____ card and _____ card are invisible.

We conclude:

> The number of perfect *inside shuffles* required to return an 8-card deck to its original order is equal to the number of perfect *outside shuffles* required to return a 10-card deck to its original order.

17. Fill in the blanks:

    a. The number of perfect *inside shuffles* required to return a deck with $2n$ cards to its original order is equal to the number of perfect *outside shuffles* required to return a deck with _____ cards to its original order.

    b. Complete the formula:

$$\begin{matrix} \textit{Number of inside shuffles necessary to return} \\ \textit{a deck with 2n cards to its original order} \end{matrix} = \underline{\hspace{2cm}}.$$

18. Use your observation from Exercise 17 to determine the following:

    a. How many perfect inside shuffles are required to return a 50-card deck to its original order?

    b. How many perfect inside shuffles are required to return a 52-card deck of playing cards to its original order?

    c. For an amateur magician who wants to learn the trick of returning a deck to its original position using perfect riffle shuffles, which would you recommend: outside shuffles or inside shuffles? Explain.

19. After 26 perfect inside shuffles, in what order will the cards of a 52-card deck be arranged?

Conjecturer's Corner (Exercises 20–24)

20. Find $\text{ord}_{13}(8)$, $\text{ord}_{13}(3)$, and $\text{ord}_{13}(8 \cdot 3)$.

21. Find $\text{ord}_{13}(12)$, $\text{ord}_{13}(8)$, and $\text{ord}_{13}(12 \cdot 8)$.

22. Find $\text{ord}_{13}(4)$, $\text{ord}_{13}(5)$, and $\text{ord}_{13}(4 \cdot 5)$.

23. Let $n \in \mathbf{N}$, and let a, b be integers that are relatively prime to n. From the preceding exercises, you might conjecture that $\text{ord}_n(ab) = \text{lcm}(\text{ord}_n(a), \text{ord}_n(b))$. However, this is not always true! Find a counterexample for $n = 13$.

24. Let $n \in \mathbf{N}$, and let a, b be integers that are relatively prime to n. Prove that $\text{ord}_n(ab) \mid \text{lcm}(\text{ord}_n(a), \text{ord}_n(b))$.

Advanced Reasoning and Proofs

EXPLORATION Computational Difficulty of Finding the Order of an Integer (Exercises 25–30)

This exploration helps you to examine an interesting computational fact: finding the order of integers modulo n is at least as difficult computationally as factoring n. That is, if you had a fast way to compute the order of any number modulo n, you would be able to factor n quickly.[3] Thus, it is believed that there is no efficient algorithm for computing $\operatorname{ord}_n(a)$ for large values of n.

Consider the following proposition:

> PROPOSITION. *Let n be a natural number, and suppose that x and y are integers such that $x^2 \equiv y^2 \pmod{n}$, but $x \not\equiv y \pmod{n}$ and $x \not\equiv -y \pmod{n}$. Then $d = \gcd(x - y, n)$ and $e = \gcd(x + y, n)$ are both nontrivial factors of n.*

Thus, if you can find x and y that satisfy the conditions of this proposition, then you can quickly find a nontrivial factor of n. (A *nontrivial factor* of n is a factor of n that is not ± 1 or $\pm n$.) In this exploration, you will discover how to find such x and y provided that you can quickly find the order of any number modulo n.

25. Prove the Proposition.

26. Suppose you wish to factor $n = 713$. Since we are assuming that you can quickly find the order of any number modulo n, you can quickly find that 3 has order 330 modulo n. Now $330 = 2 \cdot 165$ is even, and you compute

$$3^{165} \equiv 185 \pmod{n}.$$

Let $x = 185$. Explain how you know that $x^2 \equiv 1 \pmod{n}$. Use the Proposition to find a factor of n. (Make sure you use the Euclidean Algorithm to evaluate gcd's.)

27. Suppose a is an integer such that $\gcd(a, n) = 1$, and let $r = \operatorname{ord}_n(a)$. Assume that r is even, and let $x = a^{r/2}$. Show that $x^2 \equiv 1 \pmod{n}$, but $x \not\equiv 1 \pmod{n}$.

28. With notation as in the previous problem, make the additional assumption that $x \not\equiv -1 \pmod{n}$. Explain how the results of Exercises 26–27 can be used to find a factor of n.

29. In summary, assume that we evaluate $r = \operatorname{ord}_n(a)$, and two things are true:

    (*i*) $r = \operatorname{ord}_n(a)$ is even, and

    (*ii*) $a^{r/2} \not\equiv -1 \pmod{n}$.

---

[3] In fact, this observation is essential to Shor's algorithm for factoring on a quantum computer, which was mentioned in Section 9.5.

(Conditions (i) and (ii) are computationally easy to check, and it turns out that these conditions hold frequently, for at least $\frac{1}{2}$ of the possible choices of a. Hence, if we pick a few values of a, we will soon find one for which (i) and (ii) hold.)

Assume that both conditions (i) and (ii) hold. Explain how we can use ideas from the preceding exercises to factor n. Include an explanation of why this method can be done quickly on a computer.

30. **Computer Exercise** Use your method from the previous exercises to find a factor of $n = 1{,}472{,}467{,}993$. Use a computer program such as Mathematica or Maple when you need to find orders of elements modulo n. (But, of course, don't use Mathematica's or Maple's factoring function!)

31. Recall that a Mersenne prime is a prime of the form $2^p - 1$, where p is prime. In this exercise, you will prove the following theorem, which puts a restriction on the factors of a number of this form.

    THEOREM. Let p be a prime. Then any prime factor r of $2^p - 1$ satisfies $r \equiv 1 \pmod{p}$.

    a. Show that the theorem holds for $p = 2$.

    b. Now assume that the prime satisfies $p > 2$. Show that $\mathrm{ord}_r(2) = p$.

    c. Use part b to finish the proof of the Theorem.

10.2 Solving Polynomial Equations in Z_n

Your high school days

When you were in high school algebra, you were probably asked to solve many *polynomial* equations, such as

$$3x^2 + 2x + 4 = 0.$$

A **polynomial** $P(x)$ in the variable x is an expression of the form

$$P(x) = a_d x^d + a_{d-1} x^{d-1} + \cdots + a_1 x + a_0.$$

The numbers $a_d, \ldots, a_0$ are called the **coefficients** of the polynomial. If all of the coefficients of $P(x)$ are 0, then $P(x)$ is called the **zero polynomial**. For every nonzero polynomial, we may assume that $a_d \neq 0$. The integer d is called the **degree** of the polynomial $P(x)$.

A number r is called a **root** of a polynomial $P(x)$ if $P(r) = 0$.

Finding the solution of a linear equation (an equation of degree 1) was one of the first skills you learned in high school algebra. Soon after, you learned the quadratic formula, which enables you to solve any equation of degree 2. The quadratic formula expresses the roots of any quadratic polynomial in terms of the coefficients of the polynomial, using the four basic arithmetic operations and square roots. While the ancient Greeks could find positive solutions to quadratic equations using geometric methods, the quadratic formula itself was not discovered until 1100, by a Hindu mathematician named Baskhara. Formulas to solve equations of degree 3 and 4 were discovered in Italy in the 16th century. These *cubic* and *quartic* formulas, which involve the four basic operations, square roots, and cube roots, are more cumbersome than the quadratic formula (which is why you probably were not asked to learn them in high school).

For polynomial equations of degree 5 or more, the situation is even worse. Many mathematicians searched unsuccessfully for a *quintic* formula for solving polynomial equations of degree 5. In fact, it turns out that there is no quintic formula! More precisely, there is no formula that expresses the roots of a general 5th degree (or higher degree) equation using the four basic operations, square roots, cube roots, and higher roots. This surprising fact was proved by Paolo Ruffini and Niels Abel at the end of the 18th century and in the early 19th century. In 1832, Evariste Galois characterized precisely which polynomial equations can be solved using the four basic operations and nth roots.

Polynomials over $\mathbf{Z}_n$

In high school algebra, you spent a good amount of time solving *polynomials over* $\mathbf{R}$, that is, polynomials whose coefficients are real numbers. The solutions to such polynomials turn out to be complex numbers such as $2 + \sqrt{2}\,i$, and if we're lucky, the solutions are real numbers. In short, all of the numbers involved—coefficients and solutions—are real or complex numbers.

Now that we've learned about the number system $\mathbf{Z}_n$, we can also consider polynomials whose coefficients belong to this new number system. For example, let's work in $\mathbf{Z}_5$ and consider the equation

$$\overline{3}\,\overline{x}^2 + \overline{2}\,\overline{x} + \overline{4} = \overline{0}. \tag{1}$$

The expression on the left side of equation (1), $P(\overline{x}) = \overline{3}\,\overline{x}^2 + \overline{2}\,\overline{x} + \overline{4}$, is a polynomial of degree 2 whose coefficients are elements of the number system $\mathbf{Z}_5$. In this polynomial, the variable $\overline{x}$ may represent any element of the number system $\mathbf{Z}_5$. We refer to such a polynomial as a **polynomial over $\mathbf{Z}_5$**.

To see which elements of $\mathbf{Z}_5$ satisfy equation (1), we may try substituting each of the elements $\overline{0}, \overline{1}, \overline{2}, \overline{3}$, and $\overline{4}$ into the equation. For example, when $\overline{x} = \overline{2}$, we find that $P(\bar{2}) = \bar{3} \cdot \bar{2}^2 + \bar{2} \cdot \bar{2} + \bar{4} = \bar{0}$. If we calculate $P(\bar{x})$ for all $\bar{x}$ in $\mathbf{Z}_5$, we get the following results:

x	$\bar{0}$	$\bar{1}$	$\bar{2}$	$\bar{3}$	$\bar{4}$
$P(\bar{x}) = \bar{3}\bar{x}^2 + \bar{2}\bar{x} + \bar{4}$	$\bar{4}$	$\bar{4}$	$\bar{0}$	$\bar{2}$	$\bar{0}$

Thus, the equation $P(\bar{x}) = \bar{0}$ has two solutions, $\bar{x} = \bar{2}$ and $\bar{x} = \bar{4}$. (It turns out that these two solutions, which we found using trial and error, can also be found using a version of the quadratic formula. See Exercises 9–15 for details.)

EXAMPLE 1

a. Working in $\mathbf{Z}_7$, find all solutions of the polynomial equation $\bar{x}^3 + \bar{2}\bar{x} + \bar{2} = \bar{0}$.

b. Working in $\mathbf{Z}_8$, find all solutions of the polynomial equation $\bar{x}^2 - \bar{1} = \bar{0}$.

SOLUTION

a. Letting $P(\bar{x}) = \bar{x}^3 + \bar{2}\bar{x} + \bar{2}$, in $\mathbf{Z}_7$ we find the following:

x	$\bar{0}$	$\bar{1}$	$\bar{2}$	$\bar{3}$	$\bar{4}$	$\bar{5}$	$\bar{6}$
$P(\bar{x}) = \bar{x}^3 + \bar{2}\bar{x} + \bar{2}$	$\bar{2}$	$\bar{5}$	$\bar{0}$	$\bar{0}$	$\bar{4}$	$\bar{4}$	$\bar{6}$

Thus, the polynomial $P(\bar{x})$ has two roots, $\bar{2}$ and $\bar{3}$.

b. Letting $P(\bar{x}) = \bar{x}^2 - \bar{1}$, in $\mathbf{Z}_8$ we find the following:

x	$\bar{0}$	$\bar{1}$	$\bar{2}$	$\bar{3}$	$\bar{4}$	$\bar{5}$	$\bar{6}$	$\bar{7}$
$P(\bar{x}) = \bar{x}^2 - \bar{1}$	$\bar{7}$	$\bar{0}$	$\bar{3}$	$\bar{0}$	$\bar{7}$	$\bar{0}$	$\bar{3}$	$\bar{0}$

Hence, $P(\bar{x})$ has four roots: $\bar{1}, \bar{3}, \bar{5}$, and $\bar{7}$. ●

How many roots can a polynomial have?

In high school, you learned that linear equations have at most 1 solution, quadratic equations have at most 2 solutions, cubic equations have at most 3 solutions, and so forth. In the real and complex numbers, a polynomial of degree d has at most d roots. This is a very useful and fundamental fact about solving polynomial equations. Thus, while it may be difficult to find the solutions in $\mathbf{C}$ to the equation $x^3 + 3x + 1 = 0$, this principle guarantees that there will be at most 3 solutions.

Does the same principle hold for polynomials over $\mathbf{Z}_n$? Not always, as illustrated by Example 1, part b. In that example, we saw that the quadratic equation

$$\bar{x}^2 - \bar{1} = \bar{0} \text{ in } \mathbf{Z}_8 \tag{2}$$

has a total of 4 solutions, instead of the expected 2 or fewer. It is even possible for linear equations over $\mathbf{Z}_n$ to have more than 1 solution. For example, the linear equation

$$\bar{5}\bar{x} = \bar{0} \text{ in } \mathbf{Z}_{10} \tag{3}$$

has a whopping 5 solutions, namely $\bar{0}, \bar{2}, \bar{4}, \bar{6},$ and $\bar{8}$. Thus, when working with $\mathbf{Z}_n$, one is forced to live with the somewhat disturbing fact that some polynomials have "too many roots"; in other words, the number of roots may exceed the degree of the polynomial.

The reader may have noticed that in the troubling equations (2) and (3), we were working in $\mathbf{Z}_n$ where the modulus n is *composite*. Perhaps the situation will be better if we assume we are working in $\mathbf{Z}_p$ where p is *prime*. Fortunately, this is the case. That is, in $\mathbf{Z}_p$, a polynomial $P(\bar{x})$ whose degree is d can have at most d roots. This comforting fact will take some work to prove, but our efforts will be well rewarded, as we will see in the coming sections.

THEOREM 10.2.1

Let p be prime, and let $f(\bar{x})$ be a nonzero polynomial over $\mathbf{Z}_p$. Let d be the degree of $f(\bar{x})$. Then $f(\bar{x})$ has at most d roots in $\mathbf{Z}_p$.

PROOF We will prove the theorem by strong induction on the degree, d, of the polynomial.

Let d be a nonnegative integer, and suppose that the theorem is true for polynomials of degree k whenever $k < d$. **[To show: The theorem holds for polynomials of degree d.]**

First, observe that if $d = 0$, the polynomial $f(\bar{x})$ is constant: $f(\bar{x}) = \bar{c}$, for some $\bar{c} \neq \bar{0}$. Thus, $f(\bar{x})$ has no roots, and the theorem holds in this case.

Hence, we may assume that $d > 0$. Let $f(\bar{x})$ be a polynomial of degree d. We write

$$f(\bar{x}) = \bar{a}_d \bar{x}^d + \bar{a}_{d-1} \bar{x}^{d-1} + \cdots + \bar{a}_1 \bar{x} + \bar{a}_0,$$

with $\bar{a}_d \neq \bar{0}$. We must show that $f(\bar{x})$ has at most d roots. We will prove this by contradiction.

Assumption: Suppose that $f(\bar{x})$ has at least $d + 1$ distinct roots: $\bar{r}_1, \bar{r}_2, \ldots, \bar{r}_{d+1}$. Consider the polynomial

$$g(\bar{x}) = \bar{a}_d (\bar{x} - \bar{r}_1)(\bar{x} - \bar{r}_2) \cdots (\bar{x} - \bar{r}_d). \tag{4}$$

This is a polynomial of degree d, whose highest-degree term is $\bar{a}_d \bar{x}^d$. Note that this is the same as the highest-degree term of $f(x)$. Hence, if we consider the difference

$$e(\bar{x}) = f(\bar{x}) - g(\bar{x}), \tag{5}$$

the highest-degree term of $f(\bar{x})$ and the highest-degree term of $g(\bar{x})$ cancel out. Thus, all terms of the polynomial $e(\bar{x})$ have degree less than d.

We conclude that either $e(\bar{x})$ is the zero polynomial, or $e(\bar{x})$ is a nonzero polynomial of degree less than d.

Case 1 $e(\bar{x})$ is the zero polynomial.

In this case,

$$f(\bar{x}) = g(\bar{x})$$
$$= \bar{a}_d (\bar{x} - \bar{r}_1)(\bar{x} - \bar{r}_2) \cdots (\bar{x} - \bar{r}_d).$$

Substituting $\bar{r}_{d+1}$ for $\bar{x}$ in this equation and recalling that $\bar{r}_{d+1}$ is a root of $f(\bar{x})$, we get

$$\bar{0} = \bar{a}_d (\bar{r}_{d+1} - \bar{r}_1)(\bar{r}_{d+1} - \bar{r}_2) \cdots (\bar{r}_{d+1} - \bar{r}_d). \tag{6}$$

But since $\bar{a}_d \neq \bar{0}$, and all the $\bar{r}_i$ are distinct for $i = 1, 2, \ldots, d + 1$, none of the factors on the right side equals $\bar{0}$. This implies that their product cannot be $\bar{0}$ (by Theorem 8.3.5), which contradicts equation (6). $\Rightarrow\Leftarrow$

Case 2 $e(\bar{x})$ is a nonzero polynomial whose degree is less than d.

We may apply our inductive hypothesis to conclude that the number of roots of $e(\bar{x})$ is at most the degree of $e(\bar{x})$, which is less than d. Hence, $e(\bar{x})$ has fewer than d roots. However, it is not hard to check that for $i = 1, 2, \ldots, d$, each $\bar{r}_i$ is a root of $e(\bar{x})$, as follows.

Let i be in the range $1, 2, \ldots, d$. Substituting $\bar{r}_i$ into equation (4), we find

$$g(\bar{r}_i) = \bar{a}_d (\bar{r}_i - \bar{r}_1)(\bar{r}_i - \bar{r}_2) \cdots (\bar{r}_i - \bar{r}_d).$$

Note that one of the factors on the right side of this equation must be $(\bar{r}_i - \bar{r}_i)$. Hence, $g(\bar{r}_i) = \bar{0}$. Recalling that $f(\bar{r}_i) = \bar{0}$, we use equation (5) to deduce that

$$e(\bar{r}_i) = \bar{0}.$$

Thus, $\bar{r}_1, \bar{r}_2, \ldots, \bar{r}_d$ are roots of $e(\bar{x})$. However, the inductive hypothesis implies that $e(\bar{x})$ has fewer than d roots. $\Rightarrow\Leftarrow$ ∎

REMARK. Theorem 10.2.1 holds not only for $\mathbf{Z}_p$ but in fact for any field: if $P(x)$ is a nonzero polynomial over any field F, the number of roots of $P(x)$ is at most the degree of $P(x)$. The proof for an arbitrary field is almost identical to our proof of Theorem 10.2.1.

Solutions to $\bar{x}^m = \bar{1}$

Theorem 10.2.1 gives an upper bound on the number of solutions of an arbitrary polynomial equation over $\mathbf{Z}_p$, where p is prime. We now examine a particular polynomial equation,

$$\bar{x}^m = \bar{1}, \tag{7}$$

with the hope of determining exactly how many solutions it has in $\mathbf{Z}_p$.

Let's explore how many solutions this equation has in $\mathbf{Z}_{13}$ for various values of m. It will help to have the exponent table for $\mathbf{Z}_{13}$, below. (To make the $\overline{1}$'s easier to find in the table, they are all shown in **bold**.)

Exponents in $\mathbf{Z}_{13}$

$\overline{x}^1$	$\overline{1}$	$\overline{2}$	$\overline{3}$	$\overline{4}$	$\overline{5}$	$\overline{6}$	$\overline{7}$	$\overline{8}$	$\overline{9}$	$\overline{10}$	$\overline{11}$	$\overline{12}$
$\overline{x}^2$	1	4	9	3	12	10	10	12	3	9	4	1
$\overline{x}^3$	1	8	1	12	8	8	5	5	1	12	5	12
$\overline{x}^4$	1	3	3	9	1	9	9	1	9	3	3	1
$\overline{x}^5$	1	6	9	10	5	2	11	8	3	4	7	12
$\overline{x}^6$	1	12	1	1	12	12	12	12	1	1	12	1
$\overline{x}^7$	1	11	3	4	8	7	6	5	9	10	2	12
$\overline{x}^8$	1	9	9	3	1	3	3	1	3	9	9	1
$\overline{x}^9$	1	5	1	12	5	5	8	8	1	12	8	12
$\overline{x}^{10}$	1	10	3	9	12	4	4	12	9	3	10	1
$\overline{x}^{11}$	1	7	9	10	8	11	2	5	3	4	6	12
$\overline{x}^{12}$	1	1	1	1	1	1	1	1	1	1	1	1
$\overline{x}^{13}$	1	2	3	4	5	6	7	8	9	10	11	12

How many solutions does $\overline{x}^4 = \overline{1}$ have in $\mathbf{Z}_{13}$? From the table, we see that $\overline{1}^4, \overline{5}^4, \overline{8}^4$, and $\overline{12}^4$ are all equal to $\overline{1}$. Thus, there are exactly **4** solutions to the equation $\overline{x}^4 = \overline{1}$ in $\mathbf{Z}_{13}$. This corresponds to the fact that the $\overline{x}^4$ row contains the number $\overline{1}$ exactly **4** times.

How many solutions does $\overline{x}^9 = \overline{1}$ have in $\mathbf{Z}_{13}$? Since the number $\overline{1}$ appears **3** times in the $\overline{x}^9$ row, there are exactly **3** solutions to the equation $\overline{x}^9 = \overline{1}$.

By counting the $\overline{1}$'s in various rows of the table, you may notice a pattern. For any exponent m, the number of $\overline{1}$'s in the $\overline{x}^m$ row is equal to **gcd(m, 12)**. In particular, when $m \mid 12$, the number of $\overline{1}$'s in the $\overline{x}^m$ row is exactly m. These observations are captured by the following proposition, which will be crucial to our study of primitive roots (see Section 10.3). We will also revisit these ideas when we study primality tests in Chapter 12.

PROPOSITION 10.2.2

Let p be prime, and let $m \in \mathbf{N}$. Consider the equation

$$\overline{x}^m = \overline{1} \text{ in } \mathbf{Z}_p. \qquad (8)$$

(i) If $m \mid (p - 1)$, then equation (8) has exactly m solutions.

(ii) For any natural number m, equation (8) has exactly gcd(m, $p - 1$) solutions.

PROOF To prove (i), we assume $m \mid p - 1$. Thus, there exists $k \in \mathbf{Z}$ such that

$$p - 1 = km. \tag{9}$$

As the reader is invited to verify, we can then factor the polynomial $\bar{x}^{p-1} - \bar{1}$ as follows:

$$\bar{x}^{p-1} - \bar{1} = (\bar{x}^m - \bar{1})\left(\bar{x}^{(k-1)m} + \bar{x}^{(k-2)m} + \cdots + \bar{x}^{2m} + \bar{x}^m + \bar{1}\right).$$

Note that Fermat's Little Theorem (9.1.1) guarantees that every $\bar{x} \neq \bar{0}$ in $\mathbf{Z}_p$ is a root of the polynomial on the left side of this equation. Thus, each of the $p - 1$ nonzero elements of $\mathbf{Z}_p$ must be a root of one of the two factors on the right side. The second of these two factors (shown in orange) has degree $(k - 1)m$ and hence has at most $(k - 1)m$ roots by Theorem 10.2.1. Thus, the number of roots of the first factor (shown in **brown**) must be at least

$$(p - 1) - (k - 1)m.$$

This expression equals m by (9), and so we have shown that the polynomial $\bar{x}^m - \bar{1}$ has at least m roots. It then follows from Theorem 10.2.1 that this polynomial must have exactly m roots. This completes the proof of (i).

To prove (ii), we let m be an arbitrary natural number, and let $d = \gcd(m, p - 1)$.

Claim: For any $\bar{x} \in \mathbf{Z}_p$, $\bar{x}^m = \bar{1}$ if and only if $\bar{x}^d = \bar{1}$.

Proof of Claim Let $\bar{x} \in \mathbf{Z}_p$.

$[\Rightarrow]$ Assume $\bar{x}^m = \bar{1}$. **[To show: $\bar{x}^d = \bar{1}$.]**

By Fermat's Little Theorem (9.1.1), we also have

$$\bar{x}^{p-1} = \bar{1}.$$

Since $d = \gcd(m, p - 1)$, we may write d as a linear combination of m and $p - 1$ (by Corollary 5.3.3). Thus, there exist integers s and t such that

$$d = sm + t(p - 1).$$

Hence,

$$\begin{aligned} \bar{x}^d &= \bar{x}^{sm + t(p-1)} \\ &= (\bar{x}^m)^s (\bar{x}^{p-1})^t \\ &= \bar{1}^s \bar{1}^t \\ &= \bar{1}. \end{aligned}$$

$[\Leftarrow]$ Now assume $\bar{x}^d = \bar{1}$. **[To show: $\bar{x}^m = \bar{1}$.]**

Since $d \mid m$, it follows that $\bar{x}^m = \bar{1}$ (see Exercise 4).

This completes the proof of the Claim.

The Claim states that the equations $\bar{x}^m = \bar{1}$ and $\bar{x}^d = \bar{1}$ have the same solution set. By (i), the equation $\bar{x}^d = \bar{1}$ has exactly d solutions, so $\bar{x}^m = \bar{1}$ must also have exactly d solutions. ∎

EXERCISES 10.2

Numerical Problems

1. Find all solutions of each of the following equations in $\mathbf{Z}_6$.

   a. $\bar{x}^2 + \bar{x} = \bar{0}$ b. $\bar{x}^3 - \bar{x} = \bar{0}$

2. Find all solutions of each of the following equations in $\mathbf{Z}_{11}$.

   a. $\bar{x}^2 - \overline{10} = \bar{0}$ b. $\bar{x}^3 + \bar{x}^2 + \bar{1} = \bar{0}$

3. Find all solutions of each of the following equations in $\mathbf{Z}_7$.

   a. $\bar{x}^2 = \bar{2}$ b. $\bar{x}^3 = \bar{6}$ c. $\bar{x}^4 = \bar{4}$ d. $\bar{x}^5 = \bar{5}$

Reasoning and Proofs

4. Let n be a natural number, and suppose $\bar{x} \in \mathbf{Z}_n$. Let $d, m \in \mathbf{Z}$. Show that if $\bar{x}^d = \bar{1}$ and $d \mid m$, then $\bar{x}^m = \bar{1}$.

5. In Example 1, part b, we saw a polynomial of degree 2 that has 4 roots. Explain why this does not violate Theorem 10.2.1.

6. In the proof of Theorem 10.2.1, where did we used the hypothesis that p is prime?

7. For this exercise, consider the factorization $\bar{x}^{16} - \bar{1} = (\bar{x}^8 - \bar{1})(\bar{x}^8 + \bar{1})$.

   a. Explain how you know that in $\mathbf{Z}_{17}$, every reduced residue must satisfy exactly one of the following two equations: $\bar{x}^8 - \bar{1} = 0$ and $\bar{x}^8 + \bar{1} = \bar{0}$.

   b. If $\bar{a}$ satisfies $\bar{x}^8 - \bar{1} = \bar{0}$ in $\mathbf{Z}_{17}$, what are the possible values of ord($\bar{a}$)? Explain your reasoning.

   c. If $\bar{a}$ satisfies $\bar{x}^8 + \bar{1} = \bar{0}$ in $\mathbf{Z}_{17}$, what are the possible values of ord($\bar{a}$)? Explain your reasoning.

8. Let $\bar{a} \in \mathbf{Z}_{29}$, and suppose that $\bar{a}^7 \neq \bar{1}$. Prove that $\bar{a}^{21} + \bar{a}^{14} + \bar{a}^7 + \bar{1} = \bar{0}$.

EXPLORATION The Quadratic Formula in $\mathbf{Z}_p$ (Exercises 9–15)

As you know, the quadratic formula states that the solutions to the quadratic equation $ax^2 + bx + c = 0$ are given by $\dfrac{-b \pm \sqrt{b^2 - 4ac}}{2a}$. With a few small modifications, this formula can also be used to solve quadratic equations over $\mathbf{Z}_p$. The modifications are required because we have not defined square roots or division in $\mathbf{Z}_p$.

THEOREM. QUADRATIC FORMULA IN $\mathbf{Z}_p$

Let $p > 2$ be prime, and let $\bar{a}, \bar{b}, \bar{c} \in \mathbf{Z}_p$ such that $\bar{a} \neq \bar{0}$. Consider the quadratic equation

$$\bar{a}x^2 + \bar{b}x + \bar{c} = \bar{0}. \tag{10}$$

Let $\bar{d} = \bar{b}^2 - \bar{4}\bar{a}\bar{c}$.

(i) If there exists $\bar{s} \in \mathbf{Z}_p$ such that $\bar{s}^2 = \bar{d}$, then the solutions to equation (10) are given by

$$\bar{x} = (-\bar{b} + \bar{s})(\bar{2}\bar{a})^{-1} \qquad and \qquad \bar{x} = (-\bar{b} - \bar{s})(\bar{2}\bar{a})^{-1}.$$

(ii) If there does not exist $\bar{s} \in \mathbf{Z}_p$ such that $\bar{s}^2 = \bar{d}$, then equation (10) has no solutions in $\mathbf{Z}_p$.

For each of the quadratic equations in Exercises 9–12,

  a. **Compute $\bar{d} = \bar{b}^2 - \bar{4}\bar{a}\bar{c}$.**

  b. **Determine whether there exists $\bar{s} \in \mathbf{Z}_p$ such that $\bar{s}^2 = \bar{d}$.**

  c. **Use your answer to part b and the Theorem to solve the quadratic equation.**

9. $\bar{x}^2 + \bar{8}\bar{x} + \bar{4} = \bar{0}$ in $\mathbf{Z}_{11}$

10. $\bar{3}\bar{x}^2 + \bar{3}\bar{x} + \bar{1} = \bar{0}$ in $\mathbf{Z}_{11}$

11. $\bar{x}^2 + \bar{9}\bar{x} + \bar{1} = \bar{0}$ in $\mathbf{Z}_{13}$

12. $\bar{3}\bar{x}^2 + \bar{7}\bar{x} + \bar{3} = \bar{0}$ in $\mathbf{Z}_{13}$

*13. Prove the Theorem (Quadratic Formula in $\mathbf{Z}_p$). [*Hint:* It helps if you remember how the quadratic formula was derived in high school algebra.]

14. a. Explain why the statement of the Theorem would not be valid for $p = 2$.

  b. Give an example to show that the conclusion of the Theorem does not hold if p is composite.

15. Consider the equation $\bar{x}^2 - \bar{3}\bar{x} + \bar{7} = \bar{0}$.

  a. Find all solutions to this equation in $\mathbf{Z}_5$.

  b. Find all solutions to this equation in $\mathbf{Z}_7$.

  c. Find all solutions to this equation in $\mathbf{Z}_{35}$. [*Hint:* Use your answers to parts a and b.]

10.3 Primitive Roots

Definition of primitive root

If $\bar{a}$ is a nonzero element of $\mathbf{Z}_7$, then Fermat's Little Theorem (9.1.1) tells us that $\bar{a}^6 = \bar{1}$. This implies that the order of every element of $\mathbf{Z}_7$ is at most 6. At the beginning of this chapter, we saw that the element $\bar{3} \in \mathbf{Z}_7$ has order 6. (In our myth, this was the reason that Lagrange needed his prison sentence to be tripled a total of 6 times.) Elements that have the maximum possible order, such as $\bar{3} \in \mathbf{Z}_7$, are called *primitive roots*.

DEFINITION 10.3.1

Let p be prime, and let $\bar{a} \in \mathbf{Z}_p$ with $\bar{a} \neq \bar{0}$. Then $\bar{a}$ is said to be a **primitive root** *in $\mathbf{Z}_p$ if $\operatorname{ord}(\bar{a}) = p - 1$. In this case, we may also say that the integer a is a* **primitive root** *modulo p.*

REMARK. Primitive roots may also be defined for $\mathbf{Z}_n$ when n is composite. (See Exercises 15–23).

EXAMPLE 1

a. Is $\bar{2} \in \mathbf{Z}_7$ a primitive root?

b. Is $\bar{5} \in \mathbf{Z}_7$ a primitive root?

SOLUTION Computing the powers of $\bar{2}$ and $\bar{5}$ in $\mathbf{Z}_7$, we find

$$\bar{2}^1 = \bar{2}, \qquad \bar{5}^1 = \bar{5},$$
$$\bar{2}^2 = \bar{4}, \qquad \bar{5}^2 = \bar{4},$$
$$\bar{2}^3 = \bar{1}. \qquad \bar{5}^3 = \bar{6},$$
$$\bar{5}^4 = \bar{2},$$
$$\bar{5}^5 = \bar{3},$$
$$\bar{5}^6 = \bar{1}.$$

a. Thus, the order of $\bar{2}$ is 3, so $\bar{2}$ is not a primitive root.

b. We see that the order of $\bar{5}$ is 6; hence, $\bar{5}$ is a primitive root.

EXAMPLE 2

Determine which elements of $\mathbf{Z}_{11}$ are primitive roots.

SOLUTION In Example 3 of Section 10.1, we determined the orders of all of the nonzero elements of $\mathbf{Z}_{11}$:

$$\text{ord}(\overline{1}) = 1,$$
$$\text{ord}(\overline{10}) = 2,$$
$$\text{ord}(\overline{3}) = \text{ord}(\overline{4}) = \text{ord}(\overline{5}) = \text{ord}(\overline{9}) = 5,$$
$$\text{ord}(\overline{2}) = \text{ord}(\overline{6}) = \text{ord}(\overline{7}) = \text{ord}(\overline{8}) = 10.$$

Hence, the primitive roots in $\mathbf{Z}_{11}$ are $\overline{2}, \overline{6}, \overline{7},$ and $\overline{8}$.

EXAMPLE 3

Find a primitive root in $\mathbf{Z}_{17}$, if one exists.

SOLUTION For want of a better strategy, we will check the orders of the elements $\overline{1}, \overline{2}, \overline{3}, \ldots,$ until we find an element of order 16.

$$\overline{1}^{1} = \overline{1}$$

$\overline{2}^{1} = \overline{2},$	$\overline{3}^{1} = \overline{3},$
$\overline{2}^{2} = \overline{4},$	$\overline{3}^{2} = \overline{9},$
$\overline{2}^{3} = \overline{8},$	$\overline{3}^{3} = \overline{10},$
$\overline{2}^{4} = \overline{16},$	$\overline{3}^{4} = \overline{13},$
$\overline{2}^{5} = \overline{15},$	$\overline{3}^{5} = \overline{5},$
$\overline{2}^{6} = \overline{13},$	$\overline{3}^{6} = \overline{15},$
$\overline{2}^{7} = \overline{9},$	$\overline{3}^{7} = \overline{11},$
$\overline{2}^{8} = \overline{1}.$	$\overline{3}^{8} = \overline{16},$
	$\overline{3}^{9} = \overline{14},$
	$\overline{3}^{10} = \overline{8},$
	$\overline{3}^{11} = \overline{7},$
	$\overline{3}^{12} = \overline{4},$
	$\overline{3}^{13} = \overline{12},$
	$\overline{3}^{14} = \overline{2},$
	$\overline{3}^{15} = \overline{6},$
	$\overline{3}^{16} = \overline{1}.$

From the left column, we see that the element $\overline{1}$ has order **1**.
From the center column, we see that $\overline{2}$ has order **8**.

From the right column, we see that the order of $\overline{3}$ equals **16**, and we conclude that $\overline{3}$ is a primitive root.

We did not actually need to compute all of those powers of $\overline{3}$ to determine that $\overline{3}$ is a primitive root. We know that $\operatorname{ord}(\overline{3})$ must be a divisor of $17 - 1 = 16$, by Corollary 10.1.4. Once we have computed from $\overline{3}^1$ up to $\overline{3}^8$ and not yet obtained an answer of $\overline{1}$, we know that $\operatorname{ord}(\overline{3}) > 8$. Since the only divisor of 16 that is greater than 8 is 16 itself, we can then conclude that $\operatorname{ord}(\overline{3}) = 16$. (This observation is generalized in Exercise 7.) ●

In Example 3, we found that $\overline{3}$ is a primitive root in $\mathbf{Z}_{17}$. Looking again at the column of the powers of $\overline{3}$ from the example, observe that these powers cycle through all of the nonzero elements of $\mathbf{Z}_{17}$. That is, every nonzero element is equal to $\overline{3}$ raised to some power. This is a general fact about primitive roots, which we will now prove.

PROPOSITION 10.3.2

Let p be prime, and let $\overline{a} \in \mathbf{Z}_p$ be a primitive root. Then every nonzero element of $\mathbf{Z}_p$ appears exactly once in the list

$$\overline{a}^0, \overline{a}^1, \ldots, \overline{a}^{p-2}. \tag{1}$$

PROOF First, we show that the elements in list (1) are distinct. Suppose that

$$\overline{a}^j = \overline{a}^k$$

with j and k in the range $0, \ldots, p - 2$. Since $\operatorname{ord}(\overline{a}) = p - 1$, we know that two powers of $\overline{a}$ are equal if and only if the exponents are congruent modulo $p - 1$, by Proposition 10.1.3(ii). Hence,

$$j \equiv k \ (\operatorname{mod} p - 1).$$

But since j and k are restricted to the range $0, \ldots, p - 2$, it follows that $j = k$. Thus, we have shown that the elements in list (1) are all distinct.

Note that since $\overline{a} \neq \overline{0}$, each of the elements in list (1) is a nonzero element of $\mathbf{Z}_p$. There are $p - 1$ elements in list (1), and we have just shown that these elements are distinct. Since there are exactly $p - 1$ nonzero elements of $\mathbf{Z}_p$, list (1) must include every nonzero element of $\mathbf{Z}_p$. ●

⋮ Check for Understanding

In $\mathbf{Z}_{11}$, $\overline{2}$ is a primitive root. Express each of the following elements of $\mathbf{Z}_{11}$ as $\overline{2}^k$, where k is in the range $0, \ldots, 9$.

1. $\overline{5}$ 2. $\overline{9}$

The converse of Proposition 10.3.2 is also true. More precisely, if $\bar{a} \in \mathbf{Z}_p$, and if every nonzero element of $\mathbf{Z}_p$ can be expressed as a power of $\bar{a}$, then $\bar{a}$ must be a primitive root. (See Exercise 11.)

Existence of primitive roots in a prime modulus

In Example 3, we asked whether there exists a primitive root in $\mathbf{Z}_{17}$. To solve this, we did not have to search very far. Although $\bar{1}$ and $\bar{2}$ are not primitive roots, we found that $\bar{3}$ is a primitive root. If we look for primitive roots modulo other primes, we find a similar situation. For example, modulo 47, we find that 5 is a primitive root, and modulo 293 we find that 2 is a primitive root. In fact, every prime less than 100 has a primitive root that is less than or equal to 7. And every prime less than 1,000,000 has a primitive root that is less than or equal to 73.

The main result of this chapter is the Primitive Root Theorem, which states that for every prime p, there exists a primitive root modulo p. Although this is easy to state, proving it is quite tricky. So tricky, in fact, that the proof given by the great Leonhard Euler was incorrect. It was Lagrange who gave the first correct proof of the Primitive Root Theorem. (See the beginning of the chapter for a historical summary of Lagrange's life and work.)

THEOREM 10.3.3   PRIMITIVE ROOT THEOREM

Let p be prime. Then there exists a primitive root modulo p.

To prove the Primitive Root Theorem, we must prove that in $\mathbf{Z}_p$, there exists an element of order $p - 1$. Our first step is to prove that there exists an element of order q^s, where q is prime and q^s divides $p - 1$. (Numbers of the form q^s, where q is prime and $s \in \mathbf{N}$, are called *prime powers*.)

LEMMA 10.3.4

Let p be a prime, and let q^s be a prime power (where q is prime and $s \geq 1$). If $q^s \mid p - 1$, then there exists an element of order q^s in $\mathbf{Z}_p$.

Before proving this lemma, let's see a quick example.

EXAMPLE 4

Suppose that $p = 101$. Then $p - 1 = 100 = 2^2 \cdot 5^2$, which is divisible by the following prime powers:

$$2^1 = 2, \ 2^2 = 4, \ 5^1 = 5, \text{ and } 5^2 = 25.$$

Thus, Lemma 10.3.4 guarantees that in $\mathbf{Z}_{101}$, we will find elements of orders 2, 4, 5, and 25.

PROOF OF **LEMMA 10.3.4**

Let p be prime. Let q be prime and $s \in \mathbf{N}$ such that $q^s \mid p - 1$. Consider the following equations in $\mathbf{Z}_p$:

$$\overline{x}^{q^s} = \overline{1} \tag{2a}$$

$$\text{and} \quad \overline{x}^{q^{s-1}} = \overline{1}. \tag{2b}$$

By Proposition 10.2.2, equation (2a) has q^s solutions and equation (2b) has q^{s-1} solutions. Since $q^s > q^{s-1}$, there must be at least one element $\overline{a} \in \mathbf{Z}_p$ that is a solution to (2a) but not to (2b). That is,

$$\overline{a}^{q^s} = \overline{1} \tag{3a}$$

$$\text{but} \quad \overline{a}^{q^{s-1}} \neq \overline{1}. \tag{3b}$$

Claim: $\overline{a}$ has order q^s.

> **Proof of Claim** By equation (3a), it follows (using Proposition 10.1.3) that $\mathrm{ord}(\overline{a})$ divides q^s. Thus, the order of $\overline{a}$ is one of the numbers in the list
>
> $$1, \ q, \ q^2, \ldots, q^{s-1}, \ q^s. \tag{4}$$
>
> However, from equation (3b), if follows (again using Proposition 10.1.3) that $\mathrm{ord}(\overline{a})$ does not divide q^{s-1}. Since the only number in list (4) that does not divide q^{s-1} is q^s, we conclude that the order of $\overline{a}$ is q^s.

Thus, we have found an element $\overline{a} \in \mathbf{Z}_p$ with order q^s, and the lemma is proved. ∎

Suppose we want to know whether $\mathbf{Z}_{101}$ has a primitive root—that is, an element of order 100. Lemma 10.3.4 tells us that there is an element $\overline{a}$ of order **4** and an element $\overline{b}$ of order **25** (see Example 4). It turns out that the product $\overline{ab}$ will have order $4 \cdot 25 = 100$, and hence, $\overline{ab}$ is the primitive root we seek. This is guaranteed by the following lemma.

LEMMA 10.3.5

Let $n \in \mathbf{N}$, and let $\overline{a}, \overline{b} \in \mathbf{Z}_n$ be reduced residues. If $\mathrm{ord}(\overline{a})$ and $\mathrm{ord}(\overline{b})$ are relatively prime, then

$$\mathrm{ord}(\overline{ab}) = \mathrm{ord}(\overline{a}) \cdot \mathrm{ord}(\overline{b}).$$

PROOF Let $r = \mathrm{ord}(\overline{a})$ and let $s = \mathrm{ord}(\overline{b})$.

Note that

$$(\overline{ab})^{rs} = (\overline{a}^r)^s (\overline{b}^s)^r$$
$$= (\overline{1})^s (\overline{1})^r$$
$$= \overline{1}.$$

Hence,

$$\text{ord}\left(\overline{ab}\right) \le rs. \tag{5}$$

Claim: $\text{ord}(\overline{ab}) \ge rs.$

Proof of Claim Let $k = \text{ord}(\overline{ab})$. Then

$$(\overline{ab})^k = \overline{1}.$$

Raising both sides of this equation to the power r, it follows that

$$\overline{a}^{rk}\,\overline{b}^{rk} = \overline{1}. \tag{6}$$

Since $\overline{a}^r = \overline{1}$, we also have $\overline{a}^{rk} = \overline{1}$. Substitution into equation (6) yields

$$\overline{b}^{rk} = \overline{1}.$$

Thus, by Proposition 10.1.3, the order of $\overline{b}$ must divide the exponent rk. In other words, $s \mid rk$. Since s and r are relatively prime, Euclid's Lemma (5.4.2) implies that

$$s \mid k.$$

A similar argument shows that $r \mid k$. Hence, k is a common multiple of r and s. Since the least common multiple is $\text{lcm}(r, s) = rs$, we conclude that

$$k \ge rs,$$

which completes the proof of the Claim.

Combining the Claim with inequality (5), we conclude that $\text{ord}(\overline{ab}) = rs$. ●

In Example 4, we used Lemma 10.3.4 to determine that $\mathbf{Z}_{101}$ has elements of orders 4 and 25. By Lemma 10.3.6, the product of these elements has order 100, so it is a primitive root. We use this same argument, which combines Lemmas 10.3.5 and 10.3.6, to give a general proof of the Primitive Root Theorem.

PROOF OF **THE PRIMITIVE ROOT THEOREM (10.3.3)**

Let p be prime.

If $p = 2$, then $\overline{1}$ is a primitive root modulo p. Thus, we may assume that $p > 2$. Factor $p - 1$ into primes:

$$p - 1 = q_1^{a_1} q_2^{a_2} \cdots q_m^{a_m}.$$

By Lemma 10.3.4, for each $k = 1, 2, \ldots, m$, there exists an element $\overline{x}_k$ of order $q_k^{a_k}$. Let

$$\overline{x} = \overline{x}_1 \overline{x}_2 \cdots \overline{x}_m. \tag{7}$$

Applying Lemma 10.3.5 repeatedly (see Exercise 14), we find

$$\mathrm{ord}(\bar{x}) = \mathrm{ord}(\bar{x}_1) \cdot \mathrm{ord}(\bar{x}_2) \cdot \ \cdots \ \cdot \mathrm{ord}(\bar{x}_m)$$
$$= q_1^{a_1} \cdot q_2^{a_2} \cdot \ \cdots \ \cdot q_m^{a_m}$$
$$= p - 1.$$

Hence, $\bar{x}$ is a primitive root modulo p. ◼

We have now proved that $\mathbf{Z}_p$ always has at least one primitive root, but as we have seen, $\mathbf{Z}_p$ often has more than one primitive root. It turns out that once we know *one* primitive root in $\mathbf{Z}_p$, it is easy to find all of the primitive roots in $\mathbf{Z}_p$. The following lemma tells us how to do this.

Let p be prime, and let $\bar{a} \in \mathbf{Z}_p$ be a primitive root. Then for any $j \in \mathbf{Z}$, $\bar{a}^j$ is a primitive root if and only if $\gcd(j, p - 1) = 1$.

PROOF Let p be prime, and let $\bar{a} \in \mathbf{Z}_p$ be a primitive root.

By Lemma 10.1.6,

$$\mathrm{ord}(\bar{a}^j) = \frac{p - 1}{\gcd(j, p - 1)}.$$

Thus,

$$\mathrm{ord}(\bar{a}^j) = p - 1 \text{ if and only if } \gcd(j, p - 1) = 1. \qquad ◼$$

Naomi's Numerical Proof Preview: Theorem 10.3.7

Let's see whether we can use this lemma to determine how many primitive roots there are modulo 19. Let $\bar{a} \in \mathbf{Z}_{19}$ be a primitive root. Then the nonzero elements of $\mathbf{Z}_{19}$ are

$$\bar{a}^0, \ \bar{a}^1, \ \bar{a}^2, \ldots, \ \bar{a}^{17}.$$

Lemma 10.3.6 tells us which of these elements are primitive roots: precisely those in which the exponent is relatively prime to 18. Thus, the primitive roots in $\mathbf{Z}_{19}$ are

$$\bar{a}^1, \ \bar{a}^5, \ \bar{a}^7, \ \bar{a}^{11}, \ \bar{a}^{13}, \ \bar{a}^{17}.$$

We see, then, that the primitive roots in $\mathbf{Z}_{19}$ correspond to the numbers $1, 5, 7, 11, 13$, and 17, which are the positive integers less than 18 that are relatively prime to 18. The number of such integers is exactly $\varphi(18) = 6$. In general, we have the following theorem.

Let p be prime. Then $\mathbf{Z}_p$ has exactly $\varphi(p - 1)$ primitive roots.

PROOF Let $\bar{a} \in \mathbf{Z}_p$ be a primitive root. Then by Proposition 10.3.2, we know that the list of reduced residues of $\mathbf{Z}_p$ comprises the elements

$$\bar{a}^0, \bar{a}^1, \ldots, \bar{a}^{p-2}. \tag{8}$$

By Lemma 10.3.6, an element $\bar{a}^j$ in list (8) is a primitive root if and only if $\gcd(j, p - 1) = 1$. Thus, the number of primitive roots equals the number of positive integers j that are less than $p - 1$ and relatively prime to $p - 1$. By definition, this is exactly $\varphi(p - 1)$. ∎

EXERCISES 10.3

Numerical Problems

1. Find all of the primitive roots in $\mathbf{Z}_7$.

2. Find all of the primitive roots in $\mathbf{Z}_{13}$.

3. Find all of the primitive roots in $\mathbf{Z}_{17}$.

4. How many primitive roots are there in each of the following number systems?

  a. $\mathbf{Z}_{13}$ b. $\mathbf{Z}_{29}$ c. $\mathbf{Z}_{31}$ d. $\mathbf{Z}_{101}$

Reasoning and Proofs

5. If p is prime and $\bar{a} \in \mathbf{Z}_p$ is a primitive root, is $-\bar{a}$ also a primitive root? Prove it, or find a counterexample.

6. Suppose p is a prime such that $p \equiv 1 \pmod 4$. If $\bar{a} \in \mathbf{Z}_p$ is a primitive root, is $-\bar{a}$ also a primitive root? Prove it, or find a counterexample.

7. Let p be prime, and let $\bar{a}$ be a nonzero element of $\mathbf{Z}_p$. Suppose that $\mathrm{ord}(\bar{a}) > \dfrac{p-1}{2}$. Prove that $\bar{a}$ is a primitive root.

8. Let p be prime, $p > 2$, and let $\bar{a}$ be a nonzero element of $\mathbf{Z}_p$. Prove that $\bar{a}$ is a primitive root modulo p if and only if for every prime q that divides $p - 1$, we have $\bar{a}^{(p-1)/q} \neq \bar{1}$.

9. Use the result of Exercise 8 to find a primitive root modulo 31.

10. Use the result of Exercise 8 to find a primitive root modulo 101.

11. Prove the following statement, which is the converse of Proposition 10.3.2.

Let p be prime, and let $\bar{a} \in \mathbf{Z}_p$. If every nonzero element of $\mathbf{Z}_p$ is a power of $\bar{a}$, then $\bar{a}$ is a primitive root.

12. Let $\bar{a}$ be a primitive root in $\mathbf{Z}_p$, and let $j \in \mathbf{Z}$.

  a. Find a formula for the order of $\bar{a}^j$. Give a few numerical examples of your formula.

  b. Prove that your formula from part a is correct.

13. Let p be prime, and let $d \in \mathbf{N}$ such that $d \mid p - 1$. Prove that there are exactly $\varphi(d)$ elements of $\mathbf{Z}_p$ that have order d. [*Hint:* Use your result from Exercise 12.]

14. Prove the following generalization of Lemma 10.3.5.

  LEMMA. *Let $n \in \mathbf{N}$, and let $\bar{x}_1, \bar{x}_2, \ldots, \bar{x}_m \in \mathbf{Z}_n$ be reduced residues. Suppose that for all $i \neq j$, $\mathrm{ord}(\bar{x}_i)$ and $\mathrm{ord}(\bar{x}_j)$ are relatively prime. Then*

$$\mathrm{ord}(\bar{x}_1 \cdot \bar{x}_2 \cdot \ \cdots \ \cdot \bar{x}_m) = \mathrm{ord}(\bar{x}_1) \cdot \mathrm{ord}(\bar{x}_2) \cdot \ \cdots \ \cdot \mathrm{ord}(\bar{x}_m).$$

Advanced Reasoning and Proofs

EXPLORATION Primitive Roots in Composite Moduli (Exercises 15–23)

The definition of *primitive root* that we gave in the text requires that the modulus, p, be prime. However, the definition of primitive root can be generalized to composite moduli:

  DEFINITION. *Let n be a natural number, and let $\bar{a} \in \mathbf{Z}_n$ be a reduced residue. Then $\bar{a}$ is said to be a **primitive root** if the order of $\bar{a}$ is equal to $\varphi(n)$. In this case, we may also say that the integer a is a **primitive root** modulo n.*

15. **a.** Show that $\bar{5}$ is a primitive root in $\mathbf{Z}_6$.

  b. Show that $\bar{3}$ is a primitive root in $\mathbf{Z}_{10}$.

16. **a.** Find a primitive root modulo 4.

  b. Find a primitive root modulo 9.

  c. Find a primitive root modulo 22.

  d. Find a primitive root modulo 25.

17. **a.** Prove that there is no primitive root modulo 15.

  b. Prove that there is no primitive root modulo 8.

The preceding exercises are all examples of the following theorem, which is a generalization of the Primitive Root Theorem (10.3.3) to all moduli (composite or prime). For every natural number n, the theorem tells whether there is a primitive root modulo n.

  THEOREM. *Let n be a natural number. There exists a primitive root modulo n if and only if $n = 2$, $n = 4$, $n = p^k$, or $n = 2p^k$ for some prime $p > 2$ and some natural number k.*

In Exercises 18–22, you will prove one direction of this theorem (the "only if" direction of the if and only if statement).

18. Suppose $n = pq$, where p and q are distinct primes with $p, q > 2$. In this exercise you will prove that there is no primitive root modulo n. Let $a \in \mathbf{Z}$ such that $\gcd(a, n) = 1$.

   a. Prove that $a^{\varphi(n)/2} \equiv 1 \pmod{p}$ and $a^{\varphi(n)/2} \equiv 1 \pmod{q}$.

   b. Show that $a^{\varphi(n)/2} \equiv 1 \pmod{n}$.

   c. Conclude that a is not a primitive root modulo n.

19. Let n be any integer that has at least two distinct odd prime factors. In this exercise, you will show that there are no primitive roots modulo n. Let a be any integer such that $\gcd(a, n) = 1$.

   a. Let p be any prime factor of n, and let k be the number of times that p appears in the prime factorization of n. Prove that $\varphi\left(\dfrac{n}{p^k}\right)$ is even.

   b. Show that $\varphi(p^k) \mid \dfrac{\varphi(n)}{2}$.

   c. Prove that $a^{\varphi(n)/2} \equiv 1 \pmod{p^k}$.

   d. Using the fact that the result you proved in part c holds for an arbitrary prime divisor of n, show that $a^{\varphi(n)/2} \equiv 1 \pmod{n}$.

   e. Conclude that a is not a primitive root modulo n.

20. Suppose n is an integer such that $4 \mid n$, but n is not a power of 2. Follow steps a–e from Exercise 19 to show that there is no primitive root modulo n.

21. In this exercise, you will show that if $n \geq 8$ is a power of 2, then there is no primitive root modulo n. Let a be an odd integer. (That is, let a be any integer such that $\gcd(a, n) = 1$.)

   a. Prove that for any natural number $s \geq 3$, $a^{2^{s-2}} \equiv 1 \pmod{2^s}$.
      [*Hint:* Use induction.]

   b. Use your result from part a to prove that for any natural number $s \geq 3$, there is no primitive root modulo 2^s.

22. Using your results from Exercises 18–21, prove the "only if" direction of the Theorem. In other words, prove the following:

   Let $n \in \mathbf{N}$. If there exists a primitive root modulo n, then $n = 2$, $n = 4$, $n = p^k$, or $n = 2p^k$ for some prime $p > 2$ and some natural number k.

The other direction of the Theorem is harder to prove. Here we consider one particular case. (We consider other cases in an exploration in Section 12.2.)

23. Suppose $n = 2m$, where m is an odd integer. (That is, n is divisible by 2 but not by 4.)

    a. Show that $\varphi(m) = \varphi(n)$.

    b. It is a fact that 10 is a primitive root modulo 47. Explain why 10 is not a primitive root modulo $94 = 2 \cdot 47$.

    c. Use the fact that 10 is a primitive root modulo 47 to find a primitive root modulo 94.

    d. Let m be any odd integer. Prove that there exists a primitive root modulo $2m$ if and only if there exists a primitive root modulo m.

10.4 Applications of Primitive Roots

Discrete logarithms

If $\bar{a}$ is a primitive root in $\mathbf{Z}_p$, then by Proposition 10.3.2, every nonzero element of $\mathbf{Z}_p$ can be expressed as some power of $\bar{a}$. This point suggests an interesting question: Given a nonzero element $\bar{x} \in \mathbf{Z}_p$, what power of $\bar{a}$ is equal to $\bar{x}$? In other words, what number y makes the following statement true:

$$\bar{a}^y = \bar{x}?$$

If we were dealing with positive real numbers a and x instead of elements of $\mathbf{Z}_p$, logarithms would give us the answer: $y = \log_a x$. In $\mathbf{Z}_p$, the answer is known as a *discrete logarithm.*[4]

DEFINITION 10.4.1

Let p be prime and let $\bar{a}, \bar{x} \in \mathbf{Z}_p$ such that $\bar{a}$ is a primitive root and $\bar{x}$ is nonzero. The **discrete logarithm (base $\bar{a}$)** *of $\bar{x}$ is the integer y that satisfies*

$$\bar{a}^y - \bar{x}, \qquad \text{where } 0 \le y \le p - 2.$$

(By Proposition 10.3.2, there is exactly one such integer y.)

EXAMPLE 1

In $\mathbf{Z}_{17,}$ using the primitive root $\bar{3}$ as the base, find the discrete logarithm of $\overline{14}$.

SOLUTION For want of a better method, we can just make a list of all powers of $\bar{3}$, from $\bar{3}^0$ to $\bar{3}^{15}$, in $\mathbf{Z}_{17}$. We already made a list of the powers of $\bar{3}$ in Example 3 of

---

[4]*Discrete* is the opposite of *continuous*. Whereas $\mathbf{R}$ is a continuum, the set $\mathbf{Z}$ is a discrete set—pictured on the number line, the integers do not cluster (they're spaced apart). $\mathbf{Z}_n$ is also considered to be a discrete set.

Section 10.3. Referring back to that list, we see that $\overline{3}^9 = \overline{14}$. Thus, **9** is the discrete logarithm (base $\overline{3}$) of $\overline{14}$. ■

Our method for finding discrete logarithms in Example 1 is completely general. For any prime p and any primitive root $\overline{a}$ in $\mathbf{Z}_p$, we can find the discrete logarithm (base $\overline{a}$) of any nonzero element $\overline{x}$ in $\mathbf{Z}_p$ simply by raising $\overline{a}$ to every power, from $\overline{a}^0$ to $\overline{a}^{p-2}$, until we find the power of $\overline{a}$ that is equal to $\overline{x}$. Although this method will always work, it is quite slow. For a large modulus p, this method requires us to raise the base $\overline{a}$ to as many as $p - 1$ different exponents to find a discrete logarithm base $\overline{a}$. In fact, there is no known efficient (polynomial-time) algorithm for computing discrete logarithms.

Diffie-Hellman key exchange

Many methods for encrypting messages require both the sender and recipient of a message to use a shared secret key (unlike RSA, which uses a public key). This poses the following question: how can two distant parties agree on a key in such a way that someone intercepting their communications will be unable to determine the value of the key? This dilemma, known as the *key exchange problem*, was solved in 1976 by Diffie, Hellman, and Merkle.

Martin Hellman was a professor of electrical engineering at Stanford University, and Whitfield Diffie and Ralph Merkle were his graduate students. Their idea grew out of the following story: Alice plans to send Bob a confidential letter in a locked box. She wants to be sure that only Bob can open the box. If Alice were to mail the key to Bob, someone else might intercept it and open the box. Instead, she uses a trick. She locks the box with a padlock and sends the locked box to Bob. When Bob receives the box, he adds his own padlock to it and sends it back to Alice. Alice then removes her padlock and again sends it to Bob. Finally, Bob again receives the box, which he can now open since it is locked only with his own padlock.

Diffie, Hellman, and Merkle's solution to the key exchange problem enables two parties to create a shared secret key in such a way that an eavesdropper intercepting their communications has no way to figure out the key. Their solution, known as the Diffie-Hellman key exchange protocol, is believed to be secure because it relies on a *one-way function*: modular exponentiation. When the modulus, p, is very large, raising a primitive root to a power in $\mathbf{Z}_p$ can be done quickly on a computer. However, undoing this process—that is, taking a discrete logarithm in $\mathbf{Z}_p$—is practically impossible with today's computers and algorithms. Thus, the security of Diffie-Hellman key exchange depends on the computational difficulty of taking discrete logarithms.[5]

---

[5] Recall that the RSA Encryption Algorithm (9.5.1) is believed to be secure because multiplying the two large primes p and q together is easy and fast, but undoing the process (factoring their product) cannot be done in any reasonable amount of time using today's algorithms. In the same way that the security of Diffie-Hellman depends on the computational difficulty of taking discrete logarithms, the security of RSA depends on the computational difficulty of factoring large numbers.

The Diffie-Hellman key exchange protocol enables Alice and Bob, using insecure communication, to agree on the value of the key, k, in such a way that no one else knows k.

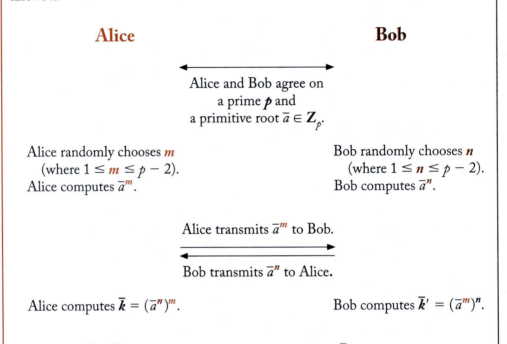

Alice   **Bob**

Alice and Bob agree on a prime p and a primitive root $\bar{a} \in \mathbf{Z}_p$.

Alice randomly chooses m (where $1 \leq m \leq p - 2$). Alice computes $\bar{a}^m$.

Bob randomly chooses n (where $1 \leq n \leq p - 2$). Bob computes $\bar{a}^n$.

Alice transmits $\bar{a}^m$ to Bob.

Bob transmits $\bar{a}^n$ to Alice.

Alice computes $\bar{k} = (\bar{a}^n)^m$.

Bob computes $\bar{k}' = (\bar{a}^m)^n$.

Note that $\bar{k} = \bar{k}'$ by the usual laws of exponents. Since $\bar{k}$ is a nonzero element of $\mathbf{Z}_p$, it corresponds to a unique natural number $k \leq p - 1$. Thus, Alice and Bob are now both in possession of the shared secret key, k.

Alice and Bob now have a common shared key, k, but is k really secret? Suppose that Evil Eve the Eavesdropper listens in on all of the messages that Alice and Bob transmit as part of the Diffie-Hellman protocol. Then Eve would know $p, \bar{a}, \bar{a}^m$, and $\bar{a}^n$ (the values on the arrows in 10.4.2). Eve can determine the value of the key, k, if she knows the value of m or n (Alice and Bob's private exponents). In other words, Eve would have to find the discrete logarithm (base $\bar{a}$) of either $\bar{a}^m$ or $\bar{a}^n$ in $\mathbf{Z}_p$. However, as discussed earlier, taking discrete logarithms is computationally intractable. Hence, Eve will be unable to find m or n, so she cannot determine k in this way.

It is perhaps possible that Eve could concoct some more efficient strategy than computing discrete logarithms, for computing k given the values of $p, \bar{a}, \bar{a}^m$, and $\bar{a}^n$. However, no such strategy is known. Thus, it is widely believed that so long as taking discrete logarithms remains computationally infeasible, the Diffie-Hellman protocol is secure.

Finding a primitive root modulo a large prime

To use the Diffie-Hellman protocol (10.4.2), Alice and Bob need to find a large prime, p, and a primitive root, a, modulo p. As mentioned in Section 9.5 and explained in detail in Chapter 12, large primes can be found quickly using a computer. Furthermore, we know by the Primitive Root Theorem (10.3.3) that there exist primitive roots modulo every prime. However, given a large prime p, it is often difficult[6] to find a primitive root modulo p. (This is because determining whether any particular number is a primitive root modulo a large prime p can take a very, very, very long time.)

There is a way around this problem. Consider the following example.

EXAMPLE 2

To communicate securely, Alice and Bob are planning to exchange keys using Diffie-Hellman (10.4.2). They wish to find a large (approximately 40-digit) prime, p, and a primitive root modulo p. If they first picked a random 40-digit prime p and then tried to find a primitive root modulo p, it could take an extremely long time for them to find a primitive root modulo p. How do they proceed?

SOLUTION Alice and Bob start by finding a random 40-digit prime q and then (using the methods of Chapter 12) test whether the number $p = 2q + 1$ is prime. If it is not, they continue generating new 40-digit primes q until they find one for which the number $p = 2q + 1$ is also prime. By the Prime Number Theorem (3.3.3), the chance that a random 40-digit number p is prime is approximately $\dfrac{1}{\log(10^{40})} \approx \dfrac{1}{92}$, which is greater than a 1% chance. Thus, after checking on the order of 100 primes q, Alice and Bob's computer will find a q for which $p = 2q + 1$ is also prime.

Alice and Bob find that for the 40-digit prime

$$q = 1768604305613921745580037409259811956979,$$

the number $p = 2q + 1$ is also prime:

$$p = 2q + 1$$
$$= 3537208611227843491160074818519623913959.$$

Alice and Bob now have a prime p for which $p - 1$ has a particularly easy factorization:

$$p - 1 = 2q.$$

---

[6]In fact, there is no known polynomial-time algorithm for finding a primitive root modulo a given prime p.

Given a natural number $a < p$, it is now computationally easy to test whether a is a primitive root modulo p, as follows.[7] By Corollary 10.1.4, $\mathrm{ord}_p(a) \mid p - 1$ or, equivalently,

$$\mathrm{ord}_p(a) \mid 2q.$$

Thus, there are only four possible values for the order of a:

$$\mathrm{ord}_p(a) = 1, 2, q, \text{ or } 2q.$$

Now Alice and Bob just need to test different values of a until they find one of order $2q$. In other words, they must find a natural number $a < p$ such that

$$a^1 \not\equiv 1 \pmod{p}, a^2 \not\equiv 1 \pmod{p}, \text{ and } a^q \not\equiv 1 \pmod{p}.$$

Alice and Bob decide to try $a = 7$. By computing $7^1, 7^2$, and 7^q, they determine that

$$7^1 \not\equiv 1 \pmod{p}, 7^2 \not\equiv 1 \pmod{p}, \text{ and } 7^q \not\equiv 1 \pmod{p},$$

They conclude that

$$\mathrm{ord}_p(7) = 2q.$$

In other words, $\mathrm{ord}_p(7) = p - 1$. Hence, $a = 7$ is a primitive root modulo p.

Alice and Bob now have their 40-digit prime p, and a primitive root modulo p. They are all set to use Diffie-Hellman.

Random and pseudorandom numbers

Random numbers are important for many applications. For example, to find the large prime numbers necessary to set up an RSA cryptosystem (9.5.1), we need a method of generating random numbers. In addition, random numbers are frequently used to create simulated data to study real-world phenomena such as weather patterns and economic trends. Such simulations often enable scientists to study the relationships between variables without having to collect actual data. However, generating a large number of random numbers is not easy, and simulations may require millions (or more) of random numbers to be used as data.

A list of numbers is said to be *random* if each number in the list is determined purely by chance. For example, random numbers can be generated using physical methods such as rolling dice, spinning a wheel, or recording the time it takes for a single atom of a radioactive substance to decay. Any list of numbers that is created

---

[7] More generally, if you know the factorization of $p - 1$, then there is a computationally efficient way to determine whether a given number is a primitive root modulo p. (See Exercises 8–10 of Section 10.3.)

using an algorithm is by definition *not* random. However, there are algorithms that generate sequences of numbers that appear to be random, can be used like random numbers, and can even pass statistical tests of randomness. Such numbers are said to be *pseudorandom*.

In 1946, John Von Neumann designed the first computer program to generate a sequence of pseudorandom numbers. His method, known as the *middle-square method*, works as follows. Suppose you wish to generate a list of 6-digit pseudorandom numbers. You start with a random 6-digit number x_0, called the seed, which will be the first pseudorandom number of your list. To get the next 6-digit pseudorandom number, you find x_0^2, which may be written as a 12-digit number, then define x_1 by selecting the middle 6 digits of this number. For example, if $x_0 = 739832$, we have

$$x_0 = 739832 \quad \Rightarrow \quad x_0^2 = 547\mathbf{351388}224.$$

$$x_1 = 351388 \quad \Rightarrow \quad x_1^2 = 123\mathbf{473526}544.$$

$$x_2 = 473526, \quad \dots$$

We can repeat this process over and over to obtain a sequence of pseudorandom numbers:

$$739832, 351388, 473526, 226872, 470904, 750577, 365832, 833052, \dots \quad (1)$$

Though this sequence is generated in a deterministic way, just eyeballing the numbers in the sequence might lead us to believe that they are random. How can we make a more precise statement about the "randomness" of sequence (1)? It turns out that precisely measuring the randomness of a sequence of numbers is a tricky business. Thus, it may be hard to say exactly how random this sequence really is.

However, sequence (1) has one property that is highly nonrandom: it cycles. This is not hard to show. Since every number in the sequence will be less than 1,000,000, we know that if we compute 1,000,000 terms of the sequence, at some point we will see a number that we saw before, and when we do, the sequence will repeat from that point on. Hence, over the long haul, sequence (1) is not very random at all! Not all is lost, however. If we continue calculating terms in the sequence, it might take a long time—possibly as many as 1,000,000 terms—before the sequence repeats. In fact, though, we find that sequence (1) begins to repeat after just 241 terms. (We find that $x_{241} = 921,000$, which is a repetition of $x_{221} = 921,000$.)

This disappointingly small number of steps before finding a repeated number is typical of the middle-squares method. Starting with an arbitrarily chosen 6-digit seed x_0, one must compute on average only 326 terms before the sequence begins to repeat. And it turns out that no matter which seed one starts with, the sequence is certain to repeat after at most 943 terms. One could also perform the middle-squares method on numbers in binary notation, instead of base 10. With 20-bit binary numbers (which are of comparable size to 6-digit decimal numbers), the sequences generated by the method will go an average of 637 steps—and at most 2537 steps—before repeating.

Fortunately, our study of primitive roots gives us a much better way to generate sequences that continue for a long time without repeating. If p is a prime number, and a is a primitive root modulo p, then we know that the sequence

$$a^1, a^2, \ldots, a^{p-1}$$

when reduced modulo p, contains every integer between 1 and $p - 1$. For example, if our modulus is the prime number $p = 1{,}000{,}003$, then the number $a = 750919$ is a primitive root, and the powers $a^1, a^2, a^3, \ldots$ reduced modulo 1,000,003 yield the following sequence:

$$750919, 652930, 71782, 305952, 480656, 640068, 780581, 345486, 223341, \ldots . \ (2)$$

This sequence, like sequence (1), has the general appearance of randomness. And while this feeling of randomness may be difficult to quantify, at least we can be absolutely certain of one thing: since $a = 750919$ is a primitive root, sequence (2) will continue for a whopping 1,000,002 terms before repeating!

The key feature of sequence (2) is that each term can be calculated from the previous term by multiplying by a and reducing modulo p. This method of generating pseudo-random numbers is called the *pure multiplicative congruential method*, or simply the *multiplicative congruential method*.

METHOD 10.4.3   MULTIPLICATIVE CONGRUENTIAL GENERATOR

Let m, a, and x_0 be natural numbers. The (pure) multiplicative congruential sequence with *modulus* m, *multiplier* a, and *seed* x_0 is defined by the recursive formula:

$$x_{n+1} \equiv a x_n \ (\mathrm{mod}\ m), \ \text{where } 0 \le x_{n+1} < m.$$

There is a simple formula for the terms x_n of this sequence generated by the multiplicative congruential method. Since we start with x_0 and repeatedly multiply by a, we find

$$x_n \equiv a^n x_0 \ (\mathrm{mod}\ m). \tag{3}$$

How many steps will it take before this sequence starts to repeat? Let

$$r = \mathrm{ord}_m(a).$$

Then $a^r \equiv 1 \ (\mathrm{mod}\ m)$, and so the rth term of sequence (3) is

$$x_r \equiv x_0 \ (\mathrm{mod}\ m).$$

It is usual to choose the multiplier a and the seed x_0 so that each is relatively prime to the modulus m. Under these conditions, the sequence x_n does not repeat until the rth term (as you will prove in Exercise 15).

Although not required for the method, it is common to choose a modulus m that is prime and to choose a multiplier a that is a primitive root modulo m. This ensures that $r = \text{ord}_m(a)$ is as large as possible; hence, the resulting pseudorandom sequence goes for as long as possible without repeating.

EXAMPLE 3

Several authors suggest using a multiplicative congruential generator whose modulus is the Mersenne prime $m = 2^{31} - 1 = 2147483647$, with multiplier $a = 16807$, a primitive root modulo m. Starting with the seed $x_0 = 123456789$ generates the following sequence of pseudorandom numbers:

$$x_0 = 123456789,$$

$$x_1 \equiv 16807 \cdot 123456789 \equiv 469049721 \pmod{m}$$

$$x_2 \equiv 16807 \cdot 469049721 \equiv 2053676357 \pmod{m}$$

$$\vdots$$

Continuing in this manner yields the sequence

$$123456789, 469049721, 2053676357, 1781357515, 1206231778,$$
$$891865166, 141988902, 553144097, 236130416, 94122056, \ldots.$$

Since a is a primitive root modulo m (i.e., $\text{ord}_m(a) = m - 1$), this sequence will produce $m - 1 = 2{,}147{,}483{,}646$ pseudorandom numbers before repeating.

A generalization of the pure multiplicative congruential method is the *linear congruential method*. Developed by D. H. Lehmer in 1949, this method is widely used today for generating pseudorandom numbers.

METHOD 10.4.4   LINEAR CONGRUENTIAL GENERATOR

Let m be a natural number, and let a, x_0, and c be nonnegative integers. The linear congruential sequence with *modulus* m, *multiplier* a, *increment* c, and *seed* x_0 is defined by the recursive formula:

$$x_{n+1} \equiv ax_n + c \pmod{m}, \text{ where } 0 \leq x_{n+1} < m.$$

Determining the period of a linear congruential generator is harder than determining the period of a multiplicative congruential generator. Since the pseudorandom numbers obtained with the linear congruential method are all in the range $0, \ldots, m - 1$, the period is at most m steps. Under certain conditions, the period turns out to be exactly m, which is as large as possible.

As in the following example, m is sometimes chosen to be a power of 2. In this case, the period turns out to be m provided that c is odd and $a \equiv 1 \pmod 4$.

EXAMPLE 4

A commonly used linear congruential generator has modulus $m = 2^{32}$, multiplier $a = 1664525$, increment $c = 1013904223$, and seed $x_0 = 123456789$. Using these values, we find

$$x_0 = \mathbf{123456789}$$

$$x_1 \equiv 1664525 \cdot \mathbf{123456789} + 1013904223 \equiv \mathbf{920370032} \pmod{2^{32}}$$

$$x_2 \equiv 1664525 \cdot \mathbf{920370032} + 1013904223 \equiv \mathbf{3761641487} \pmod{2^{32}}$$

$$\vdots$$

Continuing in this manner yields the following sequence of pseudorandom numbers:

123456789, 920370032, 3761641487, 2252023330, 1475571481, 2340457892, 1600748723, 1240767094, 2297155421, ...

This sequence continues for $m = 2^{32} = 4{,}294{,}967{,}296$ terms before repeating.

Linear congruential generators do not use a lot of memory, so they are often used for video games, as well as for statistical modeling and other applications. However, the linear congruential method is not sufficiently secure to be used for cryptography. The risk is that if someone intercepts several terms in a row, then that person can figure out the values of a, c, and m and, hence, be able to find all of the terms of the sequence. (In Exercises 20–24, you will demonstrate this lack of security by cracking a pure multiplicative congruential generator.)

Fun Facts

In 1997, Makoto Matsumoto and Takuji Nishimura developed a new method of generating pseudorandom numbers. This method, known as the *Mersenne twister*, is faster than previous methods and generates numbers that can pass more statistical tests of randomness. The name of this generator comes from the fact that its period is the Mersenne prime $2^{19937} - 1$. With such a large period, from a practical point of view, we do not have to worry about the sequence repeating itself. This method is quickly becoming a popular method for most applications of pseudorandom numbers except for cryptography.

The period of a repeating decimal

You may have wondered why some repeating decimals, such as $\frac{1}{3} = .3333\ldots = .\overline{3}$, are so simple, while some, such as $\frac{1}{7} = .\overline{142857}$, are more complicated. Contrast

$$\frac{1}{29} = .\overline{0344827586206896551724137931},$$

whose *period* (the number of digits in the repeating pattern) is 28 digits, with

$$\frac{1}{37} = .\overline{027},$$

whose period is only 3 digits. Given a fraction $\frac{1}{p}$, where p is a prime number, is there a way to tell what the period of its decimal expansion will be? Sometimes $\frac{1}{p}$ has a particularly long period: for $p = 7$ and $p = 29$, the period of $\frac{1}{p}$ is exactly $p - 1$. For which primes p does this happen?

Also observe that in all of the examples we have considered, the decimal expansion of $\frac{1}{p}$ is *purely periodic*; that is, the repeating part starts immediately after the decimal point. Is this true for every prime p?

To answer these questions, first recall that any repeating decimal, such as $12.0\overline{863}$, $.\overline{254}$, or $.\overline{7}$, can be expressed as a fraction. There is an especially easy rule for finding this fraction for purely periodic decimals, such as $.\overline{7}$ and $.\overline{254}$:

$$.\overline{7} = \frac{7}{9} \qquad .\overline{254} = \frac{254}{999}.$$

Simply place the repeating part over the number $99\ldots9$, where there is one 9 for each digit in the repeating part of the decimal. In general, if $d_1, d_2, \ldots, d_r$ is any sequence of digits, then we can express the repeating decimal $.\overline{d_1 d_2 \ldots d_r}$ as the following fraction:

$$.\overline{d_1 d_2 \ldots d_r} = \frac{d_1 d_2 \ldots d_r}{99 \ldots 9}, \tag{4}$$

where the denominator is the r-digit natural number consisting entirely of 9's. (See Exercise 17.) Now let p be a prime number. (We assume that p is not 2 or 5, since $\frac{1}{2}$ and $\frac{1}{5}$ have the finite decimal expansions 0.5 and 0.2.) We wish to find the period of the decimal expansion of $\frac{1}{p}$.

When can the fraction $\frac{1}{p}$ have a decimal expansion of the form $.\overline{d_1 d_2 \ldots d_r}$? For this to occur, by equation (4) we must have

$$\frac{1}{p} = \frac{d_1 d_2 \ldots d_r}{99 \ldots 9} = \frac{d_1 d_2 \ldots d_r}{10^r - 1}.$$

In other words,

$$10^r - 1 = p(d_1 d_2 \ldots d_r).$$

Such a number $d_1 d_2 \ldots d_r$ exists if and only if

$$p \mid 10^r - 1,$$

which is equivalent to

$$10^r \equiv 1 \pmod{p}.$$

We can summarize this discussion with the following equivalence:

$\frac{1}{p}$ has a decimal expansion of the form $. \overline{d_1 d_2 \dots d_r}$ if and only if $10^r \equiv 1 \pmod{p}$.

The period is the smallest positive integer r for which $\frac{1}{p}$ has a decimal expansion of the form $.\overline{d_1 d_2 \dots d_r}$. The smallest positive integer r for which this congruence holds is the order of 10 modulo p; that is,

$$r = \text{ord}_p(10).$$

Thus, we have shown:

OBSERVATION 10.4.5

Let p be a prime number with $p \neq 2$ and $p \neq 5$. Then the decimal expansion of $\frac{1}{p}$ is purely periodic with period $\text{ord}_p(10)$.

It follows from this observation that the period of the decimal expansion of $\frac{1}{p}$ will be a divisor of $p - 1$. Furthermore, the period of $\frac{1}{p}$ will be the particularly large number $p - 1$ exactly when $\text{ord}_p(10) = p - 1$. In other words, this will occur exactly when 10 is a primitive root modulo p. This happens relatively frequently; for primes p less than 1,000,000, more than 37% have this property. However, it is unknown whether there are infinitely many primes with this property.[8]

EXERCISES 10.4

Numerical Problems

1. In $\mathbf{Z}_{17}$, using the primitive root $\overline{3}$ as the base, find each of the following.

   a. the discrete logarithm of $\overline{13}$

   b. the discrete logarithm of $\overline{11}$

2. In $\mathbf{Z}_{13}$, find the discrete logarithm (base $\overline{2}$) of $\overline{10}$.

3. In $\mathbf{Z}_{19}$, find each of the following.

   a. the discrete logarithm (base $\overline{2}$) of $\overline{3}$

   b. the discrete logarithm (base $\overline{3}$) of $\overline{12}$

---

[8] This is a special case of *Artin's conjecture*, which asserts that given any integer $a \neq -1$ that is not a perfect square, there are infinitely many primes p for which a is a primitive root modulo p.

4. For the Diffie-Hellman Key Exchange Protocol (10.4.2), suppose Alice and Bob agree on the prime $p = 11$ and the primitive root $\bar{a} = 6$.

   a. Alice randomly chooses the exponent $m = 7$. What value does she transmit to Bob?

   b. Bob randomly chooses the exponent $n = 4$. What value does he transmit to Alice?

   c. Pretend you are Alice. Use the value that Bob transmitted to compute the key, k.

   d. Now be Bob. Use the value that Alice transmitted to compute the key, k'.

5. By eavesdropping on Alice and Bob's communications, Evil Eve overhears that they have agreed on the values $p = 47$, and $\bar{a} = 5$. Eve also intercepts that Alice transmits the number $\overline{36}$ to Bob, and Bob transmits the number $\overline{38}$ to Alice. By some method, Evil Eve figures out that the discrete logarithm (base $\bar{5}$) of $\overline{36}$ is 30. Use this to help Eve find Alice and Bob's shared key, k.

6. By hacking into Alice and Bob's instant-messaging service, Evil Eve the Eavesdropper intercepts the following chat transmissions.

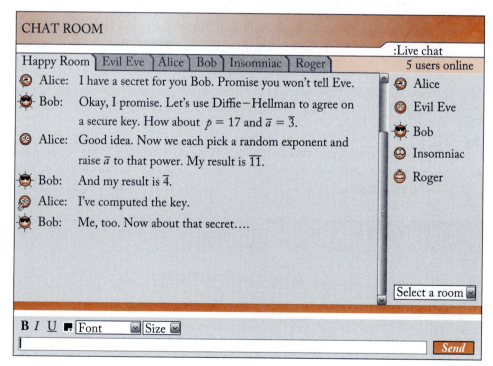

   a. Pretend you are Evil Eve. Determine the value of Alice and Bob's shared key, k.

   b. How should Alice and Bob change the numbers they used to make it much more difficult for Evil Eve to figure out the key?

7. **Computer Exercise** Using the method of Example 2, find a large (30-digit) prime, p, and a primitive root, a, modulo p. Write down the key steps of the method for your numbers.

8. **Computer Exercise** Find the first ten terms of a sequence of pseudorandom numbers generated using the middle-square method with seed $x_0 = 234818$.

9. **Computer Exercise** Find the first ten terms of a sequence of pseudorandom numbers generated using the middle-square method with seed $x_0 = 326916$.

10. **Computer Exercise** Determine the period of the sequence of pseudorandom numbers generated using the middle-square method with seed $x_0 = 722794$.

11. **Computer Exercise** Determine the period of the sequence of pseudorandom numbers generated using the middle-square method with seed $x_0 = 971582$.

12. **a.** Find the sequence of pseudorandom numbers using the pure multiplicative congruential method (10.4.3) with $x_0 = 10$, $a = 4$, and $m = 17$. What is the period of this sequence?

    b. Find the sequence of pseudorandom numbers using the pure multiplicative congruential method (10.4.3) with $x_0 = 10$, $a = 3$, and $m = 17$. What is the period of this sequence?

13. **a.** Find the sequence of pseudorandom numbers using the linear congruential method (10.4.4) with $x_0 = 5$, $a = 2$, $c = 1$, and $m = 11$. What is the period of this sequence?

    b. Find the sequence of pseudorandom numbers using the linear congruential method (10.4.4) with $x_0 = 5$, $a = 3$, $c = 1$, and $m = 11$. What is the period of this sequence?

Reasoning and Proofs

14. In the real numbers, you are familiar with the identity

$$\log(xy) = \log(x) + \log(y).$$

In this exercise, you will prove the following theorem, which is the corresponding identity for discrete logarithms.

THEOREM. *Let p be prime, and let $\bar{a}$ be a primitive root in $\mathbf{Z}_p$. (We will use $\bar{a}$ as the base for our discrete logarithms.) Let $\bar{x}, \bar{y} \in \mathbf{Z}_p$ be nonzero, let l_x be the discrete logarithm of x, let l_y be the discrete logarithm of y, and let l_{xy} be the discrete logarithm of $\bar{x} \cdot \bar{y}$. Then*

$$l_{xy} \equiv l_x + l_y \pmod{p - 1}.$$

15. Let $m, a, x_0 \in \mathbf{N}$ such that $\gcd(m, a) = 1$ and $\gcd(m, x_0) = 1$. Let $r = \operatorname{ord}_m(a)$. In this exercise, you will prove that the sequence generated by the multiplicative congruential method (10.4.3) does not repeat until the rth step.

    a. Prove that $x_r = x_0$.

    b. Prove that for all i, j in the range $0, 1, \ldots, r - 1$, if $i \neq j$ then $x_i \neq x_j$.

16. In the pure multiplicative congruential method, the nth term is given by the closed form expression of equation (3): $x_n \equiv a^n x_0 \pmod{m}$. Find a similar closed-form expression for x_n for the linear congruential method.

17. Let $d_1, d_2, \ldots, d_r$ be a sequence of decimal digits. Explain why the repeating decimal $.\overline{d_1 d_2 \ldots d_r}$ can be expressed as the following fraction:

$$.\overline{d_1 d_2 \ldots d_r} = \frac{d_1 d_2 \ldots d_r}{99 \ldots 9}.$$

    (The denominator consists of the digit 9 repeated r times.)

18. Let p be a prime number, and let a be an integer such that $1 \leq a \leq p - 1$. Consider the fraction $\frac{a}{p}$.

    a. Make a conjecture about the period of the decimal expansion for $\frac{a}{p}$.

    *b. Sometimes the decimal expansion for $\frac{a}{p}$ is a *cyclic shift* of the decimal expansion for $\frac{1}{p}$. For example, $\frac{1}{13} = .\overline{076923}$ and $\frac{3}{13} = .\overline{230769}$. However, sometimes this doesn't happen: $\frac{2}{13} = .\overline{153846}$. Make a conjecture about when this happens.

19. Artin's conjecture asserts that given any integer $a \neq -1$ that is not a perfect square, there are infinitely many primes p for which a is a primitive root modulo p. This exercise asks you to explain why the perfect squares and -1 are excluded from the conjecture.

    a. Determine all primes p for which 9 is a primitive root modulo p.

    b. Prove that if a is a perfect square, then there are only finitely many primes p for which a is a primitive root modulo p.

    c. Determine all primes p such that -1 is a primitive root modulo p. Prove that your answer is correct.

EXPLORATION Cracking a Pure Multiplicative Congruential Generator
(Exercises 20–24)

Neither the multiplicative congruential method (10.4.3) nor the linear congruential method (10.4.4) is a secure enough pseudorandom number generator to be used for cryptography. The reason is that by intercepting several terms in a row, it is possible to figure out the values of the parameters, from which you can compute all of the terms of the sequence. You will now demonstrate this lack of security by cracking a multiplicative congruential generator.

20. Consider the following multiplicative congruential generator: modulus $m = 7$, multiplier $a = 3$, and seed $x_0 = 2$.

    a. Compute $x_1, x_2, \ldots, x_7$.

    b. For each $k = 1, \ldots, 6$, compute $x_k^2 - x_{k-1} x_{k+1}$. What do you notice about all of these values?

21. Make a conjecture generalizing your observation from Exercise 20 part b. That is, if the sequence $x_0, x_1, x_2, \ldots$ is defined by $x_i \equiv a^i x_0 \pmod{m}$ for an arbitrary modulus m, multiplier a, and seed x_0, then what can you say about $x_k^2 - x_{k-1} x_{k+1}$?

22. Prove your conjecture from Exercise 21.

23. The sequence $6, 9, 8, 1, \ldots$ comes from a multiplicative congruential generator. Using the result of Exercises 21 and 22, find the modulus, m. Then find the multiplier, a.

24. The sequence $259, 480, 32, 549, 289, \ldots$ comes from a multiplicative congruential generator. Find the modulus, m, and the multiplier, a. Exhibit clairvoyance by "predicting" what the next "random" number will be.

EXPLORATION Slide Rule for Modular Multiplication (Exercises 25–28)

You can make a device for multiplying numbers modulo 13 as follows: Cut out the two circles below. Then place the small circle on top of the large circle, aligning the centers of the two circles. Fasten the two circles together at the center, but allowing the small circle to rotate.

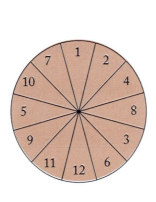
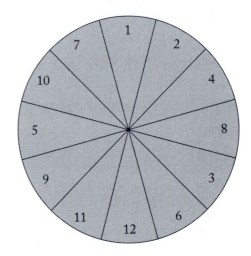

25. Figure out how to use your wheel to quickly multiply any two numbers modulo 13. Explain how to use your wheel to multiply 5 by 11 modulo 13.

26. Explain why your wheel works. (Your explanation should include the phrase "primitive root" and possibly the phrase "discrete logarithm.")

27. Create a wheel for multiplication modulo 17.

*28. Can you create a wheel for multiplication modulo 26?

Chapter 11

SPOTLIGHT ON...

Ferdinand Gotthold Max Eisenstein (1823–1852)

The mathematical genius Gotthold Eisenstein spent his short life in the city of Berlin, which at that time was a part of the Prussian Empire. Eisenstein's childhood was shaped by his father's difficulties finding satisfactory employment and by the deaths of his five brothers and sisters from meningitis. Eisenstein himself survived meningitis, as well as various other childhood illnesses, though he never completely regained his health.

Eisenstein showed unusual mathematical talent at an early age. In his autobiography, he wrote, "As a boy of six I could understand the proof of a mathematical theorem more readily than that meat had to be cut with one's knife, not one's fork." When he was 15, he taught himself calculus by reading the works of Euler and Lagrange. Eisenstein fell in love with number theory at the age of 19 as a result of reading Gauss's *Disquisitiones Arithmeticae*. Around that time, he and his family traveled to England, Wales, and Ireland, in an unsuccessful attempt to find a better job for his father. In Dublin, Eisenstein met the well-known mathematician W. R. Hamilton, who gave Eisenstein a paper to present to the Berlin Academy upon his return home. Hamilton's paper, which discussed Abel's proof that there is no general solution to a fifth-degree polynomial equation, stimulated Eisenstein to begin his own mathematical research. After returning to Berlin, not only did he present Hamilton's paper to the Academy, but he also presented a paper of his own. August Leopold Crelle, a mathematician at the Academy, offered to publish Eisenstein's paper in the journal that he edited. The following year, Eisenstein published 23 papers and two problems in Crelle's journal—some in German, some in French, and some in Latin.

The mathematician Alexander Von Humboldt was so impressed with Eisenstein's work that he asked the king, the Ministry of Education, and the Berlin Academy to give Eisenstein financial support so that he could devote himself to mathematics. Humboldt supplemented these grants with money from his own pocket. Even Gauss, who often judged young mathematicians harshly, wrote that Eisenstein's ability was such that "nature bestows on only a few in each century."

In spite of his great mathematical success, Eisenstein suffered from depression and health problems. He had very few friends in Berlin. Also, he found it demoralizing that he had to ask for money each time his government grants ran out. On top of that, Carl Gustav Jacob Jacobi, a mathematician with whom Eisenstein had been close, publicly accused Eisenstein of plagiarizing his ideas.

In 1848, there was fighting in the streets of Berlin as part of a democratic and socialist movement. Shots were fired on Prussian troops from within a house, and everyone inside the house was arrested. This included Eisenstein, who claimed that he had just stepped into the house to get away from the street fighting. The prisoners were treated badly by the soldiers, and though Eisenstein was released the following day, his emotional and physical health deteriorated substantially as a result of this experience. In addition, it became increasingly difficult for him to get government grants after his arrest. While he continued to publish papers on mathematics, during the next few years Eisenstein rarely lectured, and those lectures he delivered were from his bed. In 1852, at the age of only 29, Eisenstein died of tuberculosis.

A bit of Eisenstein's math

Eisenstein spent most of his mathematical career studying number theory. He generalized Gauss's results on Quadratic Reciprocity and on the theory of forms. He also obtained significant results on elliptic functions.

The Law of Quadratic Reciprocity is a fundamental result of number theory that is used to determine when a quadratic congruence has a solution (see Section 11.3). Euler originally stated the law in 1744. Legendre rediscovered it in 1785 and gave a partial proof. Then in 1795, when Gauss was only 18, he rediscovered the Law of Quadratic Reciprocity and proved it rigorously within a year. Gauss spoke of Quadratic Reciprocity as "the golden theorem of number theory," and during Gauss's lifetime, he published at least six different proofs of it. Many other mathematicians were fascinated by this result, and they came up with their own proofs as well. Eisenstein published four additional proofs, one of which is particularly noteworthy because it is more geometric than the previous proofs that had been given. Eisenstein's ideas play a key role in the proof of the Law of Quadratic Reciprocity that we will present in this chapter.

Chapter 11 Quadratic Residues

11.1 Squares Modulo *n*

The 19th-century mathematician Gotthold Eisenstein had a hard life and died young. What's not so well known about Eisenstein is that he was actually Hungarian royalty. His body was entombed at the Eisenstein family mansion in Transylvania, and by the late 19th century, rumors had spread throughout the region that the old abandoned mansion was haunted by the ghost of the great mathematician. Even the ruler of the county, the tyrannical Count Residulus, refused to go near the place.

On the day that old Count Residulus died, none of the county's peasants was very sad, but everyone wondered what would happen to the count's most prized possession, the exquisite square island of Quadratica. The 11 peasants of the town gathered in the town square for the reading of the count's will. Also present were the count's five daughters—Muffy, Tiffy, Buffy, Fifi, and Geraldine—arguing loudly and heatedly about which one of them deserved more of their father's inheritance.

When the town clerk finished reading the three requirements of the count's will, a murmur broke out among the crowd as everyone tried to understand the count's decree. Just then, a young peasant woman grinned broadly and began to jump up and down. "I've got it! Look at this!" The woman quickly sketched a map.

The peasant woman proposed that Quadratica would be divided into 49 equal square plots. Of these, 5 plots would be reserved for the 5 daughters' palaces. The remaining 44 plots would be apportioned equally among the 11 peasants, with each peasant receiving 4 plots for garlic farming.

All parties present applauded the pleasant peasant's proposed plan. Just then, the late count's smallest daughter, Muffy, spoke up. "Wait a minute—there's another clause in Daddy's will!"

That evening, storm clouds darkened the skies above the overgrown Eisenstein mansion, and thunder rumbled in the distance. The count's 5 daughters and the 11 peasants pushed aside cobwebs as they cautiously filed into the abandoned mansion. Just as the last peasant crossed the threshold, the wind outside picked up, causing the door to close firmly behind them with an eerie creak.

"This place is creepy," said one peasant whose face had gone pale. "I've heard it's haunted by the ghost of Dr. Eisenstein!"

"Oh, come on, there's nothing to worry about," said Muffy. "Everyone knows Eisenstein died at the early age of 29. And ever since then, his body has rested peacefully in the crypt beneath this mansion. Now let's just get this over with." Muttering under her breath, she added, "So I can get my palace!" The other daughters, standing close by, heard her and nodded enthusiastically.

At first light, the household was awakened by a shrill scream coming from the room where the count's daughters were sleeping. By the time the peasants arrived, Buffy had helped the crying Muffy to regain a small amount of composure. The other three daughters were lying on the floor, each one with two small round marks on her neck.

"Tiffy, Fifi, and Geraldine are dead!" sobbed Muffy.

"They must have been stabbed in the night by the ghost of Dr. Eisenstein," gasped one of the peasants.

Buffy knelt down to take a closer look. "These aren't knife wounds. They're teeth marks!"

"Are you suggesting that this could be the work of a vampire?" asked the peasant.

"Well," said Buffy, "it *is* well known that Eisenstein had vampiristic tendencies."

Muffy wiped the tears from her eyes. "Buffy, how can you be worried about our sisters' death at a time like this? What I want to know is, how are we going to get our palaces now? Now that there are only two daughters left, we need a new plan for dividing up the island. Where's that clever peasant woman? I'm sure she can figure it out for us."

"There's just no way to do it!" the young peasant woman apologized, stepping forward. "It's possible to divide up the island as required by the count's will if the number of daughters is 1, 3, 4, 5, or 9. But with 2 daughters, it just can't be done."

"This bites!" exclaimed Muffy. "Eisenstein must pay for his heartless murder of our sisters. If he had killed any other number of daughters, it would not have been such a tragedy. But instead that cold, calculating criminal has ruined our inheritance!"

"Yeah, that guy sucks! Let's go drive a stake through his heart!" yelled Buffy. "I'll bet he's still encrypted in the basement. Does anyone have the key?"

Buffy led her sister and the peasants down the narrow, winding staircase to the basement. She kicked open the door to reveal a dark, cavernous room. A gaunt, scholarly-looking figure was sitting upright in his coffin, lecturing. Buffy looked up at the ceiling, where a flock of bats was hanging on Eisenstein's every word.

"Herr Eisenstein!" Buffy exclaimed, stomping her foot. "You knew perfectly well that slaying three of my sisters would foil our plans and save this awful old barn of building! Why did you do it?"

"Ah, my dear," replied Eisenstein bitterly, "because I left your world with no heirs, my beautiful island became the property of the state of Transylvania, and your greedy father Count Residulus falsely claimed it as his own. But now that you cannot satisfy the conditions of his will, the island will forever remain in its natural state: a paradise for all of the indigenous species of bats, rats, and spiders."

"Well, Eisenstein, even though you've prevented us from taking your island, we can still take care of you!" Muffy cleared her throat and announced, "Gotthold Eisenstein, for your heinous murder of our sisters, we hereby condemn you under the Three Bites Law. This fair and just law carries a mandatory sentence of death by illumination. Sister, would you like to carry out the sentence?"

Buffy pulled back a heavy velvet curtain that was covering a wide window, and sunlight flooded the room. "Rise and shine, Herr Eisenstein!"

In our myth, Eisenstein ruined the day by leaving exactly 2 of the count's daughters alive. How did the peasant woman figure out that there was no way to comply with the count's will? The island would have to be divided into an x-by-x grid of square plots, with 2 plots for the living daughters' palaces and the remaining plots for garlic farming:

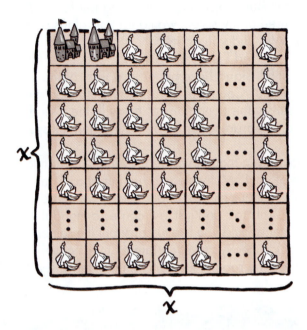

The count's will stipulates that the plots allocated for garlic farming must be divided *equally* among the **11** peasants. Thus, the number of garlic-farming plots must be equal to **11**k for some integer k. In total, the count's decree requires that the x^2 plots on the island be apportioned as **2** plots for the living daughters' palaces and **11**k plots for garlic farming:

$$x^2 = 2 + 11k$$

To follow the count's decree, we need to find a solution to this equation.

This equation is equivalent to the following congruence modulo **11**:

$$x^2 \equiv 2 \ (\text{mod } 11). \tag{1}$$

Let's make a table of all of the squares modulo **11**:

Table 11.1.1   Squares in $\mathbf{Z}_{11}$

$\bar{x}$	$\bar{0}$	$\bar{1}$	$\bar{2}$	$\bar{3}$	$\bar{4}$	$\bar{5}$	$\bar{6}$	$\bar{7}$	$\bar{8}$	$\bar{9}$	$\overline{10}$
$\bar{x}^2$	0	1	4	9	5	3	3	5	9	4	1

Notice that the only nonzero squares in $\mathbf{Z}_{11}$ are $\bar{1}, \bar{3}, \bar{4}, \bar{5}$, and $\bar{9}$. Thus, for any nonzero integer x, we see that x^2 will be congruent to 1, 3, 4, 5, or 9 modulo **11**. It follows that there is no integer x that satisfies congruence (1).

In short, because $\bar{2}$ is not a square in $\mathbf{Z}_{11}$, there is no way to comply with the count's will.

Which numbers are squares modulo n?

As we verified earlier, there are no solutions to congruence (1),

$$x^2 \equiv 2 \ (\text{mod } 11),$$

because $\overline{2}$ is not a square modulo 11. The general question of which numbers are squares modulo n opens up a gold mine of number-theoretic treasures, which will be our focus in this chapter.

Which elements of $\mathbf{Z}_7$ are squares? In other words, which elements of $\mathbf{Z}_7$ can be written as the square of some element of $\mathbf{Z}_7$? The most straightforward way to answer this question is to compute the squares of all of the elements of $\mathbf{Z}_7$:

Table 11.1.2 Squares in $\mathbf{Z}_7$

$\overline{x}$	$\overline{0}$	$\overline{1}$	$\overline{2}$	$\overline{3}$	$\overline{4}$	$\overline{5}$	$\overline{6}$
$\overline{x}^2$	0	1	4	2	2	4	1

Notice that this is simply the $\overline{x}^2$ row of the exponent table for $\mathbf{Z}_7$, which we studied in Chapter 9. From the $\overline{x}^2$ row of the table, we see that $\overline{0}, \overline{1}, \overline{2}$, and $\overline{4}$ are squares in $\mathbf{Z}_7$, while $\overline{3}, \overline{5}$, and $\overline{6}$ are not squares.

The nonzero elements of $\mathbf{Z}_n$ that are squares are called *quadratic residues*. For example, we just determined that the quadratic residues in $\mathbf{Z}_7$ are $\overline{1}, \overline{2}$, and $\overline{4}$.

DEFINITION 11.1.1

*Let $n \in \mathbf{N}$ and let $\overline{a} \in \mathbf{Z}_n$ with $\overline{a} \neq \overline{0}$. We say that $\overline{a}$ is a **quadratic residue** modulo n if there exists $\overline{b} \in \mathbf{Z}_n$ such that $\overline{b}^2 = \overline{a}$. If there is no such $\overline{b}$, then we say that $\overline{a}$ is a quadratic **nonresidue** modulo n.*

Note that $\overline{0}$ is considered to be neither a quadratic residue nor a quadratic nonresidue.

EXAMPLE 1

Find the quadratic residues modulo 9.

SOLUTION The following table shows the square of each element of $\mathbf{Z}_9$.

Squares in $\mathbf{Z}_9$

$\overline{x}$	$\overline{0}$	$\overline{1}$	$\overline{2}$	$\overline{3}$	$\overline{4}$	$\overline{5}$	$\overline{6}$	$\overline{7}$	$\overline{8}$
$\overline{x}^2$	0	1	4	0	7	7	0	4	1

Thus, the quadratic residues in $\mathbf{Z}_9$ are $\overline{1}, \overline{4}$, and $\overline{7}$. We may also express this by saying that the integers 1, 4, and 7 are quadratic residues modulo 9.

Square roots in Z_n

Having examined squares, let's turn our attention to square roots. For example, in Z_7, what is the square root of $\bar{2}$? Since

$$\bar{3}^2 = \bar{4}^2 = \bar{2},$$

we say that $\bar{3}$ and $\bar{4}$ are both square roots of $\bar{2}$. Because $\bar{4} = -\bar{3}$ in Z_7, we may also express this by saying that there are two square roots of $\bar{2}$: namely, $\bar{3}$ and $-\bar{3}$. This is a bit nicer, since we are used to the idea that any real number has the same square as its negative.

Which elements of Z_7 are square roots of $\bar{1}$? Table 11.1.2 reveals that in Z_7, $\bar{1}^2 = \bar{6}^2 = \bar{1}$, and hence, $\bar{1}$ and $\bar{6}$ are both square roots of $\bar{1}$. Since $\bar{6} = -\bar{1}$, we may say instead that $\bar{1}$ has two square roots: $\bar{1}$ and $-\bar{1}$. Who could be unhappy with this?

Looking at Z_8, however, the story is quite different, as we can tell by examining the table of squares in Z_8:

Squares in Z_8

$\bar{x}$	$\bar{0}$	$\bar{1}$	$\bar{2}$	$\bar{3}$	$\bar{4}$	$\bar{5}$	$\bar{6}$	$\bar{7}$
$\bar{x}^2$	0	1	4	1	0	1	4	1

If we ask which elements of Z_8 are square roots of $\bar{1}$, we see from the table that $\bar{1}, \bar{3}, \bar{5}$, and $\bar{7}$ all fit the bill. Thus, in Z_8, the number $\bar{1}$ has four distinct square roots. The reader may suspect that this bizarre situation can arise only when the modulus is composite, and this is indeed the case. As we will now prove, if the modulus p is a prime other than 2, then each quadratic residue in Z_p has exactly 2 square roots.

THEOREM 11.1.2

Let $p > 2$ be prime. Suppose that $\bar{a} \in Z_p$ is a quadratic residue. Then there are exactly two elements $\bar{x} \in Z_p$ that satisfy $\bar{x}^2 = \bar{a}$.

Put it in Prose, Paul!

If the modulus is prime, then any nonzero square has exactly two square roots.

PROOF Since $\bar{a}$ is a quadratic residue modulo p, there exists $\bar{b} \in \mathbf{Z}_p$ such that

$$\bar{b}^2 = \bar{a}.$$

It follows that

$$(-\bar{b})^2 = \bar{a},$$

and hence the equation $\bar{x}^2 = \bar{a}$ has the two solutions $\bar{x} = \bar{b}$ and $\bar{x} = -\bar{b}$. Furthermore, these two solutions are distinct; that is, $\bar{b} \neq -\bar{b}$. (See Exercise 18.)

At this point, we've shown that $\bar{a}$ has at least two square roots—namely, $\bar{b}$ and $-\bar{b}$. To complete the proof of the theorem, we must show that $\bar{a}$ can have no other square roots. To do this, suppose that $\bar{x} \in \mathbf{Z}_p$ satisfies

$$\bar{x}^2 = \bar{a}. \tag{2}$$

Since $\bar{a} = \bar{b}^2$, we get

$$\bar{x}^2 - \bar{b}^2 = \bar{0}$$

or, equivalently,

$$(\bar{x} + \bar{b})(\bar{x} - \bar{b}) = \bar{0}.$$

Since p is prime, one of the factors on the left side of this equation must equal $\bar{0}$ by Theorem 8.3.5. In other words,

$$\bar{x} + \bar{b} = \bar{0} \quad \text{or} \quad \bar{x} - \bar{b} = \bar{0},$$

from which it follows that

$$\bar{x} = -\bar{b} \quad \text{or} \quad \bar{x} = \bar{b}.$$

Thus, the only solutions to equation (2) are $\bar{x} = \bar{b}$ and $\bar{x} = -\bar{b}$. ●

How many quadratic residues are there?

If p is a prime number, how many elements of $\mathbf{Z}_p$ are quadratic residues? Looking at Table 11.1.2, we find that $\mathbf{Z}_7$ has 3 quadratic residues and 3 quadratic nonresidues. From Table 11.1.1, we see that in $\mathbf{Z}_{11}$, the 10 nonzero elements are also split equally, into 5 quadratic residues and 5 nonresidues. In fact, it is not difficult to show that this equal split happens in general. That is, for every prime $p > 2$, exactly half of the nonzero elements of $\mathbf{Z}_p$ are quadratic residues and half are nonresidues.

Naomi's Numerical Proof Preview: Theorem 11.1.3

To see why the number of quadratic residues is the same as the number of nonresidues, let's reexamine the squares in $\mathbf{Z}_{11}$. The list of nonzero squares in $\mathbf{Z}_{11}$ contains 10 elements:

$$\overline{1}^2, \overline{2}^2, \overline{3}^2, \overline{4}^2, \overline{5}^2, \overline{6}^2, \overline{7}^2, \overline{8}^2, \overline{9}^2, \overline{10}^2.$$

But some of these elements are equal. When we compute the values of all of the elements in the list, we get

$$\overline{1}, \overline{4}, \overline{9}, \overline{5}, \overline{3}, \overline{3}, \overline{5}, \overline{9}, \overline{4}, \overline{1}.$$

Notice that this list contains each quadratic residue exactly twice. This should not be too surprising, since we have already proved that every quadratic residue has exactly two square roots (Theorem 11.1.2). Thus, even though this list is 10 elements long, there are only $\frac{10}{2} = 5$ distinct elements in the list.

THEOREM 11.1.3

Let $p > 2$ be prime. Then there are exactly $\dfrac{p-1}{2}$ quadratic residues in $\mathbf{Z}_p$.

PROOF Consider the list of all nonzero squares in $\mathbf{Z}_p$,

$$\overline{1}^2, \overline{2}^2, \ldots, \overline{p-1}^2. \tag{3}$$

This is a list of $p - 1$ numbers that contains all of the quadratic residues of $\mathbf{Z}_p$. While list (3) appears to have $p - 1$ elements, there are some repetitions. How many *distinct* elements of $\mathbf{Z}_p$ appear in the list? Theorem 11.1.2 states that every quadratic residue has exactly two square roots. This means that every number in list (3) appears exactly twice. Hence, the number of distinct elements in the list is $\dfrac{p-1}{2}$. ∎

The Legendre symbol

The elements of $\mathbf{Z}_p$ may be divided into three classes: the **quadratic residues**, the **nonresidues**, and $\overline{0}$. The following definition assigns to these groups the numbers $+1$, -1, and 0, respectively.

DEFINITION 11.1.4

Let $p > 2$ be prime, and let $a \in \mathbf{Z}$. The Legendre symbol, $\left(\dfrac{a}{p}\right)$, is defined by

$$\left(\frac{a}{p}\right) = \begin{cases} 1 & \text{if } \overline{a} \text{ is a quadratic residue modulo } p \\ -1 & \text{if } \overline{a} \text{ is a quadratic nonresidue modulo } p \\ 0 & \text{if } \overline{a} = \overline{0} \text{ in } \mathbf{Z}_p \text{ (i.e., if } p \mid a) \end{cases}$$

EXAMPLE 2

a. Find $\left(\frac{a}{7}\right)$ for each value of a from 1 to 6.

b. Find $\left(\frac{11}{7}\right)$.

c. Find $\left(\frac{21}{7}\right)$.

d. Find $\left(\frac{-1}{7}\right)$.

SOLUTION

a. Since $\bar{1}, \bar{2},$ and $\bar{4}$ are the quadratic residues in $\mathbf{Z}_7$, we have $\left(\frac{1}{7}\right) = \left(\frac{2}{7}\right) = \left(\frac{4}{7}\right) = 1,$ while $\left(\frac{3}{7}\right) = \left(\frac{5}{7}\right) = \left(\frac{6}{7}\right) = -1.$

b. In $\mathbf{Z}_7$, we know $\overline{11} = \bar{4},$ which is a quadratic residue, and hence, $\left(\frac{11}{7}\right) = 1.$

c. Since $7 \mid 21,$ it follows that $\left(\frac{21}{7}\right) = 0.$

d. We know $\overline{-1} = \bar{6},$ which is a nonresidue in $\mathbf{Z}_7.$ Thus, $\left(\frac{-1}{7}\right) = -1.$ ●

In Example 2, part b, we saw that $\left(\frac{11}{7}\right)$ is the same as $\left(\frac{4}{7}\right)$ since $\overline{11} = \bar{4}$ in $\mathbf{Z}_7$ or $11 \equiv 4 \pmod{7}.$ This illustrates a general fact about the Legendre symbol.

THEOREM 11.1.5

Let $p > 2$ be prime, and let $a, b \in \mathbf{Z}.$ If $a \equiv b \pmod{p},$ then $\left(\frac{a}{p}\right) = \left(\frac{b}{p}\right).$

PROOF Since $a \equiv b \pmod{p},$ we have $\bar{a} = \bar{b}$ in $\mathbf{Z}_p$; that is, $\bar{a}$ and $\bar{b}$ are two different ways to represent the *same* element of $\mathbf{Z}_p.$ Hence, $\bar{a}$ is a quadratic residue (or nonresidue, or $\bar{0}$) if and only if $\bar{b}$ is a quadratic residue (or nonresidue, or $\bar{0}$). Thus,

$$\left(\frac{a}{p}\right) = \left(\frac{b}{p}\right).$$ ●

EXERCISES 11.1

Numerical Problems

1. Is $\bar{3}$ a quadratic residue in $\mathbf{Z}_{13}$?

2. Is $\overline{10}$ a quadratic residue in $\mathbf{Z}_7$?

3. List all the quadratic residues in $\mathbf{Z}_{17}.$

4. How many quadratic residues are there modulo the prime 1,000,003?

In Exercises 5–13, evaluate each Legendre symbol.

5. $\left(\frac{2}{7}\right)$

6. $\left(\frac{5}{11}\right)$

7. $\left(\frac{27}{13}\right)$

8. $\left(\frac{91}{13}\right)$

9. $\left(\frac{2}{17}\right)$

10. $\left(\frac{9}{17}\right)$

11. $\left(\frac{11}{13}\right)$

12. $\left(\frac{13}{11}\right)$

13. $\left(\frac{-5}{13}\right)$

14. Calculate the following Legendre symbols. Describe any pattern you find.

a. $\left(\frac{5}{7}\right), \left(\frac{3}{7}\right),$ and $\left(\frac{15}{7}\right)$

b. $\left(\frac{2}{11}\right), \left(\frac{6}{11}\right),$ and $\left(\frac{12}{11}\right)$

c. $\left(\frac{5}{13}\right), \left(\frac{2}{13}\right),$ and $\left(\frac{10}{13}\right)$

d. $\left(\frac{5}{17}\right), \left(\frac{3}{17}\right),$ and $\left(\frac{15}{17}\right)$

15. Find $\left(\frac{-1}{p}\right)$ for $p = 3, 5, 7, 11, 13, 17, 19$

16. Find $\left(\frac{2}{p}\right)$ for $p = 3, 5, 7, 11, 13, 17, 19$

Reasoning and Proofs

17. a. Find $\left(\frac{4}{101}\right)$.

   b. Let p be prime, and suppose $a \in \mathbf{Z}$ is a perfect square such that $p \nmid a$. Prove that $\left(\frac{a}{p}\right) = 1$.

18. Let p be prime, $p > 2$, and let $b \in \mathbf{Z}_p$ such that $\bar{b} \neq \bar{0}$. Prove that $\bar{b} \neq -\bar{b}$.

19. Let p be prime, and let $a \in \mathbf{Z}$ such that $p \nmid a$. Prove that $\left(\frac{a^2}{p}\right) = 1$.

EXPLORATION Multiplicativity of the Legendre Symbol (Exercises 20–23)

In this exploration, you will prove that the Legendre symbol is *multiplicative*. More precisely, you will prove the following theorem.

THEOREM. *Let $p > 2$ be prime, and let $a, b \in \mathbf{Z}$. Then*

$$\left(\frac{ab}{p}\right) = \left(\frac{a}{p}\right)\left(\frac{b}{p}\right).$$

(This theorem will also be proved in the next section using different techniques.)

20. Prove that the product of any two quadratic residues is itself a quadratic residue. That is, for any $a, b \in \mathbf{Z}$ and any prime $p > 2$, if $\left(\frac{a}{p}\right) = 1$ and $\left(\frac{b}{p}\right) = 1$, then $\left(\frac{ab}{p}\right) = 1$.

21. Prove that the product of a quadratic residue and a quadratic nonresidue is a quadratic nonresidue. That is, for any $a, b \in \mathbf{Z}$ and any prime $p > 2$, if $\left(\frac{a}{p}\right) = 1$ and $\left(\frac{b}{p}\right) = -1$, then $\left(\frac{ab}{p}\right) = -1$.

22. Prove that the product of any two quadratic nonresidues is a quadratic residue. That is, for any $a, b \in \mathbf{Z}$ and any prime $p > 2$, if $\left(\frac{a}{p}\right) = -1$ and $\left(\frac{b}{p}\right) = -1$, then $\left(\frac{ab}{p}\right) = 1$. [*Hint:* Consider the row of $\bar{a}$ in the $\mathbf{Z}_p$ multiplication table, and use your result from Exercise 21.]

23. Use your results from Exercises 20–22 to complete the proof of the Theorem. (Don't forget the possibility that a or b might equal 0 modulo p.)

24. Suppose that n is a natural number whose digits add up to 15. Prove that n cannot be a perfect square. [*Hint:* What is n congruent to modulo 9?]

25. Suppose that n is a natural number whose last two digits are 59. Prove that n cannot be a perfect square. [*Hint:* What is n congruent to modulo 4?]

26. Let p be prime, $p > 2$, and let a be a primitive root modulo p.

   a. Prove that $\bar{a}$ is not a quadratic residue modulo p.

   b. Let k be a natural number. Prove that $\bar{a}^k$ is a quadratic residue modulo p if and only if k is even.

Conjecturer's Corner (Exercises 27–30)

27. Find the values in this table:

p	7	11	13	17	19	23
$\left(\dfrac{p}{5}\right)$						
$\left(\dfrac{5}{p}\right)$						

28. Find the values in this table:

p	7	11	13	17	19	23
$\left(\dfrac{p}{7}\right)$						
$\left(\dfrac{7}{p}\right)$						

29. Consider the tables you completed in Exercises 27 and 28.

   a. Within each table, when is there agreement between the two cells in a column (when are both values in a column the same), and when is there disagreement?

   b. Can you make a conjecture to explain the differences you see in those agreements when comparing the two tables?

30. Calculate several $\left(\frac{p}{q}\right)$ and $\left(\frac{q}{p}\right)$, for a variety of odd primes p and q. Make a conjecture about how the two compare in general.[1]

Advanced Reasoning and Proofs

31. Let p be prime. An element $\bar{a} \in \mathbf{Z}_p$ is called a *cubic residue* if $\bar{a} \neq \bar{0}$ and there exists $\bar{b} \in \mathbf{Z}_p$ such that $\bar{b}^3 = \bar{a}$.

   a. Determine cubic residues modulo each of the primes 5, 7, 11, and 13.

   b. Make a conjecture about the number of cubic residues in $\mathbf{Z}_p$ for any prime $p > 3$.

   c. Prove your conjecture from part b.

11.2 Euler's Identity and the Quadratic Character of -1

Another way to find the Legendre symbol

In the last section, we introduced the notion of a quadratic residue and saw that in $\mathbf{Z}_p$ for any prime $p > 2$, exactly half of the nonzero elements are quadratic residues. We also noted that the quadratic residues are the elements that appear in the $\bar{x}^2$ row of the $\mathbf{Z}_p$ exponent table. For example, we see from the following table that in $\mathbf{Z}_{11}$, the quadratic residues are $\bar{1}, \bar{3}, \bar{4}, \bar{5},$ and $\bar{9}$.

Squares Modulo 11

$\bar{x}$	$\bar{1}$	$\bar{2}$	$\bar{3}$	$\bar{4}$	$\bar{5}$	$\bar{6}$	$\bar{7}$	$\bar{8}$	$\bar{9}$	$\overline{10}$
$\bar{x}^2$	1	4	9	5	3	3	5	9	4	1

There is another row of the exponent table for $\mathbf{Z}_{11}$ that can be used to determine which elements of $\mathbf{Z}_{11}$ are quadratic residues. Let's look at the $\bar{x}^5$ row of the exponent table:

5th Powers Modulo 11

$\bar{x}$	$\bar{1}$	$\bar{2}$	$\bar{3}$	$\bar{4}$	$\bar{5}$	$\bar{6}$	$\bar{7}$	$\bar{8}$	$\bar{9}$	$\overline{10}$
$\bar{x}^5$	1	10	1	1	1	10	10	10	1	10

---

[1]Note that this issue will be resolved before the end of the chapter, so do not look ahead until you have made a conjecture.

The first thing that may jump out at you is that only the numbers $\bar{1}$ and $\overline{10}$ appear in the $\bar{x}^5$ row. Since $\overline{10}$ is the same as $\overline{-1}$, we see that every element of the $\bar{x}^5$ row is either $\bar{1}$ or $\overline{-1}$.

In which columns of the table do the $\bar{1}$'s appear, and in which columns do the $\overline{-1}$'s appear? Inspection shows that $\bar{x}^5 = 1$ for $\bar{x} = \bar{1}, \bar{3}, \bar{4}, \bar{5},$ and $\bar{9}$. What distinguishes these elements from the other elements of $\mathbf{Z}_{11}$? They are exactly the quadratic residues modulo 11! For the other nonzero elements of $\mathbf{Z}_{11}$ (the quadratic nonresidues, $\bar{x} = \bar{2}, \bar{6}, \bar{7}, \bar{8},$ and $\overline{10}$), we have $\bar{x}^5 = \overline{-1}$.

What could possibly explain this unusual relationship between the $\bar{x}^2$ row of the exponent table and the $\bar{x}^5$ row of the exponent table? It may help to recall that there is another very special row of the exponent table, the $\bar{x}^{10}$ row, which consists entirely of $\bar{1}$'s by Fermat's Little Theorem (9.1.1):

10th Powers Modulo 11

$\bar{x}$	$\bar{1}$	$\bar{2}$	$\bar{3}$	$\bar{4}$	$\bar{5}$	$\bar{6}$	$\bar{7}$	$\bar{8}$	$\bar{9}$	$\overline{10}$
$\bar{x}^{10}$	1	1	1	1	1	1	1	1	1	1

The elements in the $\bar{x}^{10}$ row are the squares of the elements in the $\bar{x}^5$ row. Thus, each of the elements in the $\bar{x}^5$ row must be a square root of $\bar{1}$. These square roots are $\bar{1}$ and $\overline{-1}$. This explains why only $\bar{1}$ and $\overline{-1}$ appear in the $\bar{x}^5$ row.

Naomi's Numerical Proof Preview: Theorem 11.2.1 Part (*i*)

But in the table of 5th powers modulo 11, why should the $\bar{1}$'s appear in columns corresponding to quadratic residues, and the $\overline{-1}$'s in columns corresponding to nonresidues? To answer this, let's look at a sample quadratic residue, such as $\bar{3}$. The element $\bar{3}$ is a quadratic residue because

$$\bar{3} = \bar{5}^2.$$

Taking the 5th power of both sides, we find

$$\bar{3}^5 = \bar{5}^{10}.$$

But we know that $\bar{5}^{10} = \bar{1}$, by Fermat's Little Theorem (9.1.1). Thus,

$$\bar{3}^5 = \bar{1}.$$

A similar argument can be used to show that the 5th power of any quadratic residue in $\mathbf{Z}_{11}$ equals $\bar{1}$.

More generally, for any prime $p > 2$, we know that the $\bar{x}^{p-1}$ row consists entirely of $\bar{1}$'s by Fermat's Little Theorem (9.1.1). Hence, the $\bar{x}^{(p-1)/2}$ row will consist entirely of the square roots of $\bar{1}$, which are $\bar{1}$ and $\overline{-1}$. Furthermore, the positions of

THEOREM 11.2.1

Let $p > 2$ be prime, and let $\overline{a}$ be a nonzero element of $\mathbf{Z}_p$.

    (*i*) *If $\overline{a}$ is a quadratic residue, then $\overline{a}^{\frac{p-1}{2}} = \overline{1}$.*

    (*ii*) *If $\overline{a}$ is a quadratic nonresidue, then $\overline{a}^{\frac{p-1}{2}} = \overline{-1}$.*

We first give the proof of part (*i*), which follows the argument from the Numerical Proof Preview.

PROOF of THEOREM 11.2.1 PART (*I*)

Since $\overline{a}$ is a quadratic residue, there exists $\overline{b} \in \mathbf{Z}_p$ such that

$$\overline{a} = \overline{b}^2.$$

Raising both sides to the power $\dfrac{p-1}{2}$, we obtain

$$\overline{a}^{\frac{p-1}{2}} = \overline{b}^{p-1}.$$

But the right side of this equation equals $\overline{1}$ by Fermat's Little Theorem (9.1.1). Hence,

$$\overline{a}^{\frac{p-1}{2}} = \overline{1},$$

as desired.

Naomi's Numerical Proof Preview: Theorem 11.2.1 Part (*ii*)

The proof of part (*ii*) is more difficult, so before going into the general proof, let's consider a numerical example that illustrates the proof. In $\mathbf{Z}_{11}$, the element $\overline{6}$ is a quadratic nonresidue. Let's see if we can argue that $\overline{6}^5 = \overline{-1}$. Consider the list

$$\overline{1}, \ \overline{2}, \ \overline{3}, \ \overline{4}, \ \overline{5}, \ \overline{6}, \ \overline{7}, \ \overline{8}, \ \overline{9}, \ \overline{10}.$$

For each number $\overline{c}$ in the list, there is a number $\overline{d}$ such that

$$\overline{c} \cdot \overline{d} = \overline{6}.$$

Finding such a $\overline{d}$ is not hard—simply compute $\overline{d} = \overline{c}^{-1} \cdot \overline{6}$. In our case, one finds that

$$\overline{1} \cdot \overline{6} = \overline{6}$$

$$\overline{2} \cdot \overline{3} = \overline{6}$$

$$\overline{4} \cdot \overline{7} = \overline{6} \qquad\qquad (1)$$

$$\overline{5} \cdot \overline{10} = \overline{6}$$

$$\overline{8} \cdot \overline{9} = \overline{6}.$$

Note that it is not possible to have $\overline{d} = \overline{c}$, for then one would have $\overline{c}^2 = \overline{6}$, which is impossible since $\overline{6}$ is not a quadratic residue. Thus, the numbers from $\overline{1}$ to $\overline{10}$ can be paired up so that the product of each pair equals $\overline{6}$.

Now if we multiply these equations (1) together, we get

$$\overline{1} \cdot \overline{2} \cdot \overline{3} \cdot \overline{4} \cdot \overline{5} \cdot \overline{6} \cdot \overline{7} \cdot \overline{8} \cdot \overline{9} \cdot \overline{10} = \overline{6}^5.$$

The quantity on the left side of the equation is $\overline{10!}$, which equals $\overline{-1}$ by Wilson's Theorem (8.3.7). We thus get

$$\overline{6}^5 = \overline{-1},$$

as desired.

We now use the argument of the preceding Numerical Proof Preview to give a proof of part (*ii*).

PROOF OF **THEOREM 11.2.1 PART (*ii*)**

Suppose that $\overline{a}$ is a quadratic nonresidue modulo p. Consider the list of numbers

$$\overline{1}, \overline{2}, \overline{3}, \dots, \overline{p-1} \qquad\qquad (2)$$

Claim: For each element $\overline{c}$ in list (2), there is a unique $\overline{d}$ in the list such that $\overline{d} \neq \overline{c}$ and $\overline{c} \cdot \overline{d} = \overline{a}$.

Proof of Claim Given $\overline{c}$, we know (since $\overline{c} \neq \overline{0}$) that $\overline{c}^{-1}$ exists in $\mathbf{Z}_p$. Furthermore, the following two equations are equivalent:

$$\overline{c} \cdot \overline{d} = \overline{a} \iff \overline{d} = \overline{c}^{-1} \cdot \overline{a}.$$

Thus, there is a unique $\overline{d}$—namely, $\overline{d} = \overline{c}^{-1} \cdot \overline{a}$—that satisfies the equation $\overline{c} \cdot \overline{d} = \overline{a}$. Finally, note that $\overline{c}$ cannot equal $\overline{d}$, for then $\overline{c} \cdot \overline{d} = \overline{a}$ would imply that $\overline{c}^2 = \overline{a}$, which contradicts the hypothesis that $\overline{a}$ is a quadratic nonresidue. This completes the proof of the Claim.

It follows from the Claim that the numbers in list (2) can be paired up so that the product of each pair equals $\bar{a}$. Thus, the product of all of the numbers in list (2) equals $\bar{a}$ raised to the number of such pairs, which is $\dfrac{p-1}{2}$. We thus get

$$\bar{1} \cdot \bar{2} \cdot \bar{3} \cdot \ \cdots \ \cdot \overline{(p-1)} = \bar{a}^{\frac{p-1}{2}}.$$

The left side of this equation equals $\overline{-1}$ by Wilson's Theorem (8.3.7). Therefore,

$$\bar{a}^{\frac{p-1}{2}} = \overline{-1},$$

as desired.

■

Euler's Identity

Theorem 11.2.1 can be expressed nicely using the Legendre symbol, as follows.

11.2.2   EULER'S IDENTITY

Let $p > 2$ be prime, and let $a \in \mathbf{Z}$. Then:

$$\left(\frac{a}{p}\right) \equiv a^{\frac{p-1}{2}} \pmod{p}. \tag{3}$$

PROOF   We will prove the theorem in three cases: $\bar{a}$ is a quadratic residue modulo p, $\bar{a}$ is a quadratic nonresidue modulo p, and $\bar{a} = \bar{0}$.

Case 1 $\bar{a}$ is a quadratic residue modulo p.

Then $\left(\frac{a}{p}\right) = 1$, by definition, while Theorem 11.2.1(i) tells us that

$$a^{\frac{p-1}{2}} \equiv 1 \pmod{p}.$$

Thus, the theorem holds in this case.

Case 2 $\bar{a}$ is a quadratic nonresidue modulo p.

In this case $\left(\frac{a}{p}\right) = -1$, while Theorem 11.2.1($ii$) tells us that

$$a^{\frac{p-1}{2}} \equiv -1 \pmod{p},$$

so again the theorem holds.

Case 3 $\bar{a} = \bar{0}$ in $\mathbf{Z}_p$.

In this case, $\left(\frac{a}{p}\right) = 0$, while

$$a^{\frac{p-1}{2}} \equiv 0 \ (\text{mod } p)$$

since $a \equiv 0 \ (\text{mod } p)$. Thus, the theorem holds in this case as well. ■

EXAMPLE 1

Use Euler's Identity to find

    a. $\left(\dfrac{5}{29}\right)$

    b. $\left(\dfrac{-1}{67}\right)$

SOLUTION

a. Euler's Identity (11.2.2) tells us that

$$\left(\frac{5}{29}\right) \equiv 5^{\frac{29-1}{2}}$$

$$\equiv 5^{14} \ (\text{mod } 29).$$

The quantity 5^{14} modulo 29 can be computed efficiently using the method of repeated squaring (introduced in Section 9.3). We have

$$5^1 \equiv 5 \qquad\qquad (\text{mod } 29),$$
$$5^2 \equiv 25 \qquad\qquad (\text{mod } 29),$$
$$5^4 \equiv 25^2 \equiv 16 \quad (\text{mod } 29),$$
$$5^8 \equiv 16^2 \equiv 24 \quad (\text{mod } 29).$$

Thus,

$$5^{14} = 5^{8+4+2} \equiv 24 \cdot 16 \cdot 25 \equiv 1 \ (\text{mod } 29).$$

Hence, $\left(\dfrac{5}{29}\right) \equiv 1 \ (\text{mod } 29)$. Since by definition $\left(\dfrac{5}{29}\right)$ is $1, 0,$ or $-1,$ it follows that

$$\left(\frac{5}{29}\right) = 1.$$

b. This time Euler's Identity tells us to compute $(-1)^{\frac{67-1}{2}} \equiv (-1)^{33} \pmod{67}$.

Since -1 raised to an odd power equals -1, we have $(-1)^{33} \equiv -1 \pmod{67}$. Hence,

$$\left(\frac{-1}{67}\right) = -1. \qquad \blacksquare$$

In part a of this example, we posed the question "Is 5 a square modulo 29?" Euler's Identity (11.2.2) allowed us to answer this question by doing the arithmetical calculation $5^{(29-1)/2}$ and reducing the result modulo 29. This is one of the remarkable things about Euler's Identity. The left side of congruence (3), the Legendre symbol $\left(\frac{a}{p}\right)$, essentially asks a yes-or-no question:

$$\text{Is } a \text{ a square modulo } p? \qquad (4a)$$

In contrast, the right side of (3) is asking us to perform a modular exponentiation:

$$\text{Evaluate } a^{\frac{p-1}{2}} \text{ modulo } p. \qquad (4b)$$

It would seem on the surface that answering the yes-or-no question (4a) is quite a different task from performing the modular exponentiation (4b). However, Euler shocks us by telling us that these two tasks are identical!

Part a of Example 1 also illustrates that Euler's Identity provides an efficient method for computing the Legendre symbol $\left(\frac{a}{p}\right)$. We simply use the repeated squaring method to evaluate $a^{(p-1)/2}$ modulo p. As noted in Chapter 9 (Observation 9.3.3), the number of multiplications necessary to do this is at most

$$2\log_2\left(\frac{p-1}{2}\right),$$

which is smaller than $2\log_2 p$. This means that Euler's Identity provides a polynomial-time algorithm for computing the Legendre symbol. Using this algorithm, a computer can find $\left(\frac{a}{p}\right)$ in a fraction of a second, even if a and p are huge numbers with thousands of digits.

When is -1 a square modulo p?

For some primes p, the element $\overline{-1}$ is a quadratic residue in $\mathbf{Z}_p$, but for some primes, $\overline{-1}$ is a quadratic nonresidue. For example, in $\mathbf{Z}_{13}$, we see that $\overline{5}^2 = \overline{-1}$, so that $\overline{-1}$ is a quadratic residue modulo 13. On the other hand, $\overline{-1}$ is not a quadratic residue in $\mathbf{Z}_{11}$, as we could determine by observing that $\overline{-1} = \overline{10}$ does not appear in the $\overline{x}^2$ row of the $\mathbf{Z}_{11}$ exponent table (see Table 11.1.1). In general, for an arbitrary prime p, is there an easy way to determine when $\overline{-1}$ is a quadratic residue mod p?

Naomi's Numerical Proof Preview: Theorem 11.2.3

To answer this question, let's try using Euler's Identity (11.2.2). For example, here's what Euler's Identity tells us about $\left(\frac{-1}{13}\right)$:

$$\left(\frac{-1}{13}\right) \equiv (-1)^{\frac{13-1}{2}} \pmod{13}.$$

The right side of this congruence is $(-1)^6$, which equals 1. Thus,

$$\left(\frac{-1}{13}\right) \equiv 1 \pmod{13}.$$

Since $\left(\frac{-1}{13}\right)$ is either $-1, 0$, or 1 by definition, it follows that

$$\left(\frac{-1}{13}\right) = 1.$$

Thus, Euler's Identity has helped us to determine quickly and easily that $\overline{-1}$ is a square modulo 13.

When we determined $\left(\frac{-1}{p}\right)$ for $p = 13$, the exponent $\frac{p-1}{2}$ was even, and hence, we found that $\left(\frac{-1}{p}\right)$ was equal to -1 raised to an even power, which equals 1. This should convince us that when $\frac{p-1}{2}$ is even, then $\left(\frac{-1}{p}\right) = 1$; that is, $\overline{-1}$ is a quadratic residue modulo p. So the question then becomes, for which primes p is $\frac{p-1}{2}$ even? This happens when $p - 1$ is divisible by 4—that is, when $p \equiv 1 \pmod 4$. This completely answers our question of when $\overline{-1}$ is a square modulo p. All that is left is a formal statement and a proof of our discovery.

THEOREM 11.2.3   THE QUADRATIC CHARACTER OF -1

Let $p > 2$ be prime.

    (*i*)   *If $p \equiv 1 \pmod 4$, then $\overline{-1}$ is a quadratic residue modulo p.*

    (*ii*) *If $p \equiv 3 \pmod 4$, then $\overline{-1}$ is a quadratic nonresidue modulo p.*

The proof of this theorem will use the following lemma, whose proof you will give in Exercise 7.

LEMMA 11.2.4

Let p be an odd integer. Then

$$(-1)^{\frac{p-1}{2}} = 1 \quad \text{if } p \equiv 1 \pmod 4,$$

and

$$(-1)^{\frac{p-1}{2}} = -1 \quad \text{if } p \equiv 3 \pmod 4.$$

PROOF You will prove this lemma in Exercise 7.

We are now ready to prove the theorem.

PROOF OF **THEOREM 11.2.3**

For an arbitrary prime $p > 2$, Euler's Identity (11.2.2) tells us that

$$\left(\frac{-1}{p}\right) \equiv (-1)^{\frac{p-1}{2}} \pmod{p}.$$

The quantities on both sides of this congruence are each equal to either $+1$ or -1. Since $1 \not\equiv -1 \pmod{p}$, the two quantities must be equal. Thus,

$$\left(\frac{-1}{p}\right) = (-1)^{\frac{p-1}{2}}.$$

We now apply Lemma 11.2.4 to obtain

$$\left(\frac{-1}{p}\right) = 1 \qquad \text{if } p \equiv 1 \pmod{4},$$

and

$$\left(\frac{-1}{p}\right) = -1 \qquad \text{if } p \equiv 3 \pmod{4}.$$

This completes the proof of the theorem.

EXAMPLE 2

Determine whether $\overline{-1}$ is a quadratic residue modulo the prime $p = 103$.

SOLUTION Since $103 \equiv 3 \pmod 4$, it follows from Theorem 11.2.3 that $\overline{-1}$ is not a quadratic residue modulo 103.

The Legendre symbol is multiplicative

The next theorem is a consequence of Euler's Identity (11.2.2).

THEOREM 11.2.5 MULTIPLICATIVITY OF THE LEGENDRE SYMBOL

Let $p > 2$ be prime, and let $a, b \in \mathbf{Z}$. Then

$$\left(\frac{ab}{p}\right) = \left(\frac{a}{p}\right)\left(\frac{b}{p}\right).$$

PROOF Euler's Identity (11.2.2) tells us that

$$\left(\frac{a}{p}\right) \equiv a^{\frac{p-1}{2}} \pmod{p},$$

and

$$\left(\frac{b}{p}\right) \equiv b^{\frac{p-1}{2}} \pmod{p}.$$

Multiplying these congruences yields

$$\left(\frac{a}{p}\right)\left(\frac{b}{p}\right) \equiv (ab)^{\frac{p-1}{2}} \pmod{p}. \tag{5}$$

Euler's Identity also tells us that

$$\left(\frac{ab}{p}\right) \equiv (ab)^{\frac{p-1}{2}} \pmod{p}. \tag{6}$$

Combining congruences (5) and (6), we get

$$\left(\frac{a}{p}\right)\left(\frac{b}{p}\right) \equiv \left(\frac{ab}{p}\right) \pmod{p}.$$

Observe that the quantities on each side of this congruence equal either $+1, 0,$ or -1. Since none of these three values is congruent to any of the others modulo p, it follows that

$$\left(\frac{a}{p}\right)\left(\frac{b}{p}\right) = \left(\frac{ab}{p}\right),$$

as desired. ∎

It's easy to tell whether a product ab is a quadratic residue modulo p if we know whether a and b are quadratic residues modulo p. The following corollary of Theorem 11.2.5 tells us exactly how to do it.

COROLLARY 11.2.6

Let $p > 2$ be prime, and let $\bar{a}, \bar{b} \in \mathbf{Z}_p$, with $\bar{a}, \bar{b} \neq \bar{0}$. Then:

    (i)   *If $\bar{a}$ and $\bar{b}$ are both quadratic residues, then so is $\overline{ab}$.*

    (ii) *If $\bar{a}$ and $\bar{b}$ are both quadratic nonresidues, then $\overline{ab}$ is a quadratic residue.*

    (iii) *If one of $\bar{a}$ and $\bar{b}$ is a quadratic residue and the other is a quadratic nonresidue, then $\overline{ab}$ is a quadratic nonresidue.*

PROOF You will prove this corollary in Exercise 8.

EXERCISES 11.2

Numerical Problems

1. Evaluate each of the following Legendre symbols. (You may use the fact that the numbers 1,000,003 and 8,675,309 are prime.)

   a. $\left(\dfrac{-1}{47}\right)$
   b. $\left(\dfrac{-1}{101}\right)$
   c. $\left(\dfrac{-1}{1,000,003}\right)$
   d. $\left(\dfrac{-1}{8,675,309}\right)$

2. a. Calculate $\left(\dfrac{7}{23}\right)$ using Euler's Identity (11.2.2).

   b. Calculate $\left(\dfrac{-1}{23}\right)$ using the Quadratic Character of -1 (11.2.3).

   c. Calculate $\left(\dfrac{-7}{23}\right)$ using the Multiplicativity of the Legendre Symbol (11.2.5).

3. a. Calculate $\left(\dfrac{31}{67}\right)$ using Euler's Identity (11.2.2).

   b. Calculate $\left(\dfrac{-1}{67}\right)$ using the Quadratic Character of -1 (11.2.3).

   c. Calculate $\left(\dfrac{-31}{67}\right)$ using the Multiplicativity of the Legendre Symbol (11.2.5).

4. Let $n = 2^5 \cdot 3^2 \cdot 17^3 \cdot 23^2$. [*Hint:* For this exercise, you may want to use the Multiplicativity of the Legendre Symbol (11.2.5).]

   a. Calculate $\left(\dfrac{n}{3}\right)$.
   b. Calculate $\left(\dfrac{n}{5}\right)$.
   c. Calculate $\left(\dfrac{-n}{5}\right)$.

5. Let $n = 2^4 \cdot 11^3 \cdot 19^{17}$. [*Hint:* For this exercise, you may want to use the Multiplicativity of the Legendre Symbol (11.2.5).]

   a. Calculate $\left(\dfrac{n}{7}\right)$.
   b. Calculate $\left(\dfrac{-n}{7}\right)$.

6. Calculate $\left(\dfrac{2}{p}\right)$ for $p = 3, 5, 7, 11, 13, 17, 19, 23, 29,$ and 31, using Euler's Identity (11.2.2). Describe any pattern you find.

Reasoning and Proofs

7. Prove Lemma 11.2.4.

8. Prove Corollary 11.2.6.

9. Let $p > 2$ be prime, and let $a \in \mathbf{Z}$ such that $p \nmid a$. Explain why $\left(\dfrac{a^2}{p}\right) = 1$ using each of the following three methods:

   a. Using the definition of the Legendre symbol (11.1.4)

   b. Using Theorem 11.2.5

   c. Using Euler's Identity (11.2.2)

10. If $\bar{a}$ is a quadratic residue modulo p, then there is a computationally efficient (polynomial-time) method for finding a square root of $\bar{a}$ modulo p. This method turns out to be particularly easy in the case that $p \equiv 3 \pmod 4$, as you will see in this exercise.

    a. Let p be prime such that $p \equiv 3 \pmod 4$, and let $\bar{a}$ be a quadratic residue modulo p. Show that $\bar{b} = \bar{a}^{\frac{p+1}{4}}$ is a square root of $\bar{a}$ modulo p. That is, prove that $\bar{b}^2 \equiv \bar{a} \pmod p$.

    b. Use the result from part a to find the square roots of $\bar{2}$ modulo 23.

Advanced Reasoning and Proofs

11. Let $p > 5$ be prime. Prove that there exists $a \in \mathbf{Z}$ such that a and $a + 1$ are both quadratic residues modulo p. [*Hint:* $2 \cdot 5 = 10$.]

12. We can use Theorem 11.2.3 to prove that there are infinitely many primes of the form $4k + 1$ (where $k \in \mathbf{Z}$). Begin by assuming that there are only finitely many such primes, and call them $p_1, p_2, \ldots, p_m$. Let $N = 4 \cdot (p_1 \cdot p_2 \cdot \ \cdots \ \cdot p_m)^2 + 1$, and let q be a prime number such that $q \mid N$.

    a. Prove that $q \equiv 1 \pmod 4$. [*Hint:* Use Theorem 11.2.3.]

    b. Show that $q \neq p_j$ for every $j = 1, 2, \ldots, m$.

    c. Why does it follow from parts a and b that there are infinitely many primes of the form $4k + 1$?

11.3 The Law of Quadratic Reciprocity

Let's say you wake up one day wishing to know whether 3 is a square modulo the prime 107. Remembering that Euler's Identity (11.2.2) gives an efficient method to find the Legendre symbol, you begin computing $3^{(107-1)/2} \pmod{107}$. Just as you're about to complete your calculation, your thoughts are suddenly interrupted by a loud knock at the window, and you look up to see your dreaded friend Phil Lovett. He asks you what you're thinking about, and you reluctantly explain your calculation of $\left(\frac{3}{107}\right)$, which will now have to be redone from the beginning, thanks to his interruption.

    "It's really easy to find $\left(\frac{3}{107}\right)$," replies Phil with his usual dopey grin. "We want to determine whether 107 is a square modulo 3. Well, $107 \equiv 2 \pmod 3$, and 2 is not a square modulo 3. So 107 is also a quadratic nonresidue modulo 3. So your answer is -1. Q.E.D.! Can't get much easier than that! Dude, I'm out of here."

    "Um . . . ," you begin. But Phil has left without giving you the chance to explain what's wrong with his solution. Phil has correctly shown that 107 is not a square modulo 3. However, you wanted to know whether 3 is a square modulo 107.

Phil has figured out that $\left(\frac{107}{3}\right) = -1$, but what is the value of $\left(\frac{3}{107}\right)$? Once again, Phil has managed to screw things up.

Pondering Phil's error, you find it slightly curious that finding $\left(\frac{107}{3}\right)$ is so easy, while the value of $\left(\frac{3}{107}\right)$ seems much trickier to determine. In general, finding $\left(\frac{p}{3}\right)$ simply involves reducing p modulo 3 and checking whether the result is a square modulo 3. On the other hand, $\left(\frac{3}{p}\right)$ asks the question of whether 3 is a square mod p, an apparently unrelated question that seems more difficult to answer.

Statement of the Law of Quadratic Reciprocity

More generally, you might wonder whether there is a relationship between $\left(\frac{p}{q}\right)$ and $\left(\frac{q}{p}\right)$ for any primes p and q. In fact, there is a nice relationship between $\left(\frac{p}{q}\right)$ and $\left(\frac{q}{p}\right)$, known as the Law of Quadratic Reciprocity. This celebrated theorem is the main result of this chapter. In order to figure out the exact relationship between $\left(\frac{p}{q}\right)$ and $\left(\frac{q}{p}\right)$, let's look at a few examples.

First, let's see what happens when $q = 5$. In other words, can we find a relationship between $\left(\frac{5}{p}\right)$ and $\left(\frac{p}{5}\right)$?

EXAMPLE 1
....

Find $\left(\frac{p}{5}\right)$ and $\left(\frac{5}{p}\right)$ for $p = 13$.

SOLUTION To find $\left(\frac{13}{5}\right)$, we note that $13 \equiv 3 \pmod 5$. Since $\left(\frac{3}{5}\right) = -1$, we conclude that $\left(\frac{13}{5}\right) = -1$ as well.

To find $\left(\frac{5}{13}\right)$, we note that the squares modulo 13 are 1, 3, 4, 9, 10, and 12. Thus, 5 is not a square modulo 13, so $\left(\frac{5}{13}\right) = -1$.

Let's make a table of values of $\left(\frac{p}{5}\right)$ and $\left(\frac{5}{p}\right)$ for various primes. (If you did Exercise 27 of Section 11.1, you can check your work against this table.)

p	7	11	13	17	19	23	29	31	$\cdots$
$\left(\frac{p}{5}\right)$	-1	1	-1	-1	1	-1	1	1	$\cdots$
$\left(\frac{5}{p}\right)$	-1	1	-1	-1	1	-1	1	1	$\cdots$

It's hard to miss the similarity between the two rows of the table. The evidence suggests that even though $\left(\frac{p}{5}\right)$ and $\left(\frac{5}{p}\right)$ seem to be asking very different questions, it is always the case that $\left(\frac{5}{p}\right) = \left(\frac{p}{5}\right)$. This remarkable fact, which does indeed hold for all primes $p > 2$, is a part of the Law of Quadratic Reciprocity, which gives a general relationship between $\left(\frac{p}{q}\right)$ and $\left(\frac{q}{p}\right)$. Before stating this law, we will consider a few more examples.

Let's now compare $\left(\frac{7}{p}\right)$ with $\left(\frac{p}{7}\right)$ for various primes p.

p	11	13	17	19	23	29	31	37	$\cdots$
$\left(\frac{p}{7}\right)$	1	-1	-1	-1	1	1	-1	1	$\cdots$
$\left(\frac{7}{p}\right)$	-1	-1	-1	1	-1	1	1	1	$\cdots$

This time things are more complicated. We see that $\left(\frac{p}{7}\right)$ and $\left(\frac{7}{p}\right)$ are sometimes equal and sometimes not. Let's leave 7 for a while and move on to another prime.

Here is a table of the values of $\left(\frac{p}{13}\right)$ and $\left(\frac{13}{p}\right)$ for various primes p.

p	7	11	17	19	23	29	31	$\cdots$
$\left(\frac{p}{13}\right)$	-1	1	-1	1	-1	1	1	$\cdots$
$\left(\frac{13}{p}\right)$	-1	1	-1	1	-1	1	1	$\cdots$

That's more like it! It appears that $\left(\frac{p}{13}\right) = \left(\frac{13}{p}\right)$ for any odd prime p, just as $\left(\frac{p}{5}\right) = \left(\frac{5}{p}\right)$ for all odd primes. We have noticed a nice pattern that holds for 5 and 13, but not for 7. It turns out this pattern holds for all primes congruent to 1 modulo 4. That is, if q is prime and $q \equiv 1 \pmod{4}$, then

$$\left(\frac{p}{q}\right) = \left(\frac{q}{p}\right). \tag{1}$$

This gives us a remarkable relationship between $\left(\frac{p}{q}\right)$ and $\left(\frac{q}{p}\right)$ in the case $q \equiv 1 \pmod{4}$.

Note that equation (1), which we claim to hold for all primes $q \equiv 1 \pmod{4}$, is symmetric in the primes p and q. This symmetry implies that equation (1) must also hold whenever $p \equiv 1 \pmod{4}$. That is, if either p or q is congruent to 1 modulo 4, then $\left(\frac{p}{q}\right) = \left(\frac{q}{p}\right)$.

To complete the picture, it would be nice to find a relationship between $\left(\frac{p}{q}\right)$ and $\left(\frac{q}{p}\right)$ that is valid in the remaining case—the case in which both p and q are congruent to 3 modulo 4. Let's reexamine the table we made of values of $\left(\frac{p}{7}\right)$ and $\left(\frac{7}{p}\right)$. This time we will only include primes p that are congruent to 3 modulo 4.

p	11	19	23	31	43	47	59	67	$\cdots$
$\left(\frac{p}{7}\right)$	1	-1	1	-1	1	-1	-1	1	$\cdots$
$\left(\frac{7}{p}\right)$	-1	1	-1	1	-1	1	1	-1	$\cdots$

$\leftarrow p \equiv 3\ (\text{mod } 4)$

In this new table, a clear pattern emerges. We find $\left(\frac{p}{7}\right) = -\left(\frac{7}{p}\right)$ for any prime p that is congruent to 3 modulo 4. In general, if p and q are both congruent to 3 modulo 4, then $\left(\frac{p}{q}\right) = -\left(\frac{q}{p}\right)$.

To summarize, our calculations suggest the following theorem, known as the Law of Quadratic Reciprocity.

11.3.1   LAW OF QUADRATIC RECIPROCITY

Let p and q be odd primes with $p \neq q$.

(*i*)   *If $p \equiv 1$ (mod 4) or $q \equiv 1$ (mod 4), then* $\left(\frac{p}{q}\right) = \left(\frac{q}{p}\right)$.

(*ii*)   *If $p \equiv 3$ (mod 4) and $q \equiv 3$ (mod 4), then* $\left(\frac{p}{q}\right) = -\left(\frac{q}{p}\right)$.

Most of the remainder of this chapter is devoted to developing the machinery that we will need for the proof of the Law of Quadratic Reciprocity, which we will complete in Section 11.6. But first, let's see an example.

EXAMPLE 2

Use the Law of Quadratic Reciprocity to find

     a. $\left(\frac{11}{47}\right)$

     b. $\left(\frac{3}{107}\right)$

SOLUTION

a. Since 11 and 47 are both congruent to 3 modulo 4, the Law of Quadratic Reciprocity tells us that

$$\left(\frac{11}{47}\right) = -\left(\frac{47}{11}\right).$$

Since 47 reduces to 3 modulo 11, we know that $\left(\frac{47}{11}\right) = \left(\frac{3}{11}\right)$. Our problem is now reduced to determining whether 3 is a square modulo 11. A table of squares modulo 11 shows that 3 is a square modulo 11, and hence, $\left(\frac{3}{11}\right) = 1$. Therefore, $\left(\frac{47}{11}\right) = 1$ as well, and thus,

$$\left(\frac{11}{47}\right) = -1.$$

b. Since 3 and 107 are both congruent to 3 modulo 4, quadratic reciprocity tells us that

$$\left(\frac{3}{107}\right) = -\left(\frac{107}{3}\right)$$

$$= -\left(\frac{2}{3}\right) \quad \leftarrow \textbf{since } 107 \equiv 2 \ (\textbf{mod } 3).$$

Because 2 is not a square modulo 3, we know that $\left(\frac{2}{3}\right) = -1$, and hence,

$$\left(\frac{3}{107}\right) = 1.$$

Thus, we see that modulo 107, the number 3 is a quadratic residue.[2] ●

Restatement of the Law

The Law of Quadratic Reciprocity (11.3.1) gives a direct relationship between $\left(\frac{p}{q}\right)$ and $\left(\frac{q}{p}\right)$ for prime numbers p and q. We can restate the law in an equivalent, more compact form.

11.3.2   LAW OF QUADRATIC RECIPROCITY (Restatement)

Let p and q be odd primes with $p \neq q$. Then

$$\left(\frac{p}{q}\right)\left(\frac{q}{p}\right) = (-1)^{\frac{p-1}{2} \cdot \frac{q-1}{2}}. \tag{2}$$

---

[2]Not surprisingly, it has turned out that Phil's answer from the beginning of the section was wrong. When you tell this to Phil, he responds, "Dude, I was basically right—I was only off by a minus sign!"

When we prove the Law of Quadratic Reciprocity in Section 11.6, we will prove this second formulation (11.3.2). We now give the proof that the restatement (11.3.2) implies the original statement of the Law of Quadratic Reciprocity (11.3.1). In Exercise 12, you will prove that the original statement of the law (11.3.1) implies the restatement (11.3.2), thus completing the proof that 11.3.1 and 11.3.2 are equivalent statements.

PROOF **THAT 11.3.2 IMPLIES 11.3.1**

Let p and q be odd primes with $p \neq q$. We assume that equation (2) holds, and we must show that the statement of Theorem 11.3.1 is true.

Case 1 $p \equiv 1 \pmod 4$ or $q \equiv 1 \pmod 4$. **[To show:** $\left(\dfrac{p}{q}\right) = \left(\dfrac{q}{p}\right)$.**]**

Without loss of generality, we may assume that $p \equiv 1 \bmod 4$. Then there exists $k \in \mathbf{Z}$ such that $p = 4k + 1$. This implies that

$$\frac{p-1}{2} = 2k,$$

which is an even number. Hence, the quantity on the right side of equation (2) is -1 raised to an even power, which equals 1. Thus, equation (2) tells us

$$\left(\frac{p}{q}\right)\left(\frac{q}{p}\right) = 1.$$

Since $\left(\dfrac{p}{q}\right)$ and $\left(\dfrac{q}{p}\right)$ must each equal 1 or -1, we may conclude that either both equal 1 or both equal -1. Thus,

$$\left(\frac{p}{q}\right) = \left(\frac{q}{p}\right).$$

Case 2 $p \equiv 3 \pmod 4$ and $q \equiv 3 \pmod 4$. **[To show:** $\left(\dfrac{p}{q}\right) = -\left(\dfrac{q}{p}\right)$.**]**

Since $p \equiv 3 \bmod 4$, there exists $k \in \mathbf{Z}$ such that $p = 4k + 3$. Thus,

$$\frac{p-1}{2} = 2k + 1,$$

which is an odd number. Similarly, $\dfrac{q-1}{2}$ is odd. Hence, the right side of equation (2) is -1 raised to an odd power, which equals -1. Therefore, equation (2) tells us that

$$\left(\frac{p}{q}\right)\left(\frac{q}{p}\right) = -1. \tag{3}$$

Since the two Legendre symbols $\left(\dfrac{p}{q}\right)$ and $\left(\dfrac{q}{p}\right)$ must each equal 1 or -1, we may conclude from equation (3) that one of them equals 1 and the other equals -1. Thus,

$$\left(\frac{p}{q}\right) = -\left(\frac{q}{p}\right). \qquad\blacksquare$$

EXERCISES 11.3

Numerical Problems

In Exercises 1–8, use the Law of Quadratic Reciprocity to evaluate each Legendre symbol.

1. $\left(\dfrac{3}{71}\right)$ 2. $\left(\dfrac{5}{71}\right)$ 3. $\left(\dfrac{7}{71}\right)$ 4. $\left(\dfrac{11}{71}\right)$

5. $\left(\dfrac{3}{53}\right)$ 6. $\left(\dfrac{5}{53}\right)$ 7. $\left(\dfrac{7}{53}\right)$ 8. $\left(\dfrac{11}{53}\right)$

Reasoning and Proofs

9. Describe all primes p such that 3 is a quadratic residue modulo p.

10. Describe all primes p such that 5 is a quadratic residue modulo p.

11. Describe all primes p such that 7 is a quadratic residue modulo p.

12. Prove that Theorem 11.3.1 implies Theorem 11.3.2, thus completing the proof that these two formulations of the Law of Quadratic Reciprocity are equivalent.

Advanced Reasoning and Proofs

13. Euler originally conjectured a law of quadratic reciprocity as follows:

    THEOREM. *Let p and q be odd primes, and let a be an integer that is not divisible by p.*

    If $p \equiv q \pmod{4a}$ or $p \equiv -q \pmod{4a}$, then $\left(\dfrac{a}{p}\right) = \left(\dfrac{a}{q}\right)$.

    In this exercise, you will prove that this Theorem is equivalent to the Law of Quadratic Reciprocity.

    a. Using the Law of Quadratic Reciprocity, prove the Theorem. [*Hint:* You may also use the Quadratic Character of 2 (11.4.3).]

    b. Prove that the Theorem implies the Law of Quadratic Reciprocity.

11.4 Gauss's Lemma

In the previous section, we stated the Law of Quadratic Reciprocity, a relationship between $\left(\dfrac{p}{q}\right)$ and $\left(\dfrac{q}{p}\right)$ for any odd primes p and q. Proving this theorem will be our main goal for the rest of the chapter. Along the way, we will encounter some other interesting ways of viewing the Legendre symbol.

When we met Euler's Identity in Section 11.2, we saw that quadratic residues, which appear in the $\bar{x}^2$ row of the exponent table for $\mathbf{Z}_p$, can also be identified by looking at the $\bar{x}^{(p-1)/2}$ row of the exponent table. As it turns out, the *multiplication* table for $\mathbf{Z}_p$ also contains hidden information about quadratic residues. To discover this link, it will be useful to examine a different form of the modular multiplication tables, with negative as well as positive entries.

The lazy multiplication table

One day, your friend Phil Lovett is complaining about his remedial math class, Arithmetic in Z_{11}. "Dude, this multiplication stuff is killing me. It's cool when they ask me to multiply small numbers together, but multiplying big numbers is just too much work. Like, for homework tonight, we have to multiply $\overline{9} \cdot \overline{10}$. Multiplying those big numbers and then reducing is totally harshing my mellow. You gotta help me out, dude."

The nonzero elements of Z_{11} are

$$\overline{1}, \overline{2}, \overline{3}, \overline{4}, \overline{5}, \overline{6}, \overline{7}, \overline{8}, \overline{9}, \overline{10}.$$

If we want to be lazy like Phil and avoid working with large numbers, we could write the numbers larger than 5 as small negative numbers. For example, instead of writing the number $\overline{8}$, we could write $\overline{-3}$, which is equivalent. In this way, we find that we can write the nonzero elements of Z_{11} as

$$\overline{1}, \overline{2}, \overline{3}, \overline{4}, \overline{5}, \overline{-5}, \overline{-4}, \overline{-3}, \overline{-2}, \overline{-1}.$$

This certainly makes some arithmetic problems in Z_{11} easier. For example, Phil's homework problem of multiplying $\overline{9} \cdot \overline{10}$ is equivalent to the much simpler problem, $\overline{-2} \cdot \overline{-1}$, which equals $\overline{2}$. Table 11.4.1 shows the entire multiplication table for Z_{11} rewritten in this lazy way.

Table 11.4.1 Lazy Multiplication Table Modulo 11

$\cdot$	$\overline{1}$	$\overline{2}$	$\overline{3}$	$\overline{4}$	$\overline{5}$	$\overline{-5}$	$\overline{-4}$	$\overline{-3}$	$\overline{-2}$	$\overline{-1}$
$\overline{1}$	1	2	3	4	5	−5	−4	−3	−2	−1
$\overline{2}$	2	4	−5	−3	−1	1	3	5	−4	−2
$\overline{3}$	3	−5	−2	1	4	−4	−1	2	5	−3
$\overline{4}$	4	−3	1	5	−2	2	−5	−1	3	−4
$\overline{5}$	5	−1	4	−2	3	−3	2	−4	1	−5
$\overline{-5}$	−5	1	−4	2	−3	3	−2	4	−1	5
$\overline{-4}$	−4	3	−1	−5	2	−2	5	1	−3	4
$\overline{-3}$	−3	5	2	−1	−4	4	1	−2	−5	3
$\overline{-2}$	−2	−4	5	3	1	−1	−3	−5	4	2
$\overline{-1}$	−1	−2	−3	−4	−5	5	4	3	2	1

A close look at Table 11.4.1 reveals many patterns. When we first looked at multiplication tables in prime moduli, we noticed that every row contains each element exactly once (Theorem 8.3.6). Even more can be said about Table 11.4.1. Examining the first half of each row (the first 5 entries of each row), shown in Table 11.4.2, one sees that each number from 1 to 5 appears exactly once, provided we ignore any minus signs that appear.

Table 11.4.2 The Left Half of the Lazy Multiplication Table Modulo 11

$\cdot$	$\overline{1}$	$\overline{2}$	$\overline{3}$	$\overline{4}$	$\overline{5}$
$\overline{1}$	1	2	3	4	5
$\overline{2}$	2	4	-5	-3	-1
$\overline{3}$	3	-5	-2	1	4
$\overline{4}$	4	-3	1	5	-2
$\overline{5}$	5	-1	4	-2	3
$\overline{-5}$	-5	1	-4	2	-3
$\overline{-4}$	-4	3	-1	-5	2
$\overline{-3}$	-3	5	2	-1	-4
$\overline{-2}$	-2	-4	5	3	1
$\overline{-1}$	-1	-2	-3	-4	-5

← **Each row contains the numbers 1, 2, 3, 4, and 5 exactly once— provided signs are ignored.**

This is a general fact, which will be useful for our investigation of quadratic residues. We state and prove this fact soon, in Theorem 11.4.1. But first, we introduce a couple of new terms.

Let p be a prime greater than 2. The nonzero elements of $\mathbf{Z}_p$, which are usually expressed as

$$\overline{1}, \overline{2}, \ldots, \overline{p-2}, \overline{p-1}$$

may also be written as

$$\overline{1}, \overline{2}, \ldots, \overline{\frac{p-1}{2}}, \overline{-\frac{p-1}{2}}, \ldots, \overline{-2}, \overline{-1}$$

We will refer to the elements $\overline{1}, \overline{2}, \ldots, \overline{\frac{p-1}{2}}$ as **positive residues** and to the remaining elements, $\overline{-\frac{p-1}{2}}, \ldots, \overline{-2}, \overline{-1}$, as **negative residues**.

The following theorem expresses the pattern we observed in the left half of each row of the $\mathbf{Z}_{11}$ multiplication table (see Table 11.4.2).

THEOREM 11.4.1

Let $p > 2$ be prime, and let $\bar{a}$ be a nonzero element of $\mathbf{Z}_p$. Then in the multiplication table for $\mathbf{Z}_p$, the elements appearing in the first half of the row of $\bar{a}$—namely,

$$\bar{a} \cdot \bar{1}, \; \bar{a} \cdot \bar{2}, \ldots, \; \bar{a} \cdot \overline{\frac{p-1}{2}}$$

may be reordered to obtain the list

$$\bar{1}, \; \bar{2}, \ldots, \; \overline{\frac{p-1}{2}},$$

provided the signs of the elements are ignored.

PROOF We know from Theorem 8.3.6 that the full row of $\bar{a}$ in the multiplication table for $\mathbf{Z}_p$—namely, the list

$$\bar{a} \cdot \bar{1}, \; \bar{a} \cdot \bar{2}, \ldots, \; \bar{a} \cdot \overline{\frac{p-1}{2}}, \; \bar{a} \cdot \left(\overline{-\frac{p-1}{2}} \right), \ldots, \bar{a} \cdot (\overline{-2}), \; \bar{a} \cdot (\overline{-1}) \qquad (1)$$

contains every element of $\mathbf{Z}_p$ exactly once. Furthermore, the elements in the second half of this list are the negatives of the elements in the first half of the list: Each $\bar{a} \cdot (\overline{-j})$ is the negative of $\bar{a} \cdot \bar{j}$. For each positive residue $\bar{x}$ (from $\bar{1}$ to $\overline{\frac{p-1}{2}}$), both $\bar{x}$ and $\overline{-x}$ must appear in the list (1), and they cannot both appear in the same half of the list. It follows that exactly one of $\bar{x}$ or $\overline{-x}$ appears in the first half of the list. Thus, when signs are ignored, every positive residue appears exactly once in the first half of the list. ■

Gauss's Lemma

Our insights into the multiplication table for $\mathbf{Z}_p$ provide a new way to determine which elements of $\mathbf{Z}_p$ are quadratic residues. To see how this works, let's take as an example the element $\bar{3}$ in $\mathbf{Z}_{11}$. We will use the first half of the row of $\bar{3}$ in the $\mathbf{Z}_{11}$ multiplication table to determine whether $\bar{3}$ is a quadratic residue modulo 11.

Naomi's Numerical Proof Preview: Gauss's Lemma (11.4.2)

Consider the first half of the row of $\overline{3}$ in the $\mathbf{Z}_{11}$ multiplication table:

$\cdot$	$\overline{1}$	$\overline{2}$	$\overline{3}$	$\overline{4}$	$\overline{5}$
$\overline{3}$	3	-5	-2	1	4

Let's write these five products explicitly:

$$\overline{3} \cdot \overline{1} = \overline{3}$$
$$\overline{3} \cdot \overline{2} = \overline{-5}$$
$$\overline{3} \cdot \overline{3} = \overline{-2}$$
$$\overline{3} \cdot \overline{4} = \overline{1}$$
$$\overline{3} \cdot \overline{5} = \overline{4}.$$

If we multiply all of these equations together, we get

$$\overline{3}^5 \cdot \overline{1} \cdot \overline{2} \cdot \overline{3} \cdot \overline{4} \cdot \overline{5} = \overline{3} \cdot (\overline{-5}) \cdot (\overline{-2}) \cdot \overline{1} \cdot \overline{4}.$$

The left side of this equation is simply $\overline{3}^5 \cdot \overline{5!}$. What about the right side? The pattern we observed in the table (Theorem 11.4.1) tells us that the numbers from $\overline{1}$ to $\overline{5}$ each appear exactly once, and the only question is how many minus signs appear. In this case, there are **2** minus signs, which means that the product of the numbers on the right side is $(-\overline{1})^2 \cdot \overline{5!}$. Thus, we get

$$\overline{3}^5 \cdot \overline{5!} = (\overline{-1})^2 \cdot \overline{5!}$$

Canceling the $\overline{5!}$, we get

$$\overline{3}^5 = (\overline{-1})^2.$$

But now we can use Euler's Identity (11.2.2), which tells us that $\overline{3}^5 \equiv \left(\frac{3}{11}\right)$ (mod 11). Substituting, we get

$$\left(\frac{3}{11}\right) \equiv (-1)^2 \text{ (mod 11)},$$

and hence, $\left(\frac{3}{11}\right) = 1$, so we have determined that $\overline{3}$ is a quadratic residue modulo 11.

We have just discovered a new way to determine whether an element $\overline{a} \in \mathbf{Z}_p$ is a quadratic residue. We simply find the number g of negative residues that appear in the first half of the row of $\overline{a}$, and then $\left(\frac{a}{p}\right)$ equals $(-1)^g$.

Let $p > 2$ be prime, and let $\bar{a}$ be a nonzero element of $\mathbf{Z}_p$. Let $\mathbf{g}$ be the number of negative residues appearing in the list

$$\bar{a} \cdot \bar{1}, \; \bar{a} \cdot \bar{2}, \ldots, \; \bar{a} \cdot \overline{\frac{p-1}{2}}.$$

Then

$$\left(\frac{a}{p}\right) = (-1)^g.$$

PROOF By Theorem 11.4.1, the list

$$\bar{a} \cdot \bar{1}, \; \bar{a} \cdot \bar{2}, \ldots, \; \bar{a} \cdot \overline{\frac{p-1}{2}} \tag{2}$$

contains each of the numbers

$$\bar{1}, \; \bar{2}, \; \bar{3}, \ldots, \; \overline{\frac{p-1}{2}} \tag{3}$$

exactly once if minus signs are ignored. Thus, the product of the numbers in list (2) equals the product of the numbers in list (3) multiplied by (-1) raised to the power g, the number of negative residues in list (2). We thus have

$$(\bar{a} \cdot \bar{1}) \cdot (\bar{a} \cdot \bar{2}) \cdot \; \cdots \; \cdot \left(\bar{a} \cdot \overline{\frac{p-1}{2}}\right) = (\overline{-1})^g \cdot \bar{1} \cdot \bar{2} \cdot \; \cdots \; \cdot \overline{\frac{p-1}{2}}.$$

Canceling $\bar{1} \cdot \bar{2} \cdot \; \cdots \; \cdot \overline{\frac{p-1}{2}}$ from both sides, we obtain

$$\bar{a}^{\frac{p-1}{2}} = (\overline{-1})^g.$$

Now consider the left side of this equation. Euler's Identity (11.2.2) tells us that $a^{\frac{p-1}{2}} \equiv \left(\frac{a}{p}\right) \pmod{p}$. We thus obtain

$$\left(\frac{a}{p}\right) \equiv (-1)^g \pmod{p}. \tag{4}$$

It follows (see Exercise 17) that

$$\left(\frac{a}{p}\right) = (-1)^g, \tag{5}$$

as was to be shown. ■

EXAMPLE 1

Determine whether $\bar{2}$ is a quadratic residue modulo 11.

SOLUTION From Table 11.4.2, we see that the first half of the row of $\bar{2}$ in the $\mathbf{Z}_{11}$ multiplication table contains the numbers $\bar{2}, \bar{4}, -\bar{5}, -\bar{3}$, and $-\bar{1}$. Since there are **3** negative numbers in this list (i.e., $g = 3$), we use Gauss's Lemma (11.4.2) to get

$$\left(\frac{2}{11}\right) = (-1)^3 = -1.$$

Thus, $\bar{2}$ is not a quadratic residue modulo 11. ●

When is 2 a square mod p?

In Section 11.2, we found (Theorem 11.2.3) that if $p > 2$ is prime, then $-\bar{1}$ is a quadratic residue modulo p if and only if $p \equiv 1 \pmod 4$. This was a consequence of Euler's Identity. We use Gauss's Lemma to determine when $\bar{2}$ is a quadratic residue modulo p.

THEOREM 11.4.3   THE QUADRATIC CHARACTER OF 2

Let $p > 2$ be prime.

   (*i*)   *If $p \equiv 1 \pmod 8$ or $p \equiv -1 \pmod 8$, then $\left(\dfrac{2}{p}\right) = 1$.*

   (*ii*)   *If $p \equiv 3 \pmod 8$ or $p \equiv -3 \pmod 8$, then $\left(\dfrac{2}{p}\right) = -1$.*

Note: Conditions (*i*) and (*ii*) cover all primes $p > 2$: Every odd number is congruent to $1, 3, 5$, or $7 \pmod 8$ and hence is congruent to $1, 3, -3$, or $-1 \pmod 8$.

PROOF OF THEOREM 11.4.3

We will use Gauss's Lemma (11.4.2), which asks us to compute the number g of negative residues in the list

$$\bar{2} \cdot \bar{1}, \ \bar{2} \cdot \bar{2}, \ldots, \bar{2} \cdot \overline{\frac{p-1}{2}}. \tag{6}$$

Note that each of the products $2 \cdot 1, \ 2 \cdot 2, \ldots, 2 \cdot \dfrac{p-1}{2}$ is between 0 and $p - 1$ and hence is already reduced modulo p. Thus, for each integer k from 1 to $\dfrac{p-1}{2}$, the product $\bar{2} \cdot \bar{k}$ is a positive residue precisely when

$$2 \cdot k \le \frac{p-1}{2},$$

which is equivalent to

$$k \le \frac{p-1}{4}.$$

Hence, the number of positive residues in our list (6) is the number of positive integers k less than or equal to $\dfrac{p-1}{4}$. Another way to say this is that the number of positive residues in our list is $\dfrac{p-1}{4}$ rounded down to the nearest integer, which is denoted by $\left\lfloor \dfrac{p-1}{4} \right\rfloor$. Since there are $\dfrac{p-1}{2}$ total numbers in our list (6), the number g of negative residues is given by

$$g = \frac{p-1}{2} - \left\lfloor \frac{p-1}{4} \right\rfloor. \tag{7}$$

Case 1 $p \equiv 1 \pmod 8$.

We know that for some integer m,

$$p = 8m + 1.$$

Hence, by equation (7),

$$\begin{aligned}
g &= \frac{p-1}{2} - \left\lfloor \frac{p-1}{4} \right\rfloor \\
&= 4m - \lfloor 2m \rfloor \\
&= 4m - 2m \\
&= 2m.
\end{aligned}$$

Thus, g is even, and hence by Gauss's Lemma (11.4.2),

$$\begin{aligned}
\left(\frac{2}{p} \right) &= (-1)^g \\
&= 1.
\end{aligned}$$

Case 2 $p \equiv 3 \pmod 8$.

Now we know that for some integer m,

$$p = 8m + 3.$$

Hence, by (7),

$$\begin{aligned}
g &= \frac{p-1}{2} - \left\lfloor \frac{p-1}{4} \right\rfloor \\
&= 4m + 1 - \left\lfloor 2m + \frac{1}{2} \right\rfloor \\
&= 4m + 1 - 2m \\
&= 2m + 1.
\end{aligned}$$

Thus, g is odd, and hence by Gauss's Lemma (11.4.2),

$$\left(\frac{2}{p}\right) = (-1)^g$$

$$= -1.$$

We leave the remaining two cases, $p \equiv -1 \pmod 8$ and $p \equiv -3 \pmod 8$, as exercises.

EXAMPLE 2

Determine whether 2 is a square modulo 103.

SOLUTION Since $103 \equiv -1 \pmod 8$, we see by Theorem 11.4.3 (the Quadratic Character of 2) that $\left(\frac{2}{103}\right) = 1$. Thus, 2 is a square modulo 103. ●

Theorem 11.4.3 can be restated in a slightly slicker form. We leave the proof of this restatement as an exercise.

THEOREM 11.4.4 THE QUADRATIC CHARACTER OF 2 (Restatement)

Let $p > 2$ be prime. Then

$$\left(\frac{2}{p}\right) = (-1)^{\frac{p^2-1}{8}}.$$

PROOF You will prove this in Exercise 19.

Having figured out exactly when 2 is a square modulo p, the reader may be wondering whether we can determine a similar rule that tells us when 3 is a square modulo p. This is indeed possible, using the Law of Quadratic Reciprocity (see Exercise 9 of Section 11.3). It is also possible, though a bit more difficult, to answer this question using Gauss's Lemma (see Exercise 21).

EXERCISES 11.4

Numerical Problems

In Exercises 1–6, evaluate each Legendre symbol in three different ways:

    a. **By making a table of squares**

    b. **By using Euler's Identity (11.2.2)**

    c. **By using Gauss's Lemma (11.4.2)**

1. $\left(\dfrac{2}{5}\right)$ 2. $\left(\dfrac{2}{7}\right)$ 3. $\left(\dfrac{3}{7}\right)$

4. $\left(\dfrac{3}{11}\right)$ 5. $\left(\dfrac{4}{11}\right)$ 6. $\left(\dfrac{7}{19}\right)$

7. a. Evaluate $\left(\dfrac{2}{11}\right)$ using Theorem 11.4.3.

   b. Evaluate $\left(\dfrac{2}{11}\right)$ using Theorem 11.4.4. (Verify that your answer is the same as in part a.)

8. Use the Quadratic Character of 2 (Theorem 11.4.3 or 11.4.4) to evaluate $\left(\dfrac{2}{p}\right)$ for $p = 3, 5, 7, 11, 13,$ and 17.

9. Evaluate $\left(\dfrac{2}{p}\right)$ for each prime, p.

   a. $p = 1000003$ b. $p = 17837$ c. $p = 8675309$

10. Evaluate $\left(\dfrac{8}{47}\right)$. [*Hint:* You may want to use the Multiplicativity of the Legendre Symbol (11.2.5).]

11. Evaluate $\left(\dfrac{18}{101}\right)$. [*Hint:* You may want to use the Multiplicativity of the Legendre Symbol (11.2.5).]

Reasoning and Proofs

12. a. Let $p > 2$ be prime. Prove that $\left(\dfrac{4}{p}\right) = 1$.

    b. For which primes p is 8 a quadratic residue modulo p? Explain.

13. Is it possible to find $x \in \mathbf{Z}$ for which $x^2 \equiv -1 \pmod{55}$? If so, find such an x. If not, prove that none exists.

14. Is it possible to find $x \in \mathbf{Z}$ for which $x^2 \equiv 2 \pmod{21}$? If so, find such an x. If not, prove that none exists.

15. Is it possible to find $x \in \mathbf{Z}$ for which $x^2 \equiv -1 \pmod{85}$? If so, find such an x. If not, prove that none exists.

16. Is it possible to find $x \in \mathbf{Z}$ for which $x^2 \equiv 2 \pmod{49}$? If so, find such an x. If not, prove that none exists.

17. In the proof of Gauss's Lemma (11.4.2), explain why congruence (4) implies equation (5).

18. Complete the proof of Theorem 11.4.3 by proving the cases $p \equiv -1 \pmod 8$ and $p \equiv -3 \pmod 8$.

19. Prove Theorem 11.4.4.

20. For which primes p is 6 a quadratic residue modulo p? [*Hint:* Use the Law of Quadratic Reciprocity and the Quadratic Character of 2.]

Advanced Reasoning and Proofs

21. **a.** Consider the Legendre symbol $\left(\frac{3}{11}\right)$. What is the value of g in Gauss's Lemma? Use this to evaluate $\left(\frac{3}{11}\right)$.

   b. Answer the questions from part a for the Legendre symbols $\left(\frac{3}{13}\right)$ and $\left(\frac{3}{23}\right)$.

   c. Let $p > 3$ be prime. For the Legendre symbol $\left(\frac{3}{p}\right)$, find a formula for the value of g in Gauss's Lemma. Your formula should resemble equation (7).

   d. Use your formula from part c to answer the following question: For which primes p is $\overline{3}$ a quadratic residue modulo p?

11.5 Quadratic Residues and Lattice Points

Keeping careful count of quotients

So far, we have seen several different ways to compute the Legendre symbol $\left(\frac{a}{p}\right)$. In this section, we introduce yet another method and see that it has a nice geometric interpretation in terms of lattice points in a triangle. This interpretation will be crucial in our proof of Quadratic Reciprocity in the next section.

Naomi's Numerical Proof Preview: Eisenstein's Lemma (11.5.1)

Our new method relates closely to Gauss's Lemma (11.4.2). To compute $\left(\frac{7}{11}\right)$ using Gauss's Lemma, we compute the products

$$\overline{7} \cdot \overline{1}, \; \overline{7} \cdot \overline{2}, \; \overline{7} \cdot \overline{3}, \; \overline{7} \cdot \overline{4}, \; \overline{7} \cdot \overline{5}$$

in $\mathbf{Z}_{11}$ and see how many negative residues appear in the list. Until now, when working in $\mathbf{Z}_{11}$, we have looked only at remainders after dividing by 11, and we have ignored the quotients.

For our new approach, however, it pays to keep track of the quotients. Let's write out what happens when we use the Division Theorem to divide each of the products $7 \cdot 1, \; 7 \cdot 2, \; 7 \cdot 3, \; 7 \cdot 4$, and $7 \cdot 5$ by 11.

$$7 \cdot 1 = \mathbf{0} \cdot 11 + 7$$
$$7 \cdot 2 = \mathbf{1} \cdot 11 + 3$$
$$7 \cdot 3 = \mathbf{1} \cdot 11 + 10 \tag{1}$$
$$7 \cdot 4 = \mathbf{2} \cdot 11 + 6$$
$$7 \cdot 5 = \mathbf{3} \cdot 11 + 2$$

Recall that for the purposes of Gauss's Lemma (11.4.2), we are supposed to use lazy mod 11 arithmetic; that is, we want our remainders to be in the range $-5, \ldots, 5$. Thus, for example, in the fourth equation in the list (1),

$$7 \cdot 4 = \mathbf{2} \cdot 11 + 6,$$

which has a remainder of 6, we will replace the 6 by the expression $(-5) + \mathbf{11}$:

$$7 \cdot 4 = \mathbf{2} \cdot 11 + (-5) + \mathbf{11}.$$

Similarly, we can rewrite the first and third equations in our list (1). The other equations have remainders that are already in the range $-5, \ldots, 5$, so we'll leave them as is. Our list thus becomes

$$7 \cdot 1 = \mathbf{0} \cdot 11 + (-4) + \mathbf{11}$$
$$7 \cdot 2 = \mathbf{1} \cdot 11 + 3$$
$$7 \cdot 3 = \mathbf{1} \cdot 11 + (-1) + \mathbf{11}$$
$$7 \cdot 4 = \mathbf{2} \cdot 11 + (-5) + \mathbf{11}$$
$$7 \cdot 5 = \mathbf{3} \cdot 11 + 2.$$

Let's see what happens when we add all of these equations together. We get

$$7 \cdot (1 + 2 + 3 + 4 + 5) =$$
$$(\mathbf{0} + \mathbf{1} + \mathbf{1} + \mathbf{2} + \mathbf{3}) \cdot 11 + (-4 + 3 + -1 + -5 + 2) + \mathbf{3} \cdot \mathbf{11}. \tag{2}$$

Note that the **3** in the last term of this equation is the number of negative residues in the list; that is, **3** is the value of g in Gauss's Lemma (11.4.2). Gauss's Lemma tells us that the value of the Legendre symbol $\left(\dfrac{7}{11}\right)$ is determined by whether g is even or odd. In other words, this Legendre symbol is determined by the value of g modulo 2. So, let's see what happens when we reduce equation (2) modulo 2.

$$7 \cdot (1 + 2 + 3 + 4 + 5) \equiv$$
$$(\mathbf{0} + \mathbf{1} + \mathbf{1} + \mathbf{2} + \mathbf{3}) \cdot 11 + (-4 + 3 + -1 + -5 + 2) + \mathbf{3} \cdot \mathbf{11} \pmod 2. \tag{3}$$

Since any integer is congruent to its negative modulo 2, we may replace all of the minus signs in congruence (3) by plus signs. Furthermore, since 7 and 11 are odd, both are congruent to 1 modulo 2, so we may replace both 7 and 11 by 1. Congruence (3) then simplifies to

$$1 + 2 + 3 + 4 + 5 \equiv (0 + 1 + 1 + 2 + 3) + (4 + 3 + 1 + 5 + 2) + 3 \pmod 2.$$

Now we may subtract $(1 + 2 + 3 + 4 + 5)$ from both sides of the congruence and move the **3** to the left side. (It becomes -3, but minus signs don't matter modulo 2, as we observed earlier, so we may leave it as **3**.) We get

$$3 \equiv 0 + 1 + 1 + 2 + 3 \pmod 2. \qquad (4)$$

Now that the dust has settled, we're left with something rather striking. To determine whether the value of **g** in Gauss's Lemma (in this case **3**) is even or odd, we can just add up the quotients $0 + 1 + 1 + 2 + 3$.

Congruence (4) implies that

$$(-1)^3 = (-1)^{0+1+1+2+3}.$$

But Gauss's Lemma tells us that the left side of this equation equals the Legendre symbol $\left(\dfrac{7}{11}\right)$. Thus, we get

$$\left(\frac{7}{11}\right) = (-1)^{0+1+1+2+3}.$$

Though it took some effort, these derivations demonstrate something remarkable. We have expressed the Legendre symbol $\left(\dfrac{7}{11}\right)$ in terms of the quotients $0, 1, 1, 2,$ and 3, which arise when the numbers $7 \cdot 1$, $7 \cdot 2$, $7 \cdot 3$, $7 \cdot 4$, and $7 \cdot 5$ are divided by 11.

Eisenstein's Lemma

The ideas we have just seen give us a totally new way to compute the Legendre symbol $\left(\dfrac{a}{p}\right)$ for *odd* values of a, which we state and prove next.

LEMMA 11.5.1 **EISENSTEIN'S LEMMA**

Let $p > 2$ be prime and $a \in \mathbf{Z}$ such that $p \nmid a$. For each $k = 1, 2, \ldots, \dfrac{p-1}{2}$, let q_k be the quotient when $a \cdot k$ is divided by p. Let $T(a, p)$ be the sum of these quotients:

$$T(a, p) = q_1 + q_2 + \cdots + q_{\frac{p-1}{2}}$$

If a is odd, then

$$\left(\frac{a}{p}\right) = (-1)^{T(a, p)}.$$

Before proving Eisenstein's Lemma, let's see a couple of examples.

EXAMPLE 1
....

    a. Use Eisenstein's Lemma to compute $\left(\frac{7}{13}\right)$.

    b. Use Eisenstein's Lemma to compute $\left(\frac{33}{5}\right)$.

SOLUTION

    a. To compute $T(7, 13)$, we need to take the first $\frac{13-1}{2} = 6$ multiples of 7 and find the quotients when these are divided by 13:

$$7 \cdot 1 = \mathbf{0} \cdot 13 + 7$$
$$7 \cdot 2 = \mathbf{1} \cdot 13 + 1$$
$$7 \cdot 3 = \mathbf{1} \cdot 13 + 8$$
$$7 \cdot 4 = \mathbf{2} \cdot 13 + 2$$
$$7 \cdot 5 = \mathbf{2} \cdot 13 + 9$$
$$7 \cdot 6 = \mathbf{3} \cdot 13 + 3$$

Thus,

$$T(7, 13) = 0 + 1 + 1 + 2 + 2 + 3 = 9.$$

We conclude that $\left(\frac{7}{13}\right) = (-1)^9 = -1$.

    b. To compute $\left(\frac{33}{5}\right)$, we take the first $\frac{5-1}{2} = 2$ multiples of 33 and divide them by 5:

$$33 \cdot 1 = \mathbf{6} \cdot 5 + 3$$
$$33 \cdot 2 = \mathbf{13} \cdot 5 + 1.$$

Hence, $\left(\frac{33}{5}\right) = (-1)^{6+13} = -1$.   ●

Note that to compute $\left(\frac{33}{5}\right)$, one would normally use the fact that $33 \equiv 3 \pmod 5$ to reduce $\left(\frac{33}{5}\right)$ to $\left(\frac{3}{5}\right)$, which is easier to compute. However, as illustrated in Example 1 part b, Eisenstein's Lemma can be used to compute $\left(\frac{a}{p}\right)$ directly even when a is greater than p. This fact will be useful in our investigation of Quadratic Reciprocity in the next section.

Proof of Eisenstein's Lemma

To prove Eisenstein's Lemma (11.5.1), we will follow the reasoning from the Numerical Proof Preview that preceded the statement of the lemma.

PROOF OF **EISENSTEIN'S LEMMA (11.5.1)**

Let p be prime and $a \in \mathbf{Z}$ be an odd number such that $p \nmid a$. Let $h = \dfrac{p-1}{2}$. Since q_k is the quotient that arises when $a \cdot k$ is divided by p, we have

$$a \cdot 1 = q_1 \cdot p + r_1$$

$$a \cdot 2 = q_2 \cdot p + r_2$$

$$\vdots$$

$$a \cdot h = q_h \cdot p + r_h,$$

with each r_k in the range $0, \ldots, p - 1$.

Adding these equations together, we obtain

$$a \cdot (1 + 2 + \cdots + h) = T(a, p) \cdot p + (r_1 + r_2 + \cdots + r_h). \qquad (5)$$

We define s_k to be the "lazy" way of writing r_k; that is,

$$s_k = \begin{cases} r_k - p & \text{if } r_k > \dfrac{p-1}{2} \\[2mm] r_k & \text{if } r_k \le \dfrac{p-1}{2} \end{cases}$$

Each time r_k is a negative residue, $s_k = r_k - p$. Thus,

$$r_1 + r_2 + \cdots + r_h = (s_1 + s_2 + \cdots + s_h) + g \cdot p,$$

where g counts the number of k's for which r_k is a negative residue. Note that this is the same quantity g that appears in the statement of Gauss's Lemma (11.4.2). Substituting this into equation (5), we obtain

$$a \cdot (1 + \cdots + h) = T(a, p) \cdot p + (s_1 + \cdots + s_h) + g \cdot p.$$

We know that $a \equiv 1 \pmod 2$ and $p \equiv 1 \pmod 2$ because they are both odd. Thus, we obtain

$$1 + \cdots + h \equiv T(a, p) + (s_1 + \cdots + s_h) + g \pmod 2. \qquad (6)$$

Each number s_k is in the range $-h, \ldots, h$ and satisfies $s_k \equiv a \cdot k \pmod{p}$. It follows from Theorem 11.4.1 that the list

$$s_1, s_2, \ldots, s_h$$

may be reordered to obtain the list

$$1, 2, \ldots, h$$

if signs are ignored. Since signs may safely be ignored modulo 2, we have

$$s_1 + s_2 + \cdots + s_h \equiv 1 + 2 + \cdots + h \pmod{2}.$$

Substituting into congruence (6) gives

$$1 + 2 + \cdots + h \equiv T(a, p) + (1 + 2 + \cdots + h) + g \pmod{2}.$$

Subtracting g from both sides, using the fact that $g \equiv -g \pmod{2}$, and canceling the sum $(1 + 2 + \cdots + h)$ from both sides, we get

$$g \equiv T(a, p) \pmod{2}.$$

It follows that

$$(-1)^g = (-1)^{T(a, p)}.$$

Since $\left(\frac{a}{p}\right) = (-1)^g$ by Gauss's Lemma (11.4.2), we obtain

$$\left(\frac{a}{p}\right) = (-1)^{T(a, p)}$$

as desired. ∎

Visualizing Eisenstein's Lemma

Eisenstein's Lemma expresses the Legendre symbol $\left(\frac{a}{p}\right)$ in terms of a quantity that we called $T(a, p)$. The quantity $T(a, p)$ is defined as the sum of the quotients when the numbers $a \cdot 1, \ a \cdot 2, \ldots, a \cdot \frac{p-1}{2}$ are divided by p. (We will use this notation throughout the rest of this chapter.)

We will now see that $T(a, p)$ has a geometric interpretation as the number of lattice points in a certain triangle. Recall that a *lattice point* in the plane is a point (x, y) whose coordinates are both integers.

Naomi's Numerical Proof Preview: Proposition 11.5.2

Let's revisit the calculation of $T(7, 13)$ from Example 1, part a. In our solution, we divided the first 6 multiples of 7 by 13:

$$7 \cdot 1 = 0 \cdot 13 + 7$$
$$7 \cdot 2 = 1 \cdot 13 + 1$$
$$7 \cdot 3 = 1 \cdot 13 + 8 \tag{7}$$
$$7 \cdot 4 = 2 \cdot 13 + 2$$
$$7 \cdot 5 = 2 \cdot 13 + 9$$
$$7 \cdot 6 = 3 \cdot 13 + 3.$$

Consider the fourth equation,

$$7 \cdot 4 = 2 \cdot 13 + 2.$$

This equation states that when 28 is divided by 13, the quotient is 2. Another way to say this is that the fraction $\frac{28}{13}$, when rounded down to the nearest integer, gives 2. To express this geometrically, consider a segment of length $\frac{28}{13}$. Let's divide this segment into segments of length 1, with a smaller leftover segment at the top.

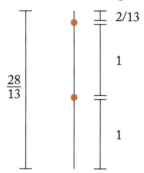

The number of orange dots we need to divide our segment in this way is 2, which equals the quotient when 28 is divided by 13. Similarly, all of the divisions we performed in (7) can be represented by segments of lengths $\frac{7}{13}, \frac{14}{13}, \frac{21}{13}, \frac{28}{13}, \frac{35}{13}$, and $\frac{42}{13}$, as shown in the following diagram:

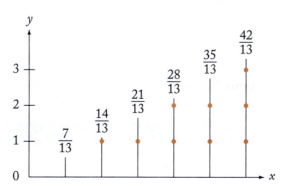

The number of orange dots needed to divide these six segments are, in order, $0, 1, 1, 2, 2$, and 3. Notice that these are exactly the quotients we found in (7).

Also note that the six segments in the diagram fit nicely into a triangle:

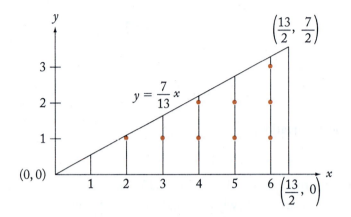

The triangle in this diagram is bounded by the three lines $y = \dfrac{7}{13}x$, $y = 0$, and $x = \dfrac{13}{2}$. The total number of lattice points that lie in the interior of the triangle is $0 + 1 + 1 + 2 + 2 + 3 = 9$, which is exactly $T(7, 13)$. (Note that we are only counting the lattice points in the *interior* of the triangle, and not the lattice points on the edges of the triangle.)

In general, the quantity $T(a, p)$ can be found geometrically by counting the number of lattice points in the interior of the triangle in Figure 1, which is denoted $\Delta(a, p)$.

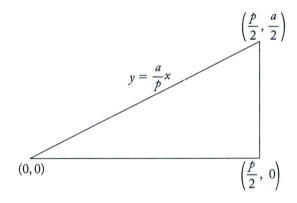

Figure 1 The triangle known as $\Delta(a, p)$.

PROPOSITION 11.5.2

Let $p > 2$ be prime, and let $a \in \mathbf{N}$ such that $p \nmid a$. Then the number of lattice points in the interior of the triangle $\Delta(a, p)$ equals $T(a, p)$.

PROOF Every lattice point in $\Delta(a, p)$ has its x-coordinate in the range $1, 2, \ldots, \dfrac{p-1}{2}$. The following claim tells us how to count the number of lattice points with a given x-coordinate.

Claim: For any $k = 1, 2, \ldots, \dfrac{p-1}{2}$, there are exactly q_k lattice points in the interior of $\Delta(a, p)$ that have x-coordinate equal to k. (As in Eisenstein's Lemma (11.5.1), we use q_k to denote the quotient when $a \cdot k$ is divided by p.)

> **Proof of Claim** For any $k = 1, 2, \ldots, \dfrac{p-1}{2}$, the point (k, y) is in the interior of $\Delta(a, p)$ if and only if

$$0 < y < \frac{a}{p} \cdot k. \tag{8}$$

We wish to count how many integers y satisfy this inequality. Since $p \nmid a$ and $p \nmid k$, the quantity $\frac{a}{p} \cdot k$ on the right side of the inequality (8) is not an integer. Thus, the number of integers y that satisfy (8) is the same as the number of integers y that satisfy

$$0 < y \le \frac{a}{p} \cdot k. \tag{9}$$

The largest such integer y is the quotient when ak is divided by p. By definition, this quotient is q_k. Thus, the integers y which satisfy inequality (9) are

$$y = 1, 2, \ldots, q_k.$$

There are q_k such integers. This completes the proof of the Claim.

Since for each $k = 1, 2, \ldots, \dfrac{p-1}{2}$, there are q_k lattice points in $\Delta(a, p)$ that have x-coordinate equal to k, the total number of lattice points in $\Delta(a, p)$ is $q_1 + q_2 + \cdots + q_{(p-1)/2}$ which equals $T(a, p)$. This completes the proof. ●

EXERCISES 11.5

Numerical Problems

In Exercises 1–6, for the given values of a and p,

    a. Find $T(a, p)$.

    b. Use Eisenstein's Lemma (11.5.1) to evaluate $\left(\dfrac{a}{p}\right)$.

1. $a = 3$, $p = 5$.   **2.** $a = 3$, $p = 7$.   **3.** $a = 25$, $p = 7$.

4. $a = 9$, $p = 19$.   **5.** $a = 17$, $p = 11$.   **6.** $a = 1$, $p = 47$.

7. Compute $\left(\dfrac{11}{19}\right)$ using Eisenstein's Lemma.

In Exercises 8–10, evaluate each Legendre symbol in four different ways:

    a. By making a table of squares

    b. By using Euler's Identity (11.2.2)

    c. By using Gauss's Lemma (11.4.2)

    d. By using Eisenstein's Lemma (11.5.1)

8. $\left(\dfrac{5}{11}\right)$ **9.** $\left(\dfrac{7}{17}\right)$ **10.** $\left(\dfrac{3}{23}\right)$

In Exercises 11–13, for the given values of a and p, do the following:

    a. Draw the triangle $\Delta(a, p)$ in the Cartesian plane. (See Figure 1.)

    b. Plot all of the lattice points in the interior of the triangle. How many are there?

    c. Use your answer from part b to find $\left(\dfrac{a}{p}\right)$.

11. $a = 5,\ p = 7$ **12.** $a = 7,\ p = 13$ **13.** $a = 13,\ p = 7$

Reasoning and Proofs

EXPLORATION Lattice Points in Polygons (Exercises 14–20)

In this section, we discovered that there is a close relationship between the Legendre symbol and the number of lattice points in a certain triangle. (As we will see in the next section, this relationship is the key to our proof of the Law of Quadratic Reciprocity.) In this exploration, you will discover a general formula for the number of lattice points in a polygon. The formula, known as *Pick's theorem*, was discovered in 1899. (Note that Pick's theorem requires that the vertices of the polygon be located at lattice points, and therefore does not apply to the triangle $\Delta(a, p)$ we studied in this section.)

For each polygon P described in Exercises 14–17, do the following:

    a. Draw the polygon P in the Cartesian plane.

    b. Find T, the number of lattice points in P (including boundary points).

    c. Find A, the area of P.

    d. Find B, the number of lattice points located on the boundary (edges and vertices) of P.

14. P_1 = the triangle with vertices $(0, 0)$, $(1, 0)$, $(1, 1)$.

15. P_2 = the triangle with vertices $(0, 0)$, $(3, 0)$, $(1, 2)$.

16. P_3 = the triangle with vertices $(0, 0)$, $(5, 0)$, $(3, 4)$.

17. P_4 = the octagon with vertices $(0, 0)$, $(1, -1)$, $(2, -1)$, $(3, 0)$, $(3, 1)$, $(2, 2)$, $(1, 2)$, and $(0, 1)$.

18. a. Summarize your results from Exercises 14–17 by making a table of the values of $T, A,$ and B for all of the polygons.

    b. Make a conjecture expressing T in terms of A and B. [*Hint:* You may want to include the values of $\frac{B}{2}$ in your table.] This relationship is known as Pick's theorem, and it holds for any polygon whose vertices are lattice points.

    c. Use Pick's theorem (your conjecture from part b) to find the number of lattice points in the triangle with vertices $(0, 0), (4, 0),$ and $(2, 0)$.

    d. Draw the triangle from part c in the Cartesian plane, and count the lattice points to verify your answer from part c.

19. Let a and b be natural numbers, and let P be the triangle with vertices $(0, 0)$, $(a, 0)$, and $(0, b)$. Prove that Pick's theorem holds for P.

20. Let $\Delta(a, p)$ be the triangle described in the text (see Figure 1), where $a \in \mathbf{N}$ and p is prime such that $p \nmid a$. Show that the formula of Pick's theorem does not always hold for $\Delta(a, p)$.

Advanced Reasoning and Proofs

21. Use the ideas of the proof of Eisenstein's Lemma (11.5.1) to give another proof of Theorem 11.4.3 (the Quadratic Character of 2).

22. [*Note:* This exercise requires Section 10.1.] Recall that a Mersenne prime is a prime of the form $2^p - 1$, where p is prime. In this exercise, you will prove the following theorem, which puts a restriction on the factors of numbers of the form $2^p - 1$. (You may wish to compare this theorem with the one in Exercise 31 of Section 10.1.)

> THEOREM. *Let $p > 2$ be prime. Then any prime factor r of $2^p - 1$ satisfies $r \equiv \pm 1 \pmod 8$.*

    a. Show that $\text{ord}_r(2) = p$.

    b. Use part a to conclude that $p \mid \frac{r-1}{2}$.

    c. Now show that 2 is a quadratic residue modulo r. [*Hint:* Use Euler's Identity (11.2.2).]

    d. Finish the proof of the Theorem.

11.6 Proof of Quadratic Reciprocity

We are now ready to prove the Law of Quadratic Reciprocity, which we stated in two equivalent forms (11.3.1 and 11.3.2).

A picture-perfect proof

The Law of Quadratic Reciprocity expresses a direct relationship between $\left(\frac{p}{q}\right)$ and $\left(\frac{q}{p}\right)$.

As we saw in the previous section, these Legendre symbols can be computed by counting the lattice points in certain triangles. A geometric relationship between these triangles is the key to the proof we will give of the Law of Quadratic Reciprocity.

Naomi's Numerical Proof Preview: Pretty Picture Proposition (11.6.1) and Law of Quadratic Reciprocity (11.3.2)

Let's consider the example $p = 13$ and $q = 7$. Eisenstein's Lemma (11.5.1) tells us that

$$\left(\frac{13}{7}\right) = (-1)^{T(13,\,7)} \text{ and } \left(\frac{7}{13}\right) = (-1)^{T(7,\,13)}.$$

Proposition 11.5.2 tells us that $T(13, 7)$ and $T(7, 13)$ count the number of lattice points in the two triangles shown here:

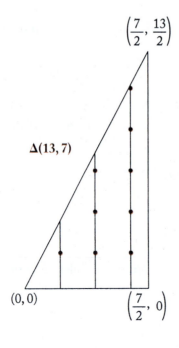

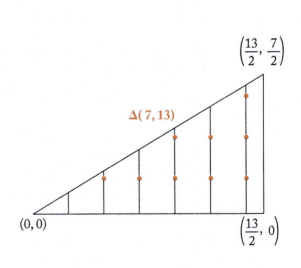

Take the triangle on the left, and reflect it about the line $y = x$. The result is

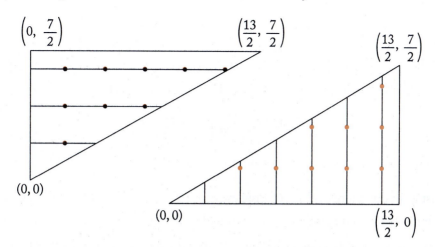

Moving the triangles together, we find that they fit together to form a perfect rectangle:

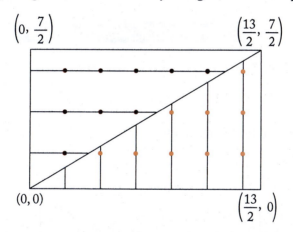

While it may not be so easy to count the number of lattice points in each triangle, it is quite easy to count the number of lattice points in the rectangle. These lattice points form a 6 by 3 grid, so there are $6 \cdot 3 = 18$ total lattice points in the rectangle.

Note that 6, the number of columns in this grid, is equal to $\frac{13 - 1}{2}$, while 3, the number of rows in the grid, equals $\frac{7 - 1}{2}$. Thus, the total number of lattice points may be expressed as

$$\text{Number of lattice points in the above rectangle} = \frac{13 - 1}{2} \cdot \frac{7 - 1}{2}.$$

Our rectangle is divided into two triangles, and the quantities $T(13, 7)$ and $T(7, 13)$ express the number of lattice points in these triangles. Hence, our equation becomes

$$T(13, 7) + T(7, 13) = \frac{13 - 1}{2} \cdot \frac{7 - 1}{2}.$$

Therefore, we find.

$$(-1)^{T(13,\,7)\,+\,T(7,\,13)} = (-1)^{\frac{13-1}{2}\cdot\frac{7-1}{2}}.$$

Now the left side of this equation can be expressed as $(-1)^{T(13,\,7)}\,(-1)^{T(7,\,13)}$, which equals $\left(\dfrac{13}{7}\right)\left(\dfrac{7}{13}\right)$ by Eisenstein's Lemma (11.5.1). Thus, we get

$$\left(\frac{13}{7}\right)\left(\frac{7}{13}\right) = (-1)^{\frac{13-1}{2}\cdot\frac{7-1}{2}},$$

which is exactly what the Law of Quadratic Reciprocity says when $p = 13$ and $q = 7$.

Just as in the Numerical Proof Preview above, the general proof of the Law of Quadratic Reciprocity will be based on the idea that $\left(\dfrac{p}{q}\right)$ and $\left(\dfrac{q}{p}\right)$ can be calculated by counting lattice points in two triangles and that these triangles fit together to form a rectangle.

PROPOSITION 11.6.1   PRETTY PICTURE PROPOSITION

Let p and q be odd primes with $p \neq q$. Then

$$T(p,q) + T(q,p) = \frac{p-1}{2}\cdot\frac{q-1}{2}.$$

PROOF Consider the following rectangle:

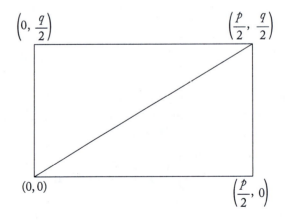

Let L be the number of lattice points in the interior of this rectangle. These lattice points form a $\dfrac{p-1}{2}$ by $\dfrac{q-1}{2}$ grid. Hence,

$$L = \frac{p-1}{2}\cdot\frac{q-1}{2}. \tag{1}$$

Claim: None of the lattice points in the interior of the rectangle lie on the diagonal segment joining $(0, 0)$ and $\left(\frac{p}{2}, \frac{q}{2}\right)$.

Proof of Claim (by contradiction). Suppose that (x, y) is a lattice point in the interior of the rectangle, and that (x, y) lies on the diagonal segment joining $(0, 0)$ and $\left(\frac{p}{2}, \frac{q}{2}\right)$. The equation for this diagonal is $y = \frac{q}{p} x$. This implies that

$$py = qx.$$

It follows that $p \mid qx$. Thus, $p \mid q$ or $p \mid x$ (by the Fundamental Property of Primes, 6.1.2). But since p and q are distinct primes, $p \nmid q$. Hence, $p \mid x$. Since (x, y) lies in the interior of our rectangle,

$$0 < x < \frac{p}{2},$$

which contradicts the fact that $p \mid x. \Rightarrow\Leftarrow$. Thus, the Claim is established.

The Claim implies that every lattice point in the interior of the rectangle lies either in the interior of the upper triangle or the interior of the lower triangle. The lower triangle is the triangle we called $\Delta(q, p)$. By Proposition 11.5.2, the number of lattice points in the interior of this triangle is $T(q, p)$.

To figure out how many lattice points are in the upper triangle, we reflect this triangle about the line $y = x$:

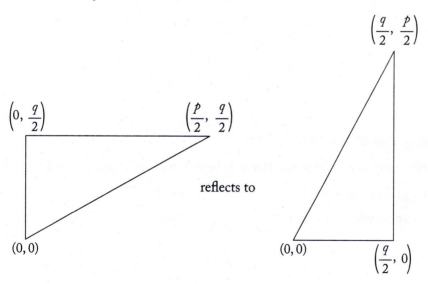

This reflection takes each lattice point (x, y) to the lattice point (y, x); hence, the reflected triangle contains exactly the same number of lattice points as the original triangle. Observe that the reflected triangle is exactly $\Delta(p, q)$. By Proposition 11.5.2, the number of lattice points in the interior of this triangle is $T(p, q)$.

Thus, the total number L of lattice points in the interior of the rectangle equals

$$L = T(p, q) + T(q, p). \tag{2}$$

Combining equations (1) and (2) gives

$$T(p, q) + T(q, p) = \frac{p-1}{2} \cdot \frac{q-1}{2}. \qquad \blacksquare$$

Proof of the Law

Having established the relationship between $T(p, q)$ and $T(q, p)$, it is now easy to prove the Law of Quadratic Reciprocity, which states that

$$\left(\frac{p}{q}\right)\left(\frac{q}{p}\right) = (-1)^{\frac{p-1}{2} \cdot \frac{q-1}{2}}.$$

PROOF OF **THE LAW OF QUADRATIC RECIPROCITY (RESTATEMENT 11.3.2)**

Let p and q be odd primes with $p \neq q$. By Eisenstein's Lemma (11.5.1),

$$\left(\frac{p}{q}\right)\left(\frac{q}{p}\right) = (-1)^{T(p,q)}(-1)^{T(q,p)}$$

$$= (-1)^{T(p,q) + T(q,p)}.$$

Hence, by the Pretty Picture Proposition (11.6.1),

$$\left(\frac{p}{q}\right)\left(\frac{q}{p}\right) = (-1)^{\frac{p-1}{2} \cdot \frac{q-1}{2}}. \qquad \blacksquare$$

The long arm of the Law

Quadratic reciprocity can be an extremely helpful tool in evaluating the Legendre symbol $\left(\frac{p}{q}\right)$. Let's look at a few examples.

Suppose we wish to evaluate $\left(\frac{11}{47}\right)$. Using quadratic reciprocity, we see that

$$\left(\frac{11}{47}\right) = -\left(\frac{47}{11}\right)$$

$$= -\left(\frac{3}{11}\right). \quad \leftarrow \textbf{since 47 reduces to 3 modulo 11}$$

At this point, we could evaluate $\left(\frac{3}{11}\right)$ by looking at the table of squares modulo 11.

However, we could instead evaluate $\left(\frac{3}{11}\right)$ by using quadratic reciprocity again:

$$\left(\frac{3}{11}\right) = -\left(\frac{11}{3}\right).$$

Since $\left(\frac{11}{3}\right) = \left(\frac{2}{3}\right)$, which equals -1 by Theorem 11.4.3 (the Quadratic Character of 2), we find that

$$\left(\frac{3}{11}\right) = 1.$$

This argument suggests that we can evaluate Legendre symbols by using the Law of Quadratic Reciprocity repeatedly. Let's look at another example.

EXAMPLE 1
. . . .

Evaluate $\left(\frac{17}{79}\right)$.

SOLUTION

$$\left(\frac{17}{79}\right) = \left(\frac{79}{17}\right) \qquad \leftarrow \textbf{by the Law of Quadratic Reciprocity, since } 17 \equiv 1 \,(\textbf{mod } 4)$$

$$= \left(\frac{11}{17}\right) \qquad \leftarrow \textbf{since 79 reduces to 11 modulo 17}$$

$$= \left(\frac{17}{11}\right) \qquad \leftarrow \textbf{again by the Law of Quadratic Reciprocity, since } 17 \equiv 1 \,(\textbf{mod } 4)$$

$$= \left(\frac{6}{11}\right) \qquad \leftarrow \textbf{since 17 reduces to 6 modulo 11.}$$

At this point it would be nice if we could use Quadratic Reciprocity once again to reduce our problem further. However, our reciprocity law does not apply since the number 6 is not prime. We can get around this by using the multiplicativity of the Legendre symbol:

$$\left(\frac{6}{11}\right) = \left(\frac{2}{11}\right)\left(\frac{3}{11}\right) \qquad \leftarrow \textbf{by Multiplicativity of the Legendre Symbol (11.2.5)}$$

$$= (-1) \cdot \left(\frac{3}{11}\right) \qquad \leftarrow \textbf{by Theorem 11.4.3 (the Quadratic Character of 2)}$$

$$= (-1) \cdot -\left(\frac{11}{3}\right) \qquad \leftarrow \textbf{again using Quadratic Reciprocity, since}$$
$$\qquad\qquad\qquad\qquad \textbf{both 3 and 11 are congruent to 3 modulo 4}$$

$$= (-1) \cdot (-1) \cdot \left(\frac{2}{3}\right) \qquad \leftarrow \textbf{since 11 reduces to 2 modulo 3}$$

$$= (-1) \cdot (-1) \cdot (-1) \qquad \leftarrow \textbf{by Theorem 11.4.3 (the Quadratic Character of 2)}$$

$$= -1.$$

Putting this all together, we find the answer to our original question: $\left(\frac{17}{79}\right) = -1.$ ●

In this example, we succeeded in evaluating a Legendre symbol using repeated applications of quadratic reciprocity. The example makes the following points clear:

- Sometimes after reducing, we encounter a Legendre symbol $\left(\frac{a}{p}\right)$, where a is not prime. We can handle this by factoring a into primes and using the multiplicativity of the Legendre symbol (Theorem 11.2.5).

- Sometimes after factoring, we encounter a Legendre symbol $\left(\frac{2}{p}\right)$. We can evaluate this using Theorem 11.4.3 (the Quadratic Character of 2).

Keeping these points in mind, our method can be used to evaluate any Legendre symbol. Let's consider another example with somewhat larger numbers.

EXAMPLE 2

Use quadratic reciprocity repeatedly to find $\left(\frac{227}{673}\right)$.

SOLUTION

$$\left(\frac{227}{673}\right) = \left(\frac{673}{227}\right) \qquad \leftarrow \text{ by quadratic reciprocity, since } 673 \equiv 1 \pmod 4$$

$$= \left(\frac{219}{227}\right) \qquad \leftarrow \text{ since 673 reduces to 219 (mod 227)}$$

$$= \left(\frac{3}{227}\right)\left(\frac{73}{227}\right) \qquad \leftarrow \text{ by Multiplicativity of the Legendre Symbol (11.2.5)}$$

$$= -\left(\frac{227}{3}\right)\left(\frac{227}{3}\right) \qquad \leftarrow \text{ by quadratic reciprocity since both 3 and 227 are congruent to 3 modulo 4, and } 73 \equiv 1 \pmod 4$$

$$= -\left(\frac{2}{3}\right)\left(\frac{8}{73}\right) \qquad \leftarrow \text{ 227 reduces to 2 (mod 3) and 227 reduces to 8 (mod 73)}$$

$$= -\left(\frac{2}{3}\right)\left(\frac{2}{73}\right)\left(\frac{2}{73}\right)\left(\frac{2}{73}\right) \leftarrow \text{ by Multiplicativity of the Legendre Symbol (11.2.5)}$$

$$= -(-1) \cdot 1 \cdot 1 \cdot 1 \qquad \leftarrow \text{ by Theorem 11.4.3 (Quadratic Character of 2)}$$

$$= 1.$$

We conclude that $\left(\frac{227}{673}\right) = 1$.

For numbers of modest size, this method is a reasonable way to compute the Legendre symbol. However, there is one step in this process that makes it impractical for large numbers. As mentioned in Chapter 9, factoring large numbers can be difficult even on today's most powerful computers. Since this method requires factoring, it is not an efficient method for finding the Legendre symbol $\left(\frac{a}{p}\right)$ when p is very large.[3] Alternatively, one could stick with Euler's Identity (11.2.2), which (as demonstrated in Section 11.2) provides a computationally efficient method for evaluating the Legendre symbol.

EXERCISES 11.6

Numerical Problems

In Exercises 1–4, evaluate each Legendre symbol using quadratic reciprocity (the method of Examples 1 and 2).

1. $\left(\dfrac{22}{37}\right)$ 2. $\left(\dfrac{31}{61}\right)$ 3. $\left(\dfrac{590}{733}\right)$ 4. $\left(\dfrac{6161}{11777}\right)$

Reasoning and Proofs

5. Let $p = 3$, and let $q > 3$ be an odd prime.

   a. Find $T(3, q)$. (Your answer will be an expression involving the floor function[4].)

   b. Find $T(q, 3)$.

   c. Using your expressions from parts a and b, prove directly that the Pretty Picture Proposition (11.6.1) holds when $p = 3$.

6. Let $p > 2$ be prime, and suppose that $q = 2p + 1$ is also prime. Determine $\left(\frac{p}{q}\right)$. (Your answer may depend on whether $p \equiv 1$ (mod 4) or $p \equiv 3$ (mod 4).)

7. Let $p > 2$ be prime, and suppose that $2p + 1$ and $9p + 4$ are also both prime. Prove that $\left(\dfrac{2p + 1}{9p + 4}\right) = 1$.

* 8. Consider the very large number $B = 3^{500001} + 1$. Prove that B is divisible by 1,000,003. Use quadratic reciprocity, but do not use a calculator. You may assume that 1,000,003 is prime.

---

[3]This method can be modified to compute the Legendre symbol without factoring, by using quadratic reciprocity for the *Jacobi symbol*. (Quadratic reciprocity for the Jacobi symbol is stated and proved in Exercise 15.)

[4]Recall that the floor function, $\lfloor x \rfloor$, denotes the greatest integer less than or equal to x.

EXPLORATION The Jacobi Symbol (Exercises 9–17)

The Legendre symbol $\left(\frac{a}{p}\right)$ is only defined when p is prime. However, we may generalize the symbol so that it does not require the number on the bottom to be prime. We make the following definition:

> DEFINITION. *Let n be an odd positive integer with prime factorization* $n = p_1^{k_1} \cdot p_2^{k_2} \cdot \ldots \cdot p_m^{k_m}$, *and let a be any integer relatively prime to n. Then we define the* **Jacobi symbol** $\left(\frac{a}{n}\right)$ *as follows:*
>
> $$\left(\frac{a}{n}\right) = \left(\frac{a}{p_1}\right)^{k_1} \cdot \left(\frac{a}{p_2}\right)^{k_2} \cdot \ldots \cdot \left(\frac{a}{p_m}\right)^{k_m},$$
>
> *where the expression on the right side is a product of powers of Legendre symbols.*

9. Explain why the Jacobi symbol is equivalent to the Legendre symbol when n is prime.

10. Use the definition of the Jacobi symbol to calculate $\left(\frac{22}{35}\right)$.

In Exercises 11–15, you will prove that in many ways, the Jacobi symbol behaves exactly the same as the Legendre symbol.

11. Let n be an odd integer, and let a and b be integers relatively prime to n.

    Prove that if $a \equiv b \pmod{n}$, then $\left(\frac{a}{n}\right) = \left(\frac{b}{n}\right)$.

12. Let n be an odd integer, and let a and b be integers relatively prime to n.

    Prove that $\left(\frac{ab}{n}\right) = \left(\frac{a}{n}\right)\left(\frac{b}{n}\right)$.

13. In this exercise, you will prove that for any odd integer n,

    $$\left(\frac{-1}{n}\right) = (-1)^{\frac{n-1}{2}}. \tag{3}$$

    To begin, let n be an odd positive integer with prime factorization

    $$n = p_1^{k_1} \cdot p_2^{k_2} \cdot \ldots \cdot p_m^{k_m}.$$

    a. Recall from the binomial theorem that

    $$p_j^{k_j} = [1 + (p_j - 1)]^{k_j}$$

    $$= 1 + k_j(p_j - 1) + \binom{k_j}{2}(p_j - 1)^2 + \binom{k_j}{3}(p_j - 1)^3 + \cdots + (p_j - 1)^{k_j}.$$

    Explain why it follows that $p_j^{k_j} \equiv 1 + k_j(p_j - 1) \pmod{4}$.

b. Use the result from part a to prove that

$$n - 1 \equiv k_1 (p_1 - 1) + k_2 (p_2 - 1) + \cdots + k_m(p_m - 1) \pmod 4.$$

c. Use the result from part b to prove that

$$\frac{n - 1}{2} \equiv k_1 \left(\frac{p_1 - 1}{2} \right) + k_2 \left(\frac{p_2 - 1}{2} \right) + \cdots + k_m \left(\frac{p_m - 1}{2} \right) \pmod 2.$$

d. Use the result from part c to prove that equation (3) is true.

14. Prove that $\left(\frac{2}{n} \right) = (-1)^{\frac{n^2 - 1}{8}}$ for any odd integer n. [*Hint:* The proof will be similar to your proof from Exercise 13.]

15. Prove that there is a reciprocity law that holds true for the Jacobi symbol; namely, if m and n are relatively prime odd integers, then $\left(\frac{m}{n} \right) \cdot \left(\frac{n}{m} \right) = (-1)^{\frac{m-1}{2} \cdot \frac{n-1}{2}}$.
    [*Hint:* Your proof from Exercise 13d may be helpful.]

16. Alas, although the Jacobi symbol behaves a lot like the Legendre symbol, it does not in fact address the question of squares in $\mathbf{Z}_n$. Show for $n = 15$ that $\overline{8}$ is not a square in $\mathbf{Z}_n$, but $\left(\frac{8}{15} \right) = 1$.

17. Although the Jacobi symbol does not give us a simple way to determine whether an integer is a square in a composite modulus, we can still prove nice results about when an integer is a square in a composite modulus. Here is an example. Prove the following theorem.

    THEOREM. *Let $a, m \in \mathbf{Z}$ such that a is odd and $m \geq 3$. Prove that the congruence*

    $$x^2 \equiv a \pmod{2^m}$$

    has exactly four solutions if $a \equiv 1 \pmod 8$ and has no solutions if $a \not\equiv 1 \pmod 8$.

Chapter 12

Paul Erdős (1913–1996)

The brilliant and colorful mathematician Paul Erdős (pronounced "AIR-dish," the name is Hungarian for *wooded*) wrote over 1500 papers, though for most of his adult life he had no home and no job. In addition to publishing more papers than any other mathematician except for Euler, Erdős spent his life traveling around the world, collaborating with hundreds of people, coming up with elegant solutions to difficult problems, posing intriguing problems for others to solve, and giving away what little money he earned from lectures and mathematical prizes.

Erdős grew up in a Jewish family in Hungary during the rise of anti-Semitism. Both of his parents were math teachers, and he showed unusual mathematical talent from an early age. Even at the age of four, Erdős enjoyed doing mathematical computations, such as how long it would take for a train to travel to the sun, or the number of seconds that a person had lived. When Erdős was only 1 year old, Erdős's father was captured by the Russian army at the beginning of World War I and spent the next 6 years imprisoned in Siberia. After having lost her two older children to scarlet fever and with her husband imprisoned, Erdős's mother was extremely overprotective. She did everything to take care of Erdős, who never learned to do his own laundry, cook, or drive a car. In fact, he never

even buttered his own toast until he was 21. Since his mother was afraid that Erdős would catch some horrible disease if he went to school, she hired a tutor to homeschool him until he was in high school.

In 1920, a law was passed in Hungary restricting the number of Jews admitted to universities to no more than 5% of all the students admitted. This made it extremely difficult for Jews to get into a university, but Erdős was able to get around this law by winning a national competition. As a result, Erdős was admitted to the University Pázmány Péter in Budapest at the age of 17. He graduated four years later with a Ph.D. in mathematics.

As a first-year student at the university, Erdős came up with his own proof of Chebyshev's Theorem, which states that for every integer $n \geq 2$, there is at least one prime number between n and $2n$. Erdős's proof of this result is considered to be more elegant than Chebyshev's original proof. In addition, he subsequently improved Chebyshev's result by showing that for every integer $n > 6$, there is at least one prime of the form $4k + 1$ and at least one prime of the form $4k + 3$ between n and $2n$.

After completing his Ph.D., Erdős found it difficult to get a job in Hungary because of anti-Semitism. He took a position as a post-doctoral fellow in England at the University of Manchester. He remained there for four years, which was the longest he ever stayed in one position. After leaving England, Erdős came to the United States, where he began his pattern of traveling from one university to another. In 1952, he accepted a position at Notre Dame. But two years later, when he left the United States to attend the International Congress of Mathematicians in Amsterdam, he was not permitted to reenter the country due to his ties to communist Hungary and his correspondence with a mathematician in China.

For the rest of his life, Erdős had no regular full-time employment. From 1954 until his death, Erdős held the position of Permanent Visiting Professor at the Technion (the Israel Institute of Technology). Until the last ten years of his life, Erdős spent about 1 month each year at the Technion, where he was paid for whatever time he spent there. But Erdős was not interested in a permanent job or in making money. Often quoted as saying "property is a nuisance," Erdős lived out of a suitcase, staying with one collaborator after another, living by the saying "another roof, another proof." He would arrive at the house of a mathematician, suitcase in hand, and declare "my brain is open," meaning that he planned to stay there and collaborate. Most mathematicians were happy to offer him food and a bed for the privilege of collaborating with him. During Erdős's lifetime, he published papers with more than 450 different collaborators. Altogether, more than 4500 mathematicians did research with him. Some of this collaborative work was published after his death, and more continues to be published to this day.

Of the money that Erdős earned, he kept little for himself. When Erdős won the Wolf Prize for Mathematics, which amounted to $50,000, he only kept $750 for his own needs; the rest he gave away, some to relatives and some to set up a scholarship in his mother's name. Erdős would also use money to try to get other people to solve problems that he couldn't solve. The rewards he offered ranged from $10 for "easy" problems to $3000 for "impossible" problems.

Erdős was known for his unique vocabulary. When talking about a truly elegant proof, he would say "this one's from The Book," referring to an imaginary book filled with the most elegant proofs, which were supposed to have been written down by God. He called children "epsilons," women "bosses," and men "slaves." He referred to alcohol as "poison," to music as "noise," and to lecturing about mathematics as "preaching." Erdős referred to God as "The Supreme Fascist," or "SF" for short.

Erdős was always very close to his mother. She began accompanying him on his travels when she was in her eighties. Her death in 1971 affected him profoundly, and his doctor prescribed Benzedrine (a type of amphetamine that is closely related to methamphetamine) for depression. Erdős had always worked at a furious pace, but after his mother died he began working 20 hours each day. The Benzedrine pills, together with caffeine tablets and large quantities of coffee, allowed Erdős to maintain this intense schedule. Erdős's friend and fellow mathematician Ron Graham was worried that Erdős had become addicted to Benzedrine, so Graham made a bet with Erdős that he couldn't give up Benzedrine for 30 days. Erdős won the bet, but said that he regretted the loss of his mathematical productivity during the 30 days. He immediately resumed his use of the amphetamine.

Erdős would often quote the mathematician Alfréd Rényi, who had said "a mathematician is a machine for turning coffee into theorems." Other mathematicians described Erdős as frail and disheveled, noting that he sometimes wore his pajama top rather than a shirt. But Erdős was able to keep up his intense pace of work until the day he died. Erdős died of a heart attack at the age of 83, while he was at a conference in Warsaw. He had just finished giving two lectures and was about to leave for Vilnius to give another.

Because Erdős had collaborated with so many people, his friends created the idea of the *Erdős number* to represent how far a mathematician was from collaborating with Erdős. The Erdős number is computed as follows. Erdős has Erdős number 0, all of his collaborators have Erdős number 1, the collaborators of Erdős's collaborators have Erdős number 2, and so on. You should ask your number theory professor what his or her Erdős number is.

In spite of his eccentricities, Erdős was a joyful person who delighted in being with children. Known to mathematicians as "Pali bácsi," or "Uncle Paul," Erdős cared deeply about his friends and colleagues. He gave money to young mathematicians, countless charities, and homeless people he encountered on the street. He was also generous with ideas, and through his example, many came to view mathematics as a social rather than a solitary activity. His life had a profound impact on his myriad friends and collaborators, and many were greatly saddened by his death in 1996.

A bit of Erdős's math

Erdős made major contributions to many areas of mathematics, including number theory, combinatorics, graph theory, set theory, analysis, probability, and geometry. Many of his results are easy to state but very hard to prove. Here are a few of his results in number theory.

Erdős and Atle Selberg surprised the mathematical world by giving an "elementary" proof of the Prime Number Theorem (3.3.3), which asserts that as n gets larger and larger, the ratio of $\pi(n)$ to $\frac{n}{\log n}$ approaches 1. This result had been proven earlier by the French mathematicians Hadamard and Vallée-Poussin, using deep results in complex analysis. The proof given by Erdős and Selberg is considered "elementary" because it does not rely on complex analysis. For over 50 years, mathematicians had believed that such a proof was not possible.

In 1939, Erdős and Mark Kac obtained another major result in number theory. They proved that the number of prime factors of a randomly chosen large integer is distributed in the form of a bell curve, according to a Gaussian distribution (for more on this distribution, see the biography of Gauss at the beginning of Chapter 7). This result marked the beginning of the field of probabilistic number theory.

Since the 18th century, mathematicians had wondered whether the product of consecutive integers could ever be a *perfect power*—that is, be equal to m^n for integers $m, n > 1$. For example, $2 \cdot 3 = 6$ is a product of consecutive integers but is not a perfect power, and $3 \cdot 4 \cdot 5 = 60$ is likewise not a perfect power. In 1975, Erdős and John L. Selfridge answered this question by proving that the product of two or more consecutive integers is never a perfect power.

In 1956, Erdős presented an informal argument that there must be infinitely many Carmichael numbers (see Sections 12.1 and 12.2 for more information about Carmichael numbers). Erdős's argument provided some of the key ideas for the formal proof of this fact that was eventually given in 1994 by the mathematicians Alford, Granville, and Pomerance.

Chapter 12   Primality Testing

12.1   Primality Testing

The problem of distinguishing prime numbers from composite numbers and of resolving the latter into their prime factors is known to be one of the most important and useful in arithmetic. It has engaged the industry and wisdom of ancient and modern geometers to such an extent that it would be superfluous to discuss the problem at length. . . . Further, the dignity of the science itself seems to require that every possible means be explored for the solution of a problem so elegant and so celebrated.

— C. F. Gauss, *Disquisitiones Arithmeticae*, Article 329 (1801)

More than 200 years ago, long before the computer age, Gauss singled out two computational problems of number theory that have turned out to be critical to modern applications:

1. *Primality*: Given a natural number n, determine whether n is prime or composite.

2. *Factoring*: Given a natural number n, find the factorization of n into primes.

At first glance, it might appear that the two problems are more or less the same. After all, how could you possibly tell whether a number is prime or composite without factoring the number? First appearances aside, however, these two problems are of very different character. As we will see in this chapter, it is possible to quickly determine whether a very large number is prime or composite without factoring the number! In contrast, it is widely believed that there is no fast algorithm for factoring a very large number. These two facts are good news for the RSA Encryption Algorithm (explained in Section 9.5), which relies on the difficulty of factoring large numbers, but whose setup requires one to find large prime numbers.

Probabilistic primality testing

In Section 12.3, we will present a primality test known as the Miller-Rabin Test. We now give a quick preview of how the test is used. The Miller-Rabin Test is a *probabilistic* test, in the following sense:

A. If the input number n is composite, then the Miller-Rabin Test returns the result **definitely composite** with probability at least $\frac{3}{4}$, and returns the result **may be prime** with probability at most $\frac{1}{4}$.

B. If the input number n is prime, then the Miller-Rabin Test always returns the result **may be prime**.

Developing this test and proving assertions A and B will take some work, which culminates in Section 12.5. We can use the Miller-Rabin Test to quickly determine whether any given number n is prime or composite with *extraordinarily* high probability, as follows.

The trick is to run the Miller-Rabin Test repeatedly—say, 100 times—with the same input number n:

A-100. If the number n is composite, then the test will return 100 results, each of which is either **definitely composite** or **may be prime**. By assertion A, the probability of getting the result **may be prime** on any given test is at most $\frac{1}{4}$, so the probability of getting **may be prime** on all 100 tests is at most $\left(\frac{1}{4}\right)^{100}$, which is less than 10^{-60}, a minuscule chance indeed!

B-100. If the number n is prime, then by assertion B, the test will return the result **may be prime** all 100 times.

So if we have a natural number n, and we are uncertain whether n is prime or composite, we go ahead and run the Miller-Rabin Test 100 times. If we receive the answer **definitely composite** at any point, it follows from B-100 that n is composite. On the other hand, if we receive the answer **may be prime** 100 times in a row, we can conclude that with overwhelming odds, the number n is indeed prime. (Were n composite, then by A-100, the odds are overwhelmingly against getting the answer **may be prime** 100 times in a row.) Thus, starting with any natural number n, we end up concluding either that n is definitely composite or that n is prime with extremely high probability.

As we will soon see, the Miller-Rabin Test is based on modular exponentiation. It follows from our discussion in Section 9.3 (see Observation 9.3.3) that the Miller-Rabin Test can be carried out efficiently. The existence of an efficient probabilistic primality test is an important theoretical result, and it is crucial for the RSA encryption algorithm (9.5.1). Using Miller-Rabin, probabilistic primes with hundreds or even thousands of digits can be found in seconds!

Probabilistic primes are great for practical work; even the most cautious person is willing to ignore a possibility whose probability is only 10^{-60}. And if not, just run the Miller-Rabin 200 times instead of 100, thereby cranking the odds down to a fantastically unlikely $\left(\frac{1}{4}\right)^{200}$, which is less than 10^{-120}. With these exponentially small odds, the chance that your probabilistic prime is actually composite is smaller than the chance that a random particle from outer space will hit your bank's computer, change your account balance, and make you an instant billionaire.

From a more mathematical perspective, it would be nice to know for *certain* that your number was prime. Thus, one could ask for a *deterministic* primality test, which can efficiently determine with absolute certainty whether a given large number is prime or composite. The first such test was discovered quite recently—in 2002! This deterministic test, known as the AKS test after discoverers Agrawal, Kayal, and Saxena, is described in Section 12.7. While provably correct and efficient, the AKS test runs slower in practice than Miller-Rabin, and therefore it is not currently used for practical applications. Still, its discovery represents a breakthrough in computational number theory, and provides a satisfying answer to the question Gauss raised more than 200 years ago.

Before discussing the details of the Miller-Rabin Test, we present a simpler test, which we call the Fermat Test, that is based on Fermat's Little Theorem (9.1.2). The Fermat Test is not as effective as the Miller-Rabin Test; however, it illustrates most of the main ideas.

Witnesses to compositeness

For many years, the number $n = 91709$ has been sneaking around, trying to pass itself off as a prime number. But if we compute $2^{n-1} \pmod{n}$, we find that $2^{91708} \equiv 12821 \pmod{91709}$. The fact that

$$2^{91708} \not\equiv 1 \pmod{91709} \tag{1}$$

reveals 91709's dirty little secret: it is composite. How do we know for sure? If 91709 were a genuine prime, then by Fermat's Little Theorem (9.1.2), 2^{91708} would be congruent to 1 modulo 91709. But as statement (1) shows, this is not the case. Thus, 91709 cannot be prime.

This is the basis of the following test for compositeness:

12.1.1   FERMAT TEST FOR COMPOSITENESS

Suppose $n \in \mathbf{N}$. If there exists an integer a in the range $1, 2, \ldots, n - 1$ such that

$$a^{n-1} \not\equiv 1 \pmod{n},$$

then n is composite. In this case, the number a is called a **Fermat witness** *to the compositeness of n.*

For example, statement (1) demonstrates that 2 is a Fermat witness to the compositeness of 91709. Curiously, although the Fermat Test (12.1.1) tells us that 91709 is composite, it tells us nothing about the factors of 91709 (which turn out to be 293 and 313).

Until 2007, RSA Laboratories was offering a $200,000 prize (see Section 9.5) for factoring the following 617-digit number:

$$n = 25195908475657893494027183240048398571429282126204032027777713783$$
$$60436620207075955626401852588078440691829064124951508218929855914917618450280848912007284499268739280728777673597141834727026189637501497182469116507761337985909570009733045974880842840179742$$
$$91006424586918171951187461215151726546322822168699875491824224336372590851418654620435767984233871847744479207399342365848238242$$
$$81198163815010674810451660377306056201619676256133844143603833904414952634432190114657544454178424020924616515723350778707749817$$
$$12577246796292638635637328991215483143816789988504044536402352738195137863656439121201039712282212072035 7$$

Let's see what the Fermat Test for compositeness (12.1.1) can tell us about this number. Choosing $a = 2$, and reducing 2^{n-1} modulo n, a task that requires only a fraction of a second on a computer, we find

$$2^{n-1} \equiv 2389876990640375833545372818643546169782664542883858277948872$$
$$14712213451886634080490185518305978025041363051076239263457001684321189064659401742441747483502385695825972772481033228020752439749454411424095050428567096982675910864550199213890449953208197505454662430091754576964618716098832521798758422473009975500008792012925696853398623528818957071180372317934229403275957715539547658421660913880376367993370855152806224965514792640583053859824196379816894126008393340915904990973872341684188911297117770433827812105999548318534714642945425389312774179462672529412586430384135922869943151699031186867950129900 55769026 \pmod{n}.$$

Since $2^{n-1} \not\equiv 1 \pmod{n}$, the Fermat Test tells us with certainty that n is composite.

Factoring the 617-digit number n is so difficult that doing so would have netted you $200,000. But with a little help from Fermat, you can prove that n is composite in an instant.

EXAMPLE 1

Let's use the Fermat Test for Compositeness to test the number $n = 341$. Choosing $a = 2$, we find

$$2^{340} \equiv 1 \pmod{341}.$$

Thus, 2 is *not* a Fermat witness for 341. What can we conclude? Does this mean that 341 is definitely prime? No, at this point we cannot conclude anything at all about 341. It may be prime, or it may be composite.

Let's see whether some other number a is willing to testify against 341. Trying $a = 3$, we find

$$3^{340} \equiv 56 \pmod{341}.$$

Thus, 3 is a Fermat witness for 341, and we conclude that 341 is a composite number after all.

Fermat fooled

Consider the number 252601. If we wish to use the Fermat Test for Compositeness (12.1.1), we might begin computing $a^{252600} \pmod{252601}$ for various values of a. We find

$$2^{252600} \equiv 1 \pmod{252601},$$
$$3^{252600} \equiv 1 \pmod{252601}, \qquad (2)$$
$$4^{252600} \equiv 1 \pmod{252601},$$
$$5^{252600} \equiv 1 \pmod{252601}.$$

We might continue by trying the same computation using random values of a between 1 and 252600. For example,

$$7853^{252600} \equiv 1 \pmod{252601},$$
$$39472^{252600} \equiv 1 \pmod{252601}, \qquad (3)$$
$$103480^{252600} \equiv 1 \pmod{252601}.$$

At this point, we might be tempted to conclude that 252601 is prime. However, we would be wrong! The number 252601 factors as $41 \cdot 61 \cdot 101$, in spite of all of the congruences in (2) and (3). But the situation is even worse than that: it turns out that for *every* integer a that is relatively prime to 252601,

$$a^{252600} \equiv 1 \pmod{252601}.$$

Though 252601 is composite, it is very difficult to get a Fermat witness to this fact. The only Fermat witnesses are numbers a that share a common factor with 252601. Thus, finding a Fermat witness for 252601 is just as hard as finding a factor of 252601. For the number 252601, Fermat's test is virtually useless. Numbers like this are called *Carmichael numbers*.

A composite number n is called a **Carmichael number** *if for every integer a,*

$$\gcd(a, n) = 1 \quad \Rightarrow \quad a^{n-1} \equiv 1 \ (\mathrm{mod}\ n).$$

That is, for a number n to be a Carmichael number, two conditions must be satisfied:

(*i*) n must be composite.

(*ii*) The only Fermat witnesses to the compositeness of n are numbers a that share a common factor with n.

Fun Facts

The smallest Carmichael number is $561 = 3 \cdot 11 \cdot 17$. Of the first one million natural numbers, only 43 are Carmichael numbers. Until recently, it was unknown whether there are infinitely many Carmichael numbers. This was settled in 1994, when mathematicians Alford, Granville, and Pomerance proved the infinitude of Carmichael numbers.

Characterizing Carmichael

If p is a prime factor of a natural number n, does it follow that $p - 1$ divides $n - 1$? Absolutely not. Counterexamples abound; for instance, $5 \mid 35$, but $4 \nmid 34$.

However, let's consider the Carmichael number 252601, which factors into primes as follows:

$$252601 = 41 \cdot 61 \cdot 101.$$

In this case, we observe that $40, 60$, and 100 are all divisors of 252600. This coincidence, it turns out, is responsible for the fact that 252601 is a Carmichael number. Let's see why.

Naomi's Numerical Proof Preview: Theorem 12.1.3

Let a be any integer that is relatively prime to 252601. Then a is not divisible by 41, so Fermat's Little Theorem (9.1.1) tells us that

$$a^{40} \equiv 1 \ (\mathrm{mod}\ 41).$$

Since $40 \mid 242600$, it follows that

$$a^{252600} \equiv 1 \ (\mathrm{mod}\ 41).$$

Similar reasoning yields

$$a^{252600} \equiv 1 \ (\mathrm{mod}\ 61),$$

and

$$a^{252600} \equiv 1 \pmod{101}.$$

By the Chinese Remainder Theorem (see Observation 7.4.5), it follows that

$$a^{252600} \equiv 1 \pmod{252601}.$$

Thus, by definition, 252601 is a Carmichael number.

This argument provides half of the proof of the following theorem, which gives a complete characterization of Carmichael numbers.

THEOREM 12.1.3　CHARACTERIZATION OF CARMICHAEL NUMBERS

Let n be a composite number. Then n is a Carmichael number if and only if the following two conditions hold:

(i)　For every prime p such that $p \mid n$, we have $p - 1 \mid n - 1$.

*(ii)　n is the product of distinct primes (i.e., n is **squarefree**).*

Before we prove this theorem, let's see an example.

EXAMPLE 2

Use Theorem 12.1.3 to show that 561 is a Carmichael number.

SOLUTION The prime factorization

$$561 = 3 \cdot 11 \cdot 17$$

demonstrates that 561 is composite, and also demonstrates that 561 is squarefree (which is condition (*ii*) of Theorem 12.1.3). To check condition (*i*), we note that 2, 10, and 16 are all divisors of 560. Hence, both conditions of Theorem 12.1.3 are satisfied, and 561 is thus a Carmichael number. ■

We now prove one direction of Theorem 12.1.3, that conditions (*i*) and (*ii*) imply that n is a Carmichael number. We will prove the other direction—that if n is a Carmichael number, then conditions (*i*) and (*ii*) hold—in the next section.

PROOF OF THEOREM 12.1.3 ([⇐] DIRECTION)

[⇐] Let n be a composite number, and assume that conditions (*i*) and (*ii*) are satisfied. **[To show: n is a Carmichael number.]**
Then

$$n = p_1 p_2 \cdots p_s,$$

where for all $j = 1, 2, \ldots, s$, the p_j are distinct primes and $p_j - 1 \mid n - 1$.

Let $a \in \mathbf{Z}$ be relatively prime to n. Then for any $j = 1, 2, \ldots, s$, we know that $p_j \nmid a$. Hence, by Fermat's Little Theorem (9.1.2),

$$a^{p_j - 1} \equiv 1 \pmod{p_j}.$$

Since $p_j - 1 \mid n - 1$, it follows that

$$a^{n-1} \equiv 1 \pmod{p_j}.$$

Since this congruence holds modulo every prime factor of n (i.e., for every $j = 1, 2, \ldots, s$), it follows from the Chinese Remainder Theorem (see Observation 7.4.5) that

$$a^{n-1} \equiv 1 \pmod{n}.$$

Hence, n is a Carmichael number. ■

EXERCISES 12.1

Numerical Problems

1. Using a calculator or computer, you find that $2^{246} \equiv 220 \pmod{247}$. From this congruence, what can you conclude about 247? What do we call 2 in this case?

2. Using a calculator or computer, you find that $2^{250} \equiv 1 \pmod{251}$. From this congruence, what does the Fermat Test (12.1.1) allow us to conclude about 251? Explain.

In Exercises 3–6, apply the Fermat Test for Compositeness (12.1.1) to the natural number n using the given value(s) of a. Explain what you can conclude about n.

3. $n = 39$; $a = 2$ 4. $n = 33$; $a = 10, 12$

5. $n = 21$; $a = 13, 16$ 6. $n = 15$; $a = 4$

Computer Exercise In Exercises 7–11, apply the Fermat Test for Compositeness (12.1.1) to the natural number n using the given value of a and at least two other values of a. Explain what you can conclude about n.

7. $n = 91$, $a = 3$ 8. $n = 645$, $a = 2$ 9. $n = 671$, $a = 2$

10. $n = 1729$, $a = 5$ 11. $n = 501001$, $a = 6$

12. Observe that $6601 = 7 \cdot 23 \cdot 41$. Is 6601 a Carmichael number? Explain your reasoning.

13. a. Compute 2^{-230} in $\mathbf{Z}_{231}$.

    b. Does your answer to part a allow you to conclude that 231 is prime? Explain your reasoning.

    c. Is 231 a Carmichael number? Justify your answer.

Reasoning and Proofs

14. Prove that for any $n \in \mathbf{N}$, the number 1 is not a Fermat witness for n.

15. Prove or disprove: 1105 is a Carmichael number.

16. Use your answer to Exercise 15 to deduce whether 14365 is a Carmichael number. [*Hint:* $14365 = 13 \cdot 1105$.]

17. Suppose the last two digits of n are 56. Can n be a Carmichael number? Explain your answer.

18. Suppose the last two digits of n are 35. Can n be a Carmichael number? Explain your answer.

19. Let p, q, and r be distinct primes. Explain why $n = q(p^3 + p)(qr^2 + q^3r^3)$ is not a Carmichael number.

20. Prove that all Carmichael numbers are odd. [*Hint:* Consider the value $a = -1$.]

21. Let $n \in \mathbf{N}$, $n > 2$. Prove that the number $n - 1$ is a Fermat witness for n if and only if n is even.

22. Let $n \in \mathbf{N}$ be odd, and let a be an integer in the range $1, 2, \ldots, n - 1$. Prove that a is a Fermat witness for n if and only if $n - a$ is a Fermat witness for n.

23. Let $n \in \mathbf{N}$ and let a be an integer in the range $1, 2, \ldots, n - 1$. Show that if $\gcd(a, n) \neq 1$, then a is a Fermat witness for n.

Advanced Reasoning and Proofs

24. Recall that $561 = 3 \cdot 11 \cdot 17$ is a Carmichael number.

    a. Show that for all $a \in \mathbf{Z}$, $a^{561} \equiv a \pmod{561}$.

    b. Let $n \in \mathbf{N}$, $n > 1$. Show that

       n is a Carmichael number if and only if for all $a \in \mathbf{Z}$, $a^n \equiv a \pmod{n}$.

25. Suppose that $n = p_1 p_2 \cdots p_s$, where the p_i are distinct primes, and the number of primes is $s \geq 2$. Find a formula for the number of Fermat witnesses to the compositeness of n. [*Hint:* Use Proposition 10.2.2 and the Chinese Remainder Theorem.]

In this Exploration, we will consider pseudoprimes, which are composite numbers that fool the Fermat Test (whose compositeness is not revealed by the Fermat Test) for some value of a. More precisely:

> **DEFINITION.** *Let $n \in \mathbf{N}$ be a composite number, and let a be an integer in the range $1, 2, \ldots, n - 1$. We say that n is a **pseudoprime (to the base a)** if a is not a Fermat witness for n.*

The smallest pseudoprime (to the base 2) is 341.

26. Recall from Section 3.3 that a *Fermat number* is a number of the form $2^{2^n} + 1$, where n is any natural number. Prove that every Fermat number that is composite is a pseudoprime (to the base 2).

In the next two exercises, you will show that there are infinitely many pseudoprimes (to the base 2).

27. Let $r, s \in \mathbf{N}$. Show that if $r \mid s$, then $2^r - 1 \mid 2^s - 1$.

28. **a.** Prove that if an odd natural number n is a pseudoprime (to the base 2), then $2^n - 1$ is also a pseudoprime (to the base 2). [*Hint:* Use the result of Exercise 27.]

    b. Explain why it follows from part a that there are infinitely many pseudoprimes (to the base 2).

12.2 Continued Consideration of Carmichael Numbers

In this section, we'll prove a number of results about Carmichael numbers, including the remaining direction of Theorem 12.1.3. Before finishing that proof, it will be useful to have a couple of lemmas.

LEMMA 12.2.1

Let p be prime, and let $x, y \in \mathbf{Z}$. If $x \equiv y \pmod{p}$, then $x^p \equiv y^p \pmod{p^2}$.

PROOF We may factor $x^p - y^p$ as follows:

$$x^p - y^p = (x - y)(x^{p-1} + x^{p-2}y + x^{p-3}y^2 + \cdots + xy^{p-2} + y^{p-1}). \tag{1}$$

Since $x \equiv y \pmod{p}$, the first factor (shown in **brown**) is divisible by p. The second factor (shown in **orange**) is also divisible by p, since it follows from the congruence $x \equiv y \pmod{p}$ that

$$x^{p-1} + x^{p-2}y + x^{p-3}y^2 + \cdots + y^{p-1} \equiv x^{p-1} + x^{p-2}x + x^{p-3}x^2 + \cdots + x^{p-1} \pmod{p}$$
$$\equiv x^{p-1} + x^{p-1} + x^{p-1} + \cdots + x^{p-1} \pmod{p}$$
$$\equiv px^{p-1} \pmod{p}$$
$$\equiv 0 \pmod{p}.$$

Thus, in equation (1), p divides both factors on the right hand side (the **brown** factor and the **orange** factor). We conclude that p^2 divides their product—that is,

$$p^2 \mid x^p - y^p.$$

Hence,

$$x^p \equiv y^p \pmod{p^2}. \qquad \blacksquare$$

To finish the proof of Theorem 12.1.3 (Characterization of Carmichael Numbers), a large part of our job will be to show that if a number is not squarefree, then it cannot be a Carmichael number. In other words, we need to show that if a number n is divisible by the square of some prime, then n must have Fermat witnesses that are relatively prime to n. The following lemma will help us do this by putting a somewhat severe restriction on numbers that are not Fermat witnesses. In particular, this lemma states that given any two numbers that are not Fermat witnesses, if they are congruent modulo p then they must be congruent modulo p^2, a much stronger condition.

LEMMA 12.2.2

Let $n \in \mathbf{N}$, and let p be prime such that $p^2 \mid n$. Suppose that $x, y \in \mathbf{Z}$ satisfy the congruences $x^{n-1} \equiv 1 \pmod{n}$ and $y^{n-1} \equiv 1 \pmod{n}$. If $x \equiv y \pmod{p}$, then $x \equiv y \pmod{p^2}$.

PROOF Let $n \in \mathbf{N}$, let p be prime such that $p^2 \mid n$, and let $x, y \in \mathbf{Z}$ such that $x^{n-1} \equiv 1 \pmod{n}$ and $y^{n-1} \equiv 1 \pmod{n}$. Assume that $x \equiv y \pmod{p}$.

Applying Lemma 12.2.1, we find that $x^p \equiv y^p \pmod{p^2}$. Raising both sides of this congruence to the power $\frac{n}{p}$ (which is an integer since $p \mid n$), we get:

$$x^n \equiv y^n \pmod{p^2}. \qquad (2)$$

From the hypothesis that $x^{n-1} \equiv 1 \pmod{n}$, we get $x^n \equiv x \pmod{n}$. Since $p^2 \mid n$, this implies (by Lemma 7.1.10) that

$$x^n \equiv x \pmod{p^2}. \tag{3}$$

By similar reasoning,

$$y^n \equiv y \pmod{p^2}. \tag{4}$$

Combining congruences (2), (3), and (4), we find that

$$x \equiv y \pmod{p^2}. \qquad \blacksquare$$

With this lemma in hand, we are now ready to complete the proof of Theorem 12.1.3.

PROOF OF THEOREM 12.1.3 ([⟹] DIRECTION)

[⟹] Let n be a Carmichael number. We must show that conditions (i) and (ii) hold.

In order to prove condition (i), let p be a prime such that $p \mid n$.
[To show: $p - 1 \mid n - 1$.]

Claim: For any $b \in \mathbf{Z}$, if $p \nmid b$ then $b^{n-1} \equiv 1 \pmod{p}$.

Proof of Claim Let $b \in \mathbf{Z}$ such that $p \nmid b$. Then there exists $a \in \mathbf{Z}$ such that $a \equiv b \pmod{p}$ and $\gcd(a, n) = 1$. (See Exercise 12.) Since n is a Carmichael number, we know that $a^{n-1} \equiv 1 \pmod{n}$, and therefore (by Lemma 7.1.10),

$$a^{n-1} \equiv 1 \pmod{p}.$$

Since $a \equiv b \pmod{p}$, this implies that $b^{n-1} \equiv 1 \pmod{p}$, which completes the proof of the Claim.

From the Claim, it follows that in $\mathbf{Z}_p$, any nonzero element $\bar{x}$ satisfies the equation

$$\bar{x}^{n-1} = \bar{1}. \tag{5}$$

Thus, equation (5) has exactly $p - 1$ solutions in $\mathbf{Z}_p$. By Proposition 10.2.2, on the other hand, we know that equation (5) has exactly $\gcd(n - 1, p - 1)$ solutions in $\mathbf{Z}_p$. Hence, we know that

$$\gcd(n - 1, p - 1) = p - 1.$$

It follows that $p - 1 \mid n - 1$, as desired.

In order to prove condition (ii), and thus complete the proof of Theorem 12.1.3, we must show that n is squarefree. We proceed by contradiction.

Assumption: Suppose n is not squarefree.

Then there exists a prime, p, such that $p^2 \mid n$. Let x be an integer such that $\gcd(x, n) = 1$. Since n is a Carmichael number,

$$x^{n-1} \equiv 1 \; (\text{mod } n).$$

Using the Chinese Remainder Theorem (see Exercise 13), there exists an integer y such that

$$y \equiv x \; (\text{mod } p), \tag{6}$$

$$y \not\equiv x \; (\text{mod } p^2), \tag{7}$$

and

$$\gcd(y, n) = 1. \tag{8}$$

Since n is Carmichael number, $y^{n-1} \equiv 1 \; (\text{mod } n)$. By Lemma 12.2.2, it now follows that $x \equiv y \; (\text{mod } p^2)$, which contradicts congruence (7). $\Rightarrow\Leftarrow$

We conclude that n is squarefree. ●

Numbers with a square factor have many Fermat witnesses

If a number n is divisible by the square of some odd prime, then n cannot be a Carmichael number. This is guaranteed by Theorem 12.1.3. Thus, there exist Fermat witnesses a to the compositeness of n that satisfy $\gcd(a, n) = 1$. In fact, as we are about to show, at least $\frac{3}{4}$ of the numbers in the range $1, 2, \ldots, n - 1$ are Fermat witnesses! Thus, a randomly chosen value of a has at least a 75% chance of being a Fermat witness to the compositeness of n. This is the subject of the following theorem, which we will also use in our analysis of the Miller-Rabin Test in Section 12.5.

THEOREM 12.2.3

Let n be a natural number and let p be an odd prime such that $p^2 \mid n$. Then the number of integers a in the range $1, \ldots, n - 1$ that satisfy the congruence

$$a^{n-1} \equiv 1 \; (\text{mod } n) \tag{9}$$

is at most $\frac{1}{4}(n - 1)$.

PROOF OF THEOREM 12.2.3

Let $n \in \mathbf{N}$, and let p be an odd prime such that $p^2 \mid n$.

Let S be the set of all solutions to congruence (9) that are in the range $0, 1, \ldots, n - 1$:

$$S = \{a \text{ in the range } 0, 1, \ldots, n - 1 \mid a^{n-1} \equiv 1 \; (\text{mod } n)\}.$$

If $a \in S$ then $a^{n-1} \equiv 1 \; (\text{mod } n)$, from which it follows (using the fact that $p \mid n$ and Lemma 7.1.10) that $a^{n-1} \equiv 1 \; (\text{mod } p)$. Hence, $a \not\equiv 0 \; (\text{mod } p)$.

Thus, every element $a \in S$ is congruent modulo p to some number in the range $1, 2, \ldots, p - 1$. We now split the elements of S into subsets according to their congruence classes modulo p. Let b be an integer in the range $1, 2, \ldots, p - 1$, and define S_b as the set of all elements of S that are congruent to $b \pmod{p}$:

$$S_b = \{a \in S \mid a \equiv b \pmod{p}\}.$$

Claim: The set S_b has at most $\dfrac{n}{p^2}$ elements.

Proof of Claim Note that any two elements of S_b are congruent modulo p (because they are both congruent to b modulo p). Hence, by Lemma 12.2.2, all of the elements of S_b are congruent modulo p^2. Thus, all of the elements of S_b are congruent to some integer $v \pmod{p^2}$. Since there are exactly $\dfrac{n}{p^2}$ integers in the range $0, 1, \ldots,$ $n - 1$ that are congruent to $v \pmod{p^2}$, it follows that the number of elements of S_b is at most $\dfrac{n}{p^2}$.

We can now use the Claim to give an upper bound on the number of elements of S. As discussed earlier, every element of S is an element of S_b for some b in the range $1, 2, \ldots, p - 1$. By the Claim, each of the sets S_b has at most $\dfrac{n}{p^2}$ elements, and thus,

$$\text{Number of elements of } S \leq (p - 1) \cdot \frac{n}{p^2}. \tag{10}$$

Number of possible   **Upper bound for the number**
values of b   **of elements in each set S_b**

On the right side of the inequality above, n is multiplied by the quantity $\dfrac{p - 1}{p^2}$. The odd prime that maximizes the value of this ratio is $p = 3$ (see Exercise 8), in which case this ratio has the value $\dfrac{3 - 1}{3^2} = \dfrac{2}{9}$. Hence, we may conclude that

$$\text{Number of elements of } S \leq \frac{2}{9}n. \tag{11}$$

From Exercise 9, it follows that $\dfrac{2}{9}n \leq \dfrac{1}{4}(n - 1)$. Hence,

$$\text{Number of elements of } S \leq \frac{1}{4}(n - 1), \tag{12}$$

as desired. ●

Carmichael numbers are divisible by at least three primes

The smallest Carmichael number, $561 = 3 \cdot 11 \cdot 17$, is the product of three distinct primes. So is our favorite Carmichael number, $252601 = 41 \cdot 61 \cdot 101$. One might wonder whether there exists a Carmichael number that is the product of only two primes. The answer is no, as the following theorem and its corollary assert.

THEOREM 12.2.4

Let p and q be distinct primes, and let $n = pq$. Then $p - 1 \nmid n - 1$ or $q - 1 \nmid n - 1$.

COROLLARY 12.2.5

Every Carmichael number is divisible by at least three distinct primes.

PROOF OF THEOREM 12.2.4

Let p and q be primes such that $p \neq q$, and let $n = pq$. Without loss of generality, we may assume that $p > q$. We will now prove by contradiction that $p - 1 \nmid n - 1$.

Assumption: Suppose $p - 1 \mid n - 1$.

Then $p - 1 \mid pq - 1$, so there exists $d \in \mathbf{N}$ such that

$$d(p - 1) = pq - 1. \tag{13}$$

Considering this equation modulo p yields the congruence $-d \equiv -1 \pmod{p}$, which implies that

$$d \equiv 1 \pmod{p}.$$

Also observe that $d \neq 1$ (otherwise, equation (13) would become $p - 1 = pq - 1$, which implies that $q = 1$). Aside from 1, the smallest natural number that is congruent to 1 (mod p) is $p + 1$. Thus,

$$d \geq p + 1.$$

Multiplying this inequality by $p - 1$, we get

$$d(p - 1) \geq p^2 - 1,$$
$$> pq - 1.$$

This contradicts equation (13). $\Rightarrow\Leftarrow$

We conclude that $p - 1 \nmid n - 1$. ●

PROOF OF COROLLARY 12.2.5

We prove this by contradiction. Let n be a Carmichael number.

Assumption: Suppose n is not divisible by at least three distinct primes.

Then by Theorem 12.1.3, n must be the product of two distinct primes, p and q:

$$n = pq.$$

It now follows by Theorem 12.2.4 that

$$p - 1 \nmid n - 1 \ \text{ or } \ q - 1 \nmid n - 1.$$

Thus, condition (i) of Theorem 12.1.3 is not satisfied, which implies that n is not a Carmichael number. $\Rightarrow\Leftarrow$

Since we have reached a contradiction, the Assumption must be false. We conclude that n is divisible by at least three distinct primes. ∎

EXERCISES 12.2

Numerical Problems

In Exercises 1–3, for each value of n, explain how many Fermat witnesses for n are guaranteed to exist by Theorem 12.2.3.

1. $n = 144$ **2.** $n = 591$ **3.** $n = 2079$

4. a. Choose two distinct primes, p and q, both larger than 10. Demonstrate that Theorem 12.2.4 holds for these values.

   b. Let $n = pq$, where p and q are distinct primes. Find an example where $p, q > 10$, and one of the two statements

$$p - 1 \nmid n - 1, \ q - 1 \nmid n - 1$$

   is true, but the other is false.

Reasoning and Proofs

5. Suppose you are given an odd natural number n, and you are trying to determine whether n is prime. You have performed the Fermat Test (12.1.1) many times, and after trying 90% of the numbers a between 1 and $n - 1$, chosen at random, you have yet to find a Fermat witness to the compositeness of n.

   a. Is it possible for n to be composite? Explain.

   b. Is it possible that n is divisible by the square of some prime, p? Explain your reasoning.

6. Let $n = 693$.

   a. What does Theorem 12.2.3 guarantee about the number of Fermat witnesses for n?

   b. In fact, you can use the ideas in the proof of Theorem 12.2.3 to do even better. What does inequality (11) imply about the number of Fermat witnesses for n?

7. Let $n \in \mathbf{N}$ such that $121 \mid n$. Prove that a randomly chosen integer a in the range $1, 2, \ldots, n - 1$ has at least a 90% chance of being a Fermat witness for n. [*Hint:* Use inequality (10).]

8. Consider the function $f(p) = \dfrac{p - 1}{p^2}$. Prove that for any integer $p \geq 3$, we have $f(p) \leq f(3)$. [*Hint:* Is f a decreasing function?]

9. Let $p, n \in \mathbf{N}$ such that p is an odd prime and $p^2 \mid n$. Show that $\frac{2}{9} n \leq \frac{1}{4}(n - 1)$. [*Hint:* What is the smallest possible value of n?]

10. Show that if we slightly alter Theorem 12.2.3 by removing the requirement that p is odd, then the theorem no longer holds.

11. Show that if we slightly alter Theorem 12.2.4 by removing the requirement that p and q are distinct, then the theorem no longer holds.

12. In the beginning of the proof of the Claim in Theorem 12.1.3, we observed that since $p \nmid b$, there exists $a \in \mathbf{Z}$ such that $a \equiv b \pmod{p}$ and $\gcd(a, n) = 1$. Prove this observation.

13. In this exercise, you will fill in the details of a step of the proof of Theorem 12.1.3. We assume that n is a natural number, p is a prime number such that $p^2 \mid n$, and x is an integer such that $\gcd(x, n) = 1$.

    a. Let n' be the integer obtained from n by dividing out the largest possible power of p. Using the Chinese Remainder Theorem, show that there exists an integer y satisfying both of the following congruences:

    $$y \equiv x + p \pmod{p^2},$$

    and

    $$y \equiv x \pmod{n'}.$$

    b. Now show that this number y satisfies the three properties (6), (7), and (8), which were required in the proof of Theorem 12.1.3.

14. Show that 561 is the smallest Carmichael number.

15. a. Find the next smallest Carmichael number after 561.

    b. Show that your answer from part a is the smallest Carmichael number that is greater than 561.

Advanced Reasoning and Proofs

EXPLORATION Primitive roots modulo p^2 (Exercises 16–22)

This exploration relies on Section 10.3, which discusses *primitive roots*. Recall that the definition of *primitive root* that we gave in the text requires that the modulus, p, be prime.

Nonetheless, it is possible to generalize the definition of primitive roots to composite moduli, as follows:

> DEFINITION. *Let n be a natural number, and let $\bar{a} \in \mathbf{Z}_n$ be a reduced residue. Then $\bar{a}$ is said to be a* **primitive root** *if the order of $\bar{a}$ is equal to $\varphi(n)$. In this case, we may also say that the integer a is a* **primitive root** *modulo n.*

(We introduced this definition before, in the exploration in Section 10.3, Exercises 15–23. In case you're wondering which composite moduli have primitive roots, see that exploration.)

In the following exercises, you will show that if p is prime, then there exists a primitive root modulo p^2.

16. Let p be prime, and let a be an integer such that $\gcd(a, p) = 1$. Show that $\mathrm{ord}_{p^2}(a) \mid p(p-1)$. [*Hint:* Use Euler's Theorem (9.3.1).]

17. Let p be prime. Show that if a is any primitive root modulo p, then $p - 1 \mid \mathrm{ord}_{p^2}(a)$.

18. Let p be prime. Use the results from Exercises 16 and 17 to show that if a is a primitive root modulo p, then $\mathrm{ord}\, p^2(a) = p - 1$ or $\mathrm{ord}_{p^2}(a) = p(p-1)$.

19. Let p be prime, and let $a, b \in \mathbf{Z}$ such that $a \equiv b \pmod{p}$ and $a^{p-1} \equiv b^{p-1} \pmod{p^2}$. Prove that $a \equiv b \pmod{p^2}$. [*Hint:* Use Lemma 12.2.2.]

20. Let p be prime, and let a be a primitive root modulo p. Consider the number $b = a + p$. Use the result from Exercise 19 to show that $\mathrm{ord}_{p^2}(a)$ and $\mathrm{ord}_{p^2}(b)$ cannot both equal $p - 1$.

21. Let p be prime. Show that if a is a primitive root modulo p, then either a or $a + p$ is a primitive root modulo p^2.

22. Explain why it follows from the result of Exercise 21 that for any prime p, there exists a primitive root modulo p^2.

12.3 The Miller-Rabin Primality Test

An improvement on Fermat

All things considered, the Fermat Test for Compositeness (12.1.1) does a pretty good job of distinguishing prime numbers from composite numbers most of the time. But as we have seen, this test has a major problem. The Fermat Test has trouble distinguishing prime numbers from Carmichael numbers, which are composite numbers that have very few Fermat witnesses to their compositeness. Ideally, we would like a test for which *every* composite number has many witnesses to its compositeness. The Miller-Rabin Test, a variant of Fermat's test that we will present shortly, is just such a test.

Let's go back to our favorite Carmichael number, $n = 252601$. For this value of n, the number $a = 5$ is not a Fermat witness. But with more thorough cross-examining, maybe we can get $a = 5$ to testify. The reason 5 is not a Fermat witness is that $5^{252600} \equiv 1 \pmod{252601}$. In other words, in $\mathbf{Z}_{252601}$,

$$\overline{5}^{252600} = \overline{1}. \tag{1}$$

Now let's take advantage of the fact that the exponent is even: $\dfrac{252600}{2} = 126300$. What can we say about the number $\overline{r} = \overline{5}^{126300}$? Well, $\overline{r}$ is a square root of $\overline{5}^{252600}$; that is, $\overline{r}$ is a square root of $\overline{1}$ in $\mathbf{Z}_{252601}$. If $n = 252601$ is prime, then by Lemma 8.3.8, the only square roots of $\overline{1}$ are $\overline{+1}$ and $\overline{-1}$. So we're off to the computer to calculate $\overline{r} = \overline{5}^{126300}$. If the result is anything other than $\overline{+1}$ or $\overline{-1}$, then we will conclude that 252601 is composite. The computer tells us... (drum roll, please)...

$$\overline{5}^{126300} = \overline{1}. \tag{2}$$

So now what? We still haven't shown that $n = 252601$ is composite. But we can go through the same process again. The exponent 126300 is itself even: $\dfrac{126300}{2} = 63150$, so $\overline{s} = \overline{5}^{63150}$ is a square root of $\overline{5}^{126300}$; that is, $\overline{s}$ is a square root of $\overline{1}$. So back to the computer to calculate $\overline{s} = \overline{5}^{63150}$, to see whether we get something other than $\overline{1}$ or $\overline{-1}$. This time we find ...

$$\overline{5}^{63150} = \overline{67772}. \tag{3}$$

Now we may conclude that 252601 is definitely composite! Equations (2) and (3) imply that

$$\overline{67772}^{2} = \overline{1},$$

so in $\mathbf{Z}_{252601}$, the number $\overline{1}$ has a square root other than $\overline{+1}$ and $\overline{-1}$. Hence, by Lemma 8.3.8, $n = 252601$ cannot be prime.

We have succeeded in our goal of coercing the evasive number 5 to witness against 252601.

Naomi's Numerical Proof Preview: Miller-Rabin Test (12.3.1)

Another way of viewing this is as follows: From equation (1), we see that in $\mathbf{Z}_{252601}$, $\overline{a} = \overline{5}$ is a solution to the polynomial equation

$$\overline{a}^{252600} - \overline{1} = \overline{0}.$$

This polynomial can be factored, using the difference of squares formula repeatedly:

$$\overline{a}^{252600} - \overline{1} = (\overline{a}^{126300} + \overline{1})(\overline{a}^{126300} - \overline{1})$$
$$= (\overline{a}^{126300} + \overline{1})(\overline{a}^{63150} + \overline{1})(\overline{a}^{63150} - \overline{1})$$
$$= (\overline{a}^{126300} + \overline{1})(\overline{a}^{63150} + \overline{1})(\overline{a}^{31575} + \overline{1})(\overline{a}^{31575} - \overline{1}). \qquad (4)$$

Since 31575 is odd, it is not possible to use the difference of squares formula to factor this polynomial any further. Observing that $252600 = 2^3 \cdot 31575$, note that we were able to apply the difference of squares formula 3 times in a row because 2^3 is the largest power of 2 that divides 252600.

Now suppose someone claims that the number 252601 is prime. Recall that in a prime modulus, if a product equals $\overline{0}$, then one of the factors must equal $\overline{0}$ (Theorem 8.3.5). Thus, since $\overline{a} = \overline{5}$ makes the left side of equation (4) equal to $\overline{0}$, we know that if the modulus 252601 were prime, then $\overline{5}$ would have to be a root of one of the factors on the right side of equation (4). In other words, the number $a = 5$ would have to satisfy one of the following congruences:

$$a^{31575} \equiv 1 \pmod{252601},$$
$$a^{31575} \equiv -1 \pmod{252601},$$
$$a^{63150} \equiv -1 \pmod{252601},$$
$$a^{126300} \equiv -1 \pmod{252601}.$$
$$(5)$$

But one computes that

$$5^{31575} \equiv 123524 \pmod{n}, \quad 5^{63150} \equiv 67772 \pmod{n}, \quad \text{and} \quad 5^{126300} \equiv 1 \pmod{n}.$$

Thus, $a = 5$ does not satisfy any of the congruences in (5). At this point, we have a contradiction. We conclude that 252601 must not be prime.

The Miller-Rabin Test, stated below, is based on these observations. In the argument above, the fact that $a = 5$ does not satisfy any of the congruences in (5) proves that 252601 is composite. Because of its special role in proving the compositeness of 252601, the number $a = 5$ is called a *Miller-Rabin witness* to the compositeness of 252601.

12.3.1 MILLER-RABIN TEST

Suppose $n \in \mathbf{N}$ is odd. Factor $n - 1$ as $n - 1 = 2^k q$, where $k, q \in \mathbf{N}$ and q is odd. Let a be an integer in the range $1, 2, \ldots, n - 1$ and consider the following congruences:

$$a^q \equiv 1 \pmod{n},$$
$$a^q \equiv -1 \pmod{n},$$
$$a^{2q} \equiv -1 \pmod{n},$$
$$a^{4q} \equiv -1 \pmod{n},$$
$$\vdots$$
$$a^{2^{k-1}q} \equiv -1 \pmod{n},$$
$$(6)$$

If none *of the congruences in this list holds, then n is composite. In this case, the* integer *a is called a* **Miller-Rabin witness** *to the compositeness of n.*

We now prove the main statement of the Miller-Rabin Test: that if none of the congruences in list (6) holds, then *n* must be composite.

PROOF OF CORRECTNESS OF THE MILLER-RABIN TEST (12.3.1)

Let $n \in \mathbf{N}$ be odd such that $n - 1 = 2^k q$, where $k, q \in \mathbf{N}$ and q is odd. Let a be a number in the range $1, 2, \ldots, n - 1$ such that none of the congruence statements in list (6) holds.

[We must show that *n* is composite. We will prove this by contradiction.]

Assumption: Suppose that *n* is prime.

We factor $\bar{a}^{n-1} - \bar{1}$ in $\mathbf{Z}_n$ by using the difference of squares formula repeatedly:

$$\bar{a}^{n-1} - \bar{1} = \bar{a}^{2^k q} - \bar{1}$$
$$= \left(\bar{a}^{2^{k-1}q} + \bar{1}\right)\left(\bar{a}^{2^{k-1}q} - \bar{1}\right)$$
$$= \left(\bar{a}^{2^{k-1}q} + \bar{1}\right)\left(\bar{a}^{2^{k-2}q} + \bar{1}\right)\left(\bar{a}^{2^{k-2}q} - \bar{1}\right)$$
$$\vdots$$
$$= \left(\bar{a}^{2^{k-1}q} + \bar{1}\right)\left(\bar{a}^{2^{k-2}q} + \bar{1}\right)\left(\bar{a}^{2^{k-3}q} + \bar{1}\right) \cdots \left(\bar{a}^q + \bar{1}\right)\left(\bar{a}^q - \bar{1}\right). \qquad (7)$$

Since *n* is prime, by Fermat's Little Theorem (9.1.1), $\bar{a}^{n-1} - \bar{1} = \bar{0}$. Hence, the product on the right side of equation (7) also equals $\bar{0}$. In a prime modulus, a product can only equal $\bar{0}$ when one of the factors equals $\bar{0}$ (Theorem 8.3.5). Thus, one of the factors on the right side of equation (7) must equal $\bar{0}$, and it follows that one of the congruences in list (6) must hold. This contradicts the hypothesis that none of the congruences in list (6) holds. $\Rightarrow\Leftarrow$

We conclude that *n* is composite. ■

The following observation states that every Fermat witness to the compositeness of *n* is also a Miller-Rabin witness for *n*. This observation guarantees that the Miller-Rabin Test is at least as good as the Fermat Test.

OBSERVATION 12.3.2

Let a and n be integers such that n is odd and composite. If a is a Fermat witness to the com-positeness of n, then a is also a Miller-Rabin witness to the compositeness of n.

PROOF Let $a, n \in \mathbf{Z}$ such that *n* is odd and composite.

Suppose that *a* is a Fermat witness to the compositeness of *n*. Then $a^{n-1} \not\equiv 1 \pmod{n}$. Hence, in $\mathbf{Z}_n$,

$$\bar{a}^{n-1} - \bar{1} \neq \bar{0}.$$

Thus, the left side of equation (7) is nonzero, so none of the factors in the product on the right side of equation (7) can be equal to $\bar{0}$. In other words, none of the congruences in list (6) hold. Thus, a is a Miller-Rabin witness to the compositeness of n. ∎

The Miller-Rabin Miracle

Observation 12.3.2 guarantees that the Miller-Rabin witnesses for any number n are at least as numerous as the Fermat witnesses for n. And as we have seen, there are also Miller-Rabin witnesses that are not Fermat witnesses. For instance, we saw in the Numerical Proof Preview for the Miller-Rabin Test that $a = 5$ is a Miller-Rabin witness to the compositeness of $n = 252601$, even though (by equation (1)) the number $a = 5$ is not a Fermat witness to the compositeness of 252601. Nonetheless, we may still wonder just how numerous the Miller-Rabin witnesses are.

Our next theorem expresses the miracle of the Miller-Rabin Test: For *every* odd composite number n, at least 75% of the integers a in the range $1, \ldots, n - 1$ are Miller-Rabin witnesses. Thus, while the Fermat Test (12.1.1) is foiled by Carmichael numbers (for which Fermat witnesses are extremely rare), the Miller-Rabin Test (12.3.1) is not foiled by any composite number n. Every composite number has plenty of Miller-Rabin witnesses!

THEOREM 12.3.3	EFFECTIVENESS OF THE MILLER-RABIN TEST

Let $n \in \mathbf{N}$ be odd and composite. Then the number of Miller-Rabin witnesses for n is at least $\dfrac{3}{4}(n - 1)$.

We are not yet ready to prove this remarkable result. The math we develop in the following two sections will culminate in the proof of this theorem.

Without Theorem 12.3.3, the Miller-Rabin Test (12.3.1), like the Fermat Test (12.1.1), is merely a test for compositeness. Sometimes it tells us that the given number n is composite, and sometimes it tells us that n may or may not be composite (which is not very helpful). In light of Theorem 12.3.3, however, the Miller-Rabin Test turns into something much more powerful, a *probabilistic primality test*. That is, Theorem 12.3.3 enables us to use the Miller-Rabin Test to quickly determine, with incredibly high probability, whether a given number n is prime.

METHOD 12.3.4	USING MILLER-RABIN AS A PROBABILISTIC PRIMALITY TEST

Starting with any odd natural number n, we run the Miller-Rabin Test (12.3.1) many times (e.g., 100 times) in a row, each time using a random value of a in the range $1, 2, \ldots, n - 1$.

(*i*) If any of these values a is a Miller-Rabin witness to the compositeness of n, conclude that n is definitely composite.

(*ii*) If none of these values a is a Miller-Rabin witness for n, conclude that with very high probability, n is prime.

Let's consider what happens when we use Method 12.3.4 to test a number n. If n is prime, then no value of a will ever be a witness to the compositeness of n. On the other hand, if n is composite, then by Theorem 12.3.3, each time we run the Miller-Rabin Test (12.3.1), the probability that a will be a witness for n is at least $\frac{3}{4}$. In other words, the probability that a will not be a witness for n is at most $\frac{1}{4}$. If we start with a composite number n and run the Miller-Rabin Test 100 times with randomly selected values of a, then it is extremely likely that in at least one of the tests, a will be a witness to the compositeness of n. To be precise, if we run the Miller-Rabin Test 100 times on a composite number n, the probability that we will never find an a that is a witness for n is less than $\left(\frac{1}{4}\right)^{100}$, an *extremely* low probability!

Miller-Rabin meets the Prime Number Theorem

As we noted in our discussion of public key cryptography in Section 9.5, the main reason we want a primality test is to generate large prime numbers. This is accomplished by picking a large (e.g., 200-digit) random odd number n, then applying the Miller-Rabin Test multiple times (Method 12.3.4) to find out whether n is prime.

Let's say we run the Miller-Rabin Test once with a random value of a, and we find that a is not a Miller-Rabin witness for n. Using Theorem 12.3.3, what is the probability that n is actually composite? You might be tempted to think the correct answer is **at most** $\frac{1}{4}$, but in fact that is not correct! The reason is that most 200-digit numbers are not prime, so when n is first selected at random, before we run any tests, it is initially highly likely that n is composite. In order to accurately estimate the probability that n is composite after we run the Miller-Rabin Test, we need an estimate of the probability that n is composite *before* we run the test. How can this initial probability, known as the *prior probability* that n is composite, be obtained? The Prime Number Theorem (3.3.3) gives us such an estimate: By the Prime Number Theorem (see Exercises 24–28), the prior probability that a random odd 200-digit number is prime is approximately $\dfrac{2}{\log(10^{200})}$, which is about $\dfrac{1}{230} \approx 0.0043$.

Thus, the prior probability that a random 200-digit odd number, n, is composite is about $1 - 0.0043 = 0.9957$. From this prior probability, we can find the probability that n is composite *after* the Miller-Rabin Test is run. To compute this after-test probability, known as the *posterior probability* that n is composite, we use a probabilistic identity known as *Bayes Rule*. (See Exercises 24–28.) If we choose a random odd 200-digit number, then run the Miller-Rabin Test once and don't find a Miller-Rabin witness, then n is still most likely to be composite, not prime. In fact, all that we may conclude from this test result is that the probability that n is composite is **at most .983**, which is a far cry from $\frac{1}{4}$! (See Exercise 26.)

How about if we run the Miller-Rabin Test 100 times on a random 200-digit number, n, and never find a Miller-Rabin witness for n. By Theorem 12.3.3, what is the probability that n is composite? You might be tempted to think the correct answer is **at most** $\left(\frac{1}{4}\right)^{100}$. Using Bayes Rule, however, we find (see Exercise 27) that the actual probability that n is composite is approximately 230 times as large as our original estimate: **at most** $230 \cdot \left(\frac{1}{4}\right)^{100}$. The good news is that $\left(\frac{1}{4}\right)^{100}$ is so tiny that the extra factor of 230 is insignificant. The probability $230 \cdot \left(\frac{1}{4}\right)^{100}$ is still mind-bogglingly small. In other words, as long as we run the Miller-Rabin Test enough times, we don't need to worry about the prior probability.

Miller-Rabin goes deterministic?

Using Method 12.3.4, the Miller-Rabin Test is an excellent method for finding large prime numbers, but it is a probabilistic test. In Section 12.6, we will discuss the AKS primality test, which is deterministic. There is also a deterministic version of the Miller-Rabin Test that is based on the following conjecture:

Every composite number n has a Miller-Rabin witness less than $2(\log_2 n)^2$. (8)

This conjecture is believed to be true, and if it is, then one gets an efficient (polynomial-time) deterministic primality test: Just check to see whether any of the numbers up to $2(\log_2 n)^2$ is a Miller-Rabin witness. If so, then n is composite. If not, then it follows from conjecture (8) that n must be prime.

One can even prove conjecture (8) if one assumes the Generalized Riemann Hypothesis. This statement and its cousin, the Riemann Hypothesis, are conjectures that are widely believed to be true but have notoriously remained unproven. The Riemann Hypothesis, a conjecture about the roots of an analytic function called the *Riemann zeta function*, is the holy grail of analytic number theory. It has myriad consequences about the distribution of prime numbers, including conjecture (8).

Fun Facts

The Clay Mathematics Institute is offering a $1 million prize for proving the Riemann Hypothesis.

EXERCISES 12.3

Numerical Problems

1. Observe that
$$5^{216} \equiv 1 \pmod{217},$$
$$5^{108} \equiv 1 \pmod{217},$$
$$5^{54} \equiv 1 \pmod{217},$$
and $$5^{27} \equiv 125 \pmod{217}.$$

What can we conclude from the Miller-Rabin Test (12.3.1)?

2. Observe that

$$3^{120} \equiv 1 \pmod{121},$$

$$3^{60} \equiv 1 \pmod{121},$$

$$3^{30} \equiv 1 \pmod{121},$$

and $\qquad 3^{15} \equiv 1 \pmod{121}.$

What does the statement of the Miller-Rabin Test (12.3.1) allow us to conclude about 121?

3. For each value of n, write $n - 1$ in the form $2^k q$, where $k, q \in \mathbf{N}$ and q is odd.

   a. $n = 341$ **b.** $n = 149$ **c.** $n = 561$ **d.** $n = 1059$

In Exercises 4–7, apply the Miller-Rabin Test (12.3.1) to determine whether a is a Miller-Rabin witness for n.

4. $n = 25, a = 7$ **5.** $n = 15, a = 4$ **6.** $n = 15, a = 11$ **7.** $n = 15, a = 14$

8. Determine all of the Miller-Rabin witnesses for $n = 9$. Justify your answer.

9. Determine all of the Miller-Rabin witnesses for $n = 7$. Justify your answer.

Computer Exercise In Exercises 10–15, first apply the Fermat Test (12.1.1), using the given value of a, to determine whether n is composite. If the Fermat Test is inconclusive, then use the Miller-Rabin Test (12.3.1) with the same value of a, and explain what the statement of the Miller-Rabin Test allows you to conclude.

10. $n = 341, a = 2$ **11.** $n = 149, a = 3$ **12.** $n = 91, a = 29$

13. $n = 561, a = 5$ **14.** $n = 1059, a = 1$ **15.** $n = 145, a = 17$

16. For each value of n, apply Theorem 12.3.3 (effectiveness of the Miller-Rabin Test) to give a lower bound on the number of Miller-Rabin witnesses for n.

   a. $n = 91$ **b.** $n = 53$ **c.** $n = 133$ **d.** $n = 101$

Reasoning and Proofs

17. Prove that for any odd $n \in \mathbf{N}$, neither 1 nor $n - 1$ is a Miller-Rabin witness for n.

18. Suppose you are using Method 12.3.4 to determine whether the odd number $n > 1$ is prime. You randomly select v different values of a in the range $1, 2, \ldots, n - 1$ to use in your test.

   a. If n is prime, what is the probability that at least one of your values of a will be a Miller-Rabin witness for n?

b. If n is composite, what (according to Theorem 12.3.3) can you say about the probability that your test will reveal that n is composite?

19. Let $n > 1$ be odd, and let a be in the range $1, 2, \ldots, n - 1$. Show that if $a^{(n-1)/2} \not\equiv 1 \pmod{n}$ and $a^{(n-1)/2} \not\equiv -1 \pmod{n}$, then a is a Miller-Rabin witness to the compositeness of n.

20. Let $n > 100$ such that $3^{(n-1)/4} \equiv 7 \pmod{n}$. Show that n is composite.

21. **Phil Lovett** Your friend Phil Lovett has just read the statement of the Miller-Rabin Test (12.3.1), and he is not impressed.

"Dude, Miller-Rabin is a really far-out idea. But it's totally impractical for big numbers. Like, in the very first step, you've got to factor $n - 1$ as $2^k q$, where $k \in \mathbf{N}$ and q is odd. My remedial math professor says factoring big numbers is practically impossible. So that means that if n is large, Miller-Rabin is totally useless."

Could Phil be right? If not, explain the error in Phil's reasoning.

Advanced Reasoning and Proofs

22. **a.** **Computer Exercise** Write a program that performs Method 12.3.4 (Using Miller-Rabin as a Probabilistic Primality Test) to test the primality of large input numbers.

    b. Use your program to find a 300-digit number that you are virtually certain is prime. You should be confident enough that you would be willing to bet your iPod that this number is in fact prime.

23. When one finds a Fermat witness or a Miller-Rabin witness for a number n, one deduces immediately that n is composite, but one does not usually learn anything about the factors of n. However, in some situations, it is possible to factor n.

    a. Let n be odd, and suppose that a is a Miller-Rabin witness for n, but a is not a Fermat witness for n. Explain how you can use this knowledge to quickly find a factor of n.

    b. Suppose that n is a Carmichael number. Explain how you can quickly find a factor of n.

EXPLORATION Bayes Rule (Exercises 24–28)

In this Exploration, you will use *Bayes Rule* and Theorem 12.3.3 to correctly compute an upper bound on the probability that a randomly chosen number is prime based on the results of the Miller-Rabin Test. By the Prime Number Theorem (3.3.3), the probability that a randomly chosen 200-digit natural number is prime is about $\dfrac{1}{\log(10^{200})}$; and thus, the probability that a randomly chosen 200-digit *odd* number is prime is $\dfrac{2}{\log(10^{200})}$.

Suppose we randomly select a 200-digit odd number, n. Let C represent the event n **is composite**. (Then $\sim C$ represents the event n **is prime**). The *prior probability* that n is prime (before the Miller-Rabin Test is run) is

$$P(\sim C) \approx \frac{2}{\log(10^{200})}.$$

24. Give an estimate of $P(C)$, the *prior probability* (before the Miller-Rabin Test is run) that n is composite. Let M be the event that we run the Miller-Rabin Test once with a random value of a and find that a is *not* a Miller-Rabin witness for n. Theorem 12.3.3 tells us that if n is composite, then the probability that the Miller-Rabin Test does not find a witness, denoted $P(M \mid C)$, is at most $\frac{1}{4}$:

$$P(M \mid C) \leq \frac{1}{4}.$$

$P(M \mid C)$ is called the *conditional probability* of M given C.

25. If n is prime, what is the probability that the Miller-Rabin Test will not find a witness for n? This probability is denoted $P(M \mid \sim C)$. After we see the result of the Miller-Rabin Test, our conclusion, known as the *posterior probability* that n is composite, is denoted $P(C \mid M)$. This posterior probability is computed using Bayes Rule:

$$P(C \mid M) = \frac{P(M \mid C)P(C)}{P(M \mid C)P(C) + P(M \mid \sim C)P(\sim C)} \qquad \leftarrow \textbf{Bayes Rule}$$

26. Suppose we choose a random 200-digit number n, then run the Miller-Rabin Test once and find that our randomly chosen a is not a Miller-Rabin witness for n. Apply Bayes Rule, and interpret your result: what can you say about the probability than n is composite?

27. Suppose we choose a random 200-digit number n, then run the Miller-Rabin Test 100 times (Method 12.3.4), and never find a Miller-Rabin witness for n.

    a. Use Theorem 12.3.3 to determine $P(M \mid C)$ for this case (in which the Miller-Rabin Test is run 100 times).

    b. Explain why in this case, $P(C)$, $P(\sim C)$, and $P(M \mid \sim C)$ are all the same as in the previous case.

    c. Apply Bayes Rule, and interpret your result: what can you say about the probability than n is composite in this case?

28. Suppose we choose a random 1500-digit number n, then run the Miller-Rabin Test 50 times (Method 12.3.4), and never find a Miller-Rabin witness for n. Apply Bayes Rule, and interpret your result: what can you say about the probability that n is composite in this case?

12.4 Two Special Polynomial Equations in Z_p

Our main goal over the next two sections is to prove Theorem 12.3.3 (Effectiveness of the Miller-Rabin Test). In order to move toward that goal, we need to learn more about the congruences in the Miller-Rabin Test, which all have one of two forms: $a^m \equiv 1$, or $a^m \equiv -1$. These congruences can also be stated as equations using barred numbers: $\bar{x}^m = \bar{1}$, and $\bar{x}^m = \overline{-1}$. In this section, we build on the results of Section 10.2 to learn more about the number of solutions to each of these two equations when the modulus is prime.

Solutions to $\bar{x}^m = \bar{1}$

Let us revisit the Z_{13} exponent table, shown in Table 12.4.1. In the table, we color in brown the entries that equal $\bar{1}$. In addition, we color in orange the entries that equal $\overline{12}$, which we write as $\overline{-1}$.

Table 12.4.1 Exponents in Z_{13}. *Entries that equal $\bar{1}$ or $\overline{-1}$ are highlighted.*

	$\bar{1}$	$\bar{2}$	$\bar{3}$	$\bar{4}$	$\bar{5}$	$\bar{6}$	$\bar{7}$	$\bar{8}$	$\bar{9}$	$\overline{10}$	$\overline{11}$	$\overline{-1}$
$\bar{x}^1$	1	2	3	4	5	6	7	8	9	10	11	-1
$\bar{x}^2$	1	4	9	3	-1	10	10	-1	3	9	4	1
$\bar{x}^3$	1	8	1	-1	8	8	5	5	1	-1	5	-1
$\bar{x}^4$	1	3	3	9	1	9	9	1	9	3	3	1
$\bar{x}^5$	1	6	9	10	5	2	11	8	3	4	7	-1
$\bar{x}^6$	1	-1	1	1	-1	-1	-1	-1	1	1	-1	1
$\bar{x}^7$	1	11	3	4	8	7	6	5	9	10	2	-1
$\bar{x}^8$	1	9	9	3	1	3	3	1	3	9	9	1
$\bar{x}^9$	1	5	1	-1	5	5	8	8	1	-1	8	-1
$\bar{x}^{10}$	1	10	3	9	-1	4	4	-1	9	3	10	1
$\bar{x}^{11}$	1	7	9	10	8	11	2	5	3	4	6	-1
$\bar{x}^{12}$	1	1	1	1	1	1	1	1	1	1	1	1
$\bar{x}^{13}$	1	2	3	4	5	6	7	8	9	10	11	-1

Of all the rows in the exponent table modulo 13, which rows contain the most $\overline{1}$'s? Looking at the table, we see that the clear winner is the $\overline{x}^{12}$ row, in which all of the entries are $\overline{1}$'s; the runner-up is the x^6 row, in which $\frac{1}{2}$ of the entries are $\overline{1}$'s.

In the Miller-Rabin Test (12.3.1), there is only one congruence in which the right side is 1, and in that congruence, $a^q \equiv 1$, the exponent q is odd. Thus, we will examine only the *odd* rows of the $\mathbf{Z}_p$ exponent table (i.e., the rows of $\overline{x}^m$ in which the exponent m is odd). Restricting our attention to the odd rows in Table 12.4.1, we find that the most $\overline{1}$'s appear in the $\overline{x}^3$ row and the $\overline{x}^9$ row, in which exactly $\frac{1}{4}$ of the entries equal $\overline{1}$. The following lemma generalizes these observations.

LEMMA 12.4.1

Let $p > 2$ be prime, and let $m \in \mathbf{N}$. Consider the equation

$$\overline{x}^m = \overline{1} \text{ in } \mathbf{Z}_p. \tag{1}$$

(i) If m is odd, then equation (1) has at most $\dfrac{p-1}{2}$ solutions.

(ii) If m is odd and $\dfrac{p-1}{2} \nmid m$, then equation (1) has at most $\dfrac{p-1}{4}$ solutions.

PROOF Let m be an odd natural number, and let S denote the number of solutions to equation (1). By Proposition 10.2.2(*ii*), $S = \gcd(m, p-1)$. Hence,

$$S \mid m \quad \text{and} \quad S \mid p-1. \tag{2}$$

Since m is odd, S must be odd. Hence, S is an odd divisor of the even number $p-1$. It follows that

$$S \left| \frac{p-1}{2} \right. . \tag{3}$$

This implies $S \le \dfrac{p-1}{2}$, which proves (*i*).

To prove (*ii*), we now make the additional assumption that

$$\frac{p-1}{2} \nmid m. \tag{4}$$

Now we cannot have $S = \dfrac{p-1}{2}$, for then by statement (2), $\dfrac{p-1}{2}$ would divide m, which contradicts hypothesis (4). Hence, it follows from statement (3) that S is at most half of $\dfrac{p-1}{2}$. That is,

$$S \le \frac{p-1}{4}.$$

Solutions to $\overline{x}^m = \overline{-1}$

The two types of congruences in the Miller-Rabin Test all have the form $a^m \equiv 1$ or $a^m \equiv -1$. We have so far focused on the first type, by considering solutions to the equation $\bar{x}^m = \bar{1}$ in $\mathbf{Z}_p$. We now concern ourselves with the second type, by considering solutions to the related equation, $\bar{x}^m = \overline{-1}$ in $\mathbf{Z}_p$.

Let's reexamine Table 12.4.1. In each row of the table, compare the number of times $\bar{1}$ appears with the number of times $\overline{-1}$ appears. A pattern becomes clear. In each row, there are two possibilities: either $\bar{1}$ and $\overline{-1}$ appear equally often, or $\overline{-1}$ does not appear at all.

The next proposition asserts that this pattern holds in general: The number of solutions to $\bar{x}^m = \overline{-1}$ in $\mathbf{Z}_p$ is either equal to the number of solutions of $\bar{x}^m = \bar{1}$ in $\mathbf{Z}_p$, which is $\gcd(m, p-1)$, or equal to 0. The proposition also tells us how to determine when each of these possibilities occurs.

PROPOSITION 12.4.2

Let $p > 2$ be prime. Let m be a natural number, and consider the polynomial equation

$$\bar{x}^m = \overline{-1} \ \ in \ \mathbf{Z}_p. \tag{5}$$

The number of solutions to this equation is

$$\begin{cases} \gcd(m, p-1) & \textit{if the factor 2 appears more times in the} \\ & \textit{prime factorization of } p-1 \textit{ than in the prime} \\ & \textit{factorization of } m. \\ \\ 0 & \textit{otherwise.} \end{cases}$$

PROOF We begin by considering the polynomial $\bar{x}^{2m} - \bar{1}$. This polynomial can be factored as a difference of squares:

$$\bar{x}^{2m} - \bar{1} = (\bar{x}^m - \bar{1})(\bar{x}^m + \bar{1}).$$

Hence, any root of the polynomial $\bar{x}^{2m} - \bar{1}$ must be a root of $\bar{x}^m - \bar{1}$ or a root of $\bar{x}^m + \bar{1}$. Furthermore, no value of $\bar{x}$ can be a solution to both of these equations (see Exercise 9). Thus

Number of roots of $\bar{x}^{2m} - \bar{1} =$
$$\textbf{Number of roots of } \bar{x}^m - \bar{1} + \textbf{Number of roots of } \bar{x}^m + \bar{1}.$$

Applying Proposition 10.2.2(*ii*), we obtain

$$\gcd(2m, p-1) = \gcd(m, p-1) + \textbf{Number of solutions to equation (5),}$$

and hence,

$$\textbf{Number of solutions to equation (5)} = \gcd(2m, p-1) - \gcd(m, p-1). \tag{6}$$

If the prime 2 appears more times in the prime factorization of $p - 1$ than in the prime factorization of m, then $\gcd(2m, p - 1) = 2 \cdot \gcd(m, p - 1)$, and thus, by equation (6), the number of solutions to equation (5) is $\gcd(m, p - 1)$.

On the other hand, if 2 appears at least as many times in the prime factorization of m as in the prime factorization of $p - 1$, then $\gcd(2m, p - 1) = \gcd(m, p - 1)$. Thus, by equation (6), the number of solutions to equation (5) is 0. ∎

In the same way that Lemma 12.4.1 limits the number of solutions to equation (1), the following lemma limits the number of solutions to the equation $\overline{x}^m = \overline{-1}$ in $\mathbf{Z}_p$. Let's take another look at the exponent table for $\mathbf{Z}_{13}$. In Table 12.4.1, which rows contain the most $\overline{-1}$'s? This time, the winner is the $\overline{x}^6$ row, in which $\frac{1}{2}$ of the entries are equal to $\overline{-1}$. Tying for second place are the $\overline{x}^3$ and $\overline{x}^9$ rows, in which $\frac{1}{4}$ of the entries are equal to $\overline{-1}$. These observations generalize to any prime modulus:

LEMMA 12.4.3

Let $p > 2$ be prime, and let $m \in \mathbf{N}$. Consider the equation

$$\overline{x}^m = \overline{-1} \ in \ \mathbf{Z}_p. \tag{7}$$

(*i*) *Equation (7) has at most $\dfrac{p - 1}{2}$ solutions.*

(*ii*) *If $\dfrac{p - 1}{2} \nmid m$, then equation (7) has at most $\dfrac{p - 1}{4}$ solutions.*

PROOF If the prime 2 appears at least as many times in the prime factorization of m as in the prime factorization of $p - 1$, then by Proposition 12.4.2, there are no solutions to equation (7), and both statements (*i*) and (*ii*) hold.

Hence, we may assume that

> The prime 2 appears more times in the prime factorization of $p - 1$ than in the prime factorization of m. $\tag{8}$

By Proposition 12.4.2, the number of solutions to equation (7) is $\gcd(m, p - 1)$. It follows from statement (8) that $\gcd(m, p - 1) = \gcd\left(m, \dfrac{p - 1}{2}\right)$, so we get

$$\text{Number of solutions to equation (7)} = \gcd\left(m, \frac{p - 1}{2}\right), \tag{9}$$

and thus statement (*i*) holds.

If $\dfrac{p - 1}{2} \nmid m$, then $\gcd\left(m, \dfrac{p - 1}{2}\right)$ must be a divisor of $\dfrac{p - 1}{2}$ other than $\dfrac{p - 1}{2}$ itself, which implies that $\gcd\left(m, \dfrac{p - 1}{2}\right) \leq \dfrac{p - 1}{4}$. Hence, by equation (9), statement (*ii*) is true. ∎

EXERCISES 12.4

Numerical Problems

In Exercises 1–3, answer parts a and b for each given value of the modulus, p.
[*Hint:* Section 9.1 contains exponent tables for each of these moduli, p.]

   a. For each m in the range $1, 2, \ldots, p$, tell how many solutions there are to the equation $\bar{x}^m = \bar{1}$ in $\mathbf{Z}_p$.

   b. For each m in the range $1, 2, \ldots, p$, tell how many solutions there are to the equation $\bar{x}^m = \overline{-1}$ in $\mathbf{Z}_p$.

  1. $p = 5$ **2.** $p = 7$ **3.** $p = 11$

In Exercises 4–7, for the given values m and p, determine whether Lemma 12.4.1 or Lemma 12.4.3 applies. If so, use the appropriate lemma to find

   a. an upper bound on the number of solutions to the equation $\bar{x}^m = \bar{1}$ in $\mathbf{Z}_p$.

   b. an upper bound on the number of solutions to the equation $\bar{x}^m = \overline{-1}$ in $\mathbf{Z}_p$.

  4. $p = 101, m = 15$ **5.** $p = 101, m = 50$

  6. $p = 103, m = 29$ **7.** $p = 103, m = 51$

  8. For the given values of m and p, use Proposition 12.4.2 to determine the exact number of solutions to the equation $\bar{x}^m = \overline{-1}$ in $\mathbf{Z}_p$.

     a. $m = 12, p = 41$ **b.** $m = 60, p = 101$ **c.** $m = 49, p = 71$

Reasoning and Proofs

  9. Let $p > 2$ be prime, and let $m \in \mathbf{N}$. Show that there is no $\bar{x} \in \mathbf{Z}_p$ that satisfies both of the polynomial equations $\bar{x}^m = \bar{1}$ and $\bar{x}^m = \overline{-1}$.

 10. Demonstrate that if we remove the hypothesis that m is odd from Lemma 12.4.1(i), the result is no longer true.

 11. Demonstrate that if we remove the hypothesis that m is odd from Lemma 12.4.1(ii), the result is no longer true.

 12. Demonstrate that if we remove the hypothesis that $\frac{p-1}{2} \nmid m$ from Lemma 12.4.1(ii), the result is no longer true.

 13. Demonstrate that if we remove the hypothesis that $\frac{p-1}{2} \nmid m$ from Lemma 12.4.3(ii), the result is no longer true.

 14. **a.** Find a prime $p > 2$ and an odd natural number m for which the equation $\bar{x}^m = \bar{1}$ has exactly $\frac{p-1}{2}$ solutions in $\mathbf{Z}_p$.

b. Let p be prime. Show that if $p \equiv 3 \pmod 4$, then there exists an odd $m \in \mathbf{N}$ for which the equation $\bar{x}^m = \bar{1}$ has exactly $\dfrac{p-1}{2}$ solutions in $\mathbf{Z}_p$.

15. a. Find a prime $p > 2$ and an odd natural number m such that $\dfrac{p-1}{2} \nmid m$ and the equation $\bar{x}^m = \bar{1}$ has exactly $\dfrac{p-1}{4}$ solutions in $\mathbf{Z}_p$.

   b. Find a prime $p > 2$ and an odd natural number m such that $\dfrac{p-1}{2} \nmid m$ and the number of solutions in $\mathbf{Z}_p$ to the equation $\bar{x}^m = \bar{1}$ is less than $\dfrac{p-1}{4}$.

16. Let p be prime, and let $\bar{a}$ be a reduced residue in $\mathbf{Z}_p$. Let m be a natural number, and consider the polynomial equation $\bar{x}^m = \bar{a}$ in $\mathbf{Z}_p$. Show that the number of solutions to this equation is either 0 or $\gcd(m, p-1)$.

17. Let p be prime such that $p \equiv 1 \pmod 4$, and let $m \in \mathbf{N}$ be odd.

   a. Show that the number of solutions in $\mathbf{Z}_p$ to the equation $\bar{x}^m = \bar{1}$ is at most $\dfrac{p-1}{4}$.

   b. Give an example to show that this maximum number of solutions can be achieved.

18. Let $p \equiv 3 \pmod 4$, and consider the equation $\bar{x}^m = -\bar{1}$ in $\mathbf{Z}_p$.

   a. If m is even, show that the equation has no solutions.

   b. If m is odd, show that the number of solutions to the equation equals $\gcd\!\left(m, \dfrac{p-1}{2}\right)$.

   c. If m is odd and $m \nmid \dfrac{p-1}{2}$, show that the number of solutions to the equation is at most $\dfrac{p-1}{6}$.

12.5 Proof that Miller-Rabin Is Effective

The goal of this section is to prove Theorem 12.3.3 (Effectiveness of the Miller-Rabin Test). Throughout this section, we will assume the hypotheses and notation from the Miller-Rabin Test (12.3.1):

Suppose $n \in \mathbf{N}$ is odd, and $n - 1 = 2^k q$, where $k, q \in \mathbf{N}$ and q is odd.

The congruences in the Miller-Rabin Test are

$$
\begin{aligned}
(\mathrm{C}_{-1}): &\quad a^q \equiv 1, \\
(\mathrm{C}_0): &\quad a^q \equiv -1, \\
(\mathrm{C}_1): &\quad a^{2q} \equiv -1, \\
(\mathrm{C}_2): &\quad a^{4q} \equiv -1, \\
&\quad \vdots \\
(\mathrm{C}_{k-1}): &\quad a^{2^{k-1}q} \equiv -1.
\end{aligned}
\tag{1}
$$

Recall that a number a in the range $1, 2, \ldots, n - 1$ is called a *Miller-Rabin witness* to the compositeness of n if *none* of the congruences in list (1) holds modulo n.

Suppose we are given a number n and apply the Miller-Rabin Test using a random value of a in the range $1, 2, \ldots, n - 1$. If we find that a is a Miller-Rabin witness for n, then we know with certainty that n is composite. On the other hand, if we find that a is not a Miller-Rabin witness for n, we can only make a probabilistic conclusion: Every number a that is not a Miller-Rabin witness makes it more probable that n is prime (see Method 12.3.4). If n is composite, then each time we get unlucky and choose an a that is not a Miller-Rabin witness, we are misled: the number a misleads us by making us believe it is more likely that n is prime, when in fact n is composite. For this reason, if a is not a Miller-Rabin witness for n, we call it a *misleader* for n.

More precisely, given a composite natural number n, we call an integer a in the range $1, 2, \ldots, n - 1$ a **misleader for n** if *at least one* of the congruences in list (1) holds modulo n. (Such an a is sometimes called a *nonwitness*, or a *Miller-Rabin liar*, or a *strong liar* for n, but the authors feel the term *misleader* is more accurate than *liar*.)

EXAMPLE 1

Let $W = 43$. Show that W is a misleader for the composite number $n = 185$.

SOLUTION Factoring the largest possible power of 2 from $n - 1$, we get $184 = 2^3 \cdot 23$. We now raise 43 to each of the powers 23, $46 = 2 \cdot 23$, and $92 = 2^2 \cdot 23$, in order to determine whether $a = 43$ satisfies any of the congruences (C_{-1}), (C_0), (C_1), or (C_2).

$$43^{23} \equiv 142 \pmod{185} \qquad \leftarrow \textbf{Congruences } (\textbf{C}_{-1}) \textbf{ and } (\textbf{C}_0) \textbf{ do not hold.}$$

$$43^{46} \equiv 184 \equiv -1 \pmod{185} \quad \leftarrow \textbf{Congruence } (\textbf{C}_1) \textit{ does } \textbf{hold!}$$

$$43^{92} \equiv 1 \pmod{185} \qquad \leftarrow \textbf{Congruence } (\textbf{C}_2) \textbf{ does not hold.}$$

Since congruence (C_1) holds, we have shown that W is a misleader for n. ∎

Before delving deeper, we observe that the congruences in list (1) have a curious property:

OBSERVATION 12.5.1

Suppose one of the congruences in list (1) holds modulo m, for some natural number $m > 2$. Then none of the other congruences in list (1) can hold modulo m.

Before we prove this observation, note that even though the congruences in the Miller-Rabin Test are congruences modulo n, this observation allows us to look at these congruences in other moduli. This will be useful shortly when we look at the congruences in list (1) modulo the prime factors of n.

PROOF OF OBSERVATION 12.5.1

Let $m \in \mathbf{N}$ such that $m > 2$.

Suppose that Congruence (C_{-1}) holds; that is,

$$a^q \equiv 1 \pmod{m}.$$

Squaring this congruence repeatedly reveals that each of the numbers $a^{2q}, a^{4q}, \ldots, a^{2^{k-1}q}$, is also congruent to 1 (mod m). Hence, none of the congruences $(C_0), (C_1), \ldots, (C_{k-1})$ holds.

On the other hand, suppose the earliest congruence in list (1) that holds modulo m is congruence (C_j), for some $j = 0, \ldots, k - 1$. Then congruence (C_j) states

$$a^{2^j q} \equiv -1 \pmod{m}.$$

Repeatedly squaring this congruence, we find that the numbers $a^{2^{j+1}q}, a^{2^{j+2}q}, \ldots, a^{2^{k-1}q}$ are all congruent to 1 (mod m). Thus, none of the congruences that come after (C_j) in the list can hold. ∎

Working one prime at a time

By definition, a misleader for a composite number n is a number a that satisfies one of the congruences in list (1) modulo n. How can we tell whether a congruence holds modulo n? The Chinese Remainder Theorem has something to say about the matter. If n is the product of distinct primes, then any congruence holds modulo n if and only if the congruence holds modulo each prime factor of n (by Observation 7.4.5). For instance, returning to our favorite composite number, $n = 252601 = 41 \cdot 61 \cdot 101$, we can say that

Any congruence holds modulo 252601 if and only if it holds modulo 41, 61, and 101.

Let's write the congruences in list (1) for $n = 252601$. Since $n - 1$ factors as $n - 1 = 252600 = 2^3 \cdot 31575$, these congruences are

$$
\begin{aligned}
(C_{-1}): && a^{31575} &\equiv 1 \\
(C_0): && a^{31575} &\equiv -1 \\
(C_1): && a^{63150} &\equiv -1 \\
(C_2): && a^{126300} &\equiv -1.
\end{aligned}
\qquad (2)
$$

Suppose you're wondering whether $a = 5$ is a misleader for n. One way to determine this is to check directly whether any of the above congruences holds modulo n. (We did this in Section 12.3 and found that $a = 5$ satisfies none of these congruences, so 5 is not a misleader for 252601.) But a snazzier method is to examine whether $a = 5$ satisfies each of the congruences in list (2) modulo 41, 61, and 101.

	Modulo 41	Modulo 61	Modulo 101
5^{31575}	$5^{31575} \equiv 32$	$5^{31575} \equiv -1$	$5^{31575} \equiv 1$
5^{63150}	$5^{63150} \equiv -1$	$5^{63150} \equiv 1$	$5^{63150} \equiv 1$
5^{126300}	$5^{126300} \equiv 1$	$5^{126300} \equiv 1$	$5^{126300} \equiv 1$

From the table, we see that

- modulo 41, $a = 5$ satisfies only Congruence (C_1);

- modulo 61, $a = 5$ satisfies only Congruence (C_0);

- modulo 101, $a = 5$ satisfies only Congruence (C_{-1}).

Thus, we see that no congruence from list (2) holds simultaneously modulo 41, 61, and 101. Hence, none of the congruences in the list holds modulo 252601. It follows that 5 is not a misleader for 252601.

From this numerical example, we begin to suspect something: It's not easy being a misleader. Let's look at this more closely.

Naomi's Numerical Proof Preview: Lemma 12.5.2

Suppose you randomly choose a number a, and you're hoping that a is a misleader for 252601. For a to be a misleader, a must satisfy one of the congruences in list (2) modulo 252601. Thus, a must satisfy one of these congruences modulo 41. There is no guarantee that a randomly selected a will satisfy *any* of the congruences in the list modulo 41, but suppose it's your lucky day, and it turns out that your number a satisfies congruence (C_1) modulo 41.

So, let's see what we can conclude if we assume that a is a misleader and

a satisfies Congruence (C_1) modulo 41.

By Observation 12.5.1,

a satisfies none of the other three congruences in the list modulo 41.

Therefore,

a satisfies none of these other three congruences modulo 252601.

Using the fact that for a to be a misleader, it must satisfy one of the four congruences in list (2) modulo 252601, this implies

a satisfies (C_1) modulo 252601.

It follows that

a satisfies (C_1) modulo 61 and modulo 101.

To summarize, we have made the following two observations:

(*i*)   In order for any number a to be a misleader for 252601, the number a must satisfy one of the congruences (C_j) modulo 41.

(*ii*) Suppose a does indeed satisfy one of the congruences (C_j) modulo 41. Then in order for a to be a misleader, a must satisfy the same congruence (C_j) modulo 61 and modulo 101.

These observations are formalized in the following lemma.

LEMMA 12.5.2

Let n be a product of distinct odd primes: $n = p_1 p_2 \cdots p_s$.

Let a be any misleader for n.

(*i*) *Then a must satisfy one of the congruences in list (1) modulo p_1.*

(*ii*) *If one of the congruences in list (1) holds modulo p_1, then the same congruence also holds modulo p_i, for all $i = 2, \ldots, s$.*

PROOF Let $n = p_1 p_2 \cdots p_s$, where the p_i are distinct odd primes, and suppose a is a misleader for n.

Proof of (*i*) Since a is a misleader for n, we know a must satisfy one of the congruences in list (1) modulo n. Since $p_1 \mid n$, we know (by Lemma 7.1.10) that a must satisfy the same congruence modulo p_1. This proves assertion (*i*).

Proof of (*ii*) Now suppose that for some $j = -1, 0, 1, \ldots, k - 1$, the congruence (C_j) holds modulo p_1. [**To show: Congruence (C_j) also holds modulo p_i, for all $i = 2, \ldots, s$.**]

Since Congruence (C_j) holds modulo p_1, Observation 12.5.1 asserts that none of the other congruences in list (1) holds modulo p_1. Since (C_j) is the only congruence from list (1) that holds modulo p_1, it follows that (C_j) is the only congruence from the list that can possibly hold modulo n (by Lemma 7.1.10). But since a is a misleader for n, we know that *some* congruence from the list must hold modulo n. Hence, Congruence (C_j) must hold modulo n. It follows from the Chinese Remainder Theorem (see Observation 7.4.5) that Congruence (C_j) also holds modulo all of the primes p_i for $i = 2, \ldots, s$. ∎

Proof of the effectiveness of the Miller-Rabin Test

We are nearing our goal of proving Theorem 12.3.3, which asserts that for any odd composite number n, at most $\frac{1}{4}$ of the numbers in the range $1, 2, \ldots, n - 1$ are misleaders for n. The next proposition will bring us tantalizingly close to that goal.

Naomi's Numerical Proof Preview: Proposition 12.5.3

To understand why misleaders are so rare, let's return to our example $n = 252601$. Suppose we have picked a random number a in the range $1, 2, \ldots, 252600$. Let's try to estimate the probability that a is a misleader. By Lemma 12.5.2(*i*), in order for a to be misleader, a must satisfy one of the congruences in list (2) modulo 41. Let's suppose we're lucky enough that this happens, and a satisfies Congruence (C_j) for some $j = -1, 0, 1, 2$. Then we can apply Lemma 12.5.2(*ii*). In order for a to be a misleader, a must also satisfy the same congruence (C_j) modulo 61 and modulo 101.

What are the chances that a random number a will satisfy (C_j) modulo 61? Well, either congruence (C_j) has the form $a^d \equiv -1$, or it has the form $a^d \equiv 1$ with an odd exponent d. We learned all about solutions to these congruences in Section 12.4. Looking back at Lemmas 12.4.1(i) and 12.4.3(i), we see that the congruence (C_j) will have at most $\dfrac{61-1}{2} = 30$ solutions modulo 61. Thus, the probability that a random number a will satisfy congruence (C_j) modulo 61 is at most $\dfrac{1}{2}$. Similarly, the probability is at most $\dfrac{1}{2}$ that a random number a will satisfy Congruence (C_j) modulo 101. Thus, the chance that our number a will satisfy (C_j) both modulo 61 and modulo 101 is at most $\dfrac{1}{2} \cdot \dfrac{1}{2} = \dfrac{1}{4}$. Hence, the probability that our number a is a misleader is at most $\dfrac{1}{4}$. This should look familiar! It is exactly the conclusion of Theorem 12.3.3.

This argument shows that for our number n that is the product of 3 distinct primes, the probability that a randomly chosen a will be a misleader is at most $\dfrac{1}{2^2}$. (We obtained this quantity by multiplying together one factor of $\dfrac{1}{2}$ for each prime after the first.) In general, if we have a number n that is the product of s distinct primes, then the probability that a randomly chosen a will be a misleader for n is at most $\dfrac{1}{2^{s-1}}$.

PROPOSITION 12.5.3

Suppose that n is the product of two or more distinct odd primes:

$$n = p_1 p_2 \cdots p_s, \text{ where } s \geq 2.$$

Then the number of misleaders for n is less than $\dfrac{1}{2^{s-1}}(n-1)$.

PROOF Let n be a product of two or more distinct odd primes: $n = p_1 p_2 \cdots p_s$, where $s \geq 2$.

Let M be the set of all misleaders for n:

$$M = \{a \mid a \text{ is a misleader for } n.\}$$

We now split the elements of M into subsets according to their congruence classes modulo p_1. Let b be an integer in the range $0, 1, \ldots, p_1 - 1$, and define M_b as the set of all misleaders that are congruent to $b \pmod{p_1}$:

$$M_b = \{a \in M \mid a \equiv b \pmod{p_1}\}. \tag{3}$$

Claim: The set M_b has at most $\dfrac{p_2 - 1}{2} \cdot \dfrac{p_3 - 1}{2} \cdot \ \cdots \ \cdot \dfrac{p_s - 1}{2}$ elements.

Proof of Claim First note that if M_b is empty, then the Claim holds. Thus, we may assume that M_b is nonempty.

If a is any element of M_b, then Lemma 12.5.2(i) tells us that a satisfies one of the congruences in list (1) modulo p_1. Thus, b must satisfy the same congruence modulo p_1. In other words, either $b^q \equiv 1 \pmod{p_1}$, or $b^{2^j q} \equiv -1 \pmod{p_1}$ for some $j = 0, 1, \ldots, k - 1$.

Case 1 $b^q \equiv 1 \pmod{p_1}$.

Every element $a \in M_b$ satisfies the congruence $a^q \equiv 1 \pmod{p_1}$, so by Lemma 12.5.2(ii),

$$a^q \equiv 1 \pmod{p_i}$$

for all $i = 2, 3, \ldots, s$. Hence, by Lemma 12.4.1(i), there are at most $\dfrac{p_i - 1}{2}$ possibilities for the value of $a \pmod{p_i}$.

Any number $a \in M_b$ is in the range $1, 2 \ldots, n - 1$, so by the Chinese Remainder Theorem (7.4.3), the value of a is uniquely determined by its value modulo each of the primes $p_1, p_2, p_3, \ldots, p_s$. Since $a \equiv b \pmod{p_1}$, there is only **1** possibility for the value of $a \pmod{p_1}$. For each of the other primes p_i, we have already shown that there are at most $\dfrac{p_i - 1}{2}$ possibilities for the value of $a \pmod{p_i}$. Thus,

$$\text{Number of elements of } M_b \ \leq \ 1 \ \cdot \ \frac{p_2 - 1}{2} \ \cdot \ \frac{p_3 - 1}{2} \ \cdot \ \ldots \ \cdot \ \frac{p_s - 1}{2}.$$

Number of possible values of a modulo p_1	Number of possible values of a modulo p_2	Number of possible values of a modulo p_3	Number of possible values of a modulo p_s

This completes the proof of Case 1.

Case 2 $b^{2^j q} \equiv -1 \pmod{p_1}$ for some $j = 0, 1, \ldots, k - 1$.

Every element $a \in M_b$ satisfies the congruence $a^{2^j q} \equiv -1 \pmod{p_1}$, so by Lemma 12.5.2($ii$),

$$a^{2^j q} \equiv -1 \pmod{p_i}$$

for all $i = 2, 3, \ldots, s$. Hence, by Lemma 12.4.3(i), there are at most $\dfrac{p_i - 1}{2}$ possibilities for the value of $a \pmod{p_i}$.

Reasoning as in Case 1, we see again that

$$\text{Number of elements of } M_b \leq \frac{p_2 - 1}{2} \cdot \frac{p_3 - 1}{2} \cdot \ \ldots \ \cdot \frac{p_s - 1}{2},$$

which completes the proof of Case 2.

This completes the proof of the Claim.

We can now give an upper bound for the number of elements of M. Observe that every element of M is an element of M_b for some b in the range $0, 1, \ldots, p_1 - 1$.

By the Claim, each of the sets M_b has at most $\dfrac{p_2 - 1}{2} \cdot \dfrac{p_3 - 1}{2} \cdot \ldots \cdot \dfrac{p_s - 1}{2}$ elements. Thus,

$$\text{Number of elements of } M \; \leq \; \underset{\substack{\nearrow \\ \textcolor{brown}{\textbf{Number of possible}} \\ \textcolor{brown}{\textbf{values of } b}}}{p_1} \cdot \underbrace{\dfrac{p_2 - 1}{2} \cdot \dfrac{p_3 - 1}{2} \cdot \ldots \cdot \dfrac{p_s - 1}{2}}_{\substack{\textcolor{brown}{\textbf{Upper bound for the number of}} \\ \textcolor{brown}{\textbf{elements in each set } M_b}}}.$$

Since $n = p_1 p_2 \cdots p_s$, it follows that

$$\text{Number of elements of } M < \frac{1}{2^{s-1}} (n - 1). \qquad \blacksquare$$

We are now ready to prove Theorem 12.3.3 (Effectiveness of the Miller-Rabin Test), which asserts that for any odd composite number n, the number of Miller-Rabin witnesses for n is at least $\frac{3}{4}(n - 1)$. To prove this, we will prove the equivalent statement that the number of misleaders for n is at most $\frac{1}{4}(n - 1)$.

Notice that Proposition 12.5.3 is tantalizingly close to Theorem 12.3.3. In fact, if n is squarefree with at least three factors ($s \geq 3$), the proposition tells us that the number of misleaders for n is less than $\frac{1}{2^{s-1}}(n - 1) \leq \frac{1}{4}(n - 1)$. Hence, in this case, we are completely done: the conclusion of Theorem 12.3.3 holds. The only other possible number of factors is $s = 2$. Even in this case, Proposition 12.5.3 tells us that there are at most $\frac{1}{2}(n - 1)$ misleaders for n, but in order to get this down to $\frac{1}{4}(n - 1)$, we will need a more careful argument.[1]

PROOF of **THEOREM 12.3.3 (EFFECTIVENESS OF THE MILLER-RABIN TEST)**

Let $n \in \mathbf{N}$ be odd and composite.

[To show: the number of misleaders for n is at most $\frac{1}{4}(n - 1)$.]

Suppose first that n is divisible by the square of some prime. Then by Theorem 12.2.3, the number of integers a in the range $1, 2, \ldots, n - 1$ that are not Fermat witnesses is at most $\frac{1}{4}(n - 1)$. Since every Fermat witness for n is also a Miller-Rabin witness for n (Observation 12.3.2), it follows that the number of misleaders for n is at most $\frac{1}{4}(n - 1)$, and we are done.

---

[1]Even with just the knowledge that at most $\frac{1}{2}$ of the possible values of a are misleaders, the worth of the Miller-Rabin Test is still solidified: It still follows that applying the test repeatedly is an efficient probabilistic primality test.

Thus, we may assume that n is squarefree. If n has 3 or more prime factors, then by Proposition 12.5.3, the number of misleaders for n is less than $\frac{1}{2^2}(n-1)$, and we are done.

The only remaining possibility is that n is the product of 2 distinct primes, so for the remainder of the proof, we assume that

$$n = p_1 p_2,$$

where p_1 and p_2 are distinct primes.

Let M be the set of all misleaders for n:

$$M = \{a \mid a \text{ is a misleader for } n\}.$$

We now split the elements of M into subsets according to their congruence classes modulo p_1. Let b be an integer in the range $0, 1, \ldots, p_1 - 1$, and define

$$M_b = \{a \in M \mid a \equiv b \pmod{p_1}\}.$$

Claim: The set M_b has at most $\dfrac{p_2 - 1}{4}$ elements.

This Claim is similar to the claim that we used to prove Proposition 12.5.3, but stronger by a factor of 2. (The 2 in the denominator has been replaced by a 4.) As before, we will use Lemmas 12.4.1 and 12.4.3, but now instead of part (i) of each lemma, we will need the stronger part (ii). In both lemmas, part (ii) requires an additional hypothesis that we will now establish.

Since $n = p_1 p_2$, Theorem 12.2.4 tells us that $p_1 - 1 \nmid n - 1$ or $p_2 - 1 \nmid n - 1$. Without loss of generality, we may assume that

$$p_2 - 1 \nmid n - 1. \tag{4}$$

Since $n - 1 = 2^k q$, it follows that for all $j = 0, 1, \ldots, k - 1$, we have $2^{j+1} q \mid n - 1$. Then from statement (4), we deduce that $p_2 - 1 \nmid 2^{j+1} q$. Equivalently,

$$\frac{p_2 - 1}{2} \nmid 2^j q \quad \text{for all} \quad j = 0, 1, \ldots, k - 1. \tag{5}$$

We are now ready to prove the Claim.

Proof of Claim As in the proof of Proposition 12.5.3, we may assume that either $b^q \equiv 1 \pmod{p_1}$, or $b^{2^j q} \equiv -1 \pmod{p_1}$ for some $j = 0, 1, \ldots, k - 1$.

Case 1 $b^q \equiv 1 \pmod{p_1}$.

For every $a \in M_b$, $a^q \equiv 1 \pmod{p_1}$, so by Lemma 12.5.2(ii),

$$a^q \equiv 1 \pmod{p_2}.$$

By statement (5), $\dfrac{p_2-1}{2}\nmid q$, so we may apply Lemma 12.4.1(ii) to conclude that there are at most $\dfrac{p_2-1}{4}$ possibilities for the value of $a \pmod{p_2}$. By the Chinese Remainder Theorem (7.4.1), the value of a is uniquely determined by its value modulo p_1 and its value modulo p_2. Thus,

$$\text{Number of elements of } M_b \;\le\; 1 \;\cdot\; \frac{p_2-1}{4}.$$

↗   ↖

Number of possible   **Number of possible**
values of a mod-   **values of a mod-**
ulo p_1   **ulo p_2**

This completes the proof of Case 1.

Case 2 $b^{2^{j}q} \equiv -1 \pmod{p_1}$ for some $j = 0, 1, \ldots, k-1$.

For every $a \in M_b$, $a^{2^{j}q} \equiv -1 \pmod{p_1}$, so by Lemma 12.5.2($ii$), $a^{2^{j}q} \equiv -1 \pmod{p_2}$. By statement (5), $\dfrac{p_2-1}{2}\nmid 2^{j}q$, so we may apply Lemma 12.4.3(ii) to conclude that there are at most $\dfrac{p_2-1}{4}$ possibilities for the value of $a \pmod{p_2}$. Reasoning as in Case 1, we see again that

$$\text{Number of elements of } M_b \le \frac{p_2-1}{4}.$$

This completes the proof of the Claim.

Every element of M is an element of M_b for some b in the range $0, 1, \ldots, p_1-1$, and by the Claim, each of the sets M_b has at most $\dfrac{p_2-1}{4}$ elements. Thus,

$$\text{Number of elements of } M \;\le\; p_1 \cdot \frac{p_2-1}{4}.$$

↗   ↖

Number of possible   **Upper bound**
values of b   **for the number of elements in**
   each set M_b

Since $n = p_1 p_2$, it follows that

$$\text{Number of elements of } M \le \tfrac{1}{4}(n-1). \qquad \blacksquare$$

Numerical Problems

1. Let $n = 341, a = 2$.

    a. Write $n - 1$ in the form $2^k q$, where q is odd.

    b. Observe that $341 = 11 \cdot 31$. Fill in every cell in the following table.

	modulo 11	modulo 31
2^{85}	$2^{85} \equiv$ _____	$2^{85} \equiv$ _____
2^{170}		

    c. Using $a = 2$, is there any congruence in list (1) that holds both modulo 11 and modulo 31?

    d. Is 2 a Miller-Rabin witness for 341? Explain.

2. Let $n = 341, a = 2$. Show that Observation 12.5.1 holds in this case by demonstrating each of the following. [*Hint:* The table from Exercise 1 may be useful.]

    a. At most one of the congruences in list (1) is satisfied modulo 11.

    b. At most one of the congruences in list (1) is satisfied modulo 31.

3. Let $a = 8$. Show that a is a misleader for the composite number $n = 9$.

4. Let $V = 7$. Show that V is a misleader for the composite number $n = 25$.

5. Let $T = 8$. Show that T is a misleader for the composite number $n = 65$.

6. Show that 9 is a misleader for 91.

7. Show that 10 is a misleader for 99.

8. Let $n = 227953 = 11 \cdot 17 \cdot 23 \cdot 53$.

    a. What does Theorem 12.3.3 guarantee about the number of misleaders for n?

    b. In fact, you can use Proposition 12.5.3 to do even better. What does Proposition 12.5.3 imply about the number of misleaders for n?

Reasoning and Proofs

9. In Observation 12.5.1, we assumed that $m \in \mathbf{N}, m > 2$. Explain why the proof would not be correct without the assumption that $m > 2$.

10. Let n be a product of r distinct odd primes: $n = p_1 p_2 \cdots p_r$. Suppose $n - 1 = 2^5 q$, where $q \in \mathbf{N}$ is odd. Let a be a misleader for n. Prove that if $a^{4q} \equiv -1 \pmod{p_r}$, then $a^{8q} \equiv 1 \pmod{p_1}$.

11. Let n be a product of 4 distinct odd primes: $n = p_1 p_2 p_3 p_4$. Suppose $n - 1 = 2^k q$, where $k, q \in \mathbf{N}$ and q is odd. Let a be a misleader for n. Prove that if $a^q \equiv -1 \pmod{p_2}$, then $a^{2q} \equiv 1 \pmod{p_1 p_3 p_4}$.

Advanced Reasoning and Proofs

EXPLORATION When 2 is a misleader (Exercise 12)

For many composite numbers n, one need look no further than $a = 2$ to find a Miller-Rabin witness. Nevertheless, it turns out—as you will show in this exploration—that there are infinitely many composite numbers n for which $a = 2$ is a misleader.

12. Let $p > 5$ be prime, and let $n = \dfrac{4^p + 1}{5}$.

    a. Show that n is an integer.

    b. Prove that $2^p + 2^{\frac{p+1}{2}} + 1$ and $2^p - 2^{\frac{p+1}{2}} + 1$ are factors of $4^p + 1$. [*Hint:* Multiply them together.] Conclude that n is composite.

    c. Show that $n - 1 = 4q$, where q is an odd integer.

    d. Prove that $p \mid q$. [*Hint:* Use Fermat's Little Theorem (9.1.1).]

    e. Show that $4^p \equiv -1 \pmod{n}$, and use this to conclude that $4^q \equiv -1 \pmod{n}$.

    f. Show that 2 is a misleader for n. Conclude that there are infinitely many composite numbers n for which 2 is a misleader.

EXPLORATION Exactly How Many Misleaders are There?
(Exercises 13–17)

In this exploration, you will use the ideas of this section to give an exact formula for the number of misleaders for any squarefree composite number n.

13. Let $n = 252601 = 41 \cdot 61 \cdot 101$.

    a. Consider the congruences in list (1) for this value of n. (These are also restated more explicitly in list (2).) For each prime divisor p of n, determine exactly how many solutions each of the congruences in the list has modulo p.

    b. Now use your results from part a to determine exactly how many solutions each of the congruences in the list has modulo n.

    c. Finally, determine exactly how many misleaders there are for $n = 252601$.

In each of Exercises 14–16, repeat all three parts of Exercise 13 for the given value of n.

14. $n = 561 = 3 \cdot 11 \cdot 17$

15. $n = 1105 = 5 \cdot 13 \cdot 17$

16. $n = 1241 = 17 \cdot 73$

17. Let $n = p_1 p_2 \cdots p_s$, where the factors p_i are distinct odd primes, and $s \geq 2$. Give a general formula for the number of misleaders for n. (In addition to the usual notation we use for Miller-Rabin, let t_i represent the number of times the factor 2 appears in the prime factorization of $p_i - 1$, and let $t = \min\{t_1, t_2, \ldots, t_s\}$.)

12.6 Prime Certificates

The two powers*

The 20th-century mathematician Paul Erdős was renowned as a mathematical wizard who spent all of his time traveling. What's not so well known about Erdős is that he was really a medieval wizard who was always time-traveling. As King Arthur's Royal Mathemagician, Erdős's most important duty was to feed the King's insatiable appetite for proofs. Although the king was generally a benevolent ruler, he was nonetheless quite rigorous when it came to mathematics. Whenever King Arthur caught wind of a new theorem or conjecture, he called upon his Royal Mathemagician to find him a proof. Giving no thought to expense, Erdős often traveled to the ends of the Earth and across time itself in a quest to find the shortest and most elegant proof.

Erdős's prized possession was an orb of unlimited computational power. Erdős had simply to wave his hand and this crystalline globe would instantly perform any numerical computation, no matter how complex.

One summer, Camelot was abuzz with rumors that King Arthur's Social Security number, which had long been suspected to be prime, was actually composite. Needing certainty, the king ordered Erdős to establish once and for all whether his Social Security number, 986-74-5901, was prime or composite.

Erdős brought out his orb of computation, and the clear crystal became dark with swirling clouds. "O, orb more brilliant than the Sun, factor 986745901." Looking deep into the orb, Erdős saw the factors of the king's number in milky white numerals glimmering amidst the coffee-colored clouds. He then proved to the king that his number was composite by giving him the factorization: $986745901 = 16273 \cdot 60637$.

The king was impressed. "Erdős, you truly are the Supreme Factorer! You know, my anniversary with Queen Guinevere is next week, and I haven't been

---

*For this myth, the authors are indebted to László Babai's lectures and writings on Arthur-Merlin proofs.

able to find a gift worthy of her magnificence. If memory serves, Guinevere's favorite number is 1-888-555-1417. Hmm. . . . That's also the number for Lancelot's chat line, the All-Knight Party Line. I wonder . . . Nah, coincidence. Anyway, if you could factor 18885551417 for me, I could give her that."

Erdős again turned to his orb of computation and intoned, "O orb of starlight in the heavens, factor 18885551417." Erdős looked into the orb, then jumped up suddenly. "Sire, the orb says that number is prime!"

King Arthur responded enthusiastically, "Even better! I can give Gwinnie a proof that her favorite number is prime. She'll love that!"

Erdős paused for a moment, taking a sip from his goblet of espresso. "No problem, Your Highness. We can just use the Miller-Rabin Test to verify with extremely high probability that her number is prime—"

"Balderdash!" interrupted the king. "A probabilistic argument works great for applications, like encrypting royal missives. But it's no kind of gift to give a lady! For goodness sake, I can't proclaim my undying love with a probabilistic argument! Proofs are forever. For a lady as pure as Guinevere, it's got to be a proof."

The next day, Erdős arrived at the king's chambers. Erdős beamed, "I've finished your gift for the queen. I hereby present Your Highness with a proof that 18885551417 is prime. Roll it in, boys!"

"I divided 18885551417 by every integer up to its square root," Erdős explained, "and wrote down all of the results. As you read this scroll, you will see that none of the integers goes in evenly. Therefore, 18885551417 is prime. Q.E.D." Erdős smiled at the king, and bowed with a flourish.

"You expect me to read this whole thing?!" bellowed the king. "I could spend the rest of my natural life and still not have enough time to verify all of these divisions! I want a beautiful, elegant proof I can give my queen! At the very least, I want a proof short enough to read. Erdős, you bring me back a short proof by noon tomorrow, or it's off with your head!" The king paused, then added cheerfully, "By the way, I like your new robes. Blue is a nice color on you. Well, see you tomorrow."

The next morning, Erdős entered the king's chambers. "Sire, I heard a rumor that some Hungarians in the 20th century had discovered a magical tome, known simply as *The Book*, that contains the most elegant proof of every theorem. I realized that if I traveled to the future and tracked down this Book, then I could return and give you the short proof you desire. So I set my time machine's dials for twentieth-century Hungary."

The king leaned forward in his throne. "Did you find The Book?"

"When I arrived in the future," Erdős replied, "I was surrounded by palm trees, and I knew something had gone wrong. I was in a place called California, and when I mentioned The Book, they thought I was joking. Anyway, I met a nice fellow named Professor Pratt. He spoke terrible Hungarian, but he was able to show me how I could use my orb to find a short proof of the primality of Guinevere's number—or indeed, of any prime number."

Erdős pulled a small scroll from his robes. "Your Majesty, I hereby present you with a proof certifying that 18885551417 is prime. To know for sure that it's prime, all you need to do is check a few modular exponentiations."

Short proofs of compositeness

In our myth, Erdős saved his hide by finding a short proof that 18885551417 is prime. But how is it possible to prove that such a large number is prime without performing an enormous number of divisions? The Miller-Rabin Test can quickly convince us that 18885551417 is prime with extremely high probability, but it cannot satisfy our mathematical itch for a rigorous proof.

Before we consider this problem, let's look at the opposite problem: proving that a number is composite. In our story, Erdős proved that 986745901 (King Arthur's Social Security number) is composite by giving a factorization.

THEOREM A. *The number 986745901 is composite.*

PROOF $986745901 = 16273 \cdot 60637$. ●

Does this proof count as a "short" proof? Well, it certainly looks short. In addition, anyone who wishes to verify that this proof is correct (King Arthur, for example), need only multiply 16273 and 60637. Since multiplication can be done quickly (in polynomial time), this proof can be considered a short proof.

Checking this proof is an easy task, but discovering it in the first place is a different matter. As mentioned before, finding the factors of a large number is believed to be computationally infeasible. But once we have obtained a factorization (using a magical orb, or by luck, or just plain hard work), there is a short proof that any composite number is composite.

OBSERVATION 12.6.1

If n is a composite number, then there is a short proof that n is composite. (That is, there is a proof that can be checked in polynomial time.) This proof may be hard to find, but it exists.

Other than factoring n, another way to prove that n is composite is to exhibit a Miller-Rabin witness for n. Checking that a given number is indeed a Miller-Rabin witness can be done quickly, since only a few modular exponentiations are required. Also, by Theorem 12.3.3 (Effectiveness of the Miller-Rabin Test), there are plenty of such witnesses, so finding one is likely to be a lot easier than finding a factor of n.

Short proofs of primality

So short proofs of compositeness always exist, but what about primality? That is, given any prime p, is there a short proof that p is prime? One way to prove that any prime number, such as 18885551417, is prime, would be to exhibit (on a giant scroll, say) a massive number of trial divisions, thus showing that 18885551417 cannot be factored. The problem with this proof is that it would be exceptionally long (i.e., exponentially long and hence not checkable in polynomial time).

In 1975, Vaughan Pratt, an MIT computer scientist who later moved to Stanford, showed that there is a better way. It is based on the fact that for any prime p, there exists an integer a such that $\mathrm{ord}_p(a) = p - 1$. (If you have read Section 10.3, recall that this is the Primitive Root Theorem (10.3.3), and such an integer a is called a *primitive root modulo p*.)

The converse of this statement is also true:

LEMMA 12.6.2

Let $n \in \mathbf{N}$. Suppose there exists an integer a such that $\mathrm{ord}_n(a) = n - 1$. Then n is prime.

PROOF The hypothesis guarantees that $\mathrm{ord}_n(a)$ is defined, and this includes the assumption that a is a reduced residue modulo n. Hence, we may apply Euler's Theorem (9.3.2) to get $a^{\varphi(n)} \equiv 1 \pmod{n}$, from which it follows that

$$\mathrm{ord}_n(a) \leq \varphi(n).$$

Since $\mathrm{ord}_n(a) = n - 1$,

$$n - 1 \leq \varphi(n).$$

But by definition of Euler's φ-function, $\varphi(n) \leq n - 1$. Hence,

$$\varphi(n) = n - 1.$$

It follows that n is prime. ∎

Thus, if you are in the business of trying to convince somebody that n is a prime number, it suffices to exhibit an integer a such that $\mathrm{ord}_n(a) = n - 1$.

Naomi's Numerical Proof Preview: Theorem 12.6.3

Suppose we wish to use Lemma 12.6.2 to prove that the number $n = 101$ is prime. We need a number a such that $\mathrm{ord}_{101}(a) = 100$. The number $a = 7$ has this property. The most naive way to verify this would be to compute the powers $7^1, 7^2, 7^3, \ldots \pmod{101}$ and see directly that 7^{100} is the first power of 7 that reduces to 1 modulo 101. Though quite straightforward, this method is slow.

Instead, let's cut right to the end and compute

$$7^{100} \equiv 1 \pmod{101}. \tag{1}$$

This tells us that $\mathrm{ord}_{101}(7)$ is a divisor of 100 (by Proposition 10.1.3). So letting $r = \mathrm{ord}_{101}(7)$, we have $r \mid 100$, but how can we be sure that $r = 100$? One way would be to eliminate all the other divisors of 100 by computing the powers $7^1, 7^2, 7^4, 7^5, 7^{10}, 7^{20}, 7^{25}, 7^{50} \pmod{101}$, and checking that none of these reduce to 1 modulo 101.

But we can do even better. Suppose we do only the two computations

$$7^{50} \equiv 100 \pmod{101} \tag{2}$$

and

$$7^{20} \equiv 84 \pmod{101}. \tag{3}$$

Then Proposition 10.1.3 tells us that $r \nmid 50$ and $r \nmid 20$. Thus, r is a divisor of 100 that does not divide 50 and does not divide 20. There is only one such divisor: $r = 100$. But that's just what we wanted! Since we have shown that $\mathrm{ord}_{101}(7) = 100$, we can now conclude by Lemma 12.6.2 that 101 is prime.

Our proof that 101 is prime only required the three modular exponentiations (1), (2), and (3), which can be rewritten as

$$7^{100} \equiv 1 \pmod{101},$$
$$7^{\frac{100}{2}} \not\equiv 1 \pmod{101},$$
$$7^{\frac{100}{5}} \not\equiv 1 \pmod{101}.$$

In general, we have the following:

THEOREM 12.6.3

Let n be a natural number. Suppose there exists an integer a such that the following two conditions hold:

(*i*) $a^{n-1} \equiv 1 \pmod{n}$,

(*ii*)   *For any prime divisor q of $n - 1$,* $a^{\frac{n-1}{q}} \not\equiv 1 \pmod{n}$.

Then n is prime.

PROOF From condition (*i*) it follows that a is a reduced residue modulo n. (See Exercise 8.)

Let $r = \mathrm{ord}_n(a)$.

By Proposition 10.1.3, condition (*i*) implies that

$$r \mid n - 1, \tag{4}$$

and condition (*ii*) implies that for any prime q such that $q \mid n - 1$,

$$r \nmid \frac{n-1}{q}. \tag{5}$$

Statement (4) implies that for some integer d,

$$n - 1 = d \cdot r.$$

If $d > 1$, then d must have a prime divisor, q, that is itself a divisor of $n - 1$, and thus

$$\frac{n-1}{q} = \frac{d}{q} \cdot r.$$

It follows that $r \mid \frac{n-1}{q}$, which contradicts statement (5). $\Rightarrow\Leftarrow$

Since the assumption that $d > 1$ led to a contradiction, we must have $d = 1$. Hence, $r = n - 1$.

Since $\mathrm{ord}_n(a) = n - 1$, we may use Lemma 12.6.2 to conclude that n is prime. ∎

We now use Theorem 12.6.3 to prove that a certain number (Queen Guinevere's favorite number from our myth) is prime.

THEOREM B. *The number* 18885551417 *is prime.*

PROOF Let $n = 18885551417$. We now show that $a = 3$ satisfies conditions (i) and (ii) of Theorem 12.6.3.

To check condition (i), one verifies by direct computation that

$$3^{18885551416} \equiv 1 \ (\text{mod } 18885551417).$$

Now we verify condition (ii). The prime factorization of $n - 1$ is

$$n - 1 = 18885551416 = 2^3 \cdot 31 \cdot 76151417. \tag{6}$$

For each of the primes $q = 2, 31$, and 76151417, we find $3^{\frac{n-1}{q}}$ by direct computation:

$$3^{\frac{18885551416}{2}} \equiv 18885551416 \ (\text{mod } 18885551417),$$

$$3^{\frac{18885551416}{31}} \equiv 5229761816 \ (\text{mod } 18885551417),$$

$$3^{\frac{18885551416}{76151417}} \equiv 5144434237 \ (\text{mod } 18885551417).$$

None of these results is congruent to 1 (mod n). Therefore, condition (ii) is satisfied. We conclude that 18885551417 is prime. ∎

We have succeeded in using Theorem 12.6.3 to prove that 18885551417 is prime. Should we consider this to be a short proof? Though not quite as short as the compositeness proof above (Theorem A), it's still pretty short. And to verify that it's correct, one only needs to check one multiplication (6) and a total of three modular exponentiations. Since both multiplication and modular exponentiation can be done quickly, the entire proof can be verified in short order.

Furthermore, the verifier of this proof, who we are assuming is familiar with Theorem 12.6.3, only needs to be told the value of a and the prime factorization of $n - 1$. The verifier then knows which operations (multiplications and modular exponentiations) need to be done, and can go off and do them on his or her own.

Boiling the above proof of Theorem B down to its bare bones, we can present all the information the verifier needs to know as follows:

THEOREM B. *The number* 18885551417 *is prime.*

ABBREVIATED PROOF

Take $a = 3$. Use the prime factorization $18885551416 = 2^3 \cdot 31 \cdot 76151417$.

Now the proof is truly a one-liner!

The reader may have noticed that in our proof of Theorem B, one important detail has been overlooked. We claimed that equation (6) is the prime factorization of 18885551416. Direct multiplication shows equation (6) to be correct, but how does the verifier know that the factors 2, 31, and 76151417 are really prime? The proof is not complete until we show that these numbers are prime. How can we prove, for example, that 76151417 is prime? Although trial division by every number up to $\sqrt{76151417}$ is an option, it would make for a *very* long proof. A much better choice is to use Theorem 12.6.3 once again. Here then is the abbreviated proof for the primality of 76151417:

THEOREM C. *The number* 76151417 *is prime.*

ABBREVIATED PROOF

Take $a = 3$. Use the prime factorization $76151416 = 2^3 \cdot 11 \cdot 865357$.

Exercise 4 asks you to verify that for $n = 76151417$, the number $a = 3$ satisfies the conditions of Theorem 12.6.3.

This proof of Theorem C begs the question of proving that 2, 11, and 865357 are prime. We keep getting more numbers whose primality we must prove! But don't despair. The good news is that these prime numbers are getting small in a hurry! Pretty soon we inevitably arrive at the smallest prime number, 2, whose primality needs no further proof.

This suggests a recursive procedure for proving that a number n is prime. The first step involves giving a value of a, and a purported prime factorization of $n - 1$. We must then prove that each purported prime factor, p, of $n - 1$ is really prime. The proof that each such p is prime involves giving a value of a, and a purported prime factorization of $p - 1$. This process continues recursively, with the prime factors getting smaller and smaller, until we reach the prime factor 2.

Using this recursive procedure, we now give a complete proof that 18885551417 is prime.

PROOF THAT 18885551417 IS PRIME

18885551417 is prime: Take $a = 3$. Use $18885551416 = 2^3 \cdot \mathbf{31} \cdot \mathbf{76151417}$.

      31 is prime: Take $a = 3$. Use $30 = 2 \cdot \mathbf{3} \cdot \mathbf{5}$.

            3 is prime: Take $a = 2$. Use $2 = 2$.

            5 is prime: Take $a = 2$. Use $4 = 2^2$.

  76151417 is prime: Take $a = 3$. Use $76151416 = 2^3 \cdot \mathbf{11} \cdot \mathbf{865357}$.

      11 is prime: Take $a = 2$. Use $10 = 2 \cdot \mathbf{5}$.

          5 is prime: Take $a = 2$. Use $4 = 2^2$.

  865357 is prime: Take $a = 2$. Use $865356 = 2^2 \cdot \mathbf{3} \cdot \mathbf{37} \cdot \mathbf{1949}$.

      3 is prime: Take $a = 2$. Use $2 = 2$.

      37 is prime: Take $a = 2$. Use $36 = 2^2 \cdot 3^2$.

          3 is prime: Take $a = 2$. Use $2 = 2$.

  1949 is prime: Take $a = 2$. Use $1948 = 2^2 \cdot \mathbf{487}$.

      487 is prime: Take $a = 3$. Use $486 = 2 \cdot 3^5$.

        3 is prime: Take $a = 2$. Use $2 = 2$.   ■

This proof is known as a *Pratt certificate* for the primality of 18885551417.

EXAMPLE 1

Give a Pratt certificate to prove that 239 is prime.

SOLUTION 239 is prime: Take $a = 7$. Use $238 = 2 \cdot 7 \cdot 17$.

        7 is prime: Take $a = 3$. Use $6 = 2 \cdot 3$.

        3 is prime: Take $a = 2$. Use $2 = 2$.

     17 is prime: Take $a = 3$. Use $16 = 2^4$.   ●

In Exercise 7, you will verify that each value of a in Example 1 satisfies the conditions of Theorem 12.6.3.

We have now seen that for a small number such as 239, and for a large number such as 18885551417, a Pratt certificate provides a very short primality proof. How about a much larger number? Here is a Pratt certificate for $10^{21} + 117$, the first prime number larger than one sextillion. For simplicity, we omit proofs of the primality of any number less than 30.

1000000000000000000117 is prime:

   Take $a = 2$. Use $1000000000000000000116 = 2^2 \cdot 3^2 \cdot 2530907 \cdot 3054967 \cdot 3592649$.

      2530907 is prime: Take $a = 2$. Use $2530906 = 2 \cdot 7 \cdot 180779$.

         180779 is prime: Take $a = 6$. Use $180778 = 2 \cdot 13 \cdot 17 \cdot 409$.

            409 is prime: Take $a = 21$. Use $408 = 2^3 \cdot 3 \cdot 17$.

      3054967 is prime: Take $a = 3$. Use $3054966 = 2 \cdot 3 \cdot 227 \cdot 2243$.

         227 is prime: Take $a = 2$. Use $226 = 2 \cdot 113$.

            113 is prime: Take $a = 3$. Use $112 = 2^4 \cdot 7$.

         2243 is prime: Take $a = 2$. Use $2242 = 2 \cdot 19 \cdot 59$.

            59 is prime: Take $a = 3$. Use $58 = 2 \cdot 29$.

      3592649 is prime: Take $a = 3$. Use $3592648 = 2^3 \cdot 239 \cdot 1879$.

         239 is prime: Take $a = 7$. Use $238 = 2 \cdot 7 \cdot 17$.

         1879 is prime: Take $a = 6$. Use $1878 = 2 \cdot 3 \cdot 313$.

            313 is prime: Take $a = 10$. Use $312 = 2^3 \cdot 3 \cdot 13$.

To verify the correctness of this 13-line proof, you only have to do 52 modular exponentiations. This calculation is a snap on a computer, and it compares quite favorably with doing the more than 30 billion trial divisions that would be necessary using the more naive method of trial division up to the square root of this number.

Thus, we find that even for huge primes, Pratt certificates are relatively short. Intuitively, the reason for this is not hard to see: When creating a Pratt certificate, the primes we need to consider decrease in a hurry. Indeed, if p is a prime, then the primes we need to consider at the next step are the prime factors of $p - 1$. But any such factor is less than half of p. This leads to an exponential decrease in the size of these factors, and it follows that the number of steps in a Pratt certificate for any prime p is at most $\log_2 p$, which is much smaller than p (see Exercises 11–13). Even for a whopping 300-digit prime, a Pratt certificate will occupy fewer than $\log_2(10^{300}) \approx 997$ lines, and it can be verified by a computer in seconds.

So verifying an existing Pratt certificate is a cakewalk. On the other hand, creating a new Pratt certificate can be a hairier problem.

EXAMPLE 2

Using the Miller-Rabin Test (see Method 12.3.4), you determine that the 301-digit number

$p = 11131348583112151972871798459201603496021853220673788011059944830$
$56000976186399427145481542842458307548523863746792360943169879666$
$01349788303684644118714619989136450197534310862153340987566152176$
$90189022025710041940459794869802586843476309597353583676184464711$
$79705367894522743146145368168123883321027$

is almost certainly prime. Can you find a Pratt certificate for the primality of p?

SOLUTION To find a Pratt certificate for p, you need to find the prime factorization of $p - 1$, as well as find an element of $\mathbf{Z}_p$ that has order $p - 1$ (if you've read Section 10.3, recall that this is called a *primitive root* modulo p). Since both of these problems are computationally infeasible, you cannot find a Pratt certificate for the primality of p during your lifetime.[2] (Unless you have help from a Mathemagician with an orb of unlimited computational power, a quantum computer, or an *extremely* long life.) ●

We conclude with

OBSERVATION 12.6.4

If n is a prime number, then a Pratt certificate for n can be used to give a short proof that n is prime. (That is, there is a proof that can be checked in polynomial time.) This proof may be hard to find, but it exists.

P versus NP

As discussed in Section 4.2, an algorithm is considered feasible if it runs in polynomial time. Computer scientists have given the name **P** to the class of all problems that can be solved by a polynomial-time algorithm. For example, computing gcd's is in **P**, since the Euclidean algorithm runs in polynomial time.[3] If your life depends on quickly finding the gcd of two large numbers, there's no need to start quaking in your boots—just calmly apply the Euclidean algorithm. The problem of factoring large numbers is a different story: there is no known polynomial-time algorithm for solving this problem.

Table 12.6.1 contains a summary of some of the topics we've studied to date, divided according to whether each is in **P** or suspected not to be in **P**. Note that it is only suspected that the problems in the right column of Table 12.6.1 are not in **P**. It has not been proved that any of these problems is not in **P**. More on this shortly.

We have just learned that if a number is prime, then there is a short proof (called a Pratt certificate) that the number is prime. Many problems in mathematics and computer science fall into this category, which computer scientists have given the name **NP** (non-deterministic polynomial time). The class **NP** consists of those problems whose solution can be *verified* in polynomial time. For a problem to be in **NP**, it is not required that *finding* a solution be achievable in polynomial time, only that the solution can be checked in polynomial time. (Roughly, this is what is meant by *non-deterministic*: the algorithm to find the certificate does not have to be specified.)

---

[2]Only the authors of this book know a Pratt certificate for this number p. How did we create such a large prime together with a Pratt certificate for p? See Exercise 10.

[3]Complexity classes like **P** are usually considered to contain only decision problems—that is, problems with a yes or no answer (rather than a numerical answer.) We ignore this technicality for our informal discussion here.

Table 12.6.1 Problems discussed in various chapters of this book, divided according to whether each is in P or suspected not to be in P.

Known to be in P	Suspected not to be in P
Basic arithmetic operations	
Division with remainder **[Ch. 3]**	
Finding the gcd (Euclidean Algorithm) **[Ch. 4]**	
Solving linear Diophantine equations **[Ch. 5]**	
	Factoring into primes **[Ch. 6]**
Modular arithmetic operations, including modular exponentiation **[Ch. 7–9]**	
	Finding the order of a number modulo n. **[Ch. 10]**
	Finding a primitive root modulo p **[Ch. 10]**
	Calculating discrete logarithms **[Ch. 10]**
Computing the Legendre symbol $\left(\frac{a}{p}\right)$ **[Ch. 11]**	
Determining whether a number is prime **[Ch. 12]**	

Thus, our discussion of Pratt certificates (Observation 12.6.4) can be summed up by saying that primality testing is in **NP**. As another example, the problem of factoring an integer is in **NP** since it is easy to verify that a given number is a factor—just do the division. Never mind that finding factors, or finding Pratt certificates, may require almost magical computational powers. Once found, their correctness can be verified in polynomial time.

The reader may be getting the idea that the class **NP** contains rather a lot of problems. In fact, all of the problems in both columns of Table 12.6.1, and many more, belong to **NP**. From the definition of these complexity classes, it follows directly that every problem in **P** belongs also to **NP**. Indeed, if you have a fast algorithm that provably finds the solution, then this same algorithm also verifies solutions quickly (without resorting to any non-deterministic magic). Thus,

$$\mathbf{P} \subseteq \mathbf{NP}.$$

Are there any problems that are in **NP** but not in **P**? A good candidate would be the factoring problem, which, as we have seen, is in **NP**. But is factoring in **P**? Nobody knows.

In fact, it is not known whether there are *any problems at all* that are in **NP** but not in **P**. The million-dollar question is

$$\text{Does } \mathbf{P} = \mathbf{NP}?$$

This is one of the greatest unanswered questions in mathematics—solving it would represent a major breakthrough in computer science.

Fun Facts

Like the Riemann hypothesis (which we mentioned in Section 12.3), solving the $\mathbf{P} = \mathbf{NP}$ question would make you not only famous but also rich. The Clay Mathematics Institute is offering a \$1 million prize for its solution.

Since many problems, such as factoring, are in **NP** but do not seem to be in **P**, it would appear that the answer is no, $\mathbf{P} \neq \mathbf{NP}$. However, if it turned out to everyone's surprise that $\mathbf{P} = \mathbf{NP}$, then we would know that factoring is in **P**. This would compromise the RSA cryptosystem (see Section 9.5), whose security depends on the intractability of factoring. In fact, if it were the case that $\mathbf{P} = \mathbf{NP}$, the consequences for cryptography would be even more dire. Intuitively, $\mathbf{P} = \mathbf{NP}$ says that anything that can be verified quickly can be computed quickly, and this would apply to the computation of decryption keys in any public key system. Thus, if it were the case that $\mathbf{P} = \mathbf{NP}$, there would be a theoretical obstacle to creating the kind of trapdoor necessary for *any* effective public key cryptosystem.

Fortunately for cryptographers, everyone suspects that $\mathbf{P} \neq \mathbf{NP}$. Proving this would establish in a formal sense that there are mathematical problems that are hard to solve, but whose solution, once found, can be verified quickly. Somehow, this intuitive idea seems inherent to the very nature of problem solving. Of course finding solutions is often much harder than verifying them! When working on a mathematical problem, have you ever had the feeling that a solution was hard to find, but then with the right additional information, the problem became easy? When asked to prove a mathematical statement, you may have to rack your brain and do lots of scratch work before you find a proof. But once found, a proof (especially a short, elegant proof like those in Erdős's fabled Book) may be easily verified to be correct. Creating proofs seems inherently harder than checking them. The Book takes ages to write, but not nearly so long to read.

EXERCISES 12.6

Numerical Problems

1. Give a Pratt certificate to prove that 11 is prime.

2. Give a Pratt certificate to prove that 43 is prime. [*Hint:* Your magical orb tells you that $\mathrm{ord}_{43}(3) = 42$. So for the first step of your Pratt certificate, take $a = 3$.]

3. Give a Pratt certificate to prove that 71 is prime. [*Hint:* Your magical orb tells you that $\text{ord}_{71}(7) = 70$. So for the first step of your Pratt certificate, take $a = 7$.]

4. **Computer Exercise** Let $n = 76151417$, and let $a = 3$. Demonstrate that conditions (*i*) and (*ii*) of Theorem 12.6.3 are both satisfied. In your demonstration of condition (*ii*), you may assume that the prime factorization of $n - 1$ that we give in the proof of Theorem C is correct.

5. **Computer Exercise** Give a Pratt certificate to prove that 52739501 is prime.

6. **Computer Exercise** Sir Perceval's beverage report can be reached at (866)GRAY-ALE. Prove that this phone number, 8664729253, is prime.

7. For each line of the proof that 239 is prime from Example 1, demonstrate that conditions (*i*) and (*ii*) of Theorem 12.6.3 are both satisfied.

Reasoning and Proofs

8. Let $n \in \mathbf{N}$, and let $a \in \mathbf{Z}$ such that condition (*i*) of Theorem 12.6.3 holds. Prove that a is a reduced residue modulo n.

9. Let $n \in \mathbf{N}$ be odd, and let $a \in \mathbf{Z}$.

   a. Show that if $a^{\frac{n-1}{2}} \equiv -1 \pmod{n}$ and $a^{\frac{n-1}{q}} \not\equiv 1 \pmod{n}$ for every odd prime divisor q of $n - 1$, then n is prime.

   b. Is the converse of your result from part a true? Explain. (If you studied *primitive roots* in Section 10.3, you may be able to modify the converse to make it a true statement.)

10. This exercise relies on Section 10.3, which discusses *primitive roots*. Suppose you have two 50-digit primes, q and r, together with Pratt certificates proving that q and r are prime. We now describe a method for getting a 100-digit prime, p, together with a Pratt certificate for the primality of p. First calculate $2qr + 1, 4qr + 1, 6qr + 1, \ldots$, and test each one using the Miller-Rabin Test (12.3.4) until you find a number (of the form $2kqr + 1$) that is prime.

    a. Explain why you do not bother testing any of the numbers $qr + 1, 3qr + 1, 5qr + 1, \ldots$ for primality.

    b. Use the Prime Number Theorem (3.3.3) to approximate how many values of k you will need to try until you find a number $2kqr + 1$ that is prime.

    c. Let $p = 2kqr + 1$ be the first prime you find. Note that p will have around 100 digits. Using Theorem 10.3.7, estimate the probability that a randomly chosen number a in the range $1, \ldots, p - 1$ will be a primitive root. (Give a formula in terms of k, but without p's and q's.)

    d. Explain how within a reasonable period of time, your computer can produce a Pratt certificate for p.

    e. Explain how to bootstrap this procedure to produce huge primes together with Pratt certificates for those primes.

In this exploration, you will prove a result about the length of the Pratt certificate for any prime p. For our purposes here, we will define the length of a certificate to be the number of lines in the certificate—that is, the total number of primes greater than 2 that are considered in the certificate. For example, the certificate in Example 1 has length 4. We use the notation $L(p)$ to denote the length of the certificate for p. Thus, looking back at the Pratt certificates from this section, we see that $L(239) = 4$, and $L(18885551417) = 14$. We define $L(2) = 0$, since in our certificates we do not verify the primality of 2.

11. Find $L(5)$, $\mathrm{L}(7)$, and $L(71)$.

12. Suppose that $p > 2$ is prime, and let $p - 1 = 2^k q_1{}^{a_1} q_2{}^{a_2} \cdots p_s{}^{a_s}$ be the prime factorization of $p - 1$. Explain why $L(p) = 1 + L(q_1) + L(q_2) + \cdots + L(q_s)$.

13. Use your result from Exercise 12 to prove that for any prime p, $L(p) < \log_2 p$. [*Hint:* Use strong induction.]

Advanced Reasoning and Proofs

14. **Pocklington's Theorem** Theorem 12.6.3 is useful for proving that a number n is prime, provided that you know the prime factorization of $n - 1$. Sometimes you might not know the complete factorization of $n - 1$, but only a partial factorization: That is, $n - 1 = f \cdot r$, where you know the prime factors of f but not the prime factors of r. In this case, if f is large enough, you can still prove that n is prime using the following theorem, which you will prove in this exercise.

> **THEOREM.** *Let $n \in \mathbf{N}$ such that $n - 1 = fr$, where $f, r \in \mathbf{N}$ and $f \geq \sqrt{n}$. Suppose there exists an integer a such that the following two conditions hold:*
>
> (*i*) $a^{n-1} \equiv 1 \pmod{n}$
>
> (*ii*) *for every prime q such that $q \mid f$, we have $\gcd(a^{\frac{n-1}{q}} - 1, n) = 1$.*
>
> *Then n is prime.*

    a. Let p be any prime factor of n. Prove that $p \equiv 1 \pmod{f}$.
    [*Hint:* Show that $\mathrm{ord}_p(a^r) = f$.]

    b. Use your result from part a to prove that n is prime.

The nth Fermat number is given by $F_n = 2^{2^n} + 1$. As discussed in Section 3.3, Fermat conjectured that for every $n \in \mathbf{N}$, F_n is prime. In 1732, Euler proved that Fermat's conjecture was incorrect by showing that F_5 is composite. In fact, $F_0, F_1, \ldots, F_4$ are the only Fermat numbers that are known to be prime. Many Fermat numbers are known to be composite, but it is still unknown whether there are infinitely many Fermat numbers that are prime, or even whether there are infinitely many Fermat numbers that are composite.

In this Exploration, you will prove the following criterion for the primality of F_n:

THEOREM. *For any $n \in \mathbf{N}$, F_n is prime if and only if $3^{\frac{F_n-1}{2}} \equiv -1 \pmod{F_n}$.*

15. In this exercise, you will prove the [$\Leftarrow$] direction of the Theorem. Assume that $3^{\frac{F_n-1}{2}} \equiv -1 \pmod{F_n}$. Use Theorem 12.6.3 to prove that F_n is prime.

16. In this exercise, you will prove the [$\Rightarrow$] direction of the Theorem.

   a. Assume that F_n is prime. Determine the value of F_n modulo 12.

   b. Use your result from part a to evaluate the Legendre symbol $\left(\dfrac{3}{F_n}\right)$.

   c. Conclude that $3^{\frac{F_n-1}{2}} \equiv -1 \pmod{F_n}$.

17. Now that you have established the Theorem, prove that it still holds when 3 is replaced by 5. That is, prove that for any $n \in \mathbf{N}$, F_n is prime if and only if $5^{\frac{F_n-1}{2}} \equiv -1 \pmod{F_n}$.

12.7 The AKS Deterministic Primality Test

The return of the king

When we last left King Arthur's court, the great mathemagician Erdős had just succeeded in finding a short proof, a Pratt certificate, of the primality of Queen Guinevere's favorite number, 18885551417. The queen couldn't stop talking about how pleased she was with her anniversary gift. Soon, all of the knights of the Kingdom wanted their own proofs of primality to present as gifts to their loved ones.

As a result, Erdős was deluged with requests for short primality proofs. Since he was the only one with the orb, he was the only one who could factor large numbers quickly and find primitive roots modulo large primes. As such, no one else had the power to quickly conjure up a proof of the primality of an arbitrary prime number.

At the next Round Table discussion, King Arthur was overwhelmed by grievances about his wizard. The knights complained that whenever they needed a primality proof and went looking for the Mathemagician, Erdős was never at home.

"I know. I know," apologized the king. "It seems that our Erdős has become obsessed with searching for a mythical Book. His quest has led him to the future, where he wanders the globe in an effort to gather any information he can about The Book.

"His single-mindedness reminds me of Sir Perceval and his obsession with fermented beverages—Percy would walk a thousand miles to find the perfect cup of gray ale. But I digress. It's simply unacceptable for Erdős to be gallivanting around when we're in need of primality proofs!"

"Indeed," boomed a voice from within a dense cloud of smoke that had suddenly materialized in the center of the Round Table. The smoke cleared to reveal the robed figure of Erdős hovering a few inches above the table's surface. "While roaming around the future in search of The Book, I found myself in far India, floating down the Ganges River. In Kanpur, I met three pleasant young mathematicians named Agrawal, Kayal, and Saxena. As luck would have it, they had just discovered a purely deterministic primality test and proved that their test is efficient! Give their AKS test any number, and it will tell you whether the number is prime or composite with 100% accuracy, guaranteed!"

The king asked, "But won't we nonmagical mortals still need you and your orb to come up with *proofs* of primality?"

"No sir," Erdős beamed. "If you want to know whether a number is prime, just run the AKS test on your number. If the test says the number is prime, you have the proof you need!"

"Huzzah!" all the knights cheered.

"And now if you don't mind," said Erdős, "it's time I returned to my search for The Book. Back to the future!" As he finished saying this, Erdős took a glowing gold ball from his pocket, popped it into his mouth, and vanished.

King Arthur and his knights would never see the great wizard again. Shortly after his return to the future, Erdős's time machine broke down in 20th-century Hungary, and he was never able to return home. Still, Erdős managed to make the best of the situation, spending many years spreading his own brand of mathematical wizardry throughout the world.

The AKS algorithm

August 2002 brought big news. Mathematicians and computer scientists worldwide were excited by the discovery of a new primality test discovered by Manindra Agrawal, Neeraj Kayal, and Nitin Saxena of the Indian Institute of Technology in Kanpur. Named for its three discoverers, the AKS Primality Test is the first primality test guaranteed to determine efficiently (in polynomial time) whether a given integer is prime or composite. It may surprise you to learn that Kayal and Saxena were undergraduate students, who completed their undergraduate thesis in 2002 under Agrawal's direction. Working over the summer, they were able to develop results from their thesis to produce their groundbreaking result.

Though we will not give a proof that the AKS test works, we will discuss several of the key ideas involved and give a statement of the test.

One of the main ideas behind the AKS test is the binomial theorem for expanding $(x + y)^n$. The coefficients appearing in this expansion come from Pascal's triangle:

				1					← row 0
			1		1				← row 1
		1		2		1			← row 2
	1		3		3		1		← row 3
1		4		6		4		1	← row 4

1	5	10	10	5	1	← row 5
1	6	15	20	15	6	1	← row 6
1	7	21	35	35	21	7	1	← row 7

$\vdots$ $\vdots$ $\vdots$ $\vdots$ $\vdots$ $\vdots$ $\vdots$ $\vdots$ $\vdots$

Fun Facts

The mathematical object that we call *Pascal's triangle* was actually discovered many centuries before Pascal. The triangle was first referred to in 200 BCE by Indian mathematician Halayuda, in a book about poetic rhythm and meter. Arabic mathematician Al-Karaji independently discovered the triangle around the year 1000. In Iran, it is referred to as *Khayyam's triangle*, since it was discovered by Persian astronomer and poet Omar Khayyam around 1100. In China, the triangle was discovered by mathematician Chia Hsian around 1050, but it is known as *Yanghui's triangle*, after the mathematician who first produced an image of the triangle in 1261.

In the 17th century, Mersenne described the triangle to French mathematician Blaise Pascal, who then used it in his development of the theory of probability. As a result, French mathematicians began referring to it as "le triangle de Pascal" (*Pascal's triangle*) in the early 18th century.

There is a difference between prime rows and composite rows of Pascal's triangle. In row 7, except for the 1's at the two ends of the row, every number is divisible by 7. In row 6, though, not every number is divisible by 6. This holds in general: in Exercises 8–9, you will show that an integer $n > 1$ is prime if and only if all of the numbers in row n of Pascal's triangle (except the 1's at the ends of the row) are divisible by n.

Naomi's Numerical Proof Preview: Theorem 12.7.1

This observation has an interesting consequence about polynomials over $\mathbf{Z}_n$. For example, the binomial theorem tells us that if we raise $(x + 1)$ to the 5th power, we get a polynomial whose coefficients come from row 5 of Pascal's triangle:

$$(x + 1)^5 = x^5 + 5x^4 + 10x^3 + 10x^2 + 5x + 1. \tag{1}$$

This equation is an identity of polynomials with real coefficients. In Section 10.2, we discussed polynomials with coefficients in $\mathbf{Z}_n$. Let's work in $\mathbf{Z}_5$, and consider $\overline{x} + \overline{1}$, a polynomial over $\mathbf{Z}_5$ in the variable $\overline{x}$. What happens if we raise this polynomial to the 5th power? Just as in equation (1), we find

$$(\overline{x} + \overline{1})^5 = \overline{x}^5 + \overline{5}\,\overline{x}^4 + \overline{10}\,\overline{x}^3 + \overline{10}\,\overline{x}^2 + \overline{5}\,\overline{x} + \overline{1}. \tag{2}$$

But since we are working in $\mathbf{Z}_5$, $\overline{5} = \overline{10} = \overline{0}$, so this equation reduces to

$$(\overline{x} + \overline{1})^5 = \overline{x}^5 + \overline{1},$$

a quite pleasing identity indeed.

Still working over $\mathbf{Z}_5$, let's try expanding $(\overline{x} + \overline{2})^5$. Here, we find

$$(\overline{x} + \overline{2})^5 = \overline{x}^5 + \overline{5}\,\overline{x}^4 \cdot \overline{2} + \overline{10}\,\overline{x}^3 \cdot \overline{2}^2 + \overline{10}\,\overline{x}^2 \cdot \overline{2}^3 + \overline{5}\,\overline{x} \cdot \overline{2}^4 + \overline{2}^5.$$

Again, all terms except the first and the last go away. We also note that by Fermat's Little Theorem (9.1.1), $\overline{2}^5 = \overline{2}$. Thus, we obtain

$$(\overline{x} + \overline{2})^5 = \overline{x}^5 + \overline{2}.$$

In general, we have the following theorem.

THEOREM 12.7.1

Let $n > 1$ be a natural number, and let a be any integer with $\gcd(a, n) = 1$. Then n is prime if and only if

$$(\overline{x} + \overline{a})^n = \overline{x}^n + \overline{a}$$

as polynomials over $\mathbf{Z}_n$.

PROOF See Exercises 8–10.

EXAMPLE 1

Use Theorem 12.7.1 to determine whether the number $n = 9167$ is prime.

SOLUTION For simplicity, we choose $a = 1$. To apply Theorem 12.7.1, we must compute the following polynomial over $\mathbf{Z}_{9167}$:

$$(\bar{x} + \bar{1})^{9167} = \bar{x}^{9167} + \overline{9167}\bar{x}^{9166} + \overline{42012361}\bar{x}^{9165} + \cdots + \overline{9167}\bar{x} + \bar{1}.$$

When we reduce the coefficients in this equation modulo 9167, many of them are equal to $\bar{0}$. After reducing the coefficients, we get

$$(\bar{x} + \bar{1})^{9167} = \bar{x}^{9167} + \overline{103}\bar{x}^{9078} + \overline{89}\bar{x}^{9064} + \overline{5253}\bar{x}^{8989} + \cdots + \overline{103}\bar{x}^{89} + \bar{1}. \quad (3)$$

Now we compare this polynomial with $\bar{x}^{9167} + \bar{1}$, and we see that they are not equal. Therefore, by Theorem 12.7.1, the number 9167 is not prime. ◼

How much time will it take to compute the right hand side of equation (3)? Since computations are done modulo 9167, the coefficients do not get very large. But there are lots of coefficients to keep track of! The polynomial $(\bar{x} + \bar{1})^{9167}$ has degree 9167, so there are thousands of coefficients. In general, the polynomial $(\bar{x} + \bar{1})^n$ has degree n, so even if n has only 30 digits or so, it would take eons to compute this polynomial, not to mention the massive amount of computer memory we would need to store all of these coefficients!

A clever shortcut

AKS to the rescue! Here's the idea that they got to work: Though it may be hard to expand $(\bar{x} + \bar{1})^{9167}$ as a polynomial over $\mathbf{Z}_{9167}$, it's easy to compute the *remainder* when this polynomial is divided by a polynomial $b(\bar{x})$ of small degree. For their algorithm, AKS chose polynomials $b(x)$ of the form $\bar{x}^r - \bar{1}$. For example, let $b(\bar{x}) = \bar{x}^3 - \bar{1}$. Then while $(\bar{x} + \bar{1})^{9167}$ has the beastly expression given in equation (3), the remainder when $(\bar{x} + \bar{1})^{9167}$ is divided by the polynomial $\bar{x}^3 - \bar{1}$ will be a polynomial of degree at most 2. We find

$$(\bar{x} + \bar{1})^{9167} \text{ leaves a remainder of } \overline{8520}\,\bar{x}^2 + \overline{8519}\,\bar{x} + \overline{8520} \quad (4)$$

when divided by $\bar{x}^3 - \bar{1}$. On the other hand,

$$\bar{x}^{9167} + \bar{1} \text{ leaves a remainder of } \bar{x}^2 + \bar{1} \quad (5)$$

when divided by $\bar{x}^3 - \bar{1}$.

From statements (4) and (5), we deduce that $(\bar{x} + \bar{1})^{9167} \neq \bar{x}^{9167} + \bar{1}$ and hence conclude that 9167 must be composite (by Theorem 12.7.1). Our new argument has a big advantage over the proof in Example 1: Unlike the massive polynomial in (3), the compact remainders in (4) and (5) can be computed efficiently.

This suggests the following method for testing whether a given number n is prime or composite. In some manner, choose an integer a, and a reasonably small value of r. Working with polynomials over $\mathbf{Z}_n$, test whether

$$(\bar{x} + \bar{a})^n \text{ and } \bar{x}^n + \bar{a} \text{ leave the same remainder when divided by } \bar{x}^r - \bar{1}. \quad (6)$$

If statement (6) fails, we conclude that n is composite. If statement (6) holds, the test is inconclusive.

So far, this may remind you of a probabilistic primality test. Perhaps if we were to repeat enough times, with different values of a and r, we could eventually conclude that n is prime with very high probability.

In fact, however, the AKS test is not just probabilistic, but deterministic! The real achievement of Agrawal, Kayal, and Saxena was to prove that if statement (6) holds for certain particular values of a and r, then n is *certainly* prime. We state their result without proof.

It is standard to use the notation

$$(x + a)^n \equiv x^n + a \pmod{n, x^r - 1}$$

to indicate that statement (6) holds in $\mathbf{Z}_n$. This is the notation we will use in our statement of the AKS test.

THEOREM 12.7.2

Let $n > 1$ be a natural number. Suppose that $r \in \mathbf{N}$ such that $\text{ord}_r(n) > (\log_2 n)^2$. Suppose that n is not a perfect power (i.e., is not expressible as a^b, for some $a, b \in \mathbf{N}$ where $b \geq 2$), and that n has no prime factor less than or equal to r. Then n is prime if and only if

$$(x + a)^n \equiv x^n + a \pmod{n, x^r - 1} \quad (7)$$

for all natural numbers $a \leq \sqrt{r} \cdot \log_2 n$.

This theorem can easily be turned into a primality test, as follows.

METHOD 12.7.3   AKS PRIMALITY TEST

Given a natural number n,

    Step 1.   Determine whether n is a perfect power. If so, output **composite** and stop.

    Step 2.   Determine a number r such that $\text{ord}_r(n) > (\log_2 n)^2$.

    Step 3.   Check to see whether n has a prime factor $\leq r$. If so, output **composite** and stop.

    Step 4.   Check whether congruence (7) holds for all natural numbers $a \leq \sqrt{r} \cdot \log_2 n$. If so, output **prime**. Otherwise output **composite**.

What makes this test so great? First,

- The check in Step 1 can be done quickly.

However, an important question arises: To find a value of r that satisfies the condition in Step 2, how many candidate values for r need to be checked? It turns out that it is always possible to find a value of r that satisfies this condition and is approximately the size of $(\log_2 n)^3$. Since $\log_2 n$ is tiny compared to n, the value of r is relatively small. This is fantastic news with several consequences:

- In Step 2, the search for r will not last very long.

- In Step 3, there are not very many trial divisions to be done.

- In Step 4, the number of values of a that must be checked is fairly small.

- Also in Step 4, the degree of $x^r - 1$ is relatively small. Thus, it will be fairly easy to compute the remainders necessary to check congruence (7).

In fact, a careful analysis reveals that this algorithm takes fewer than about $(\log_2 n)^{7.5}$ steps. Thus, AKS is a polynomial-time algorithm.

EXERCISES 12.7

Numerical Problems

In Exercises 1–6, for the given values of n, a, and r:

    a. Find the remainder when $(\bar{x} + \bar{a})^n$ is divided by $\bar{x}^r - \bar{1}$.

    b. Find the remainder when $\bar{x}^n + \bar{a}$ is divided by $\bar{x}^r - \bar{1}$.

    c. Determine whether congruence (7) holds.

    d. Based on your answers to parts a–c, which of the following can you conclude: that n is composite, that n is prime, or that n might be either prime or composite?

1. $n = 4$, $a = 2$, $r = 2$ 2. $n = 4$, $a = 1$, $r = 3$

3. $n = 5$, $a = 1$, $r = 3$ 4. $n = 6$, $a = 1$, $r = 1$

5. **Computer Exercise** $n = 11$, $a = 2$, $r = 2$

6. **Computer Exercise** $n = 12$, $a = 2$, $r = 3$

Reasoning and Proofs

7. Suppose you wish to run the AKS Test (Method 12.7.3) on the number $n = 5964481352575809641534947$ (which is approximately $6 \cdot 10^{24}$).

a. In Step 2 of the AKS Test, it is always possible to find a value of r that is no more than about $(\log_2 n)^3$. Based on this assertion, approximately what is the largest possible value of r?

b. When searching for r, you find that $r = 6779$ satisfies the conditions of Step 2 of the AKS Test. Determine how many values of a will have to be checked in Step 4.

c. In Step 4, congruence (7) will have to be checked. This involves checking the equality of certain polynomials. What is the maximum degree of these polynomials?

EXPLORATION Proof of Theorem 12.7.1 (Exercises 8–10)

In this Exploration, you will prove Theorem 12.7.1, which provides a key motivation for the AKS Primality Test. You will need to use the binomial theorem, including the formula for the binomial coefficients, $\binom{n}{k} = \dfrac{n!}{k!(n-k)!}$.

8. Let p be prime. Show that for any $k = 1, 2, \ldots, p - 1$, we have $p \mid \binom{p}{k}$.

9. Now prove the converse of the result from Exercise 8. That is, assume that $n > 1$, and suppose that $n \mid \binom{n}{k}$ for all $k = 1, 2, \ldots, n - 1$. Show that n is prime. [*Hint:* Assume that n is composite, and let q be any prime factor of n. Show that $n \nmid \binom{n}{q}$.]

10. Use your results from Exercises 8 and 9 to prove Theorem 12.7.1.

Chapter 13

Leonhard Euler (1707–1783)

No other mathematician has been as productive as the Swiss mathematician Leonhard Euler (pronounced "oiler"). As a student, Euler's father Paul had attended lectures by the mathematician Jakob Bernoulli and found them quite interesting. Paul encouraged his son's youthful enthusiasm for mathematics, though he assumed that Euler would follow him into the ministry. Since mathematics was not part of the curriculum at Euler's school, Paul hired a private tutor to give Euler math lessons. At the age of 13, Euler became a student at the University of Basel, where he studied theology, mathematics, Greek, Hebrew, medicine, astronomy, and physics. The university had fewer than 150 students and only 19 professors. Euler wrote in his autobiography about his experiences learning math from one of these professors: Johann Bernoulli, brother of the late Jakob Bernoulli:

> I soon found an opportunity to be introduced to a famous professor, Johann Bernoulli. . . . True, he was very busy and so refused flatly to give me private lessons; but he gave me much more valuable advice to start reading more difficult mathematics books on my own and to study them as diligently as I could; if I came across some obstacle or difficulty, I was given permission to visit him freely every Saturday afternoon and he kindly explained to me everything I could not understand . . . and

this, undoubtedly, is the best method to succeed in mathematical subjects.

While Euler was a student, he became friends with Johann Bernoulli's two sons, Nicolaus and Daniel, who both later became mathematicians. Through his interactions with Johann Bernoulli and his sons, Euler developed an even deeper interest in mathematics. After earning his degree from the University of Basel at the age of 15, Euler followed his father's wishes and began to study theology. Despite being quite religious, however, Euler did not do well in his study of theology because his heart was in mathematics. Fortunately for Euler, as well as for the subsequent development of mathematics, Johann Bernoulli eventually convinced Euler's father that Euler's true calling was mathematics rather than the ministry. With his father's approval, Euler quit divinity school and devoted himself entirely to mathematics.

Euler published his first research paper in mathematics when he was 18. The following year, the Paris Academy offered a prize for the best paper describing an efficient arrangement of ship masts. Euler submitted an original paper on the topic and won an honorable mention. This was impressive, not only because Euler was only 19, but also because he had lived his entire life in a country that had no access to the sea.

There were not many opportunities for mathematicians in Switzerland. In 1725, Euler's friends Nicolaus and Daniel Bernoulli left Switzerland to become mathematics professors at the St. Petersburg Academy in Russia, a prestigious institute whose faculty were paid good salaries to do scientific and mathematical research. Once they had established themselves as professors at the academy, the Bernoulli brothers convinced the administration to offer Euler a position that had become available in the Physiology Department. Before leaving for St. Petersburg, Euler remained at the University of Basel for several months to learn about physiology and its relationship to mathematics. Shortly after his arrival, Euler became an adjunct member of the Mathematics Department, and within a few years he became a professor of physics. In 1733, Euler attained the position of chief of mathematics at the St. Petersburg Academy.

Euler married in 1733 and had 13 children, only 5 of whom survived to adulthood. Although Euler spent a lot of time with his family, it did not interfere with his mathematical productivity. He had a phenomenal memory and is reputed to have done mathematics in his head while playing with his children. In describing Euler's abilities, the French mathematician François Arago wrote that he "calculated without apparent effort, as men breathe, or as eagles sustain themselves in the wind." Every time Euler completed a research paper, he would add it to the huge stack of research papers on his desk. Whenever the St. Petersburg Academy was about to print an issue of their transactions, the printer would gather some papers from the top of Euler's stack in order to

include them in the journal. As a result, many of Euler's papers were published in the opposite order from which they were written.

After an illness in 1738, Euler lost the sight in his right eye. The disease occurred just after he had spent three days computing a set of astronomical tables that people had thought would take him three months to complete. Some mathematicians have said that working too hard brought on Euler's illness, but there is no hard evidence of this. In any case, having sight in only one eye appears to have had no negative impact on Euler's mathematical research.

In 1741, King Frederick the Great offered Euler the position of director of mathematics at the Berlin Academy. The Russian political situation had become so oppressive at that time that Euler was glad to have the opportunity to leave St. Petersburg. When Euler moved to the Berlin Academy, however, he did not give up all of his administrative responsibilities at the St. Petersburg Academy. In fact, he worked for both academies simultaneously for the 25 years that he lived in Berlin. During these years, Euler was asked to assume more and more administrative responsibilities at the Berlin Academy. When the president of the Berlin Academy died in 1759, Euler took over this job and reported directly to King Frederick. Euler was not happy with the situation because he and the king had many serious intellectual and religious conflicts, and Euler felt that he was not treated as well as the other intellectuals in Frederick's court. The king did not respect pure mathematics, and he required Euler to work on a number of practical problems. For instance, he asked Euler to design the hydraulics for the fountains at the king's summer palace, Sans Souci. He also asked Euler to engineer a canal and to compute the probability that running a lottery would allow him to pay off his debts.

Because of Euler's unhappiness with his situation in Berlin, he was eager to return to the St. Petersburg Academy when he was invited by the new ruler of Russia, Catherine the Great, in 1766. Soon after his return, Euler lost most of the sight in his second eye due to a cataract, and after a failed eye operation he became completely blind. Even this blindness had no effect on the level of his mathematical productivity. His memory was sufficient that he could do all of his research in his head. He continued writing mathematics papers by dictating them to his sons, his collaborators, and even his servants. Euler proved new theorems right up until his death from a brain hemorrhage in 1783. Moments before he lost consciousness, Euler said, "I am dying." He never regained consciousness and died several hours later.

When Euler died, 50 pages of his obituary were devoted to listing his publications. During his lifetime, he published more than 700 books and articles, averaging 800 pages per year. In fact, St. Petersburg Academy did not finish publishing all of Euler's remaining papers until 47 years after his death!

A bit of Euler's math

Euler made important contributions to a wide variety of fields of mathematics including geometry, analysis, optics, differential equations, infinite series, complex numbers, graph theory, and number theory. In addition to his numerous theorems, Euler introduced mathematical notation that revolutionized the way mathematics was written and facilitated future mathematical development. In particular, Euler introduced the symbol e for the base of a natural logarithm, the symbol i for $\sqrt{-1}$, the symbol $f(x)$ for a function of the variable x, the symbol Σ for summation, and the terminology for the trigonometric functions that we still use today.

As we mentioned in the history of Fermat at the beginning of Chapter 9, although Fermat (who lived a century before Euler) is now considered the founder of modern number theory, Fermat was not well known in his own time since he published almost nothing and rarely wrote down proofs of his assertions. In contrast, Euler's work in number theory brought attention to Fermat's earlier results and increased interest in number theory in general. Euler proved many of Fermat's unproven assertions, as well as building upon them to obtain more general results. Some of Euler's many contributions to number theory appear throughout the second half of this textbook. They include the Euler φ-function that he used to generalize Fermat's Little Theorem (see Chapter 9) and the proof of Fermat's Last Theorem for $n = 3$ and $n = 4$ (see Chapter 15). While Fermat himself had proved Fermat's Last Theorem for $n = 4$, Euler proved it independently using a simpler approach. Euler also studied continued fractions (discussed in Chapter 14), which he used to find the smallest integer solutions to equations of the form $x^2 - dy^2 = 1$, where d is a natural number that is not a perfect square. (This equation, known as Pell's Equation, is studied in Chapter 15.) Above all, Euler discovered the Law of Quadratic Reciprocity (see Chapter 11). Although Euler did not produce a proof of Quadratic Reciprocity (it was proved later by Gauss), many mathematicians believe that the discovery of the Law of Quadratic Reciprocity was Euler's most significant contribution.

Chapter 13 Gaussian Integers

13.1 Definition of the Gaussian Integers

Recall that the set of complex numbers is given by

$$\mathbf{C} = \{a + bi \mid a, b \in \mathbf{R}\},$$

where i is the square root of -1. The complex number $a + bi$ has a real part, a, and an imaginary part, b.

In this chapter, we will study a subset of the complex numbers known as the *Gaussian integers*, which consists of all complex numbers whose real and imaginary parts are integers.

DEFINITION 13.1.1

*The set of **Gaussian integers**, denoted $\mathbf{Z}[i]$, read "$\mathbf{Z}$ adjoined i," is defined by*

$$\mathbf{Z}[i] = \{a + bi \mid a, b \in \mathbf{Z}\},$$

where i is the square root of -1.

Like complex numbers, Gaussian integers may be added and multiplied:

EXAMPLE 1

$$(3 + 5i)(2 - 6i) = 6 - 18i + 10i - 30i^2$$
$$= 36 - 8i. \qquad \leftarrow \textbf{since } i^2 = -1.$$

More formally, the following definition summarizes how to compute the sum and product of any two Gaussian integers.

DEFINITION 13.1.2 ADDITION AND MULTIPLICATION IN THE GAUSSIAN INTEGERS

Let $a + bi$ and $c + di$ be Gaussian integers. Their sum and product are defined as follows:

$$(a + bi) + (c + di) = (a + c) + (b + d)i$$
$$(a + bi) \cdot (c + di) = (ac - bd) + (ad + bc)i.$$

From the formulas in this definition, we see immediately that the sum and product of any two Gaussian integers are themselves Gaussian integers. In other words, $\mathbf{Z}[i]$ is closed under addition and multiplication.

The Gaussian integers form a ring

Using the definition of addition and multiplication (13.1.2), it is not hard to prove that these operations satisfy many of the usual algebraic properties (see Exercises 14–15):

- Associativity of addition and multiplication

- Commutativity of addition and multiplication

- Existence of additive and multiplicative identities

- Existence of additive inverses

- Distributivity of addition over multiplication

Recall from Theorem 8.2.6 that this list of properties can be summarized succinctly by saying that $\mathbf{Z}[i]$ is a ring.

Every nonzero complex number z has a multiplicative inverse, $\frac{1}{z}$, that is also a complex number (see Exercises 5 and 10). For this reason, we say that the ring of the complex numbers, $\mathbf{C}$, is a *field*. In contrast, there are many Gaussian integers that do not have a multiplicative inverse in the Gaussian integers (see Exercises 12 and 13). Thus, the ring $\mathbf{Z}[i]$ is not a field. In Section 13.2, we will determine exactly which elements of $\mathbf{Z}[i]$ have a multiplicative inverse in $\mathbf{Z}[i]$. (These elements are called units—see Definition 13.2.3.)

A geometric view of the Gaussian integers

Just as the set $\mathbf{R}$ of real numbers can be represented by the number line, the set $\mathbf{C}$ of all complex numbers can be represented by a plane. In the complex plane, the complex number $a + bi$ is represented by the point (a, b) in Cartesian coordinates. Every complex number can also be represented by a vector in the complex plane. For example, the following diagram shows the point representing the complex number $5 - 3i$ and the vector representing $-2 + 4i$.

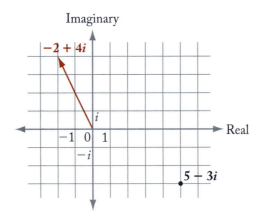

The *magnitude* of any complex number z, denoted $|z|$, is the distance from the point z to the origin. In other words, $|z|$ is the length of the vector that represents z.

We compute the **magnitude** of any complex number $a + bi$ using the following formula:

$$|a + bi| = \sqrt{a^2 + b^2}.$$

Graphically, the set of Gaussian integers $\mathbf{Z}[i]$ is represented by the set of all lattice points in the complex plane, as shown in Figure 1.

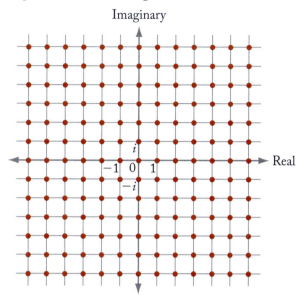

Imaginary

Real

Figure 1 The Gaussian integers, $\mathbf{Z}[i]$, are represented by lattice points in the complex plane.

The norm of a Gaussian integer

One important property of the ring $\mathbf{Z}$ is that we can compare its elements: given any two integers r and s, if they are not equal then either $r < s$ or $s < r$. (For more on the order properties of $\mathbf{Z}$, see the Appendix on the Student Companion Website.) With the Gaussian integers, the situation is not so simple. Consider the Gaussian integers $r = 4 + 7i$ and $s = 7 + 4i$. They are not equal, but it would be difficult to say that either one of them is less than the other.

Although we can't compare Gaussian integers directly, we can compare their magnitudes. For instance, $r = 4 + 7i$ and $s = 7 + 4i$ have the same magnitude: $|r| = |s| = \sqrt{4^2 + 7^2} = \sqrt{65}$. The magnitude of these Gaussian integers is $\sqrt{65}$, an irrational number, but the square of this magnitude is 65, an integer. Since integers are easier to work with, we will use the square of the magnitude to measure Gaussian integers. The square of the magnitude of a Gaussian integer is called its *norm*.

Let $z = a + bi$ be a Gaussian integer. The **norm** of z, denoted $N(z)$, is given by

$$N(z) = a^2 + b^2.$$

EXAMPLE 2

Find the norm of each Gaussian integer.

    a. $7 - 3i$ **b.** $2i$ **c.** 5

SOLUTION Use the definition of norm (13.1.3).

    a. $N(7 - 3i) = 7^2 + (-3)^2 = 58.$
    b. $N(0 + 2i) = 0^2 + 2^2 = 4.$
    c. $N(5 + 0i) = 5^2 + 0^2 = 25.$

EXAMPLE 3

Let $r = 4 - 2i$, $s = 3 + 2i$. Find $N(r), N(s),$ and $N(rs)$.

SOLUTION

$$N(r) = 4^2 + (-2)^2 = 20$$
$$N(s) = 3^2 + 2^2 = 13$$

Since $rs = (4 - 2i)(3 + 2i) = 16 + 2i$,

$$N(rs) = 16^2 + 2^2 = 260.$$

Visualizing sums and products of Gaussian integers

Graphically, the sum of two Gaussian integers (indeed, the sum of any two complex numbers) can be represented by a parallelogram. For example, the following equation on the left is represented by the diagram on the right:

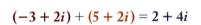

$(-3 + 2i) + (5 + 2i) = 2 + 4i$

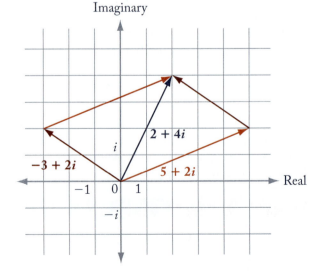

Multiplication of complex numbers also has a nice (and alliterative) geometric interpretation:

Angles Add and Magnitudes Multiply!

That is, if r and s are any two complex numbers, the product rs has an angle (measured counterclockwise from the real axis) that is the sum of the angle of r and the angle of s, and it has magnitude

$$| rs | = | r | | s |.$$

When r and s are Gaussian integers, this statement about magnitudes can also be expressed in terms of norms. You may have noticed that in Example 3, $N(rs) = N(r)N(s)$. This relationship holds not just for the particular r and s in Example 3, but for any Gaussian integers r and s.

LEMMA 13.1.4   MULTIPLICATIVITY OF THE NORM

Let r and s be Gaussian integers. Then

$$N(rs) = N(r)N(s).$$

Put it in Prose, Paul!

The norm of a product is the product of the norms.

PROOF In Exercises 7 and 8, you will prove this lemma in two different ways.

Using trigonometry, it is not hard to prove that angles add when Gaussian integers (or any complex numbers) are multiplied (see Exercise 16). Here, we will consider a few special cases of this fact that are easy to visualize.

When a Gaussian integer, z, is multiplied by a real number, c, the length of the vector z is multiplied by $| c |$. If c is positive, the direction of cz is the same as the direction of z. If c is negative, then the vector cz points in the opposite direction from z. For example, the diagram shows the points z, $2z$, $3z$, $-z$, and $-2z$, where $z = 2 + i$.

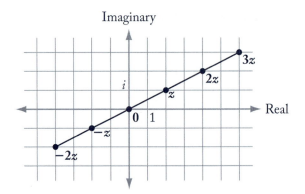

So multiplying a Gaussian integer by a real number is easy geometrically. What about multiplying by an imaginary number? For instance, what happens geometrically to a Gaussian integer when it is multiplied by i? The diagram below shows the Gaussian integer $x = 3 + 5i$, as well as the product $i \cdot x$, which is

$$i(3 + 5i) = -5 + 3i.$$

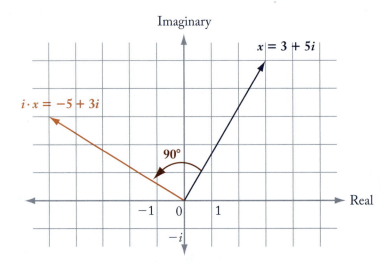

Multiplying the Gaussian integer x by i causes the vector x to be rotated 90° counterclockwise, without changing its length. This is true for any complex number (see Exercise 9).

We have just seen what happens geometrically when a Gaussian integer, z, is multiplied by a real number, and also what happens when z is multiplied by an imaginary number. What happens when a Gaussian integer, z, is multiplied by another Gaussian integer whose real and imaginary parts are both nonzero? For instance, let's examine geometrically the multiplication of a Gaussian integer z by the Gaussian integer $4 + 2i$. One way to do this is to find both $4z$ and $2iz$ in the complex plane, then add these together geometrically by drawing a parallelogram (in this case, a rectangle).

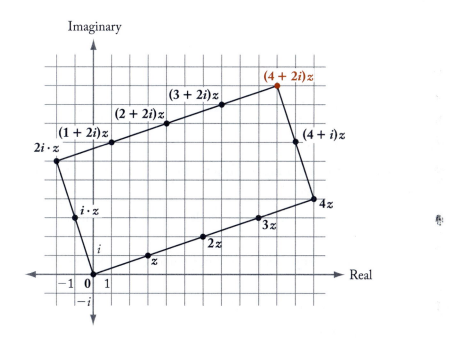

EXERCISES 13.1

Numerical Problems

1. Draw the vectors in the complex plane representing each Gaussian integer.

   a. $1 + i$ **b.** $-3 - 5i$ **c.** $5 - 2i$

2. Compute each Gaussian integer sum.

   a. $(8 + 2i) + (-3 + 2i)$ **b.** $(-1 + i) + (-2 - 4i)$

   c. $(10i) + (2 - i)$ **d.** $(a + bi) + (a - bi)$

3. Compute each Gaussian integer product.

   a. $(8 + 2i)(-3 + 2i)$ **b.** $(-1 + i)(-2 - 4i)$

   c. $(10i)(2 - i)$ **d.** $(a + bi)(a - bi)$

4. Compute the norm of each Gaussian integer.

   a. $1 - i$ **b.** $1 + 2i$ **c.** $-2 - 3i$ **d.** $10 - 3i$

5. **a.** Compute the multiplicative inverse of $z = 5 + 7i$ in the complex numbers, and express it in the form $a + bi$ (where a and b are real numbers).

   b. Show that $5 + 7i$ and your answer to part a are multiplicative inverses by demonstrating that their product equals 1.

6. Let $z = -5 - 3i$.

    a. Compute $i \cdot z$.

    b. Plot the vector z and the vector iz in the complex plane.

    c. How are the two vectors you plotted in part b related geometrically?

Reasoning and Proofs

7. Prove Lemma 13.1.4 using the formula for the product of Gaussian integers (from Definition 13.1.2).

8. For any complex number $z = a + bi$, let z^* denote its complex conjugate: $z^* = a - bi$.

    a. Prove that for any complex numbers r and s, the following identity holds: $(rs)^* = r^*s^*$.

    b. Show that for any $z \in \mathbf{Z}[i]$, $N(z) = z \cdot z^*$.

    c. Use your answers from parts a and b to give a proof of Lemma 13.1.4. [*Hint:* Use your result from part b to write $N(rs)$ using complex conjugates.]

9. The following diagram shows a vector (a, b) in the Cartesian plane, as well as the vector P that results when (a, b) is rotated 90° counterclockwise about the origin.

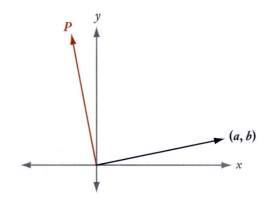

    a. What is the x-coordinate of P?

    b. What is the y-coordinate of P?

    c. Use your answers to parts a and b to show that when any complex number $a + bi$ is multiplied by i, it is rotated 90° counterclockwise about the origin.

10. Prove that every nonzero complex number $z = a + bi$ has a multiplicative inverse in the complex numbers, by expressing $\dfrac{1}{a + bi}$ in the form $x + yi$ (where x and y are real numbers), and demonstrating that $(a + bi)(x + yi) = 1$.

11. Let z be any nonzero complex number. By the result in Exercise 10, we know that z has a multiplicative inverse in $\mathbf{C}$. Prove that the multiplicative inverse of z is unique. [*Hint:* Assume that $w_1, w_2 \in \mathbf{C}$ are both multiplicative inverses of z, and show that $w_1 = w_2$.]

12. **a.** Find the multiplicative inverse (in the complex numbers) of $8 - 3i$, and express it in the form $r + si$, where r and s are rational numbers.

   b. Explain why it follows that $8 - 3i$ does not have a multiplicative inverse in the Gaussian integers. [*Hint:* Use the result from Exercise 11.]

13. Prove that $1 + i$ has no inverse in the Gaussian integers. [*Hint:* Assume that $a + bi$ is a multiplicative inverse of $1 + i$, and get a contradiction.]

14. State the theorem that $\mathbf{Z}[i]$ is a ring. [*Hint:* Model your statement on Theorem 8.2.6.]

15. Prove that $\mathbf{Z}[i]$ is a ring by proving your statement from Exercise 14.

16. Let $r = a + bi$ and $s = c + di$ be any two complex numbers. Let θ_r and θ_s respectively represent the angle of r and the angle of s (measured counterclockwise from the real axis).

   a. Write a trigonometric equation relating θ_r, a, and b.

   b. Prove that the angle of the product, rs, is equal to $\theta_r + \theta_s$.

17. If possible, find a Gaussian integer z with the given norm. If it is not possible, explain why not.

   a. $N(z) = 17$

   b. $N(z) = 11$

   c. $N(z) = 25$

   d. $N(z) = 21$

13.2 Divisibility and Primes in $\mathbf{Z}[i]$

Until this chapter, our main object of study has been the set of integers, $\mathbf{Z}$. One of the most important properties of $\mathbf{Z}$ is the Fundamental Theorem of Arithmetic (6.1.1), which states that every positive integer can be written in a unique way as a product of prime integers.

Now that we have defined the Gaussian integers, $\mathbf{Z}[i]$, we might wish to define primes in this new number system and to study how Gaussian integers factor into primes. In particular, we'd like to find an analogue of the Fundamental Theorem of Arithmetic for the Gaussian integers.

It is not hard to define primes in the Gaussian integers, and we will do so shortly. Given this, you might be tempted to assume that Gaussian integers can be factored uniquely into primes. But before you do something so rash, let's consider another number system that is similar to the Gaussian integers. The Gaussian integers $\mathbf{Z}[i]$ could be denoted $\mathbf{Z}[\sqrt{-1}]$. We can similarly define the number system $\mathbf{Z}[\sqrt{-5}]$ as follows:

$$\mathbf{Z}[\sqrt{-5}] = \{a + b\sqrt{-5} \mid a, b \in \mathbf{Z}\}.$$

In $\mathbf{Z}[\sqrt{-5}]$, we have

$$6 = \mathbf{2} \cdot \mathbf{3}$$

and

$$6 = (\mathbf{1 + \sqrt{-5}})(\mathbf{1 - \sqrt{-5}}).$$

But each of the numbers $\mathbf{2}$, $\mathbf{3}$, $(\mathbf{1 + \sqrt{-5}})$, and $(\mathbf{1 - \sqrt{-5}})$ is prime in $\mathbf{Z}[\sqrt{-5}]$. (See Exercises 20–24.) Thus, we see that in this number system, the number 6 can be factored into primes in more than one way. In $\mathbf{Z}[\sqrt{-5}]$, prime factorizations are not unique!

To explore whether prime factorizations are unique in the Gaussian integers, we first need to define divisibility and primality in $\mathbf{Z}[i]$.

Divisibility in Z[*i*]

The definition of divides for the Gaussian integers is almost identical to the definition of divides for the ordinary integers (3.1.1).

DEFINITION 13.2.1

Let $a, d \in \mathbf{Z}[i]$. We say that d **divides** a if there exists $q \in \mathbf{Z}[i]$ such that $a = qd$. We express this in symbols as $d \mid a$.

EXAMPLE 1

    a. $6 \mid 18 - 6i$ because $(3 - i)6 = 18 - 6i$.

    b. $1 + i \mid 2i$ because $(1 + i)(1 + i) = 2i$.

    c. $2 + 3i \mid -7 + 22i$ because $(4 + 5i)(2 + 3i) = -7 + 22i$.

The Linear Combination Lemma (3.1.3) has an exact analogue in the Gaussian integers.

Let d, m, n, x, y ∈ Z[i]. If d | x and d | y, then d | mx + ny.

PROOF You will prove this in Exercise 13.

Primes in Z[*i*]

Gaussian integers that have multiplicative inverses in the Gaussian integers are called *units*.

DEFINITION 13.2.3

*Let u ∈ Z[i]. We say u is a **unit** if there exists z ∈ Z[i] such that u · z = 1.*

OBSERVATION 13.2.4

The Gaussian integers have exactly four units: 1, −1, i, and −i.

PROOF

Part 1: The Gaussian integers 1, −1, *i*, and −*i* are units. (You will prove this part in Exercise 11.)

Part 2: If *u* is a unit in $\mathbf{Z}[i]$, then *u* must equal 1, −1, *i*, or −*i*.

Let *u* be a unit of $\mathbf{Z}[i]$. Then there exists $z \in \mathbf{Z}[i]$ such that

$$u \cdot z = 1.$$

By Lemma 13.1.4,

$$N(u) \cdot N(z) = N(1)$$
$$= 1.$$

Thus, $N(u) \mid 1$. Since the norm of any Gaussian integer is a nonnegative integer, this implies that $N(u) = 1$. It follows (see Exercise 12) that *u* must equal 1, −1, *i*, or −*i*. ●

We are now ready to define *prime* in the Gaussian integers.[1] The definition is very similar to the Equivalent Condition for Primality in the integers (Lemma 3.2.3).

---

[1]In fact, Definition 13.2.5 actually defines what are usually called the *irreducible* elements of a ring.

Let $p \in \mathbf{Z}[i]$ such that p is not a unit. We say p is **prime** if for every $a, b \in \mathbf{Z}[i]$, $p = ab$ implies that a is a unit or b is a unit.

EXAMPLE 2

   a. 5 is not prime as a Gaussian integer because $5 = (1 + 2i)(1 - 2i)$.

   b. $1 + 3i$ is not prime as a Gaussian integer because $1 + 3i = (1 - i)(-1 + 2i)$.

Naomi's Numerical Proof Preview: Lemma 13.2.6

Suppose we want to know whether $5 + 4i$ is prime as a Gaussian integer. Let's consider the norm of this number:

$$N(5 + 4i) = 5^2 + 4^2 = 41.$$

Now suppose $5 + 4i$ factors as a product of two Gaussian integers:

$$5 + 4i = ab.$$

By Lemma 13.1.4,

$$N(5 + 4i) = N(a)N(b),$$

or, equivalently,

$$41 = N(a)N(b).$$

Since 41 is a prime natural number, and $N(a)$ and $N(b)$ are both nonnegative integers, it must be the case that $N(w) = 1$ or $N(z) = 1$. Hence, either a is a unit or b is a unit.

Thus, by the definition of prime in $\mathbf{Z}[i]$, $5 + 4i$ is a prime Gaussian integer. This argument generalizes to any Gaussian integer whose norm is prime:

Let z be a Gaussian integer. If $N(z)$ is prime in the integers, then z is prime as a Gaussian integer.

PROOF You will prove this lemma in Exercise 15.

EXAMPLE 3

The number $9 + 4i$ is prime as a Gaussian integer, because $N(9 + 4i) = 97$ is a prime integer.

   The converse of Lemma 13.2.6 is not true (see Exercise 16). In Section 13.5, we will explore in more detail the question of precisely which Gaussian integers are prime.

Recall that two integers are relatively prime if they have no common factors other than 1 and -1. Similarly, we call two Gaussian integers *relatively prime* if they have no common factors other than the units $(1, -1, i,$ and $-i)$.

Let $a, b \in \mathbf{Z}[i]$. We say that a and b are **relatively prime** *if for all* $d \in \mathbf{Z}[i]$, $d \mid a$ and $d \mid b$ implies that d is a unit.

EXAMPLE 4

Show that the Gaussian integers $3 + 5i$ and $7 - 2i$ are relatively prime.

SOLUTION Let $d \in \mathbf{Z}[i]$ such that $d \mid 3 + 5i$ and $d \mid 7 - 2i$. The norm of d must satisfy

$$N(d) \mid N(3 + 5i) \text{ and } N(d) \mid N(7 - 2i)$$

by Lemma 13.1.4 (see Exercise 10). In other words,

$$N(d) \mid 34 \text{ and } N(d) \mid 53.$$

Since $\gcd(34, 53) = 1$, it follows that $N(d) = 1$, and hence, d is a unit. ■

The set of all multiples of a Gaussian integer

Suppose we have a Gaussian integer, z. What does the set of all Gaussian integer multiples of z look like geometrically? The following diagram shows all of the Gaussian integer multiples of $z = 3 + i$. Notice that the multiples of z are at the vertices of a square grid with side length $|z|$.

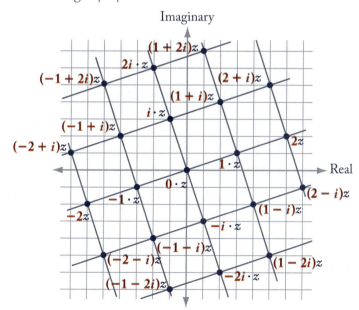

EXERCISES 13.2

Numerical Problems

1. Show that $5 + 6i$ is prime as a Gaussian integer.

2. Show that $-3 + 8i$ is prime in $\mathbf{Z}[i]$.

3. Find all of the factors of 6 in the Gaussian integers.

4. **a.** Demonstrate that 13 is not prime in $\mathbf{Z}[i]$.

   b. Demonstrate that 17 is not prime in $\mathbf{Z}[i]$.

5. Show that $2 + 5i$ and $3 + 2i$ are relatively prime.

Reasoning and Proofs

6. **a.** Find all of the factors of 2 in the Gaussian integers.
   [*Hint:* There are 12 of them.]

   b. Use norms to show that 2 has no other factors in $\mathbf{Z}[i]$.

7. In any number system R, the *units* are the elements of R that have multiplicative inverses in R:

   DEFINITION. *Let $u \in R$. We say u is a **unit** of R if there exists $z \in R$ such that $uz = 1$.*

   (In this definition, 1 represents the multiplicative identity of the number system R.)

   List or describe all of the units of each of the following number systems.

   a. $\mathbf{Z}$ **b.** $\mathbf{Q}$ **c.** $\mathbf{Z}_{10}$ **d.** $\mathbf{Z}_n$ (where n is a natural number)

8. Suggest a definition for *prime* in $\mathbf{Z}$ that allows negative integers to be prime.
   [*Hint:* You may want to base it on the definition of prime in $\mathbf{Z}[i]$ and use your answer to Exercise 7, part a.]

9. Let $a, b \in \mathbf{Z}$ such that $a \neq 0$. Prove that the integer $n = a^2 + b^2$ is not prime in $\mathbf{Z}[i]$.

10. Let d and z be Gaussian integers.

    a. Prove that if $d \mid z$ then $N(d) \mid N(z)$.

    b. Is it true that if $N(d) \mid N(z)$, then $d \mid z$? Prove it or find a counterexample.

11. Prove that $1, -1, i$, and $-i$ are units in the Gaussian integers. (This is Part 1 of the proof of Observation 13.2.4).

12. Let $r \in \mathbf{Z}[i]$. Prove that if $N(r) = 1$, then $r = 1, -1, i$, or $-i$.

13. Prove the Linear Combination Lemma for the Gaussian integers (13.2.2).

14. Assume that r and s are Gaussian integers such that $r \mid s$ and $s \mid r$. Show that $r = us$, where u is a unit.

15. Prove Lemma 13.2.6.

16. **a.** Prove that the number 3 is prime as a Gaussian integer.

  b. State the converse of Lemma 13.2.6.

  c. Show that the converse of Lemma 13.2.6 is false.

  d. Can you find a counterexample to the converse of Lemma 13.2.6 that is not real or purely imaginary? That is, can you find a Gaussian integer z that is neither real nor imaginary such that z is prime as a Gaussian integer but $N(z)$ is not a prime integer? If so, give an example; if not, explain why not.

17. **a.** Let $a, b \in \mathbf{Z}[i]$. Prove that if $\gcd(N(a), N(b)) = 1$, then a and b are relatively prime Gaussian integers.

  b. Write the converse of the result from part a. Is this converse true? Prove it, or find a counterexample.

18. Let a and b be relatively prime integers. Show that a and b are relatively prime in $\mathbf{Z}[i]$.

19. Let S be the set of all positive even integers. Note that the set S is closed under multiplication. If $n \in S$, we say n is *prime in S* if there do not exist $a, b \in S$ such that $n = ab$.

  a. Is 46 prime in S? How about 48?

  b. Determine exactly which elements of S are prime in S.

  c. Show that every element of S may be written as the product of one or more primes in S.

  d. Show that prime factorizations in S are not unique. That is, give an example of an element of S that has two distinct factorizations into primes in S.

Advanced Reasoning and Proofs

EXPLORATION Nonuniqueness of Prime Factorizations (Exercises 20–24)

At the beginning of this section, we met a number system in which unique factorization fails:

$$\mathbf{Z}[\sqrt{-5}] = \{a + b\sqrt{-5} \mid a, b \in \mathbf{Z}\}.$$

In this Exploration, you will prove rigorously that unique factorization fails in this ring. Just as in the Gaussian integers, we must first define the *units* and *primes* of $\mathbf{Z}[\sqrt{-5}]$. To do this, we use definitions that are identical to Definitions 13.2.3 and 13.2.5, except that we replace $\mathbf{Z}[i]$ by $\mathbf{Z}[\sqrt{-5}]$ in the definitions.

Also, as in the Gaussian integers, it will be useful to have a *norm* function. We define

$$N(a + b\sqrt{-5}) = a^2 + 5b^2.$$

20. Prove that for any $z, w \in \mathbf{Z}[\sqrt{-5}]$, $N(zw) = N(z)N(w)$.

21. Suppose that $z \in \mathbf{Z}[\sqrt{-5}]$ is a unit. What can you say about $N(z)$?

22. Prove that the four elements 2, 3, $1 + \sqrt{-5}$, and $1 - \sqrt{-5}$ are each prime in $\mathbf{Z}[\sqrt{-5}]$.

23. Show that the two factorizations

$$6 = (1 + \sqrt{-5})(1 - \sqrt{-5}) \text{ and } 6 = 2 \cdot 3$$

are distinct up to order and up to multiplication by units. This proves that in $\mathbf{Z}[\sqrt{-5}]$, prime factorizations are not unique.

Now consider the number system $\mathbf{Z}[\sqrt{10}] = \{a + b\sqrt{10} \mid a, b \in \mathbf{Z}\}$, with norm defined by $N(a + b\sqrt{10}) = a^2 - 10b^2$.

24. **a.** Find two factorizations of 6 in $\mathbf{Z}[\sqrt{10}]$.

   b. Using Exercises 20–23 as your guide, carefully prove that unique factorization fails in the ring $\mathbf{Z}[\sqrt{10}]$.

13.3 The Division Theorem for the Gaussian Integers

In Chapters 3–6, we proved several important theorems about the integers, each of which built on the ones that preceded it:

Division Theorem (3.5.1) $\Rightarrow$ Linear Diophantine Equations (5.1.2), solved using the Euclidean Algorithm (4.1.1) $\Rightarrow$ Euclid's Lemma (5.4.2) $\Rightarrow$ Fundamental Property of Primes (6.1.2) $\Rightarrow$ Fundamental Theorem of Arithmetic (6.1.1)

The subsequent chapters have all built upon the framework of these fundamental results.

As a number system, the Gaussian integers $\mathbf{Z}[i]$ bears many similarities to the ordinary integers, $\mathbf{Z}$. In fact, every one of the results about the integers in the chain above has an analogue in the Gaussian integers. In this section and the next, our goal is to state and prove the Gaussian integer analogues of these important results, building up a framework for the Gaussian integers that will culminate in the Fundamental Theorem of Gaussian Arithmetic (13.4.4).

As we will see, these basic results about $\mathbf{Z}[i]$ will enable us to achieve a greater depth of understanding of our old friend, $\mathbf{Z}$. In the final section of this chapter, we use our newfound knowledge of the Gaussian integers to prove Fermat's Two Squares Theorem, a substantial result about the ordinary integers.

Field of greens

Born in 18th-century Switzerland, Leonhard Euler was one of the most prolific mathematicians of all time. What's not so well known about Euler is that he was personally responsible for the Great Rabbit Plague of 1749. At that time, he was head groundskeeper at Sans Souris, the royal summer residence of King Frederick the Grim. Euler had earned his prestigious post when he rid the palace grounds of mice by releasing wild rabbits to scare them off. This worked fabulously, except for one problem. The rabbits quickly multiplied and began tearing up the beautiful, endless lawn that covered the palace grounds.

There were rabbits everywhere! Euler devised a plan to trap the pests using lettuce as bait. He put in lettuce plants in a regular square grid covering the lawn. At the center of the lawn was his shed. Euler decided to represent the locations of the lettuce plants using Gaussian integers, as shown here:

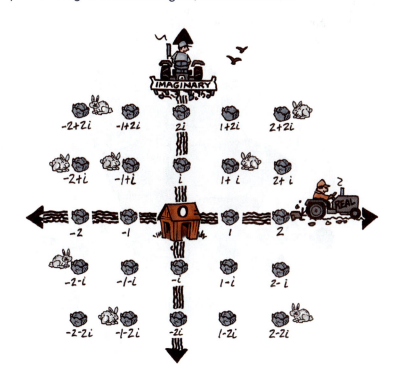

Whenever a bunny started to nibble on one of the lettuce plants, a sensor was triggered that let Euler know the location of the plant being eaten. Euler would then run from his shed to that lettuce plant and nab the offending bunny. After a while, Euler grew tired of running after so many bunnies. Once, when he was having a particularly bad hare day, Euler complained to the king that he needed a way to get around the palace grounds more quickly.

When Euler arrived at work the next morning, he was surprised to see a large chair sitting in the middle of his shed. On the arm of the chair were two large buttons marked "GO" and "ROTATE." The chair seemed to be pointed at a funny angle. Euler tried to align it with the axes of his lettuce patch, but it was too heavy for him to move. So he sat down in the chair and pushed the GO button. The chair shook violently and began to move forward. Before Euler knew it, the chair had taken him directly to the lettuce plant at $4 + 3i$.

Euler enjoyed his ride so much that he pressed the GO button again. The chair took him straight to the lettuce plant at $8 + 6i$. Just then, Euler noticed he was sitting on a sheet of fine parchment bearing the royal seal and a message written in the king's own hand.

Congratulations, Lenny. I hereby bestow upon you this special rocket-powered chair to assist you in your pursuit of rabbits. In case you haven't figured it out, the GO button propels you forward 5 units in the direction the chair is pointing. The ROTATE button causes the chair to rotate 90° counterclockwise.

Looking back at the arm of his chair, Euler pressed ROTATE, and the chair turned 90° to the left. Then he pressed GO, and the chair took him to the lettuce plant at $5 + 10i$.

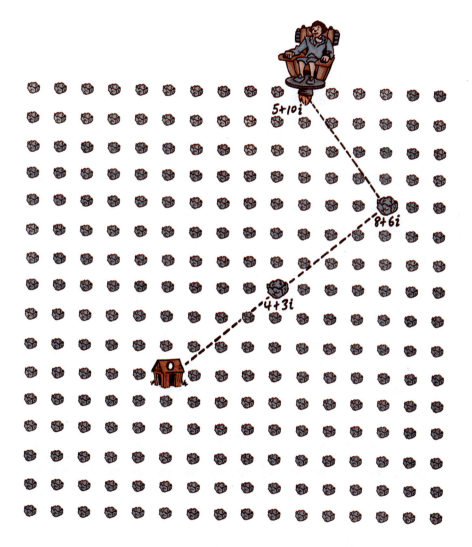

Pleased with the king's gift, Euler wondered which of the lettuce plants on the palace grounds he could reach using his rocket chair.

"Hmm, let me see," reasoned Euler. "This chair can rotate 90°, and it can move 5 units in the direction it is pointing. That means the chair travels along the edges of a square grid whose squares have sides of length 5. . . . Aha, I've got it! My new chair can reach all of the Gaussian integer multiples of $4 + 3i$."

Although the chair helped Euler to maneuver around the field quickly, there were many lettuce plants that his chair could not reach. Euler wanted to catch the rabbits nibbling these other plants without having to leave his chair.

He figured that he could attach his rabbit net to a long pole and use that to catch the bunnies. Euler went to the king and asked for a pole 5 units long.

The king furrowed his brow. "Lenny, that sounds pretty expensive. Can't you make do with anything shorter than 5 units?"

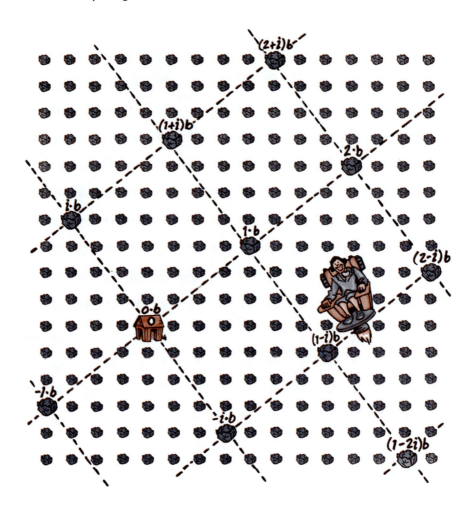

Euler's chair could reach all of the Gaussian integer multiples of $b = 4 + 3i$.

"Your Highness, I'm just not sure whether I can make do with anything less than 5. . . ." Euler well knew that it was a slippery problem, and he dug deep in search of some crude solution. Until that moment, it didn't strike Euler that he could possibly rig up a solution as slick as the one he was about to discover.

"Eureka!" Euler gushed. "The farthest I ever need to reach is to the center of a square with side length 5. I need a pole exactly $\frac{5}{\sqrt{2}}$ units long."

The Division Theorem

In our myth, Euler realized that every Gaussian integer is within $\dfrac{5}{\sqrt{2}}$ units of some multiple of $b = 4 + 3i$. This observation is the key to formulating a division theorem for the Gaussian integers. Let's take a closer look.

Naomi's Numerical Proof Preview: Geometric Proof of Theorem 13.3.1

Suppose we pick an arbitrary Gaussian integer a, which we would like to divide by $b = 4 + 3i$. (Think of a as the location of the bunny that Euler needs to catch.) The multiples of $b = 4 + 3i$ are at the vertices of a square grid with side length $|b| = 5$. Because this square grid covers the entire complex plane, we know that no matter where a is, it must be inside of (or on an edge of) one of the squares. Of the four corners of this square, the one that is closest to a is the multiple of b that is closest to a; we call this multiple $q \cdot b$.

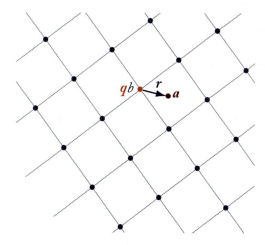

As shown in the diagram above, the vector r is the difference between a and the nearest multiple of b. In other words, $r = a - qb$. What is the longest that r could be? The worst-case scenario (the longest r) would arise if a were at the center of the square, as shown here:

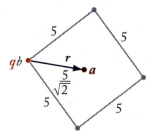

In this case, there are four multiples of b (the four corners of the square) that are equally close to a. We arbitrarily chose one of these four and labeled it qb. Since r is a vector from the corner of a square with side length 5 to the center of the square, we know its length is $|r| = \dfrac{5}{\sqrt{2}}$. In this worst-case scenario, the multiple of b that is closest to a is exactly $\dfrac{5}{\sqrt{2}}$ units away from a. Thus, any Gaussian integer a will be at most $\dfrac{5}{\sqrt{2}}$ units away from the nearest multiple of b.

As discussed earlier, we would like to have an analogue of the Division Theorem (3.5.1) for $\mathbf{Z}[i]$. What would this mean? Well, returning to the case of $\mathbf{Z}$, dividing an integer a by a natural number b means finding a multiple of b that is within b units of a. In the Gaussian integers, we want to find a multiple of b whose distance from a is less than $|b|$. In the discussion above, we were able to do even better: we saw that it is always possible to find a multiple of b whose distance from a is at most $\dfrac{|b|}{\sqrt{2}}$. We can generalize this argument to prove the following Division Theorem for $\mathbf{Z}[i]$.

THEOREM 13.3.1   THE DIVISION THEOREM FOR THE GAUSSIAN INTEGERS

Let $a, b \in \mathbf{Z}[i]$, with $b \neq 0$. Then there exist Gaussian integers q and r such that $a = qb + r$ and $N(r) < N(b)$.

We will see two different proofs of this theorem in this section. The first proof will be more geometric than most of the proofs we have encountered so far. The second proof, which comes later in the section, is an algebraic proof.

GEOMETRIC PROOF of THE DIVISION THEOREM FOR THE GAUSSIAN INTEGERS (13.3.1)

Let $a, b \in \mathbf{Z}[i]$. The Gaussian integer multiples of b are at the vertices of a square grid with side length $|b|$. Let qb be the multiple of b that is closest to a (if there are two or more multiples of b that are equally close to a, choose any one of these to be qb). Let $r = a - qb$.

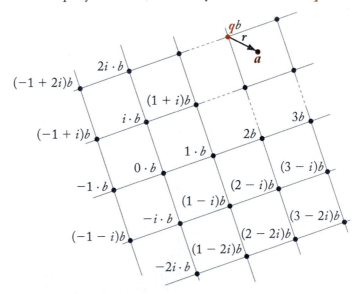

The largest possible value of $|r|$ would occur if a were at the very center of one of the $|b| \times |b|$ squares. In that case, $|r|$ would be equal to $\dfrac{|b|}{\sqrt{2}}$. Thus, no matter what the value of **a** is (no matter where **a** lies on the complex plane), we are guaranteed that

$$|r| \leq \frac{|b|}{\sqrt{2}}.$$

Squaring both sides of this inequality gives

$$N(r) \leq \frac{N(b)}{2}.$$

Since $N(b) > 0$, we have

$$N(r) < N(b). \qquad \blacksquare$$

One important difference between the Division Theorem for the Gaussian integers (13.3.1) and the Division Theorem for the ordinary integers (3.5.1) is that with the ordinary integers, **q** and **r** are unique, but with the Gaussian integers there is often more than one pair of Gaussian integers **q** and **r** that satisfy the theorem.

EXAMPLE 1

Divide $9 + 11i$ by $7 - 4i$.

SOLUTION We begin by drawing a diagram.

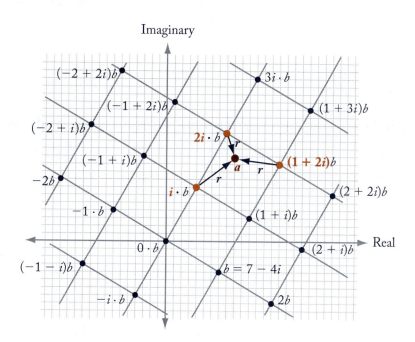

From the diagram, we see that there are three solutions that fulfill the conditions of the Division Theorem (13.3.1):

$$a = qb + r, \qquad\qquad\qquad N(r) < N(b)$$

$$9 + 11i = 2i\,(7 - 4i) + (1 - 3i), \qquad N(1 - 3i) < N(7 - 4i)$$

$$9 + 11i = (1 + 2i)(7 - 4i) + (-6 + i), \qquad N(-6 + i) < N(7 - 4i)$$

$$9 + 11i = i\,(7 - 4i) + (5 + 4i), \qquad N(5 + 4i) < N(7 - 4i). \qquad \blacksquare$$

We have just seen a geometric proof and example of the Division Theorem for Gaussian Integers (13.3.1). The geometric proof of this theorem relies on our geometric intuition, and could be considered less rigorous than the other proofs in this text. We will now see a more rigorous, algebraic proof of the Division Theorem for Gaussian Integers.

Naomi's Numerical Proof Preview: Algebraic Proof of Theorem 13.3.1

Suppose we want to divide the Gaussian integer $a = -9 + 7i$ by the Gaussian integer $b = 4 + 2i$, and we want to obtain a quotient q and remainder r that are both Gaussian integers. We may not know how to do that algebraically (without drawing a picture), but (as you may have learned when you first studied complex numbers), we do know how to find the complex quotient $\frac{a}{b}$:

$$\frac{a}{b} = \frac{-9 + 7i}{4 + 2i}$$

$$= \frac{-9 + 7i}{4 + 2i} \cdot \frac{4 - 2i}{4 - 2i} \qquad \leftarrow \textbf{Multiply numerator and denominator by}$$
$$\textbf{the complex conjugate of the denominator}$$

$$= \frac{-22 + 46i}{20}$$

$$= -\frac{11}{10} + \frac{23}{10}i.$$

Now we can obtain a Gaussian integer quotient, q, just by rounding z to the nearest Gaussian integer. To do this, we individually round the real part of $\frac{a}{b}$ $\left(\text{which is } -\frac{11}{10}\right)$ and the imaginary part of $\frac{a}{b}$ $\left(\text{which is } \frac{23}{10}\right)$ to the nearest integer:

$$q = -1 + 2i.$$

Then we let $r = a - qb$,

$$r = -9 + 7i - (-1 + 2i)(4 + 2i)$$

$$= -1 + i.$$

The reader is invited to verify (in Exercise 1) that $a = qb + r$ and that $N(r) < N(b)$.

In this discussion, the remainder r turned out to have smaller norm than b. But how do we know that no matter which Gaussian integers a and b we start with, the remainder r will always satisfy $N(r) < N(b)$? Recall that to find q, we rounded the real and imaginary parts of $\frac{a}{b}$ to the nearest integer. In the rounding process, each component of $\frac{a}{b}$ is altered by at most $\frac{1}{2}$, and as a result

$$|q - \frac{a}{b}| \leq \sqrt{\left(\frac{1}{2}\right)^2 + \left(\frac{1}{2}\right)^2} \tag{1}$$

$$= \frac{\sqrt{2}}{2}.$$

It follows that

$$|r| \leq \frac{\sqrt{2}}{2}|b|,$$

and hence, $N(r) < N(b)$, the desired conclusion. In Exercises 15–17, you will fill in the details to make this a rigorous argument.

ALGEBRAIC PROOF of **THE DIVISION THEOREM FOR THE GAUSSIAN INTEGERS (13.3.1)**

Let $a, b \in \mathbf{Z}[i]$, with $b \neq 0$.

[To show: There exist $q, r \in \mathbf{Z}[i]$ such that $a = qb + r$ and $N(r) < N(b)$.]

Let z be the complex number

$$z = \frac{a}{b}. \tag{2}$$

As you will show in Exercises 15–17, every complex number is within $\frac{\sqrt{2}}{2}$ units of a Gaussian integer, i.e., there exists $q \in \mathbf{Z}[i]$ such that

$$|z - q| \leq \frac{\sqrt{2}}{2}. \tag{3}$$

Let $r = a - qb$. Then

$$a = qb + r.$$

Furthermore,

$$
\begin{aligned}
|r| &= |a - qb| \\
&= |(z - q)b| \qquad &\leftarrow \text{by equation (2)} \\
&= |z - q||b| \\
&\leq \frac{\sqrt{2}}{2}|b| \qquad &\leftarrow \text{by inequality (3)} \\
&< |b|.
\end{aligned}
$$

We conclude that $N(r) < N(b)$. ■

Theorem 13.3.1 establishes that in the Gaussian integers, division with remainder is always possible. We will use this result to prove that any Gaussian integer can be factored uniquely into a product of prime Gaussian integers. In some sense, the hard work is done, and to achieve a version of the Fundamental Theorem of Arithmetic (6.1.1) for the Gaussian integers, all that remains is to retrace the steps we took in Chapters 4–6. This will be our project in the next section.

EXERCISES 13.3

Numerical Problems

1. Let $a, q, b,$ and r have the values stated in the Numerical Proof Preview for the algebraic proof of Theorem 13.3.1.

    a. Show that that $a = qb + r$.

    b. Show that $N(r) < N(b)$.

In Exercises 2–4, use the Division Theorem (13.3.1) to geometrically perform the division indicated (as in Example 1), following these steps:

   a. Draw a diagram.

   b. Find all solutions (q, r) that fulfill the conditions of the Division Theorem.

2. Divide $7 + 3i$ by $2 + 2i$.

3. Divide $-4 - i$ by $3i$.

4. Divide $7 + 5i$ by $3 - 2i$.

In Exercises 5–7, use the Division Theorem (13.3.1) to algebraically perform the division indicated. Use the method of the Numerical Proof Preview for the algebraic proof of Theorem 13.3.1 to find a solution (q, r) that fulfills the conditions of the Division Theorem.

5. Divide $7 - 3i$ by $2 + 3i$.

6. Divide $25 - 8i$ by $5i$.

7. Divide $71 + 15i$ by $-9 + 2i$.

In Exercises 8–10, use the Division Theorem (13.3.1) to perform the division indicated. Use either the geometric method of Example 1 or the method of the Numerical Proof Preview for the algebraic proof of Theorem 13.3.1 to find a solution (q, r) that fulfills the conditions of the Division Theorem.

8. Divide $4 + i$ by $1 - 2i$.

9. Divide $-2 + 5i$ by $1 - i$.

10. Divide $61 + 5i$ by $5 + 7i$.

Reasoning and Proofs

11. When you use the Division Theorem for the ordinary integers (3.5.1) to divide any integer by 3, there are three possible remainders: 0, 1, and 2.

    a. Draw the grid of all Gaussian integer multiples of 3 in the complex plane.

    b. List all of the possible remainders when you use the Division Theorem for the Gaussian integers (13.3.1) to divide any Gaussian integer by 3. [*Hint:* think about the grid from part a.]

12. **a.** Draw the grid of all Gaussian integer multiples of $2 + i$ in the complex plane.

    b. List all of the possible remainders when you use the Division Theorem (13.3.1) to divide a Gaussian integer by $2 + i$.

13. Consider the following assertion:

    For all $a, b \in \mathbf{Z}[i]$, if $a \mid b$, then $N(a) \leq N(b)$.

    Is this statement true? If so, prove it. If not, provide a counterexample, then alter the statement to make it true and prove your new statement.

14. Consider the following assertion:

    For all $a, b, c \in \mathbf{Z}[i]$, if $ab \mid ac$, then $b \mid c$.

    Is this statement true? If so, prove it. If not, provide a counterexample, then alter the statement to make it true and prove your new statement.

EXPLORATION A Key Step in the Algebraic Proof of the Division Theorem (Exercise 15–17)

In this exploration, you will prove the following lemma. It follows from this lemma that in the algebraic proof of the Division Theorem for Gaussian Integers (13.3.1), inequality (3) follows from equation (2).

> LEMMA. *Let $a, b \in \mathbf{Z}[i]$, with $b \neq 0$. Let z be the complex number $z = \frac{a}{b}$.*
>
> *Then there exists $q \in \mathbf{Z}[i]$ such that $|z - q| \leq \frac{\sqrt{2}}{2}$.*

Note that relying on our geometric intuition, this fact is easily justified. Rounding the real and imaginary parts of the complex number z to the nearest integer produces the element of $\mathbf{Z}[i]$ closest to z. Since the Gaussian integers form a lattice of 1×1 squares, the Gaussian integer closest to z will be at most $\frac{\sqrt{2}}{2}$ away from z. Now you will prove this lemma rigorously.

15. Let $a = s + ti$ and $b = u + vi$ be any two Gaussian integers such that $b \neq 0$. Consider the complex number $\frac{a}{b}$. Prove that the real and imaginary parts of $\frac{a}{b}$ are both rational numbers.

16. **a.** Prove the following variation of the Division Theorem (3.5.1):

    If $a \in \mathbf{Z}$ and $b \in \mathbf{N}$, there exists $q, r \in \mathbf{Z}$ such that $a = qb + r$ and $|r| \leq \frac{b}{2}$.

    b. Let $x \in \mathbf{Q}$. Show that there exists $n \in \mathbf{Z}$ such that $|x - n| \leq \frac{1}{2}$.
    [*Hint:* Use the result from part a.]

17. Prove the Lemma stated at the beginning of this exploration. [*Hint:* Apply the result from Exercise 16 part b to both the real and imaginary parts of z.]

Advanced Reasoning and Proofs

18. Let $a, b \in \mathbf{Z}[i]$, and assume that we are dividing a by b. Suppose that in the complex plane, a lies somewhere inside the following square (where k is some Gaussian integer):

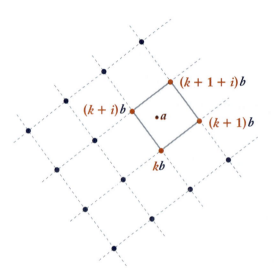

Depending on the precise location within this square where the lattice point a lies, there may be one, two, three, or four solutions (q, r) that satisfy the requirements of the Division Theorem (13.3.1).

a. Divide the square into four regions where a could lie: the region that results in exactly one solution, the region that results in exactly two solutions, the region that results in exactly three solutions, and the region that results in four solutions. Be sure to indicate what happens if a lies on an edge of the square or on a boundary between regions.

b. Assume that your bedroom wall represents the complex plane. Your roommate has painted a dark blue dot at every Gaussian integer multiple of some $b \in \mathbf{Z}[i]$. You throw a dart at your wall at random, and it lands on a brown dot that represents the Gaussian integer a. If you were to divide a by b, which number of solutions (1, 2, 3, or 4) is the likeliest? Which number of solutions is the second likeliest? Explain your answer.

19. Note that a solution (q, r) to the Division Theorem for the Gaussian Integers (13.3.1) is found by selecting any Gaussian integer multiple of b that is closer to a than the magnitude $|b|$ of the divisor. The Division Theorem for $\mathbf{Z}$ (3.5.1) is not quite analogous.

    a. State a new Division Theorem for $\mathbf{Z}$ that is analogous to the Division Theorem for $\mathbf{Z}[i]$ (13.3.1).

    b. Using the new Division Theorem that you stated in part a, divide 17 by 5. Give all solutions (q, r) in $\mathbf{Z}$ that satisfy the conditions of your new Division Theorem.

    c. Using the new Division Theorem that you stated in part a, divide 17 by -5. Give all solutions (q, r) in $\mathbf{Z}$ that satisfy the conditions of your new Division Theorem.

EXPLORATION The Euclidean Algorithm in Z[*i*] (Exercises 20–25)

In the ordinary integers, $\mathbf{Z}$, once we had proved the Division Theorem (3.5.1), we learned how to perform the Euclidean Algorithm (4.1.1), which is a series of repeated divisions, on any two natural numbers a and b. Similarly, now that we have proven the Division Theorem for $\mathbf{Z}[i]$, we can also do the Euclidean Algorithm in the Gaussian integers:

THE EUCLIDEAN ALGORITHM IN $\mathbf{Z}[i]$. *Given two Gaussian integers a and b, with $b \neq 0$, apply the Division Theorem (13.3.1) to divide a by b. If the remainder is not zero, then continue by dividing the previous step's divisor by its remainder. When a remainder of 0 is obtained, stop. The algorithm results in the following system of equations:*

$$a = q_1 b + r_1 \qquad\qquad N(r_1) < N(b)$$
$$b = q_2 r_1 + r_2 \qquad\qquad N(r_2) < N(r_1)$$
$$r_1 = q_3 r_2 + r_3 \qquad\qquad N(r_3) < N(r_2)$$
$$\vdots$$
$$r_{n-2} = q_n r_{n-1} + r_n \qquad\qquad N(r_n) < N(r_{n-1})$$
$$r_{n-1} = q_{n+1} r_n + 0,$$

where all of the q_i and r_i are Gaussian integers.

As we saw in Example 1, there is often more than one solution when applying the Division Theorem (13.3.1) to divide one Gaussian integer by another. Since the Euclidean Algorithm consists of repeated divisions, there will also often be more than one way to perform the Euclidean Algorithm. For example, here are two different ways to perform the Euclidean Algorithm on $2 + 5i$ and $3 + 2i$.

A. $2 + 5i = 1(3 + 2i) + (-1 + 3i)$ B. $2 + 5i = (1 + i)(3 + 2i) + 1$

$\quad 3 + 2i = -i(-1 + 3i) + i$ $\qquad\qquad 3 + 2i = (3 + 2i)1 + 0$

$\quad -1 + 3i = (3 + i)i + 0$

20. **a.** Verify that in each step of the Euclidean Algorithm A (at left above), the norm of the remainder is less than the norm of the divisor.

   b. Verify that in each step of the Euclidean Algorithm B (at right above), the norm of the remainder is less than the norm of the divisor.

   c. Perform the Euclidean Algorithm on $2 + 5i$ and $3 + 2i$ in a third way (get a different sequence of steps than either A or B). Verify that in each step, the norm of the remainder is less than the norm of the divisor.

21. Explain why the Euclidean Algorithm performed on any pair of Gaussian integers will halt after a finite number of steps.

22. Perform the Euclidean Algorithm once for each pair of Gaussian integers.

   a. $-2 + 5i$ and $1 + i$ **b.** $6 + 3i$ and $3 + i$ **c.** $-7 + 11i$ and $-4 + 7i$

   d. $1 + 2i$ and $16 - 5i$ **e.** $13 - i$ and $-2 - 11i$ **f.** $17 + 27i$ and $-8 - i$

23. Perform the Euclidean Algorithm twice on $11 + 3i$ and $5 + 20i$. The second time you perform the algorithm, choose a different solution to the very first division that you perform. (That is, make sure that the first line is different each time you perform the Euclidean Algorithm.)

24. **Definition of gcd in $\mathbf{Z}[i]$** Let $a, b \in \mathbf{Z}[i]$, and let r_n be the last nonzero remainder when the Euclidean Algorithm is performed on a and b.

   a. Let $d \in \mathbf{Z}[i]$. Show that d divides both a and b if and only if $d \mid r_n$.

   b. Show that a and b are relatively prime if and only if r_n is a unit.

   c. Prove that if the Euclidean Algorithm is performed on a and b in a different way, resulting in the final nonzero remainder r_n', then there exists a unit $u \in \mathbf{Z}[i]$ such that $r_n' = u \cdot r_n$.

   d. Let $d \in \mathbf{Z}[i]$ be a common divisor of a and b. Prove that $N(d) \leq N(r_n)$, and prove that if $N(d) = N(r_n)$, then $d = u \cdot r_n$ for some unit $u \in \mathbf{Z}[i]$. For this reason, we define $\gcd(a, b) = r_n$, which is well defined up to units.

25. Use the result from part b of Exercise 24, along with your answers to Exercise 22, to determine which pairs of Gaussian integers from Exercise 22 are relatively prime.

13.4 Unique Factorization in Z[i]

We are well on our way to proving the Gaussian integer version of the Fundamental Theorem of Arithmetic. Up to now, we have covered the Gaussian integer analogues of definitions and results from Chapters 3 and 4. In this section, we will state and prove the Gaussian integer analogues of key results from Chapters 5 and 6. The statements and proofs of the results about Gaussian integers in this section may look familiar—they are extremely similar to the statements and proofs of the analogous results about the ordinary integers from Chapters 5 and 6.

Linear Diophantine equations in the Gaussian integers

The following theorem guarantees that if two Gaussian integers a and b are relatively prime, then the equation $aX + bY = 1$ has a solution in the Gaussian integers. It is the Gaussian integer analogue of Theorem 5.1.2 (Solving Linear Diophantine Equations).

THEOREM 13.4.1

Let a and b be Gaussian integers. The equation $aX + bY = 1$ has a solution in the Gaussian integers if and only if a and b are relatively prime.

PROOF Let $a, b \in \mathbf{Z}[i]$.

[$\Rightarrow$] Suppose there exist $x, y \in \mathbf{Z}[i]$ that satisfy $ax + by = 1$.

[To show: *a* and *b* are relatively prime.]

Let $d \in \mathbf{Z}[i]$ be a common factor of a and b.
   Since $d \mid a$ and $d \mid b$, the Linear Combination Lemma (13.2.2) guarantees that

$$d \mid ax + by.$$

It follows that $d \mid 1$, and d is hence a unit. Therefore, by definition, a and b are relatively prime.

[$\Leftarrow$] Now assume that a and b are relatively prime.

[To show: The equation $aX + bY = 1$ has a solution in the Gaussian integers.]

Consider the set S of all linear combinations of a and b:

$$S = \{ax + by \mid x, y \in \mathbf{Z}[i]\}.$$

It follows from the Well-Ordering Principle (2.2.2) that there exists a nonzero element $d \in S$ of smallest norm. (See Exercise 8.)

Claim: $d \mid a$.

Proof of Claim Divide a by d: by the Division Theorem (13.3.1), there exist Gaussian integers q and r such that

$$a = qd + r \tag{1}$$

and

$$N(r) < N(d). \tag{2}$$

Since $r = a - qd$, and d is a linear combination of a and b, we know that r can be written as a linear combination of a and b. In other words, $r \in S$.

But d was chosen to have the smallest norm of all nonzero elements of S, so it follows from inequality (2) that $r = 0$. Thus, equation (1) becomes $a = qd$. Hence, $d \mid a$. This completes the proof of the Claim.

The Claim tells us that $d \mid a$. By similar reasoning, $d \mid b$.

Since a and b are relatively prime, we conclude that d is a unit. Since $d \in S$, there exist Gaussian integers x and y such that

$$ax + by = d.$$

Since d is a unit, it has a multiplicative inverse, $d^{-1} \in \mathbf{Z}[i]$. Multiplying the previous equation by d^{-1} gives

$$a\mathbf{x}\mathbf{d}^{-1} + b\mathbf{y}\mathbf{d}^{-1} = 1.$$

Hence, $(\mathbf{x}\mathbf{d}^{-1}, \mathbf{y}\mathbf{d}^{-1})$ is a solution in the Gaussian integers to the equation $aX + bY = 1$.

■

Now that we know that Gaussian integer linear Diophantine equations have solutions, we are ready to prove the Gaussian integer version of Euclid's Lemma (5.4.2).

LEMMA 13.4.2   GAUSSIAN EUCLID'S LEMMA

Let $d, m, n \in \mathbf{Z}[i]$ such that d and m are relatively prime. If $d \mid mn$, then $d \mid n$.

PROOF The proof is similar to the proof of Euclid's Lemma in the ordinary integers (see Exercise 7).

The Fundamental Theorem of Gaussian Arithmetic

We are nearly ready to prove the Gaussian integer analogue of the Fundamental Theorem of Arithmetic (6.1.1). But first, we must prove the Gaussian integer analogue of the Fundamental Property of Primes (Proposition 6.1.2). You may notice that the proof of the following proposition is virtually identical to the proof of Proposition 6.1.2.

PROPOSITION 13.4.3   FUNDAMENTAL PROPERTY OF PRIMES IN $\mathbf{Z}[i]$

Let p be a prime Gaussian integer, and let $a, b \in \mathbf{Z}[i]$. If $p \mid ab$, then $p \mid a$ or $p \mid b$.

PROOF Let $p, a, b \in \mathbf{Z}[i]$ such that p is prime and $p \mid ab$.

[To show: $p \mid a$ or $p \mid b$.]

Assumption: Suppose that $p \nmid a$ and $p \nmid b$.

Since p is prime and $p \nmid a$, it follows that p and a are relatively prime (see Exercise 9). Thus, p divides the product ab and is relatively prime to the factor a, so we may use Euclid's Lemma (13.4.2) to conclude that $p \mid b$. This contradicts our Assumption that $p \nmid b$. $\Rightarrow\Leftarrow$

We conclude that $p \mid a$ or $p \mid b$. ∎

The Fundamental Theorem of Arithmetic for the ordinary integers (6.1.1) asserts that prime factorizations in $\mathbf{N}$ are unique *up to order* because, for instance, we consider $3 \cdot 5$ and $5 \cdot 3$ to be the same factorization of 15. In the Gaussian integers, there is an extra wrinkle. For example, 5 can be factored into Gaussian integers as

$$5 = (2 + i)(2 - i), \tag{3}$$

but it also factors as

$$5 = (2i - 1)(-2i + 1). \tag{4}$$

Both equations (3) and (4) are factorizations of 5 into a product of prime Gaussian integers. But do we want to count these as different factorizations? Surely not. Notice that the first prime in factorization (4) is just a unit times the first prime in factorization (3):

$$(2i - 1) = i(2 + i).$$

Similarly, the second prime in factorization (4) is just a unit times the second prime in factorization (3):

$$(-2i + 1) = (-i)(2 - i).$$

Since the two factorizations of the number 5 given in (3) and (4) are identical except that you have to multiply one factorization by units in order to get the other, we want to count (3) and (4) as the *same* factorization of 5. We express this idea in the Fundamental Theorem of Gaussian Arithmetic by saying that prime factorizations are unique *up to units*.

Without further ado, we present:

THEOREM 13.4.4 | **THE FUNDAMENTAL THEOREM OF GAUSSIAN ARITHMETIC**

Let $z \in \mathbf{Z}[i]$ such that z is not zero and is not a unit. Then z may be written as a product of one or more prime Gaussian integers. Furthermore, the prime factorization of z is unique up to order and units.

PROOF The existence part of the proof (the proof that z has a prime factorization) is analogous to Theorem 3.2.5, and is proven in Exercise 11.

We now prove the uniqueness part of the Fundamental Theorem of Arithmetic: we must show that every nonzero Gaussian integer z that is not a unit has at most one prime factorization up to order and units. We will use the principle of strong induction. Since z is a Gaussian integer, rather than a natural number, we cannot perform induction on z directly. Instead, we perform our induction on the norm of z.

The condition that the nonzero Gaussian integer z is not a unit is equivalent to $N(z) > 1$.

Let $z \in \mathbf{Z}[i]$, $N(z) > 1$, and assume the inductive hypothesis that every number k that satisfies $1 < N(k) < N(z)$ has at most one prime factorization up to order and units.

[To show: z has at most one prime factorization up to order and units.]

Suppose that z can be written as the product of prime Gaussian integers in two different ways. That is, suppose

$$z = p_1 \cdot p_2 \cdot \ \cdots \ \cdot p_e = q_1 \cdot q_2 \cdot \ \cdots \ \cdot q_f, \tag{5}$$

where $p_1, p_2, \ldots, p_e$ and $q_1, q_2, \ldots, q_f$ are prime Gaussian integers.

We must show that these two prime factorizations of z are the same up to order and units.

Equation (5) implies (by the definition of divides) that

$$p_1 \mid q_1 \cdot q_2 \cdot \ \cdots \ \cdot q_f.$$

It follows from the Fundamental Property of Primes (13.4.3) that p_1 divides one of the primes $q_1, q_2, \ldots, q_f$. (See Exercise 10.) Without loss of generality, we may assume that $p_1 \mid q_1$. Since p_1 and q_1 are both prime, it follows that $q_1 = u \cdot p_1$ for some unit u (see Exercise 12). Substituting this into equation (5), we obtain

$$\frac{z}{p_1} = p_2 \cdot \ \cdots \ \cdot p_e = u \cdot q_2 \cdot \ \cdots \ \cdot q_f. \tag{6}$$

First consider the case that $N\left(\frac{z}{p_1}\right) = 1$. In this case, $\frac{z}{p_1}$ is a unit. Since units cannot be divisible by a prime (see Exercise 13), there must not be any primes at all in the two factorizations in (6). In other words, $e = f = 1$, and since p_1 and q_1 are equal up to units, the factorizations of z in (5) are the same up to units.

Having covered the case $N\left(\frac{z}{p_1}\right) = 1$, we may assume for the remainder of the proof that $N\left(\frac{z}{p_1}\right) > 1$. In addition, $N\left(\frac{z}{p_1}\right) < N(z)$, as you will show in Exercise 14. Thus, we may apply our inductive hypothesis to conclude that the two factorizations in equation (6) are the same up to order and units. But the factorizations in equation (5) are just the factorizations in equation (6) with an additional factor of p_1 (and an extra factor of u in one of the factorizations). We conclude that the two factorizations in equation (5) are the same up to order and units. Thus, we have shown that z has at most one prime factorization up to order and units.

By the Principle of Strong Induction, it follows that every Gaussian integer z with norm greater than 1 has at most one prime factorization up to order and units. ●

In Chapter 6, we proved that every ordinary integer greater than 1 can be factored uniquely as a product of primes. In this section, we have shown that every Gaussian integer with norm greater than 1 can be factored uniquely as a product of Gaussian primes. In fact, the results of this section can be viewed in a much more general context, using the framework of abstract algebra. The proofs in this section actually establish that in *any* number system in which the division theorem holds, unique factorization into primes also holds. In the language of abstract algebra, this can be stated quite succinctly: Every *Euclidean domain* (i.e., ring in which the division theorem holds) is a *unique factorization domain* (i.e., ring in which all elements have unique factorizations into primes).

EXERCISES 13.4

Numerical Problems

1. Equations (3) and (4) show two different factorizations of the number 5 into prime Gaussian integers.

   a. Give two other factorizations of 5 into prime Gaussian integers.

   b. The Fundamental Theorem of Gaussian Arithmetic states that every Gaussian integer factors uniquely into primes. Doesn't your answer to part a contradict this assertion? Explain.

2. Let $z = 5 + 2i$.

   a. Explain how you know that z is prime in the Gaussian integers.

   b. Let $w = 131 - 23i$. Show that $z \mid w$.

   c. Show that w factors as $(-12 + i)(-11 + i)$.

   d. Proposition 13.4.3 (Fundamental Property of Primes) allows you to conclude that $z \mid (-12 + i)$ or $z \mid (-11 + i)$. Demonstrate which of these two statements is true.

Reasoning and Proofs

In Exercises 3–6, factor the given Gaussian integer into a product of prime Gaussian integers.

3. $9 + 7i$ **4.** $1 - 4i$ **5.** $2 - 5i$ **6.** $-8 + 9i$

7. Prove the Gaussian Euclid's Lemma (13.4.2).

8. a. Consider the following set of Gaussian integers: $T = \{1 + 3i, 4 - 2i, 3 - i, 17i\}$. Find the norm of each element of T. This set has two elements of smallest norm. Which ones are they?

    b. Consider the set U of all Gaussian integer multiples of $5 - 2i$: $U = \{(5 - 2i)z \mid z \in \mathbf{Z}[i]\}$. How many elements of smallest norm does this set have? List them.

    c. Explain why it follows from the Well-Ordering Principle (2.2.2) that every nonempty set of Gaussian integers has an element of smallest norm.

    d. Using the result stated in part c, show that in the proof of Theorem 13.4.1, the set S has a nonzero element of smallest norm. [*Hint:* First show that a and b are not both zero.]

9. Let $q, a \in \mathbf{Z}[i]$ such that q is prime and $q \nmid a$. Show that q and a are relatively prime.

10. Prove the following corollary of the Fundamental Property of Primes (13.4.3). This is the Gaussian integer analogue of Corollary 6.1.3:

    COROLLARY. *Let p be a prime Gaussian integer. Let $a_1, a_2, \ldots, a_r \in \mathbf{Z}[i]$. If $p \mid a_1 \cdot a_2 \cdot \; \cdots \; \cdot a_r$, then $p \mid a_i$ for some $i = 1, 2, \ldots, r$.*

11. Prove the existence part of the Fundamental Theorem of Gaussian Arithmetic (13.4.4). That is, prove that any Gaussian integer n, with $N(n) > 1$, can be written as a product of one or more prime Gaussian integers.

12. Let p and q be prime Gaussian integers. Prove that if $p \mid q$, then $q = u \cdot p$ for some unit $u \in \mathbf{Z}[i]$.

13. Let $v \in \mathbf{Z}[i]$ be a unit. Show that any divisor of v in $\mathbf{Z}[i]$ is also a unit.

14. Let $p, n \in \mathbf{Z}[i]$ such that p is prime and $p \mid n$. Show that $N\!\left(\frac{n}{p}\right) < N(n)$.

15. a. Prove that if $z = a + bi$ is prime in $\mathbf{Z}[i]$, then its complex conjugate, $z^* = a - bi$, is also prime in $\mathbf{Z}[i]$.

    b. For which primes z in $\mathbf{Z}[i]$ is it the case that z and z^* are equal up to units (i.e., that $z = u \cdot z^*$ for some unit u)?

Advanced Reasoning and Proofs

EXPLORATION The Number Systems Z[√−5] and Z[√−2] (Exercises 16−18)

As proven in the Theorem 13.4.4, the Gaussian integers, $\mathbf{Z}[\sqrt{-1}]$, has the property that prime factorizations are unique. In contrast, the number system $\mathbf{Z}[\sqrt{-5}]$ does not have the property of unique factorization, as discussed in Exercises 20–24 of Section 13.2. As discussed in those exercises, in the number system $\mathbf{Z}[\sqrt{-5}]$ the norm is defined by $N(a + b\sqrt{-5}) = a^2 + 5b^2$.

16. Since our proof of unique factorization in $\mathbf{Z}[i]$ rests squarely on the Division Theorem (13.3.1), it is natural to conjecture that the Division Theorem does not hold for $\mathbf{Z}[\sqrt{-5}]$. Show directly that the Division Theorem fails in $\mathbf{Z}[\sqrt{-5}]$. That is, find $a, b \in \mathbf{Z}[\sqrt{-5}]$ such that there do not exist $q, r \in \mathbf{Z}[\sqrt{-5}]$ that satisfy $a = qb + r$ and $N(r) < N(b)$.

17. It is also natural to conjecture that the Fundamental Property of Primes fails in $\mathbf{Z}[\sqrt{-5}]$. Show that this is the case by finding a prime $p \in \mathbf{Z}[\sqrt{-5}]$ and two elements $a, b \in \mathbf{Z}[\sqrt{-5}]$ such that $p \mid ab$, but $p \nmid a$ and $p \nmid b$.

18. Consider the number system $\mathbf{Z}[\sqrt{-2}]$.

    a. State and prove a Division Theorem (analogous to Theorem 13.3.1) for this ring.

    b. Do you think prime factorizations are unique in $\mathbf{Z}[\sqrt{-2}]$? Explain.

13.5 Gaussian Primes

Factoring integers into products of Gaussian integers

By the Fundamental Theorem of Gaussian Arithmetic (13.4.4), every element of $\mathbf{Z}[i]$ (other than zero and units) can be written as the product of prime Gaussian integers. But which elements of $\mathbf{Z}[i]$ are prime?

   To answer this question, we might start by asking which of the ordinary natural number primes, 2, 3, 5, 7, and so forth, are still prime in the Gaussian integers. For instance, in Section 13.2, we saw that 5 factors as $(1 + 2i)(1 - 2i)$, and hence 5 is not a prime in $\mathbf{Z}[i]$. On the other hand, 7 *is* prime in $\mathbf{Z}[i]$, as we now show.

Naomi's Numerical Proof Preview: Proposition 13.5.1

PROOF **THAT 7 IS A GAUSSIAN PRIME.**

We prove this by contradiction.

Assumption: Suppose 7 is not prime as a Gaussian integer.

   Then there exist $w, z \in \mathbf{Z}[i]$ such that neither w nor z is a unit and

$$7 = wz.$$

By Lemma 13.1.4,

$$N(7) = N(w)\,N(z),$$

or, equivalently,

$$49 = N(w)\,N(z).$$

Since w and z are not units, their norms are both greater than 1, which implies that $N(w) = N(z) = 7$. Since w is a Gaussian integer, we can write it in the form $w = a + bi$, where $a, b \in \mathbf{Z}$, and thus,

$$N(w) = a^2 + b^2.$$

That is,

$$7 = a^2 + b^2.$$

But 7 cannot be expressed as the sum of two perfect squares. $\Rightarrow\Leftarrow$
   Thus, our Assumption must be false.

In this proof, the key step was to note that 7 cannot be expressed as the sum of two perfect squares. It turns out that all primes p that are congruent to 3 (mod 4) share this property, as guaranteed by the following proposition. As a corollary, it follows that if a prime integer p is congruent to 3 (mod 4), then p is prime in the Gaussian integers.

PROPOSITION 13.5.1

Let $n \in \mathbf{N}$. If $n \equiv 3$ (mod 4), then there do not exist $a, b \in \mathbf{Z}$ such that $n = a^2 + b^2$.

Put it in Prose, Paul!

If a number is congruent to 3 (mod 4), then it's not a sum of two squares.

Before proving Proposition 13.5.1, we state the following corollary, which you will prove in Exercise 8.

Let $p \in \mathbf{N}$ be a prime integer. If $p \equiv 3$ (mod 4), then p is prime as a Gaussian integer.

We now prove Proposition 13.5.1.

PROOF OF **PROPOSITION 13.5.1** (By contradiction.)

Let $n \in \mathbf{N}$ such that $n \equiv 3$ (mod 4).

Assumption: Suppose there exist $a, b \in \mathbf{Z}$ such that $n = a^2 + b^2$.

Consider the table of squares modulo 4:

Squares modulo 4

x	0	1	2	3
x^2 **(mod 4)**	0	1	0	1

From the table, we see that 0 and 1 are the only squares modulo 4. Thus, either $a^2 \equiv 0$ (mod 4) or $a^2 \equiv 1$ (mod 4). Similarly, either $b^2 \equiv 0$ (mod 4) or $b^2 \equiv 1$ (mod 4). Since a^2 and b^2 are each congruent to 0 or 1 (mod 4), their sum, $a^2 + b^2$, must be congruent to either $0 + 0 \equiv 0$ (mod 4), $0 + 1 \equiv 1$ (mod 4), or $1 + 1 \equiv 2$ (mod 4). Hence, it is impossible for their sum, $n = a^2 + b^2$, to be congruent to 3 (mod 4). This contradicts the hypothesis that $n \equiv 3$ (mod 4). $\Rightarrow\Leftarrow$

Thus, there do not exist integers $a, b \in \mathbf{Z}$ such that $n = a^2 + b^2$. ∎

We've learned that prime integers $p \equiv 3$ (mod 4) remain prime in $\mathbf{Z}[i]$. What about prime integers $p \equiv 1$ (mod 4)? We saw in Section 13.2 that 5 is not prime in $\mathbf{Z}[i]$, because $5 = (1 + 2i)(1 - 2i)$. After 5, the next prime integer that is congruent to 1 (mod 4) is 13, and again we see that 13 is not prime in $\mathbf{Z}[i]$, since $13 = (3 + 2i)(3 - 2i)$. Indeed, every prime $p \equiv 1$ (mod 4) can be factored in $\mathbf{Z}[i]$, as asserted by the following converse of Corollary 13.5.2.

Let $p \in \mathbf{N}$ be a prime integer. If $p \equiv 1$ (mod 4), then p is not prime as a Gaussian integer.

PROOF Let p be a prime integer such that $p \equiv 1$ (mod 4).

Using Theorem 11.2.3 (alternatively, see Exercise 10), we know that there exists $x \in \mathbf{Z}$ such that

$$x^2 \equiv -1 \ (\text{mod} \ p).$$

Thus (by the Equivalent Conditions for Congruence, 8.2.4), there exists $k \in \mathbf{Z}$ such that

$$x^2 = -1 + kp,$$

or, equivalently,

$$x^2 + 1 = kp.$$

Although $x^2 + 1$ cannot be factored as a polynomial over $\mathbf{Z}$, it *can* be factored over $\mathbf{Z}[i]$:

$$(x + i)(x - i) = kp.$$

Therefore, in $\mathbf{Z}[i]$ we have

$$p \mid (x + i)(x - i).$$

We now show by contradiction that p is not prime in the Gaussian integers.

Assumption: Suppose p is prime as a Gaussian integer.

Then by Proposition 13.4.3 (Fundamental Property of Primes),

$$p \mid x + i \quad \text{or} \quad p \mid x - i.$$

Thus, there exists a Gaussian integer $m + ni$ such that

$$p(m + ni) = x + i \quad \text{or} \quad p(m + ni) = x - i.$$

Considering only the imaginary part of each Gaussian integer equation above, we get:

$$pn = 1 \quad \text{or} \quad pn = -1. \tag{1}$$

This contradicts the hypothesis that p is a prime in $\mathbf{Z}$. $\Rightarrow\Leftarrow$
    We conclude that p is not prime as a Gaussian integer. ∎

The following theorem incorporates both Corollary 13.5.2 and Theorem 13.5.3 in a single statement:

THEOREM 13.5.4

Suppose $p \in \mathbf{N}$ is prime. Then p is prime in $\mathbf{Z}[i]$ if and only if $p \equiv 3$ mod 4.

PROOF You will prove this in Exercise 7.

Theorem 13.5.4 tells us exactly which prime natural numbers remain prime in $\mathbf{Z}[i]$. If someone gives you a natural number like 47 and asks you whether it is prime in $\mathbf{Z}[i]$, you can just use Theorem 13.5.4: Since $47 \equiv 3 \pmod{4}$, we see that 47 is prime in $\mathbf{Z}[i]$. But what if you want to know whether a general Gaussian integer, like $34 + 41i$, is prime in $\mathbf{Z}[i]$? The answer is given by the following theorem.

Let $z \in \mathbf{Z}[i]$. Then z is a prime Gaussian integer if and only if one of the following conditions holds:

(i) $N(z) = 2$,

(ii) $N(z)$ is a prime integer that is congruent to 1 (mod 4),

(iii) z is a unit times a prime integer that is congruent to 3 (mod 4).

PROOF You will prove this theorem in Exercises 11–13.

EXAMPLE 1

We can use Theorem 13.5.5 to determine whether any Gaussian integer is prime. Here are three examples:

a. $z = 1 - i$ is prime in $\mathbf{Z}[i]$ by condition (i), because $N(z) = 1^2 + (-1)^2 = 2$.

b. $z = -7 + 2i$ is prime in $\mathbf{Z}[i]$ by condition (ii), because $N(z) = (-7)^2 + 2^2 = 53$, and 53 is a prime integer that is congruent to 1 (mod 4).

c. $z = -47i$ is prime in $\mathbf{Z}[i]$ by condition (iii), because $z = (-i)47$, which is a unit times a prime integer that is congruent to 3 (mod 4).

EXERCISES 13.5

Numerical Problems

1. Each of the natural numbers shown in parts a–f below is a prime integer. In each case, tell whether or not the number is prime as a Gaussian integer. Justify your answers.

   a. 29   **b.** 31   **c.** 563

   d. 1009   **e.** 2011   **f.** $2^{61} - 1$

2. Determine whether or not each number is prime as a Gaussian integer. Justify your answers.

   a. $1 + i$ **b.** $3 - 2i$ **c.** $101i$

   d. $11 + 2i$ **e.** $-103i$ **f.** $7 + 5i$

3. List all Gaussian primes with norm less than or equal to 12. Your list should be complete up to units: If $x' = ux$ for some unit u, then only one of the Gaussian integers x and x' should be on your list.

Reasoning and Proofs

4. Factor each number into a product of prime Gaussian integers:

   a. 11   b. 14   c. 100i

   d. $2 - 6i$ e. $5 - 7i$ f. $8 + 6i$

5. Show that if $n \equiv 7 \pmod 8$, then n cannot be represented as the sum of 3 perfect squares.

6. In the proof of Theorem 13.5.3, explain why statement (1) implies that p is not prime in **Z**.

7. Prove Theorem 13.5.4.

8. Prove Corollary 13.5.2. [*Hint:* Model your proof on the Numerical Proof Preview for Proposition 13.5.1.]

9. Show that if p is a prime integer such that $p \equiv 1 \pmod 4$, then p factors as the product of two Gaussian integer primes, each of which has norm p. That is, show that there exist prime Gaussian integers z and w such that $p = zw$ and $N(z) = N(w) = p$.

EXPLORATION Is -1 a Square Modulo p? (Exercise 10)

In this exploration, you will establish the following lemma, which is needed at the beginning of the proof of Theorem 13.5.3. (As noted in the proof of Theorem 13.5.3, this lemma also follows from Theorem 11.2.3, which you may recall if you studied quadratic residues.)

> LEMMA. *Let p be a prime natural number such that $p \equiv 1 \pmod 4$. Then there exists $x \in$ **Z** such that $x^2 \equiv -1 \pmod p$.*

10. In this exercise, you will prove the lemma by completing the following steps.

    a. Show that for any prime $p > 2$,

    $$(p - 1)! \equiv (-1)^{\frac{p-1}{2}} \left(\frac{p-1}{2}!\right)^2 \pmod p.$$

    [*Hint:* Starting with the product $(p - 1)! = 1 \cdot 2 \cdot \ldots \cdot (p - 1)$, replace each of the last $\dfrac{p-1}{2}$ factors in this product by a negative number to which it is congruent modulo p.]

    b. Now assume that $p \equiv 1 \pmod 4$. Finish the proof of the lemma by showing that $x = \dfrac{p-1}{2}!$ satisfies the required congruence. [*Hint:* Use Wilson's Theorem (8.3.7).]

In this exploration, you will prove Theorem 13.5.5, which gives you a general technique for determining whether any Gaussian integer is prime.

11. Theorem 13.5.5 is an if and only if statement. Prove direction [⟸] of the theorem. That is, show that if any of the conditions (i), (ii), or (iii) holds, then z is a prime Gaussian integer.

In the following exercises, you will prove direction [⟹] of Theorem 13.5.5.

Let z be a prime Gaussian integer. [You must show that one of the conditions (i), (ii), or (iii) holds.]

12. In this exercise, you will show that there exists a prime integer $p \in \mathbf{N}$ such that $z \mid p$ in $\mathbf{Z}[i]$.

    a. Show that $z \mid N(z)$ in the Gaussian integers.

    b. Show that there exists a prime natural number p such that $z \mid p$ in the Gaussian integers. [*Hint:* Factor $N(z)$ into a product of natural numbers, then apply the Fundamental Property of Primes (13.4.3).]

13. At this point, there are three possible cases for the number p:

    Case A. $p = 2$

    Case B. $p \equiv 1 \pmod 4$

    Case C. $p \equiv 3 \pmod 4$

    a. In Case A, finish the proof by showing that $N(z) = 2$ and find all possible values of z.

    b. In Case B, finish the proof by showing that $N(z) = p$.

    c. In Case C, finish the proof by showing that z is a unit times p.

13.6 Fermat's Two Squares Theorem

So far in this chapter, we have proved a series of fundamental results about the Gaussian integers, each of which has built upon the previous results. Standing atop this edifice of fundamental results about the Gaussian integers, we are able to see farther, not only into the structure of the Gaussian integers, but also into the structure of the integers themselves. In this section, we will use the basic results we have derived about the Gaussian integers, $\mathbf{Z}[i]$, to prove Fermat's Two Squares Theorem, a substantial result about the ordinary integers $\mathbf{Z}$.

A pattern in the primes

We will now leave the Gaussian integers temporarily and once again discuss the ordinary integers. Many integers can be expressed as the sum of two perfect squares. For example, 25 and 68 are both sums of two squares:

$$25 = 3^2 + 4^2, \text{ and } 68 = 2^2 + 8^2.$$

However, many integers, such as 12, cannot be expressed as the sum of two squares: there are no integers a and b for which $12 = a^2 + b^2$. Now let's look at the case where the number is prime. For example, the prime number **13** can be written as

$$\mathbf{13} = 2^2 + 3^2.$$

However, some prime numbers, such as **11**, cannot be expressed as the sum of two squares.

Let's examine which of the prime numbers between 1 and 100 can be expressed as the sum of two squares. In Table 13.6.1, the top row contains all of the squares up to 100. The left column of the table also contains all squares up to 100. Each position inside the table contains the sum of two squares: the square at the top of the column plus the square at the left of the row.

Table 13.6.1 Integers between 0 and 100 that can be expressed as the sum of two squares. Primes are shown in orange.

+	0	1	4	9	16	25	36	49	64	81	100
0	0	1	4	9	16	25	36	49	64	81	100
1	1	2	5	10	17	26	37	50	65	82	
4	4	5	8	13	20	29	40	53	68	85	
9	9	10	13	18	25	34	45	58	73	90	
16	16	17	20	25	32	41	52	65	80	97	
25	25	26	29	34	41	50	61	74	89		
36	36	37	40	45	52	61	72	85	100		
49	49	50	53	58	65	74	85				
64	64	65	68	73	80	89	100				
81	81	82	85	90	97						
100	100										

The primes in Table 13.6.1 (i.e., all of the primes up to 100 that can be expressed as the sum of two squares) are shown in **orange**. Let's copy all of the odd primes that are the sum of two squares in **orange**. All of the other odd primes (those that are not the sum of two squares) are listed below in **brown**.

Sum of two squares: 5, 13, 17, 29, 37, 41, 53, 61, 73, 89, 97 (1)

Not the sum of two squares: 3, 7, 11, 19, 23, 31, 43, 47, 59, 67, 71, 79, 83 (2)

Besides being the sum of two squares, what do the primes in the **orange** list (1) have in common? How about the primes in the **brown** list (2)? Is there an easy way to tell them apart? Can you predict, for example, which list the primes 163 and 1000033 will be in? Every odd number is either congruent to 1 (mod 4) or congruent to 3 (mod 4). Let's look back at our two lists of odd prime numbers. Observe that the **orange** list (1) contains precisely the odd primes up to 100 that are congruent to 1 (mod 4). The **brown** list (2) contains precisely the odd primes up to 100 that are congruent to 3 (mod 4). These observations lead us to the statement of Fermat's Two Squares Theorem.

THEOREM 13.6.1 FERMAT'S TWO SQUARES THEOREM

Let $p > 2$ be a prime integer. There exist $a, b \in \mathbf{Z}$ such that $p = a^2 + b^2$ if and only if $p \equiv 1$ (mod 4).

PROOF Let $p > 2$ be a prime integer.

[$\Rightarrow$] It follows from Proposition 13.5.1 that if there exist $a, b \in \mathbf{Z}$ such that $p = a^2 + b^2$, then $p \equiv 1$ (mod 4).

[$\Leftarrow$] Assume $p \equiv 1$ (mod 4). **[To show: There exist $a, b \in \mathbf{Z}$ such that $p = a^2 + b^2$.]**

By Theorem 13.5.3, p is not prime as a Gaussian integer. Thus, there exist $y, z \in \mathbf{Z}[i]$ such that

$$p = yz,$$

where $N(y) > 1$ and $N(z) > 1$. We then know that

$$N(p) = N(y) \cdot N(z);$$

that is,

$$p^2 = N(y) \cdot N(z).$$

Since $N(y)$ and $N(z)$ are both natural numbers greater than 1, the only valid possibility for their norms is

$$N(y) = p, \ N(z) = p. \tag{3}$$

Because y is a Gaussian integer, we can write

$$y = a + bi,$$

where a and b are integers. By the definition of norm,

$$N(y) = a^2 + b^2.$$

So by (3),

$$p = a^2 + b^2. \qquad \blacksquare$$

Fermat's Two Squares Theorem tells us that any prime number that is congruent to 1 modulo 4, such as 8675309, can be written as the sum of two squares: $8675309 = a^2 + b^2$. But how would we actually find the integers a and b? To answer this question, it is natural to look back at the proof of this theorem. Looking back at the proof, we see that we can find the squares in question provided that we can find the Gaussian integer factors of the number 8675309. However, the proof does not seem to tell us how to find these factors. A closer look at the proof of the supporting results (Theorems 13.5.3 and 11.2.3), though, does lead to a method—in fact, an efficient method—for finding these factors. In Exercises 11–13, you will explore this method in detail. The end result is that even for very large primes p, the numbers a and b for which $p = a^2 + b^2$ can be found efficiently (in polynomial time).

Fermat's Two Squares Theorem states that any odd prime number $p \equiv 1 \pmod 4$ can be written as a sum of two squares: $p = a^2 + b^2$. For example, we know that $13 = 2^2 + 3^2$. Are there any other ways to express 13 as a sum of two squares? The reader may verify that if

$$13 = a^2 + b^2,$$

Then the only possible values for the squares a^2 and b^2 are

(i) $a^2 = 4$ and $b^2 = 9$

         or

(ii) $a^2 = 9$ and $b^2 = 4$.

Thus, $4 + 9$ and $9 + 4$ are the only ways to write 13 as the sum of two squares. The situation is much the same for any prime $p \equiv 1 \pmod 4$, as asserted in the following theorem.

THEOREM 13.6.2   UNIQUENESS OF FERMAT'S TWO SQUARES THEOREM

Let p be a prime integer such that $p \equiv 1 \pmod 4$. Then p can be expressed as a sum of two squares in a unique way, up to order.

PROOF By Fermat's Two Squares Theorem (13.6.1), p can be written as a sum of two squares. Suppose there are two ways to write p as a sum of two squares; in other words, suppose there exist $a, b, c, d \in \mathbf{Z}$ such that

$$p = a^2 + b^2 \quad \text{and} \quad p = c^2 + d^2.$$

Then we can factor p as a Gaussian integer in two ways:

$$p = (a + bi)(a - bi) \qquad (4)$$

and

$$p = (c + di)(c - di). \qquad (5)$$

Since

$$N(a + bi) = a^2 + b^2 = p, \qquad (6)$$

we conclude (by Lemma 13.2.6) that $a + bi$ must be prime as a Gaussian integer. Similarly, the numbers $a - bi$, $c + di$, and $c - di$ are all prime as Gaussian integers. Thus, equations (4) and (5) are two ways to factor p as a product of prime Gaussian integers.

The Fundamental Theorem of Gaussian Arithmetic (13.4.4) tells us that the two factorizations (4) and (5) must be the same, up to order and units. In particular,

$$(a + bi) = u(c + di) \quad \text{or} \quad (a + bi) = u(c - di)$$

for some unit $u \in \mathbf{Z}[i]$. Then (as you will show in Exercise 7) there are only two possibilities: either

(*i*) $a^2 = c^2$ and $b^2 = d^2$

or

(*ii*) $a^2 = d^2$ and $b^2 = c^2$.

Fun Facts

In this chapter, we've learned a lot about which numbers can be expressed as the sum of two squares. It turns out that *every* nonnegative integer can be expressed as the sum of *four* squares. This result, known as the Four Squares Theorem, was originally stated by Diophantus, but he did not give a proof. In the 17th century, Fermat asserted that he had a proof of this fact. However, as there is no written record of his proof, it is not clear whether he actually had a proof. The proof of the Four Squares Theorem was finally published in 1770 by Joseph Louis Lagrange. (See the beginning of Chapter 10 for a brief history of Lagrange's life and work.) Just as we used a two-dimensional number system (the Gaussian integers) to prove the Two Squares Theorem, one can use a four-dimensional number system, the *quaternions*, to prove the Four Squares Theorem.

Numerical Problems

1. If possible, write each integer as a sum of two squares. If it is not possible, explain why not.

   a. 41   b. 18   c. 21

2. If possible, write each integer as a sum of two squares. If it is not possible, explain why not.

   a. 145   b. 103   c. 202

3. All of the following integers are prime. In each case, tell whether or not the prime number can be written as a sum of two squares. Explain your reasoning.

   a. 149   b. 151   c. 91711

   d. 10,000,019   e. 23,456,789   f. $10^{200} + 357$

4. Many composite numbers can be written as a sum of two squares. However, unlike the prime numbers that can be written in this way, the representation of composite numbers as a sum of two squares is not always unique. Write the following composite integers as a sum of two squares in more than one way.

   a. 50   b. 65   c. 221

5. In how many different ways can you write each of the following integers as a sum of two squares? Show them all.

   a. 229   b. 125   c. 107

Reasoning and Proofs

6. a. Prove that if m and n are integers that can each be written as a sum of two squares, then mn can also be written as a sum of two squares.
   [*Hint:* Write m as $a^2 + b^2$ and n as $c^2 + d^2$. Let $r = a + bi$ and $s = c + di$, and use Lemma 13.1.4.]

   b. Let $m = 13 = 2^2 + 3^2$ and $n = 101 = 1^2 + 10^2$. Use your results from part a to express $mn = 1313$ as the sum of two squares.

   c. Use your results from part a to find a different expression for 1313 as the sum of two squares. [*Hint:* You may want to consider complex conjugates.]

7. Let $a + bi$ and $c + di$ be Gaussian integers. Prove that if $(a + bi) = u(c + di)$ or $(a + bi) = u(c - di)$ for some unit $u \in \mathbf{Z}[i]$, then either

   (*i*) $a^2 = c^2$ and $b^2 = d^2$

   or

   (*ii*) $a^2 = d^2$ and $b^2 = c^2$.

8. **a.** List all of the prime Gaussian integers of norm 101, up to multiplication by units.

   b. List all Gaussian integers z of norm $n = 1030301 = 101^3$, up to multiplication by units. [*Hint:* z factors into Gaussian primes. What are the possibilities for the factors?]

   c. Using your answer to part b, find all ways of writing 1030301 as the sum of two squares.

Advanced Reasoning and Proofs

EXPLORATION How to Tell Whether Any Natural Number is the Sum of Two Squares (Exercises 9–10)

Fermat's Two Squares Theorem (13.6.1) tells us precisely which primes $p \in \mathbf{N}$ can be expressed as the sum of two squares. But which composite numbers can be expressed as the sum of two squares? In this exploration, you will prove the following theorem, which tell us when an arbitrary natural number is the sum of two squares.

> THEOREM. *A natural number $n \geq 2$ is expressible as the sum of two squares if and only if every prime $q \equiv 3 \pmod 4$ that appears in the prime factorization of n appears with an* even *exponent.*

9. Use the result of Exercise 6 part a to prove the [⇐] direction of the Theorem. (Assume that every prime $q \equiv 3 \pmod 4$ that appears in the prime factorization of n appears with an even exponent, and prove that n can be expressed as the sum of two squares.)

10. Now you will prove the [⇒] direction of the Theorem. Assume that n can be expressed as the sum of two squares.

    a. Show that there is a Gaussian integer z such that $N(z) = n$.

    b. Consider the factorization of z into primes in $\mathbf{Z}[i]$: $z = z_1 \cdot z_2 \cdot \ \cdots \ \cdot z_r$. According to Theorem 13.5.5, what are the possibilities for the norm of each Gaussian prime factor, $N(z_i)$?

    c. Finish the proof of the Theorem. [*Hint:* Use your answer to part b and the Multiplicativity of the Norm (13.1.4).]

EXPLORATION Finding the Squares in Fermat's Two Squares Theorem (Exercises 11–13)

Suppose we have a large prime number p that is congruent to 1 modulo 4. Fermat's Two Squares Theorem (13.6.1) tells us that p can be written as the sum of two squares. But how can we find these squares? This exploration introduces an efficient method to find them.

> Let $p \in \mathbf{N}$ be prime, with $p \equiv 1 \pmod 4$. We wish to find $a, b \in \mathbf{Z}$ such that $p = a^2 + b^2$.

The proof of Fermat's Two Squares Theorem reveals that finding a and b amounts to finding the factors of p in $\mathbf{Z}[i]$, which we know exist by Theorem 13.5.3. The proof of Theorem 13.5.3 relied on Theorem 11.2.3, which asserts that there exists $x \in \mathbf{Z}$ such that

$$x^2 \equiv -1 \ (\text{mod } p). \tag{7}$$

Thus, the first step in finding a and b is to find x satisfying congruence (7).

11. **a.** Suppose that a is an integer such that $a^{\frac{p-1}{2}} \equiv -1 \ (\text{mod } p)$. Show that if $x \equiv a^{\frac{p-1}{4}} \ (\text{mod } p)$, then x satisfies congruence (7).

   b. If you studied Chapter 10 or 11, explain why it is easy with the help of a computer to find an integer a satisfying the conditions of part a.

   Having found a solution x to congruence (7), you will now show that finding a and b is as easy as calculating a single greatest common divisor in $\mathbf{Z}[i]$. Finding such a gcd can be accomplished quickly using the Euclidean Algorithm for $\mathbf{Z}[i]$. (See Exercises 20–25 of Section 13.3, where it is also shown that any two Gaussian integers have a gcd that is unique up to units.)

12. Suppose that x satisfies congruence (7). Show that the greatest common divisor of $x + i$ and p in $\mathbf{Z}[i]$ has norm p. (That is, show that $x + i$ and p share a common factor $z \in \mathbf{Z}[i]$, with $N(z) = p$, but do not share any factors in $\mathbf{Z}[i]$ with norm larger than p.)

13. **Computer Exercise** To do this exercise, you will need to use software supporting a Gaussian integer gcd command. (For example, Maple has a package called GaussInt, which contains the GIgcd command.) Or as a challenge, you could write your own Gaussian gcd program using the results in this chapter.

   Following the ideas of exercises 11 and 12, express each of the following primes as the sum of two squares:

   a. 8675309   **b.** $10^{102} + 117$

EXPLORATION How Many Ways Can a Number Be Expressed as the Sum of Two Squares? (Exercises 14–18)

In Theorem 13.6.2 (Uniqueness of Fermat's Two Squares Theorem), we proved that any prime congruent to 1 modulo 4 can be expressed as a sum of two squares in a unique way. Some composite numbers, however, may be expressed as a sum of two squares in more than one way. In this exploration, you will find a general formula for the number of ways that an arbitrary natural number can be expressed as the sum of two squares.

For the purposes of counting, we will count two expressions $n = a^2 + b^2$ as the same if they differ only by change of sign and order of the terms. For instance, $65 = 4^2 + 7^2$ and $65 = (-7)^2 + 4^2$ count as just one way of expressing 65 as a sum of two squares.

14. Let n be a natural number. Explain why the number of ways to express n as the sum of two squares is the same as the number of Gaussian integers of norm n, up to conjugation and multiplication by units.

15. According to the theorem from Exercises 9–10, in order to be expressible as a sum of two squares at all, our number n must have a prime factorization (in $\mathbf{Z}$) of the form

$$n = 2^a \cdot p_1^{b_1} \cdots p_r^{b_r} \cdot q_1^{2c_1} \cdots q_s^{2c_s},$$

where each prime $p_j \equiv 1 \pmod 4$, each prime $q_j \equiv 3 \pmod 4$, and all of the exponents $a, b_i,$ and c_i are nonnegative integers. By Fermat's Two Squares Theorem (13.6.1), each p_j can be expressed as a sum of two squares: $p_j = x_j^2 + y_j^2$. Show that the prime factorization of n in $\mathbf{Z}[i]$ has the form

$$n = u \cdot (1 + i)^{2a} \cdot (x_1 + y_1 i)^{b_1} (x_1 - y_1 i)^{b_1} \cdots (x_r + y_r i)^{b_r} (x_r - y_r i)^{b_r} \cdot q_1^{2c_1} \cdots q_s^{2c_s},$$

where u is a unit.

16. Let z be a Gaussian integer of norm n. Show that $z \mid n$ in $\mathbf{Z}[i]$. Use the Fundamental Theorem of Gaussian Arithmetic (13.4.4) to conclude that z has the form

$$z = v \cdot (1 + i)^a \cdot (x_1 + y_1 i)^{f_1} (x_1 - y_1 i)^{g_1} \cdots (x_r + y_r i)^{f_r} (x_r - y_r i)^{g_r} \cdot q_1^{c_1} \cdots q_s^{c_s},$$

where v is a unit, and the f_i and g_i are nonnegative integers such that $f_i + g_i = b_i$.

17. Using your result from Exercise 16, determine how many Gaussian integers there are of norm n, up to units. (Your formula should involve the b_i, but none of the other variables.)

18. Finally, give a formula for the number of Gaussian integers of norm n, up to units and conjugation. As noted in Exercise 14, your formula also gives the number of ways to write n as the sum of two squares. [*Hint:* Note that swapping the values of f_i and g_i produces conjugate values of z. Pay special attention to the case in which all of the b_i are even.]

Chapter 14

SPOTLIGHT ON...

Srinivasa Ramanujan

(1887–1920)

The mathematical genius Srinivasa Ramanujan grew up in the village of Erode, India, not far from Madras. His father worked as a bookkeeper in a fabric shop. Ramanujan's unusual intelligence and fascination with mathematics were clear from an early age. His abilities were so far beyond those of his schoolmates that he was promoted to high school after two years of elementary school. At the age of 13, Ramanujan derived the formulas for the sums of arithmetic and geometric series. When he was 15, shortly after being taught how to solve cubic equations (polynomial equations of degree 3), Ramanujan figured out on his own how to solve quartic equations (polynomial equations of degree 4). When Ramanujan was 16, he borrowed the book *Synopsis of Pure Mathematics*, written by G. S. Carr in 1856, which contained the statements of over 6,000 theorems. As Carr's book included almost no proofs, Ramanujan set out to prove all of the theorems in the book. As a result of this experience, Ramanujan decided to devote his life to mathematics.

After high school, Ramanujan was offered a scholarship to attend the Government College in Kumbakonam. However, because of poor performance in his English classes, Ramanujan lost his scholarship after the first year, and he was forced to drop out

of college. This gave him more time to prove the theorems in Carr's book and to read the *Journal of the Indian Mathematical Society*. Ramanujan began doing mathematics research at the age of 17, when he calculated Euler's constant, $\lim_{n \to \infty} \left(\Sigma_{k=1}^{n} \frac{1}{k} - \log(n) \right)$, to 15 decimal places (Ramanujan did not know that Euler had already calculated this value to 16 decimal places). Afterward, he proved several of his own theorems on continued fractions, divergent series, and the distribution of primes.

In 1909, Ramanujan suffered from serious health problems. While Ramanujan was recuperating from an operation, his mother arranged his marriage to a 9-year-old girl, a custom that was not unusual in India at the time. Although Ramanujan did not live with his wife until she was 12, he was expected to earn a living as soon as they were married. Without a college degree, however, it was difficult for him to find a good job. He worked for a while at the Accountant General's Office and then as a clerk in the accounts section of the Madras Port Trust. He tried unsuccessfully to find a job that would pay him to do research in mathematics. Nonetheless, he continued working on mathematics in his spare time and published several papers in the *Journal of the Indian Mathematical Society*.

After reading the book *Orders of Infinity* by British mathematician G. H. Hardy, Ramanujan wrote to Hardy to ask for his help in obtaining a mathematics position in India. He explained that although he had no university degree, he had been pursuing mathematical research on his own. In his letter, Ramanujan enclosed a list of over 100 theorems that he had proven. On first reading, Hardy did not know quite what to make of the letter, thinking that perhaps it was a hoax or written by a crank. But after three hours of careful study together with his collaborator John Littlewood, Hardy concluded that the letter was so original that it could only have been written by a mathematical genius. In fact, Hardy was so impressed with Ramanujan's results that he invited him to come to Cambridge University to work with him. Ramanujan's religion forbade him from traveling outside India, however, so Hardy arranged for Ramanujan to have a two-year fellowship at the University of Madras, all the while continuing to urge him to come to England. During Ramanujan's fellowship at the University of Madras, his mother had a dream in which she saw him in a room filled with Europeans. She concluded that the family goddess was saying Ramanujan should go to England to work with Hardy. Thus, in 1914, Ramanujan accepted Hardy's offer of a fellowship at Cambridge University.

Before going to Cambridge, Ramanujan had kept three notebooks that he filled with about 3,000 mathematical discoveries. These discoveries included brilliant new observations, as well as many results that Ramanujan discovered independently though they had been known for years, and even some results that were incorrect.

Since Ramanujan had had no access to mathematics books or European mathematics journals, he had no way of knowing the current state of mathematical knowledge. In addition, Ramanujan had never been taught how to write rigorous mathematical proofs, so the arguments that he gave in his notebooks could be described more as plausibility arguments than as proofs. For years after his death, mathematicians struggled to find proofs of some of Ramanujan's claims. About Ramanujan's notebooks, Hardy wrote, "His ideas as to what constituted a mathematical proof were of the most shadowy description. All of his results, new or old, right or wrong, had been arrived at by a process of mingled argument, intuition, and induction, of which he was entirely unable to give any coherent account."

With guidance from G. H. Hardy and John Littlewood, Ramanujan began publishing his papers in British journals. During the five years that he was a fellow at Cambridge University, Ramanujan published 21 papers, 7 of which were coauthored with Hardy. He was also given the great honor of being selected as a Fellow of the Royal Society of England.

Ramanujan had health problems on and off starting the first winter he spent in England. In 1917, he became seriously ill (possibly with tuberculosis) and spent more than a year in several British hospitals and sanatoriums. Even when he was ill, his primary focus was on mathematics. Once, when Hardy came to visit Ramanujan in the hospital, he reported to Ramanujan that the taxi he had taken had a dull license plate number: 1729. Ramanujan immediately responded that 1729 is not a dull number, since it is the smallest positive integer that can be written in two different ways as the sum of two cubes ($1729 = 1^3 + 12^3$ and $1729 = 9^3 + 10^3$).

Ramanujan's health did not improve. He blamed his illness on the cold, wet English climate and the difficulty of finding vegetarian food. Because of his religion, Ramanujan would not eat meat, and it was hard to obtain an adequate supply of vegetables in England because of rationing during World War I. In 1919, Ramanujan felt that he had no choice but to return home. Back in India, he continued to do research, even as his condition rapidly deteriorated. He died the following year at the age of 33.

In addition to the three mathematical notebooks that Ramanujan had brought to England, he wrote some of his mathematical notes in a fourth notebook. This notebook was subsequently lost and was rediscovered only in 1976. It is believed that Ramanujan wrote in this notebook during his final months in India. The notebook contains 100 pages, filled with a list of 600 formulas. As usual, however, the notebook contains no proofs.

A Bit of Ramanujan's Math

Ramanujan discovered important theorems in the areas of number theory, elliptic functions, continued fractions, and infinite series. He found many different infinite series representations for π and for $\frac{1}{\pi}$. Perhaps his most important contribution to number theory concerns the theory of *partitions* of numbers into summands, which he developed with Hardy.

A *partition* is a way to break a natural number into a sum of natural numbers. For example, $3 + 5$ and $1 + 1 + 6$ are both partitions of 8. The order of the summands does not matter; for example, $5 + 3$ and $3 + 5$ are considered to be the same partition of 8.

The function $p(n)$ counts the number of partitions of n. For example, $p(5) = 7$ because there are 7 different partitions of the number 5:

$$1 + 1 + 1 + 1 + 1, \ 1 + 1 + 1 + 2,$$
$$1 + 2 + 2, \ 1 + 1 + 3, \ 2 + 3, \ 1 + 4, \ 5.$$

As n increases, $p(n)$ gets large very quickly and becomes extremely hard to compute. Ramanujan and Hardy derived the following formula that approximates $p(n)$ for large values of n:

$$p(n) \approx \frac{e^{c\sqrt{n}}}{4n\sqrt{3}},$$

where $c = \pi\sqrt{\frac{2}{3}}$. This formula gives amazingly accurate values for $p(n)$.

Most of Ramanujan's results are in the area of analytic number theory. In this branch of number theory, theorems from analysis, particularly complex analysis, are used to prove results about the natural numbers. The connection between number theory and analysis is both surprising and deep. The field of analytic number theory includes the Riemann Hypothesis, one of the most important outstanding conjectures in mathematics today. This conjecture concerns the roots of a certain analytic function known as the Riemann zeta function. Though most mathematicians believe the Riemann Hypothesis is true, the conjecture has remained unproven for over 140 years. If the Riemann Hypothesis is in fact true, then it would give us estimates of $\pi(n)$, the number of primes less than or equal to n, that are more accurate than those given by the Prime Number Theorem.

Chapter 14 Continued Fractions
or A Cantankerous Collaboration

14.1 Expressing Rational Numbers as Continued Fractions

The good doctor

Born near the Indian city of Madras, Srinivasa Ramanujan was one of the greatest mathematical geniuses of all time. What's not so well known about Ramanujan is that he was a couples' therapist. And while he had saved countless troubled marriages, he was even more famous for repairing dysfunctional mathematical collaborations. Ramanujan's office at Madras Medical School was located in the school's main building, the Ravi D. Tajma Memorial Hall. People from all over the world came to Tajma Hall seeking treatment from the great doctor. His most famous case concerned two prominent mathematics professors from Cambridge University, G. H. Hearty and I. M. A. Littleweird. From the day they arrived, Hearty and Littleweird had a hard time adjusting to life in India. They missed England's gloomy weather, bland food, and fine imported teas.

In the medical school's Division of Arithmetical and Geometrical Disorders, Hearty and Littleweird waited in Dr. Ramanujan's consultation room. Ramanujan entered carrying one of his notebooks. "I can't find anything wrong—all of your tests are orthonormal. Your issues must be complex. Have you considered analysis?"

"In the past," Hearty replied, "analysis has been integral to the success of our collaboration. But we still don't have a satisfying functional relationship."

"Alright," said Dr. Ramanujan, "It would help if I could see how the two of you operate together. Please demonstrate for me a typical work session."

"Very well. Shall I begin?" said Littleweird, turning to his collaborator. "You'll be happy to know, Hearty, that I figured out the answer to that problem we were discussing yesterday." Littleweird approached the blackboard and wrote

$$\frac{52}{9}.$$

"Well done, old friend!" Hearty responded. "But you know I have mixed feelings about improper fractions. Let me tidy that up a bit." Hearty erased the fraction and wrote

$$5 + \frac{7}{9}.$$

"My dear Hearty, why must you always be so proper?" Littleweird quickly erased what his colleague had written and replaced it with

$$5 + \frac{1}{\frac{9}{7}}.$$

Hearty gasped. "I simply cannot abide this! Such impropriety would not befit even an Oxford freshman, let alone a Cambridge don such as yourself." Turning to Dr. Ramanujan, Hearty said, "You see what I have to put up with? I can't work with him when he acts like this!"

"I'm sorry. . . . Sometimes I just can't control myself," admitted Littleweird.

"Gentlemen," interjected Dr. Ramanujan, "allow me to make a hypothesis about where the roots of your problem lie. I'm afraid that Professor Littleweird is suffering from a case of IRS, Inverted Ratio Syndrome."

"IRS?" exclaimed Littleweird. "That sounds dreadful."

"The disease manifests itself as an uncontrollable urge to invert proper fractions."

"Can you prescribe some arithmedication for the disorder?" asked Hearty.

"Unfortunately not," the good doctor replied.

Dr. Ramanujan took Hearty aside and continued in a low voice. "Professor Hearty, you should show Littleweird that you value his contributions, by working with them rather than rejecting them outright."

"Righto," said Hearty. "Suppose I rewrite the improper fraction $\frac{9}{7}$ as a mixed number, like this." Hearty erased the previous expression and wrote

$$5 + \frac{1}{1 + \frac{2}{7}}.$$

Littleweird darted to the board and replaced the expression with

$$5 + \frac{1}{1 + \frac{1}{\frac{7}{2}}}.$$

Ramanujan looked at Hearty and said, "Now don't be anxious. Remember what we talked about."

Hearty took a few deep breaths. "Ah, I think I've got it. Now I can rewrite the improper fraction $\frac{7}{2}$ as a mixed number."

$$5 + \frac{1}{1 + \frac{1}{3 + \frac{1}{2}}}.$$

Littleweird looked at the expression on the board, scratched his head, and put down the chalk.

"There, you see," Ramanujan said to Hearty. "Since the fraction $\frac{1}{2}$ has a numerator of 1, Littleweird will not invert it."

"By George, you've done it!" Hearty shook Ramanujan's hand vigorously. "My hat's off to you, Doctor—you've saved our collaboration!"

Dr. Ramanujan had indeed saved the day, and the two honorable gentlemen went on to prove many great theorems together.

Introducing continued fractions

Let's look at the steps that **Hearty** and **Littleweird** followed in our story:

$$\frac{52}{9} = 5 + \frac{7}{9}$$

$$= 5 + \cfrac{1}{\cfrac{9}{7}}$$

$$= 5 + \cfrac{1}{1 + \cfrac{2}{7}}$$

$$= 5 + \cfrac{1}{1 + \cfrac{1}{\cfrac{7}{2}}}$$

$$= 5 + \cfrac{1}{1 + \cfrac{1}{3 + \cfrac{1}{2}}}.$$

The end result is a somewhat unwieldy expression for the rational number $\frac{52}{9}$:

$$\frac{52}{9} = 5 + \cfrac{1}{1 + \cfrac{1}{3 + \cfrac{1}{2}}}. \tag{1}$$

Expressions of this form are called *continued fractions*. Notice that above every fraction bar in this continued fraction, the numerator is always 1. In fact, all of the continued fractions we study will have only 1's as their numerators. We can write the continued fraction in equation (1) much more compactly by leaving out the numerators, and writing all of the other numbers in the expression in order inside brackets:

$$\frac{52}{9} = [5, 1, 3, 2]. \tag{2}$$

This *bracket notation* is simply a shorthand for the continued fraction in equation (1).

DEFINITION 14.1.1

Let $a_1, a_2, a_3, \ldots, a_n \in \mathbf{R}$ such that $a_1 \geq 0$, and $a_i > 0$ for $i = 2, 3, \ldots, n$. Then $[a_1, a_2, a_3, \ldots, a_n]$ represents the following finite **continued fraction**:

$$[a_1, a_2, a_3, \ldots, a_n] = a_1 + \cfrac{1}{a_2 + \cfrac{1}{a_3 + \cfrac{1}{a_4 + \cfrac{1}{+ \cfrac{\cdots}{a_{n-1} + \cfrac{1}{a_n}}}}}}. \tag{3}$$

If the terms are all integers, $a_1, a_2, a_3, \ldots, a_n \in \mathbf{Z}$, then $[a_1, a_2, a_3, \ldots, a_n]$ is called a **simple continued fraction**.

Note that this definition allows the terms in a continued fraction, a_i, to be real numbers. However, most of the continued fractions we study in this chapter will be simple continued fractions, in which all of the a_i are nonnegative integers.

To convert a rational number into the equivalent simple continued fraction, we just need to put ourselves into **Hearty** and **Littleweird's** shoes. The following example demonstrates the process.

EXAMPLE 1

Express $\frac{75}{17}$ as a continued fraction.

SOLUTION

$$\frac{75}{17} = 4 + \frac{7}{17}$$ ← **Rewrite the improper fraction as a mixed number.**

$$= 4 + \frac{1}{\dfrac{17}{7}}$$ ← **Write the proper fraction as 1 over its reciprocal.**

$$= 4 + \frac{1}{2 + \dfrac{3}{7}}$$ ← **Rewrite the improper fraction as a mixed number.**

$$= 4 + \frac{1}{2 + \dfrac{1}{\dfrac{7}{3}}}$$ ← **Write the proper fraction as 1 over its reciprocal.**

$$= 4 + \frac{1}{2 + \dfrac{1}{2 + \dfrac{1}{3}}}.$$ ← **Rewrite the improper fraction as a mixed number.**

At this point we are done, and we can express our answer in bracket notation:

$$\frac{75}{17} = [4, 2, 2, 3]. \qquad \blacksquare$$

Here is a summary of the method for converting a rational number into continued fraction notation.

- **Whenever we see an improper fraction, we rewrite it as a mixed number.**

- **Whenever we see a proper fraction (whose numerator is not 1), we write it as 1 over its own reciprocal.**

EXAMPLE 2

Express $\frac{11}{27}$ as a continued fraction in bracket notation.

SOLUTION

$$\frac{11}{27} = \frac{1}{\dfrac{27}{11}} = \frac{1}{2 + \dfrac{5}{11}} = \frac{1}{2 + \dfrac{1}{\dfrac{11}{5}}}$$

$$= \frac{1}{2 + \dfrac{1}{2 + \dfrac{1}{5}}}. \qquad (4)$$

How do we write expression (4) in bracket notation? We cannot write it as $[2, 2, 5]$, because $[2, 2, 5] = 2 + \dfrac{1}{2 + \frac{1}{5}}$, which is only part of expression (4). Compare the form of expression (4) to that of the general continued fraction (3). The integer term, a_1, is missing from expression (4). The key is to realize that the integer term must be zero:

$$\frac{11}{27} = 0 + \cfrac{1}{2 + \cfrac{1}{2 + \frac{1}{5}}}. \tag{5}$$

This is easy to write in bracket notation:

$$\frac{11}{27} = [0, 2, 2, 5]. \qquad \blacksquare$$

Continued fractions give us a new way to write rational numbers. We have just seen how to express a simple fraction (a ratio of two integers) as a simple continued fraction. By following the same steps in reverse, we can express a simple continued fraction as a simple fraction.

EXAMPLE 3
. . . .

The continued fraction $[2, 1, 7]$ represents a rational number. We can express that number as a simple fraction by simplifying the continued fraction from the bottom up.

$$[2, 1, 7] = 2 + \cfrac{1}{1 + \frac{1}{7}} = 2 + \cfrac{1}{\frac{8}{7}} = 2 + \frac{7}{8}$$

$$= \frac{23}{8}.$$

Using the method of this example, we can start with any finite simple continued fraction and find the rational number that it represents. (See Exercise 10.)

Continued fractions and the Euclidean Algorithm

When we expressed $\dfrac{75}{17}$ as a continued fraction in Example 1, the first arithmetic operation we performed was to divide 75 by 17:

$$75 = 4 \cdot 17 + 7$$

The next operation was to divide 17 by the remainder, 7, then divide 7 by the new remainder, and so on. This should have a familiar feel. When we found the continued fraction for $\dfrac{75}{17}$, whether we realized it or not, we were actually performing the Euclidean Algorithm (4.1.1) on the numbers 75 and 17.

Let's examine this relationship more closely. Consider the continued fraction expansion of $\frac{75}{17}$ from Example 1:

$$\frac{75}{17} = [4, 2, 2, 3]. \tag{6}$$

Now we perform the Euclidean Algorithm on 75 and 17, both geometrically and algebraically:

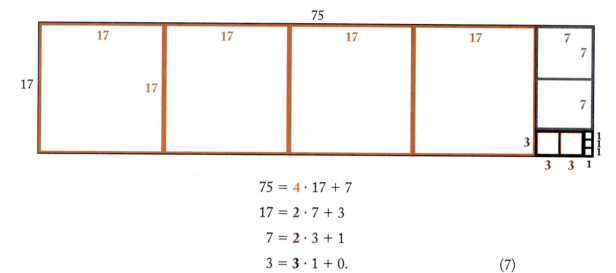

$$75 = 4 \cdot 17 + 7$$
$$17 = 2 \cdot 7 + 3$$
$$7 = 2 \cdot 3 + 1$$
$$3 = 3 \cdot 1 + 0. \tag{7}$$

Notice that the numbers 4, 2, 2, 3 in the continued fraction expansion (6) correspond to the quotients in the algebraic version of the Euclidean Algorithm (7) and to the number of boxes of each size in the geometric version of the Euclidean Algorithm. In fact, this relationship between continued fraction expansions and the Euclidean Algorithm holds for all nonnegative rational numbers (see Exercise 12).

In particular, the length of the continued fraction expansion (the number of terms inside the brackets) is equal to the number of steps required by the Euclidean Algorithm. For instance, they are both equal to 4 in equations (6) and (7). Since the Euclidean Algorithm halts when it is applied to any pair of natural numbers, the continued fraction expansion of the ratio of any two natural numbers will have finite length. In other words:

REMARK. *Every nonnegative rational number can be written as a finite simple continued fraction.*

As we saw in Example 3, it is easy to start with any finite simple continued fraction and find the rational number that it represents. Combining this with the Remark above, we have

{Nonnegative rational numbers} = {Finite simple continued fractions}.

We now have two different finite representations for nonnegative rational numbers. Any nonnegative rational can be written in the familiar manner as a ratio of two integers, or as a finite simple continued fraction. In this section, we have learned how to convert between the two representations. At this point, the idea that one might actually prefer to represent a number by its continued fraction may seem a little weird. But fear not. In the rest of this chapter, we will explore both the beauty and the usefulness of continued fractions.

EXERCISES 14.1

Numerical Problems

In Exercises 1–3, express the given rational number as a continued fraction.

1. $\dfrac{67}{29}$
2. $\dfrac{31}{264}$
3. $\dfrac{10472}{2555}$

In Exercises 4–9, express the given continued fraction as a simple fraction (as a ratio of two natural numbers).

4. $[4, 7, 3, 2]$
5. $[2, 2, 1, 1, 2, 2]$
6. $[0, 3, 4]$

7. $[0, 9, 9, 9]$
8. $[1, 1, 1, 1]$
9. $[1, 1, 1, 1, 1]$

Reasoning and Proofs

10. Prove by induction that every finite simple continued fraction $a_1, a_2, \ldots, a_n$ is a rational number. [*Hint:* Perform induction on n, the number of terms in the continued fraction.]

11. Using the recursive definition of the Fibonacci sequence (2.3.1), find and prove a formula for the simple continued fraction representation of $\dfrac{F_{k+1}}{F_k}$. Your formula should be correct for all $k \in \mathbb{N}$.

12. Prove that the terms in a continued fraction expansion correspond exactly to the quotients in the Euclidean Algorithm.

14.2 Expressing Irrational Numbers as Continued Fractions

In the previous section, we learned that every nonnegative rational number has a finite continued fraction expansion. What would happen if we tried to express an irrational number—say, π—as a continued fraction? Before we derive the continued fraction representation of π, let's look again at our method for determining continued fraction representations for rational numbers. When we found the continued fraction for $\dfrac{75}{17}$ in Example 1 of Section 14.1, the first step was converting the improper fraction $\frac{75}{17}$ into a mixed number. In other words, we rewrote $\frac{75}{17}$ as an **integer** part plus a nonnegative **fractional part** that is less than 1:

$$\frac{75}{17} = 4 + \frac{7}{17}. \tag{1}$$

The integer part, **4**, is the greatest integer that is less than or equal to $\frac{75}{17}$:

$$\left\lfloor \frac{75}{17} \right\rfloor = 4. \quad \leftarrow \textbf{The symbol } \lfloor x \rfloor, \textbf{ read "floor of } x,\textbf{" is the result of}$$
$$\textbf{rounding } x \textbf{ down to the nearest integer. More formally,}$$
$$\lfloor x \rfloor \textbf{ is defined as the greatest integer that is } \le x.$$

The next step after equation (1) in expressing $\frac{75}{17}$ as a continued fraction was to write the fractional part, $\frac{7}{17}$, as 1 over its **reciprocal**:

$$\frac{75}{17} = 4 + \frac{1}{\frac{17}{7}}.$$

To derive the continued fraction expansion for π, we will use steps that are similar to the ones we used when converting rational numbers into continued fractions. The first step is to write π as a "mixed number"—that is, as an **integer** plus a nonnegative **fractional part** that is less than 1. Since

$$\pi = \textbf{3.14159265...}, \tag{2}$$

the integer part is $\lfloor \pi \rfloor = \textbf{3}$, and the fractional part is **.14159265...**. We write

$$\pi = 3 + .14159265... \quad \leftarrow \textbf{Write } \pi \textbf{ as } \lfloor \pi \rfloor \textbf{ plus a fractional part.}$$

The second step is to write the fractional part, **.14159265...**, as 1 over its **reciprocal**:

$$\pi = 3 + \frac{1}{7.0625133...}. \tag{3}$$

Continuing in this manner, we express **7.06251331...** as an **integer** plus a nonnegative **fractional part** that is less than 1:

$$\pi = 3 + \frac{1}{7 + .0625133...}. \quad \leftarrow \textbf{Write 7.0625133...as } \lfloor 7.0625133... \rfloor$$
$$\textbf{plus a fractional part.}$$

We now write the new fractional part, **.0625133...**, as 1 over its **reciprocal**:

$$\pi = 3 + \frac{1}{7 + \frac{1}{15.996594...}}. \tag{4}$$

Continuing the procedure, we get

$$\pi = 3 + \cfrac{1}{7 + \cfrac{1}{15 + .996594\ldots}}. \qquad \leftarrow \text{Write } \mathbf{15.996594\ldots} \text{ as } \lfloor \mathbf{15.996594\ldots} \rfloor$$
$$\text{plus a fractional part.}$$

$$\pi = 3 + \cfrac{1}{7 + \cfrac{1}{15 + \cfrac{1}{1.0034172\ldots}}} \qquad (5)$$

$$= 3 + \cfrac{1}{7 + \cfrac{1}{15 + \cfrac{1}{1 + .0034172\ldots}}}$$

$$= 3 + \cfrac{1}{7 + \cfrac{1}{15 + \cfrac{1}{1 + \cfrac{1}{292.63459\ldots}}}}. \qquad (6)$$

Notice that equations (2), (3), (4), (5), and (6) can be written succinctly using bracket notation, as follows:

$$\pi = [3.14159265\ldots] \qquad (7)$$

$$= [3, 7.0625133\ldots] \qquad (8)$$

$$= [3, 7, 15.996594\ldots] \qquad (9)$$

$$= [3, 7, 15, 1.0034172\ldots] \qquad (10)$$

$$= [3, 7, 15, 1, 292.63459\ldots]. \qquad (11)$$

We can continue this procedure indefinitely, obtaining continued fraction representations for π that have more and more terms. Doing so, we obtain an *infinite* sequence of integers,

$$3, 7, 15, 1, 292, \ldots,$$

which define the *continued fraction expansion* of the irrational number π.

In fact, the same procedure that we used before to find the continued fraction expansion for π can be used to find the continued fraction expansion of any positive irrational number α. Note that this process will not terminate, for if it did, then we would have expressed α as a finite continued fraction, which would mean that α is rational.

In each step (7)–(11), we have a finite continued fraction representation of π in which the final term is irrational, and all of the preceding terms are integers.

We use the notation α_n to represent the final, irrational term of the continued fraction expansion at the nth step of the procedure.

We can then view the procedure as follows. First, define $\alpha_1 = \pi$. Let $a_1 = \lfloor \alpha_1 \rfloor$. Then π can be written as a_1 plus a **fractional part** that equals $\alpha_1 - a_1$. We define α_2 as the reciprocal of this fractional part: $\alpha_2 = \dfrac{1}{\alpha_1 - a_1}$, and let $a_2 = \lfloor \alpha_2 \rfloor$. Continuing in this manner, we get the following recursive procedure:

$$\alpha_1 = \pi = 3.14159265\ldots, \qquad\qquad a_1 = \lfloor \alpha_1 \rfloor = 3;$$

$$\alpha_2 = \frac{1}{\alpha_1 - a_1} = \frac{1}{\pi - 3} = 7.0625133\ldots, \qquad a_2 = \lfloor \alpha_2 \rfloor = 7;$$

$$\alpha_3 = \frac{1}{\alpha_2 - a_2} = 15.996594\ldots, \qquad\qquad a_3 = \lfloor \alpha_3 \rfloor = 15;$$

and so on. In this way, we can define the infinite continued fraction for any irrational number.

DEFINITION 14.2.1

Any positive irrational number α has an **infinite simple continued fraction expansion**

$$\alpha = [a_1, a_2, a_3, a_4, \ldots],$$

where $a_1, a_2, a_3, \ldots$ is an infinite sequence of integers ($a_1 \geq 0$, and $a_n > 0$ for $n = 2, 3, \ldots$), defined by the following recursive procedure:

$$\alpha_1 = \alpha,$$

$$a_n = \lfloor \alpha_n \rfloor \qquad \text{for } n \geq 1,$$

$$\alpha_{n+1} = \frac{1}{\alpha_n - a_n} \qquad \text{for } n \geq 1. \qquad (12)$$

In this procedure, α_n is called the **nth subfraction** *of α.*

In this definition, note that in equation (12), we need to take the reciprocal of a certain quantity at each step. How do we know that the denominator will never become zero? Well, if it did, then the expansion would be finite, and this would contradict the irrationality of α. The assertions that $a_1 \geq 0$, and $a_n > 0$ for $n = 2, 3, \ldots$, follow from the recursive procedure in the definition. (See Exercise 22.)

Note that equation (12) in Definition 14.2.1 can be rewritten as the following formula for the nth subfraction of α:

$$\alpha_n = a_n + \frac{1}{\alpha_{n+1}}. \qquad (13)$$

Applying this formula recursively, we see that

$$\alpha = \alpha_1$$

$$= a_1 + \frac{1}{\alpha_2}$$

$$= a_1 + \cfrac{1}{a_2 + \cfrac{1}{\alpha_3}}$$

$$= a_1 + \cfrac{1}{a_2 + \cfrac{1}{a_3 + \cfrac{1}{\alpha_4}}}$$

$$\vdots$$

In general, the following theorem tells us how to represent any positive irrational number as a finite continued fraction in which the initial terms are all integers and the final term is irrational. The theorem asserts that at each step of the recursive procedure in Definition 14.2.1, the continued fraction is equal to α.

THEOREM 14.2.2

Let α be any positive irrational number, with infinite simple continued fraction expansion $\alpha = [a_1, a_2, a_3, a_4, \ldots]$. Then for any nonnegative integer n, $\alpha = [a_1, a_2, \ldots, a_n, \alpha_{n+1}]$.

PROOF You will prove this in Exercise 23.

Definition 14.2.1 tells us how to start with any positive irrational number, α, and produce a corresponding simple continued fraction, $[a_1, a_2, a_3, a_4, \ldots]$. But now suppose we want to go in the other direction: If someone gives us a simple continued fraction, can we produce a corresponding real number? If the continued fraction is finite, we saw in Section 14.1 how to find the rational number that it represents. What about infinite continued fractions? It turns out that every infinite simple continued fraction represents a unique nonnegative real number. This fact, which we will not prove here, requires techniques from analysis. The exploration at the very end of this chapter (Section 14.4, Exercises 13–16) leads you through a proof of this correspondence.

Thus, continued fraction expansions provide an alternate representation for all nonnegative real numbers, both rational and irrational:

{Nonnegative real numbers} = {Simple continued fractions (finite and infinite)}.

Finding continued fraction expansions of irrational numbers

EXAMPLE 1

Find the continued fraction expansion of $\alpha = \sqrt{2}$.

SOLUTION Applying Definition 14.2.1, we find

$$\alpha_1 = \sqrt{2}.$$

$$a_1 = \lfloor \alpha_1 \rfloor = \lfloor \sqrt{2} \rfloor = \lfloor 1.414... \rfloor = 1.$$

$$\alpha_2 = \frac{1}{\alpha_1 - a_1} = \frac{1}{\sqrt{2} - 1} = \sqrt{2} + 1.$$

$$a_2 = \lfloor \alpha_2 \rfloor = \lfloor \sqrt{2} + 1 \rfloor = \lfloor 2.414... \rfloor = 2.$$

$$\alpha_3 = \frac{1}{\alpha_2 - a_2} = \frac{1}{\sqrt{2} + 1 - 2} = \frac{1}{\sqrt{2} - 1} = \sqrt{2} + 1.$$

At this point, we observe that α_3, the third subfraction of $\sqrt{2}$, is equal to α_2, the second subfraction of $\sqrt{2}$. Thus, following the recursive procedure in Definition 14.2.1 will be quite repetitive: We will get

$$\alpha_3 = \alpha_2, \ \alpha_4 = \alpha_3, \ \alpha_5 = \alpha_4, \ \alpha_6 = \alpha_5, \ldots$$

and

$$a_3 = a_2, \ a_4 = a_3, \ a_5 = a_4, \ a_6 = a_5, \ldots.$$

We conclude that the continued fraction expansion of $\sqrt{2}$ is

$$\sqrt{2} = [1, 2, 2, 2, 2, 2, 2, 2, 2, \ldots] = 1 + \cfrac{1}{2 + \cfrac{1}{2 + \cfrac{1}{2 + \cfrac{1}{2 + \cfrac{1}{2 + \ddots}}}}}.$$

Using an overbar to indicate the repeated portion of the continued fraction, we can write this quite succinctly:

$$\sqrt{2} = [1, \overline{2}]. \qquad \blacksquare$$

Definition 14.2.1 provides a recipe for finding the continued fraction of any irrational number. Instead of carrying out this algorithm exactly (as we just did in Example 1), we may also apply it *numerically* on a calculator or computer. Example 2 demonstrates this numerical method.

EXAMPLE 2

Using a calculator or computer, find the continued fraction expansion of $\alpha = \sqrt{6}$ numerically.

SOLUTION We apply the recursive procedure from Definition 14.2.1.

Step, n	Equation for α_n	Calculator or computer operation	α_n (Calculator display)	$a_n = \lfloor \alpha_n \rfloor$
1	$\alpha_1 = \sqrt{6}$	Evaluate $\sqrt{6}$.	2.44948974278318	$a_1 = \lfloor \alpha_1 \rfloor = 2$
2	$\alpha_2 = \dfrac{1}{\alpha_1 - a_1}$	Subtract 2 and take the reciprocal.	2.22474487139159	$a_2 = \lfloor \alpha_2 \rfloor = 2$
3	$\alpha_3 = \dfrac{1}{\alpha_2 - a_2}$	Subtract 2 and take the reciprocal.	4.44948974278315	$a_3 = \lfloor \alpha_3 \rfloor = 4$
4	$\alpha_4 = \dfrac{1}{\alpha_3 - a_3}$	Subtract 4 and take the reciprocal.	2.22474487139171	$a_4 = \lfloor \alpha_4 \rfloor = 2$
5	$\alpha_5 = \dfrac{1}{\alpha_4 - a_4}$	Subtract 2 and take the reciprocal.	4.44948974278081	$a_5 = \lfloor \alpha_5 \rfloor = 4$
$\vdots$	$\vdots$	$\vdots$	$\vdots$	$\vdots$

Do you see a pattern? The number that was showing on the display at step 4, $\alpha_4 \approx 2.22474487139171$, is almost identical to what we had already seen on the display at step 2, $\alpha_2 \approx 2.22474487139159$. It seems that these two values represent the exact same irrational number (i.e., $\alpha_4 = \alpha_2$), and they only appear different on the calculator due to numerical round-off error. If we are confident that this is the case, then we know that the numerical results in steps 5–6 will be identical (up to round-off error) to the results in steps 3–4. The pattern will repeat again for the next two steps, and so on.

Thus, it appears that the continued fraction expansion for $\sqrt{6}$ is

$$\sqrt{6} = [2, \overline{2, 4}].$$ ●

One problem with using numerical approximations to find continued fraction expansions is that numerical error can build up quickly. It is not always clear whether a small difference between the values displayed at two different steps is in fact due only to round-off error, or whether the two values are different because they actually represent two different numbers.

Periodic continued fractions

In Example 1, we used Definition 14.2.1 to find the continued fraction expansion of $\sqrt{2}$ exactly. In Example 2, we used Definition 14.2.1 with numerical approximations to find the continued fraction expansion of $\sqrt{6}$. For each of these positive irrational numbers, the continued fraction expansion was infinite and repeating:

$$\sqrt{2} = [1, \overline{2}] \qquad \text{and} \qquad \sqrt{6} = [2, \overline{2, 4}].$$

Continued fractions that are infinite and repeating, such as $[1, \bar{2}]$ and $[2, \overline{2, 4}]$, are called **periodic**.

Suppose we want to go in the other direction: What if we are given a periodic continued fraction, and need to find the unique real number that it represents?

EXAMPLE 3

Find the real number represented by $\alpha = [\overline{3, 2}]$.

SOLUTION First we write out the continued fraction expansion of α.

$$\alpha = 3 + \cfrac{1}{2 + \cfrac{1}{3 + \cfrac{1}{2 + \cfrac{1}{3 + \cfrac{1}{2 + \cfrac{1}{\ddots}}}}}} \tag{14}$$

The portion of the continued fraction shown in orange here is exactly the same as the entire continued fraction! Thus, the portion shown in orange is equal to α. Hence, we can rewrite equation (14) as follows:

$$\alpha = 3 + \cfrac{1}{2 + \cfrac{1}{\alpha}}. \tag{15}$$

(For a more rigorous justification of this equation, see Exercise 30.)

Multiplying the large fraction on the right side by $\frac{\alpha}{\alpha}$ gives

$$\alpha = 3 + \frac{\alpha}{2\alpha + 1}.$$

Multiplying both sides of this equation by $2\alpha + 1$ and collecting all of the terms on the same side yields

$$2\alpha^2 - 6\alpha - 3 = 0.$$

Now we're on familiar ground—this is just a quadratic equation. Using the quadratic formula, we find

$$\alpha = \frac{6 + \sqrt{60}}{4} \quad \text{or} \quad \alpha = \frac{6 - \sqrt{60}}{4}.$$

Since we are looking for a solution $\alpha > 0$, we may eliminate the negative solution $\frac{6 - \sqrt{60}}{4}$. Hence,

$$[\overline{3, 2}] = \frac{6 + \sqrt{60}}{4}$$
$$= \frac{3 + \sqrt{15}}{2}. \qquad \blacksquare$$

EXAMPLE 4

Find the real number represented by $\alpha = [5, \overline{4, 2}]$.

SOLUTION We expand out the continued fraction for α:

$$\alpha = 5 + \cfrac{1}{4 + \cfrac{1}{2 + \cfrac{1}{4 + \cfrac{1}{2 + \cfrac{1}{\ddots}}}}}$$

Since the number 5 only appears once in the continued fraction, α does not contain a complete copy of itself. The difficulty is that α contains both a nonrepeating part and a repeating part. However, the 2nd subfraction, α_2, is purely repeating. We have

$$\alpha = 5 + \frac{1}{\alpha_2}, \tag{16}$$

where

$$\alpha_2 = 4 + \cfrac{1}{2 + \cfrac{1}{4 + \cfrac{1}{2 + \cfrac{1}{4 + \cfrac{1}{2 + \cfrac{1}{\ddots}}}}}}$$

Notice that the continued fraction α_2 does contain a complete copy of itself. We may therefore proceed as in Example 3 to find that

$$\alpha_2 = 4 + \cfrac{1}{2 + \cfrac{1}{\alpha_2}},$$

from which it follows (as you will show in Exercise 11) that

$$\alpha_2 = 2 + \sqrt{6}.$$

Substituting this value into equation (16) gives the value of the original continued fraction, α:

$$\alpha = 5 + \frac{1}{2 + \sqrt{6}}.$$

Rationalizing the denominator and simplifying yields the final answer for the value of α:

$$[5, \overline{4, 2}] = \frac{8 + \sqrt{6}}{2}. \qquad \blacksquare$$

Classification of simple continued fractions

In Examples 3 and 4, we found the exact values of two periodic simple continued fractions. Each of those examples led to a quadratic equation, which we solved to find the real number that the continued fraction represents. As a result, the value of each repeating simple continued fraction was a **quadratic irrational**, which is a number that can be written in the form

$$\frac{a + b\sqrt{c}}{d},$$

where $a, b, c, d \in \mathbf{Z}, d > 0$, and $c > 0$ such that $\sqrt{c}$ is irrational.

It can be shown that the value of every periodic simple continued fraction is a positive quadratic irrational (see Exercise 29). It also turns out, although we will not prove it here, that every positive quadratic irrational can be represented by a periodic continued fraction.

To summarize, every real number can be expressed as a simple continued fraction that is either finite, infinite and periodic, or infinite and aperiodic. The following list summarizes which type of real number is represented by each type of continued fraction.

{Nonnegative rational numbers} = {Finite simple continued fractions},

{Positive quadratic irrationals} = {Periodic simple continued fractions},

{All other positive real numbers} = {Infinite aperiodic simple continued fractions}.

EXAMPLE 5
....

Since $\sqrt[3]{2}$ and e are positive irrational numbers that are not quadratic irrationals, their continued fraction expansions must be infinite and nonrepeating. We can find the first several terms of each of these infinite aperiodic simple continued fractions using the method of Example 2 (Definition 14.2.1 with numerical approximations):

$$\sqrt[3]{2} = [1, 3, 1, 5, 1, 1, 4, 1, 1, 8, 1, 14, 1, 10, 2 \ldots],$$

$$e = [2, 1, 2, 1, 1, 4, 1, 1, 6, 1, 1, 8, 1, 1, 10 \ldots].$$

Note that the continued fraction for e follows an interesting pattern. This pattern was discovered and proved by Roger Cotes in 1714, who was the Plumian Chair of Astronomy and Experimental Philosophy at Cambridge University. See Exercise 9 for patterns in the continued fraction expansions of other irrational numbers related to e.

EXERCISES 14.2

Numerical Problems

In Exercises 1–4, use the method of Example 1 (Definition 14.2.1, calculating the exact value of each subfraction) to find the continued fraction expansion of the given number.

1. $\sqrt{10}$ 2. $\sqrt{3}$ 3. $\sqrt{11}$ 4. $\dfrac{1 + \sqrt{5}}{2}$

In Exercises 5–8, use the method of Example 2 (Definition 14.2.1, with numerical approximations) to find the first six terms of the continued fraction expansion of the given number.

5. $\sqrt{17}$ 6. $\sqrt[3]{5}$ 7. 2π 8. $\sqrt{7} - 2$

9. Various expressions involving the irrational number e have continued fractions that are not periodic, but, surprisingly, these continued fractions follow simple patterns. Use the numerical method (the method of Example 2) to find each continued fraction expansion. Compute enough terms to discover the pattern.

   a. $e - 1$ b. $\dfrac{e - 1}{2}$ c. $\dfrac{e - 1}{e + 1}$ d. $e^{1/2}$ e. $e^{1/3}$

10. Express both roots of the polynomial $2x^2 - 4x + 1$ as continued fractions.

11. Show that the periodic continued fraction $\alpha_2 = [\overline{4, 2}]$ represents the real number $2 + \sqrt{6}$.

In Exercises 12–20, find the real number α represented by the periodic continued fraction.

12. $\alpha = [\overline{3}]$

13. $\alpha = [\overline{1}]$

14. $\alpha = [\overline{4, 5}]$

15. $\alpha = [0, \overline{7}]$

16. $\alpha = [\overline{2, 5}]$

17. $\alpha = [\overline{5, 2}]$

18. $\alpha = [1, 2, \overline{3}]$

19. $\alpha = [1, \overline{2, 3}]$

20. $\alpha = [\overline{1, 2, 3}]$

Reasoning and Proofs

21. In Example 2, we used the numerical method to obtain the continued fraction representation for $\sqrt{6}$. Verify the correctness of this particular continued fraction representation using the method of Example 4. (That is, find the real number represented by the continued fraction $\alpha = [2, \overline{2, 4}]$, and verify that $\alpha = \sqrt{6}$.)

22. Let α be a positive irrational number, and let $a_1, a_2, a_3, \ldots$ be the sequence of integers defined by the recursive procedure in the definition of infinite simple continued fractions (14.2.1).

   a. Show that $a_1 \geq 0$.   b. Use induction to prove that $a_n > 0$ for all $n \geq 2$.

23. Prove Theorem 14.2.2. [*Hint:* Use induction on n.]

24. Prove that in the continued fraction expansion for $\sqrt{2}$, we have $\alpha_n = \sqrt{2} + 1$ and $a_n = 2$ for all $n \geq 2$.

25. Let α be a positive real number with simple continued fraction expansion $\alpha = [a_1, a_2, a_3, \ldots]$. Find a formula for the simple continued fraction expansion of $\frac{1}{\alpha}$. Prove that your formula is correct.

EXPLORATION Classifying Decimal Expansions (Exercises 26–28)

In this section, we saw that different types of real numbers have different types of continued fraction expansions: finite, periodic, or infinite aperiodic. Real numbers can also be classified by how their *decimal* expansions behave: terminating (finite, such as $\frac{1}{4} = 0.25$), periodic (infinite repeating, such as $\frac{1}{3} = 0.3333333\ldots$), or infinite aperiodic (such as $\pi = 3.14159265\ldots$). In this exploration, you will classify real numbers according to how their decimal expansions behave.

26. For each of the following numbers, determine whether its *decimal* expansion terminates (i.e., is finite). Then describe, in general, what must be true if a rational number has a decimal expansion that terminates.

   a. $\frac{3}{2}$ b. $\frac{2}{3}$ c. $\frac{7}{25}$

   d. $\frac{2}{7}$ e. $\frac{3}{20}$ f. $\frac{2}{15}$

27. a. Let $\frac{a}{b}$ be a rational number written in lowest terms. Based on your answers to Exercise 26, fill in the blank:

   $\frac{a}{b}$ has a terminating decimal expansion if and only if b has no prime factors other than _____.

   b. Prove your statement from part a.

   c. Recall that the words "decimal expansion" imply that you are working in base 10 when expressing numbers. Suppose you were to work in base 12 instead. Revise your statement from part a to describe which fractions would have terminating base 12 (*duodecimal*) expansions.

28. a. Prove that if a real number has a terminating decimal expansion, then it must be a rational number.

   b. Prove that if a real number has a periodic decimal expansion, then it must be a rational number.

   c. How do the decimal expansions of irrational numbers behave?

Advanced Reasoning and Proofs

29. In this exercise, you will prove that every periodic simple continued fraction represents a quadratic irrational. (For this exercise, you may assume that every infinite simple continued fraction represents a unique nonnegative real number, which is proven in Exercises 13–16 of Section 14.4.)

Consider the periodic simple continued fraction

$$C = \left[a_1, a_2, \ldots, a_n, \overline{b_1, b_2, \ldots, b_m}\right].$$

You will prove that C is a quadratic irrational.

a. A function $S(x)$ is called a fractional linear function if it is of the form

$$S(x) = \frac{ex + f}{gx + h}$$ for real numbers e, f, g, and h. Let $T(x) = [b_1, b_2, \ldots, b_m, S(x)]$,

where $S(x)$ is a fractional linear function. (Note that since $S(x)$ is a function of x, this is not a continued fraction as defined in this chapter, but the bracket notation can still be used to interpret the meaning of this expression.) Prove by induction on m that $T(x)$ is also a fractional linear function.

b. Let

$$D = \left[\overline{b_1, b_2, \ldots, b_m}\right].$$

Use the result of part a to prove that D is a quadratic irrational, by setting up an equation to which D is the solution and solving it.

c. Prove that for any continued fraction $C = [a_1, a_2, \ldots, a_n, D]$, where each a_i is an integer and D is a quadratic irrational, C is itself a quadratic irrational. This completes the proof.

30. *Note:* This exercise requires techniques from analysis.

In Example 3, we found the real number α represented by the infinite repeating continued fraction $[\overline{3, 2}]$. The first step of the solution was to show that α satisfies equation (15). In this exercise, you will explore how to justify this step more rigorously. To do so, we need the following fact:

To find the value of an *infinite* simple continued fraction $\alpha = [a_1, a_2, a_3, \ldots]$, let r_n be the *finite* continued fraction that consists of just the first n terms:

$$r_n = [a_1, a_2, \ldots, a_n]. \tag{17}$$

Then the real number represented by the infinite continued fraction $\alpha = [a_1, a_2, a_3, \ldots]$ is given by

$$\alpha = \lim_{n \to \infty} r_n.$$

(For more detail, including a proof that this limit exists, see Exercises 13–16 of Section 14.4.)

a. Let α be the real number represented by $[\overline{3, 2}]$, and let r_n be defined by equation (17). Let $n \in \mathbf{N}$. Write an equation expressing r_{n+2} in terms of r_n.

b. Use your equation from part a to prove that equation (15) is correct—that is, prove that

$$\alpha = 3 + \cfrac{1}{2 + \cfrac{1}{\alpha}}.$$

14.3 Approximating Irrational Numbers Using Continued Fractions

Decimal approximation of $\sqrt{2}$

In Section 14.2, we learned how to express any positive irrational number as an infinite simple continued fraction. We also know another way to express any irrational number: as an infinite, aperiodic decimal. For example,

$$\sqrt{2} = 1.41421356\ldots.$$

This decimal expansion can be used to approximate $\sqrt{2}$ with rational numbers. For example, the following rational numbers are successively more accurate approximations of $\sqrt{2}$:

$1 = \dfrac{1}{1}$	← **Rounded to nearest integer (nearest 1)**
$1.4 = \dfrac{14}{10}$	← **Rounded to nearest tenth (nearest 1/10)**
$1.41 = \dfrac{141}{100}$	← **Rounded to nearest hundredth (1/100)**
$1.414 = \dfrac{1414}{1000}$	← **Rounded to nearest thousandth (1/1000)**
$1.4142 = \dfrac{14,142}{10,000}$	← **Rounded to nearest 1/10,000**
$1.41421 = \dfrac{141,421}{100,000}$	← **Rounded to nearest 1/100,000**
$1.414214 = \dfrac{1,414,214}{1,000,000}$	← **Rounded to nearest 1/1,000,000**
$1.4142136 = \dfrac{14,142,136}{10,000,000}$	← **Rounded to nearest 1/10,000,000**

$\vdots$ $\qquad\qquad\qquad\qquad$ $\vdots$

How good are these rational numbers as approximations to $\sqrt{2}$? We would like to have some idea how large the error of each approximation is.

The first approximation, $\frac{1}{1}$, is simply $\sqrt{2}$ rounded to the nearest integer. When we round an arbitrary real number x to the nearest integer, how far off could the estimate be? In the worst possible case, x will be halfway between two integers, in which case the nearest integer will be $\frac{1}{2}$ of a unit away. For example, the closest integer approximation to **7.5** is **7**, which has an error of 0.5 or $\frac{1}{2}$. (Of course, **8** is an equally good approximation to **7.5**, as it also has an error of $\frac{1}{2}$.)

Suppose we round an arbitrary real number x to the first decimal place—that is, to the nearest $\frac{1}{10}$. How large could the error of approximation be? In the worst case, x will be halfway between two tenths, such as **4.65**. In this case, the approximation to the nearest tenth will be either **4.6** or **4.7**, each of which has an error of 0.05 or $\frac{1}{20}$.

Similarly, if we round an arbitrary real number to two decimal places (to the nearest $\frac{1}{100}$), the worst-case error would be $\frac{1}{200}$.

In Table 14.3.1, we compute the error in each of our decimal approximations to $\sqrt{2}$, which is the absolute value of the difference between each approximation and $\sqrt{2}$. We then compare this actual error with the worst-case error for a decimal approximation $\left(\frac{1}{2}, \frac{1}{20}, \frac{1}{200}, \dots\right)$. To facilitate this comparison, we approximate the actual error as $\frac{1}{m}$, where m is the integer part (floor) of the reciprocal of the error.

Table 14.3.1 Decimal approximations to $\sqrt{2} = 1.414213562\ldots$

Approximation	Error of Approximation	Worst-Case Error
$1 = \frac{1}{1}$	$0.414213562\ldots \leq \frac{1}{2}$	$\frac{1}{2}$
$1.4 = \frac{14}{10}$	$0.014213562\ldots \leq \frac{1}{70}$	$\frac{1}{20}$
$1.41 = \frac{141}{100}$	$0.004213562\ldots \leq \frac{1}{237}$	$\frac{1}{200}$
$1.414 = \frac{414}{1000}$	$0.000213562\ldots \leq \frac{1}{4682}$	$\frac{1}{2000}$
$1.4142 = \frac{14,142}{10,000}$	$0.000013562\ldots \leq \frac{1}{73,733}$	$\frac{1}{20,000}$

From the table, we see that some approximations are better than others in comparison with the worst-case error. For example, the second approximation seems pretty good, since the error $\frac{1}{70}$ is more than three times smaller than the worst-case error for the same denominator, which is $\frac{1}{20}$. In contrast, the next approximation is not impressive because its error, $\frac{1}{237}$, is not much smaller than the worst-case error, $\frac{1}{200}$. Thus, it appears that not all denominators are created equal.

All of the preceding decimal approximations use a denominator that is a power of 10. However, in finding rational approximations, we do not need to restrict ourselves to powers of 10. Given any positive integer d as our denominator, it's not hard to find the integer numerator that provides the best approximation with denominator d. (See Exercise 7.) We call this method of approximation the *fixed-denominator method*.

Just as a denominator of 100 yields an approximation with worst-case error $\frac{1}{200}$, the best approximation with denominator d is guaranteed to have an error of at most $\frac{1}{2d}$. However, we might expect that if we make just the right choice of denominator d, we will get an error that is much smaller than the worst-case error $\frac{1}{2d}$. But how can we find the denominators that yield especially good approximations?

It turns out, as we will see shortly, that this is one of the areas where continued fractions really shine.

Continued fraction approximation of irrational numbers

Taking the infinite decimal expansion $\sqrt{2} = 1.4142\ldots$ and truncating (e.g., to 1.414) produces rational approximations to $\sqrt{2}$. In a similar manner, continued fractions can also be truncated to obtain rational approximations. For example, consider the continued fraction representation of $\sqrt{2}$:

$$\sqrt{2} = [1,\overline{2}] = 1 + \cfrac{1}{2 + \cfrac{1}{2 + \cfrac{1}{2 + \cfrac{1}{2 + \cfrac{1}{2 + \cfrac{1}{\ddots}}}}}}$$

Truncating the continued fraction expansion for $\sqrt{2}$ after the first 3 terms, we are left with the finite continued fraction

$$[1, 2, 2] = 1 + \cfrac{1}{2 + \cfrac{1}{2}} = \frac{7}{5}.$$

The result of truncating a continued fraction in this way is called a *convergent*.

Let α be a nonnegative real number whose simple continued fraction expansion, $\alpha = [a_1, a_2, a_3, \ldots, a_n, \ldots]$, has at least n terms. The **nth convergent** of α is $r_n = [a_1, a_2, a_3, \ldots, a_n]$, the finite continued fraction made up of the first n terms of α.

EXAMPLE 1

The fourth convergent of the infinite continued fraction $[1, 2, 2, 2, 2, \ldots]$ is the finite continued fraction $r_4 = [1, 2, 2, 2]$, which may be expressed as a ratio of two integers:

$$r_4 = [1, 2, 2, 2] = 1 + \cfrac{1}{2 + \cfrac{1}{2 + \cfrac{1}{2}}}$$

$$= \frac{17}{12}.$$

In this example, we found that the fourth convergent of the continued fraction expansion of $\sqrt{2}$ is the rational number $\frac{17}{12} = 1.416666\ldots$. This is a good approximation to $\sqrt{2} = 1.414213\ldots$. The error is $\left|\sqrt{2} - \frac{17}{12}\right| = 0.002453\ldots$. To get an intuitive feel for how small this error is, we can approximate it as $\frac{1}{m}$, where $m = \left\lfloor \frac{1}{0.002453\ldots} \right\rfloor = \lfloor 407.65 \rfloor = 407$ is the reciprocal of the error rounded down to the nearest integer. We find that the approximation $\frac{17}{12}$ has an error of less than $\frac{1}{407}$.

Let's make a table of successive convergents of $\sqrt{2}$, to determine whether they are in fact good approximations of $\sqrt{2}$. For each convergent r_n, we calculate the error of approximation, which is $|\sqrt{2} - r_n|$, and approximate that error as a fraction of the form $\frac{1}{m}$ (where m is the floor of the reciprocal of the error).

Table 14.3.2   Continued Fraction Approximations to $\sqrt{2} = 1.414213562\ldots$

n	Convergent r_n					Error $= \left|\sqrt{2} - r_n\right|$	
	Continued Fraction		*Simple Fraction*		*Decimal*		
1	$[1]$	$=$	$\frac{1}{1}$	$=$	1	$0.414213562\ldots$	$\leq \frac{1}{2}$
2	$[1, 2]$	$=$	$\frac{3}{2}$	$=$	1.5	$0.085786437\ldots$	$\leq \frac{1}{11}$
3	$[1, 2, 2]$	$=$	$\frac{7}{5}$	$=$	1.4	$0.014213562\ldots$	$\leq \frac{1}{70}$
4	$[1, 2, 2, 2]$	$=$	$\frac{17}{12}$	$=$	$1.416666666\ldots$	$0.002453104\ldots$	$\leq \frac{1}{407}$
5	$[1, 2, 2, 2, 2]$	$=$	$\frac{41}{29}$	$=$	$1.413793103\ldots$	$0.000420458\ldots$	$\leq \frac{1}{2378}$
6	$[1, 2, 2, 2, 2, 2]$	$=$	$\frac{99}{70}$	$=$	$1.414285714\ldots$	$0.000072151\ldots$	$\leq \frac{1}{13,859}$
7	$[1, 2, 2, 2, 2, 2, 2]$	$=$	$\frac{239}{169}$	$=$	$1.414201183\ldots$	$0.000012378\ldots$	$\leq \frac{1}{80,782}$
8	$[1, 2, 2, 2, 2, 2, 2, 2]$	$=$	$\frac{577}{408}$	$=$	$1.414215686\ldots$	$0.000002123\ldots$	$\leq \frac{1}{470,831}$

Consider the eighth continued fraction convergent, $\frac{577}{408}$, shown in Table 14.3.2. This fraction is an extremely good rational approximation to $\sqrt{2}$. The error is only about 0.00000212, which is less than $\frac{1}{470,000}$. Suppose we were to approximate $\sqrt{2}$ using the fixed-denominator approximation method, without using continued fractions to choose the denominator. How large a denominator d would we need in order to guarantee an error this small? Because the error is guaranteed to be at most $\frac{1}{2d}$, we would need to choose a denominator of about 235,000. This is humongous compared with 408, the denominator of the continued fraction approximation.

Tables 14.3.1 and 14.3.2 demonstrate two different methods for finding rational numbers that are close to $\sqrt{2}$. Suppose we wanted a rational approximation for $\sqrt{2}$ with denominator about 100. Would you want to use the decimal approximations from Table 14.3.1, or would you rather use the continued fraction convergents from Table 14.3.2? The decimal approximation $\frac{141}{100}$ has an error of about $\frac{1}{237}$. The convergent $r_6 = \frac{99}{70}$ has a comparable denominator but an error of only $\frac{1}{13,859}$, which is many times better. Apparently, the continued fraction convergent gives a much better approximation than a decimal approximation with a denominator of comparable size.

If we use the fixed-denominator approximation method, we must have some method of selecting our denominators (such as using powers of 10). While some chosen denominators yield good approximations, most choices of denominator will not be particularly good. On the other hand, continued fractions directly provide us with the optimal denominators for approximating irrational numbers! As we see from Table 14.3.2, using continued fractions to approximate irrational numbers gives extremely impressive results. We'll see more examples of this in the rest of this section. Then in Section 14.4, we will *prove* that continued fractions always give great approximations.

Over the course of your mathematical education, you may have learned that certain numbers, such as $\frac{22}{7}$, provide particularly good rational approximations to π. Have you ever wondered how you might go about finding your own rational approximations to π? You can do it with the convergents of the continued fraction for π, as the following example demonstrates.

EXAMPLE 2

Make a table of the first five convergents of π. Find the error of approximation of each convergent to π, and write each error approximately as $\frac{1}{m}$ (where m is the reciprocal of the error, rounded down to the nearest integer).

SOLUTION We can use the numerical method (Definition 14.2.1 with numerical approximations) to find the first several values in the continued fraction expansion of π:

$$\pi = [3, 7, 15, 1, 292, \ldots].$$

The results are shown in Table 14.3.3.

Table 14.3.3 Continued Fraction Approximations to π

n	Continued Fraction		Simple Fraction		Decimal	Error = $\left| \pi - r_n \right|$	
1	[3]	=	$\dfrac{3}{1}$	=	3	$0.141592653589\ldots \leq$	$\dfrac{1}{7}$
2	[3, 7]	=	$\dfrac{22}{7}$	=	$3.1428571428\ldots$	$0.001264489267\ldots \leq$	$\dfrac{1}{790}$
3	[3, 7, 15]	=	$\dfrac{333}{106}$	=	$3.1415094339\ldots$	$0.000083219627\ldots \leq$	$\dfrac{1}{12{,}016}$
4	[3, 7, 15, 1]	=	$\dfrac{355}{113}$	=	$3.1415929203\ldots$	$0.000000266764\ldots \leq$	$\dfrac{1}{3{,}748{,}629}$
5	[3, 7, 15, 1, 292]	=	$\dfrac{103{,}993}{33{,}102}$	=	$3.1415926530\ldots$	$0.000000000577\ldots \leq$	$\dfrac{1}{1{,}730{,}431{,}258}$

In this example, the convergents provide amazingly good approximations to π. The second convergent, the familiar approximation $\frac{22}{7}$, has a denominator of only 7 yet achieves an accuracy of about $\frac{1}{790}$. The fourth convergent, $\frac{355}{113}$, is particularly impressive because with a denominator of about 100, its error is less than $\frac{1}{3{,}000{,}000}$!

Early History of $\dfrac{355}{113}$ as an Approximation to π

In China, Zu Chongzhi (430–501) was an expert in mechanics and machinery. Around the year 480, Zu Chongzhi and his son Zu Gengzhi showed that $3.1415926 < \pi < 3.1415927$, and Chongzhi suggested using the fraction $\frac{355}{113}$ as an approximation of π.

In India, the fraction $\frac{355}{113}$ was used in the 15th century as an approximation to π.

The circle of fifths

In Chapter 6, we mentioned a peculiar consequence of the Fundamental Theorem of Arithmetic (6.1.1): It is impossible to tune a piano so that all the intervals are in their acoustically correct frequency ratios. We now explain why this is and learn what continued fractions may have to say about how Western music ended up the way it is today.

The *octave* is fundamental to music. Pluck a string on a guitar and listen to the note produced. Now put your finger down at the exact midpoint of the string, allowing only half of the string to vibrate, and pluck again. The new note is said to be an octave higher than the original note. To the human ear, two notes an octave apart sound very similar. Physically, the frequency of the sound wave produced by the higher note is exactly twice that of the lower note. The auditory system unconsciously responds to the ratio of these two frequencies, which is 2:1.

The next most basic interval is the *fifth*, which is the distance between the note C and the G above it. (Fifths can easily be recognized by thinking of the interval between the first two "twinkles" in the song "Twinkle, Twinkle, Little Star.") Fifths have a harmonious, hollow sound and are in a frequency ratio of 3:2.

If you start with the lowest C on the piano and go up a fifth, you will arrive at G. Going up another fifth gets you to D. Continuing in this manner, you end up cycling through all 12 notes of the chromatic scale in a progression known as the *circle of fifths*:

$$C \to G \to D \to A \to E \to B \to F\sharp \to C\sharp \to G\sharp \to E\flat \to B\flat \to F \to C$$

You've arrived back at C, but now you're at the very highest note of the piano. Moving by fifths, it took you **12** jumps to get from the lowest C to the highest C. But you could also have gotten there jumping up an entire octave at a time: $C \to C \to C \to \cdots \to C$. Moving in this way requires only **7** jumps to get from the lowest C to the highest C.

$$\textbf{12} \text{ jumps by fifths} = \textbf{7} \text{ jumps by octaves.} \tag{1}$$

Since a fifth has the frequency ratio 3:2, and an octave has the frequency ratio 2:1, this implies that the ratio of the frequency of the highest C to the frequency of the lowest C is

$$\left(\frac{3}{2}\right)^{12} = 2^7,$$

which means that

$$3^{12} = 2^{19}. \tag{2}$$

But a power of 3 cannot equal a power of 2 by the Fundamental Theorem of Arithmetic! Wait a minute. What's going on?

Theoretical mathematics has forced us to a practical conclusion. Since equation (2) cannot possibly be correct, we cannot tune our piano so that all the octaves are in ratio 2:1 and all the fifths are in ratio 3:2. Something has got to give.

Since octaves are more basic than fifths, we'll keep them tuned in a perfect 2:1 ratio, but this forces us to make fifths that are slightly out of tune—in other words, not exactly in the ratio 3:2. Throughout musical history, there have been many attempted solutions to this problem, leading to many different tuning systems. From these many choices, eventually a clear winner emerged, which also happened to be the simplest solution mathematically. Known as *equal temperament*, the tuning system we use today takes all fifths to be exactly equal, subject to statement (1). This tells us exactly what frequency ratio our fifths must have. Letting f represent the frequency ratio of an equal-tempered fifth, we have

$$f^{12} = 2^7,$$

and hence,

$$\begin{array}{l} \text{frequency ratio of an} \\ \text{equal-tempered fifth} \end{array} = 2^{7/12}$$

$$= 1.4983\ldots$$

Since this is very close to $\frac{3}{2}$, an equal-tempered fifth is quite close to an acoustically correct fifth. The top note of an equal tempered fifth is flat by only about 0.11% (about 2/100 of a half-step), a difference that most people do not notice.

In terms of the familiar 12-note scale (the 12 keys per octave on a piano), equal-tempered tuning corresponds to giving each half-step a frequency ratio of $2^{1/12}$.

So what does this have to do with continued fractions? Well, say you were designing your own musical scale. You haven't decided what notes to use or even how many notes there will be. But like most good-hearted people, you are fond of octaves and fifths. You realize that when you start with some note, and go up successive fifths, you are going to cycle back to your original note, but some number of octaves higher than when you started.

$$\text{Going up } A \text{ octaves} \approx \text{going up } B \text{ fifths.}$$

This implies that

$$2^A \approx \left(\frac{3}{2}\right)^B.$$

Taking the log (base 2) of both sides,

$$A \approx B \log_2\left(\frac{3}{2}\right).$$

In other words,

$$\frac{A}{B} \approx \log_2\left(\frac{3}{2}\right). \tag{3}$$

So what values should we choose for A and B? To answer this, we just need to find good rational approximations to the irrational number $\alpha = \log_2\left(\frac{3}{2}\right)$. How do you suppose we are going to do that? Using continued fractions!

The first several terms in the simple continued fraction expansion of α are

$$\log_2\left(\frac{3}{2}\right) = [0, 1, 1, 2, 2, 3, 1, 5, \ldots].$$

The convergents for α are

n	1	2	3	4	5	6	7	8	$\cdots$
r_n	0	1	$\frac{1}{2}$	$\frac{3}{5}$	$\frac{7}{12}$	$\frac{24}{41}$	$\frac{31}{53}$	$\frac{179}{306}$	$\cdots$

Consider the fifth convergent, $r_5 = \frac{7}{12}$, which corresponds to the approximation $\alpha \approx \frac{7}{12}$. Using this approximation, the octave is divided into 12 equal intervals (half-steps), and a fifth consists of 7 of these intervals. Hence, this approximation leads to the familiar 12-note scale!

Furthermore, the other convergents suggest other scales. The convergent $r_4 = \frac{3}{5}$ suggests that if we divide an octave into 5 equal intervals, 3 of these intervals will comprise an approximate fifth. This roughly corresponds to the pentatonic scale used commonly in traditional Chinese music.

Looking in the other direction, the approximation $r_6 = \frac{24}{41}$ suggests a scale with 41 notes, which might seem outrageous. But this scale would have fifths that were nearly perfectly in tune (off by less than 0.02%, or less than 1/200 of a half-step). Even more interestingly, the next approximation, $r_7 = \frac{31}{53}$, appeared in China as early as 40 BCE. Presumably without knowledge of continued fractions, King Fang observed that 53 musical fifths very nearly equal 31 octaves[1]!

EXERCISES 14.3

Numerical Problems

1. Make a table of the first eight convergents of the golden ratio, $\varphi = \frac{1 + \sqrt{5}}{2}$. Find the error of approximation of each convergent to φ, and write each error approximately as the reciprocal of an integer.

2. **a.** Create a table of the first five convergents for $\alpha = \sqrt{10}$.

    b. For each convergent, find the error of approximation to α and approximate it as $\frac{1}{m}$ for some integer m.

---

[1] See E. Dunne and M. McConnell, "Pianos and Continued Fractions," *Mathematics Magazine* 72, no. 2 (1999): 104–15.

c. For each convergent $r_n = \dfrac{h_n}{k_n}$, calculate $h_n^2 - 10k_n^2$. Describe any pattern you find.

d. For each pair of convergents r_n, r_{n+1}, calculate $h_n k_{n+1} - h_{n+1} k_n$. Describe any pattern you find.

3. **a.** Create a table of the first five convergents for $\alpha = \sqrt{3}$.

    b. For each convergent, find the error of approximation to α and approximate it as $\dfrac{1}{m}$ for some integer m.

    c. For each convergent $r_n = \dfrac{h_n}{k_n}$, calculate $h_n^2 - 3k_n^2$. Describe any pattern you find.

    d. For each pair of convergents r_n, r_{n+1}, calculate $h_n k_{n+1} - h_{n+1} k_n$. Describe any pattern you find.

4. **a.** Create a table of the first five convergents for $\alpha = \sqrt{6}$.

    b. For each convergent, find the error of approximation to α and approximate it as $\dfrac{1}{m}$ for some integer m.

    c. For each convergent $r_n = \dfrac{h_n}{k_n}$, calculate $h_n^2 - 6k_n^2$. Describe any pattern you find.

    d. For each pair of convergents r_n, r_{n+1}, calculate $h_n k_{n+1} - h_{n+1} k_n$. Describe any pattern you find.

5. Use Table 14.3.2 of the convergents for $\alpha = \sqrt{2}$.

    a. For each convergent $r_n = \dfrac{h_n}{k_n}$, calculate $h_n^2 - 2k_n^2$. Describe any pattern you find.

    b. For each pair of convergents r_n, r_{n+1}, calculate $h_n k_{n+1} - h_{n+1} k_n$. Describe any pattern you find.

6. Create a table of the first six convergents for e. Find the error of approximation of each convergent to e, and approximate it as $\dfrac{1}{m}$ for some integer m.

Reasoning and Proofs

7. Given any real number α and natural number d, describe a method for finding the best rational approximation to α that has denominator d. Explain why the error of this approximation is at most $\dfrac{1}{2d}$.

8. **a.** Write each of the first 10 convergents of the golden ratio, $\varphi = \dfrac{1 + \sqrt{5}}{2}$, as a ratio of integers.

b. Make a conjecture expressing r_n, the nth convergent of φ, in closed form using Fibonacci numbers.

c. Use induction to prove your conjecture from part b.

9. For this exercise, use Table 14.3.2 (the convergents of $\sqrt{2}$) and Table 14.3.3 (the convergents of π).

   a. In the values given in the rightmost column of Table 14.3.2 (fractions of the form $\frac{1}{m}$), demonstrate that the value of m for the nth convergent of $\sqrt{2}$ is always greater than the product of the denominators of the nth convergent and the $(n + 1)$st convergent.

   b. In the values given in the rightmost column of Table 14.3.3 (fractions of the form $\frac{1}{m}$), demonstrate that the value of m for the nth convergent of π is always greater than the product of the denominators of the nth convergent and the $(n + 1)$st convergent.

   c. Based on the pattern you found in part b, explain why the convergent $\frac{22}{7}$ gives a particularly good approximation to π.

   d. Based on the pattern you found in part b, which would you choose as an approximation for π: $\frac{333}{106}$ or $\frac{355}{113}$? Explain your reasoning.

10. An acoustically correct major third has a frequency ratio of 5:4. (You can recognize a major third as the interval between the first two notes of the song "Michael, Row Your Boat Ashore.")

    a. Show that there do not exist natural numbers m and n such that

    $$\left(\frac{5}{4}\right)^m = 2^n. \qquad (4)$$

    Thus, no number of acoustically correct major thirds can make up a whole number of octaves.

    b. Find the first four convergents of $\log_2\left(\frac{5}{4}\right)$.

    c. Use each of your answers to part b to find a pair of integers m and n for which equation (4) is approximately correct.

    d. In the equal-tempered scale, an octave consists of exactly 3 major thirds. To which convergent of $\log_2\left(\frac{5}{4}\right)$ does this correspond?

    e. Use the next convergent (beyond the one you used to answer part d) to design a scale with intervals that are very close to acoustically correct. In this scale, into how many equal steps is the octave divided? How many of these steps make up an interval with a frequency ratio of nearly 5:4? Compute this frequency ratio, and compare it to $\frac{5}{4}$.

14.4 Proving That Convergents are Fantastic Approximations

In the previous section, we saw several examples illustrating that continued fraction convergents are amazingly good approximations for irrational real numbers. In this section, we will figure out how good these approximations are in general, and prove some guarantees about the closeness of the approximations.

Finding the numerators and denominators of convergents

Let's consider the first several convergents of $\sqrt{19}$, whose continued fraction expansion is

$$\sqrt{19} = [4, \overline{2, 1, 3, 1, 2, 8}].$$

We express each convergent r_n as a ratio of two natural numbers, h_n and k_n:

$$r_n = \frac{h_n}{k_n}.$$

n	$[a_1, a_2, \ldots, a_n]$	Convergent r_n $= \dfrac{h_n}{k_n}$
1	[4]	$= \dfrac{4}{1}$
2	[4, 2]	$= \dfrac{9}{2}$
3	[4, 2, 1]	$= \dfrac{13}{3}$
4	[4, 2, 1, 3]	$= \dfrac{48}{11}$
5	[4, 2, 1, 3, 1]	$= \dfrac{61}{14}$

We would like to find a nice relationship between the terms in the two different expressions for r_n: the continued fraction expansion, $[a_1, a_2, \ldots, a_n]$, and the simple fraction expression, $\frac{h_n}{k_n}$. For instance, knowing that the next term in the continued fraction is $a_6 = 2$, can we compute the simple fraction $\frac{h_6}{k_6} = [4, 2, 1, 3, 1, 2]$ using the values of the first five convergents in the table, or do we have to compute it from scratch?

Let's write out the sequence of terms a_i as well as the sequences of numerators b_i and denominators k_i:

i	1	2	3	4	5	6
a_i	4	2	1	3	1	2
b_i	4	9	13	48	61	$\cdots$
k_i	1	2	3	11	14	$\cdots$

Can you see a pattern in the b_i and k_i values? For instance, can you determine b_6 and k_6 from the values already in the table? In general, could you determine the value of any b_i or k_i if you knew all of the values to its left ($b_1, \ldots, b_{i-1}$ and $k_1, \ldots, k_{i-1}$) and you also knew all of the terms $a_1, \ldots, a_i$?

The colors provide a hint. It turns out that any b_i term can be determined from the two terms to its left, b_{i-2} and b_{i-1}, and the term a_i. Can you see how **48**, the value of b_4, is related to the other three values that are colored in the table?

You may have noticed the relationship

$$48 = 3 \cdot 13 + 9.$$

In fact, this pattern generalizes to the other values in the table:

$$b_i = a_i \cdot b_{i-1} + b_{i-2}. \tag{1}$$

Can you find a similar pattern for k_i? As you may have observed,

$$k_i = a_i \cdot k_{i-1} + k_{i-2}. \tag{2}$$

Check for Understanding

1. Use the recurrence relations (1) and (2) to fill in the next two columns of the table for $\sqrt{19}$:

i	1	2	3	4	5	6	7
a_i	4	2	1	3	1	2	8
h_i	4	9	13	48	61	__	__
k_i	1	2	3	11	14	__	__

2. Use your answers to give the value of r_6 and r_7, the sixth and seventh convergents to $\sqrt{19}$.

Recurrence relations (1) and (2) tell us how to get the numerator h_i and denominator k_i of any convergent as long as we know the numerators and denominators of the previous two convergents. Even if we can prove that (1) and (2) always hold, they still do not give us the first two convergents' numerators, h_1 and h_2, or their denominators k_1 and k_2. To get the first two convergents using the recurrence relations (1) and (2), we cleverly (and quite artificially!) create two previous terms in the h and k sequences:

$$h_{-1} = 0, \qquad h_0 = 1,$$
$$k_{-1} = 1, \qquad k_0 = 0. \tag{3}$$

This is equivalent to adding two new columns to the left of our table for $\sqrt{19}$:

i	-1	0	1	2	3	4	5
a_i			4	2	1	3	1
h_i	0	1	4	9	13	48	61
k_i	1	0	1	2	3	11	14

The reader may verify that recurrence relations (1) and (2) now produce the correct values in the $i = 1$ and $i = 2$ columns of the table.

THEOREM 14.4.1	RECURRENCE RELATIONS FOR CONTINUED FRACTIONS

Let $n \in \mathbf{N}$, and let $\alpha = [a_1, a_2, \ldots, a_n, \ldots]$ be a continued fraction (possibly infinite) with at least n terms. Also, let $h_1, h_2, \ldots, h_n$ and $k_1, k_2, \ldots, k_n$ be defined by the following recurrence relations:

$$h_{-1} = 0, \qquad h_0 = 1$$
$$k_{-1} = 1, \qquad k_0 = 0$$
$$h_i = a_i \cdot h_{i-1} + h_{i-2}, \quad \textit{for } i = 1, \ldots, n, \tag{4}$$
$$k_i = a_i \cdot k_{i-1} + k_{i-2}, \quad \textit{for } i = 1, \ldots, n.$$

Then

$$[a_1, a_2, \ldots, a_n] = \frac{h_n}{k_n}.$$

Before proving this theorem, let's see an example. Note that Theorem 14.4.1 applies to any continued fraction (with real terms). If we restrict α to be a simple continued fraction, then Theorem 14.4.1 allows us to find the convergents of α, as the following example illustrates:

EXAMPLE 1

Use the recurrence relations (4) to find convergents $r_1, r_2, \ldots, r_5$ of

$$\sqrt{12} = [3, \overline{2, 6}].$$

SOLUTION Make a table of a_i, h_i, and k_i values. First fill in the values of $a_1, a_2, \ldots, a_5$ and the artificial values (3) of h_i and k_i for $i = 0$ and $i = -1$. Then use the recurrence relations (1) and (2) to find the values of h_i and k_i for $i = 1, 2, \ldots, 5$.

i	-1	0	1	2	3	4	5
a_i			3	2	6	2	6
h_i	0	1	3	7	45	97	627
k_i	1	0	1	2	13	28	181

Thus,

$$r_1 = \frac{3}{1} = 3, \quad r_2 = \frac{7}{2}, \quad r_3 = \frac{45}{13}, \quad r_4 = \frac{97}{28}, \quad \text{and} \quad r_5 = \frac{627}{181}.$$ ●

PROOF OF **THEOREM 14.4.1 (RECURRENCE RELATIONS FOR CONTINUED FRACTIONS)**

We prove the theorem by induction on n.

Let $P(n)$ be the statement that every continued fraction with exactly n terms, $[a_1, a_2, \ldots, a_n]$, satisfies the equation $[a_1, a_2, \ldots, a_n] = \dfrac{h_n}{k_n}$, where h_n and k_n are defined as in (4). We will use induction to show that $P(n)$ is true for all $n \in \mathbf{N}$.

Base Case For $n = 1$, we need to prove that $[a_1] = \dfrac{h_1}{k_1}$.

A continued fraction with only one term is just equal to that term, so $[a_1] = a_1$. We find the values of h_1 and k_1 using recurrence relations (4):

$$h_1 = a_1 \cdot h_0 + h_{-1} \qquad\qquad k_1 = a_1 \cdot k_0 + k_{-1}$$

$$= a_1 \cdot 1 + 0 \qquad\qquad\qquad = a_1 \cdot 0 + 1$$

$$= a_1. \qquad\qquad\qquad\qquad = 1.$$

Thus, $[a_1] = \dfrac{h_1}{k_1}$.

Inductive Step Let j be a natural number. Assume the inductive hypothesis $P(j)$.

Now let α be a continued fraction with $j + 1$ terms,

$$\alpha = [a_1, a_2, \ldots, a_{j-1}, a_j, a_{j+1}],$$

and let h_i, k_i for $i = 1, 2, \ldots, j+1$ be defined by the recurrence relations (4).

[To show: $\alpha = \dfrac{h_{j+1}}{k_{j+1}}$.]

Notice that α can be rewritten as a continued fraction with only j terms:

$$\alpha = [a_1, a_2, \ldots, a_{j-1}, x], \tag{5}$$

where

$$x = a_j + \frac{1}{a_{j+1}}.$$

By the inductive hypothesis,

$$\alpha = \frac{x \cdot h_{j-1} + h_{j-2}}{x \cdot k_{j-1} + k_{j-2}}$$

$$= \frac{\left(a_j + \dfrac{1}{a_{j+1}}\right) \cdot h_{j-1} + h_{j-2}}{\left(a_j + \dfrac{1}{a_{j+1}}\right) \cdot k_{j-1} + k_{j-2}}$$

$$= \frac{(a_j \cdot h_{j-1} + h_{j-2}) + \left(\dfrac{1}{a_{j+1}}\right) \cdot h_{j-1}}{(a_j \cdot k_{j-1} + k_{j-2}) + \left(\dfrac{1}{a_{j+1}}\right) \cdot k_{j-1}}$$

$$= \frac{h_j + \left(\dfrac{1}{a_{j+1}}\right) \cdot h_{j-1}}{k_j + \left(\dfrac{1}{a_{j+1}}\right) \cdot k_{j-1}} \qquad \leftarrow \text{by the recurrence relations (4)}$$

$$= \frac{a_{j+1} \cdot h_j + h_{j-1}}{a_{j+1} \cdot k_j + k_{j-1}}$$

$$= \frac{h_{j+1}}{k_{j+1}} \qquad \leftarrow \text{by the recurrence relations (4).}$$

Hence by the Principle of Mathematical Induction, $P(n)$ is true for all nonnegative integers n. ∎

A useful pattern in the tables of numerators and denominators

Using Theorem 14.4.1, we can find the numerators and denominators of the convergents of any simple continued fraction. For example, let's make a table of the first several convergents of $\sqrt{3} = [1, \overline{1, 2}]$.

n	−1	0	1	2	3	4	5	6	7
a_n			1	1	2	1	2	1	2
h_n	0	1	1	2	5	7	19	26	71
k_n	1	0	1	1	3	4	11	15	41

For each **×** made of **orange** and **brown** lines in the table, subtract the product of the two numbers connected by the brown line from the product of the two numbers connected by the orange line. Do you notice a pattern?

$$1 \cdot 1 - 1 \cdot 2 = \underline{\quad},$$

$$2 \cdot 3 - 1 \cdot 5 = \underline{\quad},$$

$$5 \cdot 4 - 3 \cdot 7 = \underline{\quad},$$

$$7 \cdot 11 - 4 \cdot 19 = \underline{\quad},\tag{6}$$

$$19 \cdot 15 - 11 \cdot 26 = \underline{\quad},$$

$$26 \cdot 41 - 15 \cdot 71 = \underline{\quad}.$$

You may have noticed (after filling in the blanks) that the expressions in (6) alternate in value between −1 and 1. The pattern illustrated by (6) holds not just for the continued fraction $[1, \overline{1, 2}]$ but for all continued fractions. This pattern is expressed in equation form in the following lemma.

LEMMA 14.4.2   THE CRISS-CROSS LEMMA

Let $n \in \mathbf{N}$, and let $\alpha = [a_1, a_2, \ldots, a_n, \ldots]$ be a continued fraction of length n or greater (possibly infinite). Let $h_1, h_2, \ldots, h_n \in \mathbf{R}$ and $k_1, k_2, \ldots, k_n \in \mathbf{R}$ be defined from α using the recurrence relations (4). Then

$$h_{n-1}k_n - k_{n-1}h_n = (-1)^{n+1}.$$

PROOF We prove the lemma by induction on n.

Let $P(n)$ be the statement that every continued fraction with at least n terms, $[a_1, a_2, \ldots, a_n, \ldots]$, satisfies the equation $h_{n-1}k_n - k_{n-1}h_n = (-1)^{n+1}$.

Base Case We need to prove $P(1)$. Let $[a_1, \ldots]$ be a continued fraction with at least one term. **[To show: $h_0 k_1 - k_0 h_1 = (-1)^2$; i.e., $h_0 k_1 - k_0 h_1 = 1$.]**

From the recurrence relations (4), we get

$$h_0 k_1 - k_0 h_1 = 1 \cdot 1 - 0 \cdot a_1$$
$$= 1.$$

Inductive Step Let j be a natural number, and assume the inductive hypothesis $P(j)$.

Now let α be a continued fraction with at least $j + 1$ terms,

$$\alpha = [a_1, a_2, \ldots, a_{j-1}, a_j, a_{j+1}, \ldots],$$

and let $h_1, \ldots, h_{j+1}$ and $k_1, \ldots, k_{j+1}$ be defined by the recurrence relations (4).

[To show: $h_j k_{j+1} - k_j h_{j+1} = (-1)^{j+2}$.]

By the recurrence relations (4),

$$h_j k_{j+1} - k_j h_{j+1} = h_j(a_{j+1} k_j + k_{j-1}) - k_j(a_{j+1} h_j + h_{j-1})$$
$$= a_{j+1} h_j k_j + h_j k_{j-1} - a_{j+1} h_j k_j - h_{j-1} k_j$$
$$= (-1)(h_{j-1} k_j - k_{j-1} h_j).$$

Using the inductive hypothesis to simplify the right side of this equation, we obtain

$$h_j k_{j+1} - k_j h_{j+1} = (-1)(-1)^{j+1}$$
$$= (-1)^{j+2}.$$

Thus, by the Principle of Mathematical Induction, $P(n)$ is true for all natural numbers n. ∎

Lemma 14.4.2 applies to any continued fraction, even if the terms are not integers. If α is a simple continued fraction, then (as we now prove) the h_n and k_n produced by the recurrence relations (4) are integers that are the numerator and denominator of the nth convergent of α *in lowest terms*.

LEMMA 14.4.3

Let $n \in \mathbf{N}$, and let $\alpha = [a_1, a_2, \ldots, a_n, \ldots]$ be a simple continued fraction of length n or greater (possibly infinite). For each $i = 1, 2, \ldots, n$, let $r_i = \dfrac{h_i}{k_i}$ be the ith convergent of α in lowest terms. Then the h_i and k_i satisfy the recurrence relations (4).

PROOF For $i = 1, 2, \ldots, n$, let $r_i = \dfrac{h_i}{k_i}$ be the ith convergent of α in lowest terms, and let h'_i and k'_i be given by the recurrence relations (4). We already know (by Theorem 14.4.1, Recurrence Relations for Continued Fractions) that

$$r_i = \frac{h'_i}{k'_i}.$$

Furthermore, since the terms $a_1, a_2, \ldots, a_i$ are integers, it follows from the recurrence relations (4) that h'_i and k'_i are integers. It remains to show that the fraction $\dfrac{h'_i}{k'_i}$ is in lowest terms—that is, that h'_i and k'_i are relatively prime.

Let $d = \gcd(h'_i, k'_i)$. **[To show: $d = 1$.]** By the Criss-Cross Lemma (14.4.2),

$$h'_{i-1}k'_i - k'_{i-1}h'_i = (-1)^{i+1}.$$

Since $d \mid k'_i$ and $d \mid h'_i$, d divides the left side of this equation, so

$$d \mid (-1)^{i+1}.$$

It follows that $d = 1$. ■

Convergents as rational approximations of irrational numbers

Given an infinite simple continued fraction, the following lemma expresses the difference between the continued fraction's exact value and its nth convergent, as a function of its $(n + 1)$st subfraction, α_{n+1}.

Let α be a positive irrational number with simple continued fraction expansion $\alpha = [a_1, a_2, a_3, \ldots]$. For all $i \in \mathbf{N}$, let $\dfrac{h_i}{k_i}$ be the ith convergent of α (in lowest terms), and let $k_0 = 0$. Then for all natural numbers n,

$$\alpha - \frac{h_n}{k_n} = \frac{(-1)^{n+1}}{k_n(\alpha_{n+1}k_n + k_{n-1})}.$$

PROOF Let n be a natural number.

By Theorem 14.2.2, the infinite simple continued fraction $[a_1, a_2, a_3, \ldots]$ can be written as a finite continued fraction in which the final term is not an integer, as follows:

$$\alpha = [a_1, a_2, \ldots, a_n, \alpha_{n+1}],$$

where α_{n+1} is the $(n + 1)$st subfraction of α. By Lemma 14.4.3, for all $i \in \mathbf{N}$, h_i and k_i satisfy the recurrence relations (4). (Also, let $h_0 = 1$ and $k_0 = 0$ as specified by the recurrence relations.) Then by the Recurrence Relations for Continued Fractions (14.4.1),

$$\alpha = \frac{\alpha_{n+1}b_n + b_{n-1}}{\alpha_{n+1}k_n + k_{n-1}}.$$

Thus,

$$\alpha - \frac{b_n}{k_n} = \frac{\alpha_{n+1}b_n + b_{n-1}}{\alpha_{n+1}k_n + k_{n-1}} - \frac{b_n}{k_n}$$

$$= \frac{(\alpha_{n+1}b_n + b_{n-1})k_n - (\alpha_{n+1}k_n + k_{n-1})b_n}{k_n(\alpha_{n+1}k_n + k_{n-1})}$$

$$= \frac{b_{n-1}k_n - k_{n-1}b_n}{k_n(\alpha_{n+1}k_n + k_{n-1})}.$$

Hence, by the Criss-Cross Lemma (14.4.2),

$$\alpha - \frac{b_n}{k_n} = \frac{(-1)^{n+1}}{k_n(\alpha_{n+1}k_n + k_{n-1})}.$$

This lemma leads immediately to the following theorem.

THEOREM 14.4.5   CONVERGENT APPROXIMATION THEOREM 1

Let α be a positive irrational number with simple continued fraction expansion
$\alpha = [a_1, a_2, a_3, \ldots]$. *For all $i \in \mathbf{N}$, let $\dfrac{b_i}{k_i}$ be the ith convergent of α (in lowest terms).*
Then for all natural numbers n,

$$\left| \alpha - \frac{b_n}{k_n} \right| < \frac{1}{k_n k_{n+1}}.$$

PROOF Let n be a natural number. Let $k_0 = 0$.

By Lemma 14.4.4,

$$\left| \alpha - \frac{b_n}{k_n} \right| = \left| \frac{(-1)^{n+1}}{k_n(\alpha_{n+1}k_n + k_{n-1})} \right|$$

$$= \frac{1}{k_n(\alpha_{n+1}k_n + k_{n-1})}$$

$$< \frac{1}{k_n(a_{n+1}k_n + k_{n-1})} \qquad \leftarrow \text{since } \alpha_{n+1} > a_{n+1}$$

$$= \frac{1}{k_n k_{n+1}} \qquad \leftarrow \text{by the recurrence relations (4).}$$

EXAMPLE 2

Recall from Table 14.3.3 that the third and fourth convergents of π are

$$r_3 = \frac{h_3}{k_3} = \frac{333}{106} \quad \text{and} \quad r_4 = \frac{h_4}{k_4} = \frac{355}{113}.$$

Thus, by Convergent Approximation Theorem 1, we know that the distance from r_3 to π is less than $\frac{1}{106 \cdot 113} = \frac{1}{11{,}978}$. That is, the theorem guarantees that

$$\left| \pi - \frac{333}{106} \right| < \frac{1}{106 \cdot 113}.$$

(In fact, the error is $\left| \pi - \frac{333}{106} \right| \approx \frac{1}{12{,}016}$, which is certainly less than $\frac{1}{11{,}978}$.)

We now state a corollary of Convergent Approximation Theorem 1 which is not quite as strong as that theorem. As we will see in Chapter 15, however, the following theorem is sometimes easier to use (and may be easier to remember as well).

14.4.6   **THE $\frac{1}{d^2}$ THEOREM**

Let α be a positive irrational number, and let $\frac{c}{d}$ be a continued fraction convergent for α (in lowest terms). Then

$$\left| \alpha - \frac{c}{d} \right| < \frac{1}{d^2}.$$

Put it in Prose, Paul!

When you approximate a number by a convergent whose denominator is d, the error is always less than $\frac{1}{d^2}$.

PROOF OF **THEOREM 14.4.6**

You will prove this theorem in Exercise 10.

EXAMPLE 3

As we saw in Table 14.3.2, the sixth convergent of $\sqrt{2} = [1, \overline{2}]$ is

$$r_6 = \frac{h_6}{k_6} = \frac{99}{70}.$$

The $\frac{1}{d^2}$ Theorem (14.4.6) guarantees that the error of this approximation is less than

$\frac{1}{70^2} = \frac{1}{4900}$. That is, the theorem guarantees that

$$\left| \sqrt{2} - \frac{99}{70} \right| < \frac{1}{70^2}.$$

(In fact, the error is $\left| \sqrt{2} - \frac{99}{70} \right| \approx \frac{1}{13,859}$, which is of course less than $\frac{1}{4900}$.)

The $\frac{1}{d^2}$ Theorem (14.4.6) implies that continued fraction convergents are *much* better than decimal approximations. Recall from Section 14.3 that when we approximate an arbitrary irrational number by a decimal (by a rational number whose denominator, d, is a power of 10), we are only guaranteed that the error of approximation will be less than $\frac{1}{2d}$. For instance, the decimal approximation $\sqrt{2} \approx 1.41 = \frac{141}{100}$ is guaranteed to have an error less than $\frac{1}{200}$ (and, in fact, $\left| \sqrt{2} - \frac{141}{100} \right| \approx \frac{1}{237}$, which is less than $\frac{1}{200}$). The $\frac{1}{d^2}$ Theorem shows that in contrast to the $\frac{1}{2d}$ worst-case error of decimal approximations, a continued fraction convergent with denominator d has a much smaller worst-case error of only $\frac{1}{d^2}$. For instance, in Example 3 we saw that using a denominator of only 70, a continued fraction convergent achieves an error less than $\frac{1}{70^2} = \frac{1}{4900}$.

Convergents give the best approximations

So far, we have shown that continued fractions give us great approximations to any irrational number. In fact, continued fraction convergents provide the *best possible* approximations to any irrational number using a denominator of limited size. This statement is formalized in the following theorem.

THEOREM 14.4.7   CONVERGENT APPROXIMATION THEOREM 2

Let α be a positive irrational number. For all $i \in \mathbf{N}$, let $\frac{h_i}{k_i}$ be the ith convergent of α (in lowest terms). Then for every natural number $n > 1$ and nonnegative integers a and b with $b \neq 0$,

$$\text{If } \left| \alpha - \frac{a}{b} \right| < \left| \alpha - \frac{h_n}{k_n} \right|, \text{ then } b > k_n.$$

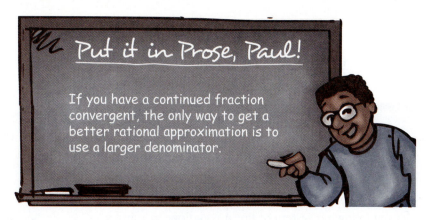

If you have a continued fraction convergent, the only way to get a better rational approximation is to use a larger denominator.

We will prove this theorem at the end of this section.

EXAMPLE 4

As we saw in Example 1, the fourth convergent of $\sqrt{12}$ is

$$r_4 = \frac{h_4}{k_4} = \frac{97}{28}.$$

Convergent Approximation Theorem 2 then makes a guarantee: If you want a rational approximation to $\sqrt{12}$, then no rational number with denominator smaller than 28 does better than $\frac{97}{28}$.

The proof of Convergent Approximation Theorem 2 relies on the following lemma. Approximating an irrational number α by a rational number, $\alpha \approx \frac{a}{b}$, is equivalent to finding integers a and b for which $\alpha b \approx a$. The goodness of $\frac{a}{b}$ as an approximation to α can be measured by how small $|\alpha b - a|$ is. The following lemma uses this measure of goodness to express how well continued fraction convergents approximate irrational numbers.

LEMMA 14.4.8

Let α be a positive irrational number. For all $i \in \mathbb{N}$, let $\frac{h_i}{k_i}$ be the ith convergent of α (in lowest terms). Then for all natural numbers n and nonnegative integers a and b with $b \neq 0$,

$$\text{If } |\alpha b - a| < |\alpha k_n - h_n|, \text{ then } b \geq k_{n+1}.$$

EXAMPLE 5

Recall (from Table 14.3.3) that the second and third convergents of π are

$$r_2 = \frac{h_2}{k_2} = \frac{22}{7} \qquad \text{and} \qquad r_3 = \frac{h_3}{k_3} = \frac{333}{106}.$$

In this case, Lemma 14.4.8 states that for any nonnegative rational number $\frac{a}{b}$,

$$\text{If} \quad |\pi b - a| < |7\pi - 22|, \quad \text{then} \quad b \geq 106.$$

PROOF of **LEMMA 14.4.8** (By contradiction.)

Assumption: Suppose there exist nonnegative integers a and b, with $b \neq 0$, such that

$$|\alpha b - a| < |\alpha k_n - h_n| \quad \text{and} \quad b < k_{n+1}.$$

Consider the following system of linear equations in x and y:

$$h_n x + h_{n+1} y = a \tag{7}$$

$$k_n x + k_{n+1} y = b \tag{8}$$

Claim 1: The solutions x and y are integers.

Proof of Claim 1 To solve this system of equations for x, we multiply equation (7) by k_{n+1}, multiply equation (8) by h_{n+1}, then subtract the second equation from the first to get

$$(h_n k_{n+1} - k_n h_{n+1})x = k_{n+1}a - h_{n+1}b.$$

By the Criss-Cross Lemma (14.4.2), this becomes

$$(-1)^{n+2}x = k_{n+1}a - h_{n+1}b,$$

from which we get

$$x = (-1)^{n+2}(k_{n+1}a - h_{n+1}b).$$

Thus, x is an integer. Similarly, y is an integer (as you will prove in Exercise 11).

Claim 2: The solutions x and y are nonzero.

Proof of Claim 2 (By contradiction.) Suppose that $x = 0$ or $y = 0$.

Case I: $x = 0$.
From equation (8), $b = k_{n+1}y$. Since b and k_{n+1} are both positive, the integer y must be positive (i.e., $y \geq 1$). Then $b \geq k_{n+1}$, which contradicts our Assumption. $\Rightarrow\Leftarrow$

Case II: $y = 0$.

From equations (7) and (8), $a = h_n x$ and $b = k_n x$. Since $b \neq 0$, we know that the integer x is nonzero. Furthermore,

$$\begin{aligned}
|\alpha b - a| &= |\alpha k_n x - h_n x| \\
&= |x(\alpha k_n - h_n)| \\
&= |x| \cdot |\alpha k_n - h_n| \\
&\geq |\alpha k_n - h_n| \qquad \leftarrow \text{since } |x| \geq 1.
\end{aligned}$$

Thus, $|\alpha b - a| \geq |\alpha k_n - h_n|$, which contradicts our Assumption. $\Rightarrow\Leftarrow$

This completes the proof of Claim 2.

Claim 3: The solutions x and y have opposite signs.

Proof of Claim 3 From equation (8),

$$k_n x = b + (-k_{n+1} y). \qquad (9)$$

By Claim 2, y is nonzero, so either $y < 0$ or $y > 0$.

Case I: $y < 0$.

Both terms on the right side of equation (9) are positive, so $k_n x$ must be positive. It follows that $x > 0$.

Case II: $y > 0$.

We know

$$b < k_{n+1} \qquad\qquad \leftarrow \textbf{by our Assumption}$$

$$\leq k_{n+1} y \qquad\qquad \leftarrow \textbf{since } y \textbf{ is a positive integer}$$

so by equation (9), $k_n x$ must be negative. It follows that $x < 0$.

Thus, Claim 3 is established.

By Lemma 14.4.4, $\left(\alpha - \dfrac{h_n}{k_n}\right)$ and $\left(\alpha - \dfrac{h_{n+1}}{k_{n+1}}\right)$ have opposite signs. Since k_n and k_{n+1} are both positive, it follows that $(\alpha k_n - h_n)$ and $(\alpha k_{n+1} - h_{n+1})$ have opposite signs. Combining this with Claim 3, we conclude that

$$x(\alpha k_n - h_n) \quad \text{and} \quad y(\alpha k_{n+1} - h_{n+1}) \text{ have the same sign.} \qquad (10)$$

Now from equations (7) and (8),

$$\alpha b - a = \alpha(k_n x + k_{n+1} y) - (h_n x + h_{n+1} y).$$

Regrouping terms and taking the absolute value of both sides gives

$$
\begin{aligned}
|\alpha b - a| &= |x(\alpha k_n - h_n) + y(\alpha k_{n+1} - h_{n+1})| \\
&= |x(\alpha k_n - h_n)| + |y(\alpha k_{n+1} - h_{n+1})| \qquad \leftarrow \textbf{by statement (10)} \\
&> |x(\alpha k_n - h_n)| \\
&= |x| \cdot |\alpha k_n - h_n| \\
&\geq |\alpha k_n - h_n| \qquad \leftarrow \textbf{since } |x| \geq 1.
\end{aligned}
$$

Thus, $|\alpha b - a| > |\alpha k_n - h_n|$, which contradicts our Assumption. $\Rightarrow\Leftarrow$

Since we have reached a contradiction, our Assumption must be false. This completes the proof of the lemma. ∎

Now that we've completed that marathon proof, we're ready for the victory lap: the proof that Convergent Approximation Theorem 2 follows from Lemma 14.4.8.

PROOF of **THEOREM 14.4.7 (CONVERGENT APPROXIMATION THEOREM 2)** We prove this indirectly.

Let α be a positive irrational number, and for all $i \in \mathbf{N}$, let $\dfrac{h_i}{k_i}$ be the ith convergent of α (in lowest terms). Suppose there exist a natural number $n > 1$ and nonnegative integers a and b, with $b \neq 0$, such that

$$\left| \alpha - \frac{a}{b} \right| < \left| \alpha - \frac{h_n}{k_n} \right| \tag{11}$$

and

$$b \leq k_n. \tag{12}$$

Multiplying inequalities (11) and (12) together, we obtain

$$|\alpha b - a| < |\alpha k_n - h_n|.$$

Then by Lemma 14.4.8, we know that $b \geq k_{n+1}$. Since $n > 1$, it follows from the recurrence relations (4) that $k_{n+1} > k_n$, and hence $b > k_n$.

But this contradicts supposition (12). $\Rightarrow\Leftarrow$

This completes the proof of the theorem. ∎

EXERCISES 14.4

Numerical Problems

In Exercises 1–5, use the method of Example 1 to find the first five convergents to each continued fraction.

1. $[1, \overline{2, 3}]$ 2. $[4, \overline{3, 1, 5}]$ 3. $[1, 2, 3, 4, 5]$ 4. $[\overline{7}]$ 5. $[0, \overline{1, 1, 3}]$

6. **a.** What does Theorem 14.4.6 tell us about $\frac{22}{7}$ as an approximation to π?

    Compare this with the actual value of the error of approximation, $\left| \pi - \frac{22}{7} \right|$.

    b. What does Theorem 14.4.6 tell us about $\frac{333}{106}$ as an approximation to π?

    Compare this with the actual error of approximation.

7. What does Convergent Approximation Theorem 1 (14.4.5) tell us about $\frac{355}{113}$ as

    an approximation to π? Compare this with the actual error of approximation.

8. What does Convergent Approximation Theorem 2 (14.4.7) tell us about $\frac{627}{181}$ as

    an approximation to $\sqrt{12} = [3, \overline{2, 6}]$?

9. **Computer Exercise**   Write a program that uses the Recurrence Relations for Continued Fractions (14.4.1) to find the continued fraction convergents of any positive real number, and computes the error of each convergent.

    a. Use your program to find the first 20 convergents of $\sqrt{e}$, and compute the error of each approximation to $\sqrt{e}$.

    b. Use your program to find the first 20 convergents of $\sqrt{31}$, and compute the error of each approximation to $\sqrt{31}$.

    c. Use your program to find the first 20 convergents of $\sqrt[3]{5}$, and compute the error of each approximation to $\sqrt[3]{5}$.

Reasoning and Proofs

10. Prove the $\frac{1}{d^2}$ Theorem (14.4.6).

11. In the proof of Lemma 14.4.8, complete the proof of Claim 1 by proving that y is an integer.

12. Let $\alpha = [a_1, a_2, a_3, \ldots]$ be an infinite simple continued fraction, let the nonnegative integers $h_{-1}, h_0, h_1, h_2, h_3, \ldots$ and $k_{-1}, k_0, k_1, k_2, k_3, \ldots$ be defined using the recurrence relations (4), and let r_n represent the nth convergent of α. Then for all natural numbers n, prove each of the following.

    a. $h_n k_{n-2} - h_{n-2} k_n = (-1)^{n+1} \cdot a_n$ **b.** $r_n - r_{n-2} = \dfrac{(-1)^n \cdot a_n}{k_n k_{n-2}}$

Advanced Reasoning and Proofs

EXPLORATION The Rigors of Continued Fractions (Exercises 13–16)

Note: This exploration requires techniques from analysis.

In Section 14.2, we stated without proof that every nonnegative real number can be expressed uniquely as a simple continued fraction. In this exploration, you will provide a rigorous justification for that statement.

Since we already established a one-to-one correspondence between rational numbers and finite continued fractions, we may focus our attention on irrational numbers.

13. Definition 14.2.1 tells us how to find the infinite simple continued fraction corresponding to a given positive irrational number. Suppose we wish to go in the other direction—from an infinite simple continued fraction, $[a_1, a_2, a_3, \ldots]$, we wish to define a real number. A natural way to do this is to take a limit of finite continued fractions. That is, for any natural number n, define

$$r_n = [a_1, a_2, \ldots, a_n]$$

(the nth convergent of the infinite continued fraction), and let

$$\alpha = \lim_{n \to \infty} r_n.$$

Prove that this limit exists. [*Hint:* Use the Criss-Cross Lemma (14.4.2) to show that the sequence r_n is a Cauchy sequence.]

14. Show that the continued fraction convergents of any positive irrational number converge to that number. That is, suppose that α is a positive irrational number with continued fraction convergents $r_n = [a_1, a_2, \ldots, a_n]$. Prove that $\lim_{n \to \infty} r_n = \alpha$.

    [*Hint:* Use the $\dfrac{1}{d^2}$ Theorem (14.4.6).]

15. Show that no two distinct irrational numbers have the same continued fraction expansion. [*Hint:* Use the result of Exercise 14.]

16. Show that any two distinct infinite simple continued fractions, $[a_1, a_2, a_3, \ldots]$ and $[b_1, b_2, b_3, \ldots]$, represent distinct irrational numbers. [*Hint:* Prove that $\alpha = \lim_{n \to \infty} [a_1, a_2, \ldots, a_n]$ lies strictly between $[a_1, a_2, \ldots, a_{n-1}, a_n]$ and $[a_1, a_2, \ldots, a_{n-1}, a_n + 1]$. Then consider the smallest natural number i for which $a_i \neq b_i$.]

Chapter 15

Sophie Germain (1776–1831)

Born in Paris in 1776, Sophie Germain is known for her results on Fermat's Last Theorem, as well as for her work on acoustics, elasticity, and curvature of surfaces. The French Revolution began when Germain was 13. Concerned with her safety, Germain's parents did not allow her to leave the house during these turbulent years. Thus, Germain spent much of her adolescence reading the books in her father's library.

At the age of 13, Germain read Montucla's account of Archimedes' death during the invasion of Syracuse by the Romans. According to Montucla, "during the invasion of his city by the Romans, Archimedes was so engrossed in the study of a geometric figure that he failed to respond to the questioning of a Roman soldier. As a result he was speared to death." This story convinced Germain that mathematics could be so absorbing that it would provide a mental escape from the horrors that were taking place in the streets of Paris. It was then that she decided to become a mathematician.

According to the Italian mathematician Count Guglielmo Libri Carucci dalla Sommaja, when Germain was an adolescent, her parents worried that her interest in math would make her unfeminine. Her father forbade her to continue learning mathematics. To prevent her from studying mathematics at night, he took away her candles, her

heating, and even her clothing. Yet Germain was so determined to study mathematics that she stashed away a supply of candles and read each night wrapped in blankets, even though at times the room was so cold that the ink in her inkwell froze. Her parents eventually came to accept her interest, and they even supported her financially when she did mathematics research as an adult.

After the French Revolution, the École Polytechnique was founded to educate future engineers and civil servants in mathematics and the sciences. This would have been the natural place for Germain to continue her mathematical studies, if she had been a man. Since that option was not open to her, Germain used the name of a former student at the École Polytechnique, Antoine-August Le Blanc, to obtain lecture notes for courses she was interested in. The administration did not know that Le Blanc was no longer a student, so his professors continued to make their notes available to him. Germain was particularly interested in Lagrange's course on analysis, and she handed in a report using the name Le Blanc. Impressed by the creativity of this report, Lagrange was eager to meet Le Blanc, who had been considered a rather mediocre student. Though Lagrange may have been surprised to discover that Monsieur Le Blanc was actually Sophie Germain, Lagrange continued to hold her in high regard, and from then on he played the role of her mathematical mentor.

Despite having no formal education in mathematics, Germain was able to learn mathematics by reading the works of the mathematicians of the day and writing to the authors about her ideas. She became interested in number theory after reading Legendre's *Théorie des Nombres*, and she corresponded at length with Legendre about some results of her own related to Fermat's Last Theorem. When Legendre published a supplement to his book, he attributed some results to Germain in a footnote; it is due to this footnote that Germain's work on Fermat's Last Theorem is well known today.

After reading Gauss's *Disquisitiones Arithmeticae* in 1804, Germain began a correspondence with Gauss using the pseudonym Monsieur Le Blanc. Gauss was impressed with Germain's work, and they corresponded for several years without Germain revealing her true identity. When Napoleon invaded Germany in 1807, Germain was concerned about Gauss's welfare. She told her friend, the French commander General Pernety, that Gauss was an important scientist whose safety must be assured. When General Pernety found Gauss and told him that Sophie Germain had sent him, Gauss responded that he knew nobody by that name. Germain eventually cleared up the resulting confusion by writing Gauss a letter in which she admitted her true identity. Rather than ridiculing the idea of a woman being interested in mathematics, as she had feared, Gauss was all the more impressed. Here is part of Gauss's response to her letter:

But how to describe to you my admiration and astonishment at seeing my esteemed correspondent M. Le Blanc metamorphose himself into this illustrious personage who gives such a brilliant example of what I would find it difficult to believe. A taste for the abstract sciences in general and, above all, for the mysteries of numbers is excessively rare: this is not surprising since the charms of this sublime science reveal themselves only to those who have the courage to go deeply into it. But when a person of the sex which, according to our customs and prejudices, must encounter infinitely more difficulties than men to familiarize herself with these thorny researches, succeeds nevertheless in surmounting these obstacles and penetrating the most obscure parts of them, then without doubt she must have the noblest courage, quite extraordinary talents, and superior genius.

Germain continued her mathematical correspondence with Gauss for several years under her own name. Although Gauss and Germain never met, in 1831 he recommended that Germain be awarded an honorary degree from the University of Göttingen based on her contributions to mathematics. Unfortunately, she died from breast cancer before receiving the degree.

A Bit of Germain's Math

Fermat's Last Theorem (FLT) states that for any integer $n > 2$, the equation $x^n + y^n = z^n$ has no solutions where x, y, and z are nonzero integers. Fermat proved FLT for $n = 4$ and observed that to prove it in general, it would be sufficient to show that it was true for all odd primes n.

Euler proved FLT for $n = 3$, and Legendre and Dirichlet independently proved FLT for $n = 5$.

Germain proved the first general result about Fermat's Last Theorem. She approached the proof of Fermat's Last Theorem by dividing the set of possible solutions to $x^p + y^p = z^p$ (where p is an odd prime) into those known as *Case I*, in which x, y, and z are not divisible by p; and those known as *Case II*, in which at least one of x, y, or z is divisible by p. Then she proved that for all odd primes $p < 100$, there is no Case I solution to $x^p + y^p = z^p$. Germain's approach to FLT continued to be used for over 150 years. In fact, in 1985, Adleman and Heath-Brown used a generalization of Germain's method to prove that for infinitely many odd primes p, there are no Case I solutions to $x^p + y^p = z^p$. (For a more detailed history of Fermat's Last Theorem and Germain's contributions, see Sections 15.2–15.5.)

In addition to her research in number theory, Germain did important work in several other areas of mathematics. In 1816, she was awarded the Grand Prize of the Mathematical Sciences by the Paris Academy of Sciences for her work on the vibrations of thin elastic plates. Germain's research on the curvature of surfaces was also a major contribution to mathematics. In particular, it was Germain who introduced the idea of mean curvature, which is still quite important in mathematics today.

Chapter 15 Some Nonlinear Diophantine Equations

15.1 Pell's Equation

Early on in our study of number theory (in Chapter 5), we learned how to solve linear Diophantine equations. In this chapter, we investigate a few Diophantine equations that are not linear.

Going straight to Pell

In Chapter 1, we proved that $\sqrt{2}$ is not a rational number. This means that there do not exist natural numbers x and y such that

$$\sqrt{2} = \frac{x}{y}.$$

If we square both sides of this equation and multiply through by y^2, we get the equation:

$$x^2 = 2y^2. \tag{1}$$

There are no natural numbers x and y that satisfy this equation. Thus, a perfect square cannot equal twice a perfect square. However, there are integers x and y for which equation (1) is *nearly* true. For example, if $x = 3$ and $y = 2$, then $x^2 = 9$ and $2y^2 = 8$. So while x^2 and $2y^2$ are not equal, they only differ by 1. That is, we have a solution to the Diophantine equation

$$x^2 - 2y^2 = 1.$$

This equation is a special case of an equation known as **Pell's equation**:

$$x^2 - dy^2 = 1,$$

where d is a natural number that is not a perfect square.

EXAMPLE 1

Consider Pell's equation with $d = 5$:

$$x^2 - 5y^2 = 1. \tag{2}$$

Does this Diophantine equation have a solution with $x, y > 0$?

SOLUTION Equation (2) can also be written in the form $x^2 = 5y^2 + 1$. Thus, we want to find a natural number y for which the quantity $5y^2 + 1$ is a perfect square. Let's make a table of values of $5y^2 + 1$, to see whether we ever get a perfect square:

y	1	2	3	**4**	5	6	$\cdots$
$5y^2 + 1$	6	21	46	**81**	126	181	$\cdots$

We find that $5 \cdot 4^2 + 1 = 9^2$. Thus, $x = 9$, $y = 4$ is a solution to $x^2 - 5y^2 = 1$. ◼

In this example, we found a solution to the Diophantine equation $x^2 - 5y^2 = 1$ by trying successive values of y until we found one that worked. If we wanted a solution to the equation $x^2 - 61y^2 = 1$, this trial-and-error method would take a long time — see the "Fun Facts".

Fun Facts

Even for relatively small values of d, the smallest solution to Pell's equation $x^2 - dy^2 = 1$ can be gigantic. When $d = 61$, Pell's equation is the Diophantine equation $x^2 - 61y^2 = 1$. In 12th-century India, Bhaskara considered this equation and found the solution $x = 1766319049$, $y = 226153980$, which is in fact the smallest solution. When $d = 3061$, Pell's equation is $x^2 - 3061y^2 = 1$, whose smallest solution involves numbers that are over 100 digits long:

$x = $ 1125301210233588984049035325007836833434570182706813305868710600 9653150953706200247914149546190456364544 9,

$y = $ 2033935240786948099791397324693815401325711604511166489710877622 42338733562573723317736495744984899734 0.

Even larger numbers appear as the solution to a trick problem proposed by Archimedes. To share knowledge with his colleagues, Archimedes would prepare lists of the statements of his discoveries, without proof. Among the true statements on these lists, Archimedes would sometimes intentionally include false statements or nearly impossible problems. This prevented his colleagues from claiming that they had already discovered all of Archimedes' theorems on their own. In a letter to Eratosthenes (whom we met in Chapter 3), Archimedes wrote:

> If thou art diligent and wise, O stranger, compute the number of cattle of the Sun, who once upon a time grazed on the fields of the Thrinacian isle of Sicily, divided into four herds of different colours, one milk white, another a glossy black, a third yellow, and the last dappled. In each herd were bulls, mighty in number according to these proportions: Understand, stranger, that the white bulls were equal to a half

and a third of the black together with the whole of the yellow, while the black were equal to the fourth part of the dappled and a fifth, together with, once more, the whole of the yellow. Observe further that the remaining bulls, the dappled, were equal to a sixth part of the white and a seventh, together with all of the yellow. These were the proportions of the cows: The white were precisely equal to the third part and a fourth of the whole herd of the black; while the black were equal to the fourth part once more of the dappled and with it a fifth part, when all, including the bulls, went to pasture together. Now the dappled in four parts were equal in number to a fifth part and a sixth of the yellow herd. Finally the yellow were in number equal to a sixth part and a seventh of the white herd. If thou canst accurately tell, O stranger, the number of cattle of the Sun, giving separately the number of well-fed bulls and again the number of females according to each colour, thou wouldst not be called unskilled or ignorant of numbers, but not yet shalt thou be numbered among the wise.

But come, understand also all these conditions regarding the cattle of the Sun. When the white bulls mingled their number with the black, they stood firm, equal in depth and breadth, and the plains of Thrinacia, stretching far in all ways, were filled with their multitude. Again, when the yellow and the dappled bulls were gathered into one herd they stood in such a manner that their number, beginning from one, grew slowly greater till it completed a triangular figure, there being no bulls of other colours in their midst nor none of them lacking. If thou art able, O stranger, to find out all these things and gather them together in your mind, giving all the relations, thou shalt depart crowned with glory and knowing that thou hast been adjudged perfect in this species of wisdom.

The fantastically complicated conditions of this problem, amounting to nine linear and quadratic equations in eight variables, eventually reduce to the Pell equation $x^2 - 4729494y^2 = 1$. The problem was first solved in 1893 by the Hillsboro, Illinois, Mathematical Club. As it turns out, the smallest solution to this equation has over 200,000 digits!

We would like a better way to determine whether solutions to $x^2 - dy^2 = 1$ exist for an arbitrary value of d and to find solutions if they do exist. If x and y satisfy $x^2 - dy^2 = 1$, then as noted, x^2 is approximately equal to dy^2, which means that the ratio $\frac{x}{y}$ is approximately $\sqrt{d}$. Thus, we expect that a solution (x, y) to Pell's equation will give us a good rational approximation $\frac{x}{y}$ to $\sqrt{d}$. In Example 1, for instance, we found the solution $(9, 4)$ for the equation $x^2 - 5y^2 = 1$, which gives us $\frac{9}{4}$ as a rational approximation to $\sqrt{5}$. This is not a bad approximation: $\frac{9}{4} = 2.25$, and $\sqrt{5} \approx 2.236$.

The problem of finding approximations to irrational numbers was discussed in Chapter 14, where we learned that continued fractions can be used to find extremely good rational approximations to irrational numbers. In fact, $\frac{9}{4} = 2 + \frac{1}{4} = [2, 4]$ is a continued fraction convergent of $\sqrt{5} = [2, \overline{4}]$. This suggests that to find solutions to Pell's equation $x^2 - dy^2 = 1$, we might try using the continued fraction convergents for $\sqrt{d}$.

EXAMPLE 2

Consider the Pell equation $x^2 - 23y^2 = 1$. Find the continued fraction for $\sqrt{23}$, and determine whether solutions to Pell's equation can be found among the convergents.

SOLUTION Using the methods of Chapter 14, we find that $\sqrt{23} = [4, \overline{1, 3, 1, 8}]$. The convergents are

$$\frac{4}{1}, \frac{5}{1}, \frac{19}{4}, \frac{24}{5}, \frac{211}{44}, \frac{235}{49}, \frac{916}{191}, \frac{1151}{240}, \frac{10124}{2111}, \ldots$$

To determine whether any of these give rise to solutions of Pell's equation, we make a table.

x	4	5	19	**24**	211	235	916	**1151**	10124	$\cdots$
y	1	1	4	**5**	44	49	191	**240**	2111	$\cdots$
$x^2 - 23y^2$	-7	2	-7	**1**	-7	2	-7	**1**	-7	$\cdots$

Thus, we find that $(\mathbf{24, 5})$ and $(\mathbf{1151, 240})$ satisfy the equation $x^2 - 23y^2 = \mathbf{1}$. ◼

Note that the values of $x^2 - 23y^2$ in this table seem to follow a repeating pattern, which suggests that every fourth convergent gives a solution to Pell's equation. From this, we might surmise that the equation $x^2 - 23y^2 = 1$ has infinitely many solutions. This is a special case of the following theorem.

THEOREM 15.1.1

Let d be a natural number that is not a perfect square. Then there exist infinitely many pairs (x, y) of natural numbers such that

$$x^2 - dy^2 = 1. \tag{3}$$

The goal of this section is to prove this theorem. In our proof of this theorem, we will see that infinitely many solutions to equation (3) can be found among the convergents for $\sqrt{d}$. In fact, it turns out (although we will not prove it here) that *every* solution to Pell's equation is found among the continued fraction convergents for $\sqrt{d}$. (See Exercises 15–18.)

One solution, many solutions

Naomi's Numerical Proof Preview: Proposition 15.1.2

If we have one solution to Pell's equation, it is not too hard to create more solutions. To see how this works, let's start with the solution $(9, 4)$ to the equation $x^2 - 5y^2 = 1$, and see if we can create some more solutions.

We have

$$9^2 - 5 \cdot 4^2 = 1,$$

which we may write as

$$(9 + 4\sqrt{5})(9 - 4\sqrt{5}) = 1. \tag{4}$$

Squaring both sides of this equation gives

$$(9 + 4\sqrt{5})^2(9 - 4\sqrt{5})^2 = 1.$$

To evaluate the left side, we must compute both

$$(9 + 4\sqrt{5})^2 = (9^2 + 5 \cdot 4^2) + (2 \cdot 9 \cdot 4)\sqrt{5} \tag{5}$$

$$= 161 + 72\sqrt{5} \tag{6}$$

and

$$(9 - 4\sqrt{5})^2 = 161 - 72\sqrt{5}.$$

Thus, when we square equation (4), the result is the equation

$$(161 + 72\sqrt{5})(161 - 72\sqrt{5}) = 1,$$

which is equivalent to

$$161^2 - 5 \cdot 72^2 = 1.$$

And voilà, we've found that $(161, 72)$ is another solution to our Diophantine equation

$$x^2 - 5y^2 = 1.$$

In general, if we have a solution to the equation $x^2 - dy^2 = 1$, this method can be used to concoct a new solution to the same equation. The formula that tells us how to do this can be seen in the right sides of equations (5) and (6). We obtained our new solution $(161, 72)$ from our old solution $(9, 4)$ using the formulas

$$161 = 9^2 + 5 \cdot 4^2,$$

$$72 = 2 \cdot 9 \cdot 4.$$

Let d be a natural number that is not a perfect square, and consider the Diophantine equation $x^2 - dy^2 = 1$. If (a, b) is a solution to this Diophantine equation, then the pair

$$(v, w) = (a^2 + db^2, 2ab)$$

is also a solution.

PROOF Since $a^2 - db^2 = 1$, we have

$$(a + b\sqrt{d})(a - b\sqrt{d}) = 1. \tag{7}$$

The square of $(a + b\sqrt{d})$ is given by

$$(a + b\sqrt{d})^2 = (a^2 + db^2) + (2ab)\sqrt{d}$$
$$= v + w\sqrt{d}.$$

Similarly,

$$(a - b\sqrt{d})^2 = v - w\sqrt{d}.$$

Thus, if we square equation (7), we obtain

$$(v + w\sqrt{d})(v - w\sqrt{d}) = 1,$$

or equivalently,

$$v^2 - dw^2 = 1.$$

Hence (v, w) is a solution to the Diophantine equation $x^2 - dy^2 = 1$. ■

This proposition tells us that given a solution (a, b) to the Diophantine equation $x^2 - dy^2 = 1$, where a and b are nonzero, we can always create the new solution $(a^2 + db^2, 2ab)$. Note that since $a^2 + db^2 > |a|$ and $|2ab| > |b|$, the absolute values of the new solution are greater than absolute values of the old solution. Hence, assuming we have a single solution, we can apply Proposition 15.1.2 repeatedly to obtain a sequence of larger and larger solutions. It follows that:

If there is a single solution to the Diophantine equation $x^2 - dy^2 = 1$, with $x, y \neq 0$, then there are infinitely many solutions.

Another approach, which generates even more solutions, is explored in Exercises 25–27. Nevertheless, the job of proving that there exists a single positive solution to Pell's equation remains; that is, we must still prove that for any natural number d that is not a perfect square, there exists a solution (x, y) to the Diophantine equation $x^2 - dy^2 = 1$.

The number system Z[√d̄]

Even though Proposition 15.1.2 is a statement about integers, we discovered it and proved it using irrational numbers. (This proposition can also be proved directly without irrational numbers; you will do so in Exercise 23.) The irrational numbers we needed for our argument have the form

$$x + y\sqrt{d},$$

where x and y are integers. (If you studied Chapter 13, this should remind you of the Gaussian integers, which would be obtained by letting $d = -1$.) We will denote the set of all numbers of this form by $\mathbf{Z}[\sqrt{d}]$, read "$\mathbf{Z}$ adjoined $\sqrt{d}$."

Let d be an integer that is not a perfect square. The number system $\mathbf{Z}[\sqrt{d}]$ is defined by

$$\mathbf{Z}[\sqrt{d}] = \{ x + y\sqrt{d} \mid x, y \in \mathbf{Z}\}.$$

The study of the number system $\mathbf{Z}[\sqrt{d}]$ is a fascinating topic that belongs to the area of *algebraic number theory*. The number system $\mathbf{Z}[\sqrt{d}]$ is a ring (see Exercise 24), and you will explore other properties of $\mathbf{Z}[\sqrt{d}]$ in the exercises. For our purposes here, however, we will only need a few properties of $\mathbf{Z}[\sqrt{d}]$.

Let d be an integer that is not a perfect square. $\mathbf{Z}[\sqrt{d}]$ is closed under addition and multiplication.

PROOF You will prove this lemma in Exercise 21.

Given any element $a = x + y\sqrt{d} \in \mathbf{Z}[\sqrt{d}]$, we define $\bar{a} = x - y\sqrt{d}$ to be the **conjugate** of a.

The number system $\mathbf{Z}[\sqrt{d}]$ has a *norm* function that assigns an integer to each element of $\mathbf{Z}[\sqrt{d}]$. (If you've read Chapter 13, this may remind you of the norm of a Gaussian integer.)

*Let d be an integer that is not a perfect square. Let $a = x + y\sqrt{d}$ be an element of $\mathbf{Z}[\sqrt{d}]$. We define the **norm** of a by*

$$N(a) = a\bar{a}$$

$$= x^2 - dy^2. \tag{8}$$

In this section, we will be interested mainly in number systems $\mathbf{Z}[\sqrt{d}\,]$ in which d is a *positive* integer. Note that in this case, the norm can be negative (unlike the norm of a Gaussian integer, which is always positive).

EXAMPLE 3

In $\mathbf{Z}[\sqrt{2}]$, consider the element $a = 5 + 4\sqrt{2}$. Find the conjugate of a and the norm of a.

SOLUTION In $\mathbf{Z}[\sqrt{2}]$, the conjugate of $a = 5 + 4\sqrt{2}$ is $\bar{a} = 5 - 4\sqrt{2}$.

$$
\begin{aligned}
N(a) &= a\bar{a} \\
&= (5 + 4\sqrt{2})(5 - 4\sqrt{2}) \\
&= 5^2 - 2 \cdot 4^2 \\
&= -7.
\end{aligned}
$$

The following lemma asserts that conjugation preserves multiplication.

LEMMA 15.1.7

Let d be an integer that is not a perfect square. Let $a, b \in \mathbf{Z}[\sqrt{d}\,]$. Then $\overline{ab} = \bar{a}\,\bar{b}$.

PROOF You will prove this in Exercise 10.

The following lemma states that the norm is multiplicative.

LEMMA 15.1.8

Let d be an integer that is not a perfect square. Let $a, b \in \mathbf{Z}[\sqrt{d}\,]$. Then

$$
N(ab) = N(a)N(b).
$$

PROOF

$$
\begin{aligned}
N(ab) &= ab \cdot \overline{ab} \\
&= ab \cdot \bar{a} \cdot \bar{b} \qquad \leftarrow \textbf{by Lemma 15.1.7} \\
&= a\bar{a} \cdot b\bar{b} \\
&= N(a)N(b).
\end{aligned}
$$

This lemma can be used to prove the following result.

Let d be an integer that is not a perfect square. Let $a, b \in \mathbf{Z}[\sqrt{d}]$. If $\frac{a}{b}$ is an element of $\mathbf{Z}[\sqrt{d}]$, *then*

$$N\left(\frac{a}{b}\right) = \frac{N(a)}{N(b)}.$$

PROOF You will prove this in Exercise 9.

Proof of existence of solutions to Pell's equation

We now prove Theorem 15.1.1, which states that for any natural number d that is not a perfect square, the Diophantine equation

$$x^2 - dy^2 = 1$$

has infinitely many solutions.

PROOF of **THEOREM 15.1.1**

Let d be a natural number that is not a perfect square. We wish to find natural numbers x and y such that $x^2 - dy^2 = 1$. To do this, it is enough to produce a pair of integers (x, y), other than $(1, 0)$ and $(-1, 0)$, that satisfy the equation. As noted earlier (Corollary 15.1.3), if we have one such solution (x, y) to the Diophantine equation $x^2 - dy^2 = 1$, then we can generate infinitely many solutions. Thus, it is enough to prove that there exists at least one such solution.

To produce this solution, we must find an element $v = x + y\sqrt{d}$ in $\mathbf{Z}[\sqrt{d}]$, where $v \neq 1$ and $v \neq -1$, such that $N(v) = 1$. The existence of such a v will mean that

$$N(x + y\sqrt{d}) = 1,$$

so

$$x^2 - dy^2 = 1,$$

and we will have found our solution to Pell's equation. (Note that y cannot equal 0, since in that case we would find $x = 1$ or -1, and hence $v = 1$ or -1.)

The heart of the proof is the existence of close approximations to $\sqrt{d}$, which we proved in Chapter 14. In particular, The $\frac{1}{d^2}$ Theorem (14.4.6) implies (since any irrational number has infinitely many convergents) that there are infinitely many positive rational numbers $\frac{x}{y}$ such that

$$\left| \sqrt{d} - \frac{x}{y} \right| < \frac{1}{y^2}. \tag{9}$$

This inequality implies that $x^2 - dy^2$ cannot be very large. In particular, we have:

Claim 1: There are infinitely many pairs (x, y) of natural numbers satisfying

$$|x^2 - dy^2| < 2\sqrt{d} + 1. \tag{10}$$

Proof of Claim 1

We will show that if a pair of natural numbers (x, y) satisfies inequality (9), then (x, y) satisfies inequality (10). We have

$$\left| x^2 - dy^2 \right| = \left| x + y\sqrt{d} \right| \left| x - y\sqrt{d} \right|$$
$$= \left| x + y\sqrt{d} \right| \left| y \right| \left| \sqrt{d} - \tfrac{x}{y} \right|$$
$$< \left| x + y\sqrt{d} \right| y \left(\tfrac{1}{y^2} \right) \qquad \leftarrow \textbf{by inequality (9).}$$

Hence,

$$\left| x^2 - dy^2 \right| < \left| x + y\sqrt{d} \right| \tfrac{1}{y}. \tag{11}$$

Since x, y, and d are positive, $\left| x + y\sqrt{d} \right| = x + y\sqrt{d}$. Substituting this into inequality (11) yields

$$\left| x^2 - dy^2 \right| < \tfrac{x}{y} + \sqrt{d}.$$

Now, how large can $\tfrac{x}{y}$ be? By inequality (9), $\tfrac{x}{y}$ is less than $\dfrac{1}{y^2}$ units away from $\sqrt{d}$. Hence, $\tfrac{x}{y} < \sqrt{d} + \dfrac{1}{y^2}$. Thus, we obtain

$$\left| x^2 - dy^2 \right| < \sqrt{d} + \dfrac{1}{y^2} + \sqrt{d}.$$

Inequality (10) now follows, which completes the proof of Claim 1.

Claim 1 tells us that infinitely many pairs of natural numbers (x, y) satisfy inequality (10). Imagine putting each of these pairs into a bin according to the value of $x^2 - dy^2$. It follows from inequality (10) that the expression $x^2 - dy^2$ is an integer with absolute value less than $2\sqrt{d} + 1$, and hence there are only finitely many possible values for this expression. So, the situation is this: Each of the infinitely many pairs (x, y) is put into one of finitely many bins. We conclude that at least one of the bins must contain infinitely many pairs (x, y). In other words, there exists an integer n for which there are infinitely many pairs of natural numbers (x, y) that satisfy

$$x^2 - dy^2 = n.$$

This implies that there are infinitely many elements $a = x + y\sqrt{d}$ in $\mathbf{Z}[\sqrt{d}]$ such that $N(a) = n$.

Next we show:

Claim 2: There exist two distinct elements $a = x + y\sqrt{d}$ and $b = z + w\sqrt{d}$ in $\mathbf{Z}[\sqrt{d}]$ such that $N(a) = N(b) = n$, where $x, y, z, w \in \mathbf{N}$ satisfy

$$x \equiv z \ (\mathrm{mod}\ n) \quad \text{and} \quad y \equiv w \ (\mathrm{mod}\ n).$$

Proof of Claim 2 We create n^2 bins labeled by pairs (i, j) of integers with $i, j = 0, 1, \ldots, n - 1$.

We have already shown that there are an infinite number of elements $a = x + y\sqrt{d}$ of norm n with $x, y \in \mathbf{N}$. We assign each such element to one of the n^2 bins, as follows. We put each such number, $x + y\sqrt{d}$, in the bin labeled (i, j) whenever $x \equiv i \pmod{n}$ and $y \equiv j \pmod{n}$. When this infinite number of elements of norm n are placed into our finite number of bins, at least one bin must contain infinitely many elements. In particular, some bin contains at least 2 elements. Let $a = x + y\sqrt{d}$ and $b = z + w\sqrt{d}$ be two distinct elements in the same bin. It follows that $x \equiv z \pmod{n}$ and $y \equiv w \pmod{n}$. This completes the proof of Claim 2.

Note that from Claim 2,

$$a - b = (x - z) + (y - w)\sqrt{d},$$

and both $x - z$ and $y - w$ are divisible by n. Thus, there exist integers r and s such that

$$a - b = n\,(r + s\sqrt{d}).$$

In other words, we have

$$a - b = nc, \tag{12}$$

where $c \in \mathbf{Z}[\sqrt{d}]$.

To complete the proof of the theorem, we make the following claim:

Claim 3: The quotient $\frac{a}{b}$ is an element of $\mathbf{Z}[\sqrt{d}]$ with norm 1.

Proof of Claim 3

From equation (12), $a = b + nc$. It follows that

$$\frac{a}{b} = 1 + \frac{n}{b}c.$$

But $N(b) = n$, and hence $b\bar{b} = n$. Thus, $\frac{n}{b} = \bar{b}$, and we may substitute to get

$$\frac{a}{b} = 1 + \bar{b}c.$$

It now follows that $\frac{a}{b} \in \mathbf{Z}[\sqrt{d}]$, since $\mathbf{Z}[\sqrt{d}]$ is closed under addition and multiplication (Lemma 15.1.5). Finally, since $N(a) = N(b)$, it follows from Corollary 15.1.9 that $N\!\left(\frac{a}{b}\right) = 1$. This completes the proof of Claim 3.

Thus, we have produced an element of $\mathbf{Z}[\sqrt{d}]$, $v = \frac{a}{b}$, with norm 1. Furthermore, since a and b are distinct and are both positive, v does not equal 1 or -1. As noted at the beginning of this proof, such an element of norm 1 gives rise to a solution to Pell's equation.

By Corollary 15.1.3, we conclude that there are infinitely many solutions to Pell's equation. ∎

EXERCISES 15.1

Numerical Problems

1. Find the norms of the following elements of $\mathbf{Z}[\sqrt{3}]$:

    a. $2 + 3\sqrt{3}$ b. $10 - \sqrt{3}$ c. 7

2. Find the norms of the following elements of $\mathbf{Z}[\sqrt{10}]$:

    a. $-2\sqrt{10}$ b. $-5 + 5\sqrt{10}$ c. $2 - 17\sqrt{10}$

3. The continued fraction for $\sqrt{6}$ is $[2, \overline{2, 4}]$. Use this to find two solutions to the Pell equation $x^2 - 6y^2 = 1$.

4. The continued fraction for $\sqrt{5}$ is $[2, \overline{4}]$. Use this to find two solutions to the Pell equation $x^2 - 5y^2 = 1$.

5. The continued fraction for $\sqrt{3}$ is $[1, \overline{1, 2}]$. Use this to find two solutions to the Pell equation $x^2 - 3y^2 = 1$.

6. Consider the following elements of $\mathbf{Z}[\sqrt{3}]$: $a = 62 + 8\sqrt{3}$, $b = 2 - 4\sqrt{3}$.

    a. Demonstrate that $\frac{a}{b} \in \mathbf{Z}[\sqrt{3}]$.

    b. Verify that Corollary 15.1.9 holds for a and b.

7. One solution to the Pell equation $x^2 - 10y^2 = 1$ is $x = 19, y = 6$. Find two more solutions. [*Hint:* Use Proposition 15.1.2.]

8. One solution to the Pell equation $x^2 - 11y^2 = 1$ is $x = 10, y = 3$. Find two more solutions.

Reasoning and Proofs

9. Prove Corollary 15.1.9.

10. Prove Lemma 15.1.7.

11. In this section, we focused mainly on $\mathbf{Z}[\sqrt{d}]$ for positive integers d, but we can also consider the number system $\mathbf{Z}[\sqrt{d}]$ when d is negative. (If you read Chapter 13, the Gaussian integers are one example of this, with $d = -1$.) Answer the following questions about $\mathbf{Z}[\sqrt{-5}]$:

    a. Find the norm of $2 + 3\sqrt{-5}$.

    b. Prove that the norm of any nonzero element is positive.

c. Which elements a (if any) have $N(a) = 1$? Justify your answer.

d. Which elements a (if any) have $N(a) = 13$? Justify your answer.

12. Consider the equation $x^2 - dy^2 = n$ when d is a square, say $d = e^2$ for some $e \in \mathbf{N}$.

    a. Factor the left side of this equation.

    b. Find all integer solutions to $x^2 - y^2 = 15$.

    c. Find all integer solutions to $x^2 - 4y^2 = 9$.

    d. Find all integer solutions to $x^2 - 9y^2 = 10$.

    e. Prove that if d is a perfect square, then the Diophantine equation $x^2 - dy^2 = n$ has only finitely many solutions for any natural number n.

13. Let d be a natural number that is a perfect square. Prove that the equation $x^2 - dy^2 = 1$ has no integer solutions with $x > 0$ and $y > 0$.

14. **a.** Prove that the Diophantine equation $x^2 - 70y^2 = -1$ has no solution. [*Hint:* Reduce modulo 7.]

    b. Let d be a natural number that is divisible by a prime p with $p \equiv 3 \pmod 4$. Prove that the Diophantine equation $x^2 - dy^2 = -1$ has no solution. [*Hint:* You may use Theorem 11.2.3, the Quadratic Character of -1.]

Conjecturer's Corner (Exercises 15–18)

Based on the period of the continued fraction for $\sqrt{d}$, you can tell which convergents will give a solution to the equation $x^2 - dy^2 = 1$. In these exercises, you will explore this pattern.

15. Each of the following continued fractions has a period of odd length. Find the first convergent that gives a solution to $x^2 - dy^2 = 1$. Which convergents give solutions to $x^2 - dy^2 = -1$?

    a. $\sqrt{5} = [2, \overline{4}]$, $d = 5$.

    b. $\sqrt{10} = [3, \overline{6}]$, $d = 10$.

    c. $\sqrt{41} = [6, \overline{2, 2, 12}]$, $d = 41$.

    d. $\sqrt{13} = [3, \overline{1, 1, 1, 1, 6}]$, $d = 13$.

16. Each of the following continued fractions has an even period. Find the first convergent that gives a solution to $x^2 - dy^2 = 1$. Which convergents give solutions to $x^2 - dy^2 = -1$?

    a. $\sqrt{3} = [1, \overline{1, 2}]$, $d = 3$.   **b.** $\sqrt{11} = [3, \overline{3, 6}]$, $d = 11$.

    c. $\sqrt{7} = [2, \overline{1, 1, 1, 4}]$, $d = 7$.   **d.** $\sqrt{59} = [7, \overline{1, 2, 7, 2, 1, 14}]$, $d = 59$.

17. Suppose the continued fraction for $\sqrt{d}$ has period n. Based on your work in Exercises 15 and 16, conjecture the answers to the following questions.

   a. Which is the first convergent to give a solution to $x^2 - dy^2 = 1$?

   b. Which is the first convergent to give a solution to $x^2 - dy^2 = -1$?

18. Trying more examples if necessary, make a conjecture by filling in the blanks in parts a–d of the following theorem statement:

   THEOREM. Let d be a natural number that is not a perfect square. Then every solution to the Diophantine equation $x^2 - dy^2 = \pm 1$ is found among the convergents of the continued fraction for $\sqrt{d}$.
   Furthermore, we can say precisely which convergents give rise to solutions of this equation. To do so, express $\sqrt{d}$ as an infinite simple continued fraction, $[a_0, a_1, a_2, a_3, \ldots]$. This expansion is periodic with period q and has the following form:

   $$\sqrt{d} = [a_0, \overline{a_1, a_2, a_3, \ldots}]$$

   CASE I: q is odd. Let x and y be relatively prime natural numbers. Then:

   a. (x, y) is a solution to the Pell equation $x^2 - dy^2 = 1$

   if and only if $\frac{x}{y} = [a_0, a_1, \ldots, a_r]$, where $r = $ _____.

   b. (x, y) is a solution to the equation $x^2 - dy^2 = -1$

   if and only if $\frac{x}{y} = [a_0, a_1, \ldots, a_r]$, where $r = $ _____.

   CASE II: q is even. Let x and y be relatively prime natural numbers. Then:

   c. (x, y) is a solution to the Pell equation $x^2 - dy^2 = 1$

   if and only if $\frac{x}{y} = [a_0, a_1, \ldots, a_r]$, where $r = $ _____.

   d. The solution set of the equation $x^2 - dy^2 = -1$ is _____.

19. Prove that no element of $\mathbf{Z}[\sqrt{10}]$ has norm 2. [*Hint:* Consider equation (8) modulo 5.]

20. Which of the following are possible norms for elements of $\mathbf{Z}[\sqrt{3}]$? If it is a valid norm, give an example of an element with that norm. If not, prove it is not a valid norm. [*Hint:* Consider equation (8) in various moduli.]

   a. 3   b. -3 c. 2   d. -2

21. Prove Lemma 15.1.5.

22. Give an example of two elements $a, b \in \mathbf{Z}[\sqrt{3}]$, with $b \neq 0$, such that $\frac{a}{b}$ is not an element of $\mathbf{Z}[\sqrt{3}]$.

23. Prove Proposition 15.1.2 by direct calculation (without using irrational numbers).

24. a. State the theorem that $\mathbf{Z}[\sqrt{d}\,]$ is a ring.
[*Hint:* Model your statement on Theorem 8.2.6.]

b. Prove that $\mathbf{Z}[\sqrt{d}\,]$ is a ring by proving your statement from part a.

EXPLORATION Generating Solutions to Pell's Equation (Exercises 25–27)

In this exploration, you will learn how to generate new solutions to Pell's equation from ones you already know.

Let d be a natural number that is not a perfect square, and consider Pell's equation:

$$x^2 - dy^2 = 1. \tag{13}$$

Suppose that we have two solutions (a, b) and (e, f) to this equation, where $a, b, e, f \in \mathbf{N}$. In $\mathbf{Z}[\sqrt{d}\,]$, consider the elements $z = a + b\sqrt{d}$ and $w = e + f\sqrt{d}$.

25. a. Show that $N(zw) = 1$.

b. Express zw in the form $g + h\sqrt{d}$ where $g, h \in \mathbf{Z}$.

c. Explain why (g, h) is a solution to Pell's equation (13).

d. Prove that (g, h) is a *new* solution to Pell's equation (13). That is, show that (g, h) is not the same as either of the solutions (a, b) or (e, f).

26. For any $n \in \mathbf{N}$, z^n can be written in the form $z^n = x_n + y_n\sqrt{d}$, where $x_n, y_n \in \mathbf{Z}$.

a. Show that for any $n \in \mathbf{N}$, (x_n, y_n) is a solution to Pell's equation (13).

b. Show that the solutions (x_n, y_n) are distinct. That is, show that for any distinct natural numbers m and n, $(x_m, y_m) \neq (x_n, y_n)$.

*27. Let the pairs (x_n, y_n) be defined as in Exercise 26. Show that if (a, b) is the smallest positive solution to Pell's equation (13), then every positive solution to (13) is one of the pairs (x_n, y_n). [*Hint:* If x, y is any solution, let $v = x + y\sqrt{d}$. Show that for some nonnegative integer n, $z^n \leq v < z^{n+1}$, and consider vz^{-n}.]

15.2 Fermat's Last Theorem

Widening the margin

The story of Fermat's Last Theorem begins with the Pythagorean Theorem. This ancient result, known to every student of geometry, states that if a and b are the lengths of the legs of a right triangle and c is the length of the hypotenuse, then $a^2 + b^2 = c^2$. It is not hard to find integers that satisfy this equation. For example, the numbers

$$a = 3, \ b = 4, \ c = 5$$

satisfy the equation. So do the triples of numbers

$$a = 5, \ b = 12, \ c = 13$$

and

$$a = 8, \ b = 15, \ c = 17.$$

Natural numbers a, b, and c that satisfy the equation $a^2 + b^2 = c^2$ are known as **Pythagorean triples**.

Around the year 250, Diophantus of Alexandria wrote the book *Arithmetica*, which summarized what was known about number theory at the time (see Chapter 5 for a historical summary of Diophantus's life and work). *Arithmetica* presented 189 problems that asked for integer or rational solutions to particular equations. Diophantus included numerical examples of solutions to these problems, but no general methods for obtaining solutions. Several of the problems in *Arithmetica* involve finding Pythagorean triples that satisfy additional conditions.

Pierre de Fermat was a French lawyer in the 17th century with a deep interest in mathematics (see Chapter 9 for a historical summary of Fermat's life and work). Reading Diophantus's *Arithmetica* inspired Fermat's mathematical work. There had not been very much progress in number theory in the Western world between the time of Diophantus and the time of Fermat, so *Arithmetica* provided Fermat with a rough account of what was known about number theory in Europe in Fermat's time. In the margins of his copy of *Arithmetica*, Fermat wrote many of his own ideas related to Diophantus's problems. Problem 8 of Book II of *Arithmetica* asks the reader "to divide a given square number into two squares"—that is, to express a given square as a sum of two squares. This problem inspired Fermat to consider the analogous problem in which squares are replaced by cubes or higher powers. In the nearby margin, Fermat wrote,

> It is impossible for a cube to be written as a sum of two cubes or a fourth power to be written as a sum of two fourth powers or, in general, for any number which is a power greater than the second to be written as a sum of like powers. I have a truly marvelous proof of this proposition which this margin is too narrow to contain.

The assertion in Fermat's marginal note has come to be known as Fermat's Last Theorem:

THEOREM 15.2.1 **FERMAT'S LAST THEOREM**

For all integers $n > 2$, there are no solutions to the equation $x^n + y^n = z^n$, where x, y, and z are nonzero integers.

Whether Fermat actually had a proof of this theorem will never be known for certain. For more than 300 years after Fermat died, many great mathematicians tried to prove his assertion, and some succeeded in proving it for particular values of n. Several important ideas in number theory grew out of various attempts to prove Fermat's Last Theorem.

Proving Fermat's Last Theorem can be reduced to the case in which the exponent is $n = 4$ and the case in which n is an odd prime:

> If Fermat's Last Theorem holds for $n = 4$ and for all odd primes n,
> then Fermat's Last Theorem holds for all integers $n > 2$.

This follows from the fact that every integer $n > 2$ is divisible by an odd prime p or is divisible by 4. (See Exercise 7.)

Fermat himself proved the $n = 4$ case of Fermat's Last Theorem, and we present a proof of this case in Section 15.3. The case in which n is an odd prime turned out to be much more difficult. In 1819, Sophie Germain had an idea of how one might prove Fermat's Last Theorem for all odd primes. Though Germain never found a full proof of Fermat's Last Theorem using her idea, she was the first to obtain general results about Fermat's Last Theorem. We present some of these results in Section 15.4. In the 18th and 19th centuries, Fermat's Last Theorem was proven for several specific values of n. In particular, Euler proved Fermat's Last Theorem for $n = 3$ in 1753, Dirichlet and Legendre proved it for $n = 5$ in 1825, Dirichlet proved it for $n = 14$ in 1832, and Lamé proved it for $n = 7$ in 1839.

In 1906, Paul Wolfskehl, a wealthy German businessman with an interest in number theory, left in his will a prize of 100,000 marks (equivalent to almost $2 million in today's dollars) to be awarded to the first person to prove Fermat's Last Theorem. Due to the magnitude of the prize offered, over 1000 false proofs of the result were submitted between 1906 and 1912. Unfortunately, inflation in Germany after World War I reduced the value of the Wolfskehl prize to a small fraction of its original value.

In 1995, more than three centuries after Fermat's original claim, Andrew Wiles, with the help of his student Richard Taylor, published a complete and correct proof of Fermat's Last Theorem for all values of n. Wiles's proof uses elliptic curves and modular forms, and it builds on results developed in the 1950s by Japanese mathematicians Taniyama and Shimura. Even with the help of these sophisticated techniques, Wiles's proof is 100 pages long and hence surely could not be contained in the margin of *Arithmetica*! If Fermat did have a proof of Fermat's Last Theorem, it must have been quite different from the one Wiles came up with.

A useful lemma

We will now prove a lemma that comes in handy when studying Fermat's Last Theorem, as we will see in the next two sections. Consider the perfect square $10^2 = 100$. This number has plenty of factors; in other words, there are lots of ways to write $100 = de$, where $d, e \in \mathbf{N}$. But let's impose the additional condition that the factors d and e must be relatively prime. Now there are only two possibilities: $100 = 1 \cdot 100$, and $100 = 4 \cdot 25$.

Note that each of these factors ($1, 4, 25, 100$) is itself a perfect square. We observe that if the perfect square 100 is factored in such a way that the factors are relatively prime, then each of the factors must itself be a perfect square. This holds for arbitrary perfect squares, and indeed for arbitrary nth powers, as asserted by the following lemma.

LEMMA 15.2.2

Let $z, n \in \mathbf{N}$. Suppose that $z^n = de$, where d and e are relatively prime natural numbers. Then there exist natural numbers x and y such that $d = x^n$ and $e = y^n$.

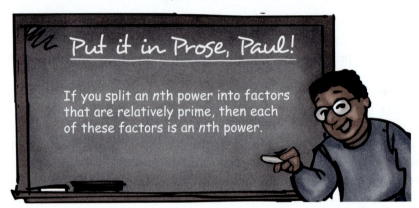

PROOF OF LEMMA 15.2.2

Let $d, e, z, n \in \mathbf{Z}$ such that $de = z^n$ and $\gcd(d, e) = 1$.

Consider the factorization of z into primes: $z = p_1^{b_1} p_2^{b_2} \cdots p_r^{b_r}$ where $p_1, p_2, \ldots, p_r$ are distinct primes. Since $de = z^n$, we have

$$de = p_1^{nb_1} p_2^{nb_2} \cdots p_r^{nb_r}. \tag{1}$$

For each $i = 1, 2, \ldots, r$, the prime p_i cannot divide both e and d, since $\gcd(d, e) = 1$. Hence, for each i, we know that $p_i^{nb_i} \mid d$ or $p_i^{nb_i} \mid e$. It follows that we may reorder the primes p_i in equation (1) so that all of the factors of d come first:

$$d = p_1^{nb_1} p_2^{nb_2} \cdots p_k^{nb_k}, \quad e = p_{k+1}^{nb_{k+1}} p_{k+2}^{nb_{k+2}} \cdots p_r^{nb_r}.$$

Hence, d is a perfect nth power, and e is a perfect nth power. ∎

A Lamé idea

One attempt to prove Fermat's Last Theorem was presented to the Paris Academy in 1847 by Gabriel Lamé. His idea was as follows: If n is a natural number, then the complex number

$$\omega = e^{\frac{2\pi i}{n}}$$

satisfies $\omega^n = 1$. In fact, the polynomial equation $x^n = 1$ has the n solutions $1, \omega, \omega^2, \ldots, \omega^{n-1}$. These numbers are called the nth *roots of unity*. If n is an odd natural number, then the solutions of the polynomial equation $x^n = -1$ are given by $-1, -\omega, -\omega^2, \ldots, -\omega^{n-1}$, and hence the polynomial $x^n + 1$ factors as

$$x^n + 1 = (x + 1)(x + \omega)(x + \omega^2) \cdots (x + \omega^{n-1}), \tag{2}$$

from which it follows (see Exercise 10) that $x^n + y^n$ can be factored as

$$x^n + y^n = (x + y)(x + \omega y)(x + \omega^2 y) \cdots (x + \omega^{n-1}y). \tag{3}$$

Hence, if we have a solution to Fermat's equation $x^n + y^n = z^n$, then we have a factorization of z^n:

$$z^n = (x + y)(x + \omega y)(x + \omega^2 y) \cdots (x + \omega^{n-1}y). \tag{4}$$

Note that the factors on the right side do not live in the integers, $\mathbf{Z}$; rather, they are elements of a larger ring known as $\mathbf{Z}[\omega]$. (This is analogous to the number system $\mathbf{Z}[\sqrt{d}]$, which we introduced in Section 15.1, and to $\mathbf{Z}[i]$, the ring of Gaussian integers, which we studied in Chapter 13.) Lamé planned to use a version of Lemma 15.2.2 for elements of $\mathbf{Z}[\omega]$ to deduce that each of the factors on the right side of equation (4) must be an nth power. Finally, he planned to use the nth roots of these factors to produce a smaller solution to the equation $x^n + y^n = z^n$, and use the Well-Ordering Principle to get a contradiction. While this seemed like a promising plan of attack, it was not wholly successful. One important reason is that Lemma 15.2.2 relies on the uniqueness of prime factorizations, and it turns out that factorizations in $\mathbf{Z}[\omega]$ are in general not unique! (We met an example of such nonuniqueness at the beginning of Section 13.2, when we considered the ring $\mathbf{Z}[\sqrt{-5}]$. For an example with roots of unity, see Exercise 17.)

In 1847, Ernst Kummer proved that Fermat's Last Theorem is true for a class of primes called *regular* primes, which includes all primes for which $\mathbf{Z}[\omega]$ has unique factorization. The regular primes include all of the primes less than 100 except for the primes 37, 59, and 67. (Kummer also gave a "proof" of Fermat's Last Theorem for the numbers 37, 59, and 67, but it was later shown that this proof contained errors.) Kummer believed that there are infinitely many regular primes, but even today it is not known whether this is the case. He was awarded the Grand Prize of the Paris Academy of Sciences for his proof of Fermat's Last Theorem for regular primes, which laid the groundwork for the field of algebraic number theory. Kummer asserted that his proof for regular primes was not part of a general program that he had for proving Fermat's Last Theorem but rather was related to his interest in generalizing the ideas of Quadratic Reciprocity (11.3.1). In fact, Kummer described his proof as "a curiosity of number theory rather than a major item."

EXERCISES 15.2

Numerical Problems

1. Two members of a Pythagorean triple are 24 and 25. What is the third?

2. Two members of a Pythagorean triple are 8 and 17. What is the third?

3. In his *Arithmetica*, Diophantus gave the following problem:

   Find a Pythagorean triple for a triangle in which the hypotenuse minus each of the legs is a cube. (In other words, the problem is to find a Pythagorean triple a, b, c in which $c - a$ and $c - b$ are perfect cubes.)

   a. Diophantus provided a solution with $a = 40$, $b = 96$. What was the value of c in his solution?

   b. Verify that the solution from part a satisfies the conditions of Diophantus's problem.

4. Consider the Diophantine equation $x^3 + y^3 + z^3 = w^3$.

   a. Find a solution where x, y, z, and w are nonzero integers.

   b. Find a different solution where x, y, z, and w are natural numbers.

Reasoning and Proofs

5. Suppose x, y, z is a Pythagorean triple. Show that for any $k \in \mathbf{N}$, kx, ky, kz is also a Pythagorean triple.

6. A Pythagorean triple is called *primitive* if $\gcd(a, b) = \gcd(b, c) = \gcd(a, c) = 1$. Show that in a primitive Pythagorean triple with $a^2 + b^2 = c^2$, the integer c must be odd.

7. In this exercise, you will show that Fermat's Last Theorem can be reduced to two cases: $n = 4$, and n is an odd prime.

   a. Prove that every integer $n > 2$ is divisible by an odd prime or divisible by 4.

   b. Let d and n be natural numbers such that $d \mid n$. Show that if there is a nonzero integer solution to $x^n + y^n = z^n$, then there is a nonzero integer solution to $x^d + y^d = z^d$.

   c. Explain why the results you proved in parts a and b imply that if Fermat's Last Theorem holds for the case $n = 4$ and the case in which n is any odd prime, then Fermat's Last Theorem holds for all $n > 2$.

8. Let $n > 2$ be an odd natural number. Prove that the Diophantine equation $x^n + y^n = z^n$ has a nonzero solution if and only if the Diophantine equation $x^n + y^n + z^n = 0$ has a nonzero solution.

9. Sometimes Fermat's Last Theorem is stated with x, y, and z being natural numbers rather than nonzero integers. In this exercise, you will prove that the two statements are equivalent. Let $n > 2$ be a natural number. Prove that the equation $x^n + y^n = z^n$ has a solution in the nonzero integers if and only if $x^n + y^n = z^n$ has a solution in the natural numbers.

10. Show that the factorization given in equation (3) is correct. [*Hint:* Start with equation (2), and substitute $\frac{x}{y}$ for x.]

11. Let x and y be integers, and let $a = 2xy$, $b = x^2 - y^2$, and $c = x^2 + y^2$.

    a. Prove that a, b, c is a Pythagorean triple.

    b. Generate five Pythagorean triples using this formula.

    c. Show that every even number greater than 2 is part of a Pythagorean triple.

    d. Show that every odd number greater than 1 is part of a Pythagorean triple.

12. Let a, b, c be a Pythagorean triple with a, b, $c \in \mathbf{N}$, and let $n > 2$ be an integer. What can you say about $a^n + b^n$ compared to c^n? Prove your answer.

13. Let p be prime, and let x, y, z be integers. Suppose $x^{p-1} + y^{p-1} = z^{p-1}$. Use Fermat's Little Theorem (9.1.1) to show that $p \mid xyz$.

14. Let p be prime, and let x, y, z be integers. Suppose $x^p + y^p = z^p$. Use Fermat's Little Theorem (9.1.1) to show that $p \mid (x + y - z)$.

15. Let $n \in \mathbf{N}$, and suppose that x, y, z are nonzero, pairwise relatively prime integers such that $x^n + y^n = z^n$. Prove that the three integers $x + y$, $z - x$, and $z - y$ are pairwise relatively prime.

16. **a.** Suppose that there exist a, b, $c \in \mathbf{Z}$ and $n \in \mathbf{N}$ such that $a^n + b^n = c^n$. Show that if a and b are relatively prime, then c is relatively prime to both a and b.

    b. Suppose that Fermat's Last Theorem is not true—that is, suppose that for some integer $n > 2$, there exist x, y, $z \in \mathbf{Z}$ such that $x^n + y^n = z^n$. Show that it follows that there exist pairwise relatively prime integers a, b, and c such that $a^n + b^n = c^n$.

Advanced Reasoning and Proofs

17. As mentioned in the text, the ring $\mathbf{Z}[\omega]$, where ω is a root of unity, does not have unique factorization in general. In this exercise, you will examine a case in which unique factorization fails.

    Consider the 23rd root of unity, $\omega = e^{2\pi i/23}$. The ring $\mathbf{Z}[\omega]$ is defined by

    $$\mathbf{Z}[\omega] = \{a_0 + a_1\omega + a_2\omega^2 + \cdots + a_{21}\omega^{21} \mid a_i \in \mathbf{Z} \quad \text{for} \quad i = 0, \ldots, 21\}$$

Define a and $b \in \mathbf{Z}[\omega]$ by

$$a = 1 + \omega^2 + \omega^4 + \omega^5 + \omega^6 + \omega^{10} + \omega^{11},$$

$$b = 1 + \omega + \omega^5 + \omega^6 + \omega^7 + \omega^9 + \omega^{11}.$$

a. Show that $\omega^{22} \in \mathbf{Z}[\omega]$. [*Hint:* Starting with the equation $\omega^{23} - 1 = 0$, factor $\omega - 1$ from the left side.]

b. Compute the product ab, and show that ab is divisible by 2 in $\mathbf{Z}[\omega]$. That is, show that there exists $d \in \mathbf{Z}[\omega]$ such that $ab = 2d$.

c. It turns out that neither a nor b is divisible by 2 in $\mathbf{Z}[\omega]$. (You might guess this, seeing that the odd number 1 appears as a coefficient of various powers of ω in the definitions of a and b.) It is also true that 2 cannot be factored in $\mathbf{Z}[\omega]$. Assuming these facts, use your result from part b to explain why unique factorization fails in the ring $\mathbf{Z}[\omega]$. (You need not try to give a fully rigorous proof; a rough explanation based on the ideas of Section 6.1 will suffice.)

15.3 Proof of Fermat's Last Theorem for $n = 4$

Fermat gave a proof of Fermat's Last Theorem for $n = 4$ that he wrote in the margin of his copy of Diophantus's *Arithmetica* (see the "Fun Facts" in Section 15.5 for more details). Independently, Euler later published proofs of Fermat's Last Theorem for $n = 3$ and $n = 4$. Since Euler's proof for $n = 4$ is simpler than that of Fermat, we will give Euler's proof. This proof uses the idea of Pythagorean triples, which we shall briefly study first.

Pythagorean Triples

As we discussed in Section 15.2, natural numbers a, b, c that satisfy the equation $a^2 + b^2 = c^2$ are called a *Pythagorean triple*.

Fun Facts

The ancient Greeks were not the first to be interested in Pythagorean triples. A Babylonian cuneiform clay tablet from 1500 BCE was found that contained a list of 15 Pythagorean triples. The fact that the tablet includes large triples such as $4961, 6480, 8161$ perhaps indicates that the Babylonians had a general method of generating Pythagorean triples, though it is also possible that they came upon these numbers by trial and error.

Taking the Pythagorean triple $50, 120, 130$ and dividing by $\gcd(50, 120)$, which is 10, produces another Pythagorean triple: $5, 12, 13$. This triple is more basic than the original one, because the numbers 5, 12, and 13 are pairwise relatively prime (that is, no two of these numbers share a common divisor). A Pythagorean triple in which the natural numbers a, b, c are pairwise relatively prime is said to be a **primitive Pythagorean triple**. If you are given a Pythagorean triple (a, b, c) that is not primitive, it is easy to find a related Pythagorean triple that is primitive: simply divide through by $\gcd(a, b)$. (See Exercise 9.) Thus, *every* Pythagorean triple is a multiple of a primitive Pythagorean triple. This means that if you want to know all about Pythagorean triples, you just need to know all about the primitive ones.

The following theorem, which completely characterizes primitive Pythagorean triples, states that every primitive Pythagorean triple can be generated by a pair of distinct relatively prime natural numbers, p and q, that have opposite parity (i.e., one is even and the other is odd). The theorem is contained in Euclid's *Elements* (though Euclid did not state it in precisely this way).

THEOREM 15.3.1 PYTHAGOREAN TRIPLE THEOREM

Let a, b, and c be natural numbers. Then a, b, c is a primitive Pythagorean triple if and only if there exist relatively prime natural numbers p and q of opposite parity such that $p > q$ and the following equations hold (after possibly interchanging a and b):

$$a = 2pq$$
$$b = p^2 - q^2 \qquad \qquad (1)$$
$$c = p^2 + q^2$$

PROOF Let a, b, and c be natural numbers.

[⇐] Suppose that p and q are relatively prime natural numbers of opposite parity, $p > q$, that satisfy the three equations (1). **[To show: a, b, c is a primitive Pythagorean triple.]**
We now compute:

$$\begin{aligned} a^2 + b^2 &= (2pq)^2 + (p^2 - q^2)^2 \\ &= 4p^2q^2 + p^4 - 2p^2q^2 + q^4 \\ &= p^4 + 2p^2q^2 + q^4 \\ &= (p^2 + q^2)^2 \\ &= c^2, \end{aligned}$$

and thus, a, b, c is a Pythagorean triple.

Claim: b and c are relatively prime.

Proof of Claim (By contradiction.)

Assumption: Suppose $\gcd(b, c) > 1$.

Then there exists a prime n such that $n \mid b$ and $n \mid c$. Thus, $n \mid b + c$, so substituting from equations (1), we get

$$n \mid 2p^2. \tag{2}$$

Since p and q have opposite parity, it follows from equations (1) that b and c are both odd, which implies that n is odd. Since n is odd, it follows from equation (2) and the Fundamental Property of Primes (6.1.2) that

$$n \mid p.$$

By similar reasoning, since $n \mid c - b$, substituting from equations (1) yields $n \mid 2q^2$, and hence,

$$n \mid q.$$

But this contradicts our hypothesis that p and q are relatively prime. $\Rightarrow\Leftarrow$. This completes the proof of the Claim.

From the Claim, it follows that a is relatively prime to both b and c. (See Exercise 7.) Therefore, a, b, c is a primitive Pythagorean triple.

[$\Rightarrow$] Now suppose that a, b, c is a primitive Pythagorean triple.
[To show: There exist relatively prime $p, q \in \mathbb{N}$ of opposite parity, with $p > q$, such that equations (1) hold.]
Since $a, b,$ and c are pairwise relatively prime, at most one of the three numbers $a, b,$ and c is even.

Claim: c is odd.

Proof of Claim (By contradiction.)

Assumption: Suppose c is even.
Then a and b are both odd. Thus, when we consider a and b modulo 4, $a \equiv 1$ or -1 (mod 4), and $b \equiv 1$ or -1 (mod 4). This implies that

$$a^2 \equiv 1 \text{ (mod 4)} \quad \text{and} \quad b^2 \equiv 1 \text{ (mod 4)}.$$

Since c is even, $c \equiv 0$ or 2 (mod 4), which implies that

$$c^2 \equiv 0 \text{ (mod 4)}.$$

Since $a^2 + b^2 \equiv c^2$ (mod 4), we have $1 + 1 \equiv 0$ (mod 4), a contradiction. $\Rightarrow\Leftarrow$. This completes the proof of the Claim.

Since c is odd, we conclude that precisely one of a or b is even and the other is odd. Without loss of generality, assume that a is even and b is odd.

Now write

$$a^2 = c^2 - b^2$$
$$= (c + b)(c - b). \tag{3}$$

Since b and c are both odd, $c + b$ and $c - b$ are both even. Thus, there are positive integers u, v, and w such that

$$a = 2u, \quad c + b = 2v, \quad \text{and} \quad c - b = 2w. \tag{4}$$

Substituting equations (4) into equation (3), we get $(2u)^2 = (2v)(2w)$. Hence,

$$u^2 = vw. \tag{5}$$

We now show that v and w are relatively prime. Let $d = \gcd(v, w)$. From equations (4), we know that $2d \mid c + b$ and $2d \mid c - b$. By adding and subtracting, it follows that $2d \mid 2c$ and $2d \mid 2b$, which implies that $d \mid c$ and $d \mid b$. Since $\gcd(b, c) = 1$, it must be that $d = 1$. Hence, v and w are relatively prime.

Thus, by equation (5) and Lemma 15.2.2, there are natural numbers p and q such that

$$p^2 = v \text{ and } q^2 = w. \tag{6}$$

Since v and w are relatively prime, p and q must also be relatively prime. Using equations (4), (5), and (6), we will now solve for a, b, and c in terms of p and q. It follows from equations (4) that $2c = 2v + 2w$, and thus, by equation (6),

$$c = v + w = p^2 + q^2. \tag{7}$$

Also from equations (4), $2b = 2v - 2w$, which in combination with equation (6) implies that

$$b = v - w = p^2 - q^2. \tag{8}$$

Finally, equations (4) and (5) imply that $a^2 = 2^2 u^2 = 2^2 vw$, so by equation (6),

$$a^2 = (2pq)^2,$$

and hence, $a = 2pq$.

We have just shown that all three equations (1) hold. Since c is odd, it follows from equation (7) that p and q have opposite parity. Furthermore, since $b > 0$, we know from equation (8) that $p > q$. This completes the proof of the theorem. ●

EXAMPLE 1

Find the values of p and q from the Pythagorean Triple Theorem (15.3.1) that produce the primitive Pythagorean triple $5, 12, 13$.

SOLUTION Since p and q have opposite parity, equations (1) imply that $a = 2pq$ is even and $b = p^2 - q^2$ is odd. Thus, for the given Pythagorean triple, 5, 12, 13, we have $a = 12$, $b = 5$, $c = 13$. (Note that this interchange of a and b is explicitly permitted in the statement of the Pythagorean Triple Theorem.)

We must find natural numbers p and q of opposite parity, with $p > q$, such that

$$12 = 2pq, \quad 5 = p^2 - q^2, \quad \text{and} \quad 13 = p^2 + q^2.$$

The second and third equations can be solved algebraically, yielding

$$p = 3 \quad \text{and} \quad q = 2.$$

(The reader is invited to verify that these values satisfy all three equations.) ∎

Euler's proof of Fermat's Last Theorem for $n = 4$

We will now use our classification of Pythagorean triples to give a proof of the $n = 4$ case of Fermat's Last Theorem: Two perfect 4th powers cannot add up to another perfect 4th power. Following Euler, we will actually show something stronger: Two perfect 4th powers cannot add up to a perfect square.

THEOREM 15.3.2

There do not exist natural numbers x, y, and w such that $x^4 + y^4 = w^2$.

To prove this theorem, we will effectively use a method that Fermat developed, known as the *method of infinite descent*. To prove that a particular statement cannot hold for any natural number, we start by assuming that the statement holds for some natural number n. The heart of the proof is to use the assumption that the statement is true for n to show that it must be true for some smaller natural number, n'. Repeating the same argument, we could use the fact that the statement is true for n' to show that it must be true for an even smaller natural number. Continuing indefinitely in this manner, we could obtain an infinite descending list of natural numbers for which the statement holds. However, we know by the Well-Ordering Principle (2.2.2) that there cannot be an infinite descending list of natural numbers. Thus, the statement must be false for every value of n.

From a modern point of view, the method of infinite descent is simply a form of proof by contradiction that makes use of the Well-Ordering Principle (2.2.2). To show that a statement cannot hold for any natural number n, we assume that it does hold for some natural number n. This means that the set of natural numbers for which the statement holds is nonempty, and hence, this set has a smallest element, s. Now we show (and again this is the real heart of the proof) that there is a natural number $s' < s$ for which the statement holds. This contradicts the minimality of s (i.e., it contradicts that s is the smallest element of the set). We conclude that the statement cannot hold for any natural number.

PROOF OF **THEOREM 15.3.2** (By contradiction.)

Assumption: Suppose that there exist natural numbers x, y, and w such that

$$x^4 + y^4 = w^2. \tag{9}$$

The Well-Ordering Principle (2.2.2) guarantees that among all solutions to this equation, there is a solution with the smallest possible value of w. Our strategy will be to produce natural numbers r, s, and t, such that $r^4 + s^4 = t^2$ and $t < w$. This will contradict the minimality of w.

Case 1 Two of the numbers x, y, and w share a common factor greater than 1.

In this case, some prime p divides two of x, y, and w. It follows that p divides all three of x, y, and w (see Exercise 12). We then have $p^4 \mid x^4$ and $p^4 \mid y^4$, and hence $p^4 \mid w^2$. Now it follows (see Exercise 8) that $p^2 \mid w$. Thus, there are integers r, s, and t, such that $x = pr$, $y = ps$, and $w = p^2 t$. We can now divide equation (9) by p^4 to obtain the equation

$$r^4 + s^4 = t^2.$$

Since $t < w$, this contradicts the minimality of w. $\Rightarrow\Leftarrow$. This completes Case 1.

Case 2 x, y, and w are pairwise relatively prime.

It follows that x^2, y^2, and w are pairwise relatively prime. Hence, by equation (9), x^2, y^2, w is a primitive Pythagorean triple. It follows from the Pythagorean Triple Theorem (15.3.1) that w is odd, and that exactly one of x^2 and y^2 is even (see Exercise 11). Without loss of generality, say y^2 is even and x^2 is odd. Hence, by the Pythagorean Triple Theorem, there are relatively prime natural numbers p and q of opposite parity such that

$$
\begin{aligned}
y^2 &= 2pq, \\
x^2 &= p^2 - q^2, \\
w &= p^2 + q^2.
\end{aligned}
\tag{10}
$$

Then $x^2 + q^2 = p^2$. Since p and q are relatively prime, it follows that x, q, p is a primitive Pythagorean triple (see Exercise 7). Hence, p is odd (by the Pythagorean Triple Theorem). Since p and q have opposite parity, q must be even. Then applying the Pythagorean Triple Theorem to the triple x, q, p, we know there are relatively prime natural numbers m and n of opposite parity such that

$$
\begin{aligned}
q &= 2mn, \\
x &= m^2 - n^2, \\
p &= m^2 + n^2.
\end{aligned}
\tag{11}
$$

By the first equation in (10), $y^2 = (2q)p$. Furthermore, $2q$ and p are relatively prime. Thus, by Lemma 15.2.2, both $2q$ and p are squares. Since $2q$ is a square and $2q = 4mn$, the number mn must be a square. Since m and n are relatively prime, we can again apply Lemma 15.2.2 to conclude that m and n are each squares. Thus, there exist natural numbers r and s such that $m = r^2$ and $n = s^2$. Substituting these into the last equation in (11) yields

$$
p = r^4 + s^4.
$$

Since p is a square, $p = t^2$ for some natural number t. Thus,

$$
r^4 + s^4 = t^2,
$$

so we have another solution to equation (9). We will now show that $t < w$. Note that

$$
t \le p < p^2 + q^2.
$$

Thus, $t < w$, which contradicts the minimality of w. $\Rightarrow\Leftarrow$. This completes Case 2. ■

COROLLARY 15.3.3

Fermat's Last Theorem is true for n = 4.

PROOF (By contradiction.) Suppose that $x^4 + y^4 = z^4$ for some nonzero integers $x, y,$ and z. Since $x, y,$ and z are each raised to an even power, replacing $x, y,$ and z with their absolute values will also yield a valid equation. Thus, without loss of generality, we may assume that $x, y,$ and z are natural numbers. We then have

$$x^4 + y^4 = (z^2)^2,$$

which contradicts Theorem 15.3.2. ∎

The proof of the following theorem is similar to the proof of Theorem 15.3.2. We will use the following theorem later in this chapter to prove Fermat's Right Triangle Theorem (15.5.1).

THEOREM 15.3.4

There do not exist nonzero integers x, y, and w such that $x^4 - y^4 = w^2$.

PROOF You will prove this in Exercise 22.

EXERCISES 15.3

Numerical Problems

1. Using the given pair of numbers as the values of p and q, apply the Pythagorean Triple Theorem (15.3.1) to generate a primitive Pythagorean triple.

   a. 4, 9   **b.** 7, 12   **c.** 15, 8

In Exercises 2–5, find the values of p and q from the Pythagorean Triple Theorem (15.3.1) that produce the given primitive Pythagorean triple.

2. 3, 4, 5   3. 39, 80, 89   4. 300, 589, 661   5. 21, 220, 221

6. Find three different solutions to the Diophantine equation $x^3 + y^3 + z^3 = 1$, such that $x, y,$ and z all have absolute value greater than 1.

Reasoning and Proofs

7. Let a, b, c be a Pythagorean triple. Prove that if any two of $a, b,$ and c are relatively prime, then all three numbers are pairwise relatively prime.

8. Let d and z be natural numbers. Prove that if $d^2 \mid z^2$, then $d \mid z$.

9. Let a, b, c be a Pythagorean triple. Let $d = \gcd(a, b)$.

   a. Show that $d \mid c$.

   b. Let $a' = \frac{a}{d}$, $b' = \frac{b}{d}$, and $c' = \frac{c}{d}$. Show that a', b', c' is a primitive Pythagorean triple.

10. In equations (4), w is defined by the equation $c - b = 2w$. Explain why w must be positive.

11. Let a, b, c be a primitive Pythagorean triple. Prove that c is odd and exactly one of the numbers a and b is even.

12. Verify the claim from Case 1 of the proof of Theorem 15.3.2. That is, assume that $x, y, w \in \mathbf{N}$ such that $x^4 + y^4 = w^2$, and p is a prime that divides two of the three numbers x, y, and w. Show that p divides all three of x, y, and w.

13. Prove that the Diophantine equation $x^4 - 2y^2 = 1$ has no solution. [*Hint:* Express $x^4 + y^4$ in terms of x alone.]

Exercises 14–18 refer to a *Pythagorean triangle*, which is a right triangle with all sides of integer length.

14. Prove that any Pythagorean triangle must have a side whose length is divisible by 3.

15. Prove that any Pythagorean triangle must have a side whose length is divisible by 4.

16. Prove that any Pythagorean triangle must have a side whose length is divisible by 5.

17. Suppose you inscribe a circle in a Pythagorean triangle. Prove that the radius of this circle is an integer.

18. Prove that a Pythagorean triangle cannot have sides whose lengths are all prime numbers.

EXPLORATION Rational Points on the Unit Circle (Exercises 19–21)

A *rational point* is a point (x, y) whose coordinates x and y are both rational numbers. The equation for the unit circle is $x^2 + y^2 = 1$.

19. Let a, b, c be a Pythagorean triple, and consider the line from the origin to (a, b). Show that this line passes through a rational point on the unit circle.

20. Let (x, y) be a rational point on the unit circle, and consider the line through the origin and (x, y). Show that this line passes through a point (a, b) such that a, b, c is a primitive Pythagorean triple for some $c \in \mathbf{N}$.

21. Show that your solutions to Exercises 19 and 20 provide a one-to-one correspondence (bijection) between the set of all primitive Pythagorean triples and the set of rational points (x, y) on the unit circle that satisfy $x \leq y$.

Advanced Reasoning and Proofs

22. Prove Theorem 15.3.4. [*Hint:* use a method similar to that used in the proof of Theorem 15.3.2.]

EXPLORATION An Elliptic Curve (Exercises 23–24)

An equation of the form $y^2 = f(x)$, where f is a cubic (third degree) polynomial, is called an *elliptic curve*. The study of rational points (points with rational coordinates x and y) on elliptic curves played a central role in Wiles's proof of Fermat's Last Theorem.

Although Fermat lived long before the general theory of elliptic curves was developed, he did consider the elliptic curve $y^2 = x^3 - x$. Fermat used his method of infinite descent to prove that the only rational points on this curve are $(1, 0)$, $(0, 0)$, and $(-1, 0)$.

23. a. Sketch a graph of the curve $y^2 = x^3 - x$.

   b. Prove that this curve contains no integer points (points with integer coordinates x and y) other than $(1, 0)$, $(0, 0)$, and $(-1, 0)$.

24. Using the following outline as a guide, write a proof that the only rational points on the elliptic curve $y^2 = x^3 - x$ are $(0, 0)$, $(1, 0)$, and $(-1, 0)$.

   I.   Assume that (x, y), with $x, y \in \mathbf{Q}$, is a solution to the equation. In lowest terms, write $x = \frac{p}{q}$ and $y = \frac{r}{s}$. Show that $\left(\frac{q^2 r}{s} \right)^2 = pq(p^2 - q^2)$.

   II. Deduce that $|p|$, $|q|$, and $|p^2 - q^2|$ are perfect squares. Let $c, d, e, \in \mathbf{Z}$ satisfy $c^2 = |p|$, $d^2 = |q|$, and $e^2 = |p^2 - q^2|$.

   III. Show that $e^2 = |c^4 - d^4|$.

   IV. Use Theorem 15.3.4 to complete your proof.

15.4 Germain's Contributions to Fermat's Last Theorem

In this section, we present some of Germain's results about Fermat's Last Theorem. An important role in these results is played by primes p for which the number $2p + 1$ is also prime. Numbers p of this type are known as **Germain primes**. These primes have been used in cryptography, particularly in the Diffie-Hellman algorithm (see Example 2 of Section 10.4). The largest known Germain prime has more than 20,000 digits. It is conjectured that there are infinitely many Germain primes, but this is not known.

EXAMPLE 1

The prime 5 is a Germain prime, since $2 \cdot 5 + 1 = 11$ is also prime.

We now state one of Germain's main results, which she proved in a letter she wrote to Legendre in the early 1820s.

Let $n > 2$ be a Germain prime, and let x, y, and z be integers. If $x^n + y^n = z^n$, then one of x, y, or z must be divisible by n.

To prove this theorem, it will be useful to have the following lemma.

LEMMA 15.4.2

Let n be a natural number such that $2n + 1$ is prime. Then for any $x \in \mathbf{Z}$, $x^n \equiv 0, 1, or -1 \pmod{2n + 1}$.

PROOF Let $n \in \mathbf{N}$ such that $2n + 1$ is prime, and let $x \in \mathbf{Z}$.

Case 1 $x \equiv 0 \pmod{2n + 1}$.
Then $x^n \equiv 0 \pmod{2n + 1}$, so we are done with Case 1.

Case 2 $x \not\equiv 0 \pmod{2n + 1}$.
Then by Fermat's Little Theorem (9.1.2), $x^{2n} \equiv 1 \pmod{2n + 1}$. Thus,

$$(x^n)^2 \equiv 1 \pmod{2n + 1}.$$

Hence, x^n is its own inverse modulo $2n + 1$, so it follows from Lemma 8.3.8 that

$$x^n \equiv 1 \quad \text{or} \quad x^n \equiv -1 \pmod{2n + 1},$$

which completes the proof for Case 2. ∎

We are now ready to prove Theorem 15.4.1.

PROOF of **THEOREM 15.4.1** (By contradiction.)

Instead of giving the general proof here, we will prove the theorem for a particular Germain prime, $n = 5$. We leave the general proof, which is essentially the same, as an exercise for the reader (Exercise 11).
    Let x, y, and z be integers such that $x^5 + y^5 = z^5$.

[To show: One of x, y, or z is divisible by 5.]

Assumption: Suppose that none of x, y, or z is divisible by 5.

    We will assume that x, y, and z are pairwise relatively prime, since otherwise we can divide out any common factors. (See Exercise 15.)
    Instead of writing the equation as $x^5 + y^5 = z^5$, we let $w = -z$ and write

$$x^5 + y^5 + w^5 = 0. \tag{1}$$

Because of the symmetry of equation (1), any argument that can be made about x has an analogous argument that can be made about both y and w.

We begin by focusing on x. Solving equation (1) for $-x^5$ and factoring, we obtain

$$-x^5 = y^5 + w^5$$
$$= (y + w)(y^4 - y^3 w + y^2 w^2 - yw^3 + w^4). \qquad (2)$$

Claim: $(y + w)$ and $(y^4 - y^3 w + y^2 w^2 - yw^3 + w^4)$ are relatively prime.

Proof of Claim (By contradiction.)

Assume there exists a prime p such that

$$p \mid (y + w) \qquad (3)$$

and

$$p \mid (y^4 - y^3 w + y^2 w^2 - yw^3 + w^4). \qquad (4)$$

It follows from (3) that $w \equiv -y \pmod{p}$, so

$$y^4 - y^3 w + y^2 w^2 - yw^3 + w^4 \equiv y^4 - y^3(-y) + y^2(-y)^2 - y(-y)^3 + (-y)^4$$
$$\equiv 5y^4 \pmod{p}.$$

The left side of this congruence is congruent to 0 modulo p by (4), so $p \mid 5y^4$, and it follows that $p = 5$ or $p \mid y$ (by Corollary 6.1.3). If $p = 5$, then by (2) and (3), $5 \mid -x^5$, which implies that $5 \mid x$, contradicting our hypothesis. If $p \mid y$, then (3) implies that $p \mid w$, which contradicts our assumption that y and z are relatively prime. This completes the proof of the Claim.

We know from equation (2) that the product of the two relatively prime integers $(y + w)$ and $(y^4 - y^3 w + y^2 w^2 - yw^3 + w^4)$ is a 5th power, $(-x)^5$, so Lemma 15.2.2 allows us to conclude that each of the two factors $(y + w)$ and $(y^4 - y^3 w + y^2 w^2 - yw^3 + w^4)$ is a 5th power (see Exercise 7).

Since equation (1) is symmetric in x, y, and w, we can use similar reasoning to show that $x + w$ and $x^4 - x^3 w + x^2 w^2 - xw^3 + w^4$, as well as $x + y$ and $y^4 - y^3 x + y^2 x^2 - yx^3 + x^4$, are all 5th powers. In other words, we can find integers a, b, c, d, e, f such that the following six equations hold:

$$y + w = a^5 \qquad y^4 - y^3 w + y^2 w^2 - yw^3 + w^4 = d^5$$
$$w + x = b^5 \qquad x^4 - x^3 w + x^2 w^2 - xw^3 + w^4 = e^5$$
$$x + y = c^5 \qquad y^4 - y^3 x + y^2 x^2 - yx^3 + x^4 = f^5$$

Since $11 = 2 \cdot 5 + 1$ is prime, we may apply Lemma 15.4.2, which tells us that x^5, y^5, and w^5 are each congruent to 0, 1, or $-1 \pmod{11}$. From equation (1), we know that $x^5 + y^5 + w^5 \equiv 0 \pmod{11}$. Also, since x, y, and z are pairwise relatively prime, x^5, y^5, and w^5 cannot all be congruent to 0 (mod 11). It follows that one of the numbers x^5, y^5, and w^5 must be congruent to each of $0, 1$, and $-1 \pmod{11}$. By the symmetry of equation (1), we may assume without loss of generality that $x^5 \equiv 0 \pmod{11}$ and, hence, that $x \equiv 0 \pmod{11}$. Now we may write the following chain of congruences:

$$0 \equiv 2x$$
$$\equiv (x + y) + (w + x) - (y + w)$$
$$\equiv c^5 + b^5 - a^5$$
$$\equiv c^5 + b^5 + (-a)^5 \pmod{11}$$

Again by Lemma 15.4.2, c^5, b^5, and $(-a)^5$ are each congruent to $0, 1$, or $-1 \pmod{11}$. Reasoning as before, we see that at least one of c^5, b^5, and $(-a)^5$ is congruent to 0 (mod 11). Thus, one of a, b, or c is congruent to 0 (mod 11).

Case 1 $b \equiv 0 \pmod{11}$.
  Since $x \equiv 0 \pmod{11}$, we have the congruence $w \equiv w + x \equiv b^5 \equiv 0 \pmod{11}$. It follows that both x and w must be divisible by 11, which contradicts our assumption that x and z are relatively prime. $\Rightarrow\Leftarrow$

Case 2 $c \equiv 0 \pmod{11}$.
  Using the same reasoning as in Case 1, we obtain a contradiction (this time to the assumption that x and y are relatively prime). $\Rightarrow\Leftarrow$

Case 3 $a \equiv 0 \pmod{11}$.
  In this case, we have $0 \equiv a^5 \equiv y + w \pmod{11}$. Hence, $w \equiv -y \pmod{11}$, and thus,

$$d^5 \equiv y^4 - y^3 w + y^2 w^2 - y w^3 + w^4$$
$$\equiv y^4 - y^3(-y) + y^2(-y)^2 - y(-y)^3 + (-y)^4$$
$$\equiv 5y^4 \pmod{11}. \tag{5}$$

Also since $x \equiv 0 \pmod{11}$, we have

$$f^5 \equiv y^4 - y^3 x + y^2 x^2 - y x^3 + x^4$$
$$\equiv y^4 \pmod{11}. \tag{6}$$

Combining congruences (5) and (6), we obtain

$$d^5 \equiv 5y^4 \equiv 5f^5 \pmod{11}.$$

By Lemma 15.4.2, d^5 and f^5 are each congruent to $0, 1$, or $-1 \pmod{11}$. But if f^5 were congruent to either 1 or -1, then d^5 would be congruent to either 5 or -5.

Hence, it must be that $f^5 \equiv 0 \pmod{11}$, and thus, $y \equiv 0 \pmod{11}$ by congruence (6), from which it follows that $w \equiv 0 \pmod{11}$, and thus, $z \equiv 0 \pmod{11}$. We have now shown that z and x are both divisible by 11. This contradicts our assumption that x and z are relatively prime. $\Rightarrow\Leftarrow$. This completes the proof of Case 3.

We conclude that if $x^5 + y^5 = z^5$, then one of x, y, or z must be divisible by 5. ■

Beyond Germain primes

After Germain proved Theorem 15.4.1, mathematicians divided the proof of Fermat's Last Theorem into two cases. To prove Fermat's Last Theorem, one must show that for any prime $n > 2$, there are no nonzero solutions to the Diophantine equation $x^n + y^n = z^n$ in the following two cases:

CASE I: None of the numbers x, y, z is divisible by n.

CASE II: One of the numbers x, y, z is divisible by n.

Using this terminology, we can restate Theorem 15.4.1 as follows (see Exercise 13):

THEOREM 15.4.3

Case I of Fermat's Last Theorem is true for every Germain prime $n > 2$.

The proof of Theorem 15.4.1 given in Germain's letter to Legendre actually implies a stronger result. Some mathematicians have come to refer to Theorem 15.4.1 as Germain's Theorem, while others (including the authors of this text) refer to the stronger result, which we will state and prove shortly (Theorem 15.4.5), as Germain's Theorem. Legendre presented this stronger result to the French Academy in 1823, and included it in a supplement to the second edition of his *Théorie des Nombres* in 1825, attributing it to Germain. In 1825, Legendre and Dirichlet each independently proved Fermat's Last Theorem for $n = 5$. Both used Germain's Theorem, so they only had to prove the result for Case II.

Theorem 15.4.3 (equivalently, Theorem 15.4.1) addresses Case I of Fermat's Last Theorem when the exponent n is a Germain prime. But what if the exponent n is not a Germain prime; that is, what if n is prime but $2n + 1$ is not?

To answer this, let's look back at the proof of Theorem 15.4.1. In this proof, our argument relied heavily on properties of the fifth powers modulo 11. More generally, Lemma 15.4.2 tells us that if n is a Germain prime, then the only nth powers modulo the prime $2n + 1$ are $-1, 0$, and 1. But what if the prime n is not a Germain prime? It turns out that we can sometimes find a prime q to play the role that $2n + 1$ played in our proof of Theorem 15.4.1. If the nth powers modulo q have just the right properties, then it is possible to use the same basic argument that was used for Germain primes. Germain isolated these properties and introduced the term *auxiliary prime* to refer to such a prime q.

Before we define *auxiliary prime*, we need just one more bit of terminology. We say that an integer a in the range $0, 1, \ldots, q - 1$ is an **nth power residue modulo q** if there is an integer b such that $b^n \equiv a \pmod{q}$. For instance, consider the 3rd power of every nonzero number modulo 13:

Table 15.4.1 The nonzero 3rd power residues modulo 13

x	1	2	3	4	5	6	7	8	9	10	11	12
$x^3 \pmod{13}$	1	8	1	12	8	8	5	5	1	12	5	12

Thus, the nonzero 3rd power residues modulo 13 are **1, 5, 8**, and **12**. Also, since no two of these four numbers are consecutive integers, we say that

$$\text{There are no consecutive nonzero 3rd power residues modulo 13.} \qquad (7)$$

DEFINITION 15.4.4

Let n be a prime number. An **auxiliary prime** *for n is a prime q that satisfies the following two conditions:*

(i) There are no consecutive nonzero nth power residues modulo q.

(ii) The number n is not an nth power residue modulo q.

The following examples illustrate this definition.

EXAMPLE 2

Show that 13 is an auxiliary prime for 3.

SOLUTION In the discussion preceding Definition 15.4.4, we determined statement (7), which fulfills part (i) of the definition of auxiliary prime for $n = 3$ and $q = 13$. To establish part (ii) of the definition, recall from Table 15.4.1 that the only 3rd power residues modulo 13 are **1, 5, 8**, and **12**. Thus, 3 is *not* a 3rd power residue modulo 13; that is, for every $x \in \mathbf{Z}$, $x^3 \not\equiv 3 \pmod{13}$.

We conclude that 13 is an auxiliary prime for 3. ∎

EXAMPLE 3

Is 19 an auxiliary prime for 3?

SOLUTION First, we compute all of the nonzero 3rd power residues modulo 19:

x	1	2	3	4	5	6	7	8	9	10	11	12	13	14	15	16	17	18
$x^3 \pmod{19}$	1	8	8	7	11	7	1	18	7	12	1	18	12	8	12	11	11	18

Thus, there are two pairs of consecutive 3rd power residues modulo 19: the pairs **7, 8** and **11, 12**. Since there exist consecutive 3rd power residues modulo 19, we conclude from part (i) of Definition 15.4.4 that 19 is *not* an auxiliary prime for 3. ●

We are now ready to state and prove Germain's Theorem. This theorem asserts that Case I of Fermat's Last Theorem is true for n, provided there exists even a single auxiliary prime for n. Note that this theorem is a generalization of Theorem 15.4.3, since if n is a Germain prime, then $2n + 1$ is an auxiliary prime for n. (See Exercise 8.)

THEOREM 15.4.5   GERMAIN'S THEOREM

Let n be an odd prime. If there exists an auxiliary prime for n, then Case I of Fermat's Last Theorem is true for n.

The proof is quite similar to the proof of Theorem 15.4.1.

PROOF OF **THEOREM 15.4.5** (By contradiction.)

Let n be an odd prime, and let q be an auxiliary prime for n. **[To show: There do not exist x, y, $z \in \mathbb{Z}$ for which $x^n + y^n = z^n$ and none of x, y, or z is divisible by n.]**

Assumption: Suppose there exist integers x, y, and z such that $x^n + y^n = z^n$ and none of x, y, or z is divisible by n.

We may assume that x, y, and z are pairwise relatively prime (see Exercise 15). By letting $w = -z$, we may rewrite the equation as

$$x^n + y^n + w^n = 0. \tag{8}$$

Solving for $-x^n$ and factoring, we obtain

$$-x^n = y^n + w^n$$

$$= (y + w)(y^{n-1} - y^{n-2}w + y^{n-3}w^2 - y^{n-4}w^3 + \cdots + w^{n-1}).$$

Claim: $(y + w)$ and $(y^{n-1} - y^{n-2}w + y^{n-3}w^2 - y^{n-4}w^3 + \cdots + w^{n-1})$ are relatively prime.

> **Proof of Claim** The proof of this Claim (which you will provide in Exercise 16) is similar to that of the Claim in the proof of Theorem 15.4.1.

We know that the product of the two relatively prime numbers $(y + w)$ and $(y^{n-1} - y^{n-2}w + y^{n-3}w^2 - y^{n-4}w^3 + \cdots + w^{n-1})$ is an nth power, $(-x)^n$, so it follows from Lemma 15.2.2 that each of the two factors $(y + w)$ and $(y^{n-1} - y^{n-2}w + y^{n-3}w^2 - y^{n-4}w^3 + \cdots + w^{n-1})$ is an nth power (see Exercise 7).

Since equation (8) is symmetric in x, y, and w, similar reasoning shows that the factors $(x + w)$ and $(x^{n-1} - x^{n-2}w + x^{n-3}w^2 - x^{n-4}w^3 + \cdots + w^{n-1})$, as well as $(x + y)$ and $(y^{n-1} - y^{n-2}x + y^{n-3}x^2 - y^{n-4}x^3 + \cdots + x^{n-1})$, are all nth powers.

In other words, we can find integers a, b, c, d, e, f such that the following six equations hold:

$$y + w = a^n \qquad y^{n-1} - y^{n-2}w + y^{n-3}w^2 - y^{n-4}w^3 + \cdots + w^{n-1} = d^n$$

$$w + x = b^n \qquad x^{n-1} - x^{n-2}w + x^{n-3}w^2 - x^{n-4}w^3 + \cdots + w^{n-1} = e^n$$

$$x + y = c^n \qquad y^{n-1} - y^{n-2}x + y^{n-3}x^2 - y^{n-4}x^3 + \cdots + x^{n-1} = f^n$$

Now we will consider our equations modulo the auxiliary prime q. From equation (8), we know that $x^n + y^n + w^n \equiv 0 \pmod{q}$. It follows by condition (i) of the definition of auxiliary prime (15.4.4) that one of $x, y,$ or z must be divisible by q (see Exercise 17). By the symmetry of equation (8), we can assume without loss of generality that x is divisible by q. Hence modulo q,

$$0 \equiv 2x$$
$$\equiv (x + y) + (w + x) - (y + w)$$
$$\equiv c^n + b^n - a^n$$
$$\equiv c^n + b^n + (-a)^n \pmod{q}$$

Again by condition (i) of the definition of auxiliary prime (see Exercise 17), one of $c, b,$ or $-a$ must be divisible by q.

Case 1 $b \equiv 0 \pmod{q}$.
Then since $x \equiv 0 \pmod{q}$, we have $w \equiv w + x \equiv b^n \equiv 0 \pmod{q}$. Thus, both x and w are divisible by q, which contradicts our assumption that x and z are relatively prime. $\Rightarrow\Leftarrow$

Case 2 $c \equiv 0 \pmod{q}$.
Using the same reasoning as in Case 1, we again find a contradiction (this time to the assumption that x and y are relatively prime). $\Rightarrow\Leftarrow$

Case 3 $a \equiv 0 \pmod{q}$.
In this case, $0 \equiv a^n \equiv y + w \pmod{q}$. Hence,

$$w \equiv -y \pmod{q}, \tag{9}$$

and thus,

$$d^n \equiv y^{n-1} - y^{n-2}w + y^{n-3}w^2 - y^{n-4}w^3 + \cdots + w^{n-1}$$
$$\equiv ny^{n-1} \pmod{q}. \tag{10}$$

Also, since $x \equiv 0 \pmod{q}$, we have

$$f^n \equiv y^{n-1} - y^{n-2}x + y^{n-3}x^2 - y^{n-4}x^3 + \cdots + x^{n-1}$$
$$\equiv y^{n-1} \pmod{q}. \tag{11}$$

Combining congruences (10) and (11), we obtain

$$d^n \equiv nf^n \pmod{q}. \tag{12}$$

If it were the case that $f \equiv 0 \pmod{q}$, then by congruence (11), $y^{n-1} \equiv 0 \pmod{q}$. Hence $q \mid y$, and thus by congruence (9), $q \mid w$. But then w and x would share the factor q, which contradicts our assumption that x and z are relatively prime.

Thus, we may assume that $\gcd(f, q) = 1$, which implies that f has a multiplicative inverse, g, modulo q. Then $f^n g^n \equiv 1 \pmod{q}$, and hence,

$$
\begin{aligned}
(dg)^n &\equiv d^n g^n \\
&\equiv n(f^n g^n) \qquad \leftarrow \textbf{by congruence (12)} \\
&\equiv n \pmod{q}.
\end{aligned}
$$

But this contradicts condition (*ii*) of the definition of auxiliary prime (15.4.4). $\Rightarrow\Leftarrow$

Since we have reached a contradiction, our Assumption must be false. It follows that Case I of Fermat's Last Theorem is true for n. ∎

After proving the above theorem, Germain proved that for every odd prime $n < 100$, there exists an auxiliary prime for n. Thus, she concluded that Case I of Fermat's Last Theorem is true for every odd prime $n < 100$.

In addition to Germain's correspondence with Legendre about Fermat's Last Theorem, she also wrote to Gauss about her results and ideas. In a letter written in 1819, she explained to Gauss her strategy for proving Fermat's Last Theorem in general. She began with the following lemma.

LEMMA 15.4.6

Suppose that there exist integers x, y, and z such that $x^n + y^n = z^n$ for some natural number n. If there exists a prime p with no consecutive nonzero nth power residues modulo p, then one of x, y, or z is divisible by p.

Note that the conclusion of Lemma 15.4.6 does not say that Case I of Fermat's Last Theorem is true for n, since the conclusion asserts that p (rather than n) divides $x, y,$ or z.

PROOF of **LEMMA 15.4.6** (By contradiction.)

Let $x, y, z \in \mathbf{Z}$ and $n \in \mathbf{N}$ such that $x^n + y^n = z^n$, and let p be prime such that there are no consecutive nonzero nth power residues modulo p.

Assumption: Suppose that none of x, y, z is divisible by p.

Then x has a multiplicative inverse, a, modulo p, and $a^n x^n \equiv 1 \pmod{p}$. Multiplying the equation $x^n + y^n = z^n$ by a^n and considering it modulo p, we obtain

$$a^n x^n + a^n y^n \equiv a^n z^n \pmod{p},$$

and thus,

$$1 + (ay)^n \equiv (az)^n \pmod{p}.$$

Furthermore, since a, y, and z are not divisible by p, neither $(ay)^n$ nor $(az)^n$ can be congruent to 0 modulo p. Thus, $(ay)^n$ and $(az)^n$ are consecutive nonzero nth power residues modulo p, which contradicts our hypothesis. $\Rightarrow\Leftarrow$

We conclude that one of x, y, or z is divisible by p. ∎

EXAMPLE 3

Let $n = 3$ and $p = 13$. We showed in Example 2 that there are no consecutive nonzero 3rd powers modulo 13. Thus, by Lemma 15.4.6, if there exist integers x, y, and z such that $x^3 + y^3 = z^3$, then one of x, y, or z must be divisible by 13.

In her 1819 letter to Gauss, Germain explained that she wanted to find a value for n such that there were infinitely many primes p with no nonzero consecutive nth power residues mod p. If there existed integers x, y, and z such that $x^n + y^n = z^n$, then by Lemma 15.4.6 each value of p would have to divide x, y, or z. Thus, at least one of x, y, or z would have infinitely many prime divisors. As this is impossible, Fermat's Last Theorem would have to be true for n. Furthermore, Germain presented a method for finding values of p that satisfied the conditions of Lemma 15.4.6 for a fixed value of n. Since p must divide x, y, or z, a large such value of p would mean that x, y, or z must be large. While Germain was not able to find an n with infinitely many values of p, she did show that for $n = 5$, if there existed integers x, y, and z such that $x^5 + y^5 = z^5$, then x, y, or z would have at least 30 digits. In a letter to Gauss, Germain wrote, "You can easily imagine, Monsieur, that I must have been able to prove that this equation is only possible for numbers whose size frightens the imagination."

EXERCISES 15.4

Numerical Problems

1. Find five Germain primes.

2. Make a table of the 5th powers modulo 7. Do there exist consecutive nonzero 5th power residues?

3. Make a table of the 3rd powers modulo 7. Do there exist consecutive nonzero 3rd power residues?

4. Make a table of the 3rd powers modulo 31. Do there exist consecutive nonzero 3rd power residues?

5. The following is a complete list of the 7th power residues modulo 29:

$$1, 12, 17, 28.$$

a. What, if anything, does Germain's Theorem (15.4.5) allow you to conclude?

b. What, if anything, does Lemma 15.4.6 allow you to conclude?

6. The following is a complete list of the 13th power residues modulo 443:

> 1, 13, 15, 18, 34, 59, 63, 67, 73, 119, 123, 169, 173, 176, 195, 209, 218, 225, 234, 248, 267, 270, 274, 320, 324, 370, 376, 380, 384, 409, 425, 428, 430, 442

a. What, if anything, does Germain's Theorem (15.4.5) allow you to conclude?

b. What, if anything, does Lemma 15.4.6 allow you to conclude?

Reasoning and Proofs

7. Prove the following variant of Lemma 15.2.2, which is needed in the proofs of Theorems 15.4.1 and 15.4.5. [*Hint:* You may use Lemma 15.2.2 in your proof.]

LEMMA. *Let $n \in \mathbf{N}$ be odd, and let $z, d, e \in \mathbf{Z}$ with $\gcd(d, e) = 1$. If $z^n = de$, then there exist $a, b \in \mathbf{Z}$ such that $d = a^n$ and $e = b^n$.*

8. Explain why Theorem 15.4.3 is a special case of Theorem 15.4.5. [*Hint:* Show that if n is a Germain prime, then $p = 2n + 1$ is an auxiliary prime for n.]

9. Let n be an odd prime.

a. Show that if q is prime such that $\gcd(n, q - 1) = 1$, then every natural number less than q is an nth power residue modulo q.

b. Show that if q is an auxiliary prime for n, or indeed if q is any prime that satisfies condition (i) of the definition of auxiliary prime, then $q \equiv 1 \pmod{n}$.

10. Prove Theorem 15.4.1 for the case $n = 2$.

11. In the text, we proved Theorem 15.4.1 for the Germain prime $n = 5$. Prove Theorem 15.4.1 for the general case in which $n > 2$ is an arbitrary Germain prime.

12. Let $n > 3$ be prime. Suppose that $p = 4n + 1$ is also prime. Prove that p is an auxiliary prime for n.

13. **a.** State the contrapositive of Theorem 15.4.1.

b. How is this related to the statement of Theorem 15.4.3?

14. Suppose you wish to know about solutions to $x^{101} + y^{101} = z^{101}$. Your computer tells you that with $n = 101$, each of the primes $p = 809, 5051, 5657, 6869, 8081$ satisfies the conditions of Lemma 15.4.6. That is, for each of these primes p, there are no consecutive 101st power residues modulo p. Use this to prove that if x, y, z are nonzero integers satisfying $x^{101} + y^{101} = z^{101}$, then at least one of x, y, z has absolute value larger than 1,000,000.

15. At the beginning of the proofs of Theorem 15.4.1 and Theorem 15.4.5, we supposed we had integers x, y, z, none of which is divisible by n, that satisfy

$x^n + y^n = z^n$. We assumed that x, y, and z are pairwise relatively prime. Justify this assumption by showing that if x, y, and z are not pairwise relatively prime, then they can be used to produce pairwise relatively prime integers x', y', and z' that satisfy $(x')^n + (y')^n = (z')^n$. [*Hint:* First explain why x, y, and z are all nonzero. Hence, x and y have a greatest common divisor, d.]

16. Prove the Claim in the proof of Germain's Theorem (15.4.5). [*Hint:* You may use our proof of the Claim in the proof of Theorem 15.4.1 as a guide.]

17. Let n be an odd prime, and let q be an auxiliary prime for n. Show that condition (*i*) of the definition of auxiliary prime (15.4.4) implies the following statement:

> For all integers x, y, and w, if $x^n + y^n + w^n \equiv 0 \pmod{q}$, then one of x, y, or w is divisible by q.

[*Hint:* Use an argument similar to the proof of 15.4.6.]

18. Look at the $\mathbf{Z}_{29}$ exponentiation table (Table 9.4.1). For which values of n does this table help us say something about possible values of x, y, and z such that $x^n + y^n = z^n$? Explain your answers.

15.5 A Geometric Look at the Equation $x^4 + y^4 = z^2$

Sophie Germain provided the first general step toward the proof of Fermat's Last Theorem. What's not so well known about Germain is that she was an inventor. Germain is credited with numerous inventions that revolutionized French society, from the snooze alarm to the flush toilet. But of all her great ideas, the invention of which she was most proud was the Walk-a-Bot, the first fully automated bipedal robot, which she invented while she was in college. In fact, Germain founded the robotics club at Paris's world-renowned École Abnormale Inférieure.

Germain's dream was to hold a college-wide robotics competition. Since batteries had not yet been invented, the electricity to power the robots needed to be transmitted through a special gold-plated conducting floor. Germain approached the university vice president, Ebenezer Scrimpsalot, to ask if he could supply the floor.

"I don't know—it's awfully expensive," said the vice president. "What if the floor were vandalized or stolen?"

"Well, I could program my Walk-a-Bot to patrol it 24 hours a day. The Walk-a-Bot is my latest invention. It can take steps that are exactly one foot long in any direction."

The vice president smiled. "In that case, I can make you a very generous offer. You can have one square sheet of the gold-plated flooring material whose sides are any whole number of feet long."

"Does the floor have to be a square?" inquired Germain. "I was thinking maybe rectangular."

"I can ask the machine shop to reshape the square sheet into a rectangle. But the material is very expensive, so I must insist that none of it go to waste."

"You mean the floor can be any size rectangle as long as its area is a perfect square?"

"That's right," replied Scrimpsalot. "And I have one more requirement." The vice president chuckled under his breath. "Your Walk-a-Bot must patrol the entire perimeter and diagonal of the floor."

"Great! I'll get right to work on the dimensions."

Back at the dorm, Germain thought about the vice president's requirements. While contemplating what size rectangular floor to ask for, she drew a picture of the floor and the robot's patrol routes.

Since the Walk-a-Bot could only take steps that were exactly one foot long, Germain knew that *a*, *b*, and *c* had to be integers in order for her robot to be able to patrol the platform.

Let's try to help Germain find the dimensions of the platform. We know by the Pythagorean Theorem that

$$a^2 + b^2 = c^2. \tag{1}$$

Also, since the area of the platform is a perfect square, we have $ab = n^2$ for some integer n. This may be rewritten as

$$a = \frac{n^2}{b}. \tag{2}$$

We substitute equation (2) into equation (1) to get

$$\left(\frac{n^2}{b}\right)^2 + b^2 = c^2.$$

Multiplying both sides of this equation by b^2, we have

$$n^4 + b^4 = (bc)^2.$$

But this contradicts Theorem 15.3.2. Thus, there is no rectangle with integer sides and integer diagonal whose area is a perfect square.

Reasoning in this manner, Germain realized that the vice president's requirements could not be satisfied by any rectangular platform. She grabbed her notes and rushed to the vice president's office, but he was gone for the day. Mustering all her courage, she knocked on the university president's door.

"Sir, I went to Vice President Scrimpsalot to ask for a platform for our robotics competition, and he said he would give us one, and all we had to do was—"

"Oh no," said the president. "Scrimpsalot didn't pull the old Euler trick on you, did he?"

"What do you mean?" asked Germain quizzically.

"Oh, he's always promising students funding under certain conditions, but the conditions invariably reduce to finding an integer solution to $x^4 + y^4 = z^2$. Why, just last year he promised the marching band new uniforms provided they could. . . . Well, that's another story for another book. In any case, don't worry. Since you were able to see through the vice president's scheme, we'll get you the gold-plated platform that you need for your competition."

The problem that was posed by the university vice president in our myth is related to a problem posed by Diophantus in *Arithmetica*. Diophantus's question was "Can the area of a right triangle be a square?" The question had the implicit assumptions that all three sides of the triangle had rational lengths and that the word "square" refers to the square of a *rational* number. This question is equivalent to the analogous question in which the numbers involved are all integers (see Exercise 4).

Fermat was the first to answer Diophantus's question, by proving that such a triangle does not exist.

Fun Facts

The only detailed proof known to have been written by Fermat is the proof of Fermat's Right Triangle Theorem (15.5.1). Although Fermat made many mathematical assertions in his letters, and it is clear from his writings and letters that he knew how to prove almost all of these assertions, most of his proofs either have been lost or were never written down.

While the margin of Diophantus's *Arithmetica* was too small to contain Fermat's "proof" of Fermat's Last Theorem, the proof of Fermat's Right Triangle Theorem was in fact found in the margin of his copy of *Arithmetica*. Since the $n = 4$ case of Fermat's Last Theorem follows from Fermat's Right Triangle Theorem (see Exercise 5), Fermat is credited with having given the first proof of the $n = 4$ case of Fermat's Last Theorem.

THEOREM 15.5.1   FERMAT'S RIGHT TRIANGLE THEOREM

The area of a right triangle with integer sides cannot be a perfect square.

PROOF (By contradiction.)

Assumption: Suppose that there is a right triangle with legs of length x and y and hypotenuse of length z, such that x, y, and z are integers and the area of the triangle is a perfect square.

Thus,

$$x^2 + y^2 = z^2, \tag{3}$$

and there exists $n \in \mathbf{Z}$ such that

$$\frac{xy}{2} = n^2. \tag{4}$$

Multiplying this equation by 4 yields

$$2xy = 4n^2. \tag{5}$$

Adding this to equation (3), we obtain

$$x^2 + y^2 + 2xy = z^2 + 4n^2,$$

or, equivalently,

$$(x + y)^2 = z^2 + 4n^2. \tag{6}$$

On the other hand, subtracting equation (5) from equation (3), we obtain

$$x^2 + y^2 - 2xy = z^2 - 4n^2$$

or, equivalently,

$$(x - y)^2 = z^2 - 4n^2. \tag{7}$$

Multiplying Equations (6) and (7) together gives

$$(x + y)^2(x - y)^2 = (z^2 + 4n^2)(z^2 - 4n^2)$$

$$= z^4 - 16n^4.$$

Thus,

$$[(x + y)(x - y)]^2 = z^4 - (2n)^4.$$

Theorem 15.3.4 implies that one of the three quantities $[(x + y)(x - y)]$, z^4, or $2n$ must be equal to zero. Since $x, y, z > 0$, we see that $(x + y)$, z^4, and $2n$ are all nonzero. Thus, the only possibility is that $(x - y) = 0$; that is, $x = y$. Equation (3) then becomes $2x^2 = z^2$. However, this is not possible since $\sqrt{2}$ is irrational. $\Rightarrow\Leftarrow$ ●

A method for solving any Diophantine equation?

In 1900, the well-known German mathematician David Hilbert gave what is considered to be the most influential mathematics lecture ever given. Speaking at the International Congress of Mathematicians in Paris, Hilbert presented his philosophy of mathematics and gave a list of what he believed to be the 23 most important problems in mathematics that were still unsolved at the end of the 19th century. Hilbert's tenth problem asks whether there is an algorithm to decide whether or not any given Diophantine equation has a solution. By "Diophantine equation," Hilbert meant a polynomial equation in one or more variables with integer coefficients.

   Imagine if there were such an algorithm. You could just go to www.diophantus.com, type in any Diophantine equation you wanted to solve, and click the "Solve" button. In seconds, the computer would tell you whether or not your equation has a solution. Type in $x^2 - 61y^2 = 1$, and the computer would respond "A SOLUTION EXISTS" (as we learned in Theorem 15.1.1 when we studied Pell's equation). Then enter $x^4 - y^4 = 49$, and soon your screen would flash "NO SOLUTION EXISTS" (as follows from Theorem 15.3.4).

At the time of Hilbert's lecture, no algorithm for solving a general Diophantine equation was known, though various methods were known for solving particular types of Diophantine equations. In 1961, Martin Davis, Hilary Putnam, and Julia Robinson made an important step. They proved that there could be no algorithm to decide whether or not a given *exponential* Diophantine equation has a solution. An exponential Diophantine equation is one in which variables are allowed to occur in the exponents.

Finally, in 1970, Russian mathematician Yuri Matiyasevich, who had learned about Hilbert's tenth problem when he was an undergraduate at City College of New York, gave a definitive answer to Hilbert's question, showing that there is no algorithm for determining whether or not a given Diophantine equation has a solution. Matiyasevich's proof combined earlier results with a very clever use of the Fibonacci numbers. The solution to Hilbert's tenth problem is considered one of the most important mathematical results of the 20th century.

While one might be initially disappointed that there is no algorithm for deciding whether or not any given Diophantine equation has a solution, mathematicians, oddly enough, also find this fact inspiring! Computers may be great for helping us to discover new patterns and theorems, but Matiyasevich's theorem shows us that no computer program can ever answer all of our questions about the natural numbers. Each new type of Diophantine equation, and each new conjecture of number theory, presents its own unique challenge. The search for truths about the natural numbers will never end.

EXERCISES 15.5

Numerical Problems

1. Find a right triangle with integer sides whose perimeter is equal to its area.

Reasoning and Proofs

2. Find every right triangle with integer sides whose perimeter is equal to its area. Prove that your list is complete.

3. Consider the curve $x^4 + y^4 = 1$.

   a. Sketch a graph of this curve.

   b. Prove that this curve contains no rational points (points with rational coordinates) other than $(0, 1)$, $(0, -1)$, $(1, 0)$, and $(-1, 0)$.

4. Prove that Fermat's Right Triangle Theorem (15.5.1) is equivalent to the assertion that if the sides of a right triangle have rational lengths, then its area cannot be the square of a rational number.

5. Prove that the $n = 4$ case of Fermat's Last Theorem follows from Fermat's Right Triangle Theorem (15.5.1).

EXPLORATION Congruent Numbers (Exercises 6–10)

A natural number n is called a *congruent number* if there exists a right triangle whose sides are all rational numbers and whose area is n.

Determining which natural numbers are congruent numbers is called the *congruent number problem*. Though this problem was originally posed in an Arab manuscript over a thousand years ago, it remains unsolved to this day. Recently, however, there has been a great deal of progress. For instance, the theory of elliptic curves has been used to show that every prime number congruent to 5 or 7 modulo 8 is a congruent number.

6. a. Show that 6 is a congruent number.

  b. Find two other congruent numbers, and demonstrate that they are congruent.

7. a. Show that 5 is a congruent number. [*Hint:* Consider a right triangle with legs of lengths $\frac{3}{2}$ and $\frac{20}{3}$.]

  b. Show, in contrast, that there is no right triangle with *integer* sides whose area is 5.

8. Show that 1 is not a congruent number. [*Hint:* Use Fermat's Right Triangle Theorem (15.5.1).]

9. This exercise introduces a connection between congruent numbers and elliptic curves. An *elliptic curve* is a curve defined by an equation of the form $y^2 = f(x)$, where $f(x)$ is a cubic (third degree) polynomial. Suppose that n is a congruent number, and $a, b, c \in \mathbf{Q}$ are the side lengths of a right triangle with area n. Thus, $a^2 + b^2 = c^2$, and $n = \frac{ab}{2}$.

  a. Let $w = \frac{c}{2}$. Prove that $w^2 + n$ and $w^2 - n$ are both squares of rational numbers.

  b. Prove that there exists $v \in \mathbf{Q}$ such that $w^4 - n^2 = v^2$.

  c. Prove that if $x = w^2$ and $y = wv$, then (x, y) lies on the elliptic curve

$$y^2 = x^3 - n^2 x.$$

  d. Conclude that if n is a congruent number, then there exists a point (x, y) with nonzero rational coordinates on the elliptic curve $y^2 = x^3 - n^2 x$. (Make sure you show that x and y are nonzero.)

10. Use your results from Exercise 9 to find nonzero rational numbers satisfying $y^2 = x^3 - 36x$.

Advanced Reasoning and Proofs

EXPLORATION Fermat's Proof of His Right Triangle Theorem (Exercise 11)

In this exploration, you will develop Fermat's original proof of his Right Triangle Theorem. (Unlike the proof we gave in the text, the proof you will give in this exploration does not rely on Theorem 15.3.4. Note also that Theorem 15.3.4 is a variant of 15.3.2, the nonexistence of solutions to $x^4 + y^4 = z^4$, for which we gave a proof due to Euler, not Fermat.)

11. Using the following outline as a guide, give another proof of Fermat's Right Triangle Theorem (15.5.1). Note that this proof uses Fermat's celebrated *method of infinite descent*.

  I. Assume that there is a triangle with legs of lengths x and y, and hypotenuse of length z, such that $x, y, z \in \mathbf{N}$ and the area is a perfect square. Among all such triangles, choose one with the shortest hypotenuse. Explain why (x, y, z) is a primitive Pythagorean triple.

  II. By the Pythagorean Triple Theorem (15.3.1), there exist $p, q \in \mathbf{N}$ of opposite parity such that $x = 2pq$, $y = p^2 - q^2$, and $z = p^2 + q^2$. Show that $p, q, p + q$, and $p - q$ are all perfect squares. Denote their square roots by a, b, c, and d, respectively.

  III. Show that $2b^2 = (c + d)(c - d)$, and that $\gcd(c + d, c - d) = 2$.

  IV. Prove that exactly one of the two numbers $c + d$ and $c - d$ is divisible by 4. This allows you to separate into two cases, $4 \mid c + d$ and $4 \mid c - d$. The remainder of this outline treats the case $4 \mid c + d$. The case $4 \mid c - d$ is similar. (Don't forget to treat that case when you're done with the first case.)

  V. Assume $4 \mid c + d$. Prove that $\dfrac{b^2}{4} = \dfrac{c + d}{4} \cdot \dfrac{c - d}{2}$, where the two factors on the right side of this equation are relatively prime integers.

  VI. Show that there exist natural numbers s and t such that $\dfrac{c + d}{4} = s^2$ and $\dfrac{c - d}{2} = t^2$.

  VII. Prove that $a^2 = \dfrac{c^2 + d^2}{2} = 4s^4 + t^4$. [*Hint:* First express c and d in terms of s and t.]

  VIII. Conclude that $(2s^2, t^2, a)$ is a Pythagorean triple. Show that the area of the corresponding right triangle is a perfect square and that the hypotenuse of this triangle is smaller than z. Derive a contradiction.

Index